NUMERICAL TECHNIQUES FOR ENGINEERING ANALYSIS AND DESIGN

NUMERICAL TECHNIQUES FOR ENGINEERING ANALYSIS AND DESIGN

Proceedings of the International Conference on Numerical Methods in Engineering: Theory and Applications, NUMETA '87, Swansea, 6-10 July 1987.

VOLUME I

Edited by

G.N. PANDE & J. MIDDLETON
University College of Swansea

1987 **MARTINUS NIJHOFF PUBLISHERS**
a member of the KLUWER ACADEMIC PUBLISHERS GROUP
DORDRECHT/BOSTON/LANCASTER

Distributors

for the United States and Canada: Kluwer Academic Publishers, 101 Philip Drive, Assinippi Park, Norwell, MA 02061, USA.
for the UK and Ireland: Kluwer Academic Publishers, MTP Press Limited, Falcon House, Queen Square, Lancaster LA1 1RN, UK.
for all other countries: Kluwer Academic Publishers Group, Distribution Center, P.O. Box 322, 3300 AH Dordrecht, The Netherlands

Library of Congress Cataloging in Publication Data

ISBN-13:978-94-010-8134-4 e-ISBN-13:978-94-009-3653-9
DOI:10.1007/978-94-009-3653-9

Proceedings of the NUMETA '87 Conference/Swansea/6-10 July 1987

PREFACE

Numerical methods and related computer based algorithms form the logical solution for many complex problems encountered in science and engineering. Although numerical techniques are now well established, they have continued to expand and diversify, particularly in the fields of engineering analysis and design. Various engineering departments in the University College of Swansea, in particular, Civil, Chemical, Electrical and Computer Science, have groups working in these areas. It is from this mutual interest that the NUMETA conference series was conceived with the main objective of providing a link between engineers developing new numerical techniques and those applying them in practice. Encouraged by the success of NUMETA '85, the second conference, NUMETA '87, was held at Swansea, 6-10 July 1987.

Over two hundred and twenty abstracts were submitted for consideration together with a number of invited papers from experts in the field of numerical methods. The final selection of contributed and invited papers were of a high quality and have culminated in the two volumes which form these proceedings. This volume contains papers on the themes of 'Numerical Techniques for Engineering Analysis and Design' and 'Developments in Engineering Software'. Many new developments on a wide variety of topics have been reported and these proceedings contain a wealth of information and references which we believe will be of great interest to theoreticians and practising engineers alike.

We are extremely grateful to all members of the Organising Committee and the Technical Advisory Panel (see p.vi) for their help and assistance in making the conference a success.

G.N. Pande
J. Middleton
(Editors)

May, 1987

NUMETA '87

CONFERENCE CO-ORDINATORS

O.C. Zienkiewicz (Chairman); G.N. Pande, J. Middleton

ORGANISING COMMITTEE

R.F. Allen
N. Bicanic
K. Board
E. Hinton
R.W. Lewis
A.R. Luxmoore
K. Morgan
D.J. Naylor
J. Peraire
J. Pittman
K.G. Stagg
C. Taylor
P. Townsend
D.R.J. Owen
R.D. Wood

TECHNICAL ADVISORY PANEL

J.H. Argyris	West Germany
S. Atluri	USA
I. Babuska	USA
J. Blaauwendraad	Netherlands
K.J. Bathe	USA
Z.P. Bazant	USA
T. Belytschko	USA
P.G. Bergan	Norway
G.F. Carey	USA
J.L. Chenot	France
M.A. Crisfield	UK
M. Crochet	Belgium
F. Darve	France
C.S. Desai	USA
J. Donea	Italy
H.R. Evans	UK
C.A. Felippa	USA
R.H. Gallagher	USA
M. Geradin	Belgium
J. Ghaboussi	USA
G.L. Goudreau	USA
J.O. Hallquist	USA
M.F. Kanninen	USA
T. Kawai	Japan
A.R. Mitchell	UK
J.T. Oden	USA
R. Ohayon	France
M.D. Olson	Canada
K.C. Park	USA
P.L. Roe	UK
A. Samuelsson	Sweden
S.F. Shen	USA
J.C. Simo	USA
I. Smith	UK
E. Stein	West Germany
G. Strang	USA
R.L. Taylor	USA
C.W. Trowbridge	UK
G.N. Vanderplaats	USA
S. Valliappan	Australia
G.B. Warburton	UK
K. Willam	USA
W. Wunderlich	Germany

CONTENTS

D10 Capturing Thermal-Stress Waves via Special Purpose Hybrid Transfinite Elements and Unified Computational Formulations
Kumar K. Tamma and Sudhir B. Railkar, *Mechanical and Aerospace Engineering, West Virginia University, Morgantown, U.S.A.*

D11 The Determination of Optimal Properties of a Layered Pavement Structure
D.F.E. Stolle and A.N. El-Bahrawy, *Dept of Civil Engineering and Engineering Mechanics, Hamilton, Ontario, Canada*

D12 A Finite Element Method for a Numerical Analysis of the Deep-Drawing Process
M. Brunet, *Laboratoire de Mécanique des Solides, Villeurbanne Cedex, France*

D13 Theory of Degenerated Curved Shell and Locking in Shell Finite Elements
Hou-Cheng Huang, *University College of Swansea, U.K.*

D14 Convergence of Hierarchical Finite Elements
Jürgen Bellmann, *Fachgebiet Elektronisches Rechnen im konstruktiven Ingenieurbau, Technische Universität München, West Germany*

D15 Numerical Methods for Three Dimensional Analysis of Buildings
Alvaro Vale e Azevedo, *LNEC - National Laboratory for Civil Engineering, Lisbon, Portugal*

D16 Matrix Mechanics to Classify Non-Linear Continua
J.D. Coleman, *Fluid Mechanics Division, Dept of Civil Engineering, The City University, London*

D17 A Mixed Eulerian-Lagrangian Contact Element to Describe Boundary and Interface Behaviour in Forming Processes
J. Huetink, J. van der Lugt, *University of Twente, The Netherlands*
J.R. Miedema, *Hoogovens IJmuiden B.V., The Netherlands*

D18 Advances in Aeroelasticity
D. Sepahy, *British Aerospace PLC, Air Weapons Division, Hatfield, England*

D19 Cylindrical Concrete Water Tanks: Analysis and Design
V. Thevendran and D.P. Thambiratnam, *Department of Civil Engineering, National University of Singapore, Singapore*

D20 A Complete Procedure for the Adjustment of a Finite Element Model from the Identified Complex Eigenmodes
Q. Zhang, G. Lallement, R. Fillod, J. Piranda, *Laboratory of Applied Mechanics, University of Franche-Comté, Besancon, France*

D21 A General Formula of the Curved Shell Elements and Adaptive Mesh Method in the Nonconservative Finite Deformation Analysis
Y.T. Zhang, H.Y. Yang and J.Y. Zhang, *Dept of Mechanics, Tianjin University, Tianjin, China*

D22 On Exact and Hierarchical Finite Elements for Frame Structures
Ragnar Larsson and Nils-Erik Wiberg, *Dept of Struc. Mech., Chalmers Univ. of Techn., Gothenburg, Sweden*

D23 Statically and Kinematically Admissible Finite Element Formulations for Elastic-Plastic Plate Analysis
J.P. Moitinho de Almeida and J.A. Teixeira de Freitas, *Departamento de Engenharia Civil, Instituto Superior Técnico, Lisboa, Portugal*

D24 Turbulence Modelling and the Effects of Directional Random Waves in Computations of Nearshore Circulation
K. Anastasiou, P. Dong and D.J. Walker, *Department of Civil Engineering, Imperial College of Science and Technology, London, U.K.*

D25 Analysis of Gravity Dam on Soft Foundation
V. Gocevski, *Hatch and Associates, Montreal, Canada*
O.A. Pekau, *Concordia University, Montreal, Canada*

D26 A Finite Element Model for Viscoelastic Foundations Supporting Plate Structures
Joseph J. Rencis, Kwo-Yih Jong and Sunil Saigal, *Center for Computer-Aided Engineering, Mechanical Engineering Department, Worcester Polytechnic Institute, Worcester, Massachusetts, U.S.A.*

D27 Computer Aided Simulation of Tooth Contact Analysis for Helical Gears with Involute Shape Teeth
C.B. Tsay, *Dept of Mechanical Engineering, National Chiao Tung University, Hsinchu, Taiwan, R.O.C.*

D28 Stress and Velocity Fields at Discharging of Silos
J. Eibl and G. Rombach, *University of Karlsruhe, IfMB, Karlsruhe, Germany*

D29 The Convergence Properties of a Series of R-Functions for Simple Polygonal Shapes
D.V. Altiparmakov, *The Boris Kidric Institute of Nuclear Sciences, Belgrade, Yugoslavia*
M.S. Milgram, *Atomic Energy of Canada Ltd, Chalk River Nuclear Laboratories, Ontario, Canada*

D30 Creation of Ship Body Form with Minimum Theoretical Resistance using Finite Element Method
Jan P. Michalski, Antti Pramila, Simo Virtanen, *Tempere Univ. of Technology, Finland*

D31 Transition Plate Bending Elements with Variable Nodes
Chang-Koon Choi and Yong-Myung Park, *Dept of Civil Engng., Korea Advanced Institute of Science and Technology, Seoul, Korea*

D32 A Simple Adaptive Scheme Based on a New Hybrid FE Model
J. Jirousek, *IREM - Swiss Federal Institute of Technology, Lausanne, Switzerland*

D33 Two-Dimensional Solidification Analysis for Twin-Roll Continuous Casting
C.G. Kang, H. Hojo, T. Saitoh and H. Yaguchi, *Dept of Mechanical Engineering II, Tohoku University, Sendai, Japan*

D34 Variational Formulation by Integral Equations for the Sound Radiation in a Non Uniform Flow
M. Ben Tahar and M.A. Hamdi, *Division Acoustique et Vibrations Industrielles, Université de Compiègne, Compiègne Cedex, France*

D35 Design and Study of Behaviour of Brush Bearing Platens by the Finite Element Method
J.M. Torrenti, *L.C.P.C., Paris, France;* and P. Royis, *E.N.T.P.E., Vaulx-en-Velin, France*

D36 Water Pollution Control using Finite Element Model
T. Oikawa and M. Kawahara, *Chuo University, Tokyo, Japan*

D37 Three-Dimensional Finite Element Analyses for a Maxwell Fluid using the Penalty Function Method
Takeo Shiojima, *Idemitsu Petrochemical Co. Ltd., Chiba, Japan*
Yoji Shimazaki, *Tokai University, Kanagawa, Japan*

D38 Optimization of Continuous Prestressed Beam for Different Construction Stages
Dragan Radic, Josip Dvornik, Vinko Candrlic, Duro Dekanovic and Josko Ozbolt, *Gradevinski Institut, Zagreb, Yugoslavia*

D39 Slope Stability Computations with Nonlinear Failure Envelope using Generalized Procedure of Slices and Optimization Techniques
Yudhbir, P.K. Basudhar and S.K. Bhowmik, *Civil Engineering Dept, Indian Institute of Technology, Kanpur, India*

D40 A Computer Code for the Stress Analysis of Cyclically Symmetric Components
Y.V.L.N. Murthy and C.P. Agrawal, *Bharat Heavy Electricals Ltd, Hyderabad, India*

D41 A Numerical Method for the Determination of the Moment-Rotation-Capacity of Thin Walled Members
Dr.-Ing. Ch. Stutzki and Prof. Dr.-Ing. G. Sedlacek, *Lehrstuhl für Stahlbau, RWTH Aachen, West Germany*

D42 Some Applications of Load-Deformation States
Dr. Ing. J. Lopetegui and Prof. Dr. Ing. G. Sedlacek, *Lehrstuhl für Stahlbau, RWTH Aachen, West Germany*

D43 A Second Generation Structural Shape Optimization Capability Employing a Boundary Element Formulation
J.H. Kane, *Worcester Polytechnic Institute, Worcester, Massachusetts, U.S.A.*

D44 An Experimental and Theoretical Study of Soil-Structure Interaction in the Case of a Shallow Foundation Model
S. Labanieh and M. Boulon, *Institut de Mecanique de Grenoble, St. Martin d'Hères Cédex, France*

D45 Dynamic Analysis of General Thin Shells
S. Naomis and P.C.M. Lau, *Civil Engineering Dept, University of Western Australia, Nedlands, Australia*

D46 A Simplified Analysis of Two Plate Bending Elements - The MITC4 and MITC9 Elements
Klaus-Jürgen Bathe, *Massachusetts Institute of Technology, Cambridge, Massachusetts* and Franco Brezzi, *Universita di Pavia, I.A.N. del C.N.R., Pavia, Italy*

D47 High-Speed Loading Analysis of Reinforced Concrete Columns
Shigekatsu Ichihashi, *Kozo Keikaku Eng. Inc., Nippon Holstein Kaikan, Honcho, Nakano-ku, Tokyo, Japan* and Akira Wada, *Dept. of Arch. and Build., Tokyo Institute of Tech., Ookayama, Meguro-ku, Tokyo, Japan*

D48 Analysis of Footing Behaviour on Homogeneous and Layered Soils
M. Al-Mukhtar, J.C. Robinet and I. Shahrour, *Université des Sciences et Techniques de Lille - EUDIL, Villeneuve d'Ascq Cedex, France*

D49 Fracture Mechanics for Delamination on Composite Structures in Compression
H. Chaouk and G.P. Steven, *The University of Sydney, N.S.W., Australia*

D50 A New Multivariable Finite Element Algorithm and a Breakable Element Algorithm for Elasto-Plastic Fracture Analysis
J.Y. Zhang, *Dept of Mechanics, Tianjin University, Tianjin, China*
T.R. Hsu, *Dept of Mech. Engng, University of Manitoba, Winnipeg, Manitoba, Canada*

D51 Preliminary Seismic Analysis and Design of Liquid Storage Tanks
R.C. Barros, *Departmento de Engenharia Civil, Universidade do Porto, Porto, Portugal*

D52 Optimal Design of Semiconductor Components
J.B. Waddell and J. Middleton, *Dept of Civil Engineering, University College of Swansea, U.K.*

D53 Reflections on Finite Element Plate Analysis
E. Hinton, *University College of Swansea, U.K.*

D54 Finite Element Model for Layered Plates/Shells
M.G. Rajendran, S. Rajasekaran, K. Jawahar Reddy and S. Valliammai, *PSG College of Technology, Coimbatore, Tamil Nadu, India*

D55 An Approach to Correct Elasto-viscoplastic Stress Predictions
T. Rodic, D.R.J. Owen, F. Damjanic, *Dept of Civil Engineering, University of Wales, Swansea, U.K.*

D56 Finite Element Analysis of a Viscoelastic Solid Sliding over Rigid Triangular Asperities
N. Purushothaman and I.D. Moore, *University of Newcastle, N.S.W., Australia*

D57 Non-Linear Analysis of Arch Dams
C. Pina and R. Camara, *Laboratorio Nacional de Engenharia Civil, Lisbon, Portugal*

D58 2-D Moving Grid FEM for Diffusion Problems with Chemical Reactions
P. Gerrekens and M. Hogge, *Aerospace Laboratory of the University of Liège, Belgium*

SECTION S - DEVELOPMENTS IN ENGINEERING SOFTWARE

S1 Are High Degree Elements Preferable? Some Aspects of the h and h-p Version of the Finite Element Method
Ivo Babuska, *Institute for Physical Sciences and Technology, University of Maryland, U.S.A.*

S2 Handicraft in Finite Elements
J. Blaauwendraad and A.W.M. Kok, *Delft University of Tech., Delft, The Netherlands*

S3 Adaptive Techniques in Finite Element Analysis
J.Z. Zhu and O.C. Zienkiewicz, *Dept of Civil Engineering, University College of Swansea, U.K.*
A.W. Craig, *Dept of Mathematical Science, University of Durham, U.K.*

S4 Aspects of Methodology for FE-Program Development
Harald Tägnfors, *Dept of Structural Mechanics, Chalmers Univ. of Technology, Göteborg, Sweden*

S5 The Significance and Practice of Rank Estimation in Structural Dynamics Identification Algorithms
John Brandon, *Dept of Mechanical and Manufacturing Systems Engineering, University of Wales Institute of Science and Tech., Cardiff*

S6 The Use of Tension Parameter in Surface Modelling
Da-Pan Chen and Tser-Liang Lin, *Dept of Mechanical Engineering National Chiao Tung University, Hsinchu, Taiwan, Republic of China*

S7 Direct Design Versus Range Selection Algorithms used in Mechanical Component Software
Dr. J. Vogwell, *University of Bath, U.K.*

S8 Analog/Hybrid and Digital Simulations in Civil Engineering
Hamdy Youssef, *Concordia University, Dept of Civil Engineering, Montreal, Quebec, Canada*

S9 Curve Design using Hierarchical Finite Element Forms
S. Virtanen, *Tampere University of Technology, Tampere, Finland*

S10 Integration of FEM, Optimization, and CAD on Microcomputers
Gu Yuanxian and Cheng Gengdong, *Research Institute of Engineering Mechanics, Dalian Institute of Technology, Dalian, China*

S11 An Adaptive HP-Version in the Finite Element Method
Ernst Rank, *A.G Siemens, Corporate Research and Technology, München, West Germany*

S12 A Finite Element Transitional Mesh Generation Technique
L. Carter Wellford, Jr. and M.R. Gorman, *University of Southern California, Los Angeles, California, U.S.A.*

S13 Expert System for Material Selection
James K. Blundell, *University of Missouri-Columbia/Kansas City*
R. Bryan Greenway, *Design Productivity Center, Kansas City, U.S.A.*

S14 A Dynamically Partitioned Out-of-Core Skyline Solver for Micro-Computers
A.J. du Toit and W.S. Doyle, *Dept of Civil Engineering, University of Cape Town, S. Africa*

S15 A Generalised A-Posteriori Analysis of the Discretization Error in Numerical Solutions to Linear Problems
D.W. Kelly, *University of New South Wales, Kensington, N.S.W., Australia*

S16 Analysis of Cantilever Sheet Piling in Stratified Cohesive Soils
Jay S. DeNatale and German A. Ibarra-Encinas, *The University of Arizona, Tucson, Arizona, U.S.A.*

S17 A Geometrical Continuous Remeshing Procedure for Application to Finite Element Calculation of Non-Steady State Forming Processes
J.P. Cescutti and J.L. Chenot, *Ecole des Mines de Paris, Centre de Mise en Forme, Valbonne Cedex, France*

S18 Improved Systolic Designs for the Iterative Solution of Linear Systems
D.J. Evans and K. Margaritis, *Dept of Computer Studies, University of Technology, Loughborough, Leicestershire, U.K.*

S19 Superconvergence and Finite Element Post Processing
G.F. Carey and R.J. MacKinnon, *The University of Texas at Austin, Austin, Texas, U.S.A.*

S20 Automatic Generation of Shape Function Routines
Peter Bettess, *School of Marine Technology, University of Newcastle-upon-Tyne, U.K.* and Jacqueline A. Bettess, *Computer Centre, University of Durham, U.K.*

S21 The Generation of Hybrid Structured-Unstructured Grids
N.P. Weatherill, *Institute for Numerical Methods in Engineering, University College of Swansea, U.K.*

SECTION D : NUMERICAL TECHNIQUES FOR ENGINEERING ANALYSIS AND DESIGN

COUPLING OF FLUID FILM LUBRICATION AND PLASTIC DEFORMATION: FINITE ELEMENT APPROACH OF PLASTOHYDRODYNAMICS IN COLD FORGING

P. MONTMITONNET, J.L. CHENOT

ECOLE DES MINES DE PARIS. CENTRE DE MISE EN FORME
SOPHIA ANTIPOLIS 06565 VALBONNE CEDEX . FRANCE

ABSTRACT

A full fluid film lubrication mechanism is analyzed and implemented in a home-made elastic-plastic finite element code dedicated to metal forming. A realistic lubricant rheology accounting for high pressure and shear-rate has been assumed. The evolution of film thickness -and hence friction stress- may be followed all along a non-stationary process. Cold upsetting between flat, overhanging dies has been chosen as a first, simple example. Several finite difference schemes have been tested to solve the lubricant conservation equation, and their stability and precision is discussed. Numerical results are compared quantitatively with analytical solution (for the case of homogeneous deformation) or qualitatively with experimental results.

1. INTRODUCTION

Friction is known to be a major item of modelling of metal forming operations. In fact, it is by now the most serious limitation of the precision of finite element codes. Many sophisticated mathematical formulations of contact exist, but they depend on friction coefficients which are generally determined empirically, and in many cases simply used as fitting parameters. Improving this situation implies an analysis of the basic phenomena involved in dry or lubricated friction, to build simple models to be implemented at low cost in finite element codes.

Forging is a forming process most sensitive to friction, which often governs the final shape of a piece (as in forward-backward extrusion for instance). Hence a precise treatment of friction is needed. Experimental measurements of normal and tangential stress profiles (PEARSALL and BACKOFEN [1]) show that classical friction laws with constant friction coefficients cannot explain observed results: a local analysis of the friction stress is necessary.

The experimental results of [1] can be explained qualitatively on account of findings of OSAKADA and OYANE [2]: thick film lubrication prevails on a great part of the interface. Namely, for axisymmetric upsetting, a thick lubricant film (up to 100µm) exists everywhere except on an outer annular zone, as seen by different surface aspects: this results in the high friction zone of [1]. The inner region is protected by a thick oil film which leads to low friction.

This is a quite favorable case where fluid mechanics allows friction stress to be calculated locally, provided a realistic rheological law is chosen .

2. ANALYSIS OF THE FLUID FILM BEHAVIOUR

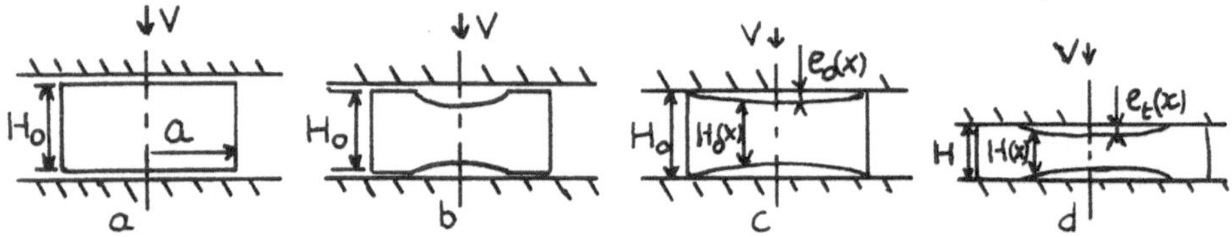

Figure 1: the successive phases of lubricant film formation and evolution

The case of upsetting of a rectangular block between flat overhanging dies has been chosen to test the coupling (figure 1). Plane strain deformation is assumed.

2.1 Isothermal case

2.1.1 lubricant trapping (figure 1a,b,c)

As lubricant is squeezed out from the interface at the beginning of die movement, it builds up a pressure profile p(x) with maximum on the symmetry axis. When this pressure peak reaches the metal yield stress, yielding begins, a depression full of lubricant is formed at the center of the workpiece and then propagates toward the edge. Integrating REYNOLDS equation between axis (x=0) and edge (x=a) gives the film thickness profile $e_0(x)$ at the end of lubricant trapping (depression closure when plastic deformation reaches the edge) [3,4]:

$$e_0(x) = \left(\frac{3\eta_0\gamma V}{1-\exp(-\gamma\sigma'_0)}\left[a - x\right]^2\right)^{1/3} \quad (1)$$

where σ'_0 is the yield stress in plane strain, V the die velocity, $\eta=\eta_0\exp(\gamma p)$ the film viscosity.

2.1.2 thickness evolution during forging [4] (figure 1d)

In the initial process described above, the lubricant imposed its pressure on the metal. After depression closure, the deforming metal will impose its pressure $p \simeq -\sigma_{yy}$ on the lubricant, which develops a velocity field u_1:

$$u_1(x,y) = \frac{1}{2\eta}\frac{dp}{dx}\,y(y-e(x)) + U_m(x)\left(1-\frac{y}{e(x)}\right) \simeq U_m(x)\left(1-\frac{y}{e(x)}\right) \quad (2)$$

The film imposes the friction stress τ on the workpiece:

$$\tau(y=0) = -(e/2)dp/dx - \eta U_m/e \quad (3)$$

U_m is the local outward velocity of the workpiece surface.

The equation of conservation of lubricant gives:

$$\partial e/\partial t = -1/2 \; d(U_m(x).e(x))/dx \quad (4)$$

Integration of (4) by the method of characteristics [4] shows that lubricant is transported outwards at an average speed of $U_m/2$; as a consequence, an unlubricated zone develops at the periphery (figure 1d), as experiments show [1,2].

2.2 Thermal and other effects

WILSON and CARPENTER [5] and WILSON and WONG [6] have shown that thermal effects are essential in high speed forging and increase the width of the unlubricated zone. On the contrary, high workpiece roughness decreases it [7]; inertial effects within the metal result in non-symmetry of top and bottom films [8].

3. LUBRICANT RHEOLOGY UNDER HIGH PRESSURE AND SHEAR RATE

Film thickness evolution is but little sensitive to rheology; however, rheology is the basis of the friction stress imposed by the lubricant as a boundary condition on the metal, which is the topic of the present paper. Hence a realistic rheological behaviour must be used in the analysis.

Many experimental data have been collected in connection with Elasto-Hydrodynamic (EHD) lubrication. Most studies used an EHD traction machine (two disks rotating at different speeds, shearing a thin oil film). A shear stress -shear rate curve results (figure 2); three parts may be distinguished:

- part 1 reflects the viscous behaviour of the lubricant; alternatively, viscoelasticity may occur [9].
- in part 3, τ decreases due to lubricant shear heating.

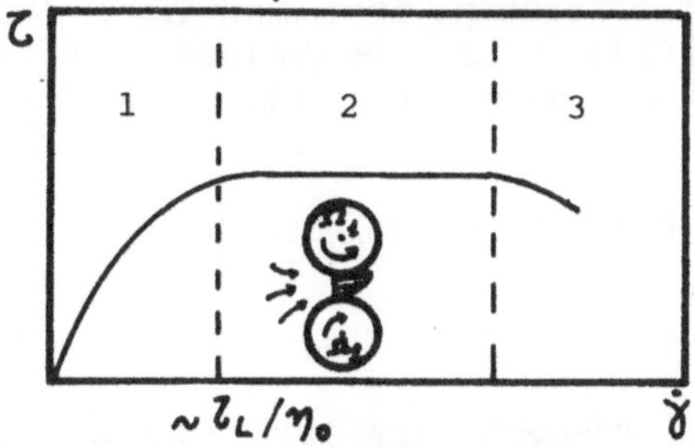

Figure 2 : flow curve under pressure from an EHD traction machine (schematic)

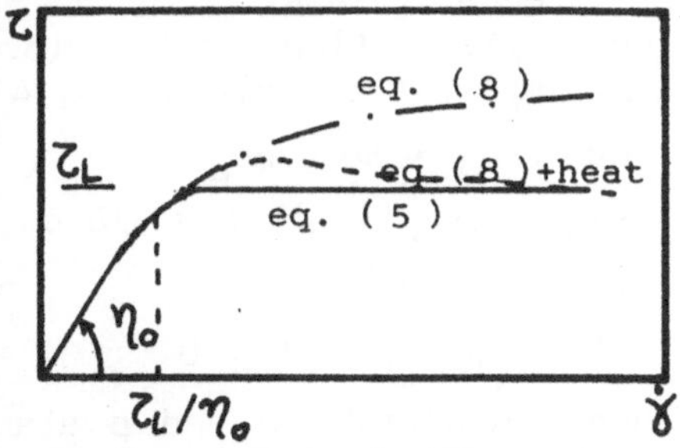

Figure 3 : intrinsic flow curve under high pressure (schematic)

- part 2 is more controversial. Its explanation generally derives from the concept of "limiting shear stress" τ_L (figure 3)[10]: according to this interpretation, a fluid under high pressure cannot sustain a shear stress greater than this limit τ_L. At shear rates higher than τ_L, the fluid behaves as a plastic solid. This behaviour change is linked with a glass transition in the fluid [11]. A model has been proposed [9]:

$$\dot{\gamma} = -(\tau_L/\eta_0) \ln(1-\tau/\tau_L) \quad \text{or} \quad \tau = \tau_L(1-\exp(-\dot{\gamma}\eta_0/\tau_L)) \qquad (5)$$

η_0 (the low shear rate viscosity) has the well known form:

$$\eta_0 = \eta_1 \exp(\gamma p - \delta\theta) \qquad (6)$$

where θ is a relative temperature. τ_L increases linearly with pressure:

$$\tau_L = \tau_{Lo} + \alpha p \qquad (7)$$

and decreases with temperature.

On the contrary, JOHNSON and GREENWOOD [12] contest the notion of limiting shear stress. By a careful thermal analysis, they show that part 2 of the traction curve may be explained by an EYRING-type viscosity (see figure 3):

$$\dot{\gamma} = (\tau_0/\eta_0) \sinh(\tau/\tau_0) \quad \text{or} \quad \tau = \tau_0 \sinh^{-1}(\eta_0\dot{\gamma}/\tau_0) \qquad (8)$$

coupled with thermal effects.

4. FORMULATION OF THE ELASTIC-PLASTIC METAL FLOW PROBLEM

4.1 Mechanical formulation and time discretization

The problem must be solved incrementally. Assuming equilibrium at time t, the Principle of Virtual Work is written at time t+Δt:

$$\forall \mathbf{V}^*, \int_{\Omega+\Delta\Omega} (\boldsymbol{\sigma}+\Delta\boldsymbol{\sigma}):\boldsymbol{\varepsilon}^* \, d\Omega = \int_{\partial\Omega(t+\Delta t)} (\mathbf{T}+\Delta\mathbf{T}).\mathbf{V}^* \, d\partial\Omega \qquad (9)$$

The unknown is the displacement field $\mathbf{\Delta X}$; hence, $\Delta\boldsymbol{\sigma}$ must be linked to $\mathbf{\Delta X}$, or to the strain increment $\Delta\boldsymbol{\varepsilon}$. The incremental flow rule $\Delta\boldsymbol{\sigma}(\Delta\boldsymbol{\varepsilon}[\mathbf{\Delta X}])$ is determined by a semi-implicit algorithm [13,14] as the solution of:

$$\Delta\boldsymbol{\varepsilon} = D_{el}^{-1}.\Delta\boldsymbol{\sigma} + \Delta\boldsymbol{\varepsilon}_p$$

$$\Delta\boldsymbol{\varepsilon}_p = \Delta\lambda_p[(1-\eta)\partial f(\boldsymbol{\sigma})/\partial\boldsymbol{\sigma} + \eta\, \partial f(\boldsymbol{\sigma}+\Delta\boldsymbol{\sigma})/\partial\boldsymbol{\sigma}] \qquad (10)$$

$$f(\boldsymbol{\sigma}, \varepsilon_p) = 0$$

$$f(\boldsymbol{\sigma}+\Delta\boldsymbol{\sigma}, \varepsilon_p+\Delta\varepsilon_p) = 0 \qquad (11)$$

where subscripts e and p stand respectively for elastic and plastic. D_{el} is the elasticity matrix, f the yield criterion, ε_p the equivalent plastic strain:

$$\varepsilon_p = \int_0^t \sqrt{(2/3)\, \boldsymbol{\varepsilon}_p:\boldsymbol{\varepsilon}_p} \, dt \qquad (12)$$

λ_p is the plastic multiplier; if $f(\boldsymbol{\sigma}, \varepsilon_p)<0$, $\lambda_p=0$

This system is solved at each integration point by a NEWTON RAPHSON method.

4.2 Spatial discretization

This continuum problem is discretized by the Finite Element Method. Hence, the equilibrium equation (9) is rewritten in terms of nodal displacements; using the incremental flow rule solution of (10) and (11), equation (9) becomes:

$$\mathbf{R}(\Delta X_i) = 0 \tag{13}$$

and is solved by a NEWTON -RAPHSON method. After each load increment, the nodal positions are actualized:

$$X_i^{n+1} = X_i^n + \Delta X_i^n \tag{14}$$

A computer code based on these principles has been built [13] for two dimensional problems (plane strain or axisymmetric). In the present study, four node linear isoparametric quadrilateral elements have been used.

4.3 Contact and friction

Contact is actualized at the beginning of each load increment. First equation (14) is used. Then the X_i^{n+1} are compared with the new position of the dies. If a point has penetrated a die, or approached it closer than a distance ε_c, it is projected onto the die surface. On the contrary, a point in contact at t may leave the die surface at t+Δt if its nodal force becomes tensile.

As for friction, a friction layer model is used hereafter:

$$\tau = -(\overline{m}\,\sigma_0(x)/\sqrt{3}).\ \Delta\mathbf{V} / |\Delta\mathbf{V}| \tag{15}$$

τ is calculated on each element, which allows variable friction along the dies.

5. COUPLING OF FLUID FILM AND PLASTIC DEFORMATION

5.1 Computation of shear stress

In this paper, the problem of lubricant trapping is not dealt with. We must hence choose an initial thickness profile $e_0(x)$. At each increment, σ is calculated within the metal; the pressure within the fluid is then deduced as $p \approx -\sigma_{yy}$. Then the resulting friction stress is given by (3), along with rheology (6); a limiting shear stress has been assumed according to (7). In this first approach, thermal and inertial effects are not accounnted for.

5.2 Solution of the fluid conservation equation

Then the film thickness is computed using (4). This is in fact the main problem in the formulation. It may first be solved analytically by the method of characteristics, yielding:

$$X = x_0 + \int_0^t U_m/2\ dt \tag{16}$$

$$e_t(x) = e_0(x_0).U_{mt}(x_0)/U_{mt}(X) \tag{17}$$

X (moving at velocity U/2 from its initial position x_0) represents the average movement of the fluid. Subscript o refers to initial values, t to values at time t. For homogeneous plane strain deformation, (17) becomes:

$$e_t(x_0\sqrt{H_0/H_t}) = e_0(x_0).\sqrt{H_t/H_0} \qquad (18)$$

This solution is not adapted to finite elements, since it implies following a virtual mesh of velocity $(U_m/2, V_m)$. Moreover, it will hardly be applicable to more complex geometries. Another solution consists in solving directly (4) at points fixed with respect to the dies. However, the explicit time integration of the Finite Element Method (14) makes it necessary to use an explicit formulation for (4), which is then highly unstable for the initial profile given by (1), mainly because of the infinite derivative at x = a.

A more satisfying solution consists in following the nodes of the FEM mesh:

$$\partial e/\partial t = 1/2 \;.\; [U_m \partial e/\partial x - e\partial U_m/\partial x] \qquad (19)$$

which is then solved by an explicit scheme.

Figure 4 compares the analytical solution (18) with the numerical film thickness for two initial profiles. The integration scheme is stable, but a rather fine mesh may be needed to obtain a good precision with initial profile (1).

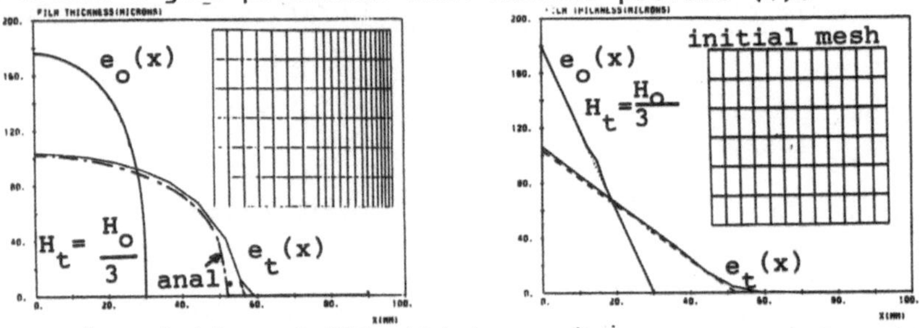

Figure 4:evolution of film thickness (homogeneous deformation)

For simpler profiles (figure 4b), stability and precision are much better, and coarser grids are satisfactory. Note that stability then resists a small irregularity on $e_0(x)$.

The key point is the computation of de/dx at each node; the average of the slopes of the two adjacent elements has to be used to obtain stability.

In figure 4, $\bar{m} = 0$ to enforce homogeneous deformation; figures 5 and 6 show a more realistic situation, where friction is higher in the unlubricated region ($\bar{m} = 0.3$). Non homogeneity ($U_m < Vx/H$) is restricted to the unlubricated region, so that the thickness profile is unchanged.

As for stresses, the normal stress σ_{yy} shows a depression near the edge, then an increase toward the axis; this is qualitatively similar to experimental results of [1] for lubricated upsetting (figure 6). The edge value is close to that predicted for high friction by HILL [15]: $\sigma = \sigma_0/\sqrt{3}\,(1+\pi/2)$.

The friction stress (figure 6) has a low value in the lubricated zone, and increases sharply in the unlubricated region, which is common sense. The aspect of the τ-profile is rather similar to the experimental results of [1].

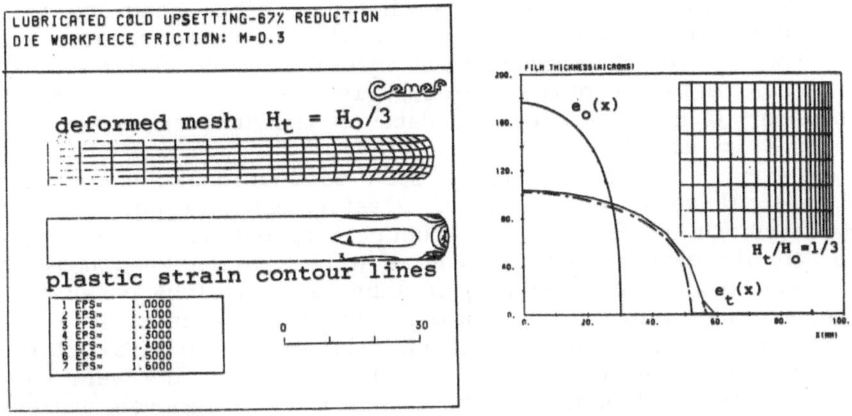

Figure 5 : effect of friction in the unlubricated zone; steel σ'_0=462 MPa; YOUNG's modulus E = 210 000 MPa; POISSON's coefficient ν=0.3

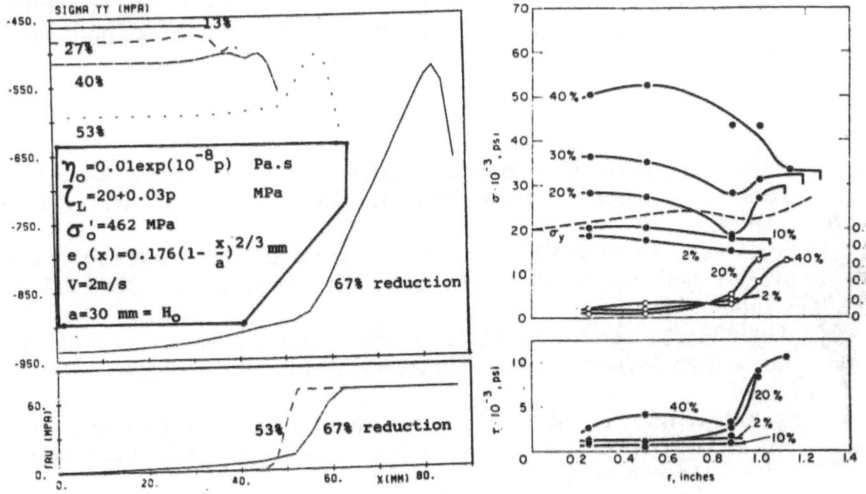

Figure 6 : qualitative comparison of calculated (a) and measured (b,[1]) stresses

6. CONCLUSION

This preliminary study proves that it is possible to implement at no cost in a FEM code a local friction stress computation based on physical arguments. Thick film lubrication of upsetting has been chosen. Most difficulties arise from solution of fluid conservation equation, because FEM imposes explicit integration. Stability can be obtained, but a good precision on lubricating action can be obtained only with grids somewhat finer than strictly necessary for plastic deformation. From this point of view, an implicit time integration in the FEM would be beneficial. As a whole, computed results are nevertheless in good qualitative agreement with available experimental data; quantitative comparison of stresses is impossible for lack of rheological data on the lubricants used in the experiments.

However, important effects such as lubricant heating due to internal shear or workpiece plastic deformation, must be accounted for if forging at industrial speeds is to be analyzed. In all cases, further studies will have to keep up with progress in the understanding of lubricant rheology under high pressure, which is far from being perfect at present.

It must be noticed that thick film lubrication is not the only case where a local friction estimate may be derived. WANHEIM and his collaborators have shown how roughness evolves during compressive plastic deformation [16]; they recently applied their results to rolling [17], where thick film lubrication may not occur.

Finally, this kind of analysis must be extended to more complex processes, such as forward [18,19] or combined forward-backward extrusion [20].

7. REFERENCES

[1] PEARSALL,G.W. and BACKOFEN,W.A. : Frictional boundary conditions in plastic compression. ASME J. Eng. Ind. 85,1 (1963) 68-76

[2] OSAKADA,K. and OYANE,M. : The effect of deformation speed on friction and lubrication in cold forging. Bull. JSME 13,66 (1970) 1504-1512

[3] OSAKADA,K. and OYANE,M. : The mechanism of lubricant trapping under dynamic compression. Bull. JSME 12,49 (1969) 149-155

[4] WILSON,W.R.D. : an isoviscous model for the hydrodynamic lubrication of plane strain forging processes with flat dies. ASME J. Lub. Tech. 95,4 (1974) 539-546

[5] WILSON,W.R.D. and CARPENTER,W.B. : a thermal hydrodynamic model for the lubrication breakdown in upsetting between overhanging dies. Wear 24 (1974) 351-360

[6] WILSON,W.R.D. and WONG,C.J. : analysis of the lubricant film formation process in plane strain forging. ASME J. Lub. Tech. 95,4 (1974) 605-610

[7] WILSON,W.R.D. and DELMOLINO,W.P. : the influence of surface roughness on lubrication breakdown in upsetting between overhanging dies. Wear 29 (1974) 1-10

[8] THOMPSON,P.J. and SYMMONS,G.R. : a plasto-hydrodynamic analysis of high speed disk forging. Proc. 17th Mach. Tool Des. Res. Conf. (1970) 587-595

[9] BAIR,S. and WINER,W.O. : A rheological model for EHD contacts based on primary laboratory data. ASME J. Lub. Tech. 101 (1979) 258-265

[10] BAIR,S. and WINER,W.O. : Shear strength measurements of lubricants at high pressure. ASME J. Lub. Tech. 101 (1979) 250-257

[11] ALSAAD,M. , BAIR,S. and WINER,W.O. : glass transition in lubricants: its relation to EHD lubrication. ASME J. Lub. Tech. 100 (1978) 404-417

[12] JOHNSON,K.L. and GREENWOOD,J.A. : thermal analysis of an EYRING fluid in EHD traction. Wear 61 (1980) 353-374

[13] BRAUDEL,H.J. : Modélisation des grandes transformations élastoplastiques d'un solide isotrope par la méthode des éléments finis- Application à la forge à froid. Thesis, Université LYON I (1986)

[14] BRAUDEL,H.J. ,ABOUAF,M. and CHENOT,J.L. : an implicit and incremental formulation for the solution of elastoplastic problems by the FEM. Comp. Struct. 22,5 (1986) 801-814

[15] HILL,R. : The mathematical theory of plasticity. (1950) Clarendon Press

[16] NELLEMAN,T. ,BAY,N. and WANHEIM,T. : real area of contact and friction stress: the role of trapped lubricant. Wear 43 (1977) 45-53

[17] WANHEIM,T. and BAY,N. : Friction and Tools, in "Design of Tools for Deformation Processes", BLAZYNSKI,T.Z. Ed. (1986) ELSEVIER

[18] WILSON,W.R.D. : the temporary breakdown of hydrodynamic lubrication during the initiation of extrusion. Int. J. Mech. Sci. 13 (1971) 17-28

[19] MAHDAVIAN,S.M. : a refined model for hydrodynamic lubrication of cold extrusion and its comparison with experiments. ASME J. Lub. Tech. 104 (1982) 46-52

[20] BLANCON,R. and FELDER,E. : Theoretical estimation of lubricant film thickness in cold forging. Proc. 5th LEEDS-LYON Symp. Tribology (LEEDS,1978)

RIGID AND FLEXIBLE MECHANISMS
A FINITE ELEMENT APPROACH BASED ON THE CONFORMAL ROTATION VECTOR

M. GERADIN * A. CARDONA **
L.T.A.S.
Dynamique des Constructions Mécaniques
Université de Liège
Rue Ernest Solvay, 21
B−4000 Liège, BELGIUM

abstract

A technique for the representation of the rotation operator in terms of only three free parameters based on the conformal rotation vector concept is described. It allows to develop efficient finite elements of mechanism components. The application to the deployment of a lightweight space structure cell demonstrates its high potential for analyzing 3-D flexible mechanisms with arbitrary topology.

1. Introduction

The finite element method has emerged in the last years as a powerful tool to treat mechanism problems. Previous approaches, i.e. the Lagrangian coordinates method [1–2] and the Cartesian coordinates method [3–5], were not ideally suited to the development of large purpose codes. In fact, the finite element method may be regarded as a powerful variant of the Cartesian coordinates approach. It is natural to think of a mechanism as an assembly of either rigid or flexible member finite elements interconnected by coupling elements such as hinges, prismatic joints, Hooke joints, etc. Various approaches have been presented in the literature to accomplish this task [6–11].

A fundamental aspect in the formalism is the representation of large finite rotations. Most approaches employ Euler parameters. They have shown to be well adapted in view of their properties of avoiding the singularities that are contained in most finite rotation representations, and leading to an algebraic description of the rotation operator and angular velocities [9–12].

However, Euler parameters have the drawback of giving a redundancy of description: in order to accurately represent any magnitude of rotation, four parameters must be carried on in the computations while the rotation operator is shown to be a function of only three free parameters. A nonlinear constraint of normality which links them should be added to the system, usually by a Lagrange multiplier technique. In this way, five degrees of freedom are employed to model rotations at each node. The final system

* Currently Visiting Professor, University of Colorado, Boulder, USA

** Becario Externo y Miembro de la Carrera del Investigador del Consejo Nacional de Investigaciones Científicas y Técnicas de la República Argentina.

to be solved includes then a large number of Lagrange multipliers which increase the computer cost and are the cause of numerical difficulties at the resolution phase.

An alternative technique for the representation of the rotation operator in terms of only three free parameters will be described. It retains all advantages of Euler parameters while using a minimal set of three free parameters to fully describe rotations. The technique is based on a conformal transformation on Euler parameters and has been referred to as the *conformal rotation vector* in the literature [13].

Finite elements of mechanism components based on this approach will be briefly presented. The application to the deployement of a lightweight space structure cell demonstrates the adequacy and high potential of the proposed technique for analyzing 3-D flexible mechanisms with arbitrary topology.

2. Conformal vector representation of finite rotations

2.1 definition

An adequate parametric representation should satisfy the following requirements:

i) Parameter singularities for particular values of rotations should not occur.
ii) Algebraic description of finite rotations is preferred, avoiding expensive calculations of trigonometric functions and lengthy analytical expressions.
iii) The set of parameters should be minimal, avoiding redundancy in description.

A particular choice that fulfills all these requirements is the conformal rotation vector (CRV) which can be introduced as follows:

Euler–Chasles theorem states that any finite rotation can be expressed as a unique rotation ϕ about a direction defined by a unit vector $\mathbf{u}$. The CRV are then defined by

$$a_i = 4\, u_i \tan\frac{\phi}{4} \qquad (i = 1,2,3) \tag{1}$$

A fourth parameter a_0 is defined in terms of the other three components

$$a_0 = \frac{1}{8}\left(16 - \mathbf{a}^{\mathrm{T}}\mathbf{a}\right) \tag{2}$$

These parameters do not present any point of singularity for all $\phi \in [-\pi, +\pi]$ since their values are bounded by

$$4 < a_i \leq 4 \qquad 0 < a_0 \leq 2 \tag{3}$$

In the sense of quaternion algebra, the CRV constitutes a quaternion [14]

$$\hat{a} = a_0 + \mathbf{a} \tag{4}$$

with non unit norm since

$$|\,\hat{a}\,|^2 = e_0^2 + \mathbf{e}^T\mathbf{e} = (4 - a_0)^2 \tag{5}$$

and it is related to the unit quaternion of Euler parameters by the conformal mapping

$$e_i = \frac{a_i}{(4 - a_0)} \qquad a_i = \frac{4e_i}{(1 + e_0)} \qquad (\mathrm{i} = 0,\ 1,\ 2,\ 3) \tag{6}$$

In terms of CRV's, the finite rotation operator takes the following form

$$R_{ij} = \frac{1}{(4 - a_0)^2}\left[(a_0^2 - a_k a_k)\delta_{ij} + 2(a_i a_j - \epsilon_{ijk} a_0 a_k)\right] \tag{7}$$

2.2 successive rotations

The law for the composition of successive rotations using CRV is derived from the quaternion multiplication rule [14]. Let the position vector $\mathbf{x}$ undergo two successive rotations $\mathbf{R}_f$ and $\mathbf{R}_g$:

$$\mathbf{z} = \mathbf{R}_g\mathbf{R}_f\mathbf{x} = \mathbf{R}\mathbf{x} \tag{8}$$

and let $\mathbf{f}$ and $\mathbf{g}$ be the conformal rotation vectors corresponding to the rotation operators $\mathbf{R}_g$ and $\mathbf{R}_f$. Then, the conformal rotation vector $\mathbf{a}$ corresponding to the total rotation matrix **R** can be expressed as

$$a_i = \frac{4\,(g_0 f_i + f_0 g_i + \epsilon_{ijk} g_j f_k)}{(4 - f_0)(4 - g_0) + f_0 g_0 - f_j g_j} \tag{9}$$

2.3 rotations of arbitrary magnitude

The definition of the conformal rotation vector was restricted to rotations in the rank $\phi = [-\pi, \pi]$. This restriction is due to the singularity of the tangent trigonometric function at $(\pi/2 + k\pi)$.

To account for rotations of arbitrary magnitude occuring during mechanism motion, the definition of CRV's can be generalized by taking account of the number k of complete revolutions about the direction $\mathbf{u}$ at a given point

$$a_i = \begin{cases} 4u_i \tan\,[(\phi + 2k\pi)/4], & \text{k even} \\ 4u_i \tan[(\phi + 2(k-1)\pi)/4], & \text{k odd} \end{cases} \tag{10}$$

$$\phi \in [-\pi, +\pi]$$

If during the time integration of motion in a given mechanism, at a certain point rotations attain a magnitude out of bounds, i.e. the angle ϕ reaches a value such that

$$\phi = \pi + \epsilon \qquad \epsilon > 0, \tag{11}$$

this condition is easily detected by the violation of the inequality

$$\mathbf{a}^T\mathbf{a} \leq 16 \tag{12}$$

and one has to update the angle and the number of turns in order to obtain values within allowed range

$$\begin{cases} \phi \rightarrow \phi' = \phi - 2\pi \\ k \rightarrow k' = k + 1 \end{cases} \tag{13}$$

Corrected values for the CRV's components are then generated as follows

$$\begin{aligned} a'_i &= 4\,u_i \tan\,[(\phi' + 2(k'-1)\pi)/4] \\ &= -\,16\,a_i/(\mathbf{a}^T\mathbf{a}) \end{aligned} \tag{14}$$

The equation for the composition of successive rotations (9) still holds. After having obtained the conformal rotation vector of the total rotation, a change of state should be performed if equation (12) is violated. Then, the conformal rotation vector of the composed rotation is finally computed by

$$a_i = \alpha\, h_i \tag{15}$$

where

$$h_i = g_0 f_i + f_0 g_i + \epsilon_{ijk} g_j f_k$$

$$\alpha = \begin{cases} -16h_0/(\mathbf{h}^T\mathbf{h}) & \text{if} \quad \mathbf{h}^T\mathbf{h} > 16h_0^2 \\ 1/h_0 & \text{otherwise} \end{cases}$$

$$h_0 = [(4 - f_0)(4 - g_0) + f_0 g_0 - f_j g_j]/4$$

3. Modelling of flexible beams

3.1 Strain energy

For the purpose of mechanism analysis, we may restrict ourselves to the case of small displacements and rotations in the dynamic frame. In such case, the strain energy is computed linearly from strain measures in the dynamic configuration. This fact allows us to develop finite elements whose matrices are simple generalizations of the usual linear analysis matrices.

Let PQ be a beam element with nodes P and Q. We will fix arbitrarily the origin of coordinates of the dynamic system at node P. We define a local coordinate system

$(P\xi_1\xi_2\xi_3)$ oriented according to the principal directions of inertia of the element PQ, and describe the beam deformation in terms of longitudinal and transverse displacements (w_1, w_2, w_3) and axial torsion α_1.

If we assume that the beam slenderness is such that the classical beam theory holds, its strain energy may be expressed as

$$\mathcal{W} = \frac{1}{2}\int_0^\ell [EA\left(\frac{\partial w_1}{\partial \xi_1}\right)^2 + EI_y\left(\frac{\partial^2 w_2}{\partial \xi_1^2}\right)^2 + EI_z\left(\frac{\partial^2 w_3}{\partial \xi_1^2}\right)^2 + GJ\left(\frac{\partial \alpha_1}{\partial \xi_1}\right)^2]\, d\xi_1 \quad (16)$$

where EI_y and EI_z are the bending stiffnesses in directions ξ_2 and ξ_3, and EA and GJ are the extensional and torsional stiffnesses.

By choosing cubic interpolation functions for the transversal displacements in the local system w_2 and w_3 and using linear interpolation functions for the axial displacement w_1 and torsion α_1, we obtain the following expression for the strain energy

$$\mathcal{W} = \frac{1}{2}\,(\mathbf{d}^T\,\mathbf{K}_e\,\mathbf{d}) \quad (17)$$

$\mathbf{K}_e$ is the elemental stiffness matrix computed in the dynamic configuration. This matrix coincides with the usual linear stiffness matrix computed at the initial configuration and remains invariant.

$\mathbf{d}$ is the vector of nodal displacements in the dynamic frame. Since we have chosen the origin of the dynamic frame at node P, it has only six non-zero components

$$\mathbf{d}^T = [\mathbf{u}_Q'^{\,T}\ \boldsymbol{\phi}_Q'^{\,T}] \quad (18)$$

and $\mathbf{K}_e$ becomes thus the corresponding (6×6) submatrix of the linear stiffness matrix.

The displacements and rotations in the dynamic frame can be computed as nonlinear functions of the positions $\mathbf{r}$ and CRV's $\mathbf{a}$ at nodes P and Q

$$\mathbf{d} = \mathbf{s}(\mathbf{r}_P, \mathbf{a}_P, \mathbf{r}_Q, \mathbf{a}_Q) = \mathbf{s}(\mathbf{q}) \quad (19)$$

where $\mathbf{q}$ is the 12-dimensional vector of elemental generalized positions

$$\mathbf{q}^T = [\mathbf{r}_P^T\ \mathbf{a}_P^T\ \mathbf{r}_Q^T\ \mathbf{a}_Q^T\ ] \quad (20)$$

Then, the final expression for the elemental strain energy reads

$$\mathcal{W} = \frac{1}{2}\mathbf{s}^T(\mathbf{q})\,\mathbf{K}_e\,\mathbf{s}(\mathbf{q}) \quad (21)$$

In order to compute the vector of generalized internal forces $\mathbf{g}$ and the tangent stiffness matrix $\mathbf{K}$, we differentiate the last equation to obtain

$$\mathbf{g} = \mathbf{B}^T\,\mathbf{K}_e\,\mathbf{s}(\mathbf{q}) = \mathbf{B}^T\,\mathbf{f} \quad (22)$$

$$\mathbf{K} = \mathbf{B}^T\,\mathbf{K}_e\,\mathbf{B} + \sum_i f_i \mathbf{B}_i^1 \quad (23)$$

where

$$\mathbf{B} = \frac{\partial \mathbf{s}}{\partial \mathbf{q}} \quad (\text{dim. } 6\times12) \qquad \mathbf{B}_i^1 = \frac{\partial^2 \mathbf{s}_i}{\partial \mathbf{q}\partial \mathbf{q}} \quad (\text{dim. } 12\times12) \quad (24)$$

and where $\mathbf{f}$ is the vector of internal forces for the element PQ in the dynamic configuration. Second derivatives of $\mathbf{s}$ can be neglected whenever internal forces remain below a certain level. They have to be included for highly flexible structures.

3.2 Kinetic energy

The kinetic energy will be computed directly in terms of velocities measured in the inertial system

$$\mathcal{K} = \frac{1}{2}\int_V \dot{\mathbf{r}}^T\,\dot{\mathbf{r}}\,d\mu \quad (25)$$

where $\dot{\mathbf{r}}$ stands for the time derivative of the position vector $\mathbf{r}$. For the beam element, we will assume that the mass is concentrated over the neutral axis, giving

$$\int_S d\mu = m \, d\xi_1 \tag{26}$$

where m is the mass per unit length. This hypothesis holds if the rotational inertia around ξ_1 is small so that we may neglect the inertia term associated with torsion.

When computing stiffness terms, local displacements were interpolated according to

$$\mathbf{w} = \mathbf{N}^*(\xi_1)\, \mathbf{y} \tag{27}$$

where $\mathbf{y} = [\mathbf{w}_Q^T \ \alpha_Q^T]$, $\eta = \xi_1/\ell$ and $\mathbf{N}^*$ is a matrix of local shape functions with cubic interpolation of transverse displacements.

For sake of simplicity, we make the choice to compute the kinetic energy directly in global coordinates using linear shape functions. This is inconsistent with the discretization of strain energy but proves to be sufficient to preserve inertia associated to rigid body motion of the beam [14]. Velocities are thus interpolated by

$$\dot{\mathbf{r}}(\xi_1) = (1 - \xi_1/\ell)\, \dot{\mathbf{r}}_P + (\xi_1/\ell)\, \dot{\mathbf{r}}_Q = \mathbf{N}(\xi_1)\, \dot{\mathbf{q}} \tag{28}$$

and we can assume the following final expression for the kinetic energy

$$\mathcal{K} = \frac{1}{2}\, \dot{\mathbf{q}}^T\, \mathbf{M}\, \dot{\mathbf{q}} \tag{29}$$

where $\mathbf{M}$ is the *constant* mass matrix

$$\mathbf{M} = \int_0^\ell \mathbf{N}^T\, \mathbf{N}\, m \, d\xi_1$$

4. Modelling of rigid joints and constraints

Rigid joints are modeled by imposing the involved kinematic constraints to the system either via an augmented Lagrangian technique or, whenever possible, by the simple Boolean identification of the concerned degrees of freedom. The *elemental Lagrangian* for a joint is then formed with the N_c constraints Φ_k of the first type

$$\mathcal{L} = \sum_{k=1}^{N_c} [-r\, \lambda_k\, \Phi_k + \frac{1}{2}\, p\, \Phi_k^2] \tag{30}$$

where

r, p : are scale and penalization factors, respectively.

k : is a Lagrange multiplier.

The scale factor is chosen so as to obtain well-conditioned matrices. In practice, we give to both factors a magnitude similar to that of the iteration matrix terms.

The elemental "internal forces" vector and "tangent stiffness" matrix can be seen as computed by differentiating the last equation with respect to the generalized degrees of freedom q_i (including within them the added Lagrange multipliers), thus giving

$$g_i = \sum_{k=1}^{N_c} (-r\lambda_k + p\, \Phi_k)\frac{\partial \Phi_k}{\partial q_i} - r\, \Phi_k\, \frac{\partial \lambda_k}{\partial q_i} \tag{31}$$

The last term in eqn (31) is, in fact, the constraint equation added to the system.

The tangent stiffness matrix then reads

$$K_{ij} = \sum_{k=1}^{N_c} -r \left(\frac{\partial \Phi_k}{\partial q_i}\frac{\partial \lambda_k}{\partial q_j} + \frac{\partial \Phi_k}{\partial q_j}\frac{\partial \lambda_k}{\partial q_i} \right) + p\, \frac{\partial \Phi_k}{\partial q_i}\frac{\partial \Phi_k}{\partial q_j} \tag{32}$$

where the contribution of second derivatives of constraints were neglected. The first term in the right-hand side is simply the addition of the gradient of Φ_k to the row and column corresponding to the Lagrange multiplier λ_k.

Reference [14] describes in detail the modelling of specific kinematic joints such as:
- spherical joint,
- revolute pair,
- universal joint,
- generalized sliding joint with curved path.

5. Assembling and solution procedures

Equations for all elements are assembled by following the standard finite element assembling procedure. According to the nature of the problem to be solved, different kinds of problems are obtained:

i) *Static structural problem:* A nonlinear system of algebraic equations is obtained. It is solved by a Newton Raphson's method.

ii) *Kinematics problem:* Motion is prescribed by imposing the position in time of certain control points of the mechanism. A succession of nonlinear systems of algebraic equations have to be solved, and Newton Raphson's method is again employed. Velocities and accelerations are computed afterwards using backward differences.

iii) *Dynamics problem:* A differential/algebraic system of equations is obtained. Various procedures are proposed to treat this kind of systems, in which special care must be taken to avoid oscillations in the response of the constraint associated terms [15,16]. In this work, we have employed the Hilber Hughes Taylor algorithm for time integrating the equations [17], introducing numerical damping to control oscillations. The iteration matrix was balanced by scaling the constraints, as indicated in reference [18].

6. Kinematic analysis of a deployable truss

The structure under consideration is a single bay of a three longeron truss, designed as a prototype component of large space structure. Each bay of the truss (fig.1.a) is made of three longerons and diagonals; two consecutive bays are interfaced by a triangle of batten members. The batten member has at each vertex a hinge body which connects two adjacent bays. Attached to each body are two longerons and diagonals.

The truss is deployable, with two bays deploying at a time. During deployment, two batten triangles are held fixed while the intermediate one rotates about the z axis. The batten members connect rigidly to hinge bodies, while longerons and diagonals are hinged to them.

To permit folding, the diagonals have mid-hinges along their length. The design is such that both fully deployed and folded configurations are nearly stress-free, while significant bending and twisting may occur during deployement.

Symmetry conditions have been used in order to limit the model to one bay. The model is made of 72 physical elements (51 beam elements, 6 rigid bodies, 15 hinge joints) and 7 additional constraints to impose the motion, giving a total of 391 DOF.

Figure 1.a displays the reference configuration (dotted line) and the initially stressed configuration obtained after assembling. Retraction is simulated in two phases:

a. in order to unlock the mechanism, mid-diagonal hinge points are moved inwards and normally to lateral faces (fig.1.b).

b. the vertical displacement of the upper batten is then controlled up to complete retraction (fig.1.b, c and d). Figure 1.e displays a vertical projection of the final configuration. Figure 1.f provides information about the evolution of bending and torsion stresses in longerons during retraction.

The kinematic analysis was made in 96 increments, with an average of 5.82 iterations per increment.

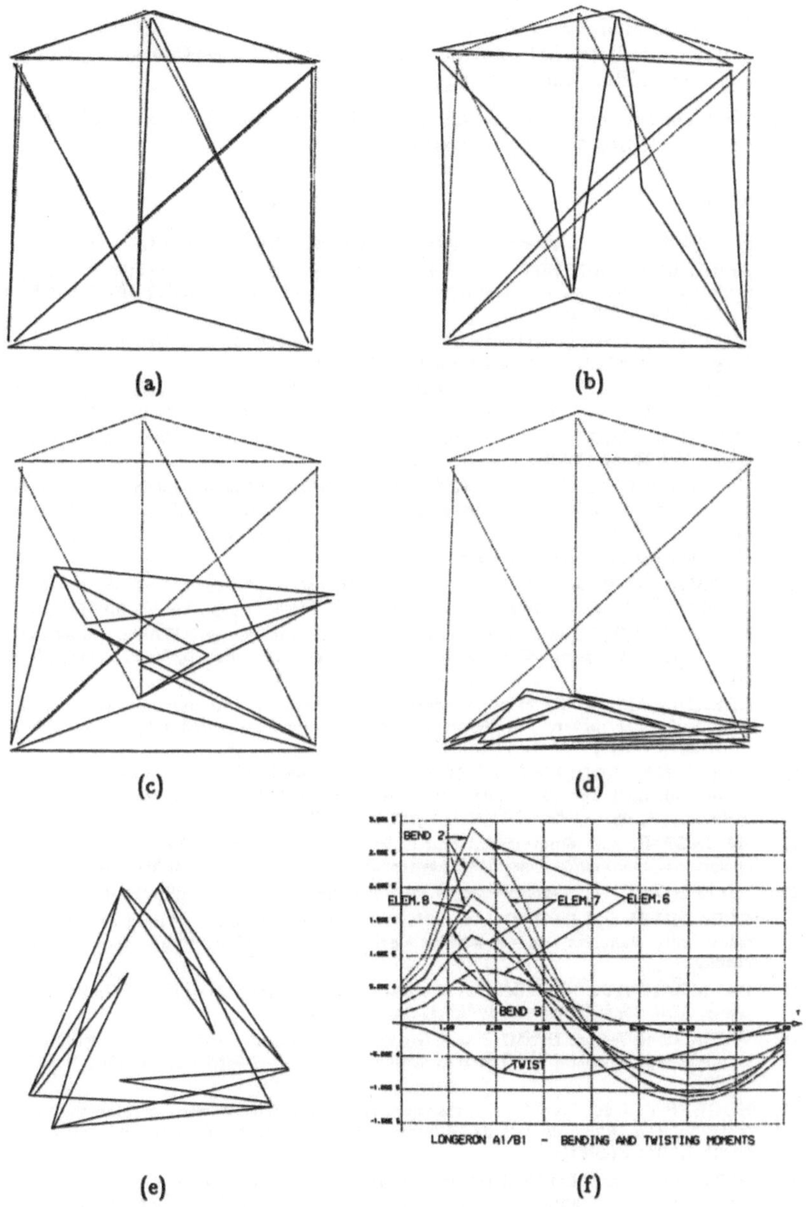

figure 1: kinematic analysis of deployable truss

(a)	reference and initial configurations
(b, (c) and *(d)*	intermediate and final configurations
(f)	top view of final configuration
(g)	bending and torsion stresses in longerons

The numerical model reproduces well the behavior of the experimental structure. In particular, it predicts that once deformations exceed a certain value, the motion proceeds without adding energy to the system.

This problem turned out to be an interesting benchmark for the MECANO software developed according to the concepts described above.

References

1. WITTENBURG J., *Dynamics of systems of rigid bodies*, Teubner, Stuttgart (1977).
2. GERADIN M., ROBERT G. and BERNARDIN, "Dynamic modelling of manipulators with flexible members", in *Advanced software in robotics*, Ed. by A. DANTHINE and M. GERADIN, North Holland (1984).
3. CHACE M., "Computer-aided engineering of large displacements dynamic mechanical systems", CAE/CAD/CAM Conference, April 1983, Salt Lake City, Utah.
4. HAUG E. J., WEHAGE R. and BARMAN N., "Dynamic analysis and design of constrained mechanical systems", Technical Report Nx 50, Div.of Eng., Univ.of Iowa, Nov. 1978.
5. SONG J. and HAUG E. J., "Dynamic analysis of mechanisms", *Comp. Meth. in Appl. Mech and Engng.* Vol 24, pp.359-381 (1980).
6. VAN DER WERFF K. and JONKER J. B., "Dynamics of flexible mechanisms", in *Computer aided analysis and optimization of mechanical system dynamics*, Ed.E. J. HAUG, Springer-Verlag (1984).
7. SHABANA A. A., "Substructure synthesis methods for dynamic analysis of multi-body systems", *Computers and Structures*, Vol.20, pp.737-744 (1985).
8. AGRAWAL O. P. and SHABANA A. A., "Application of deformable-body mean axis to flexible multibody system dynamics", *Comp. Meth. Appl. Mech. Engng.*, Vol.56, pp.217-245 (1986).
9. NIKRAVESH P., "Spatial kinematics and dynamic analysis with Euler parameters", in *Computer aided analysis and optimization of mechanical system dynamics*, Ed. by E. J. HAUG, Springer-Verlag (1984).
10. GERADIN M., "Finite element approach to kinematic and dynamic analysis of mechanisms using Euler parameters", in *Numerical methods for nonlinear problems II*, Ed. by C.TAYLOR et al, Pineridge Press, Swansea, U.K. (1984).
11. GERADIN M. ROBERT G. and BUCHET P., "Kinematic and dynamic analysis of mechanisms: a finite element approach based on Euler parameters", in *Finite element methods for nonlinear problems*, Ed. by P. BERGAN et al, Springer-Verlag (1986).
12. WEHAGE R. A., "Quaternions and Euler parameters. A brief exposition", in *Computer aided analysis and optimization of mechanical system dynamics*, Ed. by E. J. HAUG, Springer-Verlag (1984).
13. MILENKOVIC V., "Coordinates suitable for angular motion synthesis of robots".
14. GERADIN M. and CARDONA A., "Kinematics and dynamics of rigid and flexible mechanisms using finite elements and quaternion algebra", submitted for publication in *Computational Mechanics* .
15. BAUMGARTE J., "Stabilization of constraints and integrals of motion in dynamical systems", *Comp. Meth. Appl. Mech. Engng* Vol.1, pp.1-16 (1972).
16. WEHAGE R. A. and HAUG E. J., "Generalized coordinate partitioning for dimension reduction in analysis of constrained dynamic systems", *Journal of Mechanical Design*, Vol.104, pp.247-255 (1982).
17. HUGHES T. J. R., "Analysis of transient algorithms with particular reference to stability problems", in *Computational methods for transient analysis*, Ed. by T. Belytschko and T. J. R.Hughes, North Holland (1983).
18. PETZOLD L. and LOTSTEDT P.,"Numerical solution of nonlinear differential/algebraic systems from physics and engineering", in *Innovative methods for nonlinear problems*, Ed.by Liu W., Belytschko T. and Park K. C., Pineridge Press (1984).

Aknowledgment

This research has been jointly done at the University of Colorado and at the University of Liège, Belgium with support provided in part by NASA under contract NAS1-17660.

AN ANALYSIS OF DYNAMIC CRACK PROPAGATION IN A RAIL WEB WITH LONGITUDINAL RESIDUAL STRESSES

M.F. Kanninen

Southwest Research Institute
San Antonio, TX, USA

SUMMARY

To gain an improved understanding of the synergistic effects of toughness and residual stresses, a simple beam-on-elastic foundation model of a longitudinally split rail was developed. This model supplemented finite element fracture mechanics analyses that were unable to effectively address these conditions. The findings of this research specifically indicate that, if the rail does not have adequate dynamic fracture resistance, rail residual stresses can give rise to crack driving forces that are sufficient to maintain a long-running web fracture. These results are quantitatively consistent with the absence of long-running dynamic crack propagation events in standard rail and with the actual occurrence of a fracture event in a premium alloy rail.

1. INTRODUCTION

A recent train derailment resulted from catastrophic crack propagation in the web of a roller-straightened "premium-alloy" steel rail [1]. The combination of a crack initiation site at the rail end caused by torch cutting in a repair operation, high tensile residual stresses in the rail web, and the lower fracture toughness of premium alloy steel is thought to have created a near-critical fracture condition. Evidently, rapid unstable crack propagation was then triggered by the added dynamic stresses induced by the fast moving train striking a 3/16-inch rail misalignment. The reconstructed rail, showing the multiple branches that fragmented a 10 meter length of rail, is shown in Figure 1.

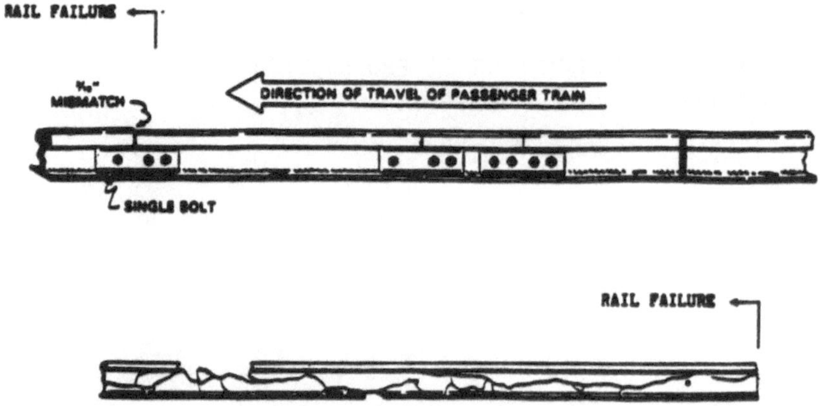

Figure 1. The Reconstruction of a Rail Involved in a Dynamic Fracture Event [2].

In view of the derailment caused by the fracture event shown in Figure 1, together with other incidents that have occurred with premium-alloy steel [1], the railroad industry must be concerned with the possible inordinate susceptibility of these steels to catastrophic fracture. On the basis that crack initiation cannot be absolutely precluded (e.g., because of fatigue cracks), the dynamic plane strain crack arrest toughness K_{Ia} becomes a key rail steel property in a safety assessment. Accordingly, this research was initiated to learn if dynamic fracture mechanics techniques can be extended to determine and apply K_{Ia} values in track applications.

Confounding the analysis problem for rail steels are (1) the presence of residual stresses, and (2) the lack of a basis for a sound two-dimensional analysis model for a rail. This research was undertaken in a heuristic spirit with a simple beam-on-elastic-foundation model being developed to overcome these difficulties. The work was aided considerably by detailed run/arrest data provided by the NKKK Steel Company of Japan [2]. The general approach is based upon the concepts of dynamic fracture mechanics as elucidated in Reference [3].

2. BASIS OF ELASTODYNAMIC FRACTURE MECHANICS

Rapid crack propagation can generally be considered to occur under the condition that the dynamically computed crack driving force is equal to the material's resistance to crack extension. Within the confines of elastodynamic behavior, this is expressed as [3]

$$K = K_{ID}(V,T) \tag{1}$$

where K_{ID}, the plane strain dynamic fracture toughness, is a function of the instantaneous crack speed V and the temperature T, while K is the dynamically calculated value of the stress intensity factor. Arrest occurs at the position and time for which K becomes less than the minimum value of K_{ID}, and remains less, for all greater times. Within this framework, the arrest criterion is $K < K_{ID}(0, T)$. It is important to note that K_{Ia}, the static crack arrest toughness, can be a good approximation to the true crack arrest toughness given by the right hand side of this inequality [3].

Experimental work performed by NKKK [2] has used the full-section impact test procedure; see Figure 2. This test is conducted on a horizontal rail with an initial blunt crack oriented to favor propagation along the mid plane of the rail web. Unstable crack propagation is initiated by dropping a weight on a wedge placed in a split pin that is located in a circular hole in the rail web. Full-section impact tests can be performed at different temperatures and drop heights, thus varying the extent and speed of the unstable crack propagation events that were observed.

3. RAIL STEEL FRACTURE TOUGHNESS AND RESIDUAL STRESS DATA

Because unequivocal values of K_{Ia} do not currently exist for rail steels, estimation procedures must be used to infer representative values. Arrest toughness values are generally comparable to, and can be estimated from, dynamic initiation values. For example, empirical relations can be used to relate K_{Id}, the dynamic initiation toughness, to the Charpy energy that in turn can be connected with K_{Ia}. Below the transition temperature (typically about 140°C for rail), K_{Ia} is approximately equal to the lower limit of K_{Id} with increasing loading rates.

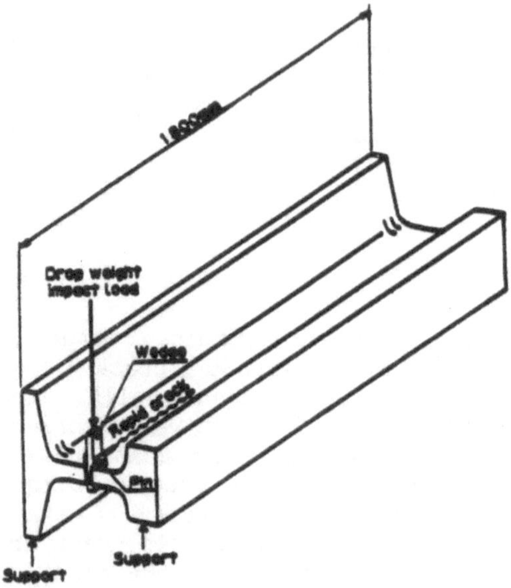

Figure 2. The Full Section Impact Test Procedure for Rail Fracture Propagation Experiments

Alternatively, Tetelman and Stone [4] suggested that a good approximation for rail steels is given by $K_{Ic} = 1.4\ K_{Id}$ where K_{Ic} is the plane strain fracture initiation toughness. It follows that K_{Ia} values for rail steel can be estimated from the alternative relation $K_{Ia} = 0.7\ K_{Ic}$. Supporting this estimate are recently obtained results showing that the ratio K_{Ia}/K_{Ic} for A533B steel is very nearly equal to 2/3 over a wide range of temperatures. In any event, K_{Ic} results from Reference [5] and other sources lead to K_{Ia} values for standard rail that range from about 20 $MPam^{\frac{1}{2}}$ to 34 $MPam^{\frac{1}{2}}$. For the rail shown in Figure 1, an estimate is $K_{Ia} = 17\ MPam^{\frac{1}{2}}$.

Of great importance in understanding rail performance are the residual stresses introduced by roller straightening operations. From the incidents cited for premium alloy rail by Orringer and Tong [1] and from other unpublicized incidents, it is clear that these rails include web tensile transverse residual stresses that assist in driving a crack longitudinally along the web. In other rails the residual stresses may be decisive in turning a crack to the flange. Figure 3 illustrates the residual stress data that are available [6].

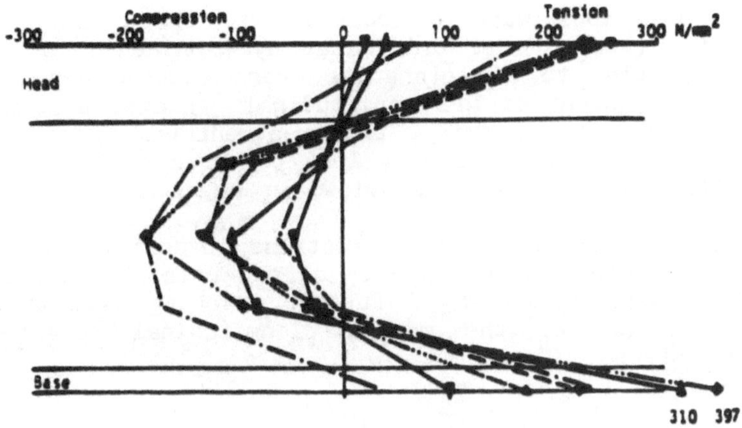

Figure 3. Example Longitudinal Direction Residual Stresses Determined on Standard Rail Steel by ORE [6].

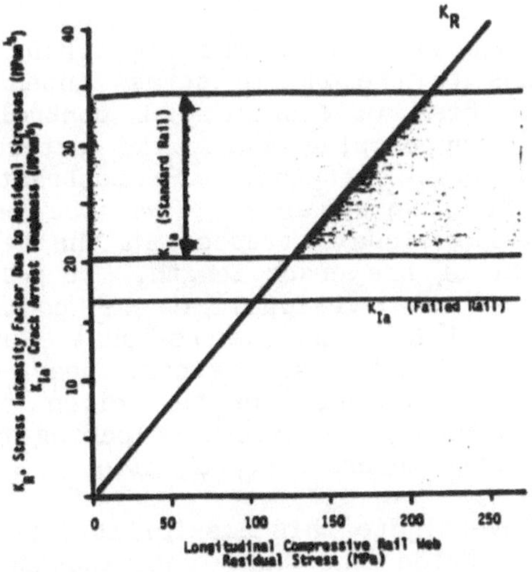

Figure 4. Tentative Basis for the Assessment of the Potential for a Long-Running Fracture in Rails with Compressive Longitudinal Web Residual Stresses.

4. A FRACTURE MECHANICS ANALYSIS FOR RAIL

To obtain a suitable stress intensity factor expression for quasi-static crack growth in a full-section impact test, two modifications were introduced into the beam-on-elastic foundation (BOEF) model originated by Kanninen [7]. First, the restriction to a rectangular cross section was removed; second, the effect of web longitudinal residual stresses was incorporated. In so doing, the treatment was specialized to that for a crack that is very long compared to the height of the rail. The details are provided in Reference [8].

The key result is for the stress intensity factor of a crack of length a acted upon by a load point displacement δ and a residual moment M_R. Further invoking beam theory to relate M_R to σ_R, the maximum longitudinal (compressive) residual stress, gives

$$K = \left(\frac{I}{b}\right)^{\frac{1}{2}} \frac{\sigma_R}{c} + 3\left(\frac{I}{b}\right)^{\frac{1}{2}} \frac{E\delta}{a^2} \qquad (2)$$

where b is the thickness of the rail web, I is the moment of inertia of the upper portion of a longitudinally split rail, and c is the distance from the crack plane to the centroid of the upper half-rail. Because the vertical constraint that exists in track is absent in a full-section impact test, this result differs from that appropriate for track.

The second term on the right hand side of Equation (2) corresponds to the static stress intensity factor for beam-like geometries in displacement control. Under a fixed or slowly varying displacement, this contribution will vanish for very long crack lengths. In contrast, the first term of Equation (2), which arises from the presence of transverse compressive residual stresses in the web, can be seen to be independent of the crack length. It follows that, for a very long crack, the residual stress contribution will become dominant. Hence, as the residual stresses are apparently constant along the rail except near the ends [1], it is apparent that cracks can be driven to virtually complete fracture without any external loading beyond that needed to initiate unstable crack propagation.

To investigate this possibility, assume that the residual stresses provide the only contribution to driving a long-running web crack. It is then only necessary to employ an expression for the dynamic stress intensity factor for steady-state dynamic crack propagation under a constant residual moment. This is as follows.

If the only external forces acting on the cracked rail are those from the residual moment M_R, a result for steady-state dynamic crack propagation at a speed V can be obtained from the beam-on-elastic foundation model. This is [8]

$$K = K_R \left(1 - \frac{V^2}{V_{cr}^2}\right) \tag{3}$$

where for typical rail dimensions and mechanical properties, $K_R = 0.159\ \sigma_R$ and $V_{cr} = 1100$ m/s.

Equation (3) provides a basis for assessing the potential for a long-running web fracture by comparing the relative values of K_R and K_{Ia}. As shown in Figure 4, using Figure 3 to estimate σ_R, the arrest toughness values of standard rail steel generally lie above K_R. Hence, consistent with track experience, long-running rail web fractures would not be expected. In contrast, for similar σ_R values, the arrest toughness of the rail in Figure 1 is clearly below K_R.

Using Equation (3), it is further possible to obtain an estimate for the crack propagation speed in a rail having a speed-independent dynamic fracture toughness value equal to its K_{Ia} value. Assuming σ_R=150 MPa for the failed rail gives V=600 m/sec. This implies that the 10 meter length of rail was fractured in about 16 ms. This crack speed can be compared with the speed of the train that initiated the accident, i.e., 32 m/sec (72 mph). That the crack speed is significantly greater than the train speed is consistent with the basic assumption that, while mechanical stresses likely served to initiate crack propagation, the long-running crack was essentially driven by residual stresses only.

5. DISCUSSION AND CONCLUSIONS

While the catastrophic rail fracture event that has focused attention on this problem arose from highly exceptional conditions, it is nevertheless prudent to assume that initiation sites always will be present; e.g., from fatigue cracks at bolt holes [9]. As shown in this paper, residual stresses are of sufficient magnitude to cause a long-running fracture. Consequently, rail safety assessments must therefore not only rely on precluding the initiation of crack growth, but should include crack arrest after a short extent of crack propagation (or abrupt crack turning) as second lines of defense against catastrophic fracture. The work reported herein also suggests that simple models can be used to reach and quantify such conclusions when finite element methods are unable to be effectively brought to bear.

6. REFERENCES

[1] Orringer, O. and Tong, P., "Investigation of Catastrophic Failure of a Premium-Alloy Railroad Rail," Fracture Problems in the Transportation Industry, edited by P. Tong and O. Orringer, American Society of Civil Engineers, New York, pp 62-79, 1985.

[2] Anon., "Crack Propagation and Arrest Test on Rail Steels," Nippon Kokan K.K. report, Kawasaki, Japan, March, 1985.

[3] Kanninen, M. F., and Popelar, C. H., Advanced Fracture Mechanics, Chapter 4, Oxford Press, New York, 1985.

[4] Tetelman, A. S., and Stone, D. H., "An Introduction to the Fracture Mechanics of Rail Material," AAR Report No. R-157, May, 1974.

[5] Jones, D. J., and Rice, R. C., "Determination of K_{Ic} Fracture Toughness for Alloy Rail Steel," Battelle Columbus Laboratories Report to Transportation Systems Center, Cambridge, MA, 15 November 1985.

[6] Anon, "Factors Influencing the Fracture Resistance of Rails in the Unused Condition," Office for Research and Experiments of the International Union of Railways, Utrecht, Report No. 1, September, 1984.

[7] Kanninen, M.F., "An Augmented Double Cantilever Beam Model for Studying Crack Propagation and Arrest," International Journal of Fracture, Vol. 9, pp. 83-92, 1973.

[8] Kanninen, M.F., Dexter, R.J., and Cardinal, J.W., "Determination of Dynamic Toughness Properties of Rail Steels," Southwest Research Institute Report to Transportation System Center, 30 September 1986.

[9] Orringer, O., Morris, J.M., and Steele, R.K., "Applied Research on Rail Fatigue and Fracture in the United States," Theoretical and Applied Mechanics, Vol. 1, pp. 23-49, 1984.

FINITE- ELEMENT-ANALYSIS AND ALGORITHMS FOR LARGE ELASTIC STRAINS

E. Stein, N. Mueller-Hoeppe
Universitaet Hannover
Hannover, FRG

SUMMARY

Elastic solids with large strains are treated in the frame of polyconvex materials so that the existence of solutions for boundary value problems can be ensured. Especially Neo-Hooke materials and a constitutive equation due to Ciarlet are treated both in material and spatial description.

A finite-element-algorithm is developed starting from the Fréchet-derivative of the principle of virtual work in material coordinates. Pushing forward this linearized form to the current configuration, performing the iteration process in this state yields a Newton method with reference to the current configuration. Isoparametric 8-node 3-D elements are most efficient for the numerical process and used for the following examples such as a bar, tensioned to nearly the double length and compressed to nearly the half length. Both examples were calculated in one increment with 4, respective 6 iteration steps.

1. INTRODUCTION

Static boundary value problems of elastic systems with large strains are treated using the concept of polyconex materials in different specialiazed forms. FE-algorithms are treated mainly using a spatial description.

2. MOTION OF A BODY AND THE DEFORMATION GRADIENT

The timeindependent triple $(X^1, X^2, X^3) = \mathbf{X} \in B_o$ is called material point. At time t it is identified by the triple $(x^1, x^2, x^3) = \mathbf{x} \in B$ which is called spatial point.

The nonlinear relation

$$\mathbf{x} = \Phi(\mathbf{X}, t) \tag{2.1}$$

is called motion or deformation.

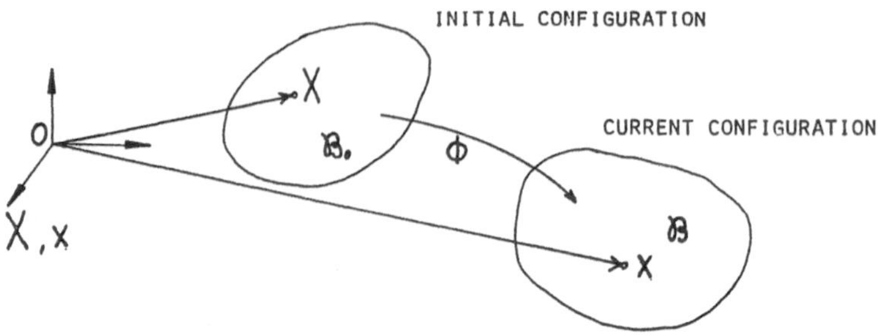

Fig. 1 Motion of a body (t fixed)

The partial derivative of the motion Φ with respect to the material coordinates is the deformation gradient

$$\mathbf{F}(\mathbf{X},t) = \frac{\partial \Phi(\mathbf{X},t)}{\partial \mathbf{X}} = Grad\, \Phi(\mathbf{X},t). \tag{2.2}$$

3. Principle of virtual work

This paper is restricted to problems where the existence of a stored energy function is ensured so that the principle of virtual work is exactly the 1. variation of the energy functional [1]

$$\pi(\Phi) = \int_{B_o} (\rho_o \psi(\mathbf{F}) - \rho_o \ddot{\mathbf{X}} \cdot \Phi)\, dV - \int_{\partial B_{o\sigma}} \hat{\mathbf{t}} \cdot \Phi\, dA. \tag{3.1}$$

The directional derivative is

$$D\pi(\Phi)\cdot\eta = \int_{B_o} (\rho_o \mathbf{g}^{-1} \frac{\partial \psi}{\partial \mathbf{F}} \cdot Grad\eta - \rho_o \ddot{\mathbf{X}} \cdot \eta)\, dV - \int_{\partial B_{o\sigma}} \hat{\mathbf{t}} \cdot \eta\, dA = 0, \tag{3.2}$$

η test function,
ρ_o density in B_o,
ψ stored energy function (free energy),
$\hat{\mathbf{t}}$ given boundary tractions,
$\mathbf{g}^{-1}$ inverse of the spatial metric tensor.

4. Material functions

A hyperelastic material is assumed so that the energy functional is known. The stored free energy function

$$\psi(\mathbf{F}) = \bar{\psi}(\mathbf{C}) = \tilde{\psi}(\mathbf{b}), \tag{4.1}$$

$\mathbf{C}$ right Cauchy-Green tensor,
$\mathbf{b}$ left Cauchy-Green tensor,

exists so that for isothermal processes the stresses [2] are defined by

$$\mathbf{P} = \rho_o \mathbf{g}^{-1} \frac{\partial \psi}{\partial \mathbf{F}}, \tag{4.2}$$

$$\mathbf{S} = 2\rho_o \frac{\partial \bar{\psi}}{\partial \mathbf{C}} = \rho_o \frac{\partial \bar{\bar{\psi}}}{\partial \mathbf{E}}, \tag{4.3}$$

$$\sigma = 2\rho \frac{\partial \tilde{\psi}}{\partial \mathbf{g}} = \rho \frac{\partial \tilde{\tilde{\psi}}}{\partial \mathbf{e}}, \tag{4.4}$$

P 1. Piola-Kirchhoff tensor,
S 2.Piola-Kirchhoff tensor,
σ Cauchy stresses,
E Green strain tensor,
e Almansi strain tensor.

Let the stored energy function be polyconvex [3] so that the minimal problem has at least one solution. The condition of polyconvexity leads to an Ogden- material, for example see [1], described by

$$\psi(\mathbf{F}) = \sum_{i=1}^{m} a_i \, tr\,(\mathbf{C}^{\alpha_i/2}) + \sum_{j=1}^{n} b_j \, tr\,((adj\,\mathbf{C})^{\beta_j/2}) + f\,(det\,\mathbf{F}) \tag{4.5}$$

with $a_i > 0,\ 1 \leq i \leq m,\ \alpha_i \geq 1;\ b_j > 0,\ 1 \leq j \leq n,\ \beta_j \geq 1$.
$f\,(det\,\mathbf{F})$ is convex and a growth condition holds.

5. Linearization

5.1 Directional derivative and linear part of a funtion

The Taylor expansion of a scalar-valued function $G(\Phi)$ leads to

$$G(\bar{\Phi} + \mathbf{u}) = G(\bar{\Phi}) + DG(\bar{\Phi}) \cdot \mathbf{u} + R, \tag{5.1}$$

$\bar{\Phi}$ the point where $G(\Phi)$ is expanded,
$DG(\bar{\Phi}) \cdot \mathbf{u}$ directional derivative of G at point $\bar{\Phi}$,
R terms of higher order.
Then

$$L(G, \mathbf{u})_{\bar{\Phi}} = G(\bar{\Phi}) + DG(\bar{\Phi}) \cdot \mathbf{u} \tag{5.2}$$

is the linear part of the function and

$$DG(\bar{\Phi}) \cdot \mathbf{u} = \frac{d}{d\epsilon}\left(G(\bar{\Phi} + \epsilon \mathbf{u})\right)_{\epsilon=0} \tag{5.3}$$

the directional derivative.

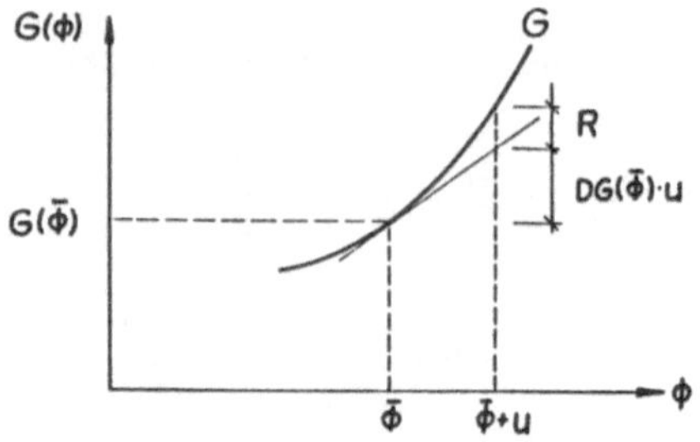

Fig. 2 Linearization of a function

5.2 Linearization of the principle of virtual work

To solve problem (3.2) with Newton's method a complete consistent linearization according to chapter 5.1 is required. Therefore $G(\Phi)$ is identified with the principle of virtual work $D\pi(\Phi)\cdot\eta$. Then $\bar{\Phi}$ is a fixed configuration during the motion Φ. All quantities with reference to this fixed configuration will be signified by "-" in the sequel.

If the stress free initial configuration is chosen as the reference configuration (5.2) yields the iterative scheme

$$DG(\bar{\Phi})\cdot \mathbf{u} = D\left(D\pi(\bar{\Phi})\cdot\eta\right)\cdot\mathbf{u} = 0. \tag{5.4}$$

The application of (5.4) to the principle of virtual work in the material description yields

$$\begin{aligned} DG(\bar{\Phi})\cdot\mathbf{u} = \int_{B_o} & tr\,(Grad\,\mathbf{u}\,\bar{\mathbf{S}}\,Grad\,\eta) \\ & + tr\left(\bar{\mathbf{F}}(\mathcal{A}\cdot(Grad^T\mathbf{u}\,\bar{\mathbf{F}} + \bar{\mathbf{F}}^T\,Grad\,\mathbf{u}))\,Grad^T\eta\right)dV \\ & = 0 \end{aligned} \tag{5.5}$$

with the elasticity tensor $\mathcal{A}$ in material description.

5.3 Push forward to the current configuration

With the help of some transformations the linearized principle of virtual work (5.5) can be pushed to the current configuration [4]. The result is

$$DG\cdot\mathbf{u} = \int_B tr\,(grad\,\eta\,\bar{\sigma}\,grad^T\mathbf{u}) + grad\,\eta\cdot(\bar{\mathbf{a}}\cdot grad^S\mathbf{u})dv = 0 \tag{5.6}$$

with the elasticity tensor a in spatial description.

5.4 Elasticity tensor

The reduction to isotropic materials yields that the free energy ψ can be described with the help of the two Lamé-constants λ and μ. In addition it is possible to describe $\bar{\psi}$ and $\tilde{\psi}$ as functions of the invariants of $\mathbf{C}$ or $\mathbf{b}$. The invariants of $\mathbf{C}$ and $\mathbf{b}$ are identical. In this context the material function [5] arises

$$\bar{\psi} = \tilde{\psi} = \frac{1}{2} a I + \frac{1}{2} b II + f(J) \tag{5.7}$$

with the invariants

$$\begin{aligned} I &= tr(\mathbf{C}) = tr(\mathbf{b}), \\ II &= \frac{1}{2}(tr^2\mathbf{C} - tr\,\mathbf{C}^2) = \frac{1}{2} tr(adj\mathbf{C}) = \frac{1}{2}(tr^2\mathbf{b} - tr\,\mathbf{b}^2), \\ III &= det\,\mathbf{C} = det\,\mathbf{b}, \\ J &= \sqrt{III}. \end{aligned}$$

The stress calculation yields in material description

$$\mathbf{S} = 2(\bar{\psi}_I + I\bar{\psi}_{II})\mathbf{G}^{-1} - 2\bar{\psi}_{II}\mathbf{C} + J\bar{\psi}_J\mathbf{C}^{-1} \tag{5.8}$$

and in spatial description

$$\sigma = (\bar{\psi}_J + \frac{2}{J} II \bar{\psi})\mathbf{g}^{-1} + \frac{2}{J}\bar{\psi}_I \mathbf{b} - 2J\bar{\psi}_{II}\mathbf{b}^{-1}, \tag{5.9}$$

$\bar{\psi}_{(\ldots)}$ partial derivative of $\bar{\psi}$ with respect to the invariants,
$\mathbf{G}^{-1}$ inverse of the material metric tensor.

To solve (5.5) or (5.6), the elasticity tensor is required to compute the tangential stiffness. The elasticity tensor can be calculated using the Lie derivative of the stresses. In the material description, the stress rate is

$$\begin{aligned} \dot{\mathbf{S}} &= 4\rho_o \frac{\partial^2 \bar{\psi}}{\partial \mathbf{C}\,\partial \mathbf{C}} \cdot \dot{\mathbf{C}} = \mathcal{A} \cdot \dot{\mathbf{C}} \\ &= \rho_o \frac{\partial^2 \bar{\bar{\psi}}}{\partial \mathbf{E}\,\partial \mathbf{E}} \cdot \dot{\mathbf{E}} = 2\mathcal{A} \cdot \dot{\mathbf{E}} \end{aligned} \tag{5.10}$$

and in the spatial description, one gets the objective rate

$$\begin{aligned} \overset{\triangledown}{\sigma} &= 4\rho \frac{\partial^2 \tilde{\psi}}{\partial \mathbf{g}\,\partial \mathbf{g}} \cdot \overset{\triangledown}{\mathbf{g}} = \mathbf{a} \cdot \overset{\triangledown}{\mathbf{g}} \\ &= \rho \frac{\partial^2 \tilde{\tilde{\psi}}}{\partial \mathbf{e}\,\partial \mathbf{e}} \cdot \overset{\triangledown}{\mathbf{e}}, \end{aligned} \tag{5.11}$$

$\overset{\triangledown}{\sigma} = \frac{1}{J}\mathbf{F}\,\dot{\mathbf{S}}\,\mathbf{F}^T$ Truesdell rate of Cauchy stresses ,
$\overset{\triangledown}{\mathbf{g}} = 2\overset{\triangledown}{\mathbf{e}} = 2\mathbf{d}$, $\mathbf{d}$ rate of deformation tensor.

5.5 The elasticity tensor of a compressible Neo-Hooke material

To calculate the elasticity tensor a special case of (5.7) with $b = 0$ is is regarded. The material function is

$$\bar{\psi} = \frac{1}{2} a\, I + f\,(J) \tag{5.12}$$

and describes a Neo-Hooke material in general. It leads to the well-known classical elasticity tensor of the linear theory if linearization is accomplished after chosing f in a special way.

In [6] the stored energy function (model 1) is suggested to

$$\rho_o\,\bar{\psi} = \frac{1}{2}\mu\,(I - 3) + f, \tag{5.13}$$

$$f = \frac{\lambda}{2}\,(ln\;J)^2 - \mu\, ln\; J. \tag{5.14}$$

This material function fulfils the following reasonable conditions of an elastic material

- the initial configuration is stressfree, and no stresses appear in the case of rigid body motion,

- the stored energy function and the strsses become $+\infty$, if the volume tends to ∞ (coercivity),

- the stored energy function becomes $+\infty$ and the stresses become $-\infty$, if the volume tends to 0.

Also the function f fullfils the growth condition but

$$f''\,(J) = \frac{\lambda}{J^2}\,(1 - ln\;J) + \frac{\mu}{J^2} \tag{5.15}$$

shows that f is not convex in general; see chapter 4. The convex domain is restricted by

$$J \geq e^{1+\frac{\mu}{\lambda}} \geq 2,718. \tag{5.16}$$

In [1] (model 2) f is assumed to

$$f = c\,J^2 - d\, ln\; J,\; c > 0,\; d > 0. \tag{5.17}$$

This function fullfils the growth condition, is convex everywhere and fullfils the resonable conditions of an elastic material.

The components of the elasticity tensor for model 1 in material description [6] are

$$\mathcal{A}^{ijkl} = \frac{\lambda}{2}\,(C^{-1})^{ij}\,(C^{-1})^{kl} + (\mu - \lambda\, ln\, J\,)\,\frac{1}{2}\big(\,(C^{-1})^{ik}\,(C^{-1})^{jl} + (C^{-1})^{il}\,(C^{-1})^{jk}\,\big) \quad (5.18)$$

with $\mathcal{A} = \mathcal{A}^{ijkl}\,\mathbf{G}_i \otimes \mathbf{G}_j \otimes \mathbf{G}_k \otimes \mathbf{G}_l$,
in spatial description

$$a^{ijkl} = \frac{\lambda}{J}\,g^{ij}\,g^{kl} + \frac{1}{J}\,(\,\mu - \lambda\, ln\, J\,)\,(\,g^{ik}\,g^{jl} + g^{il}\,g^{jk}\,) \quad (5.19)$$

with $\mathbf{a} = a^{ijkl}\,\mathbf{g}_i \otimes \mathbf{g}_j \otimes \mathbf{g}_k \otimes \mathbf{g}_l$
and for model 2 in material description with $c = \frac{\lambda}{2}$, $d = \mu$

$$\mathcal{B}^{ijkl} = \lambda\, J^2\,(C^{-1})^{ij}\,(C^{-1})^{kl} + (\,\mu - \lambda\, J^2\,)\,\frac{1}{2}\big(\,(C^{-1})^{ik}\,(C^{-1})^{jl} + (C^{-1})^{il}\,(C^{-1})^{jk}\,\big), \quad (5.20)$$

in spatial description

$$b^{ijkl} = 2\lambda\, J\, g^{ij}\,g^{kl} + \frac{1}{J}\,(\,\mu - \lambda\, J^2\,)\,(\,g^{ik}\,g^{jl} + g^{il}\,g^{jk}\,). \quad (5.21)$$

6. Finite element formulation

For computation an isoparametric 8-node 3-D element and an isoparametric 4-node 2-D element with linear interpolation functions were used. If an isoparametric element is used, there always exist perpendicular local base vectors in the parameter space. Therefore the metric coefficients of the elasticity tensor are δ_{ik}. The change of geometry of a finite element in consequence of the deformation is achieved by updating the coordinates of the nodal points. As a spatial description is used for the finite element formulation, the elasticity tensor was derived for the current configuration. For that reason it must be exactly distinguished between deformation and transformations to the unit square or cube. Therefore, updating is necessary in every iteration step to get the quadratic convergence of Newton's method.

6.1 Numerical example: Strip under compression and tension

This example was computed using an isoparametric 8-node 3-D element. With the consistent tangent tensor the compression problem was

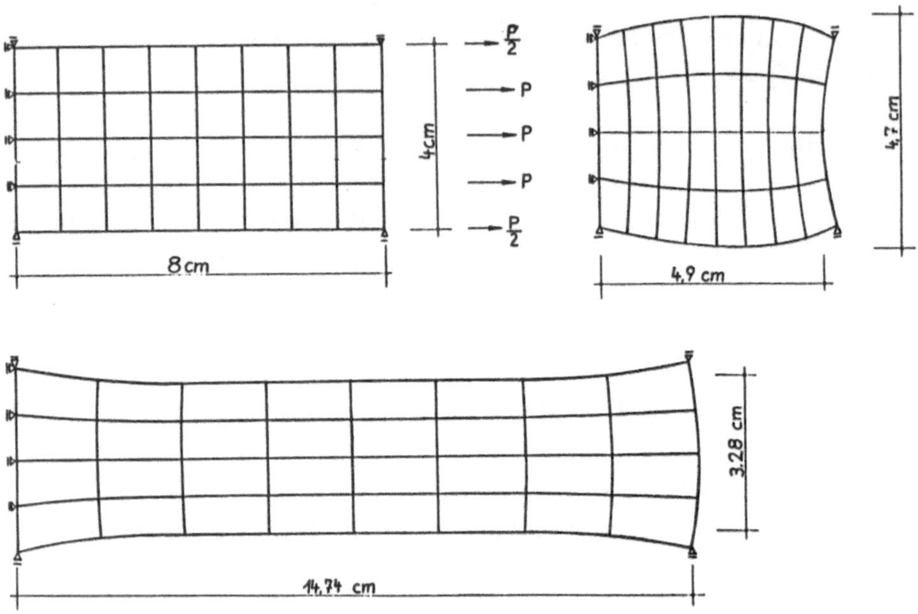

Fig. 3 Undeformed mesh, mesh under compression, mesh under tension

solved in 1 increment with 6 iterations, the tension problem in 1 increment with 4 iterations.

Material constants:
Young's modulus: $E = 21000\, kN/cm^2$, Poisson's ratio: $\nu = 0,3$
loading : $P = \mp 6000\, kN$

7. LITERATURE

1. Ciarlet, P. G. - Lectures on three-dimensional elasticity, Springer, Berlin, Heidelberg, New York, 1983.

2. Doyle, T.C. and Ericksen, J. L.,'Nonlinear elasticity', Adv. Appl. Mech. IV, Academic Press, New York, 1956.

3. Ball, J. M.,'Convexity conditions and existence theorems in nonlinear elasticity', Arch. Rational Mech. Anal. 63 (1977),337-403.

4. Marsden, J. E., Hughes, T. J. R. - Mathematical foundations of elasticity, prentice -Hall, Englewood Cliffs, N. J., 1983.

5. Ogden, R. W.,'Inequalities associated with the inversion of elastic stress-deformation relations and their implications', Math. Proc. Camb. Soc. 81 (1977),313-324.

6. Simo, J. C., Pister, K. S., 'Remarks on rate constitutive equations for finite deformation problems: computational implications', Comp. Meth. Appl. Mech. Eng. 46 (1984), 201-215.

A Consistent Finite Element Formulation of Nonlinear Frictional Contact Problems

JIANN-WEN JU ROBERT L. TAYLOR LOUIS Y. CHENG

Department of Civil Engineering

University of California, Berkeley, USA

Abstract

A *perturbed Lagrangian*-based variational formulation is proposed for the finite element solution of fully nonlinear frictional contact problems. In the spirit of an operator splitting methodology, an analogy exists between the proposed treatment for the stick-slip motion and the corresponding treatment in elastoplasticity.

Within the context of discrete formulations arising from a finite element approximation, explicit expressions for the frictional *consistent* contact tangent stiffness and residual are derived from variational equations by using a *consistent linearization* procedure for both the sliding and adhesion phases. The consistent tangent operator is *always* non-symmetric for the case of frictional sliding owing to the nature of the Coulomb's friction law employed.

1. Introduction

Frictional stick-slip contact phenomena constitute important aspects of real engineering applications. In this study, a *perturbed Lagrangian*-based formulation is proposed for the finite element solution of fully *nonlinear* frictional contact problems. The stick-slip contact phenomena is accommodated by means of a nonlinear variational formulation. In view of an analogy between the Coulomb's law of friction for the stick-slip motion and the yield criterion for classical elastoplasticity (see, e.g., Michalowski & Mroz [1978]), a two-step operator splitting methodology is employed.

In the current literature, the modification of the tangent stiffness accounting for the contribution of frictional contact often takes the form of symmetric and/or non-symmetric *rank-one* updates inherited from the *linear* theory (see, e.g., Oden & Martins [1985] and Hughes *et al.* [1976]). In the finite element solution of geometrically nonlinear frictional contact problems, however, such simple procedures are no longer adequate. In the event of frictionless contact, a consistent tangent operator has been obtained by Wriggers & Simo [1985].

Within the context of finite element discrete approximation, explicit expressions for the frictional contact tangent stiffness and the residual are derived in this paper from variational equations by using a *consistent linearization* procedure for both the sliding and adhesion phases. It is shown that for the case of frictional stick the consistent tangent operator is symmetric *only* when numerical convergence is achieved. On the other hand, the consistent tangent operator is *always* non-symmetric for the case of frictional sliding owing to the nature of the Coulomb's friction law employed. Not surprisingly, these expressions degenerate to the classical rank-one corrections of the stiffness matrices in the limiting case of infinitesimal deformations. It is emphasized that, in the presence of nonlinear contact kinematics, use of the consistent contact tangent stiffness is essential in preserving the quadratic rate of asymptotic convergence of Newton's method.

For two-dimensional applications, a three-node contact element is employed in the finite element discretization. The "perturbed Lagrangian"-based computational algorithm is capable of performing a one-pass or two-pass contact slide-line logic. A number of numerical examples are presented in Sec. 3 that illustrate the performance of the proposed variational formulation.

2. Discrete variational formulation

Within the framework of the finite element method, the governing variational equations involving fully nonlinear kinematics are considered in this section for both adhesive and sliding contact problems. In what follows, for simplicity, attention is focused on the two-dimensional (planar) geometry. The extension to three-dimensional geometry is complicated by geometric considerations only.

2.1. Finite element discretization

Throughout the remaining part of this paper, we employ the bilinear isoparametric elements for "parent" contacting bodies. Concerning the contact segment characterization, the "master-slave" slide-line contact logic is adopted (see Hallquist [1983]). In particular, a three-node contact element, consisting of two "master" nodes and one "slave" node, is used, see Figure 1. With reference to Figure 1, the tangent and normal vectors are defined as follows:

$$\mathbf{t} \equiv \frac{\mathbf{x}_2 - \mathbf{x}_1}{\| \mathbf{x}_2 - \mathbf{x}_1 \|} \tag{2.1}$$

$$\mathbf{n} \equiv \mathbf{e}_3 \times \mathbf{t} \ , \tag{2.2}$$

where $\mathbf{e}_3$ denotes the unit base vector normal to the plane of the three-node element and $\mathbf{x}_1 = \mathbf{X}_1 + \mathbf{u}_1$, $\mathbf{x}_2 = \mathbf{X}_2 + \mathbf{u}_2$ signify the *current* positions of the master nodes ($\mathbf{X}_1$, $\mathbf{X}_2$ for reference coordinates and $\mathbf{u}_1$, $\mathbf{u}_2$ for current nodal displacements). In addition, we define the current "surface coordinate" a as follows

$$a \equiv \frac{(\mathbf{x}_s - \mathbf{x}_1)}{\| \mathbf{x}_s - \mathbf{x}_1 \|} \bullet \mathbf{t} \tag{2.3}$$

in which $\mathbf{x}_s = \mathbf{X}_s + \mathbf{u}_s$ denotes the current position of the slave node. The normal and tangential gaps (penetrations) associated with a typical three-node element are defined as

$$g_n \equiv (\mathbf{x}_s - \mathbf{x}_1) \cdot \mathbf{n} \tag{2.4}$$

$$g_t \equiv (\mathbf{x}_s - \mathbf{x}_1) \cdot \mathbf{t} - a^o \, \| \mathbf{x}_2 - \mathbf{x}_1 \| \ , \tag{2.5}$$

where a^o is the (old) surface coordinate at the last time step (known). The variations (increments) of the normal and tangent vectors due to *nonlinear* kinematics can be shown to be (see Wriggers & Simo [1985])

$$\delta \mathbf{n} = \frac{-1}{\| \mathbf{x}_2 - \mathbf{x}_1 \|} (\mathbf{t} \otimes \mathbf{n}) \cdot (\boldsymbol{\eta}_2 - \boldsymbol{\eta}_1) \tag{2.6}$$

$$\delta \mathbf{t} = \frac{1}{\| \mathbf{x}_2 - \mathbf{x}_1 \|} (\mathbf{n} \otimes \mathbf{n}) \cdot (\boldsymbol{\eta}_2 - \boldsymbol{\eta}_1) \tag{2.7}$$

where $\boldsymbol{\eta}$ is the variation (increment) of $\mathbf{u}$. Furthermore, for convenience, we define the following abbreviations (operators)

$$\overline{(\bullet)} \equiv (\bullet)_s - (1-a)(\bullet)_1 - a\,(\bullet)_2 \tag{2.8a}$$

$$\overline{\overline{(\bullet)}} \equiv (\bullet)_2 - (\bullet)_1 \tag{2.8b}$$

With the above notations at hand, we now give the variational derivation.

2.2. Frictional stick

For the case of frictional stick (no-slip), we consider the following *perturbed Lagrangian* functional for bodies in contact:

$$\overline{\Pi}_\omega(\mathbf{u}\ ,\ \boldsymbol{\Lambda}_n\ ,\ \boldsymbol{\Lambda}_t) = \Pi(\mathbf{u}) + \boldsymbol{\Lambda}_n^T \mathbf{G}_n - \frac{1}{2\omega_n} \boldsymbol{\Lambda}_n^T \boldsymbol{\Lambda}_n + \boldsymbol{\Lambda}_t^T \mathbf{G}_t - \frac{1}{2\omega_t} \boldsymbol{\Lambda}_t^T \boldsymbol{\Lambda}_t \tag{2.9}$$

Here $\mathbf{u}$ designates the vector of nodal displacements, $\boldsymbol{\Lambda}_n$ ($\boldsymbol{\Lambda}_t$) the vector of normal (tangential) nodal contact forces, $\mathbf{G}_n$ ($\mathbf{G}_t$) the vector of normal (tangential) nodal gaps, and ω_n (ω_t) the normal (tangential) penalty parameters. Moreover, $\Pi(\mathbf{u})$ stands for the total potential energy of the bodies in contact.

The discrete variational equations are then obtained by taking the variations with respect to $\mathbf{u}$, $\boldsymbol{\Lambda}_n$, and $\boldsymbol{\Lambda}_t$, respectively:

$$\delta_\mathbf{u}\Pi(\mathbf{u}) + \boldsymbol{\Lambda}_n^T \delta_\mathbf{u}\mathbf{G}_n + \boldsymbol{\Lambda}_t^T \delta_\mathbf{u}\mathbf{G}_t = 0 \tag{2.10a}$$

$$\delta\boldsymbol{\Lambda}_n^T \left(-\frac{1}{\omega_n} \boldsymbol{\Lambda}_n + \mathbf{G}_n\right) = 0 \tag{2.10b}$$

$$\delta\boldsymbol{\Lambda}_t^T \left(-\frac{1}{\omega_t} \boldsymbol{\Lambda}_t + \mathbf{G}_t\right) = 0 \tag{2.10c}$$

From (2.10b,c), we obtain that $\boldsymbol{\Lambda}_n = \omega_n \mathbf{G}_n$ and $\boldsymbol{\Lambda}_t = \omega_t \mathbf{G}_t$ as the normal and tangential *penalty* contact forces.

The variation of a typical nodal normal gap $g_n \in \mathbf{G}_n$ takes the form (see Wriggers & Simo [1985])

$$\delta g_n \equiv \boldsymbol{\eta}^T \mathbf{c}_n = [\boldsymbol{\eta}_s - (1-a)\,\boldsymbol{\eta}_1 - a\,\boldsymbol{\eta}_2] \cdot \mathbf{n} \equiv \overline{\boldsymbol{\eta}} \cdot \mathbf{n} \tag{2.11}$$

where $\mathbf{c}_n \equiv \mathbf{D}_\eta(\delta g_n)$ ($\mathbf{D}$ is the directional derivative operator). Alternatively, we can write

$$\mathbf{c}_n = \bar{\mathbf{n}} \quad , \tag{2.12}$$

with the "bar" quantity defined in (2.8a). We shall give the matrix representation, within the context of three-node contact elements, of $\bar{\mathbf{n}}$ later in this section.

Similarly, the variation of a typical nodal tangential gap $g_t \in \mathbf{G}_t$ can be obtained according to

$$\begin{aligned} \delta g_t &= (\eta_s - \eta_1) \cdot \mathbf{t} + (\mathbf{x}_s - \mathbf{x}_1) \cdot \delta\mathbf{t} - a^o \, \delta \, \| \mathbf{x}_2 - \mathbf{x}_1 \| \\ &= \mathbf{t} \cdot \bar{\eta}^o + \frac{g_n}{\| \mathbf{x}_2 - \mathbf{x}_1 \|} \, (\mathbf{n} \cdot \bar{\bar{\eta}}) \equiv \eta^T \, \mathbf{c}_t \end{aligned} \tag{2.13}$$

where

$$\bar{\eta}^o \equiv \bar{\eta} \Big]_{a = a^o} = \eta_s - (1 - a^o) \, \eta_1 - a^o \, \eta_2 \quad , \tag{2.14}$$

and

$$\mathbf{c}_t \equiv \mathbf{D}_\eta(\delta g_t) = \bar{\mathbf{t}}^o + \frac{g_n}{\| \mathbf{x}_2 - \mathbf{x}_1 \|} \, \bar{\bar{\mathbf{n}}} \tag{2.15}$$

Moreover, the residual vector $\mathbf{R}_B$ and tangent stiffness $\mathbf{K}_B$ associated with the total potential energy of the contacting bodies simply read

$$\mathbf{R}_B \equiv \mathbf{D}_\eta(\Pi(\mathbf{u})) \tag{2.16}$$

$$\mathbf{K}_B \equiv \mathbf{D}_\eta(\mathbf{R}_B) \tag{2.17}$$

In the case of *inelasticity*, $\mathbf{R}_B$ and $\mathbf{K}_B$ are deduced from a (Galerkin) variational functional Π involving constitutive relations and boundary conditions.

The variational equations (2.10a,b,c) can now be stated as

$$\eta \left[\mathbf{R}_B + \mathop{\mathrm{A}}_{s=1}^{S} \left(\lambda_n^{(s)} \, \mathbf{c}_n^{(s)} + \lambda_t^{(s)} \, \mathbf{c}_t^{(s)} \right) \right] = 0 \tag{2.18a}$$

$$\delta \Lambda_n^T \left(-\frac{1}{\omega_n} \Lambda_n + \mathbf{G}_n \right) = 0 \tag{2.18b}$$

$$\delta \Lambda_t^T \left(-\frac{1}{\omega_t} \Lambda_t + \mathbf{G}_t \right) = 0 \tag{2.18c}$$

In (2.18a), the superscript (s) denotes the s-th slave node in contact and the scalars $\lambda_n^{(s)} \in \Lambda_n$, $\lambda_t^{(s)} \in \Lambda_t$. In addition, A represents an assembly operation over all three-node contact elements in consideration (S = total number of slave nodes in contact = total number of conditions of constraints). To apply the Newton's iteration scheme, consistent linearization of Eq. (2.18a,b,c) at ($\mathbf{u}$, Λ_n , Λ_t) is performed and leads to

$$
[\boldsymbol{\eta}^T , \delta\boldsymbol{\Lambda}_n^T , \delta\boldsymbol{\Lambda}_t^T] \left\{ \begin{bmatrix} \mathbf{K}_B + \mathop{\mathsf{A}}\limits_{s=1}^{S} [\mathbf{K}_n^{(s)} + \mathbf{K}_t^{(s)}] & \mathop{\mathsf{A}}\limits_{s=1}^{S} \mathbf{c}_n^{(s)} & \mathop{\mathsf{A}}\limits_{s=1}^{S} \mathbf{c}_t^{(s)} \\ \mathop{\mathsf{A}}\limits_{s=1}^{S} \mathbf{c}_n^{(s)T} & -\frac{1}{\omega_n} \mathbf{I} & \mathbf{0} \\ \mathop{\mathsf{A}}\limits_{s=1}^{S} \mathbf{c}_t^{(s)T} & \mathbf{0} & -\frac{1}{\omega_t} \mathbf{I} \end{bmatrix} \begin{Bmatrix} \Delta\mathbf{u} \\ \Delta\boldsymbol{\Lambda}_n \\ \Delta\boldsymbol{\Lambda}_t \end{Bmatrix} \right.
$$

$$
\left. = - \begin{Bmatrix} \mathbf{R}_B + \mathop{\mathsf{A}}\limits_{s=1}^{S} [\lambda_n^{(s)} \mathbf{c}_n^{(s)} + \lambda_t^{(s)} \mathbf{c}_t^{(s)}] \\ -\frac{1}{\omega_n} \boldsymbol{\Lambda}_n + \mathbf{G}_n \\ -\frac{1}{\omega_t} \boldsymbol{\Lambda}_t + \mathbf{G}_t \end{Bmatrix} \right\} \tag{2.19}
$$

where (after some algebra)

$$
\boldsymbol{\eta}^T \mathbf{K}_n^{(s)} \Delta\mathbf{u} = \frac{-\lambda_n^{(s)}}{\| \mathbf{x}_2 - \mathbf{x}_1 \|} [\Delta\bar{\mathbf{u}} \cdot (\mathbf{t} \otimes \mathbf{n}) \cdot \bar{\bar{\boldsymbol{\eta}}} + \bar{\boldsymbol{\eta}} \cdot (\mathbf{t} \otimes \mathbf{n}) \cdot \Delta\bar{\bar{\mathbf{u}}}
$$

$$
+ \frac{g_n}{\| \mathbf{x}_2 - \mathbf{x}_1 \|} \Delta\bar{\bar{\mathbf{u}}} \cdot (\mathbf{n} \otimes \mathbf{n}) \cdot \bar{\bar{\boldsymbol{\eta}}} \,]^{(s)} \tag{2.20}
$$

$$
\boldsymbol{\eta}^T \mathbf{K}_t^{(s)} \Delta\mathbf{u} = \frac{\lambda_t^{(s)}}{\| \mathbf{x}_2 - \mathbf{x}_1 \|} [\bar{\boldsymbol{\eta}}^o \cdot (\mathbf{n} \otimes \mathbf{n}) \cdot \Delta\bar{\bar{\mathbf{u}}} + \bar{\bar{\boldsymbol{\eta}}} \cdot (\mathbf{n} \otimes \mathbf{n}) \cdot \Delta\bar{\mathbf{u}}
$$

$$
- \frac{g_n}{\| \mathbf{x}_2 - \mathbf{x}_1 \|} \bar{\bar{\boldsymbol{\eta}}} \cdot (\mathbf{n} \otimes \mathbf{t} + \mathbf{t} \otimes \mathbf{n}) \cdot \Delta\bar{\bar{\mathbf{u}}} \,]^{(s)} \tag{2.21}
$$

2.2.1. Matrix representation. To facilitate finite element implementation of the above derived tangent stiffness operators and residuals, matrix formulations are given as follows. For simplicity, we will drop the superscript (s) and focus on a typical single three-node element. Let us start by introducing the following vectors

$$
\mathbf{N}_s^o \equiv [\mathbf{n} , -(1 - a^o)\, \mathbf{n} , -a^o \, \mathbf{n}]^T \tag{2.22a}
$$

$$
\mathbf{N}_s \equiv \mathbf{c}_n \equiv \bar{\mathbf{n}} \equiv [\mathbf{n} , -(1 - a)\, \mathbf{n} , -a \, \mathbf{n}]^T \tag{2.22b}
$$

$$
\mathbf{T}_s \equiv [\mathbf{t} , -(1 - a)\, \mathbf{t} , -a \, \mathbf{t}]^T \tag{2.22c}
$$

$$
\mathbf{T} \equiv \bar{\bar{\mathbf{t}}} \equiv [\mathbf{0} , -\mathbf{t} , \mathbf{t}]^T \tag{2.22d}
$$

$$
\mathbf{N} \equiv \bar{\bar{\mathbf{n}}} \equiv [\mathbf{0} , -\mathbf{n} , \mathbf{n}]^T \tag{2.22e}
$$

$$
\Delta\mathbf{u} \equiv [\Delta\mathbf{u}_s , \Delta\mathbf{u}_1 , \Delta\mathbf{u}_2]^T \tag{2.22f}
$$

$$
\mathbf{c}_t \equiv [\mathbf{t} \;,\; -(1 - a^o)\, \mathbf{t} - \frac{g_n}{\| \mathbf{x}_2 - \mathbf{x}_1 \|} \mathbf{n} \;,\; -a^o \, \mathbf{t} + \frac{g_n}{\| \mathbf{x}_2 - \mathbf{x}_1 \|} \mathbf{n}]^T \tag{2.22g}
$$

where the unified order of components in all vectors has been: slave node - master node 1 - master node 2. By using these matrix notations, Eq. (2.20) and (2.21) can be rephrased as

$$\mathbf{K}_n = -\frac{\lambda_n}{\|\mathbf{x}_2 - \mathbf{x}_1\|}\left[\mathbf{N}\,\mathbf{T}_s^T + \mathbf{T}_s\,\mathbf{N}^T + \frac{g_n}{\|\mathbf{x}_2 - \mathbf{x}_1\|}\mathbf{N}\,\mathbf{N}^T\right] \tag{2.23}$$

$$\mathbf{K}_t = \frac{\lambda_t}{\|\mathbf{x}_2 - \mathbf{x}_1\|}\left[\mathbf{N}_s^o\,\mathbf{N}^T + \mathbf{N}\,\mathbf{N}_s^T - \frac{g_n}{\|\mathbf{x}_2 - \mathbf{x}_1\|}(\mathbf{N}\,\mathbf{T}^T + \mathbf{T}\,\mathbf{N}^T)\right] \tag{2.24}$$

In addition, from Eq. (2.19) together with the fact that $\lambda_n = \omega_n\, g_n$ and $\lambda_t = \omega_t\, g_t$, the contact residual vector (due to contact only) for a single element is

$$\mathbf{R}_C = -\left[\omega_n\, g_n\, \mathbf{c}_n + \omega_t\, g_t\, \mathbf{c}_t\right] \tag{2.25}$$

The linearization of Eq. (2.25) with respect to **u** in conjunction with Eq. (2.19) then leads to the following *perturbed Lagrangian* contact tangent stiffness matrix

$$\mathbf{K}_C = \omega\left\{[\mathbf{N}_s\,\mathbf{N}_s^T + \mathbf{c}_t\,\mathbf{c}_t^T] - \frac{g_n}{\|\mathbf{x}_2 - \mathbf{x}_1\|}\left[\mathbf{N}\,\mathbf{T}_s^T + \mathbf{T}_s\,\mathbf{N}^T + \frac{g_n}{\|\mathbf{x}_2 - \mathbf{x}_1\|}\mathbf{N}\,\mathbf{N}^T\right]\right.$$
$$\left. + \frac{g_t}{\|\mathbf{x}_2 - \mathbf{x}_1\|}\left[\mathbf{N}_s^o\,\mathbf{N}^T + \mathbf{N}\,\mathbf{N}_s^T - \frac{g_n}{\|\mathbf{x}_2 - \mathbf{x}_1\|}(\mathbf{N}\,\mathbf{T}^T + \mathbf{T}\,\mathbf{N}^T)\right]\right\} \tag{2.26}$$

Here, $\omega_n = \omega_t = \omega$ (penalty parameter) has been assumed. Finally, the *total* tangent stiffness matrix and residual vector of the bodies in contact take the form

$$\mathbf{K} = \mathbf{K}_B + \mathop{\mathsf{A}}_{s=1}^{S} \mathbf{K}_C^{(s)} \tag{2.27}$$

$$\mathbf{R} = -\left[\mathbf{R}_B + \mathop{\mathsf{A}}_{s=1}^{S} \omega\,(g_n\,\mathbf{c}_n + g_t\,\mathbf{c}_t)^{(s)}\right] \tag{2.28}$$

2.3. Frictional slip

For the case of frictional slide, use of the Coulomb's law of friction renders $|\mathbf{\Lambda}_t| = \mu\,\mathbf{\Lambda}_n$, where μ denotes the coefficient of friction. Similar to the development in Sec. 2.2, characteristic variational equations (not functional) are:

$$\delta_u\Pi(\mathbf{u}) + \mathbf{\Lambda}_n^T\,\delta_u\mathbf{G}_n - \mu\,\mathbf{\Lambda}_n^T\,\delta_u\mathbf{G}_t = 0 \tag{2.29a}$$

$$\delta\mathbf{\Lambda}_n^T\left(-\frac{1}{\omega}\mathbf{\Lambda}_n + \mathbf{G}_n\right) = 0 \tag{2.29b}$$

in which ω is a penalty parameter. Note that in Eq. (2.29a) the virtual work done by the frictional force is always *negative*. Furthermore, Eq. (2.29b) yields $\mathbf{\Lambda}_n = \omega\,\mathbf{G}_n$ as the normal penalty contact force.

By taking the variations at $(\mathbf{u}\ ,\ \mathbf{\Lambda}_n)$, Eq. (2.29a,b) now read (see (2.18a,b,c))

$$\eta\left[\mathbf{R}_B + \mathop{\mathsf{A}}_{s=1}^{S}\left(\lambda_n^{(s)}\,\mathbf{c}_n^{(s)} - \mu\,\lambda_n^{(s)}\,\mathbf{c}_t^{(s)}\right)\right] = 0 \tag{2.30a}$$

$$\delta\mathbf{\Lambda}_n^T\left(-\frac{1}{\omega_n}\mathbf{\Lambda}_n + \mathbf{G}_n\right) = 0 \tag{2.30b}$$

The consistent linearization of (2.30a,b) at ($\mathbf{u}$, $\mathbf{\Lambda}_n$) then yields the following expressions (see (2.19)):

$$[\boldsymbol{\eta}^T , \delta\mathbf{\Lambda}_n^T] \begin{bmatrix} \mathbf{K}_B + \underset{s=1}{\overset{S}{\mathbb{A}}} [\mathbf{K}_n^{(s)} + \mathbf{K}_t^{(s)}] & \underset{s=1}{\overset{S}{\mathbb{A}}} [\mathbf{c}_n^{(s)} - \mu\, \mathbf{c}_t^{(s)}] \\ \underset{s=1}{\overset{S}{\mathbb{A}}} \mathbf{c}_n^{(s)T} & -\dfrac{1}{\omega} \mathbf{I} \end{bmatrix} \begin{Bmatrix} \Delta\mathbf{u} \\ \Delta\mathbf{\Lambda}_n \end{Bmatrix}$$

$$= - \begin{Bmatrix} \mathbf{R}_B + \underset{s=1}{\overset{S}{\mathbb{A}}} [\lambda_n^{(s)} \mathbf{c}_n^{(s)} - \mu\, \lambda_n^{(s)} \mathbf{c}_t^{(s)}] \\ -\dfrac{1}{\omega} \mathbf{\Lambda}_n + \mathbf{G}_n \end{Bmatrix} \tag{2.31}$$

Here, $\mathbf{K}_n$ is the same as Eq. (2.20) and (2.23), whereas $\mathbf{K}_t$ takes the following matrix form (for a single element)

$$\mathbf{K}_t = - \frac{\mu\, \lambda_n}{\| \mathbf{x}_2 - \mathbf{x}_1 \|} [\mathbf{N}_s^o\, \mathbf{N}^T + \mathbf{N}\, \mathbf{N}_s^T - \frac{g_n}{\| \mathbf{x}_2 - \mathbf{x}_1 \|} (\mathbf{N}\, \mathbf{T}^T + \mathbf{T}\, \mathbf{N}^T)] \tag{2.32}$$

It is noted that Eq. (2.32) can be obtained by simply replace λ_t in Eq. (2.24) by $[-\mu\, \lambda_n]$, as a direct consequence of the Coulomb's friction law.

Since $\lambda_n = \omega\, g_n$, the contact residual for one element is

$$\mathbf{R}_C = -\, \omega\, g_n\, [\mathbf{c}_n - \mu\, \mathbf{c}_t] \tag{2.33}$$

The linearization of (2.33) at $\mathbf{u}$ together with Eq. (2.31),(2.32),(2.23) then yield the contact tangent stiffness for frictional slip:

$$\mathbf{K}_C = \omega \Big\{ [\mathbf{N}_s\, \mathbf{N}_s^T - \mu\, \mathbf{c}_t\, \mathbf{N}_s^T] - \frac{g_n}{\| \mathbf{x}_2 - \mathbf{x}_1 \|} [\mathbf{N}\, \mathbf{T}_s^T + \mathbf{T}_s\, \mathbf{N}^T + \frac{g_n}{\| \mathbf{x}_2 - \mathbf{x}_1 \|} \mathbf{N}\, \mathbf{N}^T]$$
$$- \frac{\mu\, g_n}{\| \mathbf{x}_2 - \mathbf{x}_1 \|} [\mathbf{N}_s^o\, \mathbf{N}^T + \mathbf{N}\, \mathbf{N}_s^T - \frac{g_n}{\| \mathbf{x}_2 - \mathbf{x}_1 \|} (\mathbf{N}\, \mathbf{T}^T + \mathbf{T}\, \mathbf{N}^T)] \Big\} \tag{2.34}$$

Therefore, the *total* tangent stiffness matrix and residual vector associated with the contacting bodies are

$$\mathbf{K} = \mathbf{K}_B + \underset{s=1}{\overset{S}{\mathbb{A}}} \mathbf{K}_C^{(s)} \tag{2.35}$$

$$\mathbf{R} = -\, [\, \mathbf{R}_B + \underset{s=1}{\overset{S}{\mathbb{A}}} \omega\, g_n\, (\mathbf{c}_n - \mu\, \mathbf{c}_t)^{(s)} \,] \, . \tag{2.36}$$

3. Numerical implementation and examples

In this section, implementation of the proposed formulation within the context of the finite element method is described. Some numerical examples are also presented.

3.1. Finite element implementation

The "master-slave" slide-line contact logic is employed, which features both the "one-pass" and "two-pass" algorithms (see, e.g., Hallquist [1983]). For the planar three-node contact element under consideration, explicit vector-component expressions (6 d.o.f.) for the notations defined in Eq. (2.22a-g) can be obtained. For example,

$$\mathbf{N}_s \equiv \mathbf{c}_n = \begin{Bmatrix} -s \\ c \\ (1-a)s \\ -(1-a)c \\ a\,s \\ -a\,c \end{Bmatrix} \quad ; \quad \mathbf{c}_t = \begin{Bmatrix} c \\ s \\ -(1-a^o)c + \dfrac{g_n}{\| \mathbf{x}_2 - \mathbf{x}_1 \|}\, s \\ -(1-a^o)s - \dfrac{g_n}{\| \mathbf{x}_2 - \mathbf{x}_1 \|}\, c \\ -a^o\, c - \dfrac{g_n}{\| \mathbf{x}_2 - \mathbf{x}_1 \|}\, s \\ -a^o\, s + \dfrac{g_n}{\| \mathbf{x}_2 - \mathbf{x}_1 \|}\, c \end{Bmatrix} \tag{3.1}$$

where s , c denote $\sin\theta$, $\cos\theta$, respectively (see Fig. 1).

In the spirit of operator splitting methodology for the Coulomb's law of friction, each load (time) step is decomposed into two parts : (i) By assuming a sticking condition, a "*stick trial*" step is first performed (similar to the "elastic trial" step in classical elasto-plasticity). If the trial is successful, the contacting bodies are considered to be in a state of frictional sticking. Otherwise, (ii) a "*slip correction*" step is performed (similar to the "plastic return mapping" step in elastoplasticity) and the bodies in contact are viewed to be in a state of frictional sliding. This operator split treatment separates the no-slip and slip conditions and renders the transition from stick to slip (or vice versa) exactly the same way as the corresponding case in classical elastoplasticity. The analogy between the Coulomb's law of friction (for stick-slip contact problems) and yield criterion (for elasto-plasticity) is noted.

In all numerical examples that follow, standard Newton's method is used for solution procedure. It is emphasized that line search plays no role in numerical simulations presented in this section.

3.2. Example 1: frictional stick

This section is concerned with a (rigid or deformable) punch into an elastic foundation under the circumstance of frictional stick. See Figure 2 for (plane strain) finite element mesh and dimensions.

Case 1. Rigid punch. The material properties employed in the computation are: $E_{punch} = 10^8$ (assumed rigid), $\nu_{punch} = 0$, $E_{found} = 10^5$, $\nu_{found} = 0.3$, and $\omega = 10^7$ (penalty value). The one-pass algorithm is used in this example. The finite element solutions converge *quadratically* within 4 iterations; see Table 1 for numerical performance. The deformed mesh is displayed in Figure 3, in which the deformation is enlarged 1000 times the real scale in order to fully see the details.

Table 1. *Residual & energy norms for iterates*

Iteration	1	2	3	4
Residual	.245e+4	.299e+3	.719e–1	.173e–6
Energy	.552e+2	.151e+0	.692e–10	.727e–19

Case 2. Elastic punch. The punching block is now a deformable body. The material properties involved in the computation are: $E_{punch} = 10^4$, $\nu_{punch} = 0.3$, $E_{found} = 10^3$, $\nu_{found} = 0.3$, and $\omega = 10^7$ (penalty value). The one-pass algorithm is adopted in this example. Once again, the finite element solutions converge *quadratically* within 3 iterations; see Table 2 for numerical performance. The deformed mesh is displayed in Figure **4** (to scale).

Table 2. *Residual & energy norms for iterates*

Iteration	1	2	3
Residual	.245e+4	.528e+2	.155e–4
Energy	.682e+4	.217e–4	.116e–12

3.3. Example 2: frictional slide

Attention is now focused on the event of frictional stick-slip motion. The transition from stick to slip (or vice versa) is accounted for in this example. We once more consider an elastic punch on top of an elastic foundation made of same materials. The finite element mesh and boundary conditions are the same as Sec. **3.2** (see Fig. **2**). Moreover, the material properties used in the simulation are: $E = 10^4$, $\nu = 0$, and $\omega = 10^5$, $\mu = 0.1$ (coefficient of friction).

The punch is first vertically loaded into the elastic foundation, then move horizontally to the right by displacement controlled loading condition (vertical loads still remain). During the initial vertical loading, three bottom nodes of the punch are in contact with the foundation. In particular, the two (outer) edge nodes of the punch undergo frictional *slip* while the central node experiences frictional *stick*. The solutions converge in 7 iterations with a residual norm less than 10^{-3}. Within the proposed formulation and implementation, tangential motion across element boundaries does not impose numerical difficulties.

Before the first contact node (the rightmost contact node) of the punch reaches the right edge of the foundation, a typical iteration count for numerical convergence is 6 or 7. After the first punch element begins to overhang, the contact area is not constant and the number of contacting nodes changes. At 5% overhang (of the first punch element), the convergence takes 7 iterations. See Fig. **5** for deformed configuration (to scale). At 50% overhang, 8 iterations are taken before convergence is achieved. At 80% overhang, 11 iterations are recorded. At 90% overhang, 13 iterations are required. Finally, at 98% and 100% overhang, 15 iterations are observed. See Figures **6**, **7** for deformed meshes (to scale). After the first punch element completely overhangs, the solutions diverge which corresponds to the physical drop-off process of the punch. For this simulation, it is *crucial* to use the *two-pass* algorithm for a solution to converge. The one-pass algorithm works only before the punch overhangs. This example provides a severe test for finite element formulation of frictional contact problems.

To assess the significance of the proposed *consistent* tangent stiffness, we repeat the above numerical experiment by using the *linearized* tangent (i.e., employing only the rank-one-update terms). Before the first punch element overhangs, numerical convergence typically takes 8 or 9 iterations. At 5% overhang, the convergence takes 9 iterations. At 50% overhang, 11 iterations are observed before convergence is achieved. At 80% overhang, 19 iterations are recorded. At 90% overhang, 25 iterations are required. Finally, at 100% overhang, 33 iterations are observed. The significance of the proposed formulation versus linearized theory is clearly demonstrated.

4. Conclusion

On the basis of an operator split, the proposed formulation accommodates the frictional stick-slip motion in a variational framework. By a consistent linearization procedure, explicit expressions for the consistent contact tangent stiffness and residual have been obtained. The analogy between the proposed treatment for the stick-slip motion and the corresponding treatment in classical elasto-plasticity is noteworthy. In addition, for infinitesimal deformations (as a special case), the proposed formulation reduces to the linearized theory involving only rank-one-update terms in the tangent matrices.

To illustrate the numerical performance of the proposed formulation, some numerical examples have been presented in Sec. **3**. The significant role of the proposed tangent stiffness is fully demonstrated.

Acknowledgments. This work is sponsored by Naval Civil Engineering Laboratory with the University of California, Berkeley. This support and the continued interest of Dr. Ted Shugar are gratefully acknowledged.

References

Hallquist, J.O. [1983], "NIKE2D - A vectorized, implicit, finite deformation, finite element code for analyzing the static and dynamic response of 2-D solids," Rept. UCID-19677, Lawrence Livermore National Laboratory.

Hughes, T.R.J., Taylor, R.L., Sackman, J.L., Curnier, A. and Kanoknukulchai, W., [1976], "A finite element method for a class of contact-impact problems," *Comp. Meth. Appl. Mech. Engng.*, **8**, pp. 249-276.

Michalowski, R. and Mroz, Z. [1978], "Associated and non-associated sliding rules in contact friction problems," *Arch. of Mech.*, **30**, pp. 259-276.

Oden, J.T. and Martins, J.A.C. [1985], "Models and computational methods for dynamic friction phenomena," *Comp. Meth. Appl. Mech. Engng.*, **52**, pp. 527-634.

Wriggers, P. and Simo, J.C. [1985], "A note on tangent stiffness for fully nonlinear contact problems," *Commun. Appl. Numer. Meth.*, **1**, pp. 199-203.

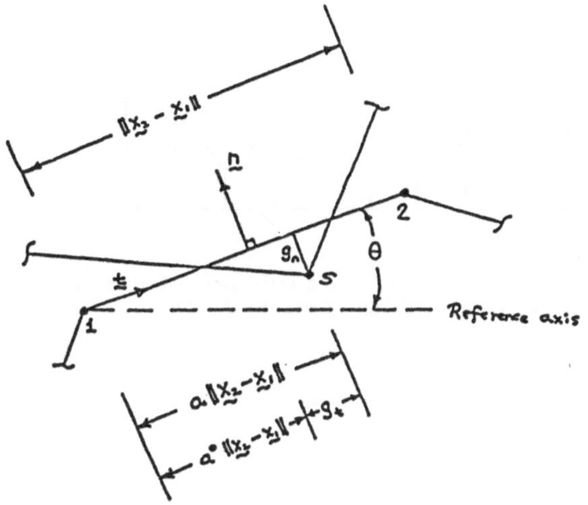

Figure 1. Geometry and definition of a typical three-node contact element.

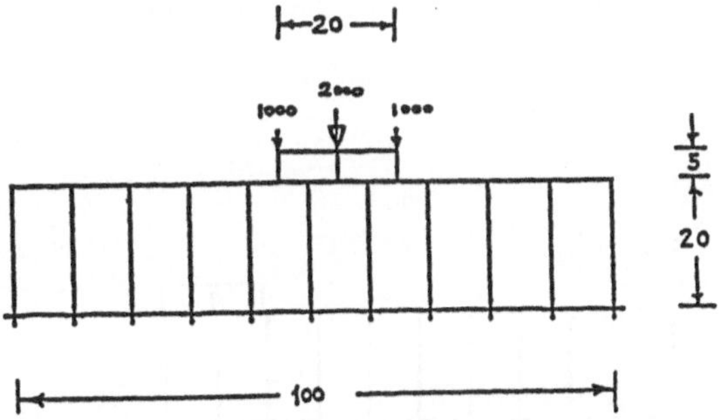

Figure 2. Finite element mesh (including boundary conditions and loads) for a punch on top of a foundation.

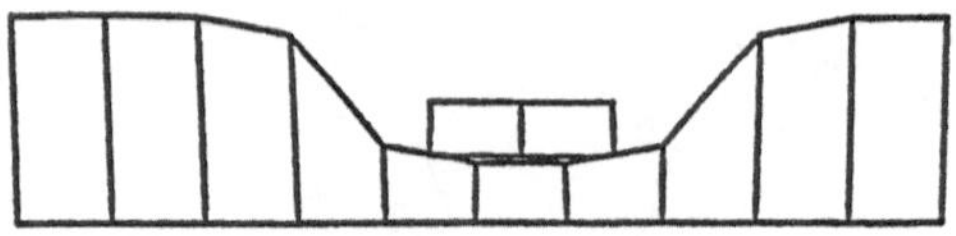

Figure 3. Deformed mesh for a rigid punch into an elastic foundation. The deformation is enlarged 1000 times for clarity.

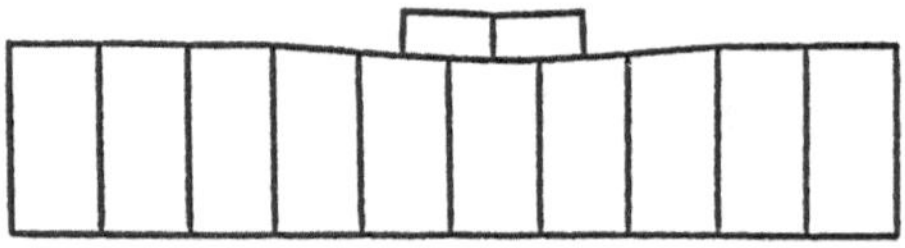

Figure 4. Deformed mesh for an elastic punch into an elastic foundation.

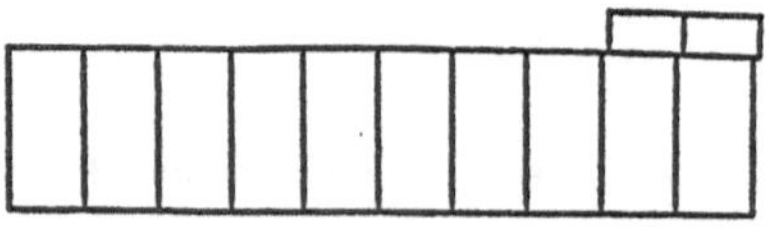

Figure 5. Deformed mesh corresponding to 5% overhang of the first (right) element of the punch.

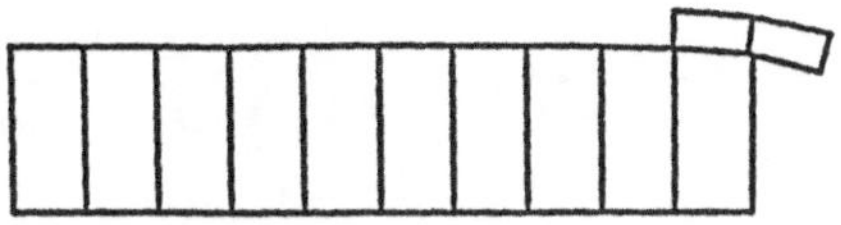

Figure 6. Deformed mesh corresponding to 98% overhang of the first punch element.

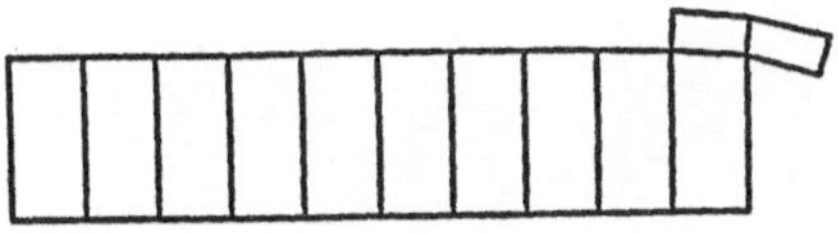

Figure 7. Deformed mesh corresponding to 100% overhang of the first punch element.

A LINEARIZATION METHOD FOR MULTILEVEL OPTIMIZATION

Michael Beers
Douglas Aircraft Company
Long Beach, California

Garrett N. Vanderplaats
University of California, Santa Barbara

ABSTRACT

A method is described wherein a large optimization problem is broken into a group of problems in a multilevel scheme. The suboptimization problems are solved simultaneously, followed by solution of a large system-level control problem. The sublevel problems involve minimizing a parameter of the designer's choice over a subset of the design variables, subject to a set of exact and approximate constraints. The system-level optimization consists of minimizing the overall system objective in an entirely linear approximate problem over all of the design variables. This problem can thus be solved using a linear programming algorithm.

THE OPTIMIZATION PROBLEM

While formal optimization has experienced increasing acceptance within industry over the past decade, its application to large engineering problems throughout the design cycle has been limited. This is due both to the cost of optimization and reoptimization as design criteria evolve, and to the unrealistic assumption that one group will design an entire large engineering system. One method for optimizing such large problems is a multilevel formulation.

In such a formulation, a design problem is broken into a number of subproblems, each of which is addressed in parallel. Several methods have been suggested to achieve this goal. In 1972, Kirsch, Reiss, and Shamir [1] presented a method wherein substructures of a large problem are optimized independently. While allowing individual substructures of a problem to be optimized, this early attempt at a multilevel scheme did not bring an optimizer to bear on the entire assembled problem. Thus, the method precluded the subsystem tradeoffs necessary to find a true system optimum. In 1978, Schmit and Ramanathan [2] published a two-level method for optimizing truss and wing structures that recognized this need. In addition, they made the significant point that interlevel coupling in a multilevel scheme must be minimized. Because subsystem design cycles may attempt to make changes counterproductive to the system-level (overall) strategy, it was suggested that a successful multilevel method should employ some methodology to reduce possible system/subsystem competition. In 1982 and 1983, Sobiezczanski-Sobieski presented papers [3,4] describing a more general method employing these ideas and incorporating optimum sensitivities. In this true multilevel formulation, the problem can be broken down into an arbitrary number of levels. The aforementioned coupling is eliminated by holding constant during the sublevel optimizations any parameters that are variables at the next higher level. In order to track lower level variables during higher level optimizations, the authors suggest the use of optimum sensitivities. These papers, then, outline the basis from which the method presented here was derived.

Linear Approximation Technique

To create linear models from the (generally) nonlinear objective and constraint functions, a first-order Taylor-series expansion will be employed. In representing an arbitrary function F(X), the approximation is (where NDV = number of design variables):

$$F(\underline{X}) = F_o + \sum_{i=1}^{NDV} \frac{\partial F(\underline{X})}{\partial X_i} (X_i - X_{i_o})$$

where $\partial F(\underline{X})/\partial X_i$ represents the partial differential of the function $F(\underline{X})$ with respect to X_i.

One-Level Optimization

The multilevel method described in this paper will be compared to a one-level formulation. Also, the bottommost level in this formulation is itself a one-level optimization. Therefore, a brief problem statement for that method is given here.

The objective is to minimize some function, F:

$$\text{Min:} \quad F(\underline{X})$$

with

$$\underline{X}^T = \{X_1, X_2, X_3, \ldots, X_{NDV}\} \tag{2}$$

$$\text{Subject to:} \quad G_J(\underline{X}) \leq 0 \qquad J = 1,NCON$$

and side constraints;

$$X_i^\ell < X_i < X_i^u \qquad i=1,NDV$$

where NCON is the number of inequality constraints, G. Equality constraints will be handled by rephrasing them as two inequality constraints, so that they can be included in $\underline{G}$. Upper and lower limits on the allowable range of X_i are prescribed by X_i^u and X_i^ℓ, respectively. Many solution techniques are available for such optimization problems [5].

Two-Level Formulation

Instead of describing a general multilevel formulation here, a two-level formulation will be detailed for simplicity and brevity. Expansion of the method described to an arbitrary number of levels follows naturally. First, the entire system to be optimized is broken down into a number of subsystems, as shown in Figure 1. In a frame structure, for instance, these subsystems may be individual frame members, which, in turn, are made up of detailed cross-sectional dimensions (flange thicknesses and widths, etc.). Each subsystem, then, has a set of design variables associated with it that are a subset of the entire system's design variable set. Vectorially:

$$\underline{X}^T = \{\underline{x}_1, \underline{x}_2, \ldots, \underline{x}_N\}$$

$$= \{(x_1,x_2,x_3,x_4),(x_5,x_6),(\ldots),(x_{NDV-2},x_{NDV-1},x_{NDV})\}$$

The objective, too, can in a sense be subdivided. In the structure example, the global goal is the minimization of material. At the subsystem level, this is also the goal, though not the primary one, as discussed previously. Since this is not an optimization governed by a system-wide strategy, it should not be allowed to dominate the problem.

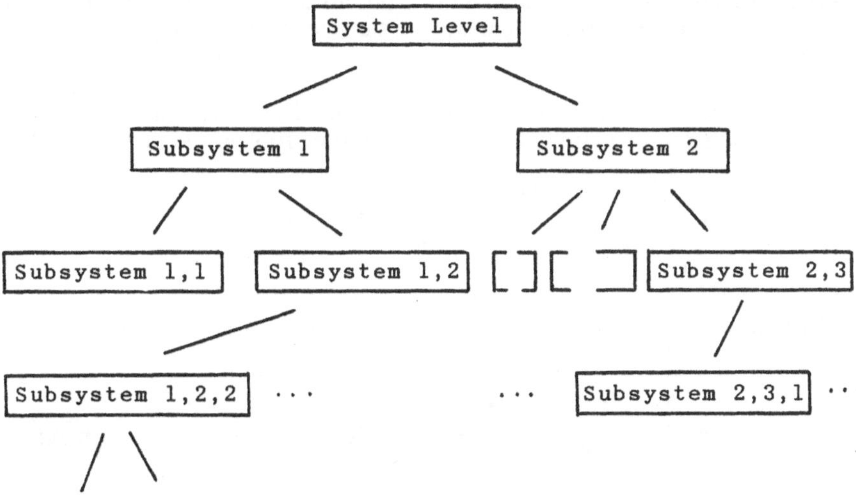

Figure 1. Hierarchical Substructuring in the Multilevel Problem Formulation

Furthermore, constraints can generally be classified into two categories. First, consider constraints that depend directly on design variables from mutliple subsystems. For instance, in a frame structure divided into subsystems by frame member, joint displacements will generally be affected by the design variable changes in multiple subproblems. Therefore, constraints of displacements should be considered at the system level, where all variables are considered. Other constraints are less strongly affected by design variables in other subsystems. In a structure, stress and local buckling constraints fall into this category. These constraints, then, can be addressed directly at the local level.

Unlike the methods discussed above, all subsystem design variables are considered together at the system level. The problem is simplified in that it is an entirely linear problem retaining only active constraints. System-level constraints are handled simply by using standard behavior sensitivities with respect to dependent design variables. Local constraints are considered in the same fashion using gradients determined at the subsystem level.

The subsystem problems are solved in their true form, employing linear approximations only for system-level constraints. The objective used is chosen by the designer. For instance, he may prefer to minimize member volume or minimize the most violated constraint in his subsystem. Constraints include local or subsystem constraints and, as mentioned above, linearized approximations of the system-level constraints that are affected by subsystem optimizations. Obviously, system-level constraints were chosen as such because they are dependent on more than one subsystem. Even so, some prediction of their change can be made at this level. Since only the "active" subsystem is modifying these constraints at any one time, the constraints are adjusted so that they appear closer to being violated or unviolated than they really are, depending on whether they are feasible or infeasible, respectively. This is done by dividing the initial value of the Taylor-series expansion by a reduction factor, R. The value used for this factor reflects the number of active subsystems that will contribute to the modification of the specific constraint. Move limits are employed to reduce the numerical difficulties characteristic of linear approximation methods. Subsystem move limits are kept to a fraction of those used at the system level so that the system-wide optimization primarily guides the design evolution.

The formulation can now be stated for a two-level optimization. At the system level, the problem is an entirely linear one, with the problem statement as follows:

$$\text{Min:} \quad F(\underline{X}) = F_o + \sum_{i=1}^{NDV} \frac{\partial F(\underline{X})}{\partial X_i} (X_i - X_{i_o})$$

$$\text{Subject to:} \quad G_J(\underline{X}) = G_{J_o} + \sum_{i=1}^{NDV} \frac{\partial G_J(\underline{X})}{\partial X_i} (X_i - X_{i_o}) \leq 0 \quad J=1,NCON$$

$$g_j(X) = g_{j_o} + \sum_{i=1}^{NDV} \frac{\partial g_j(\underline{X})}{\partial X_i} (X_i - X_{i_o}) \leq 0 \;\; 0 \quad j=1,ncon \tag{3}$$

$$X_i^{\ell} < X_i < X_i^u \quad i=1,NDV$$

where

$$X_i^u = (1 + ML) * X_i$$
$$X_i^{\ell} = (1 - ML) * X_i$$

This problem is solved within the move limits, ML, as indicated.

Following solution, the design variable values, x, for each specific subsystem are sent to the sublevel. Additionally, where system-level constraints are directly affected, they are sent to the relevant subsystem along with appropriate gradients (those with respect to the variables of that subsystem) and the reduction factor, R.

The general subsystem problem to be solved, then, is:

Min: $f(\underline{x})$ = typically nonlinear function at $\underline{x}$

Subject to: $g_j(\underline{x}) = \text{typically nonlinear function at } \underline{x}$

$$G_J(\underline{x}) = \frac{G_{J_o}}{R} + \sum_{i=1}^{ndv} \frac{\partial G_J(X)}{\partial x_i}(x_i - x_{i_o}) \leq 0 \quad J=1,NCON \quad (4)$$

$$x_i^{\ell} < x_i < x_i^{u} \quad i=1,ndv$$

$$x_i^{u} = (1 + ml) * x_i$$

$$x_i^{\ell} = (1 - ml) * x_i$$

Each of these subsystem optimizations is in turn solved within its lower level move limits, ml. Design variables, x_i, and the ncon subsystem constraint values with their gradients are then "floated" back up to the system level.

Given the problem definition above, it is apparent how the choice of the constants R, ML, and ml will significantly affect the convergence characteristics of the algorithm. The actual sequence of these steps leading to optimization is outlined in Figure 2.

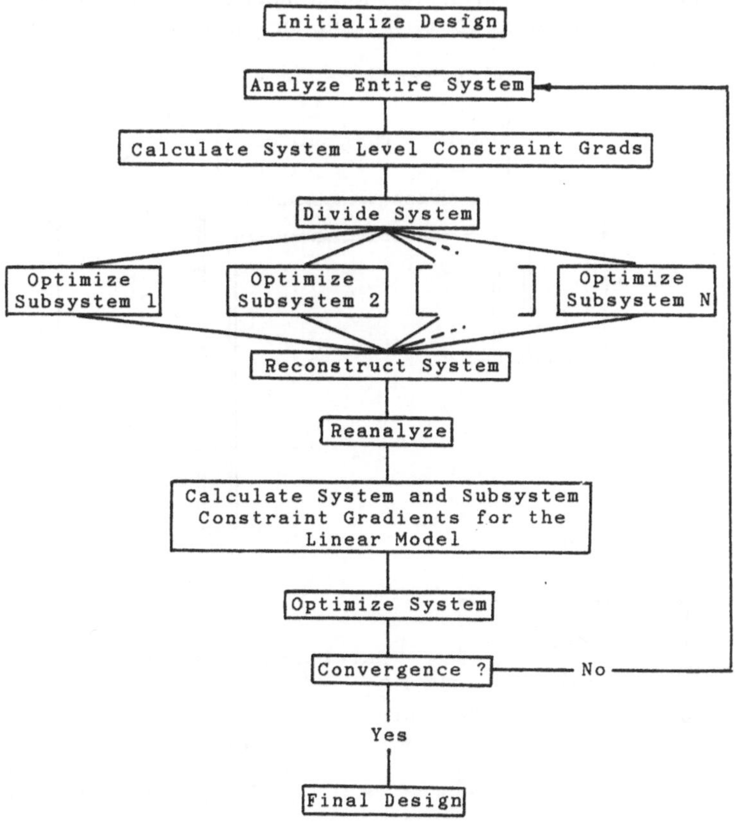

Figure 2. Multilevel Optimization Procedure

Convergence

Since the system-level or governing optimization problem for this method is effectively a sequential linear programming scheme, this formulation would be expected to have the same capability of converging to an optimum as that method so long as the system-level problem dominates the design progression. Assuming a convex design space, the method should converge eventually to an optimum design as defined by the Kuhn-Tucker conditions of optimality. As in the development of any optimization algorithm, experience will provide the true test of the method's convergence characteristics and capabilities.

Portal Frame Test Case

The problem used here to illustrate the convergence characteristics of the method is the portal frame shown in Figure 3. Each frame element, which consists of multiple cross-sectional detail dimensions, can be considered a substructure. Constraints include both those at the system level (displacement of a joint) and at the local level (stress and buckling), broken down as discussed previously.

The objective in optimizing the structure is minimizing material volume. Constraints are placed on the displacement of the upper right joint, where the forces are applied, to limit both translation and rotation at that point. Constraints are placed on the maximum normal stress for each beam, and shear stress is constrained to

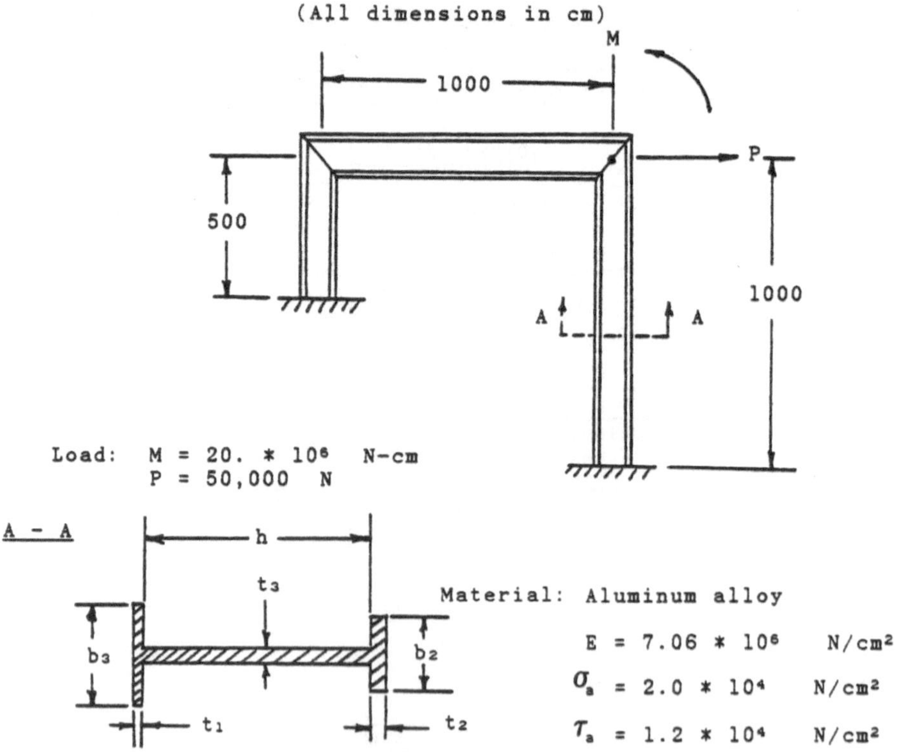

Figure 3. Portal Frame Test Problem

preclude local buckling of the beams. The problem is a nonlinear one of 18 design variables. These variables consist of flange and web widths and thicknesses for each beam. There are 58 displacement, stress, and buckling constraints.

Optimizer

Both the system and subsystem level optimizations were solved using the ADS optimization code [6]. ADS was chosen for its capability to easily compare the results and convergence rates of several optimization algorithms on the same problem. In a real application, certainly the most efficient linear problem solver, such as a Simplex scheme, would be employed at the system level.

Results for Portal Frame Example

Convergence from various initial design vectors were investigated, with the results and convergence history for one typical example shown in Figures 4 and 5. For a baseline comparison, the case was also solved by a single-level sequential linear programming method of similar sophistication.

It can be seen that the multilevel scheme converges effectively to the same solution within a similar number of iterations as the baseline method. In general, it tends to arrive at a feasible solution more rapidly than the single-level routine. This seems reasonable, as the lower level problems, while attempting to decrease component volume, move the system design toward the feasible region of the design space. In

Initial Design Feasible:

Variable Name	Initial Value, cm	Single-Level Result, cm	Multilevel Result, cm
b_1	45.0	7.43	7.25
t_1	0.95	0.378	0.498
h	95.0	85.3	83.0
t_3	0.95	0.377	0.371
b_2	45.0	7.51	6.77
t_2	0.95	0.374	0.533
b_1	45.0	5.00	5.09
t_1	0.95	0.200	0.203
h	95.0	10.0	10.1
t_3	0.95	0.200	0.202
b_2	45.0	5.00	5.06
t_2	0.95	0.200	0.201
b_1	45.0	14.3	14.5
t_1	0.95	0.633	0.642
h	95.0	72.6	71.9
t_3	0.95	0.428	0.427
b_2	45.0	14.3	14.5
t_2	0.95	0.633	0.642
Final Objective, Frame Volume, cm^3		71.34	71.57

Figure 4. Portal Frame Results

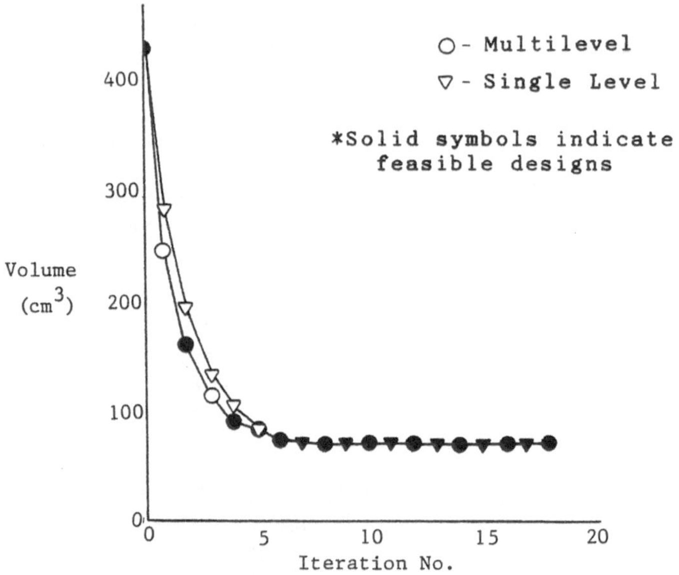

Figure 5. Portal Frame Design History — Feasible Initial Design

effect, the lower level problems tend to repair the linearization errors of the system-level optimization. This also leads to improved linearizations in subsequent system-level problems, which explains the more consistent sequence of feasible designs.

FUTURE DEVELOPMENT

The method, as it stands, is still an investigative tool, with much refinement yet to be done. Apparent areas for further development are the choice of move limits on the design variables at both the system and subsystem levels, and the method of handling the division of system-level constraints among subsystem-level problems. Work in these and other areas is continuing.

It is significant that this paper does not present an absolutely specific multilevel optimization method. Rather, it advances a multilevel problem-solving structure, which can be varied to accommodate specific details as efficiency dictates. Perhaps a purely linear system-level problem is not the best choice. For example, alternative approaches would be to use penalty functions or convex linearization. Only more work will determine whether the basic method presented here or this method modified in some manner will prove to be the most efficient.

SUMMARY

In this paper, a method is presented wherein an optimization problem is broken down in a multilevel scheme, in which separate branches of a problem can be optimized in parallel. Illustrated in a two-level format, it is shown that the method does converge efficiently while operating in a structure closely representing that which already exists in engineering design groups. Further work, including the solution of more complex test cases divided into more levels should yield a better understanding of the method's convergence characteristics and capabilities.

REFERENCES

1. Kirsch, U., Reiss, M., and Shamir, U., Optimum Design by Partitioning into Substructures, Journal of the Structural Division, Proceedings of the ASCE, January 1972, p 249.

2. Schmit, L.A., and Ramanathan, R. K., Multilevel Approach to Minimum Weight Design Including Buckling Constraints, AIAA Journal, Vol. 16, No. 2, February 1978, p 97.

3. Sobieszczanski-Sobieski, J., A Linear Decomposition Method for Large Optimization Problems — Blueprint for Development, NASA Technical Memorandum 83248, February 1982.

4. Sobieszczanski-Sobieski, J., James, B., and Dovi, A., Structural Optimization by Multilevel Decomposition, AIAA Paper No. 83-0832, presented to AIAA/ASME/ASCE/AHS 24th Structures, Structural Dynamics, and Materials Conference, Lake Tahoe, Nevada, May 2-4, 1983.

5. Vanderplaats, G. N., *Numerical Optimization Techniques for Engineering Design: with Applications*, McGraw-Hill, 1984.

6. Vanderplaats, G. N., ADS — A FORTRAN Program for Automated Design Synthesis: Version 1.00, NASA Contractor Report 172460, October 1984.

HARBOR - A PROGRAM FOR HORIZONTAL LOAD ANALYSIS OF MARINE STRUCTURES

Stanko Brčić

Civil Engineering Faculty
11000 Beograd, Yugoslavia

SUMMARY

The paper presents horizontal load analysis, both statical and dynamical, of marine structures. By marine structures we consider various piers, wharfs, dockes etc., i.e. structures consisting basically of a slab supported by a substantial number of vertical and raked piles. Equations of equilibrium and motion are formulated and special emhasis is given to the problem of impact of a ship. It is analyzed by direct evaluation of internal impulses and impact forces which are developed between the ship and the deck during impact.

FORTRAN computer code HARBOR has been developed for complete horizontal load analysis. The program has been used commercially in the main design of five river harbors in Yugoslavia, two of which are already in use, while another three are in construction.

1. INTRODUCTION

Marine structures under consideration hare are, basically, of a very simple structural form: they consist of a horizontal reinforced concrete slab supported by a number of vertical and raked piles, which may be of quite a substantial length. Such structures are located on the banks of navigable waters with the main purpuse to load or unload cargo transport vessels. Consequently, movable cranes, railways, heavy trucks, etc., are always present on the deck.

Due to a substantial free length of supporting piles, overall stability of such a structure is dominated by the action of various horizontal forces acting on the structure. Such horizontal forces are arising from various sources: namely, due to movable vehicles on the deck, wind forces, seismic forces, pull of the mooring ropes, impact of a ship, etc. Fig. 1 represents a typical cross section of considered harbor structure.

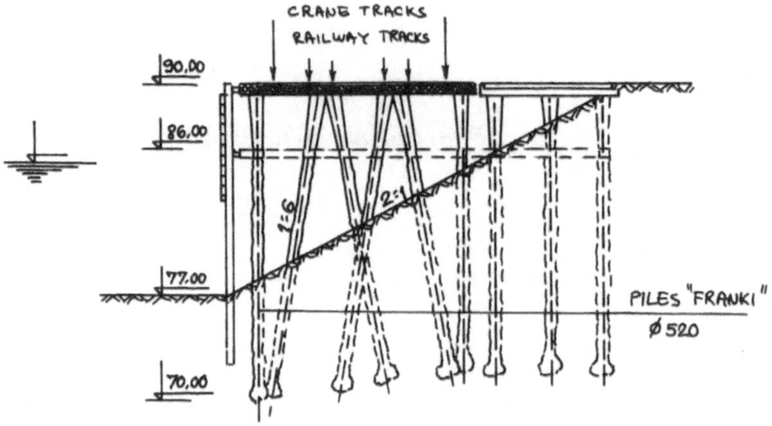

Fig. 1. Typical cross section of a harbor structure

2. STATICAL AND DYNAMICAL HORIZONTAL LOAD ANALYSIS

The main assumptions adopted in the horizontal load analysis are the following:

(1) Marine deck is infinitely stiff in its plane and it may move as a rigid body (lamina) in horizontal plane having three degrees of freedom. The planar motion of the deck is constrained by the presence of supporting piles.

(2) The piles are assumed to be built-in into the deck and into the soil at some pre-determined depth. The presence of the massive deck does not allow any rotation of the pile caps.

(3) All external forces are acting in the mid-plane of the deck.

Fig. 2 displays the plane view of marine deck with initially chosen reference coordinate system $O\bar{x}\bar{y}\bar{z}$ and corresponding three DOF: u_O, v_O, ϕ. Point P_i represents the position of one of the piles ($i=1,2,..,NP$), while B_j represents the point of application of one of the external forces $\vec{F}_j$ ($j=1,2,..,NF$).

Supporting piles are either vertical or raked, with plane of inclination being parallel to one of the principal directions of the deck: x or y , see Fig. 3.

Horizontal stiffness coefficients of vertical and raked piles, corresponding to u_i and v_i displacements of the pile caps, may be derived as follows:

- vertical piles

$$k_x = k_y = \frac{12EJ}{l^3} \tag{1}$$

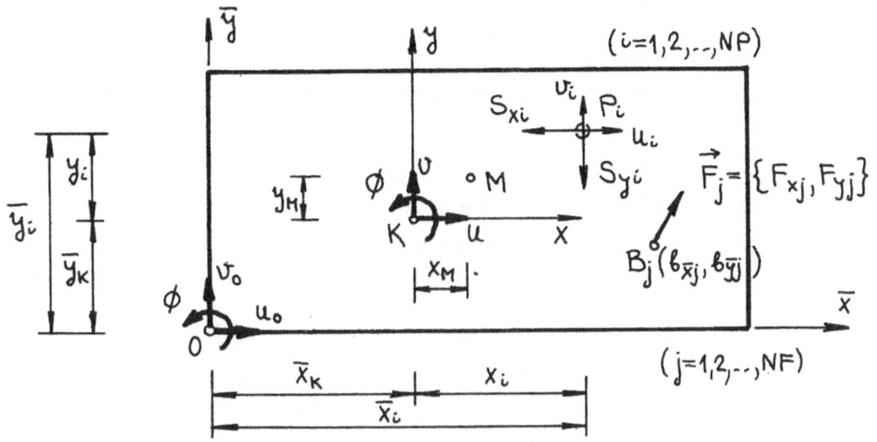

Fig.2. Definition sketch of the deck (*K* - centar of rigidity, *M* - centar of mass)

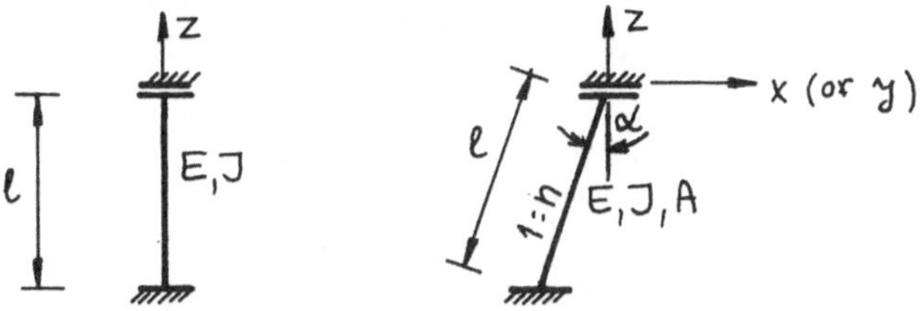

Fig. 3. Vertical and raked piles

- inclined piles (say in *xz* plane)

$$k_x = \frac{EJ}{\frac{l^3}{12}\,\theta_2 + \frac{J}{A}\,l\theta_3} \quad , \quad k_y = \frac{12EJ}{l^3} \tag{2}$$

where

$$\theta_2 = \cos^2\alpha - \theta_1 \qquad \theta_3 = \sin^2\alpha + \theta_1$$

$$\theta_1 = \sin\alpha\,\cos\alpha \cdot \frac{l^2/12 - J/A}{l^2/12\ \mathrm{tg}\alpha + J/A\ \mathrm{ctg}\alpha} \tag{3}$$

Other notations used in (1) - (3) are: E, J, A represent the pile modulus of elasticity, cross-sectional moment of inertia and area, while l and α are the pile length and angle of inclination.

Let us first consider the statical case. The deck is under the action of an arbitrary system of external forces $\vec{F}_j = \{F_{xj}, F_{yj}\}$ and forces of constraint $\vec{S}_i = \{S_{xi}, S_{yi}\}$, representing supporting piles. Forces of constraint are given by

$$S_{xi} = k_{xi} u_i \qquad S_{yi} = k_{yi} v_i \tag{4}$$

where k_{xi}, k_{yi} are the stiffness coefficients given by (1), (2), while u_i, v_i are horizontal displacements of each pile cap. Equilibrium equations with reference to initially adopted frame $O\bar{x}\bar{y}\bar{z}$ are given by:

$$[\bar{K}]\{u_o\} = \{f_o\} \tag{5}$$

The stiffness matric is obtained in the form

$$[\bar{K}] = \begin{bmatrix} \bar{k}_{11} & 0 & \bar{k}_{13} \\ 0 & \bar{k}_{22} & \bar{k}_{23} \\ \bar{k}_{13} & \bar{k}_{23} & \bar{k}_{33} \end{bmatrix} \tag{6}$$

where

$$\begin{aligned} \bar{k}_{11} &= \sum_c k_{xi} & \bar{k}_{22} &= \sum_i k_{yi} \qquad (\sum_i = \sum_{i=1}^{NP}) \\ \bar{k}_{12} &= - \sum_i \bar{y}_i k_{xi} & \bar{k}_{23} &= \sum_i \bar{x}_i k_{yi} \\ \bar{k}_{33} &= \sum_i (\bar{y}_i^2 k_{xi} + \bar{x}_i^2 k_{yi}) \end{aligned} \tag{7}$$

while displacement and force vectors are:

$$\{u_o\} = \begin{Bmatrix} u_o \\ v_o \\ \phi \end{Bmatrix} \qquad \{f_o\} = \begin{Bmatrix} F_x \\ F_y \\ M_{\bar{z}} \end{Bmatrix} \tag{8}$$

with

$$\begin{aligned} F_x &= \sum_j F_{xj} \qquad F_y = \sum_j F_{yj} \qquad (\sum_j = \sum_{j=1}^{NP}) \\ M_{\bar{z}} &= \sum_j (b_{\bar{x}j} F_{yj} - b_{\bar{y}j} F_{xj}) \end{aligned} \tag{9}$$

Equilibrium equations may be easily solved:

$$\{u_o\} = [\bar{K}]^{-1}\{f_o\} \tag{10}$$

so the forces of constraint, i.e. the forces acting on each pile cap, are given by:

$$\begin{Bmatrix} S_{xi} \\ S_{yi} \end{Bmatrix} = \begin{bmatrix} k_{xi} & 0 & -\bar{y}_i k_{xi} \\ 0 & k_{yi} & \bar{x}_i k_{yi} \end{bmatrix} \begin{Bmatrix} u_o \\ v_o \\ \phi \end{Bmatrix} \tag{11}$$

With theese forces known, normal forces (for raked piles only), shear forces and bending moments in each pile may be easily obtained.

Even trough the stiffness matrix $[\bar{K}]$ is of the order 3, it may be transformed into diagonal one by simple translation of the initial reference frame $O\bar{x}\bar{y}\bar{z}$ into the new one $Kxyz$. The origin of the new system, i.e. the center of rigidity of the deck K, is given by

$$\bar{x}_K = - \frac{\bar{k}_{23}}{\bar{k}_{22}} \qquad \bar{y}_K = \frac{\bar{k}_{13}}{\bar{k}_{11}} \tag{12}$$

Equilibrium equations corresponding to the new system $Kxyz$ are given by

$$[K]\ \{u\} = \{f\} \tag{13}$$

where $[K]$ is now diagonal matrix, with $\{u\}$ and $\{f\}$ being displacement and external force vectors corresponding to the new frame $Kxyz$. This is not really necessary in the statical analysis, since there are only three equations, but it is usefull in dynamical analysis.

If the external horizontal forces are dependent upon the time t, differential equations of motion of the deck are derived in the usual form:

$$[M]\{\ddot{u}\} + [C]\{\dot{u}\} + [K]\ \{u\} = \{f(t)\} \tag{14}$$

It is now assumed that the reference frame is in the center of rigidity, so $[K]$ is a diagonal stiffness matrix, while the mass matrix $[M]$ is given by:

$$[M] = \begin{bmatrix} m & 0 & -y_M m \\ 0 & m & x_M m \\ -y_M m & x_M m & J_z \end{bmatrix} \tag{15}$$

Coeeficient m represents the mass of the deck, while J_z is the mass moment of inertia related to Kz axis. Coordinates x_M and y_M define the position of centerof-mass related to $Kxyz$ system, see Fig.2. Finally, the damping matrix $[C]$, if desired in analysis at all, is assumed in the usual Rayleigh form.

Free vibration problem of the deck is easily solved from:

$$[M]\{\ddot{u}\} + [K]\ \{u\} = \{0\} \tag{16}$$

Seismical behaviour of the deck may be analyzed in the usual manner, with known or assumed base excitation, but it will not be elaborated here, since usual seismic design codes define some equivalent statical loading approach. Further attention will be given to the impact of the ship problem.

3. IMPACT OF A SHIP

The usual approach to deal with impact of a ship is based upon the conservation of mechanical energy. Namely, the total kinetic energy of a ship, usually including the influence of added mass of water, is equated with the maximum potential energy of the deck.

Impact of a ship is treated here in a different manner. The approach consists of direct evaluation of internal impulses exerted between the ship and the deck during impact. Two different impact modelings are considered.

In the first one, the ship is treated as a rigid rod which is moving translationally and colliding with immovable rigid boundary (obstacle) representing the deck, see Fig. 4:

We consider now the isolated ship, as a rigid rod, immidiately before and after the impact. Dynamic equations

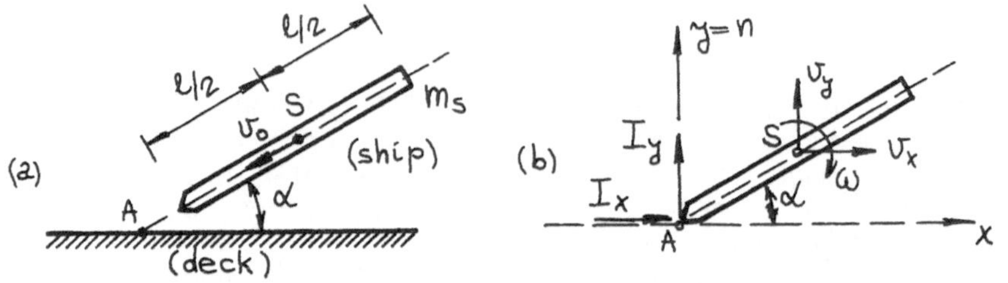

Fig. 4. Impact of a ship (model 1): (a) immidiately before and (b) immidiately after the impact (isolated ship)

defining the impact are given by

$$\vec{K}_2 - \vec{K}_1 = \vec{I}_R \qquad \vec{D}_2^{(S)} - \vec{D}_1^{(S)} = \vec{H}_R^{(S)} \tag{18}$$

where the following notation is introduced:

$\vec{K}_1, \vec{K}_2$ - momentum of a ship before and after the impact

$\vec{D}_1^{(S)}, \vec{D}_2^{(S)}$ - moment of momentum of a ship before and after the impact evaluated with regard to center-of-mass S of the ship

$\vec{I}_R, \vec{H}_R^{(S)}$ - resultant internal impuls created during impact and its moment with regard to S

Scalar form of eqs. (18) is given by, see Fig. 4:

$$\begin{aligned} m_S(v_x + v_O\cos\alpha) &= I_x \\ m_S(v_y + v_O\sin\alpha) &= I_y \\ J_S\omega &= I_y\, l/2\, \cos\alpha - I_x\, l/2\, \sin\alpha \end{aligned} \tag{19}$$

Unknown quantities in eqs. (19) are v_x, v_y, ω, which represent velocity components of center-of-mass of a ship and its angular velocity immidiately after the impact, and also I_x, I_y which represent components of internal impulse acting on a ship. Impulse I_y represents the normal component, while I_x is a consequence of assumed friction between the ship and the deck. Therefore,

$$I_x = \mu I_y \tag{20}$$

where μ represents the coefficient of friction.

The closure relation is obtained by adopting the Newton´s impact hypothesis, according to which

$$\varepsilon = \frac{v_y + \omega\, l/2\, \cos\alpha}{v_o \sin\alpha} \tag{21}$$

where ε represents the coefficient of restitution: $\varepsilon \in [0,1]$. By elimination of velocities v_x, v_y and ω from the system (19)-(21), one may obtain internal impuls I_y in the form:

$$I_y = m_S v_o (1+\varepsilon) \cdot \frac{\sin\alpha}{1+3\cos\alpha(\cos\alpha-\mu\sin\alpha)} \tag{22}$$

while I_x is given by (20). Since the impuls is defined as

$$\vec{I} = \int_{t=0}^{t+\tau} \vec{F}\, dt \tag{23}$$

where τ is some short period of time (duration of impact), if we assume duration τ, say $0.2 \div 0{,}5$ *sec*, average impact forces are given by:

$$F_x = \frac{1}{\tau} I_x \qquad F_y = \frac{1}{\tau} I_y \tag{24}$$

Now we turn our attention to the deck, Fig. 5:

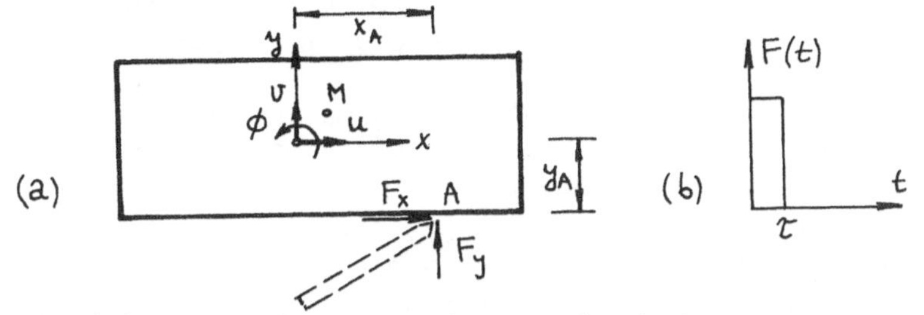

Fig. 5. (a) Impact forces acting on the deck.
(b) Assumed impulse shape

Impact forces corresponding to obtained internal impulses are clearly dynamic forces. However, if we assume some time dependence of impact forces, for instance, rectangular, triangular or sine half-wave impuls, corresponding dynamic magnification factor may be determined, so dynamical impact forces may be treated as equivalent statical forces. For instance, for rectangular shape of impulse, dynamic factor is given in the form:

$$D = \begin{cases} 2 \sin\left(\frac{\pi\tau}{T}\right) & \text{for} \quad \tau \leq 0{,}5\,T \\ 2 & \text{for} \quad \tau \geq 0{,}5\,T \end{cases} \tag{25}$$

where T represents the natural period of free vibrations (corresponding to each DOF of the deck: $u, v,$ or ϕ). Therefore, further analysis is given by eq. (13), introducing the corresponding dynamic factors:

$$\begin{bmatrix} k_{11} & & \\ & k_{22} & \\ & & k_{33} \end{bmatrix} \begin{Bmatrix} u \\ v \\ \phi \end{Bmatrix} = \begin{Bmatrix} D_x F_x \\ D_y F_y \\ D_z (F_y x_A - F_x y_A) \end{Bmatrix} \tag{26}$$

Dynamic factors are generally, different, depending upon each of the natural periods.

In the second, improved, approach, the deck is not treated

as a rigid boundary, but as a rigid body (lamina) undergoing the planar motion. Therefore, impact of a ship is treated as eccentric impact of two rigid bodies in plane motion, see Fig.6:

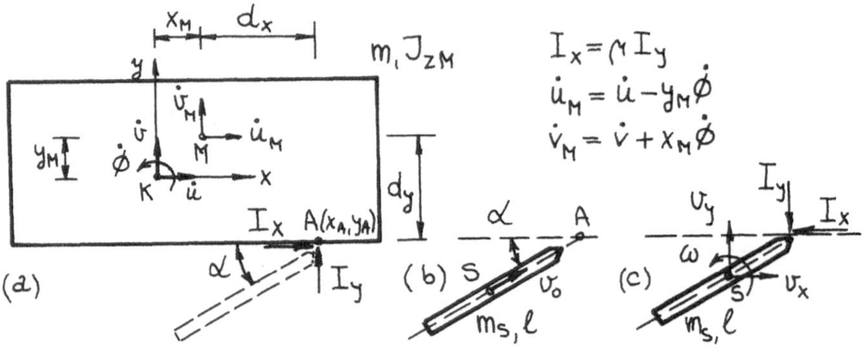

Fig.6. Impact of a ship (model 2): (a) Marine deck immidiately after the impact; (b), (c) Isolated ship immidiately before and after the impact

Eqs. (18) are written separately for isolated deck and isolated ship, which is again considered as a rigid rod. This is not a necessity, but a consequence of usual river cargo vessels. Scalar form of eqs. (18), applied to model 2 in Fig.6, is the following:

- for the deck

$$\begin{aligned} m\dot{u} - my_M\dot{\phi} &= I_x \\ m\dot{v} + mx_M\dot{\phi} &= I_y \\ J_{zM}\dot{\phi} &= I_y d_x + I_x d_y \end{aligned} \tag{27}$$

- for the ship

$$\begin{aligned} m_S v_x - m_S v_o \cos\alpha &= - I_x \\ m_S v_y - m_S v_o \sin\alpha &= - I_y \\ J_S\omega &= - I_y\, l/2 \cos\alpha + I_x\, l/2 \sin\alpha \end{aligned} \tag{28}$$

with the coefficient of restitution given by

$$\varepsilon = \frac{\dot{v} + x_A\dot{\phi} - (v_y + \omega\, l/2 \cos\alpha)}{v_o \sin\alpha} \tag{29}$$

Internal impulses I_x and I_y are correlated to each other by eq. (20). It should be mentioned that one of the commponly adopted assumptions related to impact is that any displacements during the impact are negligible. Consequently, ne reactive impulses in the center of rigidity K of the deck are created in the instant of impact. Therefore, seven eqs, (27)-(9) contain seven unknowns: velocities of the deck after the impact $\dot{u}, \dot{v}, \dot{\phi}$, velocities of the ship after impact v_x, v_y, ω and internal impuls along the normal, I_y. After elimination of all the unknowns, internal impuls component I_y is obtained in the form:

$$I_y = \frac{1}{B}(1+\varepsilon)v_o sin\alpha \tag{30}$$

where

$$B = \frac{1}{m}\{1 + \frac{m}{m_S}[1+3cos\alpha(cos\alpha - \mu sin\alpha)]\} + $$
$$+ \frac{1}{J_{zM}} d_x(d_x + \mu d_y) \tag{31}$$

with J_{zM} being the mass moment of inertia of the deck with respect to center-of-mass (Mz axis), while other notations are as before, see Fig.6. It may be easily seen that the result (30) is reduced to (22) if the deck is considered as a rigid obstacle instead of a lamina undergoing the planar motion. This is obtained in the limitting case $m\to\infty$, $J_{zm}\to\infty$, which defines the deck as a rigid obstacle. Further analysis, i.e. equivalent statical approach, is the same as before, see Fig. 5.

4. "HARBOR" COMPUTER CODE

FORTRAN computer code, called HARBOR, has been developed by the author for the purpose of commercial use related to design of marine structures. The program is based upon the presented analysis. It includes statical analysis, free vibration analysis, seismic analysis according to Yugoslav seismic design code and impact of a ship analysis as presented herein.

The main features of the present version of the HARBOR code are as follows. It may contain arbitrary number of loading cases in a single run, with up to 20 concentrated forces in each loading case. The program is executed completely in core and it uses three SSP subroutines: MINV, GMPRD, EIGEN. All the arrays are presently dimensioned to contain up to 120 piles: 60 vertical, 40 raked ones in direction along the coast and 20 raked piles normal to the coast. Such dimensions were adequate in the case of five river harbors already designed by the HARBOR code and may be easily changed if necessary.

5. CONCLUSIONS

It is unnecessary to remind that the present analysis is dealing with horizontal loads only. Therefore, complementary analysis of gravitational loads must be done additionally to obtain the necessary design information about the deck itself. As for the piles, one must combine results obtained from both horizontal and gravitational load analyses.

Experiences obtained by use of the HARBOR code and alternative calculations performed on 3D numerical models of harbor structures using SAP4 code, have resulted in complete confidence in HARBOR code. Namely, all the relevant results (displacements of the deck, bending moments, shear forces and normal forces in raked piles only) were in excellent agreement within maximum 3-4% difference (and usually less).

Preparation of input data and CPU time cannot be compared at all. For instance, for a structure with about 100 piles and 10 loading cases, including also free vibration response necessary for seismic and impact calculations, the CPU time of the HARBOR code is about 20-30 sec. Therefore, in the preliminary design stage particularly, when no definite pile arrangement is known, HARBOR code is of a great value.

SHAPE IDENTIFICATION OF A FREE SURFACE WITH A UNIFORM POTENTIAL AND FLUX

R.A.Meric
Research Institute for Basic Sciences
Gebze, Kocaeli, Turkey

ABSTRACT

The present study concerns the identification of the positic of the inner boundary surface of a hollow solid body satisfying Laplace's equation. The "inverse problem condition", providing the extra information needed for identification purposes, is given such that the inner surface has a uniform potential flux (although, an unknown quantity) with a known total value. After reformulating the problem as a shape optimization problem, the material derivative concept and adjoint variable methods are utilized in order to find a sensitivity expression for the objec tive function of optimization. The boundary element methods alor with an unconstrained minimization routine are then effectively used in an iterative numerical solution procedure.

1. INTRODUCTION

Natori and Kawarada [1] analyzed the shape identification problem of determining the outer boundary surface of a 2-dim. hollow solid body, when the unknown free boundary had a uniform potential and flux. It has been stated in Ref. [1] that this problem occured in cryogenics experiments. The same type of problem is also encountered in modelling plasma configurations in tokamak machines when the plasma is subject to skin effect [2] For the solution of this problem, Natori and Kawarada reformulated the identification problem as an optimization problem, anc then used integrated penalty methods with finite difference discretizations.

In the present study, a similar problem is attacked as in Ref [1], except that the boundary surface to be identified is tł inner surface, while the outer boundary is prescribed. After reformulating the problem as an unconstrained optimization problen the material derivative (MD) concept [3] and adjoint variable method of optimization, frequently used in structural optimization problems, are employed in order to derive shape sensitivity analysis (SSA) expressions. A slight extension in the procedure is required, however, due to the fact that the integrand of the integral objective functional of optimization involves explicitly an integral, as well. Equipped with the SSA expressions, the boundary element methods (BEM) and minimization techniques are then utilized for numerical solutions.

2. PROBLEM DEFINITION

Consider a 2-dim. isotropic hollow solid body depicted in Fig.1 satisfying Laplace's equation as

$$\text{in } D \; : \; \nabla^2 u = 0 \tag{1}$$

where D is the physical domain of unknown shape; u is the potential . Dirichlet boundary conditions are prescribed in the following form:

$$\text{on } S_1 \; : \; u = u_1(x,y) \tag{2}$$

$$\text{on } S_2 \; : \; u = u_2(\text{constant}) \tag{3}$$

where S_1 is the prescribed outer boundary; S_2 is the inner boundary to be identified; S is the total boundary with $S=S_1+S_2$; u_1 and u_2 are given quantities; x and y denote Cartesian coordinates.

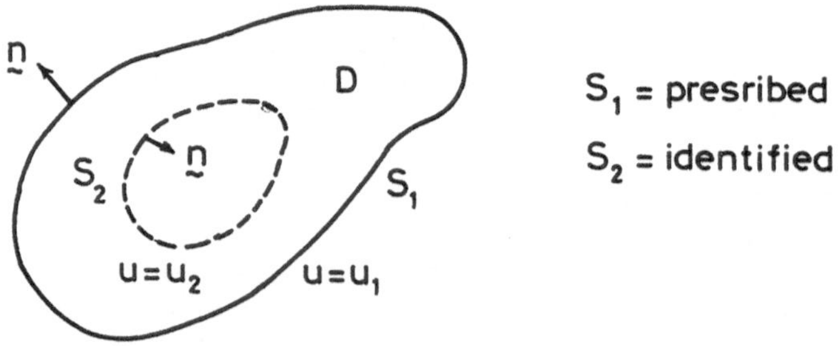

Fig.1. Schematics of problem definition

The "inverse problem conditions" (IPC) for the present shape identification problem consist of the following information:

$$\text{on } S_2 \; : \; u_{,n} = \text{constant} \tag{4}$$

$$\int_{S_2} u_{,n} \, dS = Q \; (\text{constant}) \tag{5}$$

where $u_{,n}$ is the normal component of the gradient of u on S, i.e., $u_{,n} = \nabla u . \underset{\sim}{n}$; $\underset{\sim}{n}$ is the unit vector normal to S; Q is a given problem parameter. Thus, it is noted that $u_{,n}$ is constrained to be of constant (although, unknown) value on S_2, while the total flux on S_2 is a given quantity,i.e., Q. By virtue of Eqs. (4) and (5), it may be written that

$$\text{on } S_2 : \; L\, u_{,n} = Q \tag{6}$$

where L is the total length of S_2, an unknown quantity,i.e.,

$$L = \int_{S_2} dS \tag{7}$$

The present shape identification problem requires the determination of the position of S_2 (hence D), as well as the potential u in D, while Eqs.(1)-(3) and (6) are satisfied for given u_1, u_2 and Q.

2.1. Reformulation as an optimization problem

The free boundary S_2 is assumed to be a closed convex boundary. The shape identification problem described previously may be reformulated as a shape optimization problem by choosing a suitable norm for the distance between functions. Thus, the following integral functional J, utilizing the IPC, Eq.(6), is introduced:

$$J = \frac{1}{2} \int_{S_2} (L\, u_{,n} - Q)^2 dS \tag{8}$$

where J is termed as the objective function. Hence, the original problem is transformed as a shape optimization problem, in which J will be minimized with respect to the shape configurations of S_2, subject to the system's equations (1)-(3).

In the present study, no attempts will be made for providing proofs for the existance and uniqueness of solutions of both problems. It is expected that as $J \rightarrow 0$, the optimal configuration of S_2 should approach that of the original identification problem under sufficient regularity conditions on S_1 and u_1 distributions. Otherwise, the solution of the optimization problem only constitutes an optimal shape configuration of S_2 with a minimum amount of deviation of flux from a uniform value.

3. SHAPE SENSITIVITY ANALYSIS

The total variation of J with respect to shape variation of D can be found by adopting SSA procedures in structural optimization. Thus. following Haug et al [3], the MD concept of continuum mechanics and adjoint variable method of optimization are utilized for the present shape optimization problem.

In the MD formulation of SSA, the domain D, which is varied, is treated as a continuum moving with a time-like parameter τ. The variations of a point in the domain are expressed in terms of a velocity field $\underset{\sim}{V}$, which represents a deformation velocity. The objective function J is first augmented by incorporating Eq.(1) by using an adjoint potential u^* [4,5] . After using integration by parts, the MD of the augmented functional can be taken by employing the general MD formulas given in Ref.3. Since the integrand of J involves L, which is an integral itself, the local derivative (with respect to τ) of L, Eq.(7), is evaluated through its MD formulation (cf. nonlocal calculus of variations). Integration by parts is again performed and the MD forms of the boundary conditions (2) and (3) are inserted into the resulting sensitivity expression for J.

In order to get rid of the local variations of the state variables, u^* is required to satisfy the following adjoint problem:

$$\text{in } D \ : \nabla^2 u^* = 0 \tag{9}$$

$$\text{on } S_1 \ : u^* = 0 \tag{10}$$

$$\text{on } S_2 \ : u^* = -L(Lu_{,n} - Q) \tag{11}$$

If the primary problem, Eqs.(1)-(3) and the adjoint problem, Eqs.(9)-(11), are satisfied in an iterative solution procedure, the total variation (i.e.,MD) of J, indicated by $\dot{J}$, is given at the current iterative step as follows:

$$\dot{J} = \int_{S_2} \{u_{,n}\, u^*_{,n} + H[u^*(u_{,n} + \frac{u^*}{2L^2}) - \frac{1}{L}\int_{S_2} \bar{u}_{,n}\bar{u}^* d\bar{S}]\} V_n dS \tag{12}$$

where H is the curvature of the boundary S_2; the superpositioned bar indicates that the relevant quantity is evaluated at the field point $(\bar{x},\bar{y})$, i.e., dummy variables of integration; V_n is the normal component of boundary perturbation on S_2, i.e., $V_n = \underset{\sim}{V}.\underset{\sim}{n}$.

It is noted that $\dot{J}$ is expressed in terms of only the normal component of boundary deformation, i.e., V_n. In other words, the tangential component of $\underset{\sim}{V}$ does not contribute to any variation in J. For the numerical computation of $\dot{J}$ for a given V_n distribution on S_2, a double integral is also evaluated.

4. NUMERICAL METHOD OF SOLUTION

For the numerical solution of the primary and adjoint problems the BEM with constant elements is employed. As there are no source terms in the potential problems, and the objective function J and its MD, $\dot{J}$, are expressed in terms of boundary integrals only, no internal cells in the domain D are required. This fact greatly increases the efficiency of an iterative minimisation solution procedure for J, since the boundary elements on S_2 only have to be updated at each step.

The radial lengths b^k between the kth vertex of the boundary line element on S_2 (see Fig.2) and the origin are chosen as the decision parameters (called the ray fanctions). The angle which the ray function b^k makes with the x-axis is denoted by α^k and is held fixed. Hence, as the variation of the boundary proceeds, the vertices of the boundary line elements will always move along

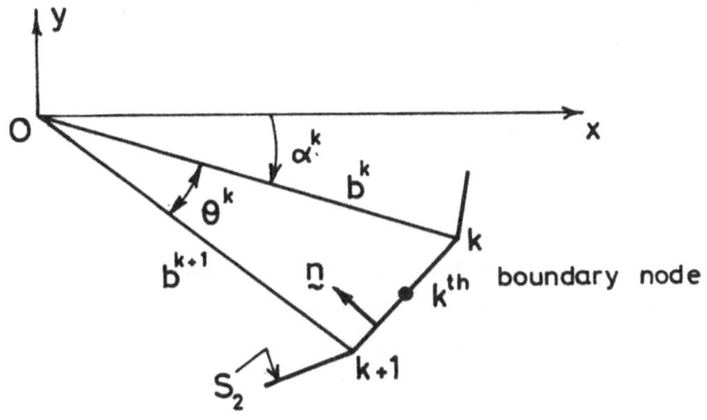

Fig.2. Ray functions b^k and b^{k+1}

the lines of the ray functions. The angle between the ray functions b^k and b^{k+1} is also a constant and is denoted by θ^k.

Starting with an initial guess for the values of the ray functions b^k, solution of the u- and u*-problem proceeds in a standard way by using the BEM [6]. It is noted that the potentials and their derivatives are directly given by the BEM solutions.

The evaluation of J, Eq.(8), is straightforward. For the calculation of $\dot{J}$, Eq.(12), however, the curvature H of each boundary element, and the normal boundary perturbation V_n have to be computed.

In order to find H^k for the kth element, the unit vectors $\underset{\sim}{n}^{k-1}$, $\underset{\sim}{n}^k$ and $\underset{\sim}{n}^{k+1}$, the tangential unit vector $\underset{\sim}{t}^k$ and the element length l^k are evaluated first (refer to Fig.3). The unit vectors $\underset{\sim}{m}^k$ and $\underset{\sim}{m}^{k+1}$ (at the element extreme points k and k+1) are then found

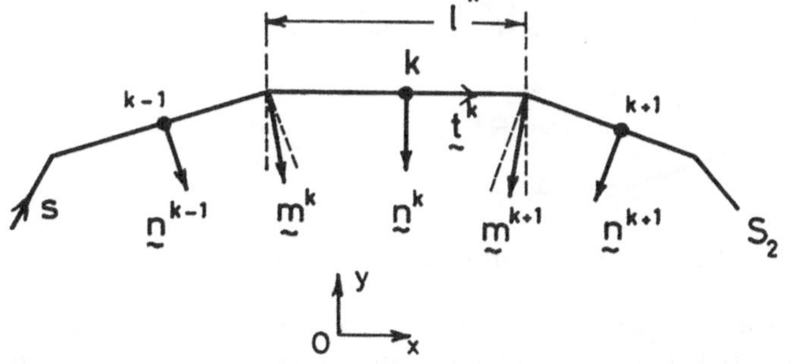

Fig.3. Definition of unit vectors $\underset{\sim}{m}^k$ and $\underset{\sim}{m}^{k+1}$

such that they are directed towards the direction of the half-angle between the direction of two respective neighbouring unit normal vectors. The kth element curvature H^k is thus approximated as follows:

$$H_k = t_x^k \frac{m_x^{k+1} - m_x^k}{l^k} + t_y^k \frac{m_y^{k+1} - m_y^k}{l^k} \qquad (13)$$

since $H = \underset{\sim}{t} . d\underset{\sim}{n}/ds$.

The integral of V_n over a boundary line element S_2^k is given in terms of the variations of the decision parameters, i.e., δb^k, by the following first-order approximation [4,5]:

$$\int_{S_2^k} V_n \, dS = \frac{1}{2} \sin\theta^k (b^{k+1} \delta b^k + b^k \delta b^{k+1}) \qquad (14)$$

Thus, it is an easy manner to evaluate J and $\dot{J}$ at each iteration step of a minimization routine for J. Since gradient information for J, i.e., $\dot{J}$, is available, a sequential quadratic programming algorithm is utilized in the present study.

5. NUMERICAL RESULTS

Numerical experiments were performed by using 64 constant boundary elements in total. For a specific run, S_2 was prescribed as an ellipse with major and minor axis lengths 1.6 and 0.6, respectively. On this surface, u_1 was given nonuniformly as $1+|\cos\phi|$, where ϕ is the angle which the line connecting the boundary node with the origin makes with the x-axis (see Fig.4). The potential was set as zero on S_2, thus $u_2=0$. The total flux Q was also prescribed as -7.5.

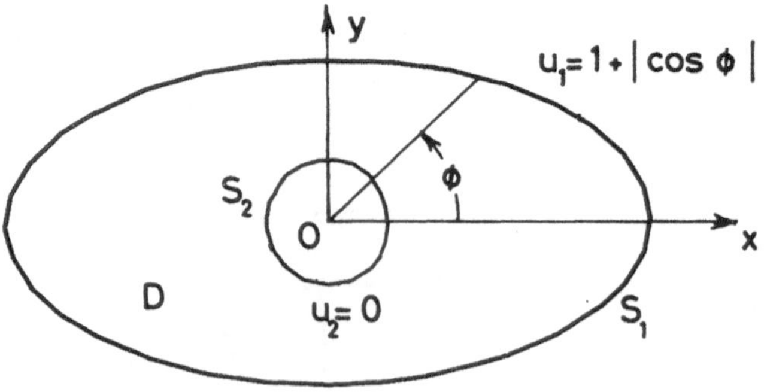

Fig.4. Identified inner boundary S_2

Twenty eight major iterations were needed for convergence of the iterative solution procedure. which resulted in $J=0.209 \times 10^{-13}$. The identified S_2 inner boundary is depicted in Fig.4, on which the normal potential derivative was found as $u_{,n}=-3.8668$, and the total length as L=1.9396. The maximum b^k was 3.149 and the minimum 3.050. Thus, the identified surface S_2 slightly deviated from a circle in order that $u_{,n}$ be uniform on the same surface.

6. CONCLUSION

During the calculations no internal cells are needed (with no mesh regriddings, except for the boundary elements) and no internal values of potentials are computed. As such, the present procedure using the BEM may be used effectively for other shape optimization and/or identification problems.

REFERENCES

1. Natori, M. and Kawarada, H. 'An Application of the Integrated Penalty Method to Free Boundary Problems of Laplace Equation', Numer.Func.Anal. and Optimiz. 3 (1981), 1-17.

2. Demidov, A.S., 'The Form of a Steady Plasma Subject to the Skin Effect in a Tokamak with Non-Circular Cross-Section', Nuclear Fusion 15 (1975).

3. Haug, E.J., Choi, K.K. and Komkov, V., *Design Sensitivity Analysis of Structural Systems,* Academic Press, New York, 1986.

4. Meric , R.A., 'Shape Design Sensitivity Analysis for Nonlinear and Anisotropic Heat Conducting Solids and Shape Optimization by the BEM', Int.J.Num.Meths.Engng. (to appear).

5. Meric, R.A.,'Shape Optimization and Identification of Solid Geometries Considering Discontinuities', Trans. ASME, J.Heat Transfer (to appear).

6. Banerjee, P.K. and Butterfield, R., *Boundary Element Methods in Engineering Science,* McGraw-Hill, New York, 1981.

A C^0 ELASTOPLASTIC SHELL ELEMENT BASED ON ASSUMED COVARIANT STRAIN INTERPOLATIONS

Peter M. Pinsky
and
Junho Jang

Department of Civil Engineering
Stanford University, Stanford, California 94305

Summary

A curved 9-node C^0 shell finite element for elastoplastic analysis is proposed which is free from serious locking problems, does not possess hourglass modes and provides solutions which are quite insensitive to mesh distortion. The element is based on the use of modified strain fields which are obtained from assumed interpolations of covariant (non-physical) strains referred to the element natural coordinate system. The linear elastic shell formulation is described first and this is then extended for an elastoplastic constitutive model. A return mapping algorithm is introduced for integration of the rate constitutive equations under the *zero normal stress* hypothesis. Some numerical results, illustrating the good convergence characteristics of the element, are reported.

1. Introduction

Although 9-node elements are complex compared to 4-node elements, they are generally considered to be advantageous in some applications such as problems involving curved boundaries or dominant inextensional bending. However, as is well known, the 9-node C^0 Lagrange isoparametric elements suffer from serious deficiencies. The 9-node selectively reduced integrated element (9-SRI) exhibits poor convergence for some problems, whereas the 9-node uniformly reduced integrated element (9-URI) has a rank deficiency problem, even though this element may be convergent.

In order to avoid the above noted deficiencies of the C^0 isoparametric elements, a number of authors have developed elements based on the concept of assumed strain interpolations, for example [1-8]. Some of these elements produce solutions which are also quite insensitive to severe mesh distortion. These elements employ the same nodal degrees of freedom as the standard C^0 Lagrange elements but are not isoparametric elements.

A 9-node C^0 curved shell element, based on interpolation of *non-physical* covariant strains computed with respect to the element natural coordinate system has recently been proposed [7,8]. For all test problems analyzed, this element exhibits excellent convergence in displacements without any serious locking or rank deficiency problems.

Solutions are also stable with respect to mesh distortion. Numerical results obtained for plate problems demonstrate that not only in-plane stresses but also transverse shear stresses converge to quite accurate values. General conditions for the convergence of shell elements based on assumed strain interpolations have been stated in [8] where it is also shown that this element generally satisfies these conditions.

In the first part of this paper, the elastic 9-node shell element based on assumed interpolations of covariant strains is briefly described. The second part of the paper then extends the formulation to include an elastoplastic constitutive model with nonlinear isotropic and linear kinematic hardening and a von Mises yield condition. An unconditionally stable return mapping algorithm is introduced for the integration of the rate constitutive equations under the ***zero normal stress*** hypothesis. Finally, an elastic element convergence study for a standard problem is reported as well as numerical results for elasto-plastic analysis of plates.

2. Shell Kinematics

We consider a curved shell finite element and introduce a global Cartesian coordinate system $\{x, y, z\}$ and an element natural coordinate system $\{\xi, \eta, \zeta\}$. Fig. 1 shows the coordinate systems, fiber vectors and nodal numbering scheme used for the 9-node element.

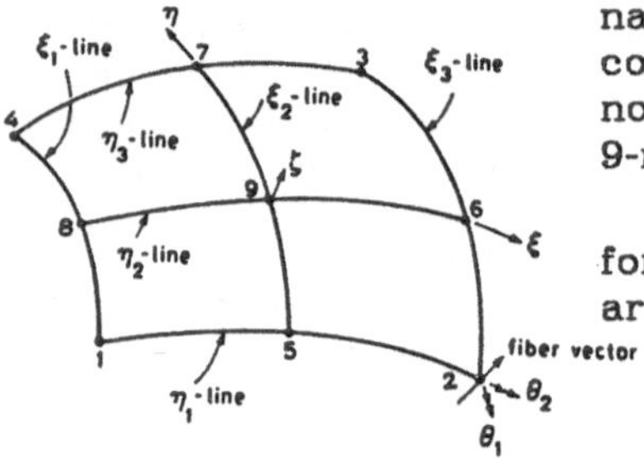

Fig. 1

Position and displacement vectors for a typical point in the shell element are introduced as follows

$$\mathbf{x} = \bar{\mathbf{x}}(\xi,\eta) + \frac{t}{2}\zeta\,\hat{\mathbf{x}}(\xi,\eta) \qquad (2.1)$$

$$\mathbf{u} = \bar{\mathbf{u}}(\xi,\eta) + \frac{t}{2}\zeta\,\hat{\mathbf{u}}(\xi,\eta) \qquad (2.2)$$

where t is the shell thickness measured along the fiber vector, $\bar{\mathbf{x}}$ denotes the position vector of a point on the shell mid-surface, $\hat{\mathbf{x}}$ denotes the unit fiber vector, $\bar{\mathbf{u}}$ denotes the displacement of the shell mid-surface and $\hat{\mathbf{u}}$ denotes the displacement of the tip of the unit fiber vector relative to the mid-surface. Finite element approximations for the position and displacement vectors may be introduced for the 9-node element as follows

$$\mathbf{x} = \sum_{a=1}^{9} N_a(\xi,\eta)\,\bar{\mathbf{x}}_a + \sum_{a=1}^{9} N_a(\xi,\eta)\,\frac{t_a}{2}\zeta\,\hat{\mathbf{x}}_a \qquad (2.3)$$

$$\mathbf{u} = \sum_{a=1}^{9} N_a(\xi,\eta)\,\bar{\mathbf{u}}_a + \sum_{a=1}^{9} N_a(\xi,\eta)\,\frac{t_a}{2}\zeta\,\hat{\mathbf{u}}_a \qquad (2.4)$$

where $N_a(\xi,\eta)$ are the element Lagrange shape functions and a subscript a indicates the nodal value of a quantity.

In practice, five nodal degrees of freedom are employed: three translations corresponding to the global Cartesian frame plus two rotations about the nodal fiber coordinate axes, see Fig. 1. In this case, the displacement components $\hat{u}_a, \hat{v}_a, \hat{w}_a$ become functions of the fiber vector direction and two local rotations.

3. Structure of Covariant Strains

The covariant components of the (linearized) strain tensor $\varepsilon_{\alpha\beta}$ defined with respect to the element natural coordinate basis may be obtained from the Cartesian strain components e_{ij} by use of the transformation

$$\varepsilon_{\alpha\beta} = \frac{\partial x^i}{\partial \xi^\alpha}\frac{\partial x^j}{\partial \xi^\beta} e_{ij} \qquad\qquad e_{ij} = \frac{1}{2}\left(\frac{\partial u_i}{\partial x^j} + \frac{\partial u_j}{\partial x^i}\right) \tag{3.1}$$

where repeated indices imply summation over the range 1 to 3. Denoting the components of $\mathbf{x}$ and $\mathbf{u}$ relative to the Cartesian coordinate system as x, y, z and u, v, w, respectively, expressions for the covariant strain components may be found from (3.1) as

$$\begin{aligned}
\varepsilon_{\xi\xi} &= u_{,\xi}\, x_{,\xi} + v_{,\xi}\, y_{,\xi} + w_{,\xi}\, z_{,\xi} \\
\varepsilon_{\eta\eta} &= u_{,\eta}\, x_{,\eta} + v_{,\eta}\, y_{,\eta} + w_{,\eta}\, z_{,\eta} \\
\varepsilon_{\zeta\zeta} &= u_{,\zeta}\, x_{,\zeta} + v_{,\zeta}\, y_{,\zeta} + w_{,\zeta}\, z_{,\zeta} \\
\gamma_{\xi\eta} &= u_{,\xi}\, x_{,\eta} + v_{,\xi}\, y_{,\eta} + w_{,\xi}\, z_{,\eta} + u_{,\eta}\, x_{,\xi} + v_{,\eta}\, y_{,\xi} + w_{,\eta}\, z_{,\xi} \\
\gamma_{\xi\zeta} &= u_{,\xi}\, x_{,\zeta} + v_{,\xi}\, y_{,\zeta} + w_{,\xi}\, z_{,\zeta} + u_{,\zeta}\, x_{,\xi} + v_{,\zeta}\, y_{,\xi} + w_{,\zeta}\, z_{,\xi} \\
\gamma_{\eta\zeta} &= u_{,\eta}\, x_{,\zeta} + v_{,\eta}\, y_{,\zeta} + w_{,\eta}\, z_{,\zeta} + u_{,\zeta}\, x_{,\eta} + v_{,\zeta}\, y_{,\eta} + w_{,\zeta}\, z_{,\eta}
\end{aligned} \tag{3.2}$$

where $\gamma_{\alpha\beta} \equiv 2\,\varepsilon_{\alpha\beta}$. The covariant strain components given by (3.3) are not *physical* strain components which is evidenced by the fact that they have dimension of length squared. Substituting (2.1) and (2.2) into (3.1), the following *shell* covariant strains are obtained

$$\begin{aligned}
\varepsilon_{\xi\xi} &= \varepsilon^m{}_{\xi\xi}(\xi,\eta) + \zeta\, \varepsilon^b{}_{\xi\xi}(\xi,\eta) + O(\zeta^2) \\
\varepsilon_{\eta\eta} &= \varepsilon^m{}_{\eta\eta}(\xi,\eta) + \zeta\, \varepsilon^b{}_{\eta\eta}(\xi,\eta) + O(\zeta^2) \\
\gamma_{\xi\eta} &= \gamma^s{}_{\xi\eta}(\xi,\eta) + \zeta\, \gamma^b{}_{\xi\xi}(\xi,\eta) + O(\zeta^2) \\
\gamma_{\xi\zeta} &= \gamma^s{}_{\xi\zeta}(\xi,\eta) + O(\zeta,\zeta^2) \\
\gamma_{\eta\zeta} &= \gamma^s{}_{\eta\zeta}(\xi,\eta) + O(\zeta,\zeta^2)
\end{aligned} \tag{3.3}$$

where

$$\begin{aligned}
\varepsilon^m{}_{\xi\xi} &= \bar{u}_{,\xi}\, \bar{x}_{,\xi} + \bar{v}_{,\xi}\, \bar{y}_{,\xi} + \bar{w}_{,\xi}\, \bar{z}_{,\xi} \\
\varepsilon^m{}_{\eta\eta} &= \bar{u}_{,\eta}\, \bar{x}_{,\eta} + \bar{v}_{,\eta}\, \bar{y}_{,\eta} + \bar{w}_{,\eta}\, \bar{z}_{,\eta} \\
\varepsilon^b{}_{\xi\xi} &= \frac{t}{2}[\, \bar{u}_{,\xi}\, \hat{x}_{,\xi} + \bar{v}_{,\xi}\, \hat{y}_{,\xi} + \bar{w}_{,\xi}\, \hat{z}_{,\xi} + \hat{u}_{,\xi}\, \bar{x}_{,\xi} + \hat{v}_{,\xi}\, \bar{y}_{,\xi} + \hat{w}_{,\xi}\, \bar{z}_{,\xi}\,] \\
\varepsilon^b{}_{\eta\eta} &= \frac{t}{2}[\, \bar{u}_{,\eta}\, \hat{x}_{,\eta} + \bar{v}_{,\eta}\, \hat{y}_{,\eta} + \bar{w}_{,\eta}\, \hat{z}_{,\eta} + \hat{u}_{,\eta}\, \bar{x}_{,\eta} + \hat{v}_{,\eta}\, \bar{y}_{,\eta} + \hat{w}_{,\eta}\, \bar{z}_{,\eta}\,] \\
\gamma^s{}_{\xi\eta} &= \bar{u}_{,\xi}\, \bar{x}_{,\eta} + \bar{v}_{,\xi}\, \bar{y}_{,\eta} + \bar{w}_{,\xi}\, \bar{z}_{,\eta} + \bar{u}_{,\eta}\, \bar{x}_{,\xi} + \bar{v}_{,\eta}\, \bar{y}_{,\xi} + \bar{w}_{,\eta}\, \bar{z}_{,\xi} \\
\gamma^b{}_{\xi\eta} &= \frac{t}{2}[\, \bar{u}_{,\xi}\, \hat{x}_{,\eta} + \bar{v}_{,\xi}\, \hat{y}_{,\eta} + \bar{w}_{,\xi}\, \hat{z}_{,\eta} + \hat{u}_{,\xi}\, \bar{x}_{,\eta} + \hat{v}_{,\xi}\, \bar{y}_{,\eta} + \hat{w}_{,\xi}\, \bar{z}_{,\eta} \\
&\qquad + \bar{u}_{,\eta}\, \hat{x}_{,\xi} + \bar{v}_{,\eta}\, \hat{y}_{,\xi} + \bar{w}_{,\eta}\, \hat{z}_{,\xi} + \hat{u}_{,\eta}\, \bar{x}_{,\xi} + \hat{v}_{,\eta}\, \bar{y}_{\xi} + \hat{w}_{,\eta}\, \bar{z}_{,\xi}\,] \\
\gamma^s{}_{\xi\zeta} &= \frac{t}{2}[\, \bar{u}_{,\xi}\, \hat{x} + \bar{v}_{,\xi}\, \hat{y} + \bar{w}_{,\xi}\, \hat{z} + \hat{u}\, \bar{x}_{,\xi} + \hat{v}\, \bar{y}_{,\xi} + \hat{w}\, \bar{z}_{,\xi}\,] \\
\gamma^s{}_{\eta\zeta} &= \frac{t}{2}[\, \bar{u}_{,\eta}\, \hat{x} + \bar{v}_{,\eta}\, \hat{y} + \bar{w}_{,\eta}\, \hat{z} + \hat{u}\, \bar{x}_{,\eta} + \hat{v}\, \bar{y}_{,\eta} + \hat{w}\, \bar{z}_{,\eta}\,]
\end{aligned} \tag{3.4}$$

where superscripts m, b and s refer to membrane, bending and shear components, respectively. In the above strain expressions, terms involving ζ^2 have been neglected. For the transverse shear given by (3.4g,h), terms involving ζ have also been neglected. The discarding of these terms is appropriate for thin shells.

4. Assumed Covariant Strain Interpolations

In order to remove the possibility locking problems in the thin shell limit, each of the strain components in (3.4) are modified by (i) evaluating the strains at certain points in the element in *terms of the nodal degrees of freedom* and (ii) extrapolating these values over the element using special functions which then define the modified strain. This concept has been employed, for example, in [2-9].

In detail, each strain component is identified with three natural coordinate lines (six in the case of in-plane shear and bending). On each line, the strain is evaluated in terms of the nodal degrees of freedom at the two *reduced* points and linearly extrapolated. These line values of the strain are then interpolated over the element with certain functions that are choosen according to the strain component. This process is fully described in Table 1.

Table 1. Structure of Assumed Covariant Strain

Interp. direction	Strain Components	Interp. Functions
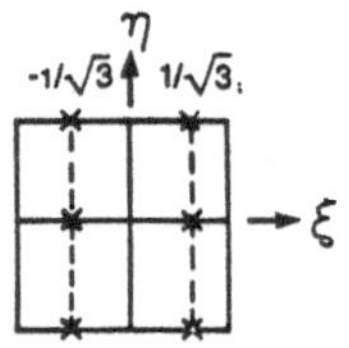	$\tilde{\varepsilon}^m_{\xi\xi},\ \tilde{\varepsilon}^b_{\xi\xi},\ \tilde{\gamma}^s_{\xi\zeta}$	$\bar{N}_1,\ \bar{N}_2,\ \bar{N}_3$
	$\frac{\tilde{\gamma}^b_{\xi\eta}}{2},\ \frac{\tilde{\gamma}^s_{\xi\eta}}{2}$	$\tilde{N}_1,\ \tilde{N}_2,\ \tilde{N}_3$
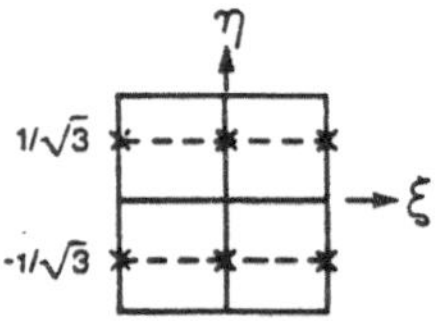	$\tilde{\varepsilon}^m_{\eta\eta},\ \tilde{\varepsilon}^b_{\eta\eta},\ \tilde{\gamma}^s_{\eta\zeta}$	$\bar{S}_1,\ \bar{S}_2,\ \bar{S}_3$
	$\frac{\tilde{\gamma}^b_{\xi\eta}}{2},\ \frac{\tilde{\gamma}^s_{\xi\eta}}{2}$	$\tilde{S}_1,\ \tilde{S}_2,\ \tilde{S}_3$

In Table 1, the interpolation functions have the form

$$\bar{N}_1 = \frac{\eta(\eta-1)}{2}, \quad \bar{N}_2 = 1-\eta^2, \quad \bar{N}_3 = \frac{\eta(\eta+1)}{2} \tag{4.1}$$

$$\tilde{N}_1 = \frac{1}{6} - \frac{\eta}{2}, \quad \tilde{N}_2 = \frac{2}{3}, \quad \tilde{N}_3 = \frac{1}{6} + \frac{\eta}{2} \tag{4.2}$$

Similar expressions, replacing η with ζ, hold for $\bar{S}_i$ and $\tilde{S}_i$.

The convergence of shell elements based on the use of these *assumed* covariant strains requires that they be capable of representing rigid body and constant strain states. The satisfaction of this requirement by the above interpolation scheme is considered in detail in [8], where general conditions for convergence of elements based on assumed strain interpolations are also stated.

5. Elastic Shell Element

Using a simple procedure, it is possible to find expressions for all the interpolated strains given in Table 1 in terms of the nodal degrees of freedom and nodal geometry. Details may be found in [7,9,10]. In order to define the strain-displacement matrix **B**, a vector of assumed (continuum) strain interpolations is introduced

$$\tilde{\varepsilon} = < \tilde{\varepsilon}_{\xi\xi}, \tilde{\varepsilon}_{\eta\eta}, \tilde{\gamma}_{\xi\eta}, \tilde{\gamma}_{\xi\zeta}, \tilde{\gamma}_{\eta\zeta} >^T \tag{5.1}$$

and relative to this ordering of strains, we let

$$\mathbf{B} = \mathbf{B}^i(\xi,\eta) + \zeta\, \mathbf{B}^d(\xi,\eta) \tag{5.2}$$

In order to obtain the elasticity matrix for the strains in (5.1), we note that $\sigma^{\alpha\beta} = C^{\alpha\beta\gamma\delta}\, \varepsilon_{\gamma\delta}$ where, for isotropic elasticity, [7]

$$\mathbf{C}^{\alpha\beta\gamma\delta} = \lambda\, g^{\alpha\beta}\, g^{\gamma\delta} + \mu\, (g^{\alpha\gamma}\, g^{\beta\delta} + g^{\alpha\delta}\, g^{\beta\gamma}) \tag{5.3}$$

and where the components of the metric tensor are given by

$$g^{\alpha\beta} = \frac{\partial \xi^{\alpha}}{\partial x^i} \frac{\partial \xi^{\beta}}{\partial x^j} \delta^{ij} \tag{5.4}$$

The *zero normal stress* assumption may be imposed directly (without the need for lamina coordinates) by requiring [7] $\sigma^{33} \equiv 0$. Using this condition, ε_{33} can be eliminated from the elastic constitutive equation resulting in a condensed elasticity tensor. The matrix form of this condensed tensor is denoted $\tilde{\mathbf{C}}$ and its components, which in general depend on ζ, are given explicitly in [7].

Noting (5.1), the element stiffness matrix $\mathbf{k}^e$ is obtained in the usual way

$$\mathbf{k}^e = \int_{\Omega^e} \int_{-1}^{1} [\, (\mathbf{B}^i)^t <J\tilde{\mathbf{C}}> (\mathbf{B}^i) + (\mathbf{B}^d)^t <\zeta^2 J\tilde{\mathbf{C}}> (\mathbf{B}^d)$$

$$+ (\mathbf{B}^i)^t <\zeta J\tilde{\mathbf{C}}> (\mathbf{B}^d) + (\mathbf{B}^d)^t <\zeta J\tilde{\mathbf{C}}> (\mathbf{B}^i)\,]\, \frac{t}{2} d\zeta\, d\Omega^e \tag{5.5}$$

Since only the $<\cdot>$ terms are functions of ζ, we can pre-integrate through ζ for those terms. We have used a 3-point integration rule for this pre-integration, although for a highly distorted mesh or a problem with variable thickness, 4-point integration might be advisable. Full 3×3 integration was used over the shell mid-surface. Note that the above shell stiffness matrix was obtained *without* the need to rotate any quantities to a shell lamina coordinate system.

6. Elastoplastic Shell Element

In this Section, rate constitutive equations for nonlinear isotropic and linear kinematic hardening elastoplasticity are introduced and numerically integrated using a return mapping algorithm under the *zero normal stress* constraint. The algorithm is similar to that proposed for plane stress elastoplasticity [12], except that in the present case the effects of transverse shear are included.

6.1. Elastoplastic Rate Constitutive Equations

Rate constitutive equations, incorporating the zero normal stress constraint, for nonlinear isotropic and linear kinematic hardening are introduced in *matrix* form with reference to a *local Cartesian lamina coordinate system* as follows:

$$
\begin{aligned}
\boldsymbol{\varepsilon} &= \boldsymbol{\varepsilon}^{e} + \boldsymbol{\varepsilon}^{p} \\
\boldsymbol{\sigma} &= \bar{\mathbf{C}}\, \boldsymbol{\varepsilon}^{e} \\
\dot{\boldsymbol{\varepsilon}}^{p} &= \dot{\lambda}\, \mathbf{P} \boldsymbol{\eta} \\
\boldsymbol{\eta} &= \boldsymbol{\sigma} - \boldsymbol{\alpha} \\
\dot{\boldsymbol{\alpha}} &= \frac{2}{3} H \dot{\lambda} \boldsymbol{\eta} \\
\dot{\bar{e}}^{p} &= \dot{\lambda} \left[\frac{2}{3} \boldsymbol{\eta}^{t} \mathbf{P} \boldsymbol{\eta} \right]^{1/2} \\
\varphi &= \frac{1}{2} \boldsymbol{\eta}^{t} \mathbf{P} \boldsymbol{\eta} - \frac{1}{3} k^{2}(\bar{e}^{p}) \le 0
\end{aligned}
\qquad
\mathbf{P} = \frac{1}{3}\begin{bmatrix} 2 & -1 & 0 & 0 & 0 \\ -1 & 2 & 0 & 0 & 0 \\ 0 & 0 & 6 & 0 & 0 \\ 0 & 0 & 0 & 6 & 0 \\ 0 & 0 & 0 & 0 & 6 \end{bmatrix}
\tag{6.1}
$$

where $\boldsymbol{\sigma} = \langle \sigma_{11}, \sigma_{22}, \sigma_{12}, \sigma_{13}, \sigma_{23} \rangle^{T}$, $\boldsymbol{\alpha} = \langle \alpha_{11}, \alpha_{22}, \alpha_{12}, \alpha_{13}, \alpha_{23} \rangle^{T}$ and $\boldsymbol{\varepsilon} = \langle \varepsilon_{11}, \varepsilon_{22}, 2\varepsilon_{12}, 2\varepsilon_{13}, 2\varepsilon_{23} \rangle$ and where $\boldsymbol{\alpha}$ is the back stress, H is the kinematic hardening parameter and $\kappa(\bar{e}^{p})$ defines the (nonlinear) isotropic hardening model. Note that these constitutive equations are exactly consistent with the zero normal stress constraint.

6.2. Return Mapping Algorithm

It is assumed that the values of $\boldsymbol{\varepsilon}_{n}$, $\boldsymbol{\varepsilon}_{n}^{p}$, $\bar{e}^{p}$ and $\boldsymbol{\alpha}_{n}$ at time $t = t_{n}$ are known. It is also assumed that $\boldsymbol{\varepsilon}_{n+1}$ is known at time $t = t_{n+1}$.
Then the generalized mid-point rule applied to the three rate equations (6.1c,e,f) results in [9,10]

$$
\begin{aligned}
\boldsymbol{\varepsilon}_{n+1}^{p} &= \boldsymbol{\varepsilon}_{n}^{p} + \bar{\lambda} \mathbf{P} \boldsymbol{\eta}_{n+\alpha} \\
\boldsymbol{\alpha}_{n+1} &= \boldsymbol{\alpha}_{n} + \bar{\lambda} \frac{2}{3} H \boldsymbol{\eta}_{n+\alpha} \\
\bar{e}_{n+1}^{p} &= \bar{e}_{n}^{p} + \bar{\lambda} \left[\frac{2}{3} \boldsymbol{\eta}_{n+\alpha}^{t} \mathbf{P} \boldsymbol{\eta}_{n+\alpha} \right]^{1/2}
\end{aligned}
\tag{6.2}
$$

where $\bar{\lambda} = \lambda_{n+1} - \lambda_{n}$ and

$$
\boldsymbol{\eta}_{n+\alpha} = \frac{1}{1 + \frac{2}{3} \alpha \lambda H} \Pi(\bar{\lambda})\, \bar{\mathbf{C}}^{-1} \boldsymbol{\eta}_{n+\alpha}^{E}
$$

$$
\Pi(\bar{\lambda}) = \left[\bar{\mathbf{C}}^{-1} + \frac{\alpha \bar{\lambda}}{1 + \frac{2}{3} \alpha \bar{\lambda} H} \mathbf{P} \right]^{-1}, \qquad \boldsymbol{\eta}_{n+\alpha}^{E} = \bar{\mathbf{C}} [\, \boldsymbol{\varepsilon}_{n+\alpha} - \boldsymbol{\varepsilon}_{n}^{p} \,] - \boldsymbol{\alpha}_{n} \tag{6.3}
$$

Finally, the updated stress is given by (6.1b) as $\boldsymbol{\sigma}_{n+1} = \bar{\mathbf{C}}(\boldsymbol{\varepsilon}_{n+1} - \boldsymbol{\varepsilon}_{n+1}^{p})$. The update procedure (6.2-3) is complete except that the value of the incremental plastic strain parameter $\bar{\lambda}$ is not yet determined. This value is found from satisfaction of the plastic yield condition. Note that the formula (6.3) defines a ***return mapping algorithm*** expressed in terms of an elastic predictor $(\boldsymbol{\eta}_{n+\alpha}^{E})$ and plastic corrector $(\boldsymbol{\eta}_{n+1})$.

Evaluating the yield condition (6.1g) at time $t = t_{n+\alpha}$, we have

$$\varphi = \frac{1}{2}\eta^t_{n+\alpha}\mathbf{P}\eta_{n+\alpha} - \frac{1}{3}\kappa^2(\bar{e}^p{}_{n+\alpha}) = 0 \tag{6.4}$$

where both $\eta_{n+\alpha}$ and $\bar{e}^p{}_{n+\alpha}$ are functions of $\bar{\lambda}$, see (6.2) and (6.3). Using the spectral decomposition of $\mathbf{P}$ and $\bar{\mathbf{C}}$, and noting that they have identical eigenvectors, a procedure described by Simo and Taylor [12] may be followed to obtain the roots ($\bar{\lambda}$) of this equation [9,10].

Finally, the shell elastoplastic tangent operator is defined by consistent linearization of the updated stress as given by the return mapping algorithm. The resulting operator must then transformed to correspond to the element natural coordinate basis. Further details may be found in [9,10].

7.1 Numerical Results: Pinched Hemisphere

A hemispherical shell subjected to self-equilibrating radial point forces that alternate in direction at 90 degree intervals along its free edge is analyzed. The pole is fixed. The analytic solution is given in [13]. Results are given in Fig. 2 where ACS - present element, LAG - Lagrangian, HET - Heterosis, SER - serendipity, STG - STAGS C^1 quadrilateral. Only the 16-LAG and 9-ACS elements converge with a crude mesh. All other elements converge slowly or lock. The 4-LAG element even converges to a value higher than the exact solution. For this problem, it is known that rigid body rotation constitutes a significant portion of the total deformation. The success of the 9-ACS element with a crude mesh is attributable in part to its capability of correctly describing rigid body rotations [8].

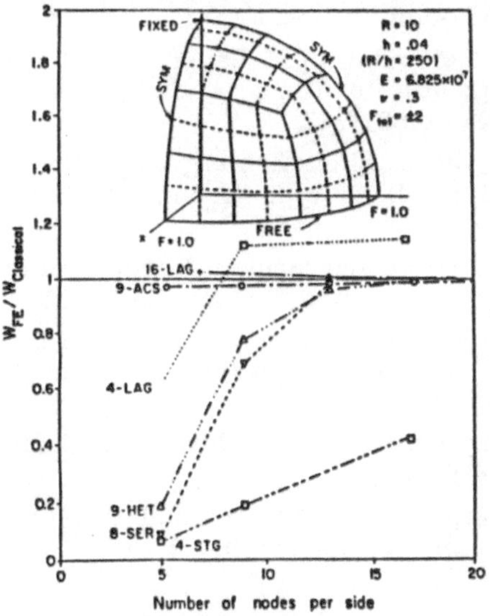

Fig. 2

7.2 Numerical Results: Elastoplastic Stretching of a Perforated Plate

An isotropic hardening rule with linear and saturation laws of the exponential type [12] is used

$$\kappa(\bar{e}^p) = K\bar{e}^p + \kappa_o + (\kappa_\infty - \kappa_0)[1 - e^{-\gamma\bar{e}^p}]$$

where $K = 2.24$ and $\kappa_\infty = \kappa_0 = 0.243$. In addition, elastic constants were taken as $E = 70$ and $\nu = 0.2$. Kinematic hardening is not considered.

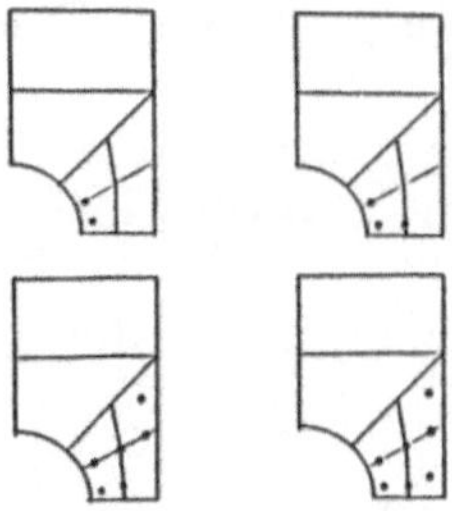

Fig. 3

Using symmetry, only a quarter of the plate, was discretized. Fig. 3 shows the mesh of only three elements. The upper boundary of the plate was given a prescribed uniform displacement of 0.04, followed by three increments of 0.01. Fig. 3 shows the yielded zone for each displacement increment, the solid dots denote a yielded quadrature point. These results appear to compare closely with those of Simo and Taylor [12], which were obtained with 164 4-node quadrilaterals.

Acknowledgements

The research reported in this paper was supported by ONR grant N00014-84-K-0715, NR-064-733 and NSF grant CEE-8411977.

References

[1] R.H. MacNeal, "Derivation of Element Stiffness Matrices by Assumed Strain Distributions," Nuclear Engineering and Design 70 (1982) pp.3-12.

[2] K.J.Bathe, E.Dvorkin, "A Four-node Plate Bending Element Based on Mindlin/Reissner Plate Theory and a Mixed Interpolation," Int. J. Num. Meth. Engr. 21 (1985) pp.367-383.

[3] K.J. Bathe and E.N. Dvorkin, "A Formulation of General Shell Elements - The Use of Mixed Interpolation of Tensorial Components," Int. J. Num. Meth. Engr. 22 (1986) pp.697-722.

[4] K.C.Park and G.M.Stanley, "A Curved C^0 Shell Element Based On Assumed Natural-Coordinate Strains," J. of Appl. Mech., 53 (1986), pp.278-290.

[5] H.C.Huang and E.Hinton, "An Improved Lagrangian 9-node Mindlin Plate Element," Proc. Numeta '85 Conference/Swansea Jan 1985.

[6] H.C.Huang and E.Hinton, "A New Nine Node Degenerated Shell Element with Enhanced Membrane and Shear Interpolation," Int. J. Num. Meth. Engr vol. 22, (1986) pp.73-92.

[7] J.Jang and P.M.Pinsky, "A Covariant Strain Based 9-Node Shell Element," To appear, Int. J. Num. Meth. Engr.

[8] J.H.Jang and P.M.Pinsky, "Convergence of Curved Shell Elements Based on Assumed Covariant Strains," To appear, Int. J. Num. Meth. Engr.

[9] J.Jang, "Curved Shell Finite Elements Based on Assumed Covariant Strain Interpolations," Ph.D. Thesis, Stanford University, Stanford, California (1987).

[10] P.M. Pinsky and J. Jang, "An Elastoplastic Curved Shell Finite Element Based on Assumed Covariant Strain Interpolations," Submitted to Engineering Mechanics, ASCE.

[11] P.M. Pinsky and J. Jang, "Convergence of Transverse Shear Stress in Plate Elements Based on Assumed Covariant Strain Interpolations," Submitted to Computers and Structures.

[12] J.C. Simo and R.L.Taylor, "A Return Mapping Algorithm for Plane Stress Elastoplasticity," Int. J. Num. Meth. Engr., 22, (1986) pp.649-670.

[13] L.S.D. Morley and A.J.Morris, "Conflict Between Finite Elements and Shell Theory," Royal Aircraft Establishment Report, London (1978).

CAPTURING THERMAL-STRESS WAVES VIA SPECIAL PURPOSE HYBRID TRANSFINITE ELEMENTS AND UNIFIED COMPUTATIONAL FORMULATIONS

Kumar K. Tamma
Associate Professor and Director-ICAD

and

Sudhir B. Railkar
Graduate Research Assistant

Mechanical and Aerospace Engineering
West Virginia University
Morgantown, WV 26506, USA

SUMMARY

The present paper represents an attempt to apply extensions of a hybrid transfinite element computational approach for accurately predicting thermoelastic stress waves. A unique feature of the proposed formulations for applicability to the Danilovskaya's problem of thermal stress waves in elastic solids lies in the hybrid nature of the unified formulations and the development of special purpose transfinite elements in conjunction with the classical Galerkin techniques and transformation concepts. Numerical test cases validate the applicability and superior capability to capture the thermal stress waves induced due to boundary heating.

1. INTRODUCTION

The prediction of propagating thermal stress waves in elastic solid media is of considerable practical importance in many mechanical, civil, and aerospace engineering problems. Typically, conventional computational approaches for transient stress wave and/or thermal-stress wave propagation problems are mostly based on space-time integration techniques. The governing partial differential equations are first discretized in space, and for the time derivatives typical finite difference approximations are employed.

The present paper represents an attempt to apply extensions of a hybrid transfinite element computational methodology recently being used for transient thermal-structural problems by Tamma et. al. [1-3]. The applicability of the present transfinite element formulations for accurately capturing the thermal stress waves induced by boundary heating for the Danilovskaya's problems [4-5] is demonstrated.

A unique feature of the present formulations for applicability to the Danilovskaya's problems [4,5] lies in the hybrid nature of the formulations and the development of special purpose finite elements in conjunction with Galerkin techniques and transformation concepts. Related applications using transform techniques with numerical schemes appears in Refs. [6-8]. In this paper, technical details of the proposed methodology is first described with emphasis on applications to thermal-stress wave propagation problems. Next, numerical test cases are presented to validate the applicability and superior ability to accurately predict the thermal- stress waves induced due to boundary heating.

2. FORMULATIONS

After dividing the physical domain of interest into finite elements appropriate for a given problem, the first step involves applying the selected transform to the governing transient equations. The next step involves the selection of interpolation functions in the transform domain which is arbitrary. Although this selection can be achieved via the general solution of the transformed differential equations or assuming appropriate approximating functions, for certain class of problems (such as the numerical test cases presented in this paper) the former is advantageous in that the corresponding basis functions generally yield more accurate results for the same element degree of freedom. This fact has been used to advantage in the hybrid formulations presented in the remainder of this paper. Once the type of elements and their interpolation functions have been selected, the finite element formulations in the transform domain are derived envoking the classical Galerkin procedures. The formulations in the transform domain are referred to as transfinite formulations and the corresponding elements are referred to as transfinite elements. The system equations are then assembled following the usual procedures of assembly and then solved in the transform domain itself. Through an appropriate inversion technique [9], the solution response is then obtained.

Typical field problems can be solved by discretizing a given region Ω into finite elements. The dependent variable $\overline{\Phi}$

(e.g. temperature or structral displacement) within each element Ω_e and the associated gradients $\bar{\Phi}_{,i}$ can be approximated in the transform domain in the form

$$\bar{\Phi} = \bar{\Phi}_\alpha N_\alpha \qquad (1)$$

$$\bar{\Phi}_{,i} = \bar{\Phi}_\alpha N_{\alpha,i} \qquad (2)$$

where N_α denotes typical element thermal or structural interpolation functions, and $\bar{\Phi}_\alpha$ denotes a vector of element unknown nodal temperatures or displacements in the transform domain respectively.

In order to demonstrate the basic advantages for accurately capturing the thermal stress waves, for the numerical test cases presented in this paper, we have used at advantage the special finite elements constructed via the selection of the interpolation functions from the transformed general solution. The formulations are developed in the transform domain in the following manner. Exact heat transfer interpolation functins are first derived for a typical thermal element for evaluating the matrices associated with the thermal model. Such formulations will yield exact solutions to the temperature problem. Therein, these temperature interpolation functions are used to derive the exact structural element interpolation functions and consequently the associated structural element matrices. It should be noted that the heat transfer and structural problems are interfaced in the transform domain itself. And, for evaluating the thermally induced stress wave response, it is necessary to invert only in the final structural formulation.

3. GOVERNING EQUATIONS AND DISCRETIZED MODELS

The governing transient heat conduction equation in domain Ω enclosed by S for the cases considered in this paper are described by

$$\rho c_v \dot{T} = (k_{ij}\ T,_j) + Q \qquad \text{in } \Omega \qquad (3a)$$

$$T = T_s \text{ on } S_1; \quad (k_{ij}\ T,_j)\ n_i = -h\ (T-T_\infty) \quad \text{on } S_2 \qquad (3b)$$

where $S = S_1 + S_2$.

For the associated structural problem, the governing equations are described by

$$\rho\ddot{u}_i = \sigma_{ji,j} + \rho f_i \qquad \text{in } \Omega \tag{4a}$$

$$e_{ij} = 1/2(u_{i,j} + u_{j,i}); \ \sigma_{ij} = E_{ijkl}\, e_{kl} - \beta_{ij}(T-T_{in}) \tag{4b}$$

with boundary conditions

$$u_i = u_g \text{ on } S_1 \ ; \qquad \sigma_{ji}n_i = u_h \qquad \text{on } S_2 \tag{4c}$$

Envoking the classical Bubnov-Galerkin scheme after applying the Laplace transform in time to eqs. 3-4, the transformed discretized element equations for each of the thermal and structural models can be represented as

$$[\bar{K}]_e \ \{\bar{\Phi}\}_e = \{\bar{F}\}_e \tag{5}$$

Using index notation, the element definitions in the transform domain for the heat conduction equations (eqs.3) and the equations of motion (eqs.4) are derived as:

Thermal Model:

$$[\bar{K}]_e = \int_{\Omega_e} k_{ij} N_{\alpha,i} N_{\beta,j} d\Omega_e + \int_{\Omega_e} \bar{\rho}\bar{c}_v N_\alpha N_\beta d\Omega_e + \int_{S_2} hN_\alpha N_\beta d\Omega_e \tag{6a}$$

$$\{\bar{F}\}_e = \int_{\Omega_e} \bar{Q} N_\alpha d\Omega_e + \int_{\Omega_e} \rho c_v T_{in} N_\alpha d\Omega_e + \int_{S_1} (\bar{q}_{ij} n_j) N_\alpha dS_1 + \int_{S_2} h\bar{T}_\infty N_\alpha dS_2 \tag{6b}$$

Structural Model:

$$[\bar{K}]_e = \int_{\Omega_e} E_{ijkl}\, N_{\alpha,k}\, N_{\beta,l}\, d\Omega_e + \int_{\Omega_e} \bar{\rho}\, N_\alpha\, N_\beta\, d\Omega_e \tag{7a}$$

$$\{\bar{F}\}_e = \int_{\Omega_e} (\bar{\rho} u_i + \rho \dot{u}_i + \rho \bar{f}_i) N_\alpha\, d\Omega_e$$

$$+ \int_{\Omega_e} \beta_{ij}\, (\bar{T} - \bar{T}_i) N_{\alpha,j}\, d\Omega_e + \int_{S_2} \bar{u}_h\, N_\alpha\, dS_2 \tag{7b}$$

For evaluating the element matrices for the numerical test cases in this paper, the following interpolation functions are proposed for each of the thermal and structural models respectively.

Thermal Model: $\bar{T} = [N_1\ N_2\ N_p]\ \{\bar{T}\}_e$ (8a)

$$N_1 = \mathrm{Sinh}[\lambda(\ell - x)]/\mathrm{Sinh}[\lambda\ell]; \ N_2 = \mathrm{Sinh}[\lambda x]/\mathrm{Sinh}[\lambda\ell] \tag{8b}$$

$$N_p = 1-N_1 - N_2 \ ; \ \text{and} \ \lambda = (\rho c_v s/k)^{1/2} \tag{8c}$$

Structural Model: $\bar{u} = [N_1 \ N_2 \ \ N_p] \ \{\bar{u}\}_e$ (9a)

$$N_1 = \text{Sinh}[\lambda^*(\ell-x)]/\text{Sinh}[\lambda^*\ell]; \ N_2 = \text{Sinh}[\lambda^* x]/\text{Sinh}[\lambda^*\ell] \tag{9b}$$

$$N_p = (1-N_1 - N_2)[\bar{u}_i + \dot{\bar{u}}_i] - \alpha_\ell \left[\frac{\lambda}{(\lambda^2-\lambda^{*2}) \ \text{Sinh}[\lambda\ell]}\right](p + q) \tag{9c}$$

$$\lambda^* = (\rho s^2/E)^{1/2} \tag{9d}$$

$$p = \{N_1 + \text{Cosh}\ [\lambda\ell]\ N_2 - \text{Cosh}\ [\lambda x]\}\ (\bar{T}_2 - \bar{T}_i) \tag{9e}$$

$$q = \{N_2 + \text{Cosh}\ [\lambda\ell]\ N_2 - \text{Cosh}\ [\lambda(\ell-x)]\}\ (\bar{T}_1 - \bar{T}_i) \tag{9f}$$

4. NUMERICAL TEST STUDIES

The problem concerns an elastic half space (x>0) with the bounding plane x = 0 subjected to sudden variations in temperature. The bounding plane is assumed to be traction free at all times, and the medium is constrained so that there is only uniaxial motion. Thus

$$u_x = u_x\ (x,t) \ ; \ \text{and} \ u_y = u_z = 0 \tag{10}$$

The initial and boundary conditions are given by

$$T(x,o) = T_i \ ; \ u_x(x,o) = u_{xt}(x,o) = 0; \ \sigma_x(o,t) = 0 \tag{11}$$

The bounding plane (x = 0) is assumed to be exposed to two types of temperature environments: (1) Sudden surface heating, (2) Convective surface heating.

<u>Danilovskaya's first problem</u>:

$$T(0,t) = \begin{cases} T_i & \text{for } t < 0 \\ T_1 & \text{for } t \geq 0 \end{cases} \tag{12}$$

<u>Danilovskaya's second problem</u>:

$$T(o,t) = T_i \ \text{for } t < o; \ k \left.\frac{dT}{dx}\right|_{x=o} = h(T\big|_{x=o} - T_\infty) \ \text{for } t \geq o \tag{13}$$

where k is the thermal conductivity, h is the convective heat transfer coefficient, and T_∞ is the convective exchange temperature.

The above problems were solved using the proposed formulations to demonstrate the applicability for capturing the thermal stress waves. The numerical data assumed is taken from the reference by Ting and Chen[10] who attempt to solve these cases using the concept of heat displacement and a variational formulation in Lagrangian form. Ting and Chen [10] model the semi-infinite elastic medium as a finite medium and introduce nondimensional variables. A nondimensional characteristic length (L=4) and mesh having 248 degrees of freedom was used by Ting and Chen[10]. Exact solutions for these problems are also available [11].

Using the proposed unified hybrid formulations in conjunction with the special purpose transfinite elements derived earlier, these models were analyzed with a mesh containing only two elements for each of the thermal and structural models respectively. Figure 1 shows the comparative time history results for temperature, structural displacement, and stress for a point $\xi=1.0$ for the first Danilovskaya's problem. Note that $\xi=1.0$ is the nondimensionalized location of the elastic thermal stress wave front at a nondimensionalized time $\tau=1.0$. The superior capability of the present formulations to predict accurate results and also capture the propagating thermal stress waves is demonstrated clearly in Fig. 1. Figure 2 depicts the comparative time histories for the temperature, structural displacement, and thermal stress for the second Danilovskaya's problem at a similar location as the previous problem. Again, the results clearly demonstrate the capability to accurately predict and therein capture the detailed responses without any fluctuations. This is because of the special purpose hybrid transfinite elements formulated in this paper. Using linear elements via the present approach [3] although the results agree well, the responses were not as accurate as the present formulations. The methodology via the special purpose elements formulated in this paper demonstrates the capability to accurately predict the thermally induced stress waves.

5. CONCLUDING REMARKS

The present paper described a unified computational methodology for accurately predicting the propagation of thermal stress waves in elastic solids via hybrid formulations which combine finite elements, Galerkin techniques, and transform approaches. A unique feature of the proposed formulations lies in the hybrid nature as well as the development of special purpose transfinite elements for modeling the propagation of thermally induced stress waves. Comparative results demonstrated the superior capability and ability of the proposed formulations to accurately predict the thermal stress waves induced due to boundary heating.

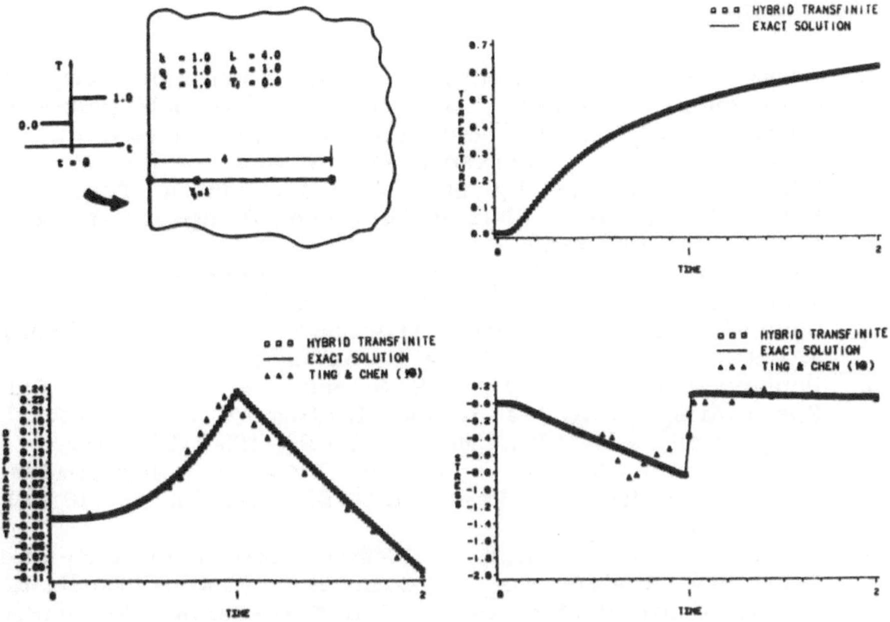

Fig. 1 Comparative temperature/displacement/stress histories for Danilovskaya's first problem.

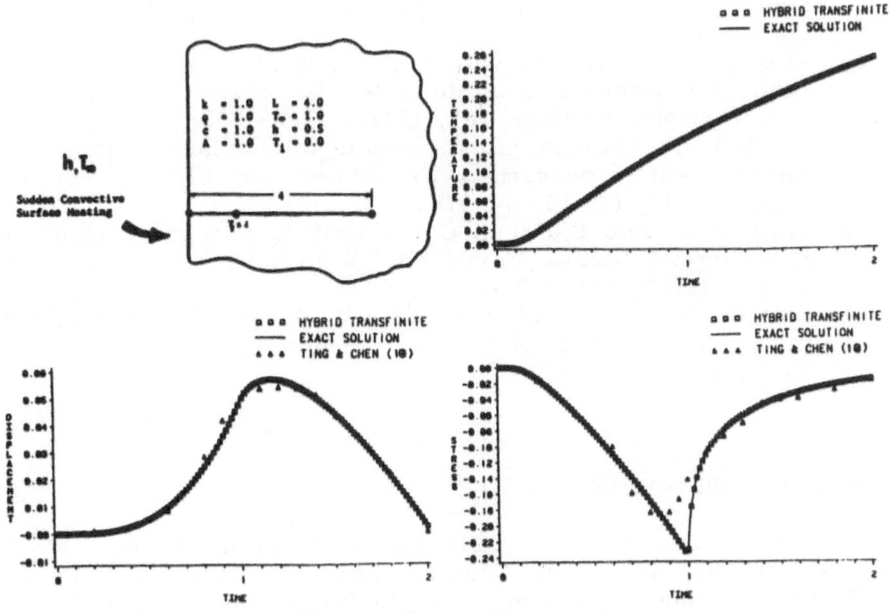

Fig. 2 Comparative temperature/displacement/stress histories for Danilovskaya's second problem.

REFERENCES

1. Tamma, K. K. and Railkar, S. B., 'Transfinite Element Methodology Towards a Unified Thermal/Structural Analysis', J. Computers and Structures (to appear).
2. Tamma, K. K., Spyrakos, C. C. and Lambi, M. A., 'Thermal-Structural Dynamic Analysis Via a Transform Method Based Finite Element Approach', J. Spacecrafts and Rockets (to appear).
3. Tamma, K. K. and Railkar, S. B., 'A Generalized Hybrid Transfinite Element Computational Approach for Nonlinear/Linear Unified Thermal/Structural Analysis', J. Computers and Structures (to appear).
4. Danilovskaya, V. I., 'Thermal Stresses in an Elastic Half Space Arising After a Sudden Heating of Its Boundary', (in Russian), Prikl. Mat. Mekh., 14 (3), (1950) 316-318.
5. Danilovskaya, V. I., 'On a Dynamical Problem of Thermoelasticity', (in Russian), Prikl. Mat. Mekh., 16 (3), (1952) 341-344.
6. Weeks, G. E. and Cost, T. L., 'Complex Stress Response and Reliability Analysis of a Composite Elastic-Viscoelastic Missile Configuration Using Finite Elements', Mechanics Research Communications, 7(2), (1980) 59-63.
7. Beskos, D. E. and Boley, B. A., 'Use of Dynamic Influence Coefficients in Forced Vibration Problems with the Aid of Laplace Transforms', J. Computers and Structures, 5, (1975) 263-269.
8. Aral, M. M. and Gulcat, U., 'A Finite Element Laplace Transform Solution Technique for the Wave Equation', Int. J. Num. Meth. in Engr., 11, (1977) 1719-1732.
9. Durbin, F., 'Numerical Inversion of Laplace Transofrms: An Efficient Improvement to Dubner and Abate's Method', Comp. J., 17, (1979) 371-376.
10. Ting, E. C. and Chen, H. C., 'A Unified Numerical Approach for Thermal Stress Waves', J. Computers and Structures, 15 (2), (1982) 165-175.
11. Sternberg, E. and Chakravorty, J. G., 'On Intertia Effects in a Transient Thermoelastic Problem', J. Appl. Mech., 26(4), (1959) 503-509.

ACKNOWLEDGEMENTS

This research was conducted with support in part by NASA-Langley Research Center, Hampton, Virginia, and the Flight Dynamics Laboratory, Wright Patterson Air Force Base, Ohio. The authors greatly appreciate the continued encouragement and support.

THE DETERMINATION OF OPTIMAL PROPERTIES OF A LAYERED PAVEMENT STRUCTURE

D.F.E. Stolle and A.N. El-Bahrawy
Department of Civil Engineering and Engineering Mechanics
Hamilton, Ontario, Canada

ABSTRACT

A two-dimensional axisymmetric model formulated within the weighted residual finite element framework and combined with least squares optimization method for finding optimal model parameters, is presented The model incorporates analytical solution in terms of Bessel functions in radial direction, thereby reducing the boundary-valued discretization from two- to one-dimension. Advantages of using this procedure over the conventional finite element method, rests in its ability to model thin layers of different material types, as in a pavement structure, without having to resort to fine discretization and the resulting large system of equations which must be inverted. An example is presented to illustrate the use of this model in conjunction with nondestructive field test data provided by a falling weight deflectometer (FWD), for establishing in situ properties of a layered pavement structure.

1. INTRODUCTION

In recent years non-destructive testing techniques involving interpretation of surface deflection measurements via rational methods have gained popularity for characterization of pavement structure integrity. These tests have the ability to locate soft spots and voids, measure structural strength, and establish pavement uniformity. While evaluation of the test data requires overall knowledge on pavement materials and an understanding of how pavements behave under in situ loading, a rational model is required to discern the in situ properties from the surface data. This subject has been addressed in several publications over the last ten years; see e.g.[1], [2].

The purpose of this paper is to present a discrete layer model for determining stresses, strains and displacements within multilayered media, and to describe a technique based on least squares method for evaluating the pseudo linear elastic in situ properties given a surface deflection basin and the geometry of the pavement structure. While the discrete layer model is most appropriate for the static loading case, the applicability of the model for discerning in situ properties from data provided by nondestructive impact devices such as the falling weight deflectometer (FWD) is also discussed.

2. DEFINITION OF PROBLEM

Let us consider the three-layer system shown in Fig. 1. The real pavement and subgrade materials exhibit nonlinear elastoplastic stress-strain behaviours which are sensitive to temperature and moisture content and are not necessarily uniform and homogeneous; e.g. subgrade may contain soft spots or the pavement wearing surface may be discontinuous. However, most analyses of pavement structures are completed by assuming linear elastic theory where the upper two layers extend to infinity in the horizontal plane, and the bottom one is semi-infinite. Consequently the properties which are used by these models to match the measured in situ behaviour are not true material properties but are effective pseudo-elastic model parameters which reflect the average bulk properties at the prevailing in situ temperature, moisture content and stress states. Since the loads which are applied to pavement structures are well below yield levels and generally of short duration, the assumption of pseudo-elastic response is considered to be reasonable.

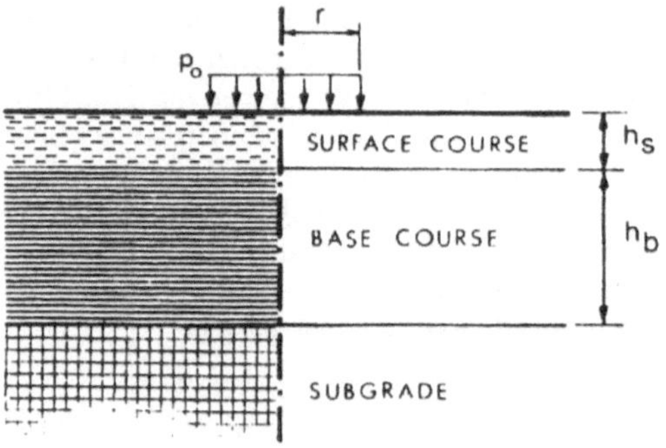

Fig. 1: Geometry for a three-layer system

3. DYNAMIC VERSUS STATIC MODELLING

The pavement properties which are discerned from the field data are sensitive to the equipment used to measure the deflection basins and the model adopted for the interpretation of data. Non-destructive testing equipment include static (Benkelman beam), harmonic (Road Rater) and impact (Falling weight deflectometer) loading devices [1]. In practice static analysis is generally used to establish the pseudo-elastic properties regardless of the nature of the loading. As a result it is not unusual to overestimate subgrade stiffness, see, e.g.[2]. While dynamic analysis is most appropriate for interpreting field data from the harmonic and impact devices, assumptions regarding the interaction between measuring equipment and structure, material properties and location of groundwater table and boundary can undermine the improved accuracies associated with this type of analysis when compared to those obtained assuming a static response.

The method expounded in this paper, although based on static analysis, may be used to interpret data from the FWD. Fig. 2 which is reproduced from Ref [2] compares measured surface displacements with those predicted by static and dynamic analyses. From this figure it may be noted that better agreement exists between static and measured deflection bowl shapes, even though the dynamic predictions have less error The shape of the deflection bowl is just as important as the deflection values themselves for establishing pseudo-elastic properties. This figure and others presented in Ref. [2] suggest that static analysis can be used to determine the in situ properties from deflection bowl data, provided that the measured displacements are adjusted via a translation, which for the data shown in Fig. 2 amounts to an extra 0.025-0.05 mm displacement.

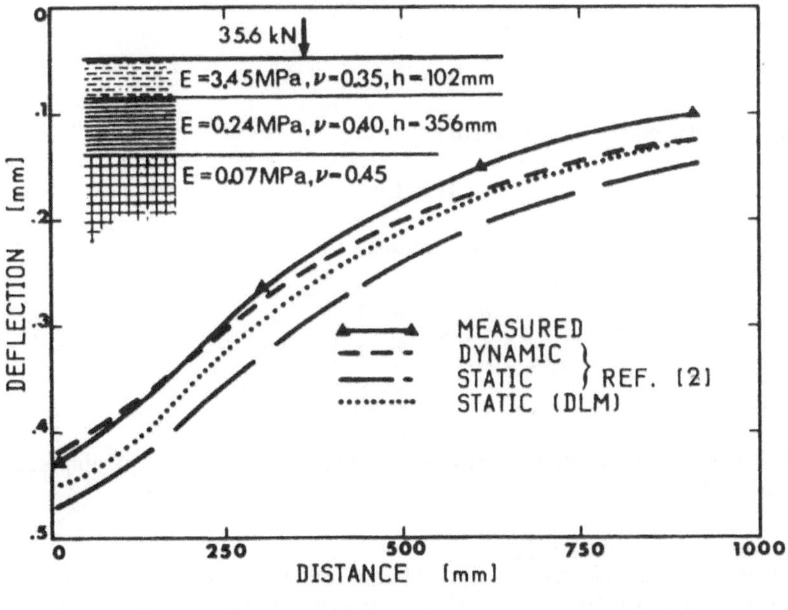

Fig. 2: Measured a predicted surface deflections, Sherrad section

4. DISCRETE LAYER MODEL

4.1 Kinematic Relationship

An examination of Burmister's [3] work suggests that the displacement field within a discrete layer element, due to an axisymmetric surface loading $t_z = c_i J_0(m_i r)$, may be approximated as

$$\mathbf{u} = \mathbf{N}\mathbf{a} \tag{1a}$$

$$\mathbf{u} = \langle u_r \, u_z \rangle^T \qquad \mathbf{a} = \langle u_{r_1} \, u_{z_1} \, u_{r_2} \, u_{z_2} \ldots \rangle^T \tag{1b}$$

with

$$\mathbf{N} = \begin{bmatrix} J_1(m_i r) & 0 \\ 0 & J_0(m_i r) \end{bmatrix} \begin{bmatrix} N_1 & 0 & N_2 & 0 & \ldots \\ 0 & N_1 & 0 & N_2 & \ldots \end{bmatrix} \tag{1c}$$

where u_r and u_z are radial and axial displacements, respectively, $J_n(m_i r)$ are Bessel functions of first kind and nth order corresponding to wave number m_i, N_i are polynomial interpolation functions for axial direction and **a** contains the generalized displacement amplitudes. Subscripts r, z and θ refer to cyclindrical coordinates in radial, axial, and circumferential directions, respectively.

Now that the displacement field is defined, we can relate **ε** to **a** assuming infinitesimal strain approximation,

$$\boldsymbol{\varepsilon} = \mathbf{B}\mathbf{a} \tag{2a}$$

where kinematic matric **B** is given by

$$\mathbf{B} = \mathbf{L}\mathbf{N} \tag{2b}$$

with L being the linear differential operator,

$$\mathbf{L} = \begin{bmatrix} \frac{\partial}{\partial r} & 0 & \frac{\partial}{\partial z} & \frac{1}{r} \\ 0 & \frac{\partial}{\partial z} & \frac{\partial}{\partial r} & 0 \end{bmatrix}^T \tag{2c}$$

For the examples presented in this paper linear interpolation of displacement amplitudes is adopted.

4.2 Constitutive Relationship

Linear elastic, isotropic material behaviour is adopted for each layer. The corresponding constitutive equation between stress **σ** and strain **ε**, is given by

$$\boldsymbol{\sigma} = \mathbf{D}\,\boldsymbol{\varepsilon} \tag{3}$$

where **D** is the constitutive matrix which is a function of the elastic modulus E and Poisson's ratio ν.

4.3 Equilibrium

The heart of the finite element methods rests in the ability to convert the differential equations for equilibrium to an integrated equivalent. This is usually accomplished by minimization of an appropriate functional or via, the more general, weighted residual procedure which is adopted within this study; see. e.g. [4]. It can be shown that an appropriate weighted residual form for the class of problems addressed in this paper is

$$\int_V \overline{B}^T \sigma \, dV - \int_{S_t} w^T T \, dS = 0 \qquad \text{in } V \tag{4}$$

where T are surface tractions on boundary S_t of the form $t_z = c_i J_o(m_i r)$,w are weighting functions,

$$\overline{B} = Lw \qquad \text{in } V \tag{5}$$

and the other terms are the same as defined previously.

In order to simplify integrations involving Bessel functions and to account for the natural radial decay associated with the displacement field, the following weighting function is appropriate.

$$w = e^{-\lambda r} N \tag{6}$$

where λ is a decay parameter and **N** contains the interpolation functions for the axial direction. It should be noted that identical interpolation was adopted in this study for the weighting function and displacement amplitudes in the axial direction. A nonsymmetric matrix equation is obtained after appropriate substitution of constitutive, kinematic and weighting relationships into equation (4). For a more realistic load distribution, e.g. circular, the surface traction can be approximated by a truncated Fourier-Bessel series, $t_z = \Sigma \, c_i \, J_o \, (m_i r)$. The solution of the boundary-valued problem consists of completing a separate analysis for each series term, and then linearly superimposing the displacement fields.

In order to model a semi-infinite half space, a spring-type boundary condition which makes use of Burmister's solution can be used for the bottom layer. Such a boundary can model the semi-infinite domain exactly and avoids having to approximate a semi-infinite medium by introducing several layers of elements for the purpose of removing the boundary from the area of interest.

5. OPTIMIZATION TECHNIQUE

Unlike the prediction of the deflection bowl given the elastic properties and thickness of each layer, the inverse procedure involving the calculation of the elastic moduli from the deflection bowl data does not guarantee convergence to a unique solution. In fact, several combinations of properties can yield identical deflection bowls as characterized by a finite number of observation points.

It was found during the early stages of this study that powerful direct search procedures like Rosenbrock's method [5] could not converge to the known moduli of the test problems. While subgrade predictions were generally consistent, that is within 10% of the known modulus, surface stiffnesses were either greatly underestimated or overestimated, depending on the initial starting values for optimization. This insensitivity of the optimal solution to surface stiffness is not surprising since generally 70-90% of the measured surface deflection is due to subgrade deformation. Owing to the inability of the direct search method to converge to the elastic moduli for the test problems a gradient approach based on a least squares technique was adopted.

Let us define the surface deflection bowl by n observation points such that $\mathbf{w}_{ob} = <w_1, w_2 \ldots, w_n>^T$. For sufficiently small variations in elastic moduli $\Delta E_k = <\Delta E_1\ \Delta E_2, \ldots, \Delta E_m>^T$ where m is the number of layers, the corresponding change in deflection bowl is given by

$$\Delta \mathbf{w}_k = \frac{\partial \mathbf{w}}{\partial \mathbf{E}} \Delta \mathbf{E}_k \tag{7}$$

which can be rewritten as

$$\left(\frac{\partial \mathbf{w}}{\partial \mathbf{E}}\right)^T \left(\frac{\partial \mathbf{w}}{\partial \mathbf{E}}\right) \Delta \mathbf{E}_k = \left(\frac{\partial \mathbf{w}}{\partial \mathbf{E}}\right)^T \Delta \mathbf{w}_k \tag{8}$$

for the purpose of finding ΔE_k given $\Delta w_k = (w_{ob} - w_k)$ where $\mathbf{w}_k$ defines the deflection bowl corresponding to $\mathbf{E}_k$. It is assumed that Poisson's ratio can be estimated to a reasonable accuracy, a priori. The gradient $\partial \mathbf{w}/\partial \mathbf{E}$ must be evaluated numerically.

To overcome the problem of converging to a wrong solution, or even diverging, a modified definition for Δw_k is suggested

$$\Delta \mathbf{w}_k = \frac{i}{n_s}(\mathbf{w}_{ob} - \mathbf{w}_o) + \mathbf{w}_o - \mathbf{w}_k \tag{9}$$

where $\mathbf{w}_o$ are the displacements corresponding to starting values $\mathbf{E}_o$, and i and n_s represent increment number and number of displacement increments, respectively. We can proceed with the optimization in a manner similar to that followed when solving elastoplasticity problems where n_s would correspond to the number of load increments and eq. (9) would reflect a residual load vector. As with incremental elastoplasticity approaches we may use the initial stiffness technique, tangential stiffness technique, or a combination of the two. For the example presented in this paper the initial stiffness technique was adopted where $(\partial \mathbf{w}/\partial \mathbf{E})^T\ (\partial \mathbf{w}/\partial \mathbf{E})$ was evaluated only once; i.e., for initial $\mathbf{E}_o$. To help expedite convergence, the initial estimates of **E** for each layer should be a lower bound.

6. NUMERICAL EXAMPLE

The discrete layer model (DLM) presented herein was tested by making comparisons of its solutions with those effected by using Burmister's [3] equation for several problems involving multilayered media. The agreement of solutions for each test case was excellent [6].

An example presented in Ref. [2] and shown in Fig. 2 was analysed with DLM. The simulation was completed by using 10 and 20 elements to approximate the surface and base layers, respectively with the spring-type bound condition to model the semi-infinite subgrade. The 35.6 kN load was uniformly distributed over a 305 mm diameter flexible plate by using a 40 term Fourier-Bessel series. It is clearly illustrated in Fig. 2 that the DLM gives a reasonable prediction of the surface deflections when using the properties that are also given in the figure. It should be noted that similar predictions were obtained when using both courser and finer discretizations. All simulations were completed with a Texas Instruments Professional computer incorporating an 8087 co-processor.

In order to illustrate the use of the DLM for estimating the elastic modulus of each layer, optimization simulations were completed using the geometry, Poissons' ratios and the measured surface deflection bowl shown in Fig. 2. The deflection bowl which reflects the response due to a FWD-type loading was adjusted by translating each measured displacement down by the same amount. The initial assumed moduli for the surface, base and subgrade layers were 1.73, 0.138 and 0.035 MPa, respectively. Convergence, as measured by the root mean square error in moduli between two successive iterations (<0.001) was achieved for all simulations within 20 iterations. The results of the simulations are given in Table 1.

		Displacement Adjustment (mm)		
		0	0.013	0.025
E (MPa)	Surface	3.189	2.977	2.784
	Base	0.229	0.249	0.267
	Subgrade	0.084	0.076	0.069

Table 1: Predicted Elastic Moduli as Function of Deflection Bowl Correction

While the base and subgrade moduli given in Ref. [2] could be reproduced through the proper selection of deflection bowl adjustment, the predicted surface stiffnesses were consistently lower than the suggested asphalt concrete modulus of 3.45 MPa. Owing to the limited number of case histories studied to date, it is not clear at this time whether the discrete layer model underestimated the surface modulus due to model limitations or whether the reported laboratory modulus overestimates the inside value.

7. CONCLUDING REMARKS

The discrete layer model presented herein is capable of yielding reasonable predictions for multilayered media response to static, axisymmetric surface loading. It is possible to estimate a range of reasonable moduli for each layer given the measured surface deflection bowl by completing two to three optimizations, each incorporating a different adjustment to the measured deflections.

ACKNOWLEDGEMENTS

The authors wish to thank the Natural Sciences and Engineering Research Council of Canada for the financial support which was provided to this project.

References

1. Hoffman, M.S. and Thompson, M.R., "Comparative Study of Selected Nondestructive Testing Devices", Transportation Research Record 852, TRB, National Research Board, Washington, D.C. (1982), 32-41.

2. Sebaaly, B.E., Mamlouk, M.S., and Davies, T.G., "Dynamic Analysis of Falling Weight Deflectometer Data", Transportation Research Board 1070, TRB, National Research Board, Washington, D.C. (1986), 63-68.

3. Burmister, D.M., "The General Theory of Stresses and Displacements in Layered Soil Systems", J. Applied Physics 16 (1945) 89-94 , 126-127, 296-302.

4. Zienkiewicz, O.C. - The Finite Element Method, McGraw-Hill, New York, 1977.

5. Rosenbrock, H.H., "An Automatic Method for Finding the Greatest or Least Value of a Function", Comp. J., 3 (1960), 174-184.

6. Stolle, D.F.E., "Axisymmetric Analysis of Multilayered Media", submitted to Engineering Analysis (1986).

A FINITE ELEMENT METHOD FOR A NUMERICAL ANALYSIS OF THE DEEP-DRAWING PROCESS

M. BRUNET
Laboratoire de Mécanique des Solides
INSA 304, 69621 VILLEURBANNE CEDEX, FRANCE

SUMMARY

For the purpose of studying the deep drawing process and related metal-forming processes, an efficient finite element method has been developed based on the updated Lagrangian approach with large elasto-plastic strain. The main object of the paper is to show a procedure which maintains equilibrium overall the process including contact and friction boundary conditions and which is also able to handle large strain increments.

1. INTRODUCTION

In recent years, considerable progress has been made in the development of solution methods for large strain and geometrically non-linear problems but practical applications in metal-forming processes often use an explicit forward integration scheme eventually augmented with correction after each step. Therefore, it may be questionable how equilibrium is maintained or tool-force predicted after sometimes thousand steps and it appears that numerical results are quite sensitive to the increment size [1].

In this paper, we show that it is possible to handle large strain increments. Also, the quasi-Newton Raphson method works on the virtual work equation at the end of the increment where the boundary constraints are applied by an incremental penalty method. The incremental objectivity of the rate constitutive equation using the corotational Kirchhoff stress rate (or eventually the Green-Nagdi stress rate when kinematic hardening is needed) is maintained over a finite step by using the mid-step configuration for evaluating both the strain increment and the rotation increment. In the mid-step configuration the constitutive equation is integrated using the so-called first order radial return algorithm where the values in the equilibrium iterations are calculated based solely on the converged values at the beginning of the step. This procedure ensures path independency and permits unloading. In order to improve the rate of convergence, the elasto-plastic tangent stiffness is consistent with the radial return algorithm.

2. GOVERNING EQUATIONS

In the deep drawing process, the surface traction acting on the sheet, are dependent on the deformation. Consequently, the work done by the external loads is path-dependent. Furthermore, using a flow theory of plasticity, the current stress-strain relation is dependent on the past stress history of the material points. To account for these non-conservative effects it is necessary to employ an incremental solution procedure of some kind. Also, the present work makes use of the finite element method based on an updated Lagrangian formulation. The principle of virtual work at time t + Δt referred to the current configuration at time t can be written in the following way [2] :

$$\int_V \{\Delta S_{ij}\delta u_{i,j} + \sigma_{ij}\Delta u_{k,i}\delta u_{k,j}\} dV = \int_{S_t} (t_i + \Delta t_i)\delta u_i dS - \int_V \sigma_{ij}\delta u_{i,j} dV \tag{1}$$

where σ_{ij} is the Cauchy stress and ΔS_{ij} the convected increment of Kirchhoff stress. This permits us to calculate a certain state of equilibrium based on the knowledge of the previous equilibrium state together with the actual increments in the external loads or the prescribed displacements. To avoid error accumulation and to permit unloading, the new equilibrium state is calculated by an iterative technique where the term :

$$R = \int_{S_t} t_i \delta u_i dS - \int_V \sigma_{ij}\delta u_{i,j} dV \tag{2}$$

is often called the residual which is expected to vanish in the increment. For history-dependent elasto-plastic materials the constitutive relations are readily incorporated in the updated Lagrangian formulation since these equations are functions of variables referred to the current state.

3. BOUNDARY CONDITIONS

In the present formulation of the deep-drawing process, it is assumed that the tool parts are completely rigid bodies, whose relative movements are forced on a deformable elastic-plastic solid. Then we can use parametric functions to describe the rigid surface of the punch and of the die.

In an incremental approach, we determine an appropriate prescribed displacement to just bring a or some material points on the surface Sc to be incontact with the rigid surface. Thus for any two in-contact points we have :

$$\Delta u_n - \Delta g_n = 0 \text{ on Sc} \qquad (3)$$

where Δu_n is the increment of the normal displacement constrained by the given rigid surface motion Δg_n. To resolve this boundary constraint, we apply the exterior penalty method. Physically the constraint (3) can be wiewed as to replace the rigid surface by a set of very stiff springs along the contact surfaces. Thus, we define :

$$\overset{\triangledown}{\Delta f}_n = - k_n(\Delta u_n - \Delta g_n) \text{ on Sc} \qquad (4)$$

with a very large penalty parameter k_n (>0). The point will stay in contact as long as the resultant contact pressure f_n updated by its material increment Δf_n remains negative. In (4), $\overset{\triangledown}{\Delta f}_n$ may be the corotational increment in order to preserve the material frame indifference of the constitutive relation (4) when large deformation occurs. We assume that there exists a Coulomb's friction law, so that no relative motion is observed if :

$$|f_t| < \mu|f_n| \qquad (5)$$

Thus, a same relation that (4) holds between the tangential increment of the friction force and the tangential increment of displacement :

$$\overset{\triangledown}{\Delta f}_t = - k_t (\Delta u_t - \Delta g_t) \text{ on Sc} \qquad (6)$$

when the state of stress of a point on Sc is such that

$$|f_t| = \mu|f_n| \quad \text{and} \quad |\overset{\triangledown}{df}_t| = \mu|\overset{\triangledown}{df}_n| \qquad (7)$$

the point is said to be sliding. Frictional effects play an important role in most of the practical applications. In general an attempt is made toward constructing an incremental friction law in analogy with the elastoplasticity theory [1].

4. FINITE ELEMENT APPROXIMATION

Using the corotational increment of Kirchhoff stress in the constitutive law, the principle of virtual work at time t + Δt referred to the current configuration at t (1), including the virtual work done by the pressure contact forces can be rewritten as :

$$\int_V \{L'_{ijkl}\Delta u_{k,l} + \sigma_{jk}\Delta u_{i,k}\}\, \delta u_{i,j} dV + \int_{Sc} [k_n du_n \delta u_n + k_t \Delta u_{ti} \delta u_{ti} - \frac{1}{2}(\Delta u_{i,j} - \Delta u_{j,i}) f_j \delta u_i] dSc = \int_{S_t} (t_i + \Delta t_i)\delta u_i dS_t + \int_{S_c} [(f_n + k_n dg_n)\delta u_n + (f_{ti} + k_t dg_{ti})\delta u_{ti}] dSc - \int_V \sigma_{ij}\delta u_{i,j} dV \qquad (8)$$

with $\forall\ \delta u_i$ such that $\delta u_i = 0$ on S_u and $S = S_c \cup S_u \cup S_c$
where : $L'_{ijkl} = C^{e-p'}_{ijkl} - \frac{1}{2}(\delta_{ik}\sigma_{jl} + \delta_{il}\sigma_{jk} + \delta_{jk}\sigma_{il} + \delta_{jl}\sigma_{ik})$ (9)

and the form of the numerical elasto-plastic moduli $C^{e-p'}_{ijkl}$ is discussed later. The spatial discretization of the integral form (8) follows the standard finite element procedures. Within each element, interpolations are made to the variables $\Delta \underset{\sim}{u}$, $\Delta \underset{\sim}{g}$, $\Delta \underset{\sim}{t}$ and $\delta \underset{\sim}{u}$ by properly chosen shape functions $N_\alpha(\underset{\sim}{x})$ which lead to an algebraic matrix equation :

$$\{\sum_e K^{ve}_{ij\alpha\beta} + \sum_e K^{Sce}_{ij\alpha\beta} - \sum_e \overset{\nabla}{K}{}^{Sce}_{ij\alpha\beta}\}\Delta u_{j\alpha} = \sum_e (f^{Ste}_{i\beta} + \Delta f^{Ste}_{i\beta}) + \sum_e (f^{Sce}_{i\beta} + \Delta f^{Sce}_{i\beta})$$

$$- \sum_e \int_{Ve} \sigma_{ij} N_{\beta,j} dVe \quad (10)$$

where

$$K^{ve}_{ij\alpha\beta} = \int_{Ve} (L'_{ijkl} + \delta_{ij}\sigma_{kl}) N_{\alpha,l} N_{\beta,k}\ dVe \quad (11)$$

$$K^{Sce}_{ij\alpha\beta} = \int_{Sce} \{k_n n_j n_i + k_t(\delta_{ij} - n_i n_j)\}\ N_\alpha N_\beta\ dSce \quad (12)$$

$$\overset{\nabla}{K}{}^{Sce}_{ij\alpha\beta} = \int_{Sce} \{N_{\alpha,k}\ f_k \delta_{ij} - N_{\alpha,i} f_j\}\ N_\beta/2\ dSce \quad (13)$$

are respectively the elastoplastic element stiffness matrix and the penalty stiffness matrices. The last one, due to the corotational derivative of the pressure contact is not symmetric and may be put on the right hand side of (10) as a load correction factor.

$$\Delta f^{Ste}_{i\beta} = \int_{Ste} \Delta t_i N_\beta dS_{te} \quad (14)$$

$$\Delta f^{Sce}_{i\beta} = \int_{Sce} \{k_n \Delta g_n n_i + k_t \Delta g_{tj}(\delta_{ij} - n_i n_j)\}\ N_\beta\ dSce \quad (15)$$

are respectively the element exterior surface load vector and the element penalty parameter load vector. Since the result of an incremental calculation is not known in advance, it is necessary to do a few iterations to assure consistent boundary conditions, equilibrium and path independency as it will be shown in the next section.

5. SOLUTION PROCEDURE

In the procedure presented here, it is attempted to satisfy equilibrium in the current state of the body. Morever, during iteration, the nodal imbalance is formulated with respect to the state obtained as a result of the previous iteration. Hence upon calculation of the latest estimate for the displacement increment, the geometry of the element will be updated corresponding to this latest estimate.

Using the classical linear expression for deformation rate and spin, but taken in the state at the middle of the total displacement increment, the incremental rotation tensor is calculated first as [2], [3] :

$$\Delta\underline{R} = e^{\Delta\underline{\omega}(t+\Delta t/2)}.\underline{I} \tag{16}$$

where $\Delta\underline{\omega}(t+\Delta t/2)$ is the incremental spin tensor evaluated at the middle of the increment. The Cauchy stress at the end of the increment then follows from :

$$\underline{\sigma}(t+\Delta t) = \Delta\underline{R}.\underline{\sigma}(t).\Delta\underline{R}^T + \Delta\underline{R}^{1/2}.\Delta\overset{\nabla}{\underline{S}}.\Delta\underline{R}^{1/2T} \tag{17}$$

If we use the Green-Nagdi stress rate of Cauchy stress instead of the corotational rate of Kirchhoff stress in the rate constitutive law, the incremental rotation tensor may be evaluated as :

$$\Delta\underline{R} = \underline{R}(t + \Delta t).\underline{R}^T(t) \tag{18}$$

$$\Delta\underline{R}^{1/2} = \underline{R}(t + \Delta t).\underline{R}^T(t + \Delta t/2) \tag{19}$$

where $\underline{R}$ is the orthogonal rotation tensor of the polar decomposition $\underline{F} = \underline{R}.\underline{U}$ of the gradient deformation tensor.

The evaluation of (17) requires an integration of the constitutive rate equations in the middle of the increment in order to obtain the corotational stress increment $\Delta\overset{\nabla}{\underline{S}}$. For Von Mises plasticity we have found that the elastic predictor-radial corrector one order implicit method is both simple and accurate [4]. This approach assures path independency and incremental objectivity.

The next aspect demands that in an efficient solution procedure the elastoplastic stiffness matrix must be consistent with the implicit integration procedure of the constitutive rate equations in order to improve the rate of convergence. A derivation of this consistent moduli $\underline{C}^{e-p'}$ can be found in [4] and [5]. For Von Mises plasticity and based on the elastic predictor-radial corrector method, it can be shown that :

$$C^{e-p'}_{ijkl} = G'(\delta_{ik}\delta_{jl}+\delta_{il}\delta_{jk})+\{2(1+\nu)G/(3(1-2\nu))-2G'/3\}\delta_{ij}\delta_{kl} - 3G'/((1+H'/3G)\bar{\sigma}^2)s_{ij}s_{kl} \tag{20}$$

In this equation, G is the elastic shear modulus, the Poisson's ratio and G' and H' are the modified shear modulus and plastic hardening coefficient related to the actual G and H by :

$$G' = G\bar{\sigma}/\bar{\sigma}_e \tag{21}$$

$$H' = H/[\bar{\sigma}(H/3G + 1)/\bar{\sigma}_e - H/3G] \tag{22}$$

In these relations, $\bar{\sigma}e$ is the equivalent elastic stress of the elastic predictor-radial corrector method. It is seen that the shear modulus and the hardening coefficient are modified accordingly such that the effective shear modulus G' is lower than the actual shear modulus G. Thus, use of the moduli $\underline{C}^{e-p'}$ improve considerably the rate of convergence.

Prior to the solution of practical problems, a number of one element tests were carried out to check the correctness of the implementation. The calculation were carried out with different increment sizes and the attempted independence of increment size was obtained.

6. NUMERICAL RESULTS

An experimental and numerical analysis of the axisymmetric deep drawing process has been carried out respectively by WOO [6] and ANDERSEN [7] using the same data of geometry, friction and material. ANDERSEN [7] used a purely incremental procedure in his calculation and he observed that the force-displacement curve tends to deviate more and more as the deformation increases.

In the present study, the element used is an axisymmetric quadrilateral and isoparametric element with 12 nodes. The area integration are performed using a (3 x 3) Gaussian scheme and also for the line integration on the contact boundary where the contact pressure and the penalty parameters are tested and evaluated at the Gauss points. After each increment, the distance from each of the tool parts to each of the Gauss points is calculated. Figure (1) shows the deformed mesh at two stages of the deep drawing analysis. We simulate the material as :

$$\sigma = 82 + 423\ \varepsilon^{p^{\,0.504}} \qquad (\mathrm{N/mm}^2)$$

in true stress and plastic logarithmic strain. The friction coefficient is 0.04 against both the punch part and the die-part of the tools.

The complete deep-drawing analysis has needed 1120 increments of displacement of about 5 % of the initial thickness of the sheets and an average of 3 iterations per increment. The equilibrium is maintained up to 1 % of the nodal forces for the euclidian norm of the residual forces. The comparison between the measured punch force and the calculated one, shown in figure 2, is in resonable good agreement, the experimental force is not given by WOO [6] after its maximum value. Other numerical results of strain and stress, not presented here, are also in a good agreement.

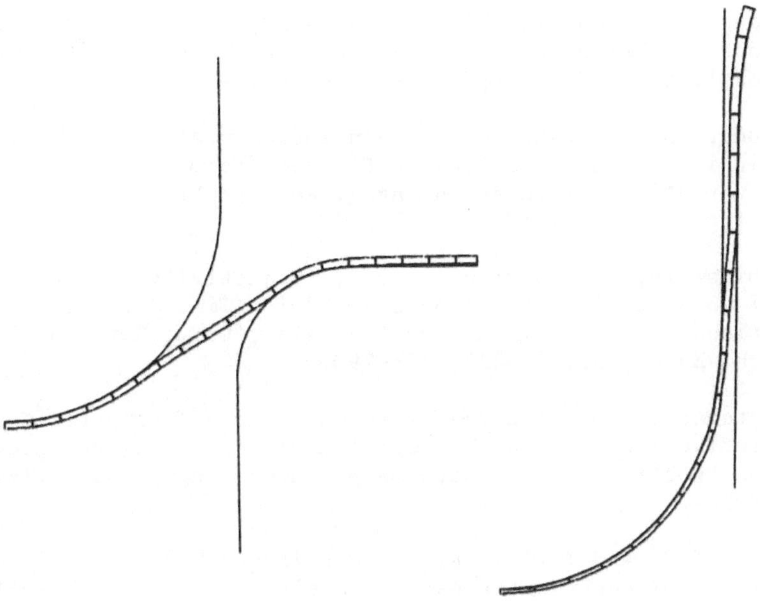

Figure 1 : F.E.M. deep-drawing

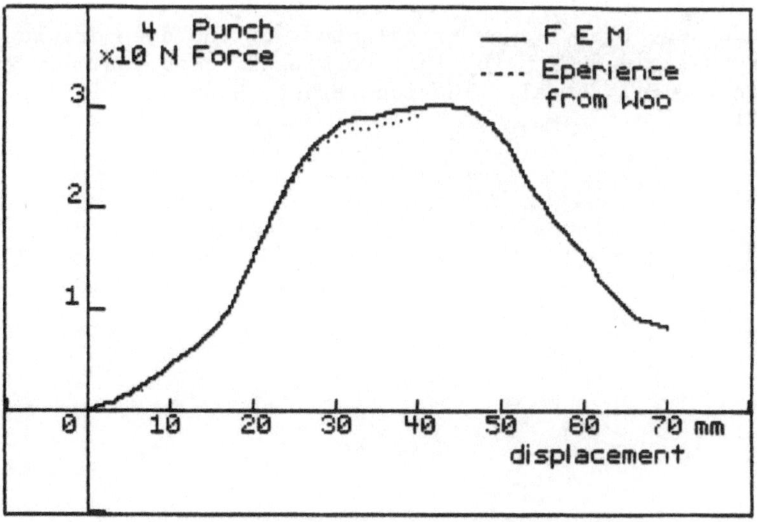

Figure 2 : Punch force versus displacement

REFERENCES

1. Cheng, J.H. and Kikuchi, N., 'An analysis of Metal Forming Processes Using Large Deformation Elastic-Plastic Formulations', Comp. Meth. Appl. Mech. Engng, 49, (1985), 71-108

2. Brunet, M. and Bahuaud, J., 'An incremental Variational Principle in Elastic-Plastic Finite Deformation Analysis' in Variational Methods in Engng. ed. Brebbia, C.A., Springer Verlag, (1985), 1-12

3. Rubinstein, R. and Atluri, S.N., 'Objectivity of incremental Constitutive Relations over Finite Time Steps in Computational Finite Deformation Analysis', Comp. Meth. Appl. Mech. Engng, 36, (1983), 277-290.

4. Nagtegaal, J.C., 'On the implementation of Inelastic Constitutive Equations with Special Reference to large Deformation Problems', Comp. Meth. Appl. Mech. Engng., 33, (1983) 469-484

5. Simo, J.C. and Taylor, R.L., 'Consistent Tangent Operators for rate-Independent Elastoplasticity', Comp. Meth. Appl. Mech. Engng, 48, (1985), 101-108

6. Woo, D.M., 'On the complete solution of the deep-drawing Problem', Int. J. Mech. Sci.,10, (1968), 83-94

7. Andersen, B.S., 'A numerical Study of the deep-drawing Process' in Num. Meth. Ind. Forming Process (ed.) by Pittmann, J.F.T. et al. Pineridge Press, Swansea, (1982), 709-721.

THEORY OF DEGENERATED CURVED SHELL AND LOCKING IN SHELL FINITE ELEMENTS

Hou-Cheng Huang
Tianjin University, China
presently: University College of Swansea, U.K.

SUMMARY: In this paper, Mindlin plate theory is extended to cater for curved shell structures. It can be considered as Mindlin type shell theory. From this theory, the C(0) continunity formulation of shell elements could be derived directly. The shear and membrane locking behaviour in degenerated shell elements is investigated and may be avoided by the assumed strain approach.

1. INTRODUCTION

Mindlin plate theory can be extended to cater for curved shell structures, if a local coordinate system is introduced at points of the shell midsurface (see Figure 1). The theory can be exprresed as that normals to the shell mid-surface before deformation remain straight but not necessarily normal to the shell mid-surface after deformation.

In the local coordinate system (x',y',z'), z' is taken perpendicular to the shell midsurface and the x'-y' coordinate plane is then taken as being tangential to the midsurface of the shell. Using this assumption, the C(0) continunity formalation of shell elements, which introduced the transverse shear deformation, could be derived directly. The element formulation is the same as that developed by Ahmad and called as the degenerated shell element.

The original Mindlin plate element and the degenerated shell element perform reasonably well for moderately thick plate and shell situations. However, for thin plates and shells when full integration is used to evaluate the stiffness matrix, overstiff solutions are often produced due to shear and membrane locking. Attempts have been made to correct this behaviour by use of reduced or selective integration techniques. However, when reduced integration of the shear and membrane stiffness matrix is carried out, apart from the obligatory zero energy modes associated with rigid body movements extra zero spurious energy modes are introduced. In order to avoid this serious defect, assumed strain shell elements were developed by the author [1,2]. In the formulation of this element, both assumed transverse shear strains and assumed membrane strains (AMS) are used to overcome the conflict between locking and mechanism problems.

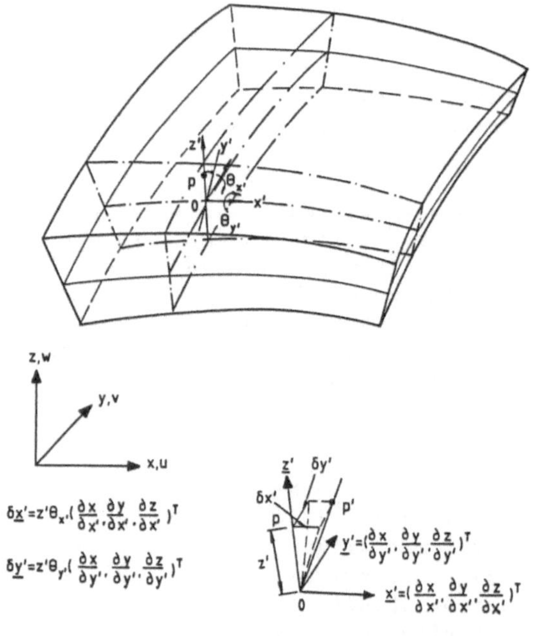

Fig. 1 Degenerated curved shell theory

2. THREE DIMENSIONAL DEGENERATED CURVED SHELL

It is assumed that the displacements of points at the midsurface are u'_o, v'_o and w'_o in the local coordinate directions x', y' and z' respectively. If the rotations $\theta_{x'}$ and $\theta_{y'}$ of the midsurface normals in the x'z' and y'z' plane are available then the following relations can be obtained at a typical material point P (Figure 1)

$$u' = u'_o + z'\theta_{x'} \; , \; v' = v'_o + z'\theta_{y'} \; , \; w' = w'_o \tag{1}$$

If the displacements u'_o, v'_o and w'_o can be transferred to the global coordinate system as u_o, v_o and w_o then it is possible to write that

$$u = u_o + z'(\theta_{x'} \frac{\partial x}{\partial x'} + \theta_{y'} \frac{\partial x}{\partial y'})$$
$$v = v_o + z'(\theta_{x'} \frac{\partial y}{\partial x'} + \theta_{y'} \frac{\partial y}{\partial y'})$$
$$w = w_o + z'(\theta_{x'} \frac{\partial z}{\partial x'} + \theta_{y'} \frac{\partial z}{\partial y'}) \tag{2}$$

It is noted that if the axes of the local coordinate system are parallel to those of the global coordinate system at all points in the shell midsurface, then

$$u = u_o + z\theta_x \qquad v = v_o + z\theta_y \qquad w = w_o \tag{3}$$

that is, the Mindlin plate expression is obtained. Therefore, the Mindlin plate is considered as a special case of the degenerated shell.

Since the stress perpendicular to the shell mid-surface is neglected, the strain components of interest are

$$\underset{\sim}{\varepsilon}' = \begin{bmatrix} \underset{\sim}{\varepsilon}'_f \\ -- \\ \underset{\sim}{\varepsilon}'_s \end{bmatrix} = \begin{bmatrix} \varepsilon_{x'} \\ \varepsilon_{y'} \\ \gamma_{x'y'} \\ \hline \gamma_{x'z'} \\ \gamma_{y'z'} \end{bmatrix} = \begin{bmatrix} \frac{\partial u'}{\partial x'} \\ \frac{\partial v'}{\partial y'} \\ \frac{\partial u'}{\partial y'} + \frac{\partial v'}{\partial x'} \\ \hline \frac{\partial u'}{\partial z'} + \frac{\partial w'}{\partial x'} \\ \frac{\partial v'}{\partial z'} + \frac{\partial w'}{\partial y'} \end{bmatrix} \tag{4}$$

where $\underset{\sim}{\varepsilon}'_f$ is the in - plane strain vector defined in the local coordinates, $\underset{\sim}{\varepsilon}'_s$ is a transverse shear strain vector, and u', v' and w' are the displacement components in the local system x'_i. $\underset{\sim}{\varepsilon}'_f$ can be divided into two parts one associated with membrane behaviour and one associated with bending behaviour. The global derivatives of the displacements u, v, and w are transformed into local derivatives of the local displacements u', v' amd w' by the standard operation.

The constitutive relationship between the five relevant stress and strain components in the local system may be written in partitioned form

$$\begin{bmatrix} \sigma_f \\ \sigma_s \end{bmatrix} = \begin{bmatrix} \underset{\sim}{D}_f & \underset{\sim}{0} \\ \underset{\sim}{0} & \underset{\sim}{D}_s \end{bmatrix} \begin{bmatrix} \varepsilon_f \\ \varepsilon_s \end{bmatrix} \tag{5}$$

in which the in-plane strains are

$$\underset{\sim}{\sigma}_f = [\sigma_{x'}, \sigma_{y'}, \tau_{x'y'}]^T \tag{6}$$

$$\underset{\sim}{\sigma}'_s = [\tau_{x'z'}, \tau_{y'z'}] \tag{7}$$

For an isotropic, homogeneous linear elastic material

$$\underset{\sim}{D}_f = E/(1-v^2) \begin{bmatrix} 1 & v & 0 \\ v & 1 & 0 \\ 0 & 0 & (1-v)/2 \end{bmatrix} \tag{8}$$

$$\underset{\sim}{D}_s = Ek/2(1+v) \begin{bmatrix} 1 & 0 \\ 0 & 1 \end{bmatrix} \tag{9}$$

where E is the elastic modulus, v is Poisson's ratio and k is a shear correction factor taken as equal to 5/6.

From the three dimensinal degenerated shell theory described above the C(0) continunity formulation of shell elements could be derived directly [3].

3. ESSENCES OF LOCKING BEHAVIOUR

According to Mindlin-type theories, the rotations are independent of the displacements in both plates and shells. The transverse shear effects should gradually diminish as the plate/shell thickness becomes extremely thin.

If we could solve the governing differential equations directly, shear locking behaviour should not appear. However, the shear strains obtained using a numerical method, such as the finite element method, may be inaccurate locally, though average shear strains over a particular region are reasonable. As we know, when the finite element method is used in conjunction with Mindlin-type formulations, the shear strain energy is always included and must be positive. Consequently, as the thickness of the plate and shell becomes extremely thin, the shear strain energy can be magnified unreasonably even though the average value of the shear strains over the area tends to zero. Therefore, when a full integration scheme is employed, the shear strain energy contribution will dominate the Total Potential Energy and the shear locking problem cannot be avoided [4].

In order to describe the essence of membrane locking let

us consider a shallow curved cylindrical shell (Figure 2). According to the Marguerre shallow shell theory [5,6] the membrane strain may be expressed as follows

$$\varepsilon_m = \frac{\partial u}{\partial x} + \frac{\partial w_o}{\partial x}\frac{\partial w}{\partial x} \qquad (10)$$

Here u and w are the x and z components of the displacemenet of the the mid-surface. w_o is the distance of the mid-surface from the x-y plane and ε_m is the membrane strain.

As in shear locking, membrane locking arises in curved elements due to the inability of most finite element formulations to adequately simulate the membrane strains. For example, in Figure 3 if the shell is subjected to pure bending, the membrane strains should equal zero. Unfortunately, isoparametric element formulations cannot represent this state, since, in (10) the two terms describing the membrane strains have different orders of polynomial. For a quadratic, isoparametric element, $u_{,x}$ is linear while $w_{o,x}w_{,x}$ is quadratic. Therefore, they do not match and cannot vanish in the state of pure bending.

4. ELIMILATIONS OF LOCKING BEHAVIOUR

According to [3, 4], it is possible to find some points at which values of the shear strains may represent the average distribution of the shear strain fields at the particular region. Assuming the shear strains tend to zero in a certain area of the 9-node Lagrangian element, it is obtained that, if $\xi = \pm a$ (where $a = (3)^{-1/2}$) which are the two Gauss points in the ξ direction, then $\gamma_{\xi\tau} = 0$ for very thin plates. Similarly, if $\eta = \pm a$ which are the two Gauss points in the η direction, then $\gamma_{\eta\tau} = 0$ for very thin plates.

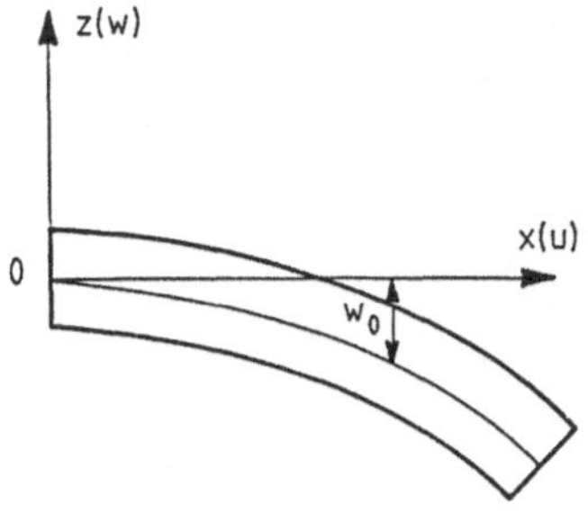

Fig. 2 Cross section of a Cylindrical shell

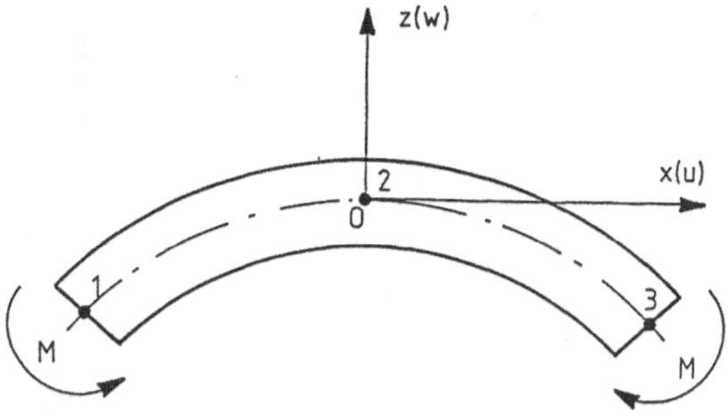

Fig. 3 Single cylindrical Shell element

As mentioned in [1, 2], six sampling points are required for the representation of the assumed shear strains. For $\gamma_{\xi\tau}$ the six sampling points are $\xi = 1$, $\xi = 0$ and $\xi = -1$ on each of lines $\xi = a$ and $\xi = -a$. Similarly, the six sampling points for $\gamma_{\eta\tau}$ are the three points $\xi = 1$, $\xi = 0$ and $\xi = -1$ at the two lines $\eta = a$ and $\eta = -a$.

It is known that the membrane strains can only be separated from the bending strains in the local Cartesian coordinate system (x', y', z') in which the x'-y' plane is tangent to the shell mid-surface. Therefore, it is expected that the use of assumed membrane strains in an orthogonal curvilinear coordinate system (r, s, t) can help to eliminate membrane locking behaviour. To find the location of the sampling points for membrane strain interpolations we consider a single shallow cylindrical shell element with 9 nodes. The element is subjected to pure bending (see Figure 3). Considering one section for simplicity we have:

$$u = \sum_{i=1}^{3} N_i u_i, \qquad w = \sum_{i=1}^{3} N_i w_i, \tag{11}$$

and

$$w_o = \sum_{i=1}^{3} N_i w_{oi}, \qquad x = \sum_{i=1}^{3} N_i x_i \tag{12}$$

where

$$N_1 = \xi(\xi - 1)/2, \quad N_2 = 1 - \xi^2, \quad N_3 = \xi(\xi + 1)/2 \tag{13}$$

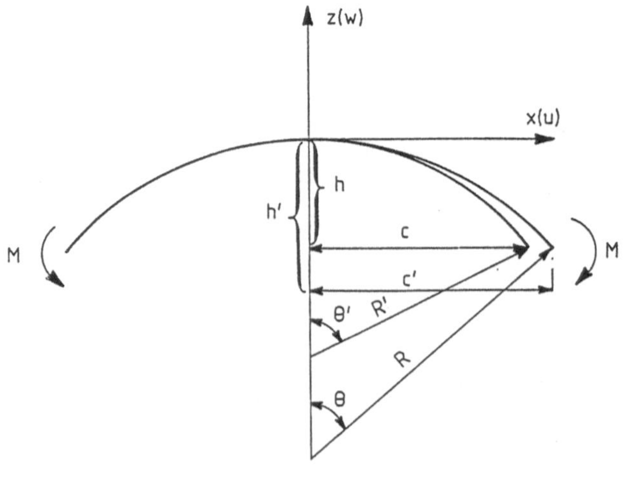

Fig. 4 The description of a deformed single element

Substituting (11)-(13) into (10) and noting that $u_2 = 0$, $u_1 = -u_3$, $w_2 = 0$, $w_1 = w_3$ and $x_2 = 0$, $x_1 = -x_3$ we have

$$\varepsilon_m = \frac{u_3}{c} + \frac{4hw_3}{c^2}\xi^2 \tag{14}$$

where $x_3 = c$, $w_{o2} = h$. Now, we try to find the value of ξ at which the membrane strain ε_m equals zero and it is assumed that an incremental rotation $\Delta\Theta$ is produced due to the moment (Figure 4). Since, in the linear case $\Delta\Theta$ is very small then we obtained that, if

$$\xi = \pm\frac{1}{2}\left[\frac{\sin\Theta(\sin\Theta-\Theta\cos\Theta)}{(1-\cos\Theta)(\cos\Theta+\Theta\sin\Theta-1)}\right]^{\frac{1}{2}} \tag{15}$$

then $\varepsilon_m = 0$. It is interesting to note that when Θ is small the value of (15) is approximately $1/\sqrt{3}$ which is the location for two Gauss points. Since (10) is true only for very shallow shells, it can be deduced that the membrane strains at lines $\xi = 1/\sqrt{3}$ and $\xi = -1/\sqrt{3}$ may tend to zero.

Therefore, for shells with a single non-zero curvature, assumed membrane strains can be interpolated from the values of the membrane strains at lines $\xi = a$, $\xi = -a$ and $\eta = a$, $\eta = -a$ (where $a = 3^{-1/2}$). For general shells, in the formulation of the the new 9-noded degenerated shell element the membrane and shear strains are interpolated from identical sampling points, but they are expressed in different coordinate systems - the membrane strains are expressed in the orthogonal curvilinear coordinate system, whereas the transverse shear strains are expressed in the natural coordinate system [2].

5. NUMERICAL TESTS

The shear locking test and other benchmark tests have been conducted in [3]. Here, only the membrane locking test are considered in which QUAD9**, the new ASM 9-node element and QUAD9S, the 9-node element with selective integration, are used. To check membrane locking behaviour in the shallow cylindrical shells a curved cantilever subjected to a tip bending moment is considered (Figure 5). When a single QUAD9** is used to represent the cantilever, fields of constant bending moment and zero membrane forces are obtained. However, when a single QUAD9S is used, though fields of constant moment is obtained, the value of bending moment is smaller than the applied value (80.0% of the applied value) and a parasitic membrane deformation occurs. This means that the membrane locking exists in QUAD9S. Certainly, the 9-node Lagrangian element with the completely reduced integration may provide fields of constant bending moment and zero membrane forces. But, for a single element, results cannot be obtained due to mechanisms unless more constriants are enforced.

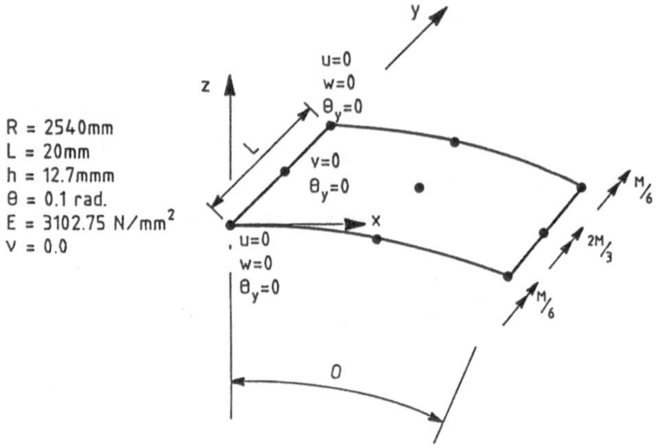

Fig. 5 A curved cantilever subjected to a tip bending moment

6. CONCLUSIONS

In this paper the three dimensional degenerated shell theory is developed. An explanation of locking behaviour and the location of sampling points for the assuumed strain fields is also described. Therefore, the behaviour of Mindlin plate elements and degenerated shell elements has been explicitly described.

7. REFERENCES

1. Huang, H.C. and Hinton, E., 'A nine node Lagrangian Mindlin plate element with enhanced shear interpolation', Eng. Comput., 1 (1984), 369–379.

2. Huang, H.C. and Hinton, E., 'A new nine node degenerated shell element with enhanced membrane and shear interpolation' Int. J. Num. Meth. Engng., 22 (1986), 73-92.

3. Huang, H.C., 'Defect-free shell elements', Ph.D. thesis, University College of Swansea, C/Ph/89/86

4. Huang, H.C., 'Implementation of assumed strain degenerated shell elements', Comput. and Struct., 25 (1987), 147-155.

5. Belytschko, Stolarski, Liu, Carpenter and Ong, ' Stress projection for membrane and shear locking in shell finite elements', Comp. Meth. Appl. Mech. Engng. 51 (1985), 221-258

6. Huang, H.C., 'Membrane locking and assumed strain shell elements', to appear in Computers and Structures

Convergence of Hierarchical Finite Elements

by Jürgen Bellmann

Fachgebiet Elektronisches Rechnen im konstruktiven Ingenieurbau
Technische Universität München, D-8000 München 2

1. Introduction

In finite-element-analysis, a new generation of programs is arrising, helping the engineer to obtain solutions with higher accuracy. The aim of these programs is to reduce the approximation error automatically. Thus it becomes necessary to get aposteriori information on the magnitude of the error and its distribution over the mesh. Among the two different types of adaptive procedures (h- and p- version), the variation of the polynomial order (p-version) for plate problems is discussed in this paper. In particular the behavior of the hierarchical elements noticeably deviating from the rectangular shape is examined.

2. Adaptive hierarchical finite elements

2.1 General construction

The formulation of the hierarchical finite elements described in this paper is similar to the formulation of isoparametric elements [1]. The real element is transformed to a standard element with the local coordinates r and s from -1 to 1. The transformation is given by the Jacobian matrix and uses the 4 up to 8 nodes of the element.
The 'ansatz' for the deformation is made separately of the geometry and uses hierarchical functions in r- and s-direction.

2.2 Hierarchical functions

The hierarchical functions are built of polynomials up to degree p. The highest actual used degree is 8.
Three different types of polynomial functions have been tested. A very simple form is given by [2]

$$N_p = (\xi^p - b) / p! \qquad (1)$$

with ξ lokal coordinates (-1 to +1)
p polynomial order ($p \geqq 2$)
b = 1 if p even
= ξ if p odd

The second form is taken from [3] and is built of integrals of the Legendre polynomials. The third form is also built of integrals of the Legendre polynomials, but integrated explicitely for every degree p.
In picture 1 the three different variants are plotted in opposite to the functions of isoparametric elements (quadratic).

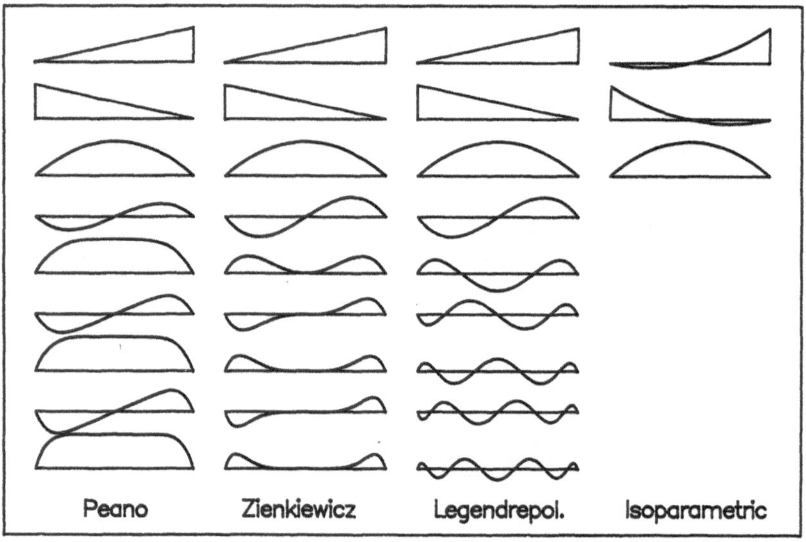

figure 1

Using an iterativ equation solver, the conditon of the stiffness matrix corresponds with the number of iteration steps.

step	max p	NDOF	iterations for equation solver Peano	Zienk.	Legendrepol.
1	2	16	16	16	16
2	3	35	48	48	49
3	4	58	98	89	84
4	5	95	249	157	144
5	6	145	465	232	204
6	7	202	1265	291	247

table 2

Table 2 shows, that for higher polynomial order, the application of the third form produces the best condition (least iteration steps).

2.3 Two dimensional shape functions

Applying the one dimensional functions in the two directions r and s, the hierarchical shape functions are obtained. They can be divided into three different types.
The linear modes describe the displacement (and the rotation) in the nodes of the structure.
The higher hierarchical modes use higher hierarchical functions ($p \geqq 2$) along one edge and a linear slope vertical to the edge.
The bubble modes use hierarchical functions in r- and s-direction with $p \geqq 2$. So the value of the shape function is zero at every point of the boundary.
Figure 3a,3b and 3c show the different modes, in figure 3d a single quadratic displacement mode on one edge of a plate is plotted separately on a mesh of 2*4 elements.

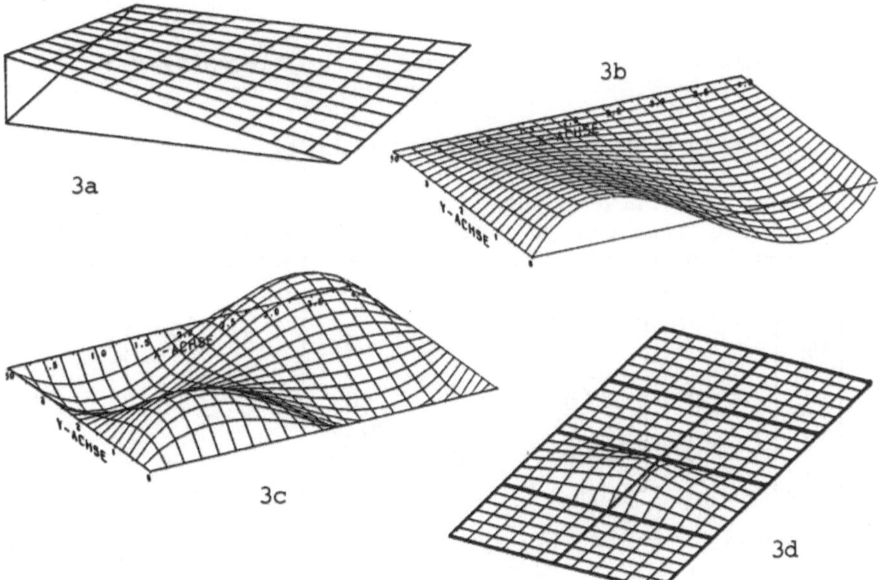

figure 3

Varying the distribution of r- and s- polynomial functions for the bubble modes, different types of elements can be obtained. A Serendipity element is built without bubble modes, a Lagrange element arises by applying the maximum number of bubble modes (up to the maximum degree of the modes at the edges). Though the Lagrange element produces best results, it wastes too many degrees of freedom. The Serendipity element needs only few degrees of freedom but does not satisfy in the results.

An effective compromise exists by taking the bubble modes with the condition

$$p_r + p_s \leq p_{max} \qquad \text{(elementtype 3).}$$

Figure 4 shows the distribution of the polynomial functions for the three different types of elements with a maximum degree of 5 in r direction and of 3 in s direction.

Serendipity	Lagrange	elementtype 3
xy	xy	xy
$x^2y \quad xy^2$	$x^2y \quad xy^2$	$x^2y \quad xy^2$
$x^3y \quad xy^3$	$x^3y \quad x^2y^2 \quad xy^3$	$x^3y \quad x^2y^2 \quad xy^3$
x^4y	$x^4y \quad x^3y^2 \quad x^2y^3$	$x^4y \quad x^3y^2 \quad x^2y^3$
x^5y	$x^5y \quad x^4y^2 \quad x^3y^3$	x^5y
	$x^5y^2 \quad x^4y^3$	
	x^5y^3	

figure 4

2.4 Formulation of the Reissner/Mindlin plate problem

The connection between the displacement w, the rotation θ_x and θ_y and the stresses of the element are as follows (θ_x is defined as the rotation around the x-axis, σ_x arises of stresses in x-direction)

$$\begin{bmatrix} m_x \\ m_y \\ m_{xy} \\ q_x \\ q_y \end{bmatrix} = D \cdot \begin{bmatrix} 0 & +\partial/\partial x & 0 \\ -\partial/\partial y & 0 & 0 \\ -\partial/\partial x & +\partial/\partial y & 0 \\ 0 & +1 & \partial/\partial x \\ -1 & 0 & \partial/\partial y \end{bmatrix} \begin{bmatrix} \theta_x \\ \theta_y \\ w \end{bmatrix} \tag{2}$$

The material matrix is equivalent to the usual form [4]. Shear-locking of the elements can be avoided by setting the polynomial order of the deflection w one degree higher than the order of the rotation θ_x and θ_y ($p_w = \max p_\theta + 1$).
In this way, also very thin elements up to t/l = 1/5000 can be calculated without reduced integration. Of course, the condition of the stiffness matrix becomes worse and the number of iteration in the equation solver increases.

3. Error computation and adaptive refinement

For adaptive refinement only the jumps of the stress values at the interface of two elements are considered to establish an error indicator [3,6,7,8] representing the lokal energy error. For the plate problem, the contribution of one stress component s at edge k to the error indicator of element i is given by equation 3.

$$\bar{\lambda}_{iks}{}^2 = AREA \,/\, (24 \cdot p_m \cdot DIAG_s) \cdot 1/k \cdot \int_{\Gamma k} (\Delta Z_{iks})^2 \, d\Gamma \qquad (3)$$

with
- ΔZ_s jump of one of the 5 stress components (m_x, m_y, m_{xy}, q_x, q_y)
- k length of the edge
- AREA area of the element
- $DIAG_s$ diagonalelement of the material-matrix
- p_m polynomial order $(p_r + p_s)/2$
- Γ edge

Separating the local error indicator $\bar{\lambda}_{iks}{}^2$ according to the stress components at edge k we get an indication for individually rising the polynomial degree of θ_x or θ_y. The degree of the displacement w is not set up explicitely by the error indicator of the stresses q_x or q_y but is given by the shear locking condition ($p_w = \max p_\theta + 1$). The error indicators of q_x or q_y however lead to a higher degree of θ_y or θ_x.
Summing the local error indicators an energy estimator for measuring the global accuracy of the finite element results is obtained.

4. Non rectangular elements

Quadrilateral elements deviating significantly from the rectangular shape often cause problems due to their overestimated stiffness. Increasing the polynomial order, this effect is reduced significantly. Figure 5a shows the displacement at the endpoint for different meshes (5b) plotted over the number of degrees of freedom. In this example, the rectangular mesh was the optimal one for low polynomial order, but for higher order nearly no differences can be observed.
For a problem with singularities it can be necessary to refine the finite element mesh near the singularity. The question arises: how far can the loss of accuracy in the non rectangular elements be covered by the profit of accuracy due to the mesh refinement?
In the following example of an L-shaped plate with a significant singularity at the reentrant corner the profit is proved to be dominant. The optimal factor of a geometrical mesh refinement towards the singular point turns out to be between 0.15 and 0.20. In figure 6a the true error in the energy norm

is shown for four different meshes (6b) with 3,9,12 and 15 elements under constant load. Figure 7 shows the bending moment m_x for the fourth mesh after 5 steps of p-refinement.

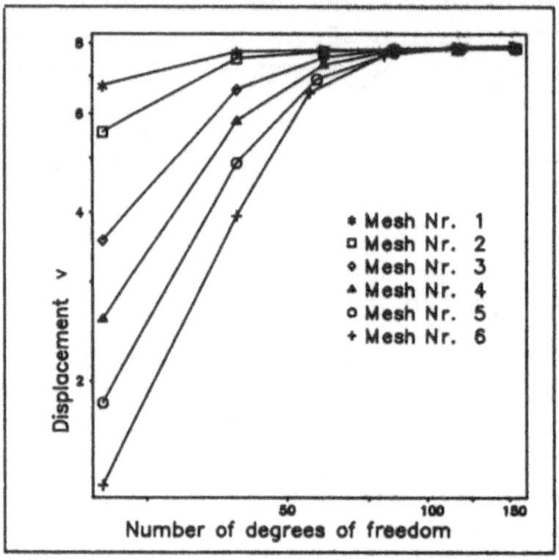

figure 5a

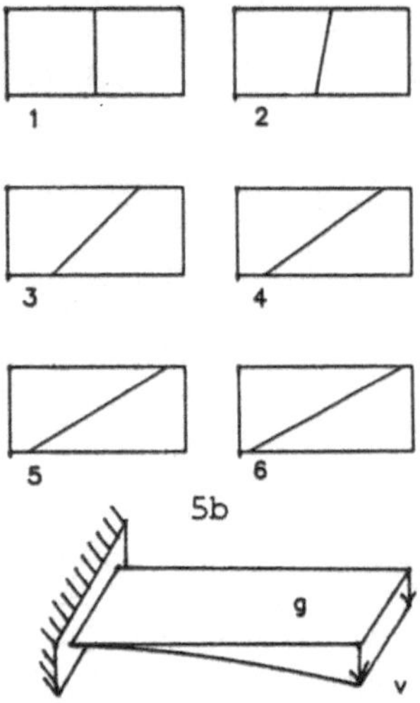

5b

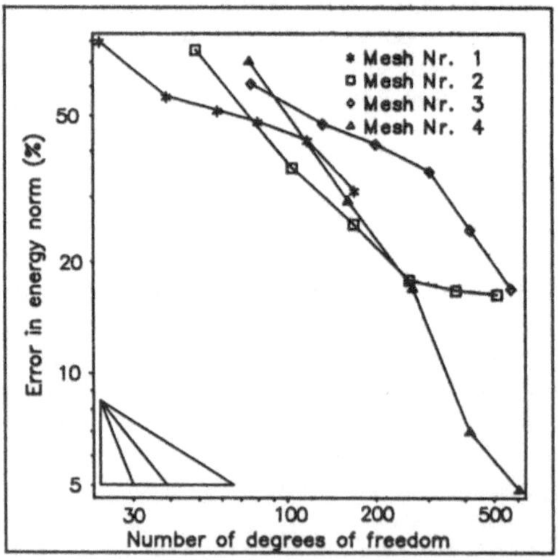

figure 6a

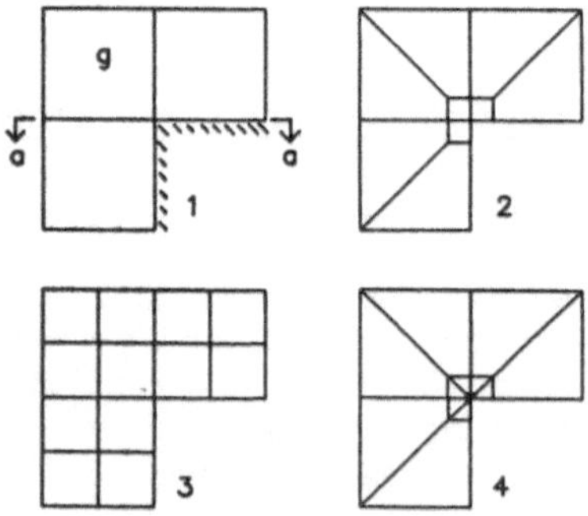

6b

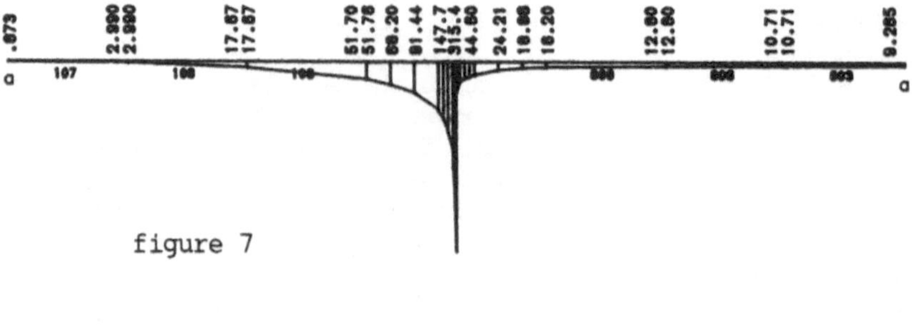

figure 7

References

/ 1/ The Finite Element Method
O. C. Zienkiewicz
Mc Graw-Hill Book Company 1977

/ 2/ Adaptive Approximation in Finite Element Structural Analysis
Peano, Pasini, Riccioni and Sardella
ISMES-Conference Bergamo Italy 1978

/ 3/ Hierarchical finite element approaches, error estimates and adaptive refinement
O. C. Zienkiewicz, D. Kelly, J. Gago and I. Babuska

/ 4/ Transient and pseudo-transient analysis of mindlin plates
A. Pica and E. Hinton
International journal for numerical methods in engineering, Vol. 15 189-208 1980

/ 5/ Shear-constraints and folded-plated structures
M. A. Crisfield
Engineering computations Vol 2 Sept. 1985

/ 6/ A posteriori error analysis and adaptive processes in the finite element method part I and part II
D. Kelly, J. Gago, O. C. Zienkiewicz and I. Babuska
International journal for numerical methods in engineering, Vol. 19 1593-1619 1983

/ 7/ A posteriori error indicators for the p-version of the finite element method
D. Dunavant and B. Szabo
International journal for numerical methods in engineering, Vol. 19 1851-1870 1983

/ 8/ A-posteriori-Fehlerabschätzungen und adaptive Netzverfeinerung für Finite-Element- und Randintegralelement-Methoden
E. Rank Mitteilungen aus dem Institut für Bauing.wesen I
Technische Universität München 1985 Heft 16

NUMERICAL METHODS FOR THREE DIMENSIONAL ANALYSIS OF BUILDINGS

Álvaro Vale e Azevedo

LNEC - National Laboratory for Civil Engineering, Lisbon, Portugal

SUMMARY

This paper describes some numerical techniques used on analytical three dimensional model that performs the structural analysis of buildings due to static actions in an integrated form with the dynamic analysis. The assembly of the floors stiffness matrix, as well as the substructuring techniques used for the solution of the system of equations, saves computation time and allows the analysis of buildings in which the floors have a large number of degrees of freedom. The techniques used for the solution of the system of equations of the overall buildings, as well as the techniques which allow to perform the dynamic analysis are presented.

1 - INTRODUCTION

There are several general computer programs which can carry out static or dynamic two and three dimensional analysis of buildings. Nevertheless, their use presents some disadvantages and sometimes makes it complex both modelling the structure and the loads acting on it. Usually the dynamic analysis programs require a large amount of static analysis to prepare data and to process results. On the other hand, the parameters of structural response to earthquake action, determined through dynamic analysis, must be combined with the values that these parameters take for other actions (static loads). These combinations of actions cannot be always made by automatic means since, generally, static analysis programs are not interconnected with dynamic analysis programs.

We have developed at LNEC an analytical three dimensional model that performs an integrated static and dynamic analysis of buildings {1}. In this model the floors are considered as diaphragms rigid in their own plane and the building structure is considered to be divided in horizontal sub-structures consisting of floors, with three degrees of freedom per nodal point (vertical displacement and rotations about the horizontal axes), and vertical substructures consisting of columns and shear walls, with six degrees of freedom per nodal point (three displacements and three rotations). Therefore, in this model the vertical displacements and the rotations about the horizontal axes are full compatible. The structural elements considered are beamelements, plate bending finite elements, column elements and shear wall finite elements.

2 - FLOOR STIFFNESS MATRIX

In building structures the dimension of the system of equations is generally very high and can reach some thousands of unknowns. The direct stiffness matrix leads to a symmetric stiffness matrix with a lot of zero coefficients.Taking into account the cost and limitations of computer central memory it is very important to optimize the computer storage of this matrix.

Considering that in building structures there are several floors structurally equal, we define the type-floor concept as a set of one or more storeys equal to one another. For the calculation of the overall stiffness matrix it will be enough to calculate the stiffness matrix of each type-floor and subsequently to assemble it suitably in the overall stiffness matrix of the building, taking into account which floors will be concerned by the type-floor in question.

In the floors we will individualize main and secondary nodal points. The main nodal points are the nodes of the structural system in the floor that are connected to another floor through columns or shear walls, or which provide support to the structure. The secondary nodal points are those that belong to the structural discretization of the floor, though they are not directly connected to other floor nor provide support to the structure (fig. 1).

The stiffness matrix of each type-floor

$$\underline{K}_{pt} \cdot \underline{\delta}_{pt} = \underline{F}_{pt} \tag{1}$$

is rearranged and partitioned in such a may that vectors $\underline{\delta}_s$ represent the displacements of the secondary nodes and vectors $\underline{\delta}_n$ represent the displacements of the main nodes.

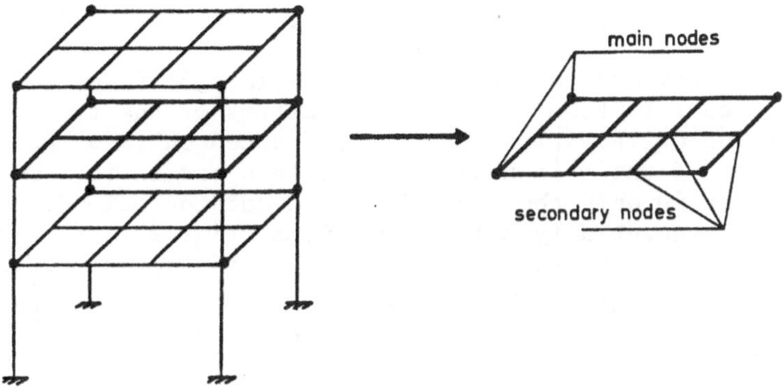

Figure 1 - Floor discretization. Main and secondary nodes.

$$\begin{bmatrix} \underline{K}_{ss} & \underline{K}_{sn} \\ \underline{K}_{sn}^{T} & \underline{K}_{nn} \end{bmatrix} \begin{bmatrix} \underline{\delta}_{s} \\ \underline{\delta}_{n} \end{bmatrix} = \begin{bmatrix} \underline{F}_{s} \\ \underline{F}_{n} \end{bmatrix} \qquad (2)$$

By static condensation of the degrees of freedom which are uncoupled in the structural system (secondary nodes), we can obtain the stiffness matrix of the floor reduced to the degrees of freedom of the main nodes

$$\begin{bmatrix} \underline{K}'_{ss} & \underline{K}'_{sn} \\ 0 & \underline{K}'_{nn} \end{bmatrix} \begin{bmatrix} \underline{\delta}_{s} \\ \underline{\delta}_{n} \end{bmatrix} = \begin{bmatrix} \underline{F}'_{s} \\ \underline{F}'_{n} \end{bmatrix} \qquad (3)$$

where $\underline{K}'_{ss}$ is an upper triangular matrix. Using this technique we can carry out the global analysis of the building considering only the main nodes of each type-floor. Once we have obtained the main nodes displacements of all the floors, the secondary nodes displacements of each floor can easily be obtained by a backsubstitution process.

$$\underline{K}'_{ss} \cdot \underline{\delta}_{s} = \underline{F}'_{s} - \underline{K}'_{sn} \cdot \underline{\delta}_{n} \qquad (4)$$

The computer program that implements the analytical model described herein performs an automatic renumbering of the nodes in secondary and main. One of the main concerns regarding this model was the optimization of computer storage of arrays. Considering that the stiffness matrix of the type-floor is a symmetric matrix with a large number of zero coefficients, and considering that the renumbering of the nodes in secondary and main can increase very much the bandwidth, the stiffness matrix of the type-floor $\underline{K}_{pt}$ is stored using the skyline technique.

Once performed the static condensation (3) it becomes necessary to separate from the stiffness matrix of the type-floor the portion of the stiffness matrix reduced only to the degrees of freedom of the main nodes $\underline{K}'_{nn}$. As this matrix is symmetric we will retain only the upper triangle and the main diagonal in a vector, sequentially column by column (fig. 2). It is also necessary to separate the portion of the load vectors reduced to the main nodes $\underline{F}'_n$ (fig. 2). Once separated, matrices $\underline{K}'_{nn}$ and $\underline{F}'_n$ are saved to computer auxiliary memory (disk), for further use when assembling the overall stiffness matrix of the building.

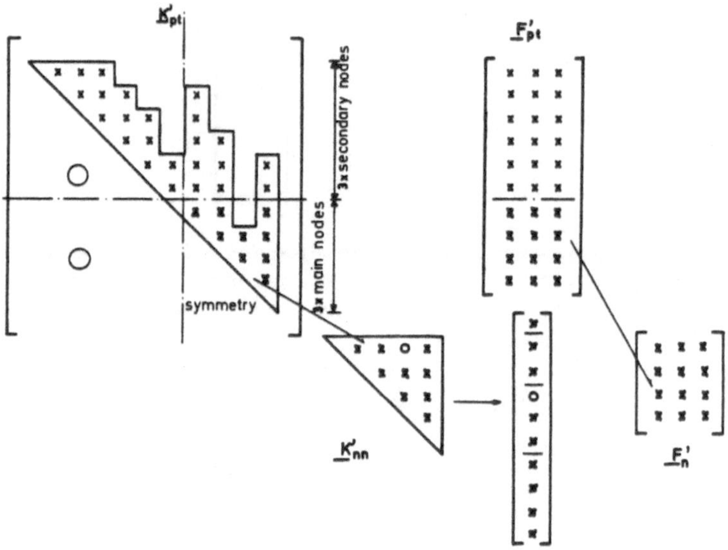

Figure 2 - Type-floor stiffness matrix

Since the stiffness matrix of the type-floor $\underline{K}_{pt}$ is stored by skyline completely in central memory, the static condensation is carried out using the method of Martin et al {5}, which is very appropriate because it works sequentially by columns and leads to a minimum number of arithmetical operations.

This substructuring technique leads to an analysis with less arithmetical operations, thus saving computation time.

3 - BUILDING STIFFNESS MATRIX

Once the stiffness matrix of each type-floor obtained and reduced to the degrees of freedom of the main nodes and once the stiffness matrix of the vertical elements (columns and shear walls) calculated, the overall stiffness matrix of the building is obtained assembling the matrices of all the substructures, floor by floor, from the top to the bottom of the building.

The system of equations that expresses the static equilibrium of the building structure is rearranged in order to individualize the degrees of freedom referring to the displacements of the floors main nodes $\underline{\delta}_n$ (vertical displacement and two rotations about the horizontal axes) and the degrees of freedom referring to the rigid body displacements of the floors $\underline{\delta}_p$ (two horizontal displacements and rotation about the vertical axis).

$$\begin{bmatrix} \underline{K}_{nn} & \underline{K}_{np} \\ \underline{K}_{np}^T & \underline{K}_{pp} \end{bmatrix} \begin{bmatrix} \underline{\delta}_n \\ \underline{\delta}_p \end{bmatrix} = \begin{bmatrix} \underline{F}_n \\ \underline{F}_p \end{bmatrix} \qquad (5)$$

By static condensation of the degrees of freedom referring to the floors main nodes displacements we obtain the stiffness matrix of the structure reduced to the rigid body displacements of the floors

$$\begin{bmatrix} \underline{K}'_{nn} & \underline{K}'_{np} \\ 0 & \underline{K}'_{pp} \end{bmatrix} \begin{bmatrix} \underline{\delta}_n \\ \underline{\delta}_p \end{bmatrix} = \begin{bmatrix} \underline{F}'_n \\ \underline{F}'_p \end{bmatrix} \qquad (6)$$

where $\underline{K}'_{nn}$ is an upper triangular matrix.

With the reduced stiffness matrix $\underline{K}'_{pp}$ it becomes possible to perform the dynamic analysis of the building, in which are considered only the degrees of freedom related with the floors rigid body displacements (two horizontal displacements and rotation about the vertical axis).

On the other hand, for static analysis, the floors rigid body displacements $\underline{\delta}_p$ are calculated solving the system of equations derived from (6).

$$\underline{K}'_{pp} \cdot \underline{\delta}_p = \underline{F}'_p \qquad (7)$$

Then, the displacements of the main nodes of all the floors are easily obtained from (6) by a process of backsubstitution.

$$\underline{K}'_{nn} \cdot \underline{\delta}_n = \underline{F}'_n - \underline{K}'_{np}\, \underline{\delta}_p \qquad (8)$$

Finally, the displacements of the secondary nodes are calculated by expression (4) as described earlier.

The schematic shape of the stiffness matrix and of the load vectors matrix of the structure is represented in fig. 3, where the broken line represents the stiffness matrix symmetric part. The submatrix $\underline{K}_{nn}$ exhibits a stair step shape.

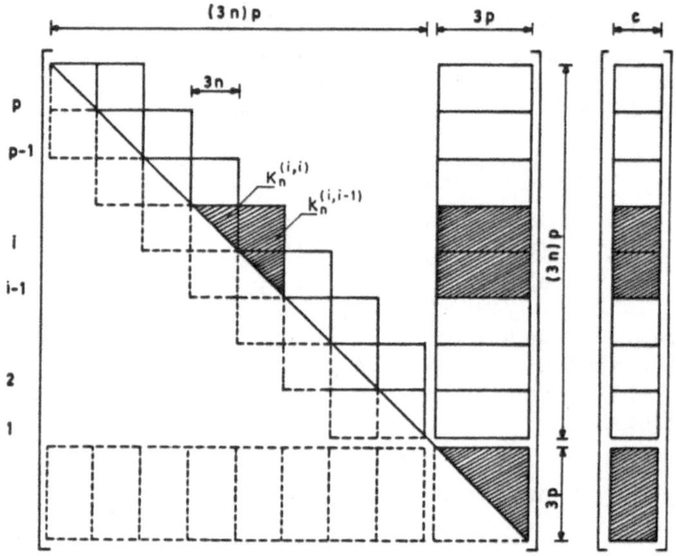

p = n. of floors
n = n. of main nodes of each floor
c = n. of static load cases

Figure 3 - Schematic shape of the system of equations

One of the main concerns regarding this model was the optimization of computer storage of arrays, as well as the minimization of the amount of computation and moderate use of transfers to auxiliary memory (disk). Therefore, the system of equations is partitioned in blocks in such a may that, in different stages of the analysis, only the submatrices needed for the solution of the system are stored in computer central memory, while the other submatrices are stored in computer auxiliary memory (disk). In order to reduce the number of transfers to auxiliary memory the system of equations assembly is carried out simultaneously with the solution. The shadow part of figure 3 shows the submatrices that are stored in computer central memory in each stage of the analysis.

For the submatrices $K_n^{(i,i)}$, which are symmetric, we only store its upper triangle and the main diagonal in a vector, sequentially column by column as described in section 2. The submatrices $K_n^{(i,i-1)}$ are stored in square matrices. Note that, in case we don't have shear wall finite elements the submatrices

$K_n^{(i,i-1)}$ give to each stair step of submatrix $\underline{K}_{nn}$ the additional characteristic of bandwidth. Submatrix $\underline{K}_{pp}$ is always present in central memory and due to its symmetry we only store its upper triangle and the main diagonal in a vector sequentially column by column.

The Gauss elimination method is used for the solution of the system of equations.

4 - DYNAMIC ANALYSIS

In the analytical model described herein the dynamic behaviour of the building is studied considering three degrees of freedom per floor (two horizontal displacements and rotation about the vertical axis), due to the fact that the floors are considered as diaphragms rigid in their own plane.

The earthquake is specified as samples of a vectorial gaussian stationary stochastic process, with a given duration and with which a power spectral density function is associated. The structural response to earthquake action is determined through modal analysis.

With the stiffness matrix reduced to the rigid body displacements of the floors, matrix $\underline{K}'_{pp}$ of expression (6), and with the mass matrix, the natural frequencies and mode shapes are determined. The modal displacements of the main nodes of the floors are calculated by expression (8) and the modal displacements of the secondary nodes are calculated by expression (4). Then the maximum dynamic displacements for each node of each floor can be determined with its six components (three displacements and three rotations) by stochastic analysis. Once determined the modal displacements of all nodes we can calculate the internal moments and forces associated to each mode in all the structural elements and by stochastic analysis we can calculate the maximum dynamic internal moments and forces, which can be combined with the values due to static actions.

5 - CONCLUSIONS

The substructuring technique used in the type-floors, reordering the nodal points in secondary and main, makes it possible to consider the slabs of the floors discretized by plate bending finite elements or to improve the floors discretization, increasing the numbers of secondary nodal points, without excessive increase of computation time.

The storage of the type-floor stiffness matrix, using the skyline, is very appropriate because economizes computer central memory and allows a great flexibility of the nodal points numbering, which is very important since the computer program renumbers automatically the nodal points in secondary and main.

The rearrangement of the overall building system of equations, in order to individualize the displacements of the floors main nodes and the rigid body displacements of the floors, allows to obtain, after static condensation, the stiffness matrix of the building reduced to the rigid body displacements of the floors, with which the dynamic analysis can be performed.

The partitioning of the system of equations in blocks, retaining only in computer central memory the submatrices needed for its solution, has the advantage of not restricting the number of floors of the building. The fact that the assembly of the system of equations is done simultaneously with its solution reduces the number of transfers to auxiliary memory.

The techniques briefly described in this paper provide an optimization of computer storage of arrays as well as a minimization of number of arithmetical operations performed and transfers to auxiliary memory. Therefore, this analytical model for buildings static and dynamic analysis is a valuable tool that makes it possible to analyse structures with a large number of degrees of freedom.

REFERENCES

1. Azevedo, A. Vale - Análise Estática e Dinâmica de Estruturas de Edifícios, Thesis, LNEC, Lisbon, 1986.

2. Azevedo, A.Vale, 'A Model for Dynamic Analysis of Buildings', Proc. 8th European Conference on Earthquake Engineering, Lisbon, 1986.

3. Bathe, K.J. and Wilson, E.L. - Numerical Methods on Finite Element Analysis, Prentice-Hall, Inc., Englewood Cliffs, New Jersey, 1976.

4. Bathe, K.J.; Wilson, E.L.; Peterson, F.E. - SAP IV - A Structural Analysis Program for Static and Dynamic Analysis of Linear Systems, Rep. EERC 73-11, University of California, Berkeley, 1973.

5. Martin, R.S.; Peters, G.; Wilkinson, J.H., 'Symmetric Decomposition of a Positive Matrix', Numeriche Mathematik, N.7, 1965.

6. Wilson, E.L.; Hollings, J.P.; Dovey, H.H. - Three Dimensional Analysis of Building Systems (extended version), Rep. EERC 75-13, University of California, Berkeley, 1975.

MATRIX MECHANICS TO CLASSIFY NON-LINEAR CONTINUA

J.D.Coleman

Senior Visiting Fellow
Fluid Mechanics Division
Department of Civil Engineering
The City University, London.

SUMMARY

Matrix Mechanics is a procedure to classify linear, semi-linear and quasi-linear continua from the system of E first order partial differential equations in E dependent variables which define them. Purpose is to index problems, seek analogues and determine boundary conditions for higher order systems. The type assists choice of stable, efficient and convergent numerical procedures for the continuum. Dummy variables are used to reduce any second order terms to first order-one more variable-one more equation. System is arranged with each dependent variable in vertical rows. Each u_x say is replaced by C_x etc., where C(x y z t)=const. is a characteristic for the system. The equation det A=0 then defines all the E characteristics. The paper gives ground rules for determining characteristic classes. Then examples are systematically presented in eight different selected classes.

1. GROUND RULES FOR CLASSIFICATION

The equation det A = 0 usually splits up into a chain of homogeneous first and second order factors in the C_x, C_x^2 etc Each first order factor gives one characteristic, each second order factor, two. If say the original set was in x,y,z,t, then a factor containing all the C_x, C_y, C_z, C_t is non-degenerate. If one or more are missing - degenerate. Factors should be reduced to a sum of squares, but in some 30 practical examples so far, most have appeared already in that form. If any second order factor is (non-degenerate);

1) Positive or negative definite- it is elliptic, [1] .
2) Indefinite with one odd sign - it is hyperbolic,
3) Indefinite with two odd signs- it is ultrahyperbolic,
4) If the factor is degenerate - it is parabolic

Courant is quite clear on these points. So far, second order factors have appeared without multipliers to the C_x^2 etc. and the sign problem is also unambiguous. Thus, a sum of squares, positive definite, in x,y,z is elliptic in x,y,z but is parabolic in x,y,z,t. Rules for first order factors must be consistent with second order; it emerges that for consistency a non-degenerate first order factor is hyperbolic. If degenerate , it is parabolic . When all factors have been assigned to class, the class of the system is their sum in terms of E,H and P. Numbers give the numbers of each type; thus 2H2P etc. We show multiplicity(overlap) of characteristics by brackets; thus 2H(2H)2P etc. Notably, characteristics can be found directly from the principal part of any eliminant of the system, but a theorem of Courant shows that that method may not give them all; the set may be incomplete . The set from the A matrix is complete . There is also an even and odd order theorem. Any totally elliptic, or elliptic(multiple) system has to be of even order; the roots are in pairs. Conversely, any odd order system has to have at least one real root; it is hyperbolic, or parabolic. Systems with only real roots, H or P, can be of any order. Results so far show that elliptic and related systems are relatively few, hyperbolic and parabolic more common, but the most numerous engineering class is mixed hyperbolic/parabolic systems, mostly with multiple characteristics. A previous paper enumerated all possible non-degenerate classes for systems of order one to eight, [2] . Degenerate classes are legion. We now consider examples arranged systematically in eight selected classes: these examples are in the main drawn from the mechanics of the earth, ocean, air, sun and space. That is; environmental civil engineering. Probably infinitely complex, we approximate.

2. EXAMPLES

2.1 Equations which change type. Classification in a non-linear system only has meaning at a given point, time, and for a given problem. Usually, systems do not change type over a usable field, but some do. These we now consider. To save space, we do not set out the bulky A matrix for each, but just give results. They can be checked using techniques given in the previous paper, [2], and in Courant and Hilbert [1] . Two examples are:

$$y \cdot u_{xx} + u_{yy} = 0 \quad \text{(Tricomi's equation)} \quad (1)$$

and

$$(a^2 - (\phi_x)^2) \cdot \phi_{xx} - 2 \cdot \phi_x \cdot \phi_y \cdot \phi_{xy} + (a^2 - (\phi_y)^2) \cdot \phi_{yy} = 0 \quad (2)$$

both from gas dynamics-subsonic/transonic/supersonic gas flow, adiabatic and frictionless. In (1) the gas flow is vertical, downwards. Positive upper half plane, subsonic and elliptic. Lower half plane, supersonic and hyperbolic.

At y=0 the flow is sonic-the x axis is the"parabolic line ". (2) is similar but in two dimensions. When-ϕ_x and-ϕ_y, the fluid velocity components, are small- elliptic. When one is large- hyperbolic - the characteristics (Mach lines) are the familiar shock diamonds seen in a rocket motor, or plane using reheat. When both components are equal to a(the sonic velocity) divided by root two(so their vector sum is a), then the equation is ;

$$\phi_{xx} - 2. \phi_{xy} + \phi_{yy} = 0 \qquad (3)$$

and it forms a perfect square - parabolic. So, at the sonic surface, the Mach lines coincide into one set, at right angles to the sonic streamlines. This loss of a characteristic root, and orthogonality, are features of the second order parabolic equation. However - the parabolic diffusion equation goes parabolic in an even more degenerate way - the perfect square is made by both B and C zero in $B^2 - A.C = 0$ .The lost characteristic is taken as arbitrary. So- characteristics for slender aircraft at the speed of sound - straight lines orthogonal to flight path. With a thick wing , small sonic areas appear at some Mach 0.8 over it, marked by vapour clouds, which also mark the line vortex from each wingtip (Joukowski circulation), and the cone vortex over each wing of delta aircraft at landing angle. Tests in a 4,000 H.P. supersonic tunnel without boundary layer suction showed a parallel bundle of rather unstable shocks, hard to photograph without a spark, with the tunnel running close to the speed of sound. But they were more or less plane and at right angles to tunnel axis. A schlieren showed them clearly - they were too unstable for the Mach-Zehnder interferometer. Similar type change in principle occurs at the Nadai roof surface between the elastic (elliptic) and plastic(hyperbolic) zones in elastic/plastic continua. Most engineering geometrical non-linearity is due to large strain/ strain rate, and constitutive n/l due to n/l material properties. Thus, we have l/l , l/nl , nl/l and nl/nl continua. Examples ; small strain elasticity, heat conduction with vapour transfer, post buckling elastic behaviour, and plasticity/ soil mechanics/high speed gas flow. Naturally , nl/nl problems are the most difficult.

2.2 Totally elliptic systems All factors non-degenerate, all roots complex, all separate. Most of the well known second order systems are of this type - Laplace, Helmholtz, Poisson and Schrodinger equations - the latter from quantum mechanics. They are 2E systems. Some others of this class;

$$(1 + (u_y)^2).u_{xx} - 2.u_x.u_y.u_{xy} + (1 + (u_x)^2).u_{yy} = 0 \quad (4)$$

which is Lagrange's equation for a minimal surface(bubble), and;

$$h_{xx}((h_x)^2 + m.(h_y)^2) + 2(1 - m)h_x h_y h_{xy} + h_{yy}((h_y)^2 + m(h_x)^2) = 0 \quad (5)$$

which describes the total head h for non-darcy flow in a rock fill spillway. Flow is defined by ; [3]

$$q = - k A (i)^{(1/m)} \qquad (6)$$

so m = 1 for darcy flow.In the fluid mechanics laboratory, m = 1.87 was found at a hydraulic gradient of about unity for flow in a horizontal pipe, nine inches(230mm.) in diam. It was 3 metres long, filled with 3-4 cm. diam. gravel.m=2 has been found by Soviet workers. Flow went non-linear, m greater than one, for very small hydraulic gradients - as low as 1/300. A log/log plot was made, the slope showing m, and the transition from darcy to non darcy flow could be seen. It had been hoped to use a soap bubble to solve (5) but no m value was found to make(4) and (5) identical; it can be solved with finite elements using a functional, or Lagrangian density, given in the previous paper at the last NUMETA[3]. It expresses minimum dissipation rate per unit volume integrated over the flow volume. For m = 1 (darcy), (5) becomes identical to Laplace's equation as expected. It has to be clearly stated that mathematicians do not appear to agree on the degree of sub-division of the various class of equations. It is generally agreed that degenerate and mixed equations are so numerous that systematic grouping is hardly possible. It is also , in the main , agreed that totally hyperbolic and hyperbolic classes be distinguished, although some authors group them as just "hyperbolic". In this paper we have chosen to distinguish totally elliptic from elliptic, totally parabolic from parabolic,but most writers do not make those distinctions. They are simply "elliptic " and "parabolic" . It will be seen that all those that attract the term " totally " are in fact second order, and in that context the term " totally " is unusual. We have retained the more detailed classes on the grounds that the definitions are at least unambiguous, that future examples may demand it, and that those who do not prefer this procedure can easily group say "totally elliptic" and "elliptic" together under the heading "elliptic" for example. So- our eight classes could be made six, or five, according to preference.

2.3 <u>Elliptic systems</u> Characteristics all complex, all non-degenerate, some multiple. Common example is ;

$$u_{xxxx} + 2.u_{xxyy} + u_{yyyy} = 0 \qquad (7)$$

for the stress function in a bent flat plate, elastic systems with axial symmetry, slow viscous flow. Often-it appears in cylindrical coordinates for axial symmetry. For viscous flow - rectangular only.For biharmonic wave and diffusion equations,characteristics are degenerate.For (7),

$$(C_x^2 + C_y^2)^2 = 0 \qquad (8)$$

showing the double complex characteristics. Other examples are the cylindrical shell (6th order) and the double curve shell (8th order). A limited but important class.

2.4 Totally hyperbolic systems   Sometimes termed "properly hyperbolic"; all real, non-degenerate and separate characteristics. Simple examples are ;

$$u_x + (1/v)\, u_t = 0 \qquad (9)$$

and

$$U.\, u_x + V.\, u_y + W.\, u_z + u_t = 0 \qquad (10)$$

for the constant velocity kinematic wave (flood waves, waves in traffic flow) and the surface in the air moving along with local velocities U, V and W parallel to the three coordinate axes. The latter is quasi-linear. Another first order example is the non-linear ;(here u = light flight time);

$$(u_x)^2 + (u_y)^2 + (u_z)^2 = n^2(x,y,z)/c^2 \qquad (11)$$

where c is velocity of light and n the refractive index. (11) defines the least time flight of light rays in optical engineering e.g. a 35mm. camera lens, binoculars, levels, telescopes, microscopes and rangefinders. The track of the light rays are the characteristics ; clearly problem dependent , but real, but in a single family of lines at 90° to wave fronts, u = const. Now computer optimised, formerly (11) was solved by laborious ray tracing. It is one of the very few truly non-linear equations in design. A second order example is Heaviside's wave equation of telephony ; the characteristic equation is ;(standard symbols);

$$(L.C.C_t^2 - C_x^2) = 0 \qquad (12)$$

from the equation ;

$$\phi_{xx} = L.C.\,\phi_{tt} + (L.G + R.C)\phi_t + R.G.\phi = 0 \qquad (13)$$

showing the two separate characteristics and the odd order term in d/dt typical of dissipation ; the system generates entropy and will not run backwards. In a sense all continua possess d/dt , as all processes have a beginning, an end, and all in reality are dissipative. Odd order d/dt, in principle , is present in every process, including biophysics. The A matrix, arbitrarily set down, either can , or cannot, be symmetrised by row/row, column/column interchange and multiplication of any row by an arbitrary number. If it can, all roots real- hyperbolic or possibly, parabolic. If **very** slightly **asymmetric**(skew component small) - the same. If very skew - almost certainly elliptic. If skew symmetric, all roots plus or minus i, or zero. If symmetric, and diagonalised, we see clearly how the value of det A is the product of the roots, and Trace A is their sum. Clearly any zero root (factor) renders det A identically zero(singular).

We also perceive that an engineering tension structure, of taut wires connected at nodes, could be seen as a system of vibrators (the nodes). If each node is seen as moving from a zero not affected by its neighbours, then each dependent variable (the displacements) are in the same number of ordinary differential equations. That is, E in E. The A matrix is then an E x E diagonal matrix - symmetric- and hyperbolic. Wave suppression on these structures is important. Often E is as much as ten thousand. With interaction - off diagonal terms appear. The wider the range of neighbour interaction, the broader the matrix band.

2.5 <u>Hyperbolic systems</u> All real roots, all non-degenerate but some multiple. An example (possible) is Maxwell's equations of radio engineering. = 0 (14)

$$(\mu/c)\,H_t = -\,\mathrm{curl}\,E\ ,\quad (K/c)\,E_t = \mathrm{curl}\,H\ ,\quad \mathrm{div}\,H = \mathrm{div}E$$

= 0; c is light velocity, mu magnetic permeability, K the dielectric constant. Courant gives the characteristic matrix and det A = 0 gives ;

$$C_t^2\ (\ C_t^2\ (\ 1\ /\ c^2\ )\ -C_x^2\ -\ C_y^2\ -C_z^2\ )^2 = 0 \quad (15)$$

; we see that the sixth order system in fact is mixed parabolic hyperbolic. A truly hyperbolic system is given by Dirac's equations (first order equations of quantum theory). Again , Courant gives ;

$$(\ C_x^2\ +\ C_y^2\ +\ C_z^2\ -\ (\ 1/c^2\ )\ C_t^2\ )^2 = 0 \quad (16)$$

a fourth order hyperbolic system - Dirac used four dep. variables to describe the electron, in x y z t. If now we truncate the system by making the electromagnetic compliance of space zero - wave velocity is infinite (like an elastic medium with infinite shear and bulk modulus). The systems then reduce to ;

$$(\ C_x^2\ +\ C_y^2\ +\ C_z^2\ )^2 = 0 \quad (\ 17)$$

of electrostatics. Another fourth order elliptic system. With ;

$$\mathrm{curl}\ E\ =\ \mathrm{curl}\ H\ =\ \mathrm{div}\ E\ =\ \mathrm{div}\ H = 0 \quad (18)$$

and we deduce (17) directly. Other hyperbolic systems are the sixth order homogeneous equation for surface waves on an elastic solid (involved in rock blasting) and the seventh order hydromagnetic wave equation for Alfven waves within the sun and the shocks in the solar wind. They are plasma waves ; in a gas, equal parts positive and negative ions. 98 per cent of matter in the universe is plasma ; including most of the ten million million million stars.

2.6 <u>Totally parabolic systems</u> Roots all real , all degenerate , all separate. The set of second order diffusion equations approaches this most closely; type $C_x^2 = 0$.

In x,t one characteristic is $C_x = 0$, that is, t = const.The other is arbitrary, usually taken as x = const. They do not overlap.Examples from mechanics of groundwater and heat in the earth are ; [3], (the first is semi-linear),

$$k\,(h.\,h_{xx} + (h_x)^2) = n\,.\,h_t \tag{19}$$

$$k(T,z).\,T_{zz} + k(T,z)_T.(T_z)^2 = \rho.c_p.\,T_t \tag{20}$$

an equation solved in situ with a " molecule" of thermo - meters. Quasi-linear forms are ;

$$k.(h/m).(h_x)^{((1/m)-1)}.h_{xx} + k.(h_x)^{((1/m)+1)} = n.h_t \tag{21}$$

for the free surface for non-darcy flow in a gravel bed [3]. All are of form ;(g_1 positive,g_1,g_2 finite,continuous,s.v. functions),

$$u_t = g_1(x,t,u,u_x,u_t).u_{xx} + g_2(x,t,u,u_x,u_t) \tag{22}$$

the most general quasi-linear parabolic form in x, t.

2.7 Parabolic systems Roots all real, all degenerate, at least some multiple - the"parabolic" definition of some other authors, who exclude the non-multiple class. Examples from marine dynamics are ;

$$C_x^4 = 0,\ [2]\ , \tag{23}$$

which includes Prandtl's boundary layer operator, and the Ekman double diffusion equation for the wind and water velocity spirals as the wind drives the currents on the upper ocean - the current is at an angle to the wind. Another example of greater difficulty is ; (24)

$$nu\ .\ del^4\ \emptyset = (del^2\ \emptyset)_t - \emptyset_y.(del^2\emptyset)_x + \emptyset_x.(del^2\emptyset)_y = 0$$

for simultaneous advection and diffusion of vorticity , $del^2\ \emptyset$, in a Navier-Stokes fluid, incompressible. Four different methods gave a singular A matrix - a zero root. At risk of losing roots, the eliminant gives ;

$$(C_x^2 + C_y^2)^2 = 0 \tag{25}$$

which is certainly degenerate and multiple. However , they are not real but complex - it may be a new class. It would be parabolic/elliptic.Not defined in any known text. Value is for the shear drag on a fixed oil rig circular pile. Inviscid theory (elliptic) gives normal forces. Lab. tests are ruined by the huge Reynold's number disparity. Notably, (24) can be split into two second order equations, one parabolic, one elliptic.Other similar examples will be sought.

2.8 Mixed parabolic/hyperbolic systems All roots real, some degenerate, some not, and some multiple. Our examples are all from the dynamics of the air. Numerical weather prediction systems, first started about 1925. A team of humans were used for these early efforts.

They are; [2],(All with symmetric A matrix), (26)

$$(u.C_x + v.C_y + C_t)(C_x^2 + C_y^2) = 0 \text{(Sutton,1962)}$$

$$(u.C_x + v.C_y + w.C_z + C_t)^2(C_x^2 + C_y^2 + C_z^2) = 0 \text{ (Crank, also 1962), (27)}$$

and

$$(u.C_x + v.C_y + w.C_z + C_t)^3 (C_z)^2 = 0 \text{ (Cullen, Norbury and Purser, in press).} \quad (28)$$

The latter is the current source of the nightly weather forecast on U.K. television. A global grid of side 150 km. has 15 heights at each node, five current variables at each of these points. A central difference scheme based directly on the five first order equations is computed by CYBER 205 at several million floating point calculations per second - a 24 hour forecast takes 4 1/2 minutes . Next machine (planned) will be 25 times faster. Oil rigs at sea can receive notice of approaching storms.A useful item, some storms can reach windspeeds of 130 m.p.h. Civil engineers on the floating rig are at great risk. (28) has three hyperbolic (kinematic) terms; three discontinuities mark a warm,or cold,or occluded,front in the air. Density, and x and y velocities (but not pressure) suffer a sharp change as the front passes over. Cold air lies below the warm and at a cold front the air suddenly chills. At a warm front - the opposite. Thunderstorms and hail often mark a cold front - sometimes called " rainbow weather ". There has been controversy about (28) but it is within our class. (26) and (27) are more doubtful -hyperbolic / parabolic/ elliptic. Engineers could well take (28) as a free example of a powerful way ahead in engineering mechanics, the solution of which is above our heads for all to see. When mechanical (storms) and thermal(infra-red) air/sea interaction is allowed for - CYBER 205 will seem like an abacus from the dark ages. Slow secular weather trends will be predictable; the marine relaxation time is several years.

3. REFERENCES

1. Courant,R. - Methods of Mathematical Physics - 2 , Partial Differential Equations ,Wiley Eastern, New Delhi, 1975.

2. Coleman,J.D., Dynamics of a Transient : Atmosphere / Upper Ocean/ Towed Rig : Systematic Method of Characteristics, Int. Conf. on Reliability of Methods of Engineering Analysis. Edit. K.J.Bathe, D.R.J.Owen. Pineridge Press, Swansea , 1986 , 641 - 658.

3. Coleman,J.D. Three non-linear diffusion equations of sub-surface mechanics, Int. Conf. on Numerical Methods in Engineering. Edit. J.Middleton, G.N.Pande. Balkema, Rotterdam, 1985, 161 - 165.

A Mixed Eulerian-Lagrangian Contact Element to describe Boundary and Interface Behaviour in Forming Processes

J. Huetink, J. van der Lugt
University of Twente
PO Box 217, 7500 AE Enschede, The Netherlands
J.R. Miedema
Hoogovens IJmuiden B.V.
PO Box 10000, 1970 CA IJmuiden, The Netherlands

SUMMARY

A special element has been developed to describe boundary and interface behaviour in forming processes. A mixed Eulerian-Lagrangian finite element method has been applied by which nodal point locations may be adapted independently of material displacement. Movement of free and contact surfaces can be taken into account by adapting nodal surface point locations in a way that they remain on the moving surface. A smoothing procedure, related to the Hu-Washizu principle, was introduced to avoid numerical instabilities. The description of the contact element was based on the assumption of a thin layer in the tool-product interface, the constitutive behaviour of this layer was taken analogous to elasto-plasticity with non-associated flow. Large relative displacements and rotations were taken into account. An application is shown by the simulation of a cold rolling process.

1. INTRODUCTION

The analysis of forming processes cannot be restricted to product deformations only. One has to include into the analysis the product-tool contact region, friction at the product-tool interface and sometimes even the tool itself. This calls for a finite element approach to model the contact conditions, especially in the multi-body case. The host finite element program "DIEKA" uses a mixed Eulerian-Lagrangian formulation in order to prevent grave mesh distortions and to properly analyze (semi-)continuous forming processes [1,2]. This formulation also proves convenient to deal with large relative surface displacements; meshes of product and tool can be kept properly aligned.

The contact element presented here was inspired on a friction layer approach [3]. Stick-slip constitutive behaviour is introduced as an analogon to elasto-plasticity with non-associated flow.

2. GEOMETRY AND KINEMATICS CONCERNING THE CONTACT REGION

Consider two bodies A and B and a (candidate) contact region C [4]. Define a reference surface S between the bodies, having equal normal or directional distances to either body:

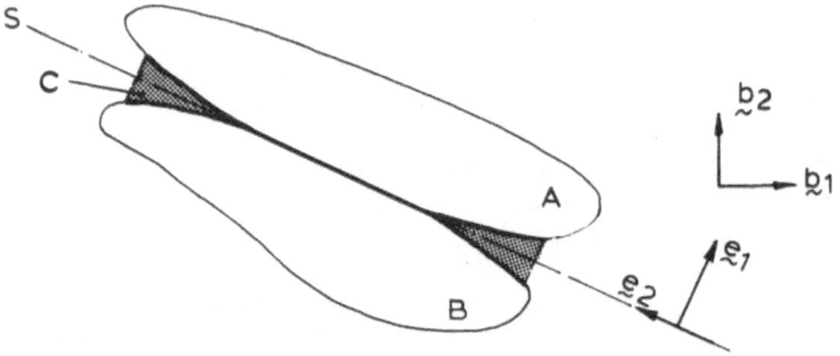

Fig. 1 Contacting bodies A and B with (candidate) contact region C and reference surface S.

Referring to Fig. 1, two coordinate systems are defined: a global, fixed, coordinate system $(\underset{\sim}{b}_1, \underset{\sim}{b}_2)$ and a local, corotating, coordinate system $(\underset{\sim}{e}_1, \underset{\sim}{e}_2)$:

$$\underset{\sim}{e}_i = Q_{i\alpha}\underset{\sim}{b}_\alpha \tag{1}$$

Taking time derivatives:

$$\dot{\underset{\sim}{e}}_i = \dot{Q}_{i\alpha}\underset{\sim}{b}_\alpha \;\Rightarrow\; \dot{\underset{\sim}{e}}_i = -\Omega_{ik}Q_{k\alpha}\underset{\sim}{b}_\alpha \tag{2}$$

where Ω_{ik} denotes rate of rotation of the local basis, which in this paper is defined to equal the rate of rotation of reference surface S.

3. STRESS IN THE CONTACT REGION

Since contact stresses are considered to be

tractions applying to only one surface (interface) S, it can be denoted as a first order tensor $\underset{\sim}{\sigma}$:

$$\underset{\sim}{\sigma} = \sigma_i \underset{\sim}{e}_i \tag{3}$$

Taking time derivatives:

$$\dot{\underset{\sim}{\sigma}} = \dot{\sigma}_i \underset{\sim}{e}_i + \sigma_i \dot{\underset{\sim}{e}}_i \quad \Leftrightarrow \quad \dot{\underset{\sim}{\sigma}} = \dot{\sigma}_i \underset{\sim}{e}_i + \Omega_{ij} \sigma_j \underset{\sim}{e}_i \tag{4}$$

Alternatively $\dot{\underset{\sim}{\sigma}}$ can be split up into a corotational derivative $\overset{\nabla}{\underset{\sim}{\sigma}}$ and a rotational change $\underline{\Omega} \cdot \underset{\sim}{\sigma}$ [5]:

$$\dot{\underset{\sim}{\sigma}} = \overset{\nabla}{\underset{\sim}{\sigma}} + \underline{\Omega} \cdot \underset{\sim}{\sigma} \tag{5}$$

Hence, tensor components for $\overset{\nabla}{\underset{\sim}{\sigma}}$ must be:

$$\overset{\nabla}{\underset{\sim}{\sigma}} = \dot{\sigma}_i \underset{\sim}{e}_i \tag{6}$$

4. RATE OF DEFORMATION c.q. SLIP AND CONSTITUTIVE BEHAVIOUR

The momentaneous situation in the contact region is shown in Fig. 2:

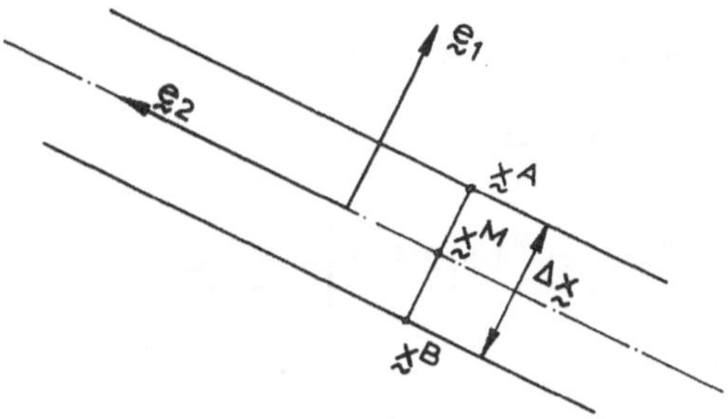

$$\underset{\sim}{x}^{A} = \underset{\sim}{x}^{M} + \tfrac{1}{2}\Delta \underset{\sim}{x}$$

$$\underset{\sim}{x}^{B} = \underset{\sim}{x}^{M} + \tfrac{1}{2}\Delta \underset{\sim}{x}$$

Fig. 2 Position of related points.

The rate of deformation is defined by:

$$d_i = \dot{x}_i^{A} - \dot{x}_i^{B} = \Delta \dot{x}_i \tag{7}$$

Thus tensor $\underset{\sim}{d}$ denotes relative surface velocity, corrected for rotation of the reference surface:

$$\underset{\sim}{d} = \Delta \underset{\sim}{v} - \underline{\Omega} \cdot \Delta \underset{\sim}{x} \tag{8}$$

Constitutive behaviour is taken analogous to elasto-plasticity. The rate of deformation is assumed to consist of a reversible (elastic) part and a non-reversible (plastic, c.q. slip) part. A linear relation is assumed between corotational stress derivative and the elastic part of rate of deformation, hence:

$$\overset{\nabla}{\underset{\sim}{\sigma}} = \underline{E}\cdot(\underset{\sim}{d} - \underset{\sim}{d}^{sl}) \quad \text{with} \quad \underline{E} = \begin{bmatrix} E_{11} & \emptyset \\ \emptyset & E_{22} \end{bmatrix} \tag{9}$$

It is assumed that normal relative displacement is strictly elastic. The non-elastic rate of deformation is assumed to have the direction of the tangential stress. The slip criterion is defined for Coulomb friction: $-\mu\sigma_1 + |\sigma_2| = \emptyset$ (10)

From (9) and (10):

$$\overset{\nabla}{\underset{\sim}{\sigma}} = \underline{E}\cdot\underset{\sim}{d} - \underline{Y}\cdot\underset{\sim}{d}\ , \quad \text{where} \quad \underline{Y} = \begin{bmatrix} \emptyset & \emptyset \\ \mu\,\mathrm{sgn}(\sigma_2)E_{11} & E_{22} \end{bmatrix} \tag{11}$$

The above relation is valid in case of slip (plastic) deformation. For the elastic situation: $\underline{Y} = \emptyset$, for the open situation: $\underline{Y} = \underline{E} = \emptyset$, $\underline{E}$ and $\underline{Y}$ are generally combined into a deformation stiffness tensor $\underline{D}$ by stating $\underline{D} = \underline{E} - \underline{Y}$, and contact behaviour becomes:

$$\overset{\nabla}{\underset{\sim}{\sigma}} = \underline{D}\cdot\underset{\sim}{d} \qquad \text{alternatively: } \dot{\sigma}_i = D_{ij}d_j \tag{12}$$

5. VIRTUAL POWER AND FINITE ELEMENT FORMULATION

For the two bodies A, B (Fig. 1) the virtual power yields:

$$\int_{V_B+V_A}\delta d_{ij}\sigma_{ij}dV + \int_{S_{contact}}(\delta v_i^A - \delta v_i^B)\sigma_i dS + \int_{S_{free}}\delta v_i T_i dS = \emptyset \ \forall \delta v_i \tag{13}$$

note: d_{ij} and σ_{ij} represent the rate of deformation and the Cauchy stress for the body interior respectively.

The description of the contact element is derived from the rate of change of the contact surface integral in (13). The real velocity distribution is approximated by interpolation of nodal point velocities v_i^N:

$$v_i = \psi^N v_i^N \tag{14}$$

Consequently:

$$d_i = B_{ik}^N v_k^N \qquad \text{and} \qquad \Omega_{ik} = \overset{*}{B}{}_{ikl}^M v_l^M \tag{15}$$

Substitution of (14) and (15) into the rate of change of the contact surface integral results into a stiffness matrix equation with components:

$$K^{NM}_{lm} = \int_{S_{contact}} \Big(B^{N}_{il} D_{ij} B^{M}_{jm} + \overset{*}{B}{}^{N}_{kil} \Delta x_k D_{ij} B^{M}_{jm} + B^{N}_{kl} \sigma_i \overset{*}{B}{}^{M}_{kim} + \overset{*}{B}{}^{N}_{kjl} \Delta x_j \sigma_i \overset{*}{B}{}^{M}_{kim} \Big) dS \tag{16}$$

6. INCREMENTAL STRESS FORMULATION

Nonlinearity and path-dependence necessitate an incremental solving procedure. After solving incremental displacements stresses are calculated in integration points [6]. For a 4-noded contact element two integration points are defined each of which is chosen at a point $\underset{\sim}{x}^{M}$ (Fig. 2) halfway the corresponding nodal points. This position enables a more accurate determination of the open-close condition in the nodal points.

Uncoupling of material and grid point displacement implies that in addition to the incremental calculation as in the Updated method, convection must be taken into account in order to be able to update the state at the grid points. The incremental stress formulation is given by [1,2]:

$$\Delta \underline{\sigma} = \hat{\underline{\sigma}} \Delta t + (\Delta \underset{\sim}{x} - \Delta \underset{\sim}{u}) \cdot \vec{\underset{\sim}{\nabla}} \underline{\sigma} \tag{17}$$

The first term on the righthand side equals the stress increment as in the Updated method, the second term represents the convective stress increment.

The determination of the convective stress increment complicates the solution procedure considerably because the stress gradient is needed and this cannot be calculated directly at element level. The convective terms are calculated from the differences between the values in adjacent elements of each material-associated quantity respectively. To avoid numerical instabilities a smoothing procedure, introduced by J. Huetink [2], has been used. This smoothing is based on taken mean values of nodal stresses in adjacent elements resulting in a continuous stress field. However, this field will generally not satisfy nodal point equilibrium. Nodal point equilibrium can be achieved using a method related to the Hu-Washizu principle [6].

The stresses are related to a coordinate system

defined by the orientation of the contact element. Consequently it is not possible to calculate nodal stresses as mean values of all adjacent elements. Besides, the smoothing procedure would give incorrect stress predictions. Therefore a procedure has been used by which the calculation of mean nodal stresses is restricted to elements of the same type.

7. CONTACT SURFACE MOVEMENT

Movement of contact surface points can be taken into account by adapting nodal surface point locations in such a way that they remain on the moving surface. The procedure is illustrated in Fig. 3 . It is observed that the new position of nodal surface points is not exactly on the surface found by the element boundaries if the material displacement increments are followed. However, if the new position of the nodal points would have been chosen on these element boundaries, material is lost at every increment. When using a spline, the amount of lost material is more or less in equilibrium with the added material.

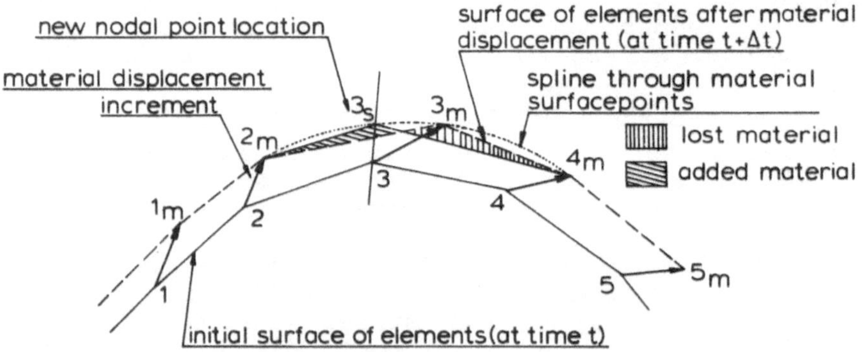

Fig. 3 Adaptation of nodal surface point location.

8. SIMULATION OF A COLD ROLLING PROCESS

The finite element simulation was carried out for stand 3 in a 5-stand cold rolling mill at HOOGOVENS IJmuiden BV., The Netherlands. The geometry and finite element mesh are shown in Fig. 4 and Fig. 5 respectively. Only half of the problem is modelled because of symmetry. The elements used are 4-noded linear isoparametric plane strain elements with constant dilatation [7,8].

The rolled material is low carbon steel, the roll

is taken into account as an elastic structure in order to predict roll deformation. The contact elements are located between roll and rolled material, the coefficient of friction is 0.07.

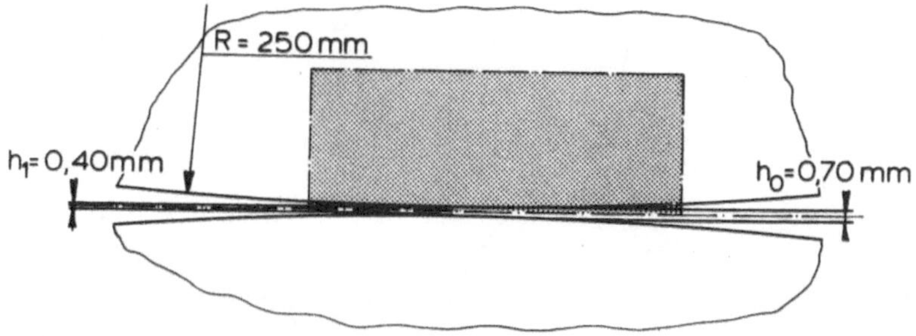

Fig. 4 Geometry of the cold rolling process.

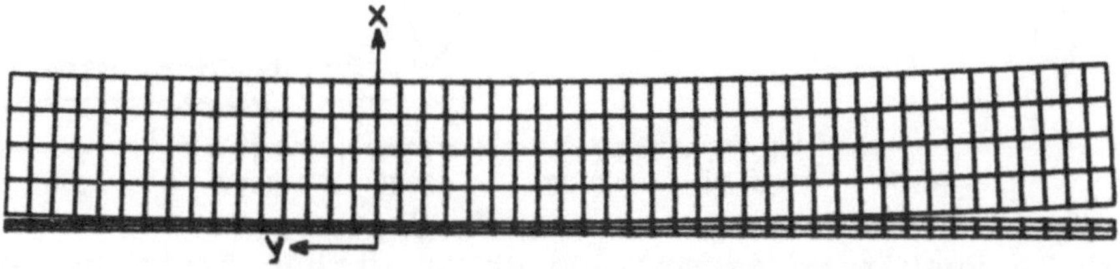

Fig. 5 Element mesh of the cold rolling simulation.

The simulation has been performed in 425 steps. In the first 15 steps an entry and exit tension of 110 N/mm^{-2} was applied to the rolled material. In each following step a displacement increment of the roll of 0.0001R was prescribed until a steady state was achieved.

The resulting normal and shear stress distribution on the interface roll and rolled material is given in Fig. 6. The point where the shear stress reverses is called the neutral point ; roll velocity equals rolled material velocity. From entry towards neutral point the roll is moving faster than the rolled material, from neutral point towards exit the roll is moving more slowly than the rolled material. This is characteristic for cold rolling processes.

Fig. 7 shows the undeformed and the deformed roll shape. The deformed roll deviates from a circular shape and the roll deformation has a relatively strong effect on the total reduction. In this simulation only a small part of the roll was modelled, if a greater part of the roll is modelled the effect of roll deformation will be stronger. This is a point of investigation in the future.

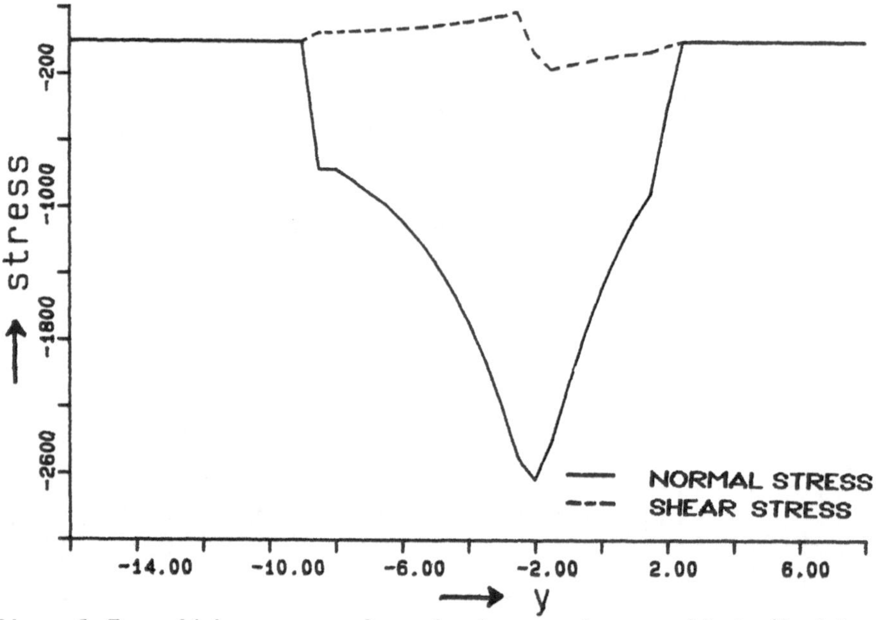

Fig. 6 Resulting normal and shear stress distribution

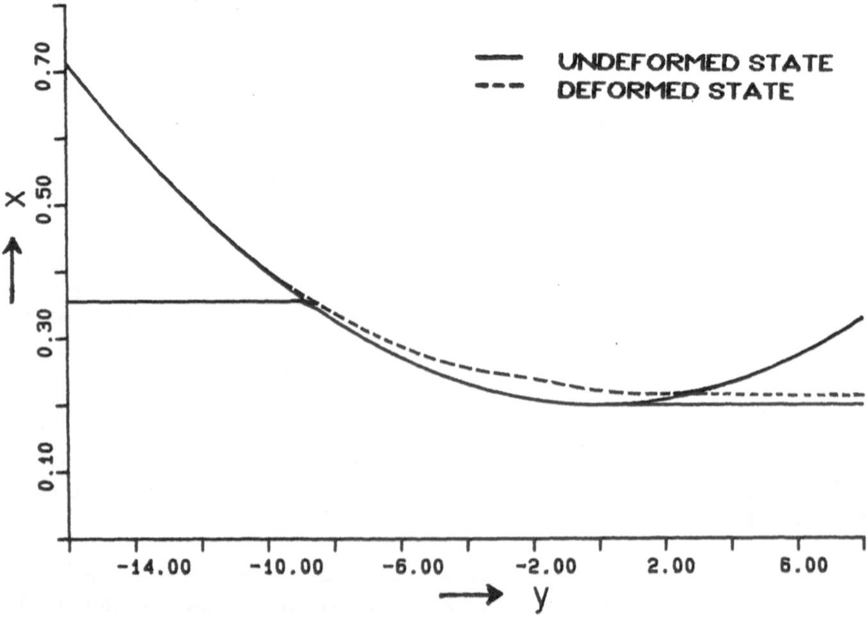

Fig. 7 Undeformed and deformed roll shape.

REFERENCES

1. Huetink, J., 'Analysis of metal forming processes based on a combined Eulerian-Lagrangian finite element formulation', Proc. Int. Conf. Num. Meth. Industr. Forming Processes, 501-509, 1982.
2. Huetink, J., 'On the simulation of thermo-mechanical forming processes',Thesis, University of Twente, The Netherlands, 1986.
3. Lugt, J. van der and Huetink, J.,'Thermal-mechanically coupled finite element analysis in metal forming processes', Comp. Meth. Appl. Mech. Eng., 54 (1986), 145-160.
4. Baaijens, F.P.T., Veldpaus, F.E. and Brekelmans, W.A.M., 'On the simulation of contact problems in forming processes', Proc. 2nd Int. Conf. Num. Meth. Industr. Forming Processes, 85-90, 1986.
5. Prager, W.-<u>Introduction to Mechanics of Continua</u>, Ginn, New York, 1961.
6. Zienkiewicz, O.C. -<u>The Finite Element Method</u>, McGraw-Hill, New York, 1983
7. Nagtegaal, J.C., Parks, D.M. and Rice, J.C., 'On numerically accurate finite element solutions in the fully plastic range', Comp. Meth. Appl. Mech. Eng., 4 (1974), 153-177.
8. Nagtegaal, J.C. and Jong, J.E. de, 'Some computational aspects of elastic plastic large strain analysis', Int. J. Num. Meth. Eng., 17, (1981), 15-41.

ADVANCES IN AEROELASTICITY

D. Sepahy

British Aerospace PLC,
Air Weapons Division,
Hatfield, England

Abstract

This paper reviews aeroelastic analysis from the earliest days of flight up to the present and anticipates future developments. An overview of the theoretical background is provided, which deals with structural and aerodynamic modelling, model superposition and transformation and the solution of the dynamic response equation. The emphasis is on the techniques used in industry to meet the needs of successive air vehicle designs and the account also includes research and development in progress. Likely future developments occuring in response to changes in flight technology and the ongoing need to produce an optimum design to maximize the economics of flight are also included.

1. HISTORICAL BACKGROUND

Aeroelasticity is the branch of engineering mechanics which is concerned with the motion of deformable bodies through fluid. Aeroelasticity plays an important role in the design and analysis of air vehicles and the essential work required to meet airworthiness authorities' standards.

Earlier work in the area of aeroelastic instability, i.e. flutter investigation of the aircraft, goes back to the 1920s with the work of Frazer & Duncan at the National Physical Laboratory in the UK; their matrix method and numerical techniques are well known [1].

Hand operated desk calculating machines were used to carry out matrix operations to six significant figures. Therefore a considerable amount of work was involved, even for limited degrees of freedom, to solve the equation of motion of a semi rigid aircraft (which provided a good representation of early low speed aircraft with a large wing aspect ratio) employing quasi steady state aerodynamic loads. However the method becomes unsuitable at higher speeds and unsteady aerodynamics, which are a function of the structural response frequency, should be used as well as consideration of air compressibility when approaching the speed of sound.

With the advent of high speed flight and low aspect ratio lifting surfaces, two dimensional unsteady aerodynamic formulation was no longer adequate and the solution of three dimensional unsteady aerodynamics in subsonic and supersonic flow was made possible with the appearance of 1st and 2nd generation digital computers. A paper by Broadbent in 1956 [2] dealt with aeroelastic problems in connection with high speed flight and gave a comprehensive review to that date. At a later date a paper by Baldock and Skingle "A survey of flutter technology in the United Kingdom" [3] argued that the reduction in flutter and vibration accidents to aircraft indicated that aeroelastic response prediction was adequately accurate.

The 54 accidents recorded for the period 1944 to 1960 were reduced to 7 mild incidences for the period 1960 to 1972, and this also led to a decrease in the proportion of prototype aircraft needed to iron out aeroelastic problems in spite of the increase in flight speeds and the number of supersonic aircraft in service.

Extensive work in aeroelasticity has been produced in the United States since the early days of powered flight and this is reviewed in the classical text by Bisplinghoff et al on aero-elasticity [4] published in 1955, giving full analytical and experimental treatment of aeroelasticity with comprehensive references for each topic. Although three decades have elapsed since publication it is still highly relevant and used throughout industry.

A manual on aeroelasticity in six volumes was published in 1968 by AGARD [5], containing contributions from a panel of experts in the field. Detailed background to aeroelasticity is given, with comprehensive surveys and references, dealing with the state of the art in the formulation of the problem, structural modelling, aerodynamic modelling, various phenomena in aeroelasticity and factual information on flutter charactistics and experimental methods. However the manual stopped short of numerical and computational techniques for digital computers and left the user to develop his own computer programme selecting from a wide choice of analytical techniques.

Over the last two decades we have seen the rapid development of digital computers with an increase in computer power and a reduction in computing costs. In parallel, numerical techniques in structural and fluid mechanics have developed with an ever-increasing speed. Special and general computer software has been developed and marketed and is readily available to industry, ranging from general purpose finite element software for large scale mainframe computers, e.g. Nastran [6], to specialized programs for a small home computer [7].

A more recent review of the state of the art in aeroelasticity is given by Garrick & Reed [8] which mainly reviews the activities and developments in unsteady aerodynamics which are used in flutter analysis.

A comprehensive review of the subject was also given by Collar [9] in the late 1970s, which covers theoretical and experimental developments with emphasis on activities in the UK.

In spite of extensive past activities, achievements and successes in computational techniques in aeroelasticity, further technology transfer is still required in the aerospace industries to solve the problems related to new challenges, e.g. dynamic response of computer controlled agile unstable fighters, highly flexible optimally designed aircraft with a computer controlled gust alleviation system, or in dealing with reusable flexible space shuttles for a range of subsonic and supersonic flight conditions.

2. THEORETICAL BACKGROUND

The term aeroelasticity includes phenomena involving interactions among inertial, aerodynamic and elastic forces, therefore in general it covers dynamic and static phenomena. The equation of motion of a deformable body through fluid can be derived using Lagrange's dynamical equations for a small perturbation of δq_r in a generalized dynamical coordinate

$$\frac{d}{dt}\left(\frac{\partial T}{\partial \dot{q}_r}\right) - \frac{\partial T}{\partial q_r} + \frac{\partial U}{\partial q_r} = Q_r \qquad \ldots\ldots\ldots\ldots (1)$$

Where T is the kinetic energy of the system, U is the strain energy of conservative forces and Q is the generalized force. For a linear system without damping:

$$T = \tfrac{1}{2}\{\dot{q}_r\}'[M]\{\dot{q}_r\} \quad \ldots (2) \qquad \{Q\} = \{F\} + \{F_i\} \quad \ldots\ldots\ldots\ldots (4)$$

$$U = \tfrac{1}{2}\{q_r\}'[K]\{q_r\} \quad \ldots (3) \qquad \{F_i\} = [\bar{B}]\{q_r\} + [\bar{C}]\{q_r\} \quad \ldots (5)$$

Where

M is a matrix of structural inertial coefficients
K is a matrix of structural stiffness coefficients
$\underline{F}$ is a vector of external aerodynamic forces
$\underline{B}$ is a matrix of aerodynamic damping coefficients
$\underline{C}$ is a matrix of aerodynamic stiffness coefficients

The latter part of equation (4) represents induced dynamic forces due to the motion of the body through the fluid, ignoring the inertia terms, since they are very small for air. For a body moving through the fluid with a forward speed of V, the equation of motion is given by

$$[M]\ \{\ddot{q}_r\} + \left[\sigma^{\frac{1}{2}} V[\bar{B}] + [D]\right]\{\dot{q}_r\} + \left[V^2[\bar{C}] + [K]\right]\ \{q_r\} = \{F\} \quad \ldots (6)$$

where

D is structural damping matrix
σ is relative air density

For an oscillating body motion can be written as

$$q = \bar{q}\ e^{pt}$$

$$p = \omega\ (\alpha \pm i)$$

Therefore the equation of oscillating bodies can be written as

$$\left[p^2.[M] + \sigma.^{\frac{1}{2}} V.p.[\bar{B}] + p.[D] + V^2.[C] + [K]\right]\{\bar{q}\}.e^{pt} = \{F\} \quad (7)$$

The matrices B and C are implicit functions of Mach number and reduced frequency, i.e. they depend on forward air speed and frequency of the oscillating body. To obtain a consistent solution of the equation of motion lengthy iterative techniques are required.

In general the structural matrices in equation (7) are constant, and are either derived directly for linear structural models or linearized equivalent structural matrices are obtained for a nonlinear structural model by consideration of the energy, or some other, method.

The frequency dependent external vector force F will describe atmospheric turbulence, noise, etc. or it can be equated to zero when the stability (flutter divergence) of the system is under consideration. Therefore equation (7) is solved in the frequency domain and the steady state response of the system is obtained for the desired range of frequencies. The stochastic processing of the result can proceed after achieving the frequency response solution and hence the fatigue life of the design can be determined. The stability of the system with or without autopilot is investigated for the specified flight envelope by obtaining characteristic roots of equation (7). In recent years equation (7) has also been used, directly or indirectly, in structured optimization. Computer programs have been developed over many years either to solve aeroelastic problems using equation (7) for a general complex three dimensional structure in subsonic and supersonic potential flow, or aeroelastic solutions are aimed at a specific problem.

Finite element techniques to generate the necessary structural matrices are now well established [10,11] and, more recently, finite element modelling has also gained some ground in fluid flow [12,13,14,15]. For linear systems the concept of superposition has been applied to derive the oscillatory aerodynamic matrices for a rigid structure and finally added to structural matrices in equation (7).

A convenient, accurate and cost-effective generalized coordinate system is that of structural free vibration modal coordinate system, i.e. equation (7) is set up in normal mode coordinates and a generalized solution is computed in modal coordinates and finally response characteristics are transferred to the physical coordinate systems. The second-order differential simultaneous equations, as symbolically represented in equation (6), can be solved in the time domain using finite difference or finite element techniques in time. For linear and nonlinear systems, numerically stable finite difference techniques have gained popularity and are generally available in most commercial finite element software packages. One of the main problems still remaining to be tackled is to set up unsteady aerodynamic coefficients in the time domain to be coupled in equation (6), on the other hand it is straight forward enough to introduce structural nonlinearity in equation (6) since structural coefficients can easily be updated after each time increment. However oscillatory aerodynamic coefficients can be computed in the frequency domain and transformed into the time domain using inverse Fourier transformation (fast Fourier transformation routine). Nevertheless the cost of the computational process is prohibitive for complex three dimensional structures and is only practical at present for a simple structural model which can be represented by a few hundred degrees of freedom. Although for nonlinear coupled structures fluid software packages are becoming available [16,17] for general use, they are not suitable for dealing with aeroelastic problems in a practical and cost effective manner, since they are specifically written for impact and penetration which are, by several orders of magnitude, in a different response time scale to that of structural response.

3. RECENT DEVELOPMENTS

An area of activity which has recently gained in popularity is that of aeroelastic optimization [18] where the structures are optimized with respect to weight for displacement frequency or flutter speed constraints. Software packages are developed by bringing together established techniques; finite element structural analysis, aeroelasticity and mathematical optimization tools. The success of the techniques in industry largely depends on the reduction in computation cost and development of expert software systems with the inclusion of rules and interactive user help for model synthesis and heuristic analysis. With the availability of fibre composites, aeroelastic design can provide wider scope in aeroelastic tailoring [19] in the early stages of aircraft design.

Recent developments in the area of unsteady aerodynamics have been in transonic viscous flow; a great number of papers have been published on the subject and a recent review can be found in the paper by McCroskey et al [20]. A number of special purpose codes [21] have been developed for transonic small-disturbance aerodynamic equations. A complete and reliable numerical simulation of the unsteady, transonic viscous flow around an aircraft could become possible within the next decade, with an increase in computer capabilities and advances in algorithm and solution methods, grid generation, turbulence modelling, vortex modelling, data processing and coupling of the aerodynamic and structural dynamic analysis. Fatigue damage of a structure may occur within the flight envelope due to the limit cycle brought about by structural nonlinearity. Structural nonlinearity has been investigated by Hauenstein et al [22] using an asymptotic expansion technique to analyse the limit cycle response of aerodynamic surfaces. Three simulation techniques were compared for response of a nonlinear system. A two degree of freedom nonlinear aeroelastic structure was examined by Ibrahim [23] using the Fokker-Planck equation technique.

A recent review of servoaeroelastic developments is given by Newsom et al [24] and coupled problems in the region of transonic flow are dealt with by Batina et al [25] who investigated three degrees of freedom stability in transonic flow, using a transonic small disturbance finite difference computer code, and demonstrated the application of the Pade aeroelastic model and time marching analysis of flutter suppression using active control.

Recent progress in aeroelasticity in research and development at NASA is given by Hanson [26] with particular emphasis on unsteady aerodynamic flow and a review of a range of recent computer codes available for solution of idealized potential flow to Navier-Stokes separated flow.

4. CONCLUSION

From the earliest days of flight aeroelastic analysis relied on advances in mathematical modelling and the use of numerical techniques to solve the dynamic response of the aircraft, to suppress aeroelastic instabilities and estimate fatigue life of the structures and hence ensure realistic design. With the present state of the art in aeroelastic analysis, numerical techniques and computer capabilities, semi-automatic design and optimization procedures can be introduced at the preliminary design phase. As a result, improvements in design and the introduction of new materials, such as fibre composite materials, are made possible as well as improvements in construction techniques.

There has also been extensive achievement and success in computational methods for solution of aeroelastic problems. Further computer tools are required in industry to meet the present and anticipated advances in flight technology, e.g. dynamic response of computer controlled agile unstable high speed fighters, highly flexible optimally designed aircraft with a computer controlled gust alleviation system, and particularly to tackle the aeroelastic problems of reuseable flexible space shuttles for a wide range of flight conditions.

In the following areas further practical economical computer software is required in industry for aeroelastic analysis:

1. Transonic and supersonic body/wing interaction

2. Accelerated flight.

3. High angle of incidence and separated flow.

4. Nonlinear numerical techniques in the solution of nonlinear aeroelastic analysis.

General enhancement in computer software technology is required in automatic aeroelastic optimization, particularly in the area of graphics for adaptive mesh generation and expert systems.

Acknowledgement

Thanks are due to British Aerospace PLC for permission to publish this paper.

5. REFERENCES

1. Frazer, R.A., Duncan W.J. and Collar A.R. Elementary Matrices, Macmillan, New York, 1946

2. Broadbent, E.G. 'Aeroelastic problems in connection with high speed flight', J. of Royal Aero.Soc., Vol.60, July 1956

3. Baldock, J.C.A. and Skingle, C.W. 'Flutter technology in the United Kingdom - a survey', Royal Aircraft Estab. Tech. Memorandum, Struc. 831 1973

4. Bisplinghoff, R.L., Ashley, A. and Hoffman, R.L. Aeroelasticity, Addison-Wesley, USA, 1955

5. Mazet, R. Gen.ed. Manual on Aeroelasticity, AGARD. 6v, 1968

6. Bellinger, D. Aeroelasticity Supplement, MacNeal Schwendler Corp., 1980

7. Simpson, A. 'The solution of large flutter problems on small computers', Aero. J., April 1984

8. Garrick, I.E. and Reed, W.H. 'Historical development of flutter', J. of Aircraft, Vol.18, No. 11, 1981

9. Collar, A.R. The first fifty years of aeroelasticity, Aerospace. Vol.5., 1978

10. Zienkiewicz, O.C. The Finite Element Method, 3rd Ed., McGraw Hill, London, 1977

11. NAFEMS A Finite Element Primer, Nat. Agency for Finite Element Methods and Standards, 1986

12. Albano, E. and Rodden, W.P. 'A doublet-lattice method for calculating lift distribution on an oscillating surface in subsonic flow', AIAA J., Vol.7, 1963

13. Pines. S., Dugundji, J. and Neuringer, J. 'Aerodynamic flutter derivatives for a flexible wing with supersonic and subsonic edges', J. of Aero. Science., Oct. 1955

14. Giesing J.P., Kalman, T.P. and Rodden, W.P. 'Subsonic steady and oscillatory aerodynamics for multiple interfering wings and bodies', J. of Aircraft, Vol. 9, Oct. 1972

15. Morino, L., Chen, L.T. and Sucin, E.O. 'Steady and oscillatory subsonic and supersonic aerodynamics around complex configurations', AIAA J., Vol.13, Mar. 1975

16. Hallquist, J.O. DYNA3D User's Manual, Lawrence Livermore Nat. Lab. April 1984

17. Hancock, S. PICES - 2 DELK Theoretical Manual, PICES International, Aug. 1985

18. Sobieski, J. 'Recent experience in multidisciplinary analysis and optimization', Conf.Proc. NASA Tech. Memorandum 87600, 1984

19. Schneider, G. and Zimmermann, H. 'Static aeroelastic effects on high performance aircraft', Messerschmitt-Boelkow-Blohm, Munich, AD-A167.595, Jan. 1986

20. McCroskey, W.J., Kulter, P. and Bridgeman, J.O. - 'Status and prospects of computational fluid dynamics for unsteady transonic viscous flows'. NASA Tech. Memorandum, 86018 Oct. 1984

21. Guruswamy, G.P., Goorjian, P.M. and Merritt, F.J. 'ATRAN35: an unsteady transonic code for clean wings'. NASA Tech. Memorandum, 86783

22. Hauenstein, A.J., Laurenson, R.M. and Gubser, J.L. 'Investigation of an asymptotic expansion technique to analyse limit cycle response of aerodynamic surfaces with structural non-linearity'. Airforce Office of Scientific Res. Tech. Report, 86-0288, July 1985

23. Ibrahim, R.A. 'Stochastic nonlinear flutter of aeroelastic structures'. Airforce Office of Scientific Res. Tech. Report 85-1076, Oct 1985

24. Newsom, J.R., Adam Jr., W.M. and Mukhopadhyay, V. 'Active control: a look at analytical methods and associated tools'. NASA Tech. Memorandum, 86269, July 1984

25. Batina, J.T. and Ynag, T.Y. 'Transonic calculation of aerofoil stability and response with active control'. NASA Tech. Memorandum, 85770, March 1984

26. Hanson, P.W. 'Aeroelasticity at the NASA Langley Research Centre, recent progress, new challenges'. NASA Tech. Memorandum, 87660, December 1985

CYLINDRICAL CONCRETE WATER TANKS: ANALYSIS AND DESIGN

V Thevendran and D P Thambiratnam

Senior Lecturers
Department of Civil Engineering
National University of Singapore
10 Kent Ridge Crescent, Singapore 0511

Summary: The minimum weight design of circular cylindrical concrete water tanks is studied in this paper. The internal radius and height of a tank are maintained constant while the thickness of the wall is varied along the axis so that the bending (tensile) and hoop stresses attain values as close as possible to their respective allowable values. Only the piecewise linear variations of wall thickness are considered. For the analysis a numerical procedure which combines the Runge-Kutta method of solution of ordinary differential equations with a numerical minimization method is used. This procedure is then imbedded into a minimization routine to deal with the design problems.

1. INTRODUCTION

The paper considers the minimum weight designs of circular cylindrical concrete water tanks with fixed internal radius and height. Minimum weight designs may not always be the most economical ones. However such a design is the preliminary step towards seeking the most economical one. The designs studied are subject to constraints that the bending (tensile) stresses and hoop stresses should not exceed certain prescribed values and that the thickness of the wall should not be less than a prescribed value. Water tanks are usually designed as having uniform wall thickness. In such a design, it is very unlikely that both critical bending and hoop stresses attain their maximum allowable values simultaneously. Only one of these two critical stresses would govern the final design. Since these stresses vary along the length of the tank, the thickness of the shell may be varied so that the stress distributions are as close as possible to the respective maximum allowable values. In the present study, only piecewise linear variations are considered as any other type of variation is considered to be impracticable, in general. No discontinuities in the values of the thickness along the length of the shell are allowed. Optimal design of such tanks poses problems as the governing differential equations and corresponding boundary conditions are difficult to solve exactly. For cylindrical tanks with constant wall thickness, as well as for those with constant mean radius and with thickness varying linearly from top to bottom, Timoshenko [1] has given approximate analytical solutions to the governing equations. However, these solutions cannot be used in the present study as the design conditions are qualitatively different from those considered by Timoshenko. The numerical methods used so far in analysis of cylindrical water tanks are the ones based on finite difference methods [2] or finite element methods [3]. In order to achieve better convergence of results, finer grids or meshes have to be used in these methods. But the optimal design using a numerical optimization procedure is iterative in nature and therefore requires a numerical analysis procedure which gives faster convergence. Thevendran [4] has presented a numerical approach, which is different from these two methods, to the analysis of circular cylindrical water tanks. The approach is based on the Runge-Kutta method of solving initial value problems involving ordinary differential equations. In the problems studied, there are boundary conditions to be satisfied at both ends of the solution domain. To solve such problems, Runge-Kutta method can be combined with a numerical optimization method and used. This procedure is then imbedded in an appropriate minimization routine to deal with optimal design of the tanks.

2. FORMULATION OF PROBLEM

2.1 PART I : Analysis

The present study deals with circular cylindrical water tanks with constant inner radius and varying outer radius. In such a case the middle surface of the tank would not be parallel to the axis of symmetry. If the slope of the middle surface with the axis of symmetry is very small, then the governing equations can be derived in the same way as those for a tank with middle surface of constant radius.

Figure 1 shows an element of the cylindrical tank with usual notation. Only the nonzero stress resultants in the axisymmetric problem studied herein are indicated on the element. Symbols M, N and Q denote bending moment, direct

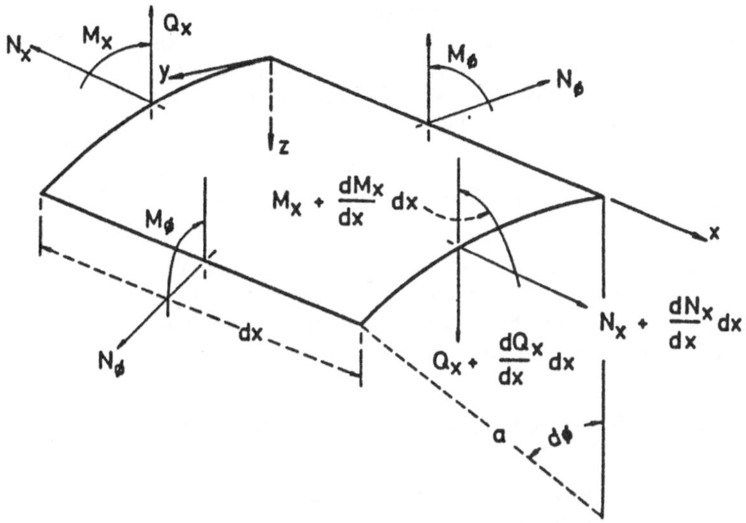

Fig. 1 Typical element of a cylindrical shell

force and shear force resultants respectively. The only equations of equilibrium are [1]

$$N_x' = 0 \tag{1}$$

$$Q_x' + (N_\phi/a) = -Z \tag{2}$$

$$M_x' - Q_x = 0 \tag{3}$$

where Z = external pressure normal to the surface, ()' indicates differentiation with respect to x and a is the radius of the middle surface. Equation (1) indicates that N_x is constant and is taken to be zero in the subsequent discussion. If they are different from zero, the deformations and stresses corresponding to such constant forces can be evaluated and superimposed on deformations and stresses produced by lateral loads. Hence, only eqns (2) and (3) are to be considered for solution. These two equations contain three unknowns N_ϕ , and Q_x and M_x. To solve the problem, the displacement of points in the middle surface have to be considered. Using the appropriate strain-displacment and stress-strain relations, it can be shown that

$$N_\phi = -Ehw/a \; ; \; M_x = -Eh^3w''/\{12(1-\nu^2)\} \tag{4}$$

where E and ν are, respectively, Young's Modulus and Poisson's ratio of the material, h the thickness of the tank, w the radial displacement and a the radius of the middle surface of the tank. From eqns (1) - (4), the following governing equation can be derived

$$(h^3 w'')'' + \{[12(1 - \nu^2)hw]/a^2\} = \{12(1 - \nu^2)/E\}Z \quad (5)$$

in which $Z = -\gamma x$, where γ is the specific weight of the water. The tank is considered to be clamped at the bottom ($x = d$, where d is the depth of the tank) and free at the top ($x = 0$). The corresponding boundary conditions are therefore

$$w(d) = 0 = w'(d) \; ; \; w''(0) = 0 = w'''(0) \quad (6)$$

To solve the above system represented by eqns (5) - (6) using the Runge-Kutta method, the following functions are defined:

$$u_1 = w; \; u_2 = w'; \; u_3 = (h^3 w''); \; u_4 = (h^3 w'')' \quad (7)$$

Thus

$$u_1' = u_2 \;\; ; \;\; u_2' = [u_3/h^3] \;\; ; \;\; u_3' = u_4 \; ;$$

$$u_4' = - 12(1 - \nu^2) \{[hu_1/a^2] + [\gamma x/E]\} \quad (8)$$

with

$$u_1(d) = 0 = u_2(d) \quad (9a)$$

$$u_3(0) = 0 = u_4(0) \quad (9b)$$

In order to apply the Runge-Kutta method $u_3(d)$ and $u_4(d)$ are treated as variable quantities which have to be chosen such that eqns (9b) are satisfied. The values of $u_3(d)$ and $u_4(d)$ are denoted by v_1 and v_2 respectively, i.e.

$$u_3(d) = v_1; \;\; u_4(d) = v_2 \quad (10)$$

Obviously $u_1(0)$, $u_2(0)$, $u_3(0)$ and $u_4(0)$ are all functions of v_1 and v_2. The solutions to eqn (8) with boundary conditions (9) is accomplished by considering the problem as the choice of v_1 and v_2 which minimizes the function

$$f(v_1, v_2) = [u_3(0)]^2 + [u_4(0)]^2 \tag{11}$$

f attains its minimum value of zero only when $u_3(0) = 0 = u_4(0)$. For practical purposes, the eqns (9) are deemed to be satisfied when f has attained a small enough value. The minization of f is done numerically.

2.2 PART II : Optimal design

The minimum weight design of a water tank with inner radius r and depth d is considered. the volume V of the construction material is minimized subject to (a) material property constraints that the critical bending tensile stress f_b and the critical hoop stress f_h should not exceed the respective allowable values F_b and F_h, and (b) a geometrical constraint that the thickness h should not be less than a prescribed value h_{min}. The design with n piecewise linear variations of the outer surface consists of minimizing.

$$V = \sum_{i=1}^{m} \pi d_i \{r(h_i + h_{i+1}) + \frac{1}{3}(h_i^2 + h_i h_{i+1} + h_{i+1}^2)\} \tag{12}$$

subject to

$$f_b < F_b \quad ; \quad f_h < F_h \quad ; h > h_{min} \tag{13}$$

where $h_i = h(x_i)$ and $h_{i+1} = h(x_{i+1})$ are the thicknesses at the ends of the i-th segment (see Figure 2); $d_i = (x_{i+1} - x_i)$.

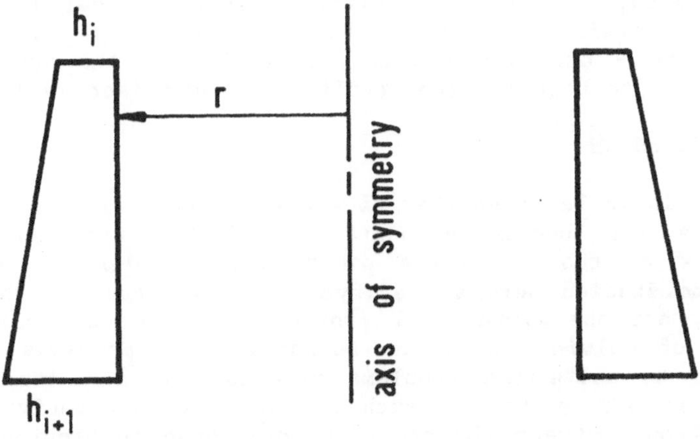

Fig. 2 Cross-section of typical i-th element

The problems studied are constrained nonlinear minimization problems. A constrained minimization problem is converted into an equivalent unconstrained problem using the 'exterior point' method of the Sequential Unconstrained Minimization Techniques (SUMT) developed by Fiacco and McCormick [6]. Accordingly, a problem of minimizing $f(\underline{x})$ subject to $g_j(\underline{x}) \geq 0$, $j = 1,2,...,m$ is solved by considering the problem of minimizing

$$F(\underline{x}, r_k) = f(\underline{x}) + r_k \sum_{j=1}^{m} [g_j(\underline{x}) - |g_j(\underline{x})|]^2 \qquad (14)$$

over a monotonically increasing sequence of r_k. For the minization of an unconstrained problem, a direct search method based on Rosenbrock's method [7] is used.

3. EXAMPLES

As examples of optimal wall shape design of tanks with constant inner wall radius, water tanks with the following dimensions are studied. All tanks are 10 m high but of different internal radii. The internal radii considered are 20 m, 10 m and 5 m. The critical hoop and bending tensile stress are not to exceed 1.44 N/mm^2 and 2.02 N/mm^2 respectively for the grade 30 concrete [8]. Young's Modulus E and Poisson's ratio ν of concrete are taken as 28 kN/mm^2 and 0.167 respectively. The specific weight of water is 9.81 kN/m^3 and that of concrete is 24 kN/m^3. The minimum acceptable thickness is 50 mm.

For each problem, three types of thickness variations are studied, viz. **Type A:** uniform wall thickness; **Type B:** a linear variation from top to bottom; and **Type C:** two linear variations - one from top to mid-height and the other from mid-height to bottom. The results are summarised in Table 1.

4. DISCUSSION

It is to be noted that nowadays finite element method of analysis has become standard method for the structural analysis of two and three dimensional problems. But the problems studied here are axisymmetric problems and thus, in effect, are one dimensional problems. Hence, any numerical method of solving two point boundary value problems can be adopted to solve the problems studied herein. The method described herein is one such method. A comparative study carried out between the procedure developed in Section 2 and the finite element method has shown that the procedure presented in this paper is more efficient computationally

Table 1: Summary of results

Inner Radius (m)	Type of Design	Thickness (m) Top / Mid-ht / Bottom	Volume V (m^3)	Savings in V %
20	A	1.006 / 1.006 / 1.006	1310.5	-
	B	0.128 / 0.654 / 1.179	847.3	35.3
	C	0.092 / 0.663 / 1.182	842.9	35.7
10	A	0.640 / 0.640 / 0.640	419.6	-
	B	0.051 / 0.378 / 0.705	246.0	41.4
	C	0.051 / 0.376 / 0.706	245.4	41.5
5	A	0.359 / 0.359 / 0.359	118.1	-
	B	0.154 / 0.265 / 0.375	86.4	26.8
	C	0.140 / 0.258 / 0.381	84.7	28.3

(see [4] for further details and comparison). This procedure is imbedded into a numerical minimization routine based on the SUMT and employed to study the optimal wall shapes of the tanks.

In solving the design problems, optimization has been carried out starting from a set of initial values for the design variables (thickness values at selected points), using 50 equal sub-intervals in the Runge-Kutta algorithm. The optimal values are then increased by 5% and optimization has been carried out using 100 equal sub-intervals. The results presented in Table 1 pertain to this procedure. When the procedure has been repeated with 300 equal sub-intervals, no appreciable differences over the earlier ones have been observed. The results show that the reduction in volume of concrete for tanks with a single slope of the outer wall are approximately 35%, 41% and 27% for tanks with internal radius of 20 m, 10 m and 5 m respectively. With two slopes of the outer walls, these values increased to approximately 36% and 28% for the tanks with internal radius of 20 m and 5 m respectively. The intermediate shell of 10 m internal radius does not show an increased saving in material. Thus it appears that the major saving in material is achieved by one slope of the outer wall itself. While two slopes may improve the saving in material for some tanks, it does not seem reasonable to increase the number of slopes beyond two. If the number of slopes is increased to three or more, the increased difficulties posed by construction may outweigh the marginal savings in material. In all the cases studied, the critical bending stresses seem to govern the final optimal design. The critical hoop stresses remain below the allowable maximum values (1.44 N/mm^2) except in the cases of the two larger tanks with two slopes.

In the derivation of the governing equations, it has been assumed that the inclination of the middle surface to the axis of symmetry can be disregarded. It can be seen from the results, the largest inclination occurs in the case of tank with 20 m inner radius and is about $\tan^{-1} 0.026$ which is small enough to be disregarded.

REFERENCES

1. Timoshenko, S.P., and Woinowsky, S., *Theory of Plates and Shells*, 2nd Edn, McGraw Hill, New York, 1957.

2. Ghali, A, *Circular Storage Tanks and Silos*, E & F.N. Spon Ltd., London, 1979.

3. Zienkiewicz, O.C. and Morgan, K., *Finite Elements and Approximations*, John Weley, New York, 1983.

4. Thevendran, V., 'A numerical approach to the analysis of circular cylindrical water tanks', *Computers and Structures*, **23** (1986), 379-383.

5. Kreyzig, E., *Advanced Engineering Mathematics*, 5th Edn, John Wiley, New York, 1983.

6. Fiacco, A.V., and McCormick, G.P., *Nonlinear Programming: Sequential Unconstrained Minimization Techniques*, John Wiley, New York, 1968.

7. Rosenbrock, H.H., 'An automatic method for finding the greatest or the least value of a function', *Computer Journal*, 3, 175-184, 1960.

8. *BS5337 : 1976, The Structural use of Concrete for Retaining Aqueous Liquids* (formerly CP2007), British Standard Institution, London (1976).

A COMPLETE PROCEDURE FOR THE ADJUSTMENT OF A FINITE ELEMENT MODEL FROM THE IDENTIFIED COMPLEX EIGENMODES

Q. Zhang, G. Lallement, R. Fillod, J. Piranda

Laboratory of Applied Mechanics, Associated to CNRS, UFR Sciences and Technology, University of Franche-Comté, 25030 BESANÇON CEDEX (France)

Abstract

In this paper, a complete procedure is presented for the adjustment of the mathematical model of a self-adjoint mechanical structure from the identified complex modes. The mathematical model considered here is a finite element model with symmetric, positive definite matrices. This procedure is constituted of the following three steps : 1) Transformation of the identified complex eigensolutions into the real eigensolutions of the associated conservative structure. 2) Localisation of the dominant errors in the finite element model. 3) Parametric correction of the finite element model by an iterative process. The two last steps take advantage of the sensitivity matrix as a function of the design parameters of the structure. This procedure is illustrated by a simulated example.

Introduction

The objective is to improve the precision of the discrete models of the sort used in finite elements for preliminary of a structure's dynamic behavior. This improvement is based on the minimization of the "distance" between the output calculated from the model and the output observed on the structure. What is sought is :

a) To treat models in which the state matrices and their constitutive submatrices have a clear physical significance, both initially and after the parametric correction.

b) To construct a corrected model converging toward the "best" discrete representation (of a given order N) of the physical structure.

The essential characteristics of the proposed procedure are:

First step : Exploitation of the calculated and observed eigensolutions (frequencies, eigenvectors, and generalized masses) of the Associated Conservative Structure (abb. : ACS).

Second step : Localization of the regions presenting the dominant modeling errors.

Third step : Correction of the design variable belonging to the dominant error regions by minimization of the distance between pairs of calculated and observed eigensolutions.

The proposed procedure is :
- related to the "local" methods [1-5],
- distinctly different from the "global" methods [6-7].

Passage from the identified complex eigensolutions to the real eigensolutions of the associated conservative structure

This passage is effected in the following two successive steps:
1) Construction of an initial estimation for the real eigenmodes of the associated conservative structure.
2) Iterative corrections of this initial estimation, followed by a verification of the quality of the obtained real eigenmodes.

1) Construction of an initial estimation : Given a complete set of N identified complex eigensolutions $\hat{Y}$; $\hat{S}$, the matrix $M^{-1} K$ can be expressed in the form :

$$M^{-1} K = - [I_m(\hat{Y}\hat{S}\hat{Y}^{-1})][I_m(\hat{Y}\hat{S}^{-1}\hat{Y}^{-1})]^{-1} \tag{1}$$

Or, as a function of the real eigensolutions $\hat{Z}$; $\hat{\Lambda}$, in the form :

$$M^{-1} K = \hat{Z}\hat{\Lambda}\hat{Z}^{-1} \tag{2}$$

M and K are the mass and stiffness matrices ; both are symmetric and positive definite , I_m is the unit matrix of order m.

The complex modal matrix $\hat{Y}$ can be expressed on the base $\hat{Z}$ of the associated conservative structure as :

$$\hat{Y} = \hat{Z}\hat{T} \qquad \text{where} \qquad \hat{T} \in \mathbb{C}^{N,N} \tag{3}$$

Substituting (3) in (1) and taking into account (2) yields :

$$\hat{\Lambda} = [I_m (\hat{T}\hat{S}\hat{T}^{-1})] [I_m(\hat{T}\hat{S}^{-1}\hat{T}^{-1})]^{-1} \tag{4}$$

This relation is used later to construct the initial estimation $Z°$ of the modal sub-matrix $Z \in \mathbb{R}^{c,m}$. In fact, the identified complex sub-base $Y \in \mathbb{C}^{c,m}$ can be represented by the sub-base $Z \in \mathbb{R}^{c,m}$ with a sufficient precision for technical applications.

Thus :

$$Y \cong ZT \quad \text{where} \quad T \in \mathbb{C}^{m,m} \tag{3'}$$

A good approximation of $\Lambda \in \mathbb{R}^{m,m}$ is then given by :

$$\Lambda \cong [I_m\,(TST^{-1})][I_m(TS^{-1}T^{-1})]^{-1} \tag{4'}$$

Initial estimation $\Lambda^{(o)}$ of Λ : The eigenvalues $s_\nu = -a_\nu\omega_\nu + j\omega_\nu$, $\nu = 1$ to m, have been identified on the structure. Since ω_ν^2 is often very close to λ_ν, the real eigenvalue of the ACS, an initial estimation $\Lambda^{(o)}$ of Λ is taken as : $\Lambda^{(o)} = \text{diag}\,\{\lambda_\nu^{(o)} = \omega_\nu^2\}$ (5)

Initial estimation $Z^{(o)}$ of Z : The initial estimation $Z^{(o)}$ is defined by (3') as : $Z^{(o)} = YT^{(o)}$ where $T^{(0)} \in \mathbb{C}^{m,m}$ is factored in the form :

$$T^{(0)} = HQ \tag{6}$$

where : $H \triangleq Y'^{+}\,Y \in \mathbb{C}^{m,m}$ is a known matrix ; $Q \in \mathbb{C}^{m,m}$ is an unknown matrix whose form is established in what follows. Taking into account (5) ; (6), equation (4') can be written :

$$\Lambda^{(0)} = Q^{-1}\,RQ \tag{7}$$

where : $R \triangleq [I_m(HSH^{-1})]\,[I_m\,(HS^{-1}H^{-1})]^{-1}$.

From (7), Q is obtained by : $RQ = Q\,\Lambda^{(0)}$ (8)

Having determined the matrix Q, the initial estimation $Z^{(0)} = YHQ$ can be immediatly calculated.

2) Iterative corrections of the initial estimations $\Lambda°$; $Z°$ and verification of the quality of the solution. These procedure are presented in detail in [8]. Only the principe will be summarized below :

a) First, construct an initial estimation $\beta°$ of the generalized damping matrix, using the matrices $\Lambda°$; $T°$.

b) From the eigenvalue problem :

$$\left(\begin{bmatrix} -\Lambda^{(0)} & 0 \\ 0 & I_m \end{bmatrix} - s_\nu^{(1)} \begin{bmatrix} (0) & I_m \\ I_m & 0 \end{bmatrix} \right) \begin{bmatrix} t_\nu^{(1)} \\ s_\nu t_\nu^{(1)} \end{bmatrix} = 0, \quad \nu = 1 \text{ to } m, \tag{9}$$

determine : $T^{(1)} = [\ldots\,;\,t_\nu^{(1)}\,;\,\ldots]$; $S^{(1)} = \text{diag}\,\{s_\nu^{(1)}\}, \in \mathbb{C}^{m,m}$, then $Y^{(1)} = [\ldots\,;\,y_\nu^{(1)}\,;\,\ldots] \in \mathbb{C}^{c,m}$ in which $y_\nu^{(1)} = Z^{(0)}\,t_\nu^{(1)}$.

c) The difference matrices : $\Delta Y^{(1)} \triangleq Y - Y^{(1)}$; $\Delta S^{(1)} \triangleq S - S^{(1)}$ are then used to correct the initial estimations $Z^{(0)}$; $\Lambda^{(0)}$, followed by the construction of the new estimations $T^{(1)}$; $\beta^{(1)}$ of the matrices T and β.

d) This procedure is iterated until convergence, such that :

$$\alpha_1 \; ||Y-Y^{(i)}|| \; + \varepsilon_2 \; ||S-S^{(i)}|| \leq \varepsilon \quad , \tag{10}$$

where α_1 ; α_2 ; are given positive scalars.

e) The verification of the quality of the eigensolutions Λ ; Z and of the generalized damping matrix β is effected simultaneously using (10). Other methods of determining the eigensolutions of the ACS are proposed in [9-12].

Localization of the dominant errors

An initial estimation is available and is constituted by a finite element model :

$M^e = \sum_{i=1}^{\ell} M_i^e$; $K^e = \sum_{i=1}^{\ell} K_i^e \in \mathbb{R}^{N,N}$, where M_i^e ; K_i^e are the mass and stiffness matrices of the ith substructure.

The results of a modal identification effected on the real structure are also available. The data set is constituted of :

. m eigenvectors of the ACS, $z_\nu \in \mathbb{R}^c$, sub-vector of $\hat{z}_\nu \in \mathbb{R}^N$, $\nu = 1,2,\ldots,m$

. m eigenvalues of the ACS λ_ν ; $\nu = 1,\ldots,m$

. n eigenvectors calculated using c dof, $z_\nu^e \in \mathbb{R}^c$, sub-vector of $\hat{z}_\nu^e \in \mathbb{R}^N$, $\nu = 1,2,\ldots,n$

. n calculated eigenvalues, λ_ν^e ; $\nu = 1,\ldots,n$.

The following inequalities are often satisfied : $N \gg c; n > m; c > m$.

A localization technique was developed in 1976 in the context of a study under the CNES [13]. Another idea proposed by Sidhu has recently been taken up [14]. A third technique is proposed here which is considered to be more robust and better adapted to complicated or large-scale structure.

1) Localization indicators :

The localization is effected by macro-elements. Define the qualitative localization indicators m_i ; k_i, $i = 1,..\ell$ intervening linearly in the mass and stiffness matrices of the macro-elements such that :

$$M = \sum_{i=1}^{\ell} M_i = \sum_{i=1}^{\ell} m_i M_i^e \; ; \; K = \sum_{i=1}^{\ell} K_i = \sum_{i=1}^{\ell} k_i K_i^e \tag{11}$$

2) <u>Qualitative estimation of localization indicators</u>:

Let : $m_i = 1 + \Delta m_i$; $k_i = 1 + \Delta k_i$, $i = 1,2,\ldots, \ell$.

The identified modal deformations $z_\nu \in \mathbb{R}^c$ are linearized with respect to the indicators m_i and k_i by a first order Taylor series expansion.

Hence : $\Delta z_\nu = z_\nu - z_\nu^e \cong S_\nu \ \Delta p$, $\nu = 1,\ldots,m$, (12)

$$\Delta p = [\Delta k_1 \ ;.; \ \Delta k_\ell \quad \Delta m_1 \ ;.; \ \Delta m_\ell \ ]^T \in \mathbb{R}^{2\ell}$$

Similarly for the eigenvalues λ_ν , the Taylor series expansion leads to :

$$\Delta\lambda_\nu = \lambda_\nu - \lambda_\nu^e = \underset{\sim}{S}_\nu \quad \Delta p \quad , \tag{13}$$

Regrouping (12) and (13), yields : $\underset{(c+1)mx1}{\Delta w} = \underset{(c+1)mx2\ell}{\tilde{S}} \ . \ \underset{2\ell x1}{\Delta p}$ (14)

It is solved with respect to $\Delta p \in \mathbb{R}^{2\ell}$ and is restricted to the over-determined case (ℓ of the order $(c+1)m/4$). In practice, the solution of (14) is effected after the introduction of dimensionless quantities, ie. after the introduction of a diagonal weithting matrix W^{-1} : $W^{-1} \ \Delta w = W^{-1} \tilde{S} \ \Delta p$ (15)

3) Solution of equation (15) :

A first method consists of solving (15) directly in the sense of Moore-Penrose pseudo-inverse : $\Delta p = [W^{-1}\tilde{S}]^+ \ W^{-1}\Delta w$ (16)

A second method consists of searching for the best sub-space of parameters of a given dimension (Δm_i ; Δk_i) which best reduces the error $||E||$: $||E|| = ||\Delta w - \tilde{S}^m \ \Delta p^m||$ where : $\tilde{S}^m$ is a sub-matrix of $\tilde{S}$; Δp^m is the corresponding sub-vector of Δp.

An analysis of the errors obtained with sub-spaces of increasing dimension permits the selection of the most probable dominant errors.

Equation (14) (or 15) implies the pairing of identified and calculated eigensolutions. In many cases only the m (m= 4 to 6) first eigensolutions can be paired in this way. For these m eigensolutions, two cases can be distinguished.

- the frequency spectrum is sparse in the band of interest : pairing could then be directly effected by the usual procedures[15]

- the frequency spectrum includes some zones of high density : the pairing is effected after frequency separation by numerical modifications introduced simultaneously in the mathematical model M^e ; K^e and in the structure [16].

Parametric correction of the model M^e ; K^e

After localization of the sub-domains presenting the dominant modeling errors in the initial estimation M^e ; K^e, a set p_j, j=1 to r, of physical parameters can be defined in these sub-domains. The correction of these parameters by an iterative calculation using the Gauss-Newton method leads to the minimization of the model-structure "distance" represented by $||\Delta w||$. The kth iteration leads the solution of a problem having the same form as(14):

$$\text{Min}/\Delta p_{(k)} \; j = ||\Delta w^{(k)} - \tilde{S}^{(k)} \Delta p^{(k)}||$$

where : $\Delta_p^{(k)} = \{\Delta p_1^{(k)}, \ldots, \Delta p_j^{(k)}, \ldots, \Delta p_r^{(k)}\}^T \in \mathbb{R}^r$

At each iteration, a check of the norm of the solution $\Delta p^{(k)}$ is performed with respect to the constraint equations : $\Delta p_j^{(k)} = 0$, introduced by the intermediary of the weighting coefficients α_j, j = 1 to r. The solution $\Delta p_{(k)}$ is effected by the intermediary of a singular value decomposition, after having determined the scalar γ such that $||\Delta p^{(k)}|| = \beta^{(k)}$, where $\beta^{(k)}$ is an arbitrarily chosen positive scalar.

Numerical simulation example :

The structure used in the simulation is represented by Figure 1. The perturbed model used to simulate the physical structure is obtained by introducing stiffness modifications in zones 3' and 5', which are contained in the sub-domains 3 and 5 respectively. In this example : $k_3' = 10\, k_3$; $k_5' = 10\, k_5$. The first 5 eigenfrequencies of the models are presented in Table 1.

Eigenfrequencies	N° 1	2	3	4	5
Calculated	24.01	58.41	71.51	103.50	199.08
Identified	26.22	62.21	75.72	120.08	202.91

Tab.1

Figure 2 represents the 8 sub-domains used in the localization. The localization results are reported in Figure 3. The stiffness of macro-elements 3 and 5 are localized by limiting the dimension of the sub-space in the sensibility matrix $\tilde{S}$ to 2.

The parameters k_3' and k_5' are then treated as parameters to be adjusted. Since the distance between the eigensolutions of the initial and perturbed models is significant , the adjustment is effected in two steps. Figure 4 shows the evolution of the relative distances $\Delta\lambda_\nu/\lambda_\nu$ and $||\Delta y_\nu||/||y_\nu||$ during the first adjustment step.

As the eigensolutions in each iteration are calculated using a reduced initial modal base, the distance between the adjusted and identified eigensolutions does not tend to zero but becomes stationary after 10 iterations. The results obtained are : $k^{\circ}_{3'} = k^{\circ}_{5'} = 6$. By taking $k^{\circ}_{3'}$; $k^{\circ}_{5'}$ as new parameter values, the new exact eigensolutions can be calculated and a second parametric adjustment can be effected. The convergence of the relative distances $\Delta\lambda_\nu/\lambda_\nu$ and (fig. 5) $||\Delta y_\nu||/||y_\nu||$ is excellent and the final results are : $k_{3'} = k_{5'} = 9.7$ (The exact values are : $k_{3'} = k_{5'} = 10$).

Conclusion

The quality of the parametric correction is conditioned by the localization step. Only a localization of the dominant modeling errors permits a reduction of the degree of indetermination and avoids the choice of a particular arbitrary solution. Only a localization can lead to the "best neighboring solution" to the "exact solution", taking into account a limited frequency window of m eigenmodes.

References

1. Chrostowski, J.D., Evensen ,D.A., Hasselman, T.K., Model Verification of mixed dynamic systems, Trans. ASME, Vol. 100, April 1978, pp. 266-273
2. Natke,H.G., Deliberations on the improvement of the computational model with measured eigenmagnitudes, Rev. Roum. Sci. Tech. Mech. Appl., Tome 28, N° 2, 1983, pp. 159-173
3. Chan, J.C., Garba, J.A., Analytical model improvement using modal test data, AIAA Journal 18, 1980, pp. 684-690
4. Berger, H., Chaquin, J.P., Ohayon, R., Finite element model adjustement using experimental vibration data, 2nd IMAC, Orlando, Florida, USA, 6-9 Feb. 1984
5. Boutin, D., Petiau, C., Dynamic identification on finite elements model, Euromech 168, Manchester, July 1983
6. Baruch, M., Methods of reference basis for identification of linear dynamic structures, Technical report TAE n° 458, Technion Israel Institute of Technology, Haifa, Israel, Sept. 1981
7. Berman, A., Nagy, E.J., Improvement of a large analytical model using test data, AIAA Journal, 21, 1983, pp. 1168-1173
8. Zhang, Q., Lallement, G., Simultaneous determination of normal modes and generalized damping matrix from complex modes, Proc. of 2nd ISASD, Aachen, W. Germany, April 1-3, 1985
9. Zhang, Q., Lallement, G., Comparison of normal eigenmodes calculation methods based on identified complex eigenmodes, Jour. Spacecraft and Rockets, vol.24, 1, Jan-Feb.87, 69-73
10. Ibrahim, S.R., Determination of normal modes from measured complex modes, Shock Vib. Bull., 52, 5, pp. 13-17, 1982
11. Natke, H.G., Rotert, D., Determination of normal modes from identified complex modes, Z. Fulgwiss.. Weltrantorsch, 9, 1985, Helt 2
12. Niedbal, N., Advance in group vibration testing using a combination of phase resonance and phase separation methods, Proc. 2nd ISASD, Aachen, Germany, April 1-3, 1985
13. Bugeat, L., Fillod, R., Lallement, G., Piranda, J., Adjustement of a conservative non gyroscopic mathematical model from measurement, Shock and Vib. Bull., 48, Part 3, Sept. 1978, pp. 71-81
14. Zhang, Q., Lallement, G., Dominant error localization in a finite element model of a mechanical structure,Mechanical Systems and Signal Processing, 1987, 1(1), 1-9
15. Bugeat, L.P., Lallement, G., Methods of matching of calculated and identified eigensolutions, Strojnicky Casopis, Rocnik 32, Cislo 5, 1981, pp. 513-523
16. Zhang, Q., Identification modale et paramétrique de structures mécaniques autoadjointes et non autoadjointes, Thèse Doct. ès Sciences, Université de Franche-Comté, Janvier 1987

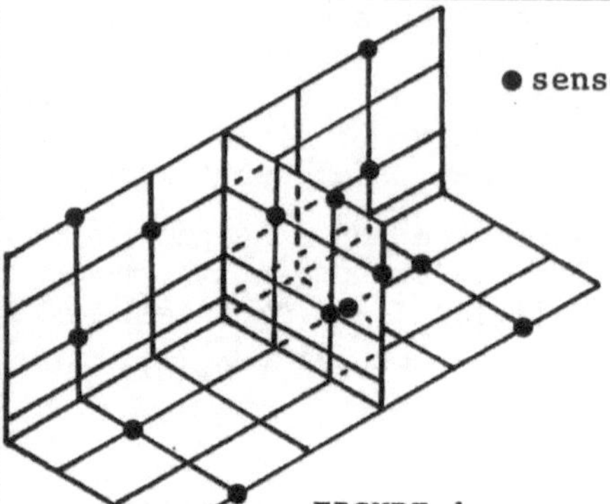

FIGURE 1

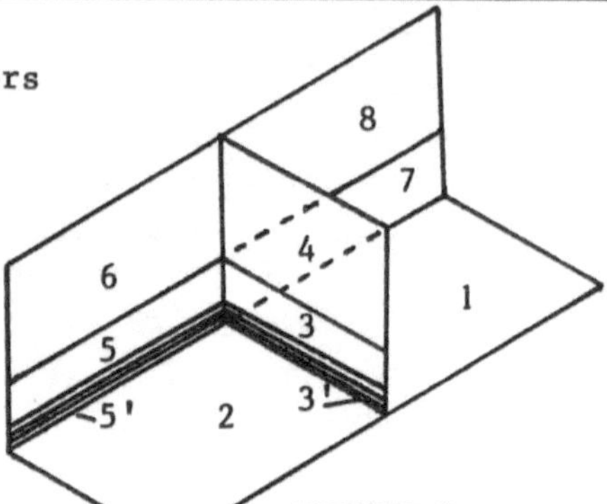

FIGURE 2

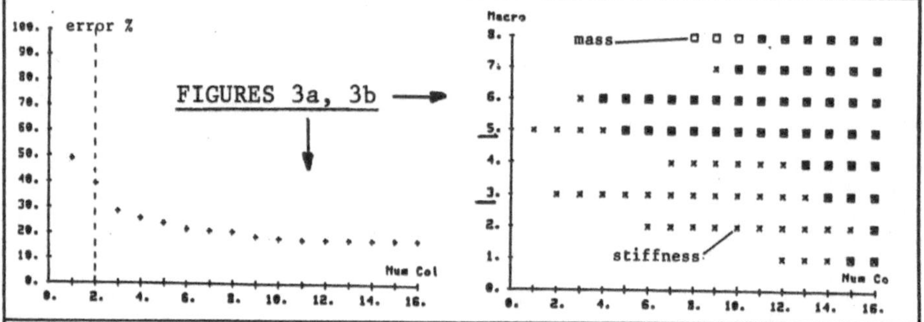

FIGURES 3a, 3b

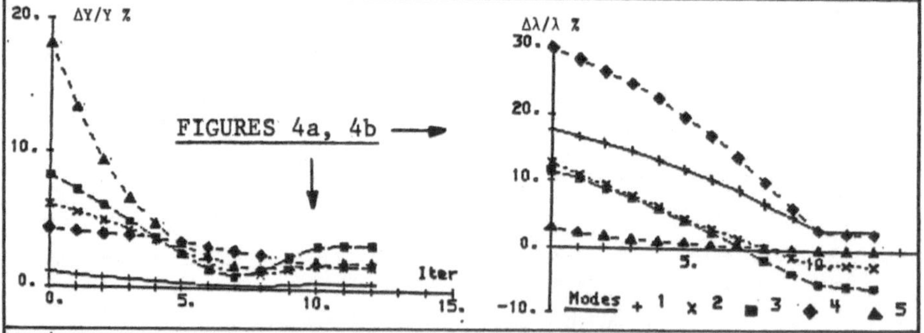

FIGURES 4a, 4b

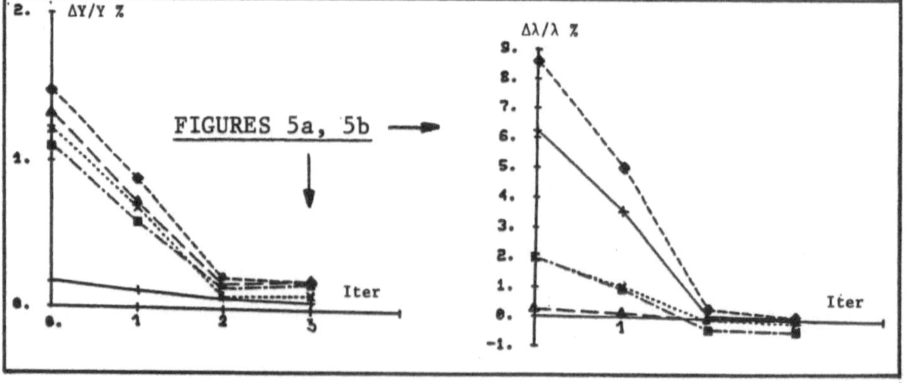

FIGURES 5a, 5b

A GENERAL FORMULA OF THE CURVED SHELL ELEMENTS AND ADAPTIVE MESH METHOD IN THE NONCONSERVATIVE FINITE DEFORMATION ANALYSIS

Y.T.Zhang, H.Y.Yang and J.Y.Zhang

Department of Mechanics, Tianjin University, Tianjin, China

SUMMARY

This paper presents a general formula of the curved shell in finite deformation analysis. The displacement field at a point $P(\xi,\eta,\zeta)$ within a element is discribed using the vector E_i and the F_i. The E_i is the displacement vector of node i in the mid-surface of shell, the vector F_i is the difference of the two unit normal vectors at node i in deformed and initial configurations. This formula is a general form and simpler than the existing ones[1],[2]. That is on the basis of simple and important physical concept that in the finite deformation the displacement vectors can be added and the rotation ones can't. It is convenient to apply this formula to some nonconservative finite deformation analysis.

1. DISPLACEMENT FIELD OF THE ELEMENT

In this paper the displacement field of element is given by

$$\begin{Bmatrix} u \\ v \\ w \end{Bmatrix} = \sum_{i=1}^{n} N_i\ (\xi,\eta)\ [\ E_i + \frac{B_i \zeta}{2}\ F_i] \qquad (1)$$

i.e.

$$\begin{Bmatrix} u \\ v \\ w \end{Bmatrix} = \sum_{i=1}^{n} N_i \left(\begin{Bmatrix} u_i \\ v_i \\ w_i \end{Bmatrix} + \frac{B_i \zeta}{2} \begin{Bmatrix} F_{ix} \\ F_{iy} \\ F_{iz} \end{Bmatrix} \right) \qquad (2)$$

Where the B_i represents the elenment thickness at node i.

In the small deformation analysis we can write

$$F_i = [\, V_{1i},\ -V_{2i}\,] \begin{Bmatrix} \alpha_i \\ \beta_i \end{Bmatrix} \tag{3}$$

Where the unit vector V_{1i}, V_{2i} and V_{3i} form the orthogonal local axes at the shell node i, and the V_{3i} is along the outward normal of shell at node i, and the α_i, β_i are known as the rotations about V_{2i} and V_{1i}. The linear function (3) is not available for the finite deformation analysis. The author of Reference 1 made a pioneering attempt and successfully gave some nonlinear formulations of the F_i in three cases. However, it is no matter whether the F_i is linear or nonlinear. It is important that the rotations can't be added together as vectors in the large deformation. This is not an approximation problem. In static analysis, in fact, what is important for us is the final equilibrium state, not the deforming process. The authors of Reference 2 made progress further. They gave a formulation that contains all the cases in the Reference 1. That formulation can be regarded as the further expression of the general formula in this paper.

Let $U_i=[l_i\ m_i\ n_i]^T$, a unit vector, be the outward normal vector of shell at node i on the final configuration, we can write

$$F_i = U_i - V_{3i} \tag{4}$$

where $l_i+m_i+n_i=1.0$
Substituting (4) into (1) we obtain the general formula of cureved shell displacement field in the finite deformation analysis as follows

$$\begin{Bmatrix} u \\ v \\ w \end{Bmatrix} = \sum_{i=1}^{n} N_i \left[\begin{Bmatrix} u_i \\ v_i \\ w_i \end{Bmatrix} + \frac{B_i \zeta}{2} \left(\begin{Bmatrix} l_i \\ m_i \\ n_i \end{Bmatrix} - \begin{Bmatrix} V_{3ix} \\ V_{3iy} \\ V_{3iz} \end{Bmatrix} \right) \right] \tag{5}$$

It is easy to show that the U_i+dU_i is also a unit vector.

Let α_i be the angle between plane(V_{2i}, V_{3i}), which passes the vector V_{2i} and V_{3i} and so on, and place(U_i, V_{2i}), β_i be the angle between plane(V_{1i}, V_{3i}) and plane(U_i, V_{1i}). When $\cos^2\alpha_i \sin^2\beta_i + \cos^2\beta_i \neq 0$, we can get

$$U_i = \begin{Bmatrix} l_i \\ m_i \\ n_i \end{Bmatrix} = \frac{1}{\sqrt{\cos^2\alpha_i \sin^2\beta_i + \cos^2\beta_i}} [V_{1i}, -V_{2i}, V_{3i}] \begin{Bmatrix} \sin\alpha_i \cos\beta_i \\ \cos\alpha_i \sin\beta_i \\ \cos\alpha_i \sin\beta_i \end{Bmatrix} \quad (6)$$

It seems to be better to regard α_i and β_i as the angles between planes than as rotations in Reference 2. The function (6) is complicated. In lot of practical constructions we can use the formula (5) directly taking the $\mathrm{sign}(V_{3iz})\sqrt{1-l_i^2-m_i^2}$ as n_i, for example, in the numerical examples of this paper.

2. INCREMENTAL EQUATION OF EQUALIBRIUM

We can write

$$\begin{Bmatrix} du \\ dv \\ dw \end{Bmatrix} = \sum_{i=1}^{n} [N_i] \begin{Bmatrix} du_i \\ dv_i \\ dw_i \\ dl_i \\ dm_i \end{Bmatrix} \quad (7)$$

where

$$[N_i] = \begin{bmatrix} N_i & 0 & 0 & N_i' & 0 \\ 0 & N_i & 0 & 0 & N_i' \\ 0 & 0 & N_i & -N_i'\frac{l_i}{n_i} & -N_i'\frac{m_i}{n_i} \end{bmatrix} \quad (8)$$

and

$$N'_i = \frac{B_i \zeta}{2} N_i$$

Taking all the same expressions as in the Reference 1, we can write

$$d\{\varepsilon\} = d\{\varepsilon^{\circ}\} + d\{\varepsilon^{1}\} = ([H] + [A]) [G] \{d\delta\} \quad (9)$$

In this paper

$$\begin{bmatrix} N_{i,x} & 0 & 0 & N'_{i,x} & 0 \\ 0 & N_{i,x} & 0 & 0 & N'_{i,x} \\ 0 & 0 & N_{i,x} & -N'_{i,x}\frac{l_i}{n_i} & -N'_{i,x}\frac{m_i}{n_i} \end{bmatrix} \quad (10)$$

$$[G_i] = \begin{bmatrix} N_{i,y} & 0 & 0 & N'_{i,y} & 0 \\ 0 & N_{i,y} & 0 & 0 & N'_{i,y} \\ 0 & 0 & N_{i,y} & -N'_{i,y}\dfrac{l_i}{n_i} & -N'_{i,y}\dfrac{m_i}{n_i} \\ N_{i,z} & 0 & 0 & N'_{i,z} & 0 \\ 0 & N_{i,z} & 0 & 0 & N'_{i,z} \\ 0 & 0 & N_{i,z} & -N'_{i,z}\dfrac{l_i}{n_i} & -N'_{i,z}\dfrac{m_i}{n_i} \end{bmatrix}$$

In the equation of static equilibrium

$$\{\phi\} = \{R\} - \int_V [B]^T \{\sigma\} \, dv \tag{11}$$

and the incremental equation of equilibrium

$$d\{\phi\} = d\{R\} - [K^T] d\{\delta\} \tag{12}$$

all the matrix are simplified in this paper because

$$\frac{\partial F_{ix}}{\partial l_i} = \frac{\partial F_{iy}}{\partial m_i} = 1, \qquad \frac{\partial F_{ix}}{\partial m_i} = \frac{\partial F_{iy}}{\partial l_i} = 0$$

$$\frac{\partial F_{iz}}{\partial l_i} = -\frac{l_i}{n_i}, \qquad \frac{\partial F_{iz}}{\partial m_i} = -\frac{m_i}{n_i}$$

$$\frac{\partial^2 F_{ix}}{\partial l^2_i} = \frac{\partial^2 F_{ix}}{\partial l_i \partial m_i} = \frac{\partial^2 F_{ix}}{\partial m^2_i} = \frac{\partial^2 F_{iy}}{\partial l^2_i} = \frac{\partial^2 F_{iy}}{\partial l_i \partial m_i} = \frac{\partial^2 F_{iy}}{\partial m^2_i} = 0$$

$$\frac{\partial^2 F_{iz}}{\partial l^2_i} = \frac{-1+m^2_i}{n^3_i}, \qquad \frac{\partial^2 F_{iz}}{\partial m^2_i} = \frac{-1+l^2_i}{n^3_i} \tag{13}$$

$$\frac{\partial^2 F_{iz}}{\partial l_i \, m_i} = \frac{-l_i m_i}{n^3_i} .$$

3. THE NONCONSERVATIVE FINITE DEFORMATUON ANALYSIS OF SHELL

It is convenient to apply the general formula of the curved shell element in this paper to solve some nonconservative problem. The $U_i=[l_i \;\; m_i \;\; n_i]^T$ is the normal at node i on the deformed configuration. The meshes of the curved shell elements are restricted to passing the normals of the nodes, the directions of the follower forces or the follower distributed forces are always along the directions of the normals of shell in many nonconservative constructions, for example, the Bourdon tube, which is used widely in pressure gauges. We can adapt the meshes conveniently by substituting U_i for V_{3i} in calculating processes.

The theorem of virtual work in finite deformation of elasticity is suitable for the both the conservative and the nonconservative analysis [3]. But in the nonconservative analysis a load correction matrix K^{nc}, which is, in general, nonsymetrical, is added to the tangent stiffness matrix if the magnitude of the load does't change[4]. In this paper, for an element,

$$K_{ij}^{nc} = q \int_s \frac{1}{\left|\sum_{i=1}^{n} N_i U_i\right|} [N_i]^T [L_j] ds \tag{14}$$

where the scalar q is the load parameter, $[N_i]$ is writen in the equation (8) and the $[L_j]$ can be writen as

$$[L_j] = \begin{bmatrix} 0 & 0 & 0 & N_j & 0 \\ 0 & 0 & 0 & 0 & N_j \\ 0 & 0 & 0 & -N_j\dfrac{l_j}{n_j} & -N_j\dfrac{m_j}{n_j} \end{bmatrix} \tag{15}$$

4. NUMERICAL EXAMPLES

Example 1

We consider a cylindrical shell shown in Fig. 1. The longitudinal boundaries of the shell are hinged and are immovable, whereas the circular edges are complitely free. The load is applied by incrementing displacement at point A.

The numerical results are presented in Table 1.

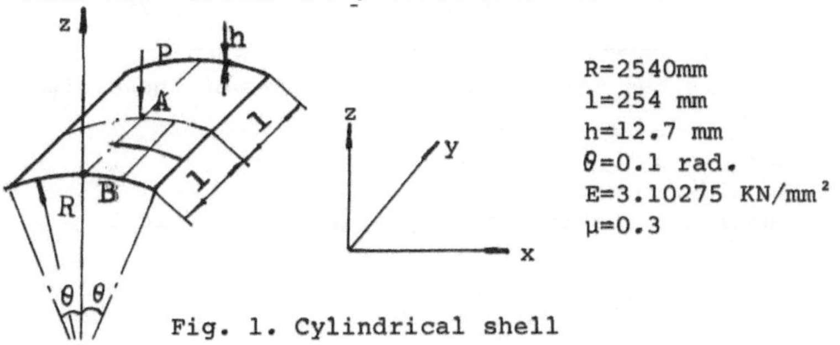

Fig. 1. Cylindrical shell

Table 1 The Results of Cylindrical Shell

	Result of this paper	Result of Reference1
A (mm)	-2.5 -5.0	-2.5 -5.0
Total load P(KN)	-0.8782 -1.5071	-0.8859 -1.5186
B (mm)	-0.5510 - 1.2531	-0.5409 -1.2672
No. of iterations	3 4	4 4

Example 2

Consider the unformly loaded circular plate with large deflection. The circular edges of the plate is clamped and the radial displacement on it are zero. A quarter of plate is shown in Fig. 2. The results are given in Table 2.

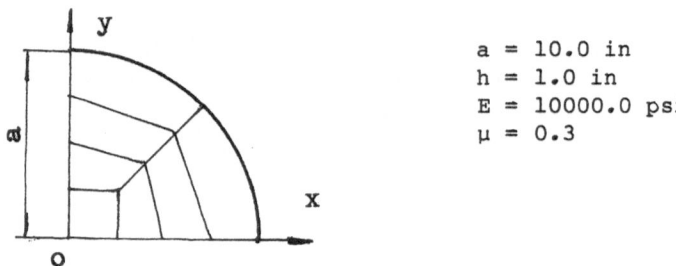

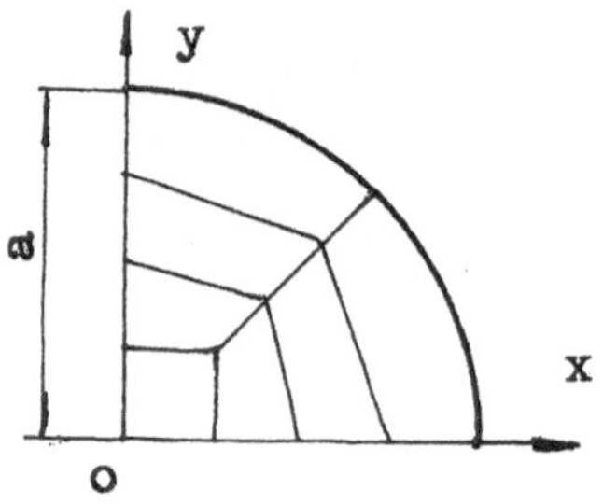

a = 10.0 in
h = 1.0 in
E = 10000.0 psi
μ = 0.3

Fig. 2 Circular Plate with Large Deffection

Table 2 The result of the uinformly loaded circulor plate

q	W_o/h	
	The result of this paper	The result of Reference [2]
15.0	1.305	1.312
30.0	1.783	1.796

where W_o is the defletion at the center of the plate.

Example 3

Consider Bourdon tubes shown in Fig. 3. The sizes of the tubes and results are represented in Table 3.

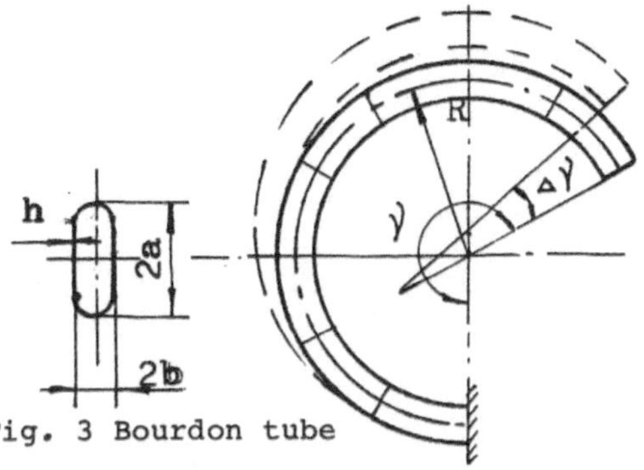

Fig. 3 Bourdon tube

Table 3. The results of the Bourdon tube

γ (deg)	μ	b (mm)	R/b	a/b	h/b	$\frac{\Delta\gamma}{\gamma}\frac{E}{P}$ The result of this paper	$\frac{\Delta\gamma}{\gamma}\frac{E}{P}$ The experimentale result of Reference [5]
241	0.28	3.05	17.2	3.40	0.125	7013	7200
241	0.28	3.17	16.5	3.28	0.168	3575	3800

Acknowlegement — The authors would like to thank the Science Foundation Committee of The Academy of Sciences of China for her financial support.

REFERENCES

1. Surana, K.S., 'Geometrically Nonlinear Formulation For the Curved Shell Element', Int. J. Num. Meth. Engng, Vol. 19(1983), 581-615.
2. Zhang, L. and Xia, Y.R., 'The Geometrically Nonlinear Analysis on the Curved Shell elements', 2nd Chinese National Conference on Calculating Mechanics', No. 83681, Aug. 1986.
3. Xiong, Y.X., 'Variational Principles of Finite Deformation of Elesticity in Nonconservative System', Acta Mechanica Sinica, Vol. 15, 1(1983), 86-90.
4. Argyris, J.H. and Symeonidis, Sp., 'Nonlinear Finite Element Analysis of Elastic Systems Under Nonconservative Loading-Natural Formulation, Part 1 Quastatic Problem', Comp. Meths. Appl. Mech. Engng., Vol. 26, 1(1981), 75-123.
5. Weng, S.C. and Lin, Y.D. etc.,- The Foundation on Designing Elastic Elements in Gauges, Beijing, 1982.

ON EXACT AND HIERACRHICAL FINITE ELEMENTS FOR FRAME STRUCTURES

Ragnar Larsson and Nils-Erik Wiberg

Dep. of Struc. Mech., Chalmers Univ. of Techn., S-412 96 Gothenburg, Sweden.

Summary

If the approximation exactly fulfills the differential equation the FE-method is said to be exact. A single element then represents a continuous part of the structure and only a few elements are sufficient for an accurate result. The paper gives a unified discussion of the exact FE-method for frame structures and also shows that the load term can be handled by use of one 'exact' hierarchical base function.

1 INTRODUCTION

The analysis of frame structures is commonly based on uncoupled 1-D boundary value problems to which exact solutions can be found for many important structure problems according to the theory of ordinary differential equations. In order to obtain an accurate and economic result the exact solutions, if available, should be incorporated in the FE analysis. Many papers concerning the exact FE analysis of frame structures have been written but the problem still continues to recieve attention. A recent paper by Banerjee and Williams[5)] describes an exact method for a range of tapered beams where Bessel functions were used to obtain the exact static stiffness matrix. The exact stiffness matrix for a beam on elastic foundation according to the first order theory has been given by Eisenberger and Yankelevsky[3)] . In Larsson and Wiberg[4)] the soil-structure interaction problem has been analysed with the aid of exact solutions for both the elements of the frame and of the foundation with special emphasis on the elastic stability of the entire structure.

There are many cases when the exact solutions are not available. For these cases the FE technique is well established. In Delpak and Peshham[1)] the hierarchical formulation has been investigated for parametric beam elements showing major improvements in the solution when hierarchical functions were added.

The aim of this paper is to formulate a unified solution technique for static elastic frames based on exact solutions and solutions obtained from a FE-method. It will be shown that the hierarchical formulation of the exact solution is equivalent to dividing the exact solution into a homogeneous and a particular part. The exact basis functions obtained from the homogeneous solution are the ordinary ones and the basis functions obtained from the particular solution are the hierarchical ones improving the solution due to the load.

2 BASIC EQUATIONS

Static-elastic analysis of 2-D or 3-D frame structures is commonly based on uncoupled differential equations regarded as 1-D boundary value problems defined in a bounded interval I with length L. The differential equations can be summarized with the aid of a differential operator L^2 for second order differential equations and with the operator L^4 for fourth order differential equations. Without loss of generality we consider homogeneous boundary conditions for the following boundary value problems:

$$\begin{cases} L_i^2 u = f_i \quad in \quad I \in (0,L), \quad i = 1,2 \\ u(0) = u(L) = 0 \end{cases} \tag{2.1a}$$

$$\begin{cases} L_i^4 u = f_i \quad in \quad I \in (0,L), \quad i = 3,4,5,6 \\ u(0) = u'(0) = u(L) = u'(L) = 0 \end{cases} \tag{2.1b}$$

The operators expanded as

$$L_i(\lambda_1, \lambda_2, \lambda_3) = \frac{d^2}{dx^2}\lambda_1\frac{d^2}{dx^2} - \frac{d}{dx}\lambda_2\frac{d}{dx} + \lambda_3, \quad L_i^2 = L_i(0, \lambda_2, \lambda_3), \quad L_i^4 = L_i(\lambda_1, \lambda_2, \lambda_3)$$

are defined in Table 1 together with their physical interpretations. The load terms f_i are interpreted as an appropriate load for each equation.

By introducing some basic concepts from Functional analysis the differential equation (2.1a,b) is recast into its variational form by multiplying with a test function w and then integrating over the length L of the interval I.

$$a(w,u) = F(w) \quad \forall w \in V \quad , F(w) = \int_0^L fwdx \tag{2.2}$$

where V is a Hilbert space, with the scalar product a(·,·) and the corresponding norm $||w||_V$, specially designed for the operators L_i; F(w) is a linear functional on V. The bilinear forms a(·,·) corresponding to each operator L_i are shown in Table 1.

The solution u can be proved to be unique provided that the bilinear form a(·,·) is V-elliptic, i.e. positive definite. The conditions on the stiffness coefficients λ_i that will ensure V-ellipticity of a(·,·) can roughly be estimated by the priori inequalities

$$\int_0^L w^2 dx \le L^2 \int_0^L w'^2 dx \le L^4 \int_0^L w''^2 dx \quad \forall w \in V_0 \tag{2.3}$$

The results obtained for the bilinear forms $a_4(\cdot,\cdot)$ are shown in Table 1. The result obtained for operator a (·,·) is in fact a crude estimate of the buckling load for a column with constant bending stiffness clamped at the base. It can also be observed that the result gives a limit of the error for the buckling load that can be obtained for any function on V.

Table 1. Differential operators and bilinear forms

Physical interpretation	Operators $L_i(\lambda_1,\lambda_2,\lambda_3) = \frac{d^2}{dx^2}\lambda_1\frac{d^2}{dx^2} - \frac{d}{dx}\lambda_2\frac{d}{dx} + \lambda_3$				Bilinear form $a_i(w,w) = \int_0^L (w''\lambda_1 w'' + w'\lambda_2 w' + w\lambda_3 w)\,dx$	Estimated stiffness coefficients $a(w,w) \geq m\,\|w\|_V^2 \quad \forall w \in V,\ m \geq 0$
	Type	λ_1	λ_2	λ_3	$a_i(\lambda_1,\lambda_2,\lambda_3)$	
1-D elasticity	L_1^2	0	EA	0	$a_1(0,\lambda_2,0)$	$\lambda_2 \geq 0$
Torsion	L_1^2	0	GJ	0	$a_1(0,\lambda_2,0)$	$\lambda_2 \geq 0$
1-D elasticity + load transfer	L_2^2	0	EA	S_s	$a_2(0,\lambda_2,\lambda_3)$	$\lambda_2 + \lambda_3 L^2 \geq 0$
Beam theory	L_3^4	EI	0	0	$a_3(\lambda_1,0,0)$	$\lambda_1 \geq 0$
Beam theory, 2nd order	L_4^4	EI	N	0	$a_4(\lambda_1,\lambda_2,0)$	$\lambda_1 + \lambda_2 2L^2 \geq 0,\ \lambda_1 \geq 0$
Torsion + prevented warping	L_4^4	EC_w	-GJ	0	$a_4(\lambda_1,\lambda_2,0)$	$\lambda_1 + \lambda_2 2L^2 \geq 0,\ \lambda_1 \geq 0$
Beam theory + load transfer	L_5^4	EI	0	S_w	$a_5(\lambda_1,0,\lambda_3)$	$\lambda_1 + \lambda_3 L^4 \geq 0$
Beam theory, 2nd order + load transfer	L_6^4	EI	N	S_w	$a_6(\lambda_1,\lambda_2,\lambda_3)$	$\lambda_1 + \lambda_2 L^2 + \lambda_3 L^4 \geq 0$

In Table 1 EI is the bending stiffness, EA is the axial stiffness, GJ is the torsional stiffness, EC_w is the warping stiffness, S_s and S_w are Winkler moduli and and N is the axial force (positive in tension).

3 EXACT SOLUTIONS

The exact solutions of the inhomogeneous boundary value problems (2.1a,b) can be obtained for non-homogeneous boundary conditions by dividing the total solution u into a homogeneous and a particular part as

$$u = u_H + u_P \tag{3.1}$$

The homogeneous part u_H satisfies (2.1a,b) with non-homogeneous boundary conditions as

$$\begin{cases} L_i^2 u_H = 0 \quad in \quad I \in (0,L), \quad i = 1,2 \\ u_H(0) = n_1, \quad u_H(L) = n_2 \end{cases} \tag{3.2a}$$

$$\begin{cases} L_i^4 u_H = 0 \quad in \quad I \in (0,L), \quad i = 3,4,5,6 \\ u_H(0) = t_1, \quad u'_H(0) = m_1, \quad u_H(L) = t_2, \quad u'_H(L) = m_2 \end{cases} \tag{3.2b}$$

The particular part satisfies (2.1a,b) with $u = u_P$.

The *homogeneous solution* of (3.2ab) can, for constant coefficients λ_i, be obtained by solving the characteristic equation arising when $u_H = e^{ri}$ is inserted or by simply direct integrating. The solution is of the form

$$u_H = \sum a_i e^{r_i x} = \Psi \tilde{a} \; ; \; i = 1,2 \;\; or \;\; i = 1, \dots, 4 \tag{3.3}$$

Where r_i are the roots of the characteristic equation. Finally, by incorporating the non-homogeneous boundary conditions the unknown coefficients $\tilde{a}$ can be determined for second and fourth order operators, respectively, as

$$C_n^2 \tilde{a} = \bar{n} \; ; \quad \tilde{a} = [a_1, a_2]^T, \quad \bar{n} = [n_1, n_2]^T \tag{3.4a}$$

$$C_n^4 \tilde{a} = \bar{n} \; ; \quad \tilde{a} = [a_1, a_2, a_3, a_4]^T, \quad \bar{n} = [t_1, m_1, t_2, m_2]^T \tag{3.4b}$$

When the coefficients are non-constant the concept of solutions based on the characteristic equation will fail. However, some classes of second order equations has solutions expressible, over a certain interval, in power series. This method becomes successful to some well known equations, e.g. the Bessel equation used by Banerjee and Williams[5] for the calculation of buckling loads of a range of tapered beams.

The *particular solution* u_P of (2.1ab) can always be reduced to find one solution u_{P2} provided that the homogeneous solution u_{P1} is known. For this purpose the method of undetermined coefficients can be used when the load term f_i has only a finite number of linearly independant derivates, e.g. when it is a polynom. The particular solution is then obtained as a sum $u = u_{P1} + u_{P2}$. Similarily as for (3.4a,b) the unknown coefficients $\tilde{a}$ can be determined by incorporating the homogeneous boundary conditions, which gives

$$C_n^2 \tilde{a} = \tilde{u}_{P2} \; ; \quad \tilde{u}_{P2} = [u_{P2}(0), u_{P2}(L)]^T \tag{3.5a}$$

$$C_n^4 \tilde{a} = \tilde{u}_{P2} \; ; \quad \tilde{u}_{P2} = [u_{P2}(0), u'_{P2}(0), u_{P2}(L), u'_{P2}(L)]^T \tag{3.5b}$$

If the particular solution is interpreted as a generalized function one may allow u_P be a weak solution. A weak solution can be allowed if the following relation holds:

$$(Lu_P, \phi) = (f, \phi) \tag{3.6}$$

where ϕ is an arbitrary test function belonging to a certain class of functions.

By insertion of u_H and u_P into (2,2) while considering non-homogeneous boundary conditions gives the corresponding relation between the element forces.

$$\bar{N} = \bar{N}_H + \bar{N}_P \tag{3.7}$$

Where N, N_H and N_P are vectors containing the end forces corresponding to the total solution, the homogeneous solution and the particular solution. The homogeneous part N can be expressed with the stiffness relation

$$\bar{S}^e \bar{n} = \bar{N}_H, \quad \bar{S}^e = C_N C_n^{-1} \tag{3.8}$$

where $n = n_H$ is a vector containing the end displacements and C_N is a matrix which determines the end forces N in terms of $\tilde{a}$ as

$$C_N \tilde{a} = \bar{N}_H \tag{3.9}$$

Finally, the *exact element stiffness relation* becomes as

$$\overline{S}^e \overline{n} = \overline{N} - \overline{N}_P \tag{3.10}$$

No distinction between the second and fourth order problems has been made here.

4 THE HIERARCHICAL FE-METHOD

The variational problem (2.2) is solved by construction of a finite dimensional subspace V_h to V. A subdivision of the interval I into subintervals, $I_j = (x_{j-1},x_j)$, $j = 1,\dots,n$ with the length $l_j = x_j - x_{j-1}$, is made. The basis functions are defined as a set of piecewise continuous functions on I so that a function $w \in V_h$ can be represented as a linear combination of the basis functions as

$$w = \sum w_k \varphi_k + \sum w_l \psi_l \tag{4.1}$$

where φ_k are the ordinary ones and ψ_l are the hierarchical ones. The solution can now be obtained as: Find $u_h \in V_h$ so that

$$a(w,u_h) = F(w) \quad \forall w \in V_h \tag{4.1}$$

The linearity of a(•,•) gives the equation system

$$\begin{bmatrix} A_{kk} & A_{kl} \\ A_{lk} & A_{ll} \end{bmatrix} \begin{bmatrix} \tilde{u}_k \\ \tilde{u}_l \end{bmatrix} = \begin{bmatrix} \tilde{b}_k \\ \tilde{b}_l \end{bmatrix}, \quad A\tilde{u} = \tilde{b} \tag{4.2}$$

The introduction of hierarchical functions thus gives a coupled partioned system. This approach was used by Jennings[2)] for approximate elastic stability analysis of plane frames. The similarity between the hierarchical FE (HFE) formulation and the exact solution technique, described in section 3, is apparent when the load term f_i is taken into account. In the HFE-formulation the solution is represented as a linear combination of basis functions of local character. We may, of course, also interpret the exact solution $u = u_H + u_P$ as a function represented by basis functions pertinent to the exact solution as

$$u = \sum u_k \varphi_k^e + \sum u_l \psi_l^e, \quad \sum u_k \varphi_k^e = \Psi C_n^{-1} \overline{n} \tag{4.3}$$

where φ^e are basis functions related to the homogeneous solution and ψ^e are basis functions related to the particular solution. Due to the fact that the solutions u_H and u_P are independant the element equations becomes uncoupled so (4.2) gets the form

$$\begin{bmatrix} A_{kk} & 0 \\ 0 & A_{ll} \end{bmatrix} \begin{bmatrix} \tilde{u}_k \\ \tilde{u}_l \end{bmatrix} = \begin{bmatrix} \tilde{b}_k \\ \tilde{b}_l \end{bmatrix} \tag{4.4}$$

In order to demonstrate the HFE-formulation when 'exact' basis functions are used we consider a uniform bar subjected to a distributed load of constant magnitude, case a, and a point load located at $x = L/2$, case b, as shown in Figure 4.1. In Figure 4.1 we also define the exact basis functions φ^e and ψ^e .

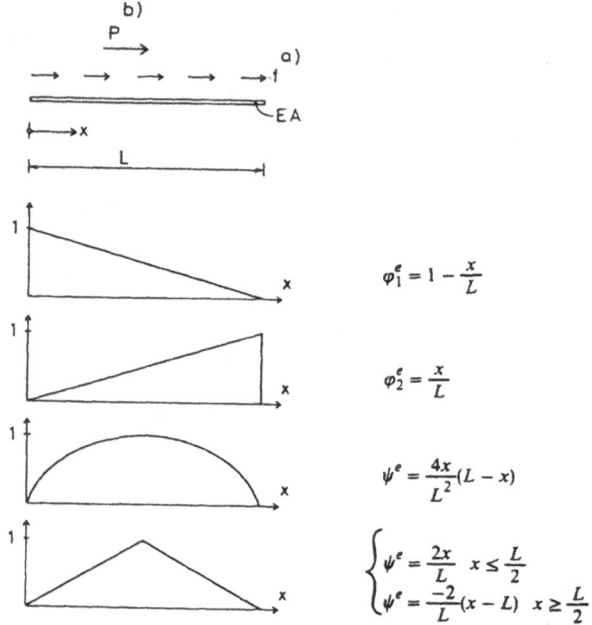

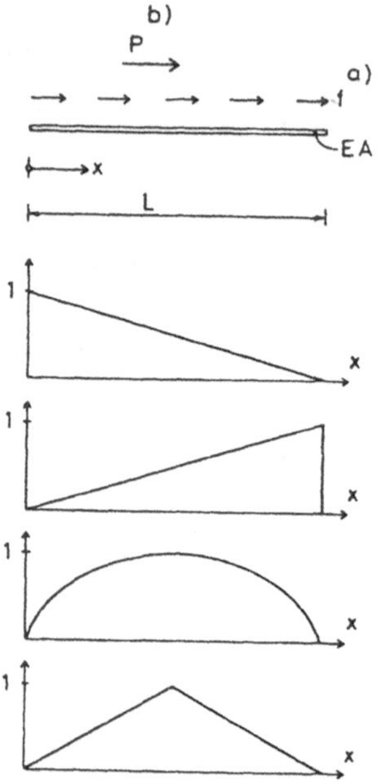

$$\varphi_1^e = 1 - \frac{x}{L}$$

$$\varphi_2^e = \frac{x}{L}$$

$$\psi^e = \frac{4x}{L^2}(L - x)$$

$$\begin{cases} \psi^e = \dfrac{2x}{L} & x \le \dfrac{L}{2} \\ \psi^e = \dfrac{-2}{L}(x - L) & x \ge \dfrac{L}{2} \end{cases}$$

Figure 4.1. Analysed uniform bar

Applying the HFE-formulation procedure for non-homogeneous boundary conditions to the problem in Figure 4.1 gives

$$\frac{EA}{L}\begin{bmatrix} 1 & -1 & 0 \\ -1 & 1 & 0 \\ 0 & 0 & 16/3 \end{bmatrix}\begin{bmatrix} u_1 \\ u_2 \\ u_p \end{bmatrix} = \begin{bmatrix} N_1 \\ N_2 \\ 0 \end{bmatrix} + \begin{bmatrix} 1/2 \\ 1/2 \\ 4/6 \end{bmatrix} fL \; ; \; \frac{EA}{L}\begin{bmatrix} 1 & -1 & 0 \\ -1 & 1 & 0 \\ 0 & 0 & 4 \end{bmatrix}\begin{bmatrix} u_1 \\ u_2 \\ u_p \end{bmatrix} = \begin{bmatrix} N_1 \\ N_2 \\ 0 \end{bmatrix} + \begin{bmatrix} 1/2 \\ 1/2 \\ 1 \end{bmatrix} P$$

which is equivalent to the element stiffness relation (4.4) for case a and b. It can be observed that the hierarchical base function ψ^e for both cases are orthonormal to the basis functions φ^e pertinent to the homogeneous solution. An interesting point is that case a is in fact equivalent to the p-version of the HFE-formulation since the hierarchical base function is of a higher order. But case b is equivalent to the hierarchical h-refinement technique as the same result would have been obtained if a hierarchical base function of the same order as the ordinary basis functions was added.

5 NUMERICAL EXAMPLE

The effectiveness in the exact solution technique is verified by the analysis of the beam on elastic foundation shown in Figure 5.1; the beam is analysed according to second order theory, by use of the computer program SITU-FRAME, Larsson et al.[6] . This problem has also been analysed, according to first order theory, by Eisenberger and Yankelevsky[3] . For the exact solution technique it is sufficient to use four elements to analyse the beam in order to take the discontinuities due to the load into account, see mesh A in Figure 5.1. The exact solution technique was compared to an approximate method where the Winkler stiffness S_w is lumped at distinct points. For the approximate method two different subdivisions of the beam, i.e. mesh B and C, was considered to study the convergence towards the exact solution.

Figure 5.2 shows the distribution of displacements, moments and shear forces for the exact and the approximate method according to second order theory. The calculations show a good agreement between the exact and the approximate method. It can be observed that for increasing order of differentiation of the solution the approximate method becomes more inexact. The example verifies the effectiveness in the exact solution technique as the exact solution was obtained with the aid of only four elements, i.e. mesh A.

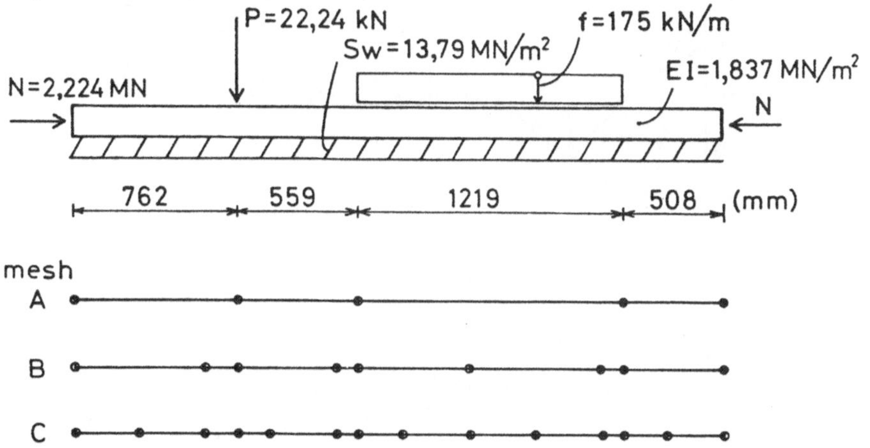

Figure 5.1. Analysed beam on elastic foundation

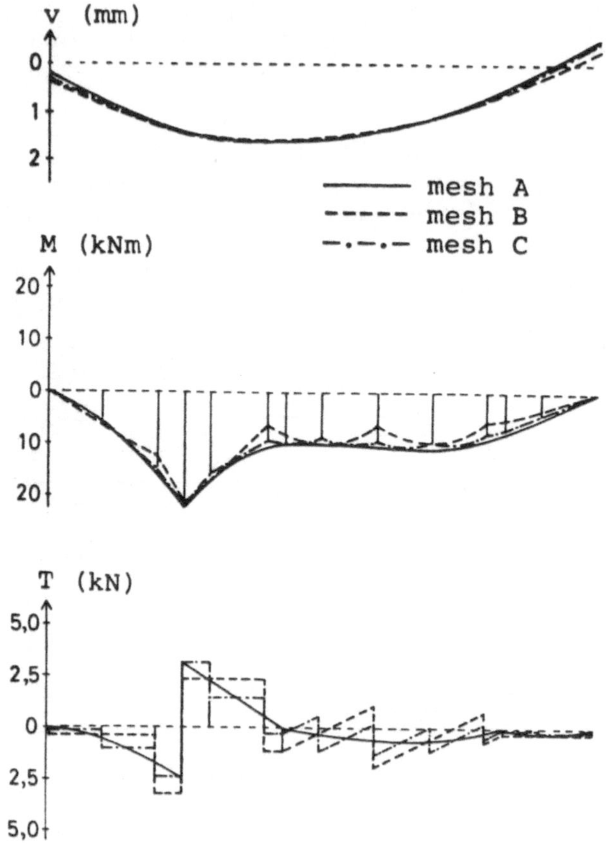

Figure 5.2. Resulting distribution of displacements, moments and shear forces.

6 CONCLUSIONS

It was demonstrated that the exact method is equivalent to the hierarchical finite element (HFE) method when the exact solution is divided into a homogeneous and a particular part.

The properties that will ensure a unique a solution was derived with reference to the uniqueness of a solution belonging to a certain Hilbert space, which is important when the HFE-method is included in an adaptive procedure.

In the numerical example the effectiveness in the exact method was verified.

7 REFERENCES

1. R. Delpak and V. Peshkam, 'A study of the influence of hierarchical nodes on the performance of selected parametric elements, Int. J. Num. Meth. Eng., 22, 153-171 (1986)

2. A. Jennings, 'The elastic stability of rigidly jointed frames', Int. J. Mech. Sci., 5, 99-113 (1963)

3. M. Eisenberger and D. Z. Yankelevsky, 'Exact stiffness matrix for beams on elastic foundation', Comp. and Struc., 21, No. 6, 1355-1359 (1985)

4. R. Larsson and N-.E. Wiberg, 'Stability of plane frames including soil structure interaction', Chalmers University of Techn., Dep. of Struc. Mech., Report 87:1, Sweden (1987)

5. J. R. Banerjee and F. W. Williams, 'Exact Bernoulli-Euler static stiffness matrix for a range of tapered beam-columns', Int. J. Num. Meth. Eng., 23, 1615-1628 (1986)

6. R. Larsson, N.-E. Wiberg, K. Runesson and H. Tägnfors, SITU-FRAME-E2 - A computer program for second order analysis of plane frames subjected to static load, Chalmers University of Techn., Dep. of Struc. Mech., Report 87:7, Sweden (1987)

STATICALLY AND KINEMATICALLY ADMISSIBLE FINITE ELEMENT FORMULATIONS FOR ELASTIC-PLASTIC PLATE ANALYSIS

J.P. Moitinho de Almeida , J.A. Teixeira de Freitas
Departamento de Engenharia Civil, Instituto Superior Técnico
1096 LISBOA PORTUGAL

1. SUMMARY

Hybrid formulations designed to generate either statically or kinematically admissible finite element solutions are presented. Consistency is ensured through Static-Kinematic Duality. The elastic-plastic analysis of a deep, simply supported beam is used to illustrate the performance of the formulations being suggested.

2. INTRODUCTION

In structural analysis, particularly in the case of plastic limit analysis, it is important to know how the formulation being adopted simulates the conditions for static and kinematic admissibility.

A solutions is said to be statically {kinematically} admissible if equilibrium {compatibility} and the yeld {flow} rule of plasticity hold throughout the domain and the static {kinematic} boundary conditions are locally satisfied.

In classical finite element formulations the displacement functions are so chosen as to guarantee interelement continuity through the assembly of nodes shared by adjacent elements. A class of finite element formulations designed to satisfy equilibrium locally has developed from the works of Veubeke [9]. Preference has been given to mixed and hybrid formulations [6, 3] which are easier to implement and tend to generate better solutions.

The formulation to be presented is hybrid in the sense that basic functions are used to discretize independently the displacement and the stress fields developing in the finite elements.

Weighted residuals equations are used to associate corresponding fields. Strains and displacements are weighted using the basic functions that describe the stress or the applied force fields. Conversely, the stresses and the applied forces are weighted using the basic functions that describe the strain or the displacement fields. The interelement boundary fields are also treated independently, using a similar procedure. This choice allows the user to select which interelement continuity conditions must be satisfied.

Formulations in which either the equilibrium or the compatibility conditions are partially relaxed can also be generated using this approach. They will in general give better approximations than the "pure" formulations, but can no longer be used to bound the solution.

As this formulation is independent of the element shape and basic functions, it can also be applied to perform hierarchical analyses.

3. EQUILIBRIUM AND COMPATIBILITY CONDITIONS

The exact solution of the problems to be solved consists of functions $\mathbf{u}$ and σ that must satisfy the following set of equations:

$$\epsilon = \mathbf{D}\,\mathbf{u} \quad \text{in } V\,, \tag{1}$$

$$\mathbf{D}^{*}\sigma + \bar{\mathbf{f}} = \mathbf{0} \quad \text{in } V\,, \tag{2}$$

$$\begin{aligned} \mathbf{u} &= \bar{\mathbf{u}} \quad \text{in } S_u\,, \\ \mathbf{N}\sigma &= \bar{\mathbf{t}} \quad \text{in } S_\sigma\,. \end{aligned} \tag{3}$$

In the compatibility condition (1), vector $\mathbf{u}$ defines the displacement field and ϵ the associated state of strain. Similarly, in the equilibrium condition (2) vector $\bar{\mathbf{f}}$ denotes the prescribed body forces associated with the state of stress σ. For planar problems these arrays can be organized as follows:

$$\mathbf{u} = \begin{bmatrix} u_x \\ u_y \end{bmatrix}, \quad \epsilon = \begin{bmatrix} \epsilon_{xx} \\ \epsilon_{yy} \\ 2\,\epsilon_{xy} \end{bmatrix}, \quad \sigma = \begin{bmatrix} \sigma_{xx} \\ \sigma_{yy} \\ \sigma_{xy} \end{bmatrix}, \quad \mathbf{f} = \begin{bmatrix} f_x \\ f_y \end{bmatrix}, \quad \mathbf{D} = \begin{bmatrix} \partial_x & 0 \\ 0 & \partial_y \\ \partial_y & \partial_x \end{bmatrix}$$

For first order problems $\mathbf{D}^{*} = \mathbf{D}^{T}$. According to the boundary conditions (3), $\bar{\mathbf{u}}$ and $\bar{\mathbf{t}}$ represent the displacement and traction fields prescribed at boundaries S_u and S_σ, respectively.

4. DEFINITION OF THE STRESS AND DISPLACEMENT FIELDS

The domain to be analised is divided in N_e subdomains. Due to this discretization an additional, interelement, boundary is be introduced in the problem. This boundary is termed $\mathbf{S}_b$.

In each element the displacement is defined as a linear combination of β_e linearly independent functions, continuous within the subdomain:

$$\mathbf{u} = \mathbf{U}\,\hat{\mathbf{u}}\,. \tag{4}$$

Similarly the stress field in each element is also defined as a linear combination of α_e linearly independent functions:

$$\sigma = \mathbf{S}\,\hat{\mathbf{s}}\,. \tag{5}$$

The dual variables of the displacement and stress parameters defined in (4) and (5), are:

$$\hat{\mathbf{e}} = \int_V \mathbf{S}^T \epsilon \, dV\,, \tag{6}$$

$$\hat{\bar{\mathbf{f}}} = \int_V \mathbf{U}^T \mathbf{f} \, dV. \tag{7}$$

Equations (1) and (2) can be written in a weighted residual form, using the functions described above, to yield:

$$\hat{\mathbf{e}} = \int_V \mathbf{S}^T \mathbf{D}\,\mathbf{U}\, dV\, \hat{\mathbf{u}}, \qquad \hat{\mathbf{e}} = \mathbf{A}\,\hat{\mathbf{u}}, \tag{8}$$

$$\int_V \mathbf{U}_T\,(\mathbf{D}^*\,\mathbf{S})\, dV\, \hat{\mathbf{s}} + \hat{\bar{\mathbf{f}}} = \mathbf{0}, \qquad \mathbf{B}\,\hat{\mathbf{s}} + \hat{\bar{\mathbf{f}}} = \mathbf{0}\,. \tag{9}$$

5. KINEMATIC MODEL

In the formulation presented herein interelement continuity is enforced by introducing auxiliary variables $\mathbf{r}$ which define the relative displacement at the boundaries,

$$\mathbf{r} = \sum_{i=1}^{N_e} \mathbf{M}_i\,\mathbf{u}_i \; - \mathbf{M}_u \bar{\mathbf{u}}\,, \text{ in } S_u, \qquad \mathbf{r} = \sum_{i=1}^{N_e} \mathbf{M}_i\,\mathbf{u}_i\,, \text{ in } S_b, \tag{10}$$

where $\mathbf{M}_i$ is the rotation matrix corresponding to the outward normal matrix of element i. Kinematic interelement continuity is enforced by setting $\mathbf{r} = \mathbf{0}$.

Associated with this variable there is an outward stress flow g, which can be described as a linear combination of basic functions $\mathbf{G}$:

$$\mathbf{g} = \mathbf{G}\,\hat{\mathbf{g}}\,, \qquad \text{in } S_u,\ S_b. \tag{11}$$

Equilibrium in the boundary is satisfied when

$$\begin{aligned} \mathbf{0} &= \mathbf{N}\,\sigma - \mathbf{M}^T \mathbf{g}, \quad \text{in } S_u,\ S_b \text{ and} \\ \mathbf{0} &= \mathbf{N}\,\sigma - \bar{\mathbf{t}}, \quad \text{in } S_\sigma. \end{aligned} \tag{12}$$

To obtain the kinematic model, equation (9) is integrated by parts, resulting in:

$$\int_S \mathbf{U}^T \mathbf{N}\,\mathbf{S}\, dS\, \hat{\mathbf{s}} - \int_V (\mathbf{D}\,\mathbf{U})^T\, \mathbf{S}\, dV\, \hat{\mathbf{s}} + \hat{\bar{\mathbf{f}}} = \mathbf{0}. \tag{13}$$

The boundary term $\mathbf{N}\,\mathbf{S}\,\hat{\mathbf{s}}$ is replaced by the interelement stress flow $\mathbf{g}$, or the applied tractions $\bar{\mathbf{p}}$ as defined in (12):

$$\int_S \mathbf{U}^T \mathbf{N}\,\mathbf{S}\,dS\,\hat{\mathbf{s}} = \int_{S_u,S_b} \mathbf{U}^T \mathbf{M}^T \mathbf{G}\,dS\,\hat{\mathbf{g}} + \int_{S_\sigma} \mathbf{U}^T \bar{\mathbf{p}}\,dS. \tag{14}$$

Then the equilibrium equation takes the form:

$$\begin{gathered}\int_{S_u,S_b} \mathbf{U}^T \mathbf{M}^T \mathbf{G}\,dS\,\hat{\mathbf{g}} + \int_{S_\sigma} \mathbf{U}^T \bar{\mathbf{p}}\,dS - \int_V (\mathbf{D}\,\mathbf{U})^T \mathbf{S}\,dV\,\hat{\mathbf{s}} + \hat{\bar{\mathbf{f}}} = 0\,, \\ \mathbf{C}_u\,\hat{\mathbf{g}}_u + \mathbf{C}_b\,\hat{\mathbf{g}}_b - \mathbf{A}^T\,\hat{\mathbf{s}} + \hat{\bar{\mathbf{t}}} + \hat{\bar{\mathbf{f}}} = 0.\end{gathered} \tag{15}$$

Interelement continuity and the kinematic boundary conditions are enforced by the weighted residual form of equations (10). The transpose of the stress flow shape functions are used as weights, to preserve static-kinematic duality [5].

$$\begin{gathered}\int_{S_u} \mathbf{G}^T \mathbf{M}\,\mathbf{U}\,dS\,\hat{\mathbf{u}} - \int_{S_u} \mathbf{G}^T \mathbf{M}\,\bar{\mathbf{u}}\,dS = \mathbf{0}\,, \quad \int_{S_b} \mathbf{G}^T \mathbf{M}\,\mathbf{U}\,dS\,\hat{\mathbf{u}} = \mathbf{0}\,, \\ \mathbf{C_u}^T\,\hat{\mathbf{u}} - \hat{\bar{\mathbf{r}}} = \mathbf{0}\,, \qquad \mathbf{C_b}^T\,\hat{\mathbf{u}} = \mathbf{0}\,.\end{gathered} \tag{16}$$

6. STATIC MODEL

Static interelement continuity is obtained by the introduction of boundary forces ϕ defined by:

$$\phi = \sum_{i=1}^{N_e} \mathbf{N}_i\,\sigma_i - \bar{\mathbf{p}}\,, \ \text{ in } S_\sigma, \qquad \phi = \sum_{i=1}^{N_e} \mathbf{N}_i\,\sigma\,, \ \text{ in } S_b. \tag{17}$$

If the force and stress fields are in equilibrium, and there are no interelement pressures, then $\phi = 0$.

A boundary displacement field $\mathbf{v}$ is associated with the auxiliary force ϕ. This field is also defined as a linear combination of basic functions $\mathbf{V}$:

$$\mathbf{v} = \mathbf{V}\,\hat{\mathbf{v}}\,, \ \text{ in } S_\sigma,\ S_b. \tag{18}$$

Compatibility at the boundary is satisfied when:

$$\begin{gathered}\mathbf{0} = \mathbf{u} - \mathbf{v}, \quad \text{ in } S_\sigma,\ S_b \text{ and} \\ \mathbf{0} = \mathbf{u} - \bar{\mathbf{u}} \quad \text{ in } S_u.\end{gathered} \tag{19}$$

To obtain the static model, equation (8) is integrated by parts, resulting in:

$$\hat{\mathbf{e}} = \int_S \mathbf{S}^T \mathbf{N}^T \mathbf{U}\,dS\,\hat{\mathbf{u}} - \int_V (\mathbf{D}^* \mathbf{S})^T \mathbf{U}\,dV\,\hat{\mathbf{u}} \tag{20}$$

The boundary term of the displacement $\mathbf{u}$ is replaced by the boundary displacement $\mathbf{v}$, as defined in (19):

$$\int_S \mathbf{S}^T \mathbf{N}^T \mathbf{U}\, dS\, \hat{\mathbf{u}} = \int_{S_\sigma, S_b} \mathbf{S}^T \mathbf{N}^T \mathbf{V}\, dS\, \hat{\mathbf{v}} + \int_{S_u} \mathbf{S}^T \mathbf{N}^T \bar{\mathbf{u}}\, dS\,. \tag{21}$$

Then the compatibility equation can be writen in the following form:

$$\hat{\mathbf{e}} = \int_{S_\sigma, S_b} \mathbf{S}^T \mathbf{N}^T \mathbf{V}\, dS\, \hat{\mathbf{v}} + \int_{S_u} \mathbf{S}^T \mathbf{N}^T \bar{\mathbf{u}}\, dS - \int_V (\mathbf{D}^* \mathbf{S})^T \mathbf{U}\, dV\, \hat{\mathbf{u}}\,,$$
$$\hat{\mathbf{e}} = \mathbf{D}_\sigma \hat{\mathbf{v}}_\sigma + \mathbf{D_b} \hat{\mathbf{v}}_b - \mathbf{B}^T \hat{\mathbf{u}} + \hat{\bar{\mathbf{u}}}\,. \tag{22}$$

Interelement equilibrium and the static boundary conditions are enforced by the weighted residual form of equations (17). The transpose of the boundary displacement shape functions are used as weights to yield:

$$\int_{S_\sigma} \mathbf{V}^T \mathbf{N}\,\mathbf{S}\, dS\, \hat{\mathbf{s}} - \int_{S_\sigma} \mathbf{V}^T \bar{\mathbf{t}}\, dS = \mathbf{0}\,, \qquad \int_{S_b} \mathbf{V}^T \mathbf{N}\,\mathbf{S}\, dS\, \hat{\mathbf{s}} = \mathbf{0}\,,$$
$$\mathbf{D}_\sigma{}^T \hat{\mathbf{s}} - \hat{\bar{\phi}} = \mathbf{0}\,, \qquad \mathbf{D_b}{}^T \hat{\mathbf{s}} = \mathbf{0}\,. \tag{23}$$

7. ELASTICITY RELATIONS

Let the elastic constitutive relations be writen in form:

$$\epsilon_E = F\,\sigma, \qquad \text{where} \qquad F = \frac{1}{E}\begin{bmatrix} 1 & -\nu & 0 \\ -\nu & 1 & 0 \\ 0 & 0 & 2(1+\nu) \end{bmatrix}, \tag{24}$$

for plane stress problems. The weighted residual form of this relation is:

$$\hat{\mathbf{e}}_E = \int_V \mathbf{S}^T F\, \mathbf{S}\, dV\, \hat{\mathbf{s}}\,, \quad \hat{\mathbf{e}}_E = \mathbf{F}\, \hat{\mathbf{s}}\,, \tag{25}$$

resulting in an unique relation between stresses and strains, as long as the stress basic functions are linearly independent. The reciprocal relation takes the form:

$$\hat{\mathbf{s}} = \mathbf{K}\, \hat{\mathbf{e}}_E, \qquad \mathbf{K} = \mathbf{F}^{-1}\,. \tag{26}$$

8. PLASTICITY RELATIONS

Maier's [4] matrix representation of plasticity is used herein. The following conditions must be satisfied locally:

$$\mathbf{y}_* = \sigma_* - \mathbf{N}_*^T \sigma\,, \quad \epsilon_P = \mathbf{N}_* \mathbf{p}\,, \quad \mathbf{y}_* \geq 0\,, \quad \mathbf{y}_*^T \mathbf{p} = 0\,, \quad \mathbf{p} \geq 0\,. \tag{27}$$

Matrix $\mathbf{N}_*$ collects the gradients of the plastic yield functions $\mathbf{y}_*$. The plastic capacities are collected in array σ_* and the plastic multiplier vector

$\mathbf{p}$ lists the contribution to ϵ_P of each yield mode associated with the yield function $\mathbf{y}_*$.

At finite element level the plastic parameters $\mathbf{p}$ are defined as a linear combination of basic functions $\mathbf{P}$, as proposed by Corradi [2]:

$$\mathbf{p} = \mathbf{P}\hat{\mathbf{p}} . \tag{28}$$

The weighted residual form of equations (27) is:

$$\begin{gathered}\int_V \mathbf{P}^T \mathbf{y}_* \, dV = \int_V \mathbf{P}^T \sigma_* - \int_V \mathbf{P}^T \mathbf{N}_*^T \mathbf{S} \, dV \, \hat{\mathbf{s}} , \\ \int_V \mathbf{S}^T \epsilon_P \, dV = \int_V \mathbf{S}^T \mathbf{N}_* \mathbf{P} \, dV \, \hat{\mathbf{p}} ,\end{gathered} \tag{29}$$

$$\hat{\mathbf{y}}_* = \hat{\sigma}_* - \mathbf{A}_*^T \hat{\mathbf{s}} , \quad \hat{\mathbf{e}}_P = \mathbf{A}_* \hat{\mathbf{p}} , \quad \hat{\mathbf{y}}_* \geq 0 , \quad \hat{\mathbf{y}}_*^T \mathbf{p} = 0 , \quad \hat{\mathbf{p}} \geq 0.$$

It is important to stress that these conditions do not guarantee that the plasticity conditions are satisfied locally. They represent average values on each element, and if the results are to be used to bound solutions they must be checked locally.

9. GOVERNING EQUATIONS

Two different sets of equations can be obtained for a displacement based model.

Using equations (8, 15, 16, 26, 29), the kinematic model is obtained:

$$\begin{bmatrix} \mathbf{A}^T\mathbf{K}\mathbf{A} & \mathbf{C}_u & \mathbf{C}_b & -\mathbf{A}^T\mathbf{K}\mathbf{A}_* & \mathbf{0} & -(\hat{\bar{\mathbf{f}}} + \hat{\bar{\mathbf{t}}}) \\ \mathbf{C}_u^T & \mathbf{0} & \mathbf{0} & \mathbf{0} & \mathbf{0} & \mathbf{0} \\ \mathbf{C}_b^T & \mathbf{0} & \mathbf{0} & \mathbf{0} & \mathbf{0} & \mathbf{0} \\ -\mathbf{A}_*^T\mathbf{K}\mathbf{A} & \mathbf{0} & \mathbf{0} & \mathbf{A}_*^T\mathbf{K}\mathbf{A}_* & -\mathbf{I} & \mathbf{0} \end{bmatrix} \begin{bmatrix} \hat{\mathbf{u}} \\ -\hat{\mathbf{g}}_u \\ -\hat{\mathbf{g}}_b \\ \hat{\mathbf{p}} \\ \hat{\mathbf{y}} \\ \lambda \end{bmatrix} = \begin{bmatrix} \mathbf{0} \\ \hat{\bar{\mathbf{r}}} \\ \mathbf{0} \\ -\hat{\sigma}_* \end{bmatrix}$$

$$\hat{\mathbf{u}}, \, \hat{\mathbf{g}}_u, \, \hat{\mathbf{g}}_b \gtrless 0, \qquad \hat{\mathbf{p}}, \, \hat{\mathbf{y}}_* \geq 0, \qquad \hat{\mathbf{y}}_*^T \hat{\mathbf{p}} = 0. \tag{30}$$

Using equations (9, 22, 23, 26, 29), and eliminating the stress components, the static model is obtained.

$$\begin{bmatrix} \mathbf{B}\mathbf{K}\mathbf{B}^T & -\mathbf{B}\mathbf{K}\mathbf{D}_\sigma & -\mathbf{B}\mathbf{K}\mathbf{D}_b & \mathbf{B}\mathbf{K}\mathbf{A}_* & \mathbf{0} & -\hat{\bar{\mathbf{f}}} \\ -\mathbf{D}_\sigma^T\mathbf{K}\mathbf{B}^T & \mathbf{D}_\sigma^T\mathbf{K}\mathbf{D}_\sigma & \mathbf{D}_\sigma^T\mathbf{K}\mathbf{D}_b & -\mathbf{D}_\sigma^T\mathbf{K}\mathbf{A}_* & \mathbf{0} & -\hat{\bar{\phi}} \\ -\mathbf{D}_b^T\mathbf{K}\mathbf{B}^T & \mathbf{D}_b^T\mathbf{K}\mathbf{D}_\sigma & \mathbf{D}_b^T\mathbf{K}\mathbf{D}_b & -\mathbf{D}_b^T\mathbf{K}\mathbf{A}_* & \mathbf{0} & \mathbf{0} \\ \mathbf{A}_*^T\mathbf{K}\mathbf{B}^T & -\mathbf{A}_*^T\mathbf{K}\mathbf{D}_\sigma & -\mathbf{A}_*^T\mathbf{K}\mathbf{D}_b & \mathbf{A}_*^T\mathbf{K}\mathbf{A}_* & -\mathbf{I} & \mathbf{0} \end{bmatrix} \begin{bmatrix} \hat{\mathbf{u}} \\ \hat{\mathbf{v}}_\sigma \\ \hat{\mathbf{v}}_b \\ \hat{\mathbf{p}} \\ \hat{\mathbf{y}}_* \\ \lambda \end{bmatrix} = \begin{bmatrix} \mathbf{B}\mathbf{K}\hat{\bar{\mathbf{u}}} \\ -\mathbf{D}_\sigma^T\mathbf{K}\hat{\bar{\mathbf{u}}} \\ -\mathbf{D}_b^T\mathbf{K}\hat{\bar{\mathbf{u}}} \\ \mathbf{A}_*^T\mathbf{K}\hat{\bar{\mathbf{u}}} - \hat{\sigma}_* \end{bmatrix}$$

$$\hat{\mathbf{u}}, \, \hat{\mathbf{v}}_\sigma, \, \hat{\mathbf{v}}_b \gtrless 0, \qquad \hat{\mathbf{p}}, \, \hat{\mathbf{y}}_* \geq 0, \qquad \hat{\mathbf{y}}_*^T \hat{\mathbf{p}} = 0. \tag{31}$$

Systems (30) and (31) can be solved using the Wolfe-Markowitz-Smith algorithm [8].

10. APPLICATIONS

The formulations presented allow the user to choose wich conditions are to be satisfied in the solution of a given problem.

In order to guarantee that the displacement field is continuous in the kinematic model, the order of functions $\mathbf{G}$ must be such that equation (16) implies local continuity.

In the static model the stress field is found to be in equilibrium at the interelement boundary when the order of function $\mathbf{V}$ satisfies a condition similar to that mentioned for the stress flow basic functions. To guarantee equilibrium within the elements, either the stress basic functions are such that equation (2) is automatically satisfied (in which case the order of $\mathbf{U}$ is irrelevant because $\mathbf{B} = \mathbf{0}$), or the order of $\mathbf{U}$ is such that equation (9) implies (2) locally.

The relation between the order of the displacement and the stress basic functions is also relevant.

In the kinematic model, the solution is not improved when the stress basic function is of an order higher than that which corresponds to the displacement function.

When the stress basic function is of lower order, the stiffness matrix is truncated (it may be interpreted as a form of reduced integration) and the solution cannot be considered kinematically admissible (though the displacements may be continuous).

The exact meaning of the relation between the order of the displacement and the stress basic functions is left for later discussion.

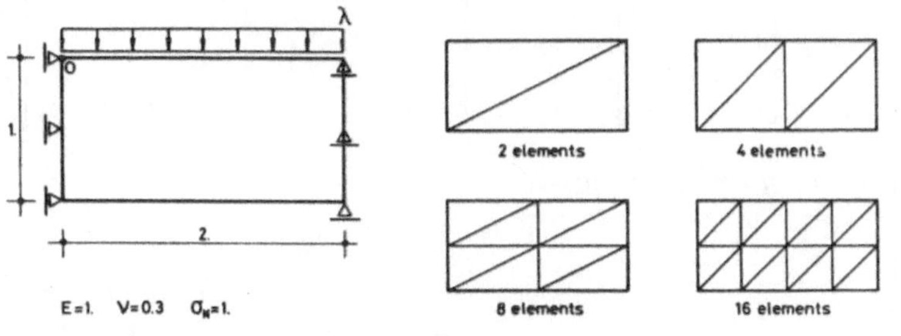

FIGURE 1

The problem studied to illustrate the method being suggested consists of a simply supported deep beam, shown in figure 1, studied by *Prager and Hodge* [7] using the beam theory, and by *Anand et al* [1] using the finite element method.

Four different meshes have been used, with both models. Typical load displacement diagrams for these meshes, using both the static and the kinematic models are shown in figure 2.

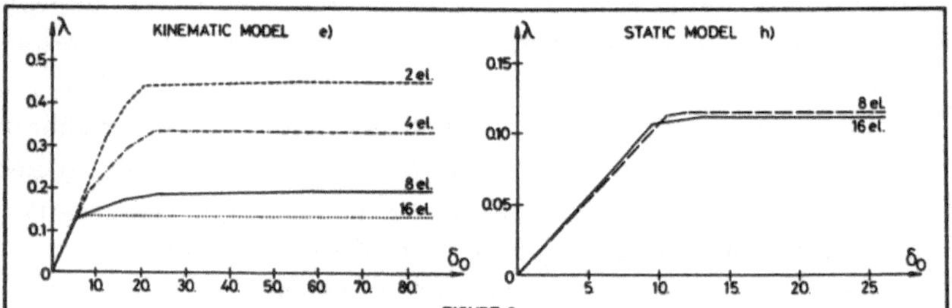

FIGURE 2

Let the elastic stiffness $\mathbf{k}$ be measured as the ratio between the load and the displacement of point O ($\mathbf{k}_{exact}$=0.02214). Given in the next tables, is the ratio $\mathbf{k}/\mathbf{k}_{exact}$ found for some combinations of the following basic functions:

$$\mathbf{U}_6 = \begin{bmatrix} 1 & x & y & & & \\ & & & 1 & x & y \end{bmatrix} \qquad \mathbf{U}_8 = \begin{bmatrix} 1 & x & y & xy & & & & \\ & & & & 1 & x & y & xy \end{bmatrix}$$

$$\mathbf{U}_{12} = \begin{bmatrix} 1 & x & y & x^2 & xy & y^2 & & & & & & \\ & & & & & & 1 & x & y & x^2 & xy & y^2 \end{bmatrix}$$

$$\mathbf{S}_3 = \begin{bmatrix} 1 & & \\ & 1 & \\ & & 1 \end{bmatrix} \qquad \mathbf{S}_9 = \begin{bmatrix} 1 & x & y & & & & & & \\ & & & 1 & x & y & & & \\ & & & & & & 1 & x & y \end{bmatrix}$$

$$\mathbf{S}_{18} = \begin{bmatrix} 1 & x & y & x^2 & xy & y^2 & & & & & & & & & & & & \\ & & & & & & 1 & x & y & x^2 & xy & y^2 & & & & & & \\ & & & & & & & & & & & & 1 & x & y & x^2 & xy & y^2 \end{bmatrix}$$

$$\mathbf{S}_{12} = \begin{bmatrix} 1 & & & x & & y & & x^2 & & xy & & y^2 \\ & 1 & & & x & & y & y^2 & x^2 & & xy & \\ & & 1 & -y & & & -x & -2xy & & -y^2/2 & -x^2/2 & \end{bmatrix}$$

$$\mathbf{G}_4 = \mathbf{V}_4 = \begin{bmatrix} 1 & r & & \\ & & 1 & r \end{bmatrix} \qquad \mathbf{G}_6 = \mathbf{V}_6 = \begin{bmatrix} 1 & r & r^2 & & & \\ & & & 1 & r & r^2 \end{bmatrix}$$

Unsafe and safe linearizations of the Tresca yield criterion are used respectively for the kinematic and static models. Matrix $\mathbf{P}$ is always constant.

Kinematic model - $\mathbf{k}/\mathbf{k}_{exact}$

	elements	2	4	8	16
a)	$\mathbf{U}_6\ \mathbf{S}_3\ \mathbf{G}_4$	5.48	3.24	2.44	1.74
b)	$\mathbf{U}_8\ \mathbf{S}_3\ \mathbf{G}_4$	–	–	0.84	0.81
c)	$\mathbf{U}_8\ \mathbf{S}_9\ \mathbf{G}_4$	2.08	1.26	1.40	1.10
d)	$\mathbf{U}_8\ \mathbf{S}_9\ \mathbf{G}_6$	2.92	1.37	1.44	1.11
e)	$\mathbf{U}_{12}\ \mathbf{S}_9\ \mathbf{G}_4$	0.75	0.88	0.98	0.99
f)	$\mathbf{U}_{12}\ \mathbf{S}_9\ \mathbf{G}_6$	1.15	1.04	1.01	1.00

Static model - $\mathbf{k}/\mathbf{k}_{exact}$

	elements	2	4	8	16
g)	$\mathbf{U}_6\ \mathbf{S}_9\ \mathbf{V}_4$	–	–	0.57	0.60
h)	$\mathbf{U}_{12}\ \mathbf{S}_9\ \mathbf{V}_4$	–	–	0.48	0.51
i)	$\mathbf{U}_6\ \mathbf{S}_{18}\ \mathbf{V}_4$	1.93	1.56	1.38	1.20
j)	$\mathbf{U}_6\ \mathbf{S}_{18}\ \mathbf{V}_6$	0.92	0.97	0.98	0.99
k)	$\mathbf{S}_{12}\ \mathbf{V}_6$	1.11	1.02	1.01	0.99

In the examples presented, some functions correspond to "exact" formulations and consequently give "true" bounds to the elastic stiffness, namely examples a), d) and f) for the kinematic model, and example h) for the static model. These elastic solutions can be considered respectively kinematically and statically admissible.

The plastic limit load obtained for the kinematic model is also kinematically admissible, and consequently corresponds to an upper bound to the plastic colapse load. This is not true for the static model for the reasons given in section (8). To obtain a lower bound the most negative potential $\mathbf{y}_*$ should be calculated using equation (27.a) and all results scaled so that (27.c) is satisfied locally.

ACKNOWLEDGMENTS

This research has been sponsored by the National (Portuguese) Institute of Scientific Research (INIC) through the Mechanics and Structural Engineering Centre (CMEST), Technical University of Lisbon.

REFERENCES

1. Anand, S.C., Lee, S.L. and Rossow, E.C. 'Finite Element Analysis of Elastic-Plastic Plane Stress Problems Based upon Tresca Yeld Criterion', Ingenieur Archiv **39**(1970), 73-86.

2. Corradi, L., 'A Displacement Formulation for the Finite Element Elastic-Plastic Problem', Meccanica **18**(1983), 77-91.

3. Atlury, S.N., *et al* (editors), *Hybrid and Mixed Finite Element Methods*, J. Willey & Sons, 1983.

4. Maier, G., 'Linear Flow Laws of Elastoplasticity: A Unified General Approach', Rc. Accad. Naz-Lincei **142A**(1968).

5. Munro, J. and Smith, D.L., 'Linear Programming Duality in Plastic Analysis and Synthesis', Int. Symp. Computer-Aided Structural Design, Conventry (1972).

6. Pian, T.H.H. 'Derivation of Element Stiffness Matrices by Assumed Stress Distributions', A.I.A.A. Journal **2**(1964), 1333-1335

7. Prager, W. and Hodge, P.G., *Theory of Perfectly Plastic Solids*, J. Willey & Sons, New York, 1951.

8. Smith, D.L. 'The Wolfe-Markowitz algorithm for nonholonomic elastoplastic analysis', Eng. Structures **1**(1978), 8-16.

9. de Veubeke, B.M. Fraeijs, 'Upper an Lower Bounds in Matrix Structural Analysis', AGARDograph, **72**(1964), Pergamon Press, 165-201.

TURBULENCE MODELLING AND THE EFFECTS OF DIRECTIONAL RANDOM WAVES IN COMPUTATIONS OF NEARSHORE CIRCULATION.

K. Anastasiou, P. Dong, D.J. Walker.
Department of Civil Engineering,
Imperial College of Science and Technology,
London SW7 2BU, U.K.

SUMMARY

The computation of nearshore, wave induced, circulation is currently receiving considerable attention owing to its importance in studies of sediment and pollutant transport in coastal environments. However, the modelling of turbulence in the surf zone is not generally as advanced as that found in other branches of applied fluid mechanics. Moreover, the randomness and directionality of the wave motion is not usually taken into account, the solutions corresponding to monochromatic and uni-directional waves. This paper implements three models of turbulence in the surf zone and presents results which show the degree to which the computed features of the flow depend on the type of turbulence model used. Similar comparisons are made with regard to the distribution of the alongshore current over a plane beach. By using linear superposition the random and three-dimensional character of the waves is also taken into account, and it is demonstrated that the computed flow fields differ significantly from those corresponding to monochromatic waves.

1. INTRODUCTION

The computation of nearshore, wave induced, circulation is normally accomplished in two stages. The wave field is first calculated and then the equations of motion are solved. Certain terms in the equations of motion, the radiation stresses, are due solely to the presence of the waves and constitute the link between the two stages. The whole cycle can be repeated a number of times if the interaction between the waves and the mean motion is to be taken into account. Owing to the computational difficulties associated with a full three-dimensional model of the flow, the established procedure is for the solution to be concerned with the two-dimensional, depth integrated and time averaged equations of motion.

The computation of the wave field is nowadays a routine task and a wide range of models are available which describe wave propagation in areas of varying bathymetry and in the presence of obstacles. Solutions based on the 'mild slope' equation [5,6,8,14] and the Boussinesq equations [1] hold a prominent position. Solutions are generally related to monochromatic waves and the combined effects of refraction, diffraction, reflection, wave breaking and energy dissipation can be taken into account.

Some effort has also been expended in an attempt to take into account the randomness and directionality of the waves [2,3,4,7]. Such solutions consider the combined effects of some of the above phenomena but not the complete ensemble. A major source of difficulty is the correct modelling of wave energy dissipation, through wave breaking, within the surf zone which is very important for the generation and subsequent behaviour of the mean currents.

Modelling of turbulence in the surf zone using one and two equation models has been reported recently [16,17]. However, a full study of the effects of using different turbulence models on the circulation pattern has not yet been undertaken. It is the main purpose of this paper to explore further the role and nature of turbulence in computations of nearshore circulation. Results from zero, one and two equation models are presented for typical coastal configurations, including a system of groynes. The structure of the alongshore current distribution is also investigated as influenced by turbulence and mixing induced by wave breaking. The circulation model is based on a Alternating Direction Implicit (ADI) finite-difference scheme.

2. THEORY

2.1. The wave field

In the present work the wave field is described by the mild slope equation

$$\nabla(G\ \nabla\Phi) + k^2 G\ \Phi = 0 \tag{1}$$

with

$$G = CC_g\ /\ g \tag{2}$$

where Φ is the wave potential at the still water level, C, C_g and k are wave celerity, group velocity and wave number as determined by linear wave theory. It has been shown [8,9] that Eq. (1) can be recast as a set of first order equations of the form

$$\nabla Q + \frac{C_g}{C}\ \frac{\partial \zeta}{\partial t} = 0 \tag{3}$$

and

$$\frac{\partial Q}{\partial t} + CC_g\ \nabla\zeta = 0 \tag{4}$$

where ζ is the free surface elevation and **Q** is a dummy vector variable defined by

$$Q = G\ \nabla\Phi \tag{5}$$

The computation of the wave field will not be dealt with in detail here. It should be mentioned that care is needed when obstacles are present in the flow so that reflections are allowed to travel out of the model. Details may be found in [2,8,9].

In the case of a random wave field which may be uni-directional or multi-directional, a number of components are present, each being associated with a phase which is usually chosen to be uniformly distributed between 0 and 2π. Eqs. (3) and (4) are equally valid for a random wave field as long as it is understood that they must be applied to each individual component in turn and, for each time step and point over the whole solution domain, values from all components are summed. As was demonstrated by Battjes [3], the radiation stresses associated with such a field can also be calculated by applying the principle of linear superposition. Such a scheme is likely to incur a heavy computational penalty but the availability nowadays of cheap computer power offsets the run time penalties involved. However, care is needed when accounting for wave breaking in shallow water. In the case of monochromatic waves it is normally checked whether

$$|\zeta| < \zeta_b \quad , \quad \text{where} \quad \zeta_b = \gamma\, h, \quad \gamma \cong 0.4 \tag{6}$$

and h is the local water depth. If $|\zeta| > \zeta_b$ then $|\zeta|$ is set equal to ζ_b and the calculation advances to the next point. Implicit in this procedure is the assumption that the waves break as spilling breakers, thus maintaining their sinusoidal form. When N components are present, where $N \geq 1$, the total instantaneous contribution to the local variance is

$$\zeta^2 = \sum_{i=1}^{N} \zeta_i^2 \tag{7}$$

In order to ensure that energy is conserved (after accounting for loss due to breaking) it is necessary to check ζ^2 against the local limit ζ_b^2. If $\zeta^2 > \zeta_b^2$, then

$$\zeta_i = \zeta_i \sqrt{(\zeta_b^2 / \zeta^2)} \tag{8}$$

It is understood, of course, that this technique does not take into account transfer of energy between components at different frequencies.

2.2. The current field

The depth integrated and time averaged equations of motion for the present problem are

$$\frac{\partial \bar{\zeta}}{\partial t} + \frac{\partial}{\partial x}(Uh) + \frac{\partial}{\partial y}(Vh) = 0 \tag{9}$$

$$\frac{\partial}{\partial t}(Uh) + \frac{\partial}{\partial x}(U^2h) + \frac{\partial}{\partial y}(UVh) = -\frac{1}{\rho}\left[\frac{\partial S_{xx}}{\partial x} + \frac{\partial S_{xy}}{\partial y} + \rho g h \frac{\partial \bar{\zeta}}{\partial x}\right] + \frac{1}{\rho}\left[\frac{\partial T_{xx}}{\partial x} + \frac{\partial T_{xy}}{\partial y}\right] - \frac{1}{\rho}\overline{\tau_x^b} \tag{10}$$

In eqns. (9) and (10) U, V are the depth-averaged mean current velocities in the alongshore and offshore directions respectively, S_{xx}, S_{yy} and S_{xy} are the radiation stresses, and variables with an overbar denote time mean values. The depth integrated Reynolds stresses T_{ij} and the shear stress $\tau^{b}{}_{i}$ at the sea bed play very important role in this work, and for this reason they will be examined in detail. Using the concept of eddy viscosity, the general expression for T_{ij} can be written as

$$T_{ij} = \nu_t h \left(\frac{\partial U_i}{\partial x_j} + \frac{\partial U_j}{\partial x_i} \right) - \frac{2}{3} h\, k\, \delta_{ij} \tag{11}$$

where ν_t is the depth averaged eddy viscosity, k is the turbulent kinetic energy and δ_{ij} is the Kronecker delta. In order to determine ν_t, the commonly adopted method is to use a zero equation model of the form

$$\nu_t = K\, u\, l \tag{12}$$

where u and l are characteristic velocity and length scales, and K is an empirical coefficient [12].

This type of turbulence model has been used extensively in a wide range of flow problems. It has been generally found that it is quite good for flows without strong recirculatory features [15]. The one and two equation models are to be preferred for cases where there is strong recirculation.

In the one equation model the turbulent kinetic energy k is used as the characteristic velocity scale and is found by solving the depth averaged transport equation, including the effects of wave breaking. This equation can be written as

$$\frac{\partial kh}{\partial t} + U \frac{\partial kh}{\partial x} + V \frac{\partial kh}{\partial y} = \frac{\partial}{\partial x}\left[\frac{\nu_t}{\sigma_k} h \frac{\partial k}{\partial x}\right] + \frac{\partial}{\partial y}\left[\frac{\nu_t}{\sigma_k} h \frac{\partial k}{\partial y}\right]$$
$$+ P_h + \frac{D_w}{\rho} - c_D \frac{k^{3/2}}{L} h \tag{13}$$

with turbulence production by shear P_h and eddy viscosity ν_t as

$$P_h = \nu_t h\left[2 \left[\frac{\partial U}{\partial x}\right]^2 + 2 \left[\frac{\partial V}{\partial y}\right]^2 + \left[\frac{\partial U}{\partial y} + \frac{\partial V}{\partial x}\right]^2 \right] \tag{14}$$

$$\nu_t = c\mu \sqrt{k}\, L \tag{15}$$

where D_w is calculated from the wave energy equation

$$\frac{\partial E}{\partial t} + \frac{\partial}{\partial x_i} \{E(U_i + C_{gi})\} + S_{ij} \frac{\partial U_j}{\partial x_i} = - D_w \tag{16}$$

The more complicated two-equation model involves solving for both the turbulent kinetic energy k and its rate of dissipation ϵ. The model equations used are [10]

$$\frac{\partial kh}{\partial t} + U\frac{\partial kh}{\partial x} + V\frac{\partial kh}{\partial y} = \frac{\partial}{\partial x}\left[\frac{\nu_t}{\sigma_k}h\frac{\partial k}{\partial x}\right] + \frac{\partial}{\partial y}\left[\frac{\nu_t}{\sigma_k}h\frac{\partial k}{\partial y}\right] + P_h + \frac{D_w}{\rho} - \epsilon h \qquad (17)$$

$$\frac{\partial \epsilon h}{\partial t} + U\frac{\partial \epsilon h}{\partial x} + V\frac{\partial \epsilon h}{\partial y} = \frac{\partial}{\partial x}\left[\frac{\nu_t}{\sigma_\epsilon}h\frac{\partial \epsilon}{\partial x}\right] + \frac{\partial}{\partial y}\left[\frac{\nu_t}{\sigma_\epsilon}h\frac{\partial \epsilon}{\partial y}\right] + c_{1\epsilon}\frac{\epsilon}{k}P_h + P_{\epsilon w} - c_{2\epsilon}\frac{\epsilon^2}{k}h \qquad (18)$$

$$\nu_t = c_\mu k^2/\epsilon \qquad (19)$$

in which σ_k, σ_ϵ, $c_{1\epsilon}$, $c_{2\epsilon}$ are standard empirical coefficients[15] and c_μ was taken as 2.5 [10]. $P_{\epsilon w}$ is an extra term due to wave breaking which presently is modelled by

$$P_{\epsilon w} = c_B\ C/h\ (D_w/\rho) \qquad (20)$$

where c_B is an empirical coefficient and can be related to $c_{2\epsilon}$ and c_μ [10]. Appropriate boundary conditions must be imposed for this type of problem. At both offshore and shoreline boundaries small values for k and ϵ are specified. At lateral boundaries either periodic or zero gradient conditions are used. Neither is strictly applicable but constitutes, nevertheless, an acceptable approximation as long as the model boundaries are chosen sufficiently far away from the area of interest.

With reference to the bottom friction term the following representations are used:

$$\overline{\tau b}_x = \frac{4\rho}{\pi} C_f\ u_{max}\ U \qquad (21)$$

$$\overline{\tau b}_y = \frac{2\rho}{\pi} C_f\ u_{max}\ V \qquad (22)$$

where C_f is the bottom friction coefficient and u_{max} is the maximum wave orbital velocity at the bottom.

Bottom friction poses a formidable problem. Eqs. (21) and (22) assume that the mean current is weak compared with the wave orbital velocity and propagates at almost 90^0 degrees to the waves. Random, directional waves do not satisfy this requirement. It seems reasonable to suggest that the formulation adopted by Ebersole and Darlymple [11] which does not make any explicit assumptions about the relationship between the waves and the mean flow might be more appropriate. However, the results presented herein are related to small directional spreading of the wave energy and, given the considerable computational effort required to implement this general formulation, it has been decided to retain the simple formulation as described in [12].

3. RESULTS AND DISCUSSION

Figures 1–4 show some results from the numerical tests. The beach slope was taken as 0.025 and the length and spacing between groynes was 100 meters. In the case of monochromatic waves a wave period of 10 sec and wave height of 1.50 m were used. For the multi-component case the period varied between 8 – 12 secs and 5 components with a directional spread of 20° were considered. The total wave energy was maintained equal to that in the monochromatic wave case.

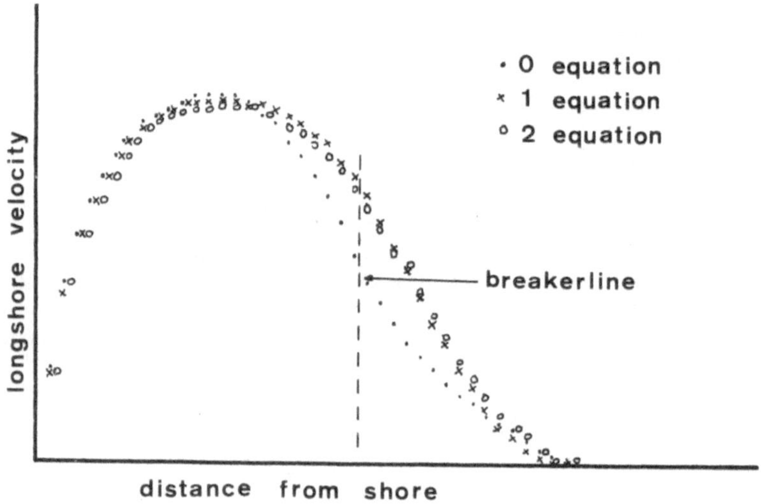

Fig.1 Longshore current distribution on a plane beach from zero, one and two equation turbulence models

Inspection of Fig. 1 indicates that in the case of a plane beach the three turbulence models produce almost identical flow patterns over most of the surf zone. This suggests that Longuet-Higgins' formulation [12] for the eddy viscosity gives satisfactory answers and constitutes a very reasonable model for turbulence and mixing in the surf zone for simple bottom geometries. Differences in the offshore region are due to the unrealistic increase of eddy viscosity outside the breaker line in the zero equation model.

The results shown in Figures 2–4 constitute a more severe test of the three turbulence models. Here, owing to the more complicated geometry and the importance of the turbulence convection and diffusion it is natural to expect that the one and two equation models would be able to predict more accurately the features of the mean flow and, especially, the character of turbulent mixing which is important in sediment and pollutant transport studies. Inspection of Figures 2 and 3 reveals that although the free longshore currents from both models are identical, slightly smaller and weaker circulations are predicted by the two equation model. Also the rip currents from two equation model change more gradually in both amplitude and direction. The reason for the lack of pronounced difference is that the variation in eddy viscosities obtained from the zero and two equation models do not affect the mean flow significantly. However suspended sediment transport is much more sensitive to variations in eddy viscosity [13], and it can be expected that differences between the two models would be pronounced in such applications.

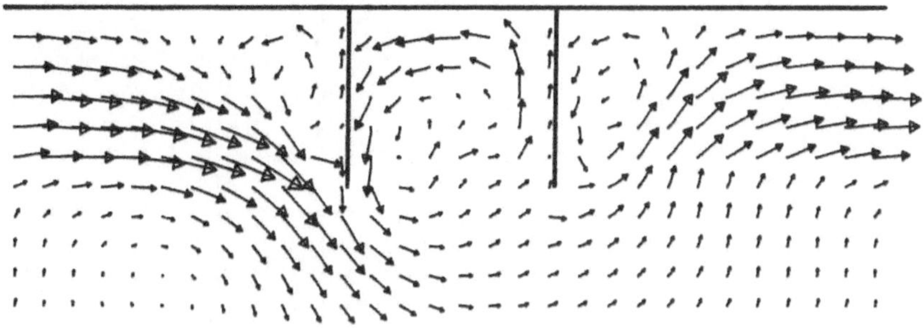

Fig.2 Circulation due to monochromatic waves with zero-equation turbulence model

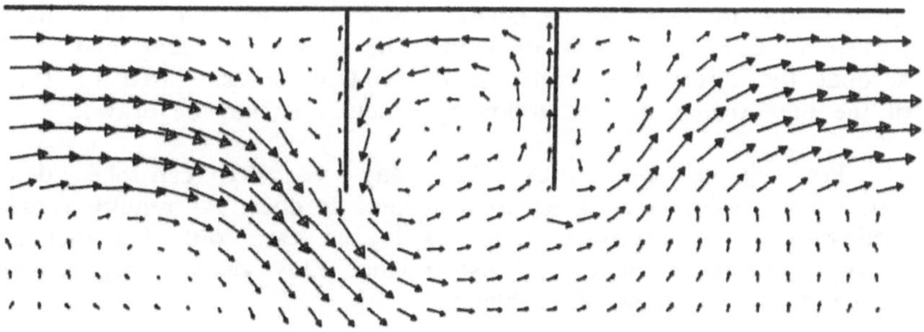

Fig.3 Circulation due to monochromatic waves with two-equation turbulence model

Figure 4 shows the circulation pattern resulting from the directional waves. Marked differences are evident between these results and those corresponding to monochromatic waves (Fig. 2). In the present case the computed flow features are weaker which may be partly attributed to the spreading of the incoming wave energy and partly to the larger magnitude of the bottom shear stress terms.

The main obstacle in properly analysing and comparing results of this nature is the lack of reliable experimental data on the surf zone flow patterns. More full scale information is required on the characteristics of turbulence and the dependence of the mean currents on wave directionality.

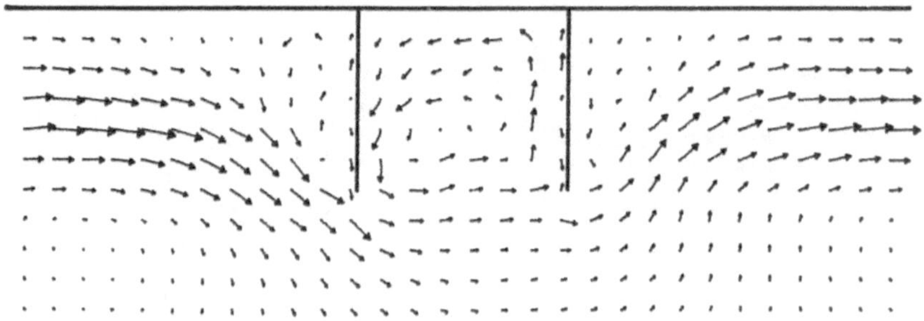

Fig.4 Circulation due to directional random waves with zero–equation turbulence model

4. CONCLUSIONS

Computations of nearshore, wave induced, circulation have been performed using a number of models of turbulent mixing and taking into account the directional properties of random waves. Based on the results of the numerical tests, the following conclusions may be drawn

a) For simple geometries the eddy viscosity can be determined satisfactorily using a zero equation model. It does not require sophisticated modelling of characteristic scales as is the case with the one and two equation models. Longuet–Higgins' original eddy viscosity representation is sufficient for this class of problems.

b) Regions with strong recirculation caused by coastal structures or complex sea bed topography require one or two equation models in order to achieve detailed description of eddy viscosity.

c) The directional properties of full scale waves are important for the circulation pattern. The computed flow fields are generally not as strong as in the case when this factor is not taken into account.

d) The computational effort required to include these directional effects is considerable. However this drawback is expected to become less significant as computers become more powerful.

5. REFERENCES

1. Abbott, M.B., Petersen, H.M. and Skovgaard, O., ' On the numerical modelling of short waves in shallow water', Jnl. Hyd. Res., Vol. 16, No.2 (1978), 173–203.
2. Anastasiou, K., 'Nearshore, random wave induced circulation', Proc. 3rd Indian Conf. on Ocean Engng, Vol A, A203–A214, 1986.
3. Battjes, J.A., 'Computation of set–up, longshore currents, run–up and overtopping due to wind generated waves', Communications in Hydraulics, Report No. 74–2, Dept. of Civil Engng., Delft University of Technology, 1974.
4. Battjes, J.A. and Jansen, J.P.F.M., 'Energy loss and set–up due to breaking of random waves', Proc. 16th Coastal Engng. Conf., 569–487, 1978.
5. Berkhoff, J.C.W., 'Computation of combined refraction and diffraction', Proc. 13th Coastal Engng Conf., Vol I, 471–490, 1972.
6. Bettess, P. and Zienkiewicz, O.C. 'Diffraction and refraction of surface waves using finite and infinite elements', Int. Jnl. for Num. Methods in Engng, Vol. II (1977), 1271–1290.
7. Booij, N., Holthuisjen, L.H. and Herbers, T.H.C., 'A numerical model for wave boundary conditions in port design', Proc. Int. Conf. Num. Hyd. Modelling of Ports and Harbours, 263–269, 1985.
8. Copeland, G.J.M., 'A practical alternative to the mild slope equation', Coastal Engng 9 (1985), 125–149
9. Copeland, G.J.M., 'Practical radiation stress calculations connected with equations of wave propagation', Coastal Engng 9 (1985), 195–219.
10. Dong, P., Walker, D.J. and Anastasiou, K., 'The modelling of turbulent mixing in the surf zone', In preparation.
11. Ebersole, B.A. and Darlymple, R.A., 'Numerical modelling of nearshore circulation', Proc. 17th Coastal Engng Conf., Chapter 163, 1980.
12. Longuet–Higgins, M. S., 'Longshore currents generated by obliquely incident sea waves,' Jnl. of Geophy. Res., Vol. 75 (1970), 6778–6801.
13. Nielsen, P. 'Some Basic Concepts of Wave Sediment Transport' Paper No. 20 ISVA, Tech Univ Denmark.(1979)
14. Radder, A.C., 'On the parabolic equation method for water wave propagation', Jnl. Fluid Mech. 95 (1979), 159–176.
15. Rodi, W., 'Turbulence models and their application in Hydraulics', Int. Assoc. for Hyd. Res., Delft, The Netherlands, 1980.
16. Visser, P. J., 'A mathematical model of uniform longshore currents and the comparison with laboratory data', Report no. 84–2, Department of Civil Engineering, Delft University of Technology, 1984.
17. Wind, H. G. and Vreugdenhil, C. B., 'Rip–current generation near structures', J. Fluid Mech., Vol. 171. (1986), 459–476.

ANALYSIS OF GRAVITY DAM ON SOFT FOUNDATION

V. Gocevski

Hatch and Associates
Montreal, Canada H3B 1S6

O.A. Pekau

Concordia University
Montreal, Canada H3G 1M8

1. SUMMARY

This paper reports the analysis of a gravity dam founded on a soft subsoil and subjected to loading phases consisting of the construction stages and subsequent fluctuation of the reservoir water level. An elasto-plastic constitutive model is employed for the soil, incorporated in the MIXDYN computer program. Examined is the importance of the accumulation of plastic deformation in the soil on the movement of the dam and the stresses in the dam and in the supporting medium.

2. INTRODUCTION

Gravity dams founded on "soft" subsoils such as weak rock or cemented granular deposits possessing poor shear strength characteristics present a major design problem. Such foundations have been encountered in many parts of the world [1] and simple force equilibrium analysis [1, 2] as well as finite element and photoelastic methods [3] have been employed in the past for evaluation of the general behaviour of dams under these conditions. However, the importance of permanent (plastic) deformations of the weak subsoil under monotonic, and even more so under cyclic, loading conditions has not been examined adequately to date.

Reported in this paper is the behaviour of a concrete gravity dam located on a cemented alluvial sand and gravel foundation. The principal parameters examined in this study are the effects of the construction process and cyclic loading due to fluctuation of the retained water level.

3. CONSTITUTIVE SOIL MODEL

The above problem is examined in this paper by applying a

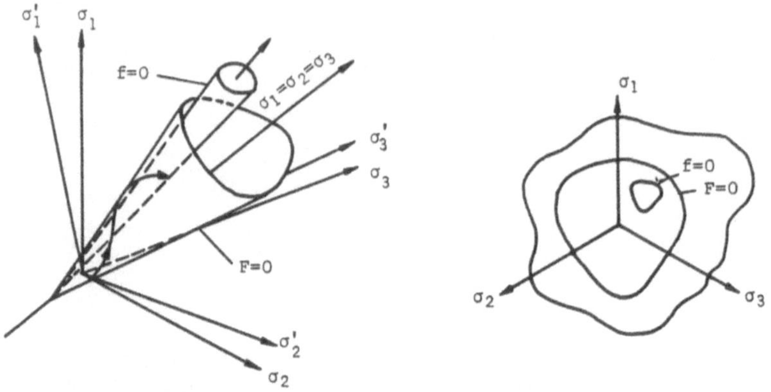

Fig. 1 Yield (f=0) and bounding surface (F=0) in principal stress space

recently developed constitutive model for granular soil with some cohesion [4]. The model is based on the theory of bounding surface plasticity (see Fig. 1) incorporating a non-associated flow rule and the concept of reflected plastic potential. The formulation constitutes an extension of the work by Poorooshasb and Pietruszczak [5] and differs from it in a number of aspects; for example, the forms of the yield surface and plastic potential, the mode of accumulation of plastic distortions and the generalized form of the local plastic potential. Also, it is more detailed as it provides the explicit analytical expressions for the gradient tensors and the location of the conjugate stress point, which are required for implementation in a computer code. A quadrilateral finite element incorporating the constitutive model was added to the existing element library of the program MIXDYN [6].

4. DETAILS OF DAM AND ANALYTICAL PROCEDURE

The configuration of the prototype concrete gravity dam on a soft foundation is illustrated in Fig. 2(a). The width of the dam is 50 m. To prevent sliding a 10 m deep shear key is incorporated at the base of the dam. If required a water barrier and a drainage system may be installed under the dam.

The following properties of the foundation medium, typical for cemented granular deposits, were assumed: Young's modulus $E = 335{,}000\ kN/m^2$; Poisson's ratio $\nu = 0.25$, frictional angle $\phi = 47.5^o$, cohesion $c = 20\ kN/m^2$. The cementation is simulated by introducing tensile strength characteristics through the addition of cohesion. The depth of the cemented deposit is assumed to be 70 m; it possesses low permeability and at the bottom is bounded by impermeable rock.

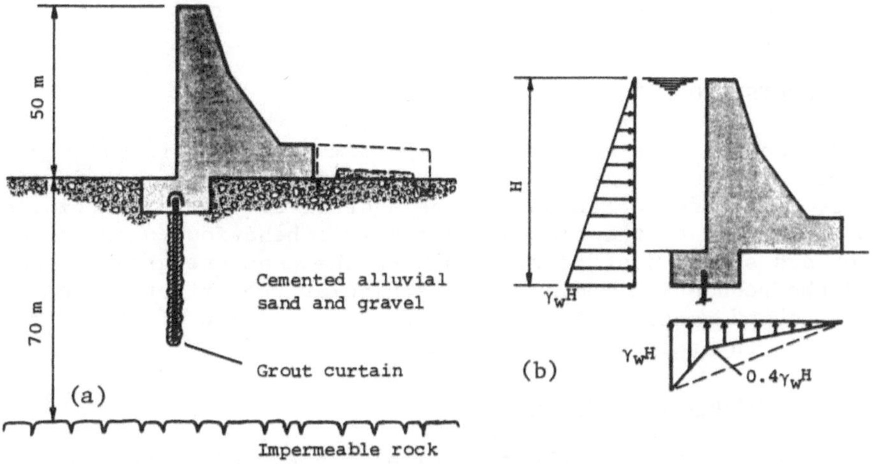

Fig. 2 Concrete gravity dam: (a) dimensions and subsoil; (b) water pressure diagram

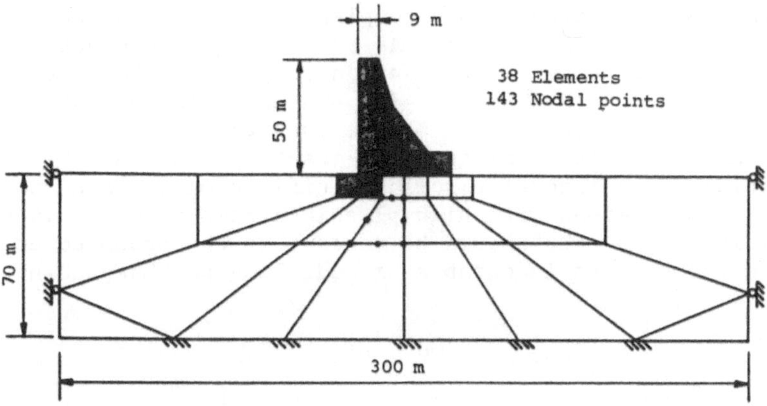

Fig. 3 Finite element model

The water pressure load considered in the analysis is shown in Fig. 2(b). In relation to the uplift pressure a pattern with cut-off based on normal design practice is assumed.

The finite element mesh is shown in Fig. 3. Both the gravity dam and the foundation medium are analysed using two-dimensional, plane-strain, 8-node isoparametric elements. A 2 x 2 Gauss integration rule is used for the stiffness evaluation. The model base is assumed to be fixed in both horizontal and vertical directions, and side boundaries are represented by vertical rollers, i.e. restrained in the horizontal direction.

The calculated stresses are the effective stresses in all steps of the analytical procedure.

5. RESULTS AND DISCUSSION

The design of a concrete gravity dam on a soft foundation involves a large number of important aspects to be considered. Here, only a few of these will be addressed. They are related to the effect of the elasto-plastic soil behaviour on deformation and stress distribution in both the concrete gravity dam and the foundation medium. In particular the effect of (a) the sequence of construction and (b) the fluctuation of the water level in the reservoir are discussed.

5.1 Construction phase effects

The first phase of the analytical process simulates the construction of the gravity dam. The entire gravity load from the dam was applied in forty increments. It should be noted that the initial stresses in the foundation medium were calculated prior to the analysis. The deformation of the concrete dam and the induced settlements upon completion of the construction phase is shown in Fig. 4(a). As a result of these settlements, regions of high tensile stresses in the dam developed through the loading process. Figure 4(b) presents contours of these stresses at the termination of the construction phase and indicates areas of potential concrete cracking. During the construction phase intensive soil plastification is found to take place. The regions with high plastic strain accumulations are shown in Fig. 5. It should be noted that the proposed constitutive formulation is capable of modelling path dependent soil

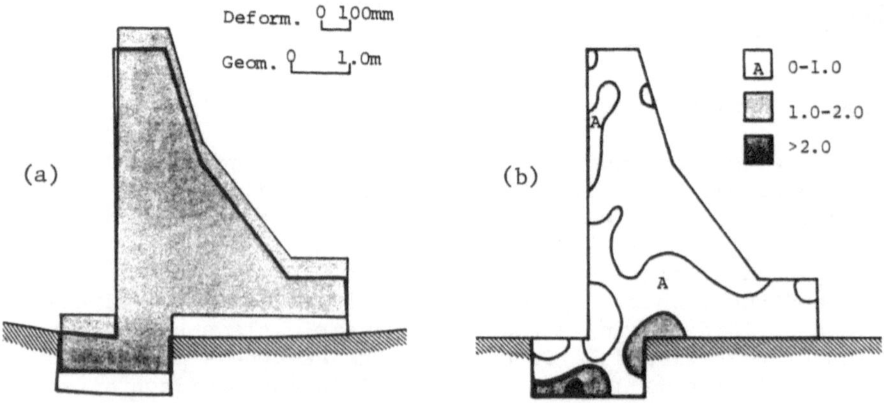

Fig. 4 Construction completed: (a) dam movement; (b) zones of tensile stresses (MPa)

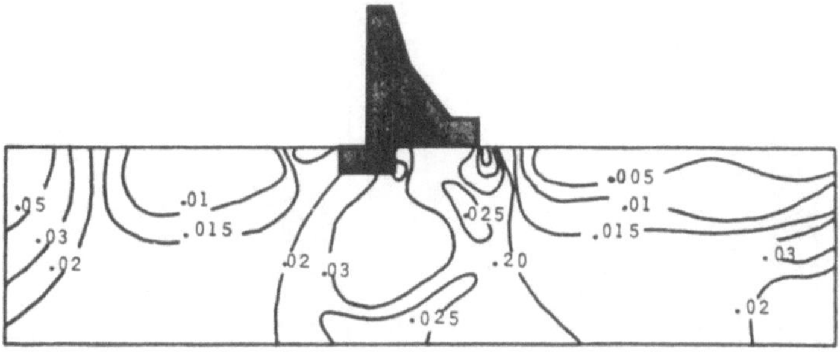

Fig. 5 Construction completed: contours of accumulated plastic strain (%)

behaviour. Therefore, another construction procedure (loading program) will obviously develop a different plastic strain distribution.

5.2 Behaviour due to water fluctuation

Consider the dam subjected to a rise in the water level to EL. 50 m. Due to the fluctuation of the water level in the reservoir, it is apparent that the dam will be subjected to loading of a cyclic nature. For convenience the rise and the subsequent fall in water level is defined as one load cycle. The magnitude and the number of loading cycles vary and depend, in the case of an actual structure, on the location, climate and purpose of the reservoir (hydro-electric, irrigation, water supply for a populated region, etc.). In the present analysis it is assumed that the magnitude of the variation of the water elevation in the accumulation is 20 m. It can be expected that the effect of plastification of the soil is more pronounced if the amplitude of the load cycles is increased.

The accumulated permanent plastic displacements of the concrete dam after the first, fifth, tenth and twelfth loading cycle are shown in Fig. 6(a), and the load-displacement history of the crest of the dam is presented in Fig. 6(b). The results presented demonstrate the importance of plastic deformation of the foundation medium even under relatively low variation of load (stresses).

The water fluctuation in the reservoir affects also the redistribution of zones with tensile stresses in the concrete gravity dam. After the first cycle a maximum tensile stress of 1140 kPa is developed at the middle of the base of the dam (see Fig. 7(a)), whereas after the twelfth cycle a maximum tensile stress of 3548 kPa is observed at the heel of the dam (see Fig. 7(b)). The latter indicates the real possibility of cracking

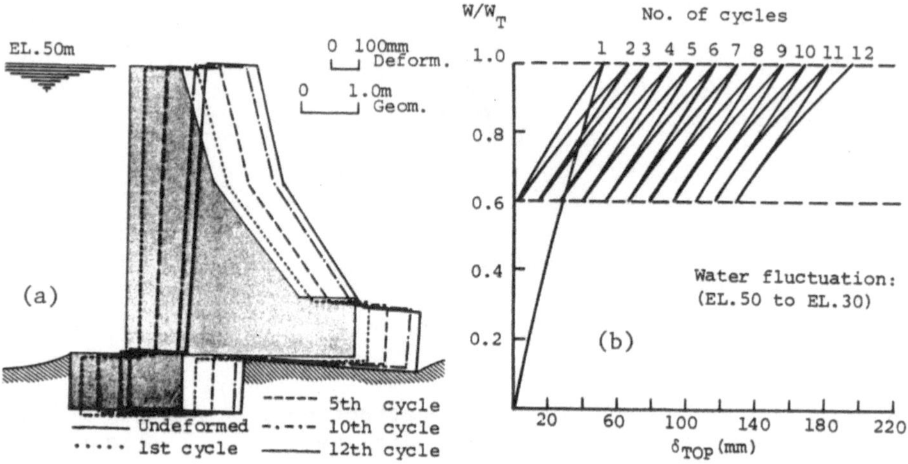

Fig. 6 Cyclic deformation of dam (a) and (b) load-displacement relation for crest

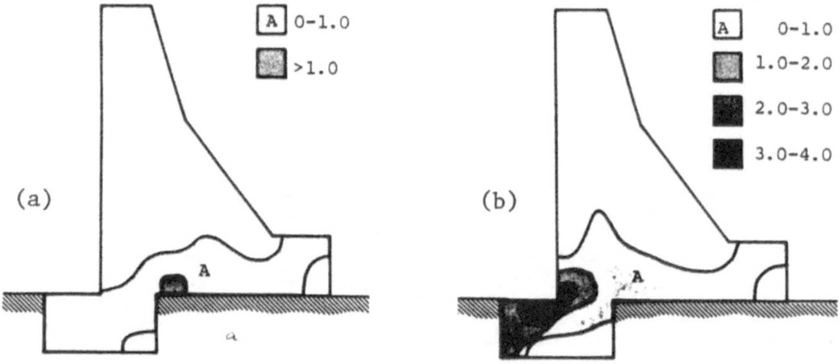

Fig. 7 Tension zones in dam (MPa): (a) after 1st load cycle; (b) after 12th load cycle

of the concrete in the affected regions.

The sliding stability of a gravity dam is directly related to the shear strength of the foundation medium. Figure 8(a) shows the contours of maximum shear stresses in the subsoil subsequent to the first loading cycle. As shown in Fig. 8(b), after twelve loading cycles, the zones with high percentage of mobilized shear strength are clearly marked, indicating possible sliding planes.

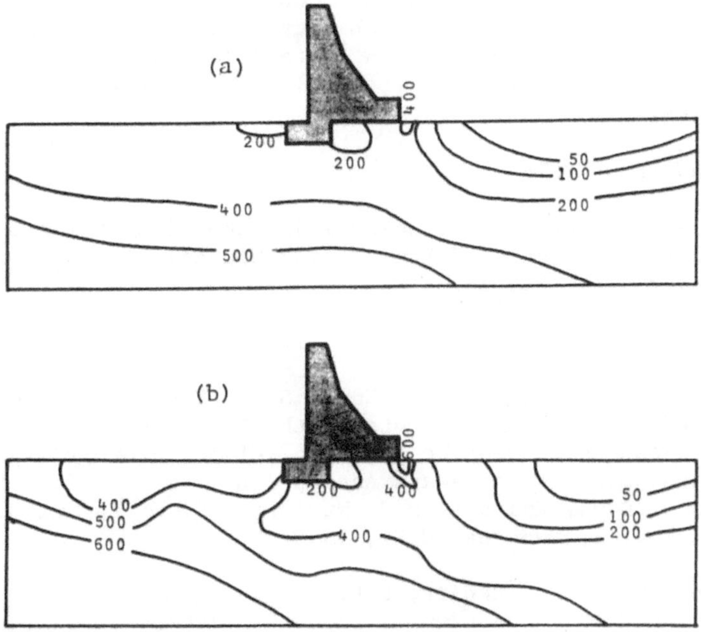

Fig. 8 Contours of maximum shear stresses (kPa): (a) after 1st load cycle; (b) after 12th load cycle

6. CONCLUSIONS

The behaviour of a concrete gravity dam on an elasto-plastic soil deposit was examined. Two important aspects, namely (a) the influence of the construction phase and (b) and influence of water fluctuation in the reservoir, were examined. From the results obtained in the analysis the following conclusions are noted in relation to the simulated elasto-plastic dam behaviour:

1. Depending upon the compressibility of the foundation medium, considerable settlement of the dam may occur during the construction period. The relatively low state of stress under the dam during and after completion of the construction phase results in formation of zones with a considerable magnitude of plastic flow.

2. The fluctuation of the water in the reservoir is found to have significant influence on both (a) the permanent dam deformations and (b) the magnitude and distribution of stresses in the dam and the foundation. Zones with tensile stresses develop in the concrete dam after a few loading cycles. These tensile stresses exceed the rupture strength of concrete and hence cracking was predicted. Also, the

magnitude of shear stresses in the foundation which relates directly to the sliding stability of the dam increases with successive load cycles. An average increase of 50% in the maximum shear stress in the soil occurs after twelve load cycles.

7. ACKNOWLEDGEMENTS

The authors gratefully acknowledge the financial support by the Natural Sciences and Engineering Council of Canada for this work under Grant No. A8258.

8. REFERENCES

1. Dant, B., Mokhashi, L.S., Knot, S.A., Thomas, B., and Mathew, V.P. "Some Aspects of the Design of Shear Keys for Dams Founded on Weak Rock", Proc. of an International Symposium on Criteria and Assumptions for Numerical Analysis of Dams, Editors, D.J. Naylor, K.G. Stagg and O.C. Zienkiewicz, Swansea, U.K., 1975.

2. Carns, C.F. and Nesbitt, R.H., "Sliding Stability of Three Dams on Weak Foundations", Proc. 9th ICOLD, Vol. 1, p. 463.

3. Kaneko, K., "Finite Element Methods Applied for Foundation Analysis of Three Dams in Azusagava Project", Discussion on Q. 37, Proc. 10th ICOLD Vol. 6, p. 507.

4. Gocevski, V., "Elasto-Plastic Two Surface Soil Model and Its Finite Element Formulation and Application", Thesis in the Department of Civil Engineering, presented in partial fulfillment of the requirements for the Degree of Doctor of Philosophy at Concordia University, Montreal, Canada, 1985.

5. Poorooshasb, H.B. and Pietruszczak, S., "On Yield and Flow of Sand; a Generalized Two-Surface Model", Computers and Geotechnics, Vol. 1, 1985, pp. 35-58.

6. Owen, D.R.J. and Hinton, E., "Finite Elements in Plasticity; Theory and Practice", Pineridge Press Limited, Swansea, U.K., 1980.

A FINITE ELEMENT MODEL FOR VISCOELASTIC FOUNDATIONS SUPPORTING PLATE STRUCTURES

Joseph J. Rencis
Kwo-Yih Jong
Sunil Saigal

Center for Computer-Aided Engineering
Mechanical Engineering Department
Worcester Polytechnic Institute
Worcester, Massachusetts 01609, U.S.A.

SUMMARY

In this paper a numerical approach for the analysis of elastic plate on viscoelastic foundations is developed. The effects of foundation responses are introduced by an analogy to mechanical plate vibrations. The differential and integral types of viscoelastic foundation responses are formulated by particular forms of finite difference equations to analyze the time effects. The structural responses are modeled by the finite element displacement method. The present method combines these two formulations for the response of elastic plates on visco- elastic foundations. The technique is validated by comparing the finite element results to the analytical solution of an infinite elastic beam on a Kelvin foundation. Numerical results are also included for a simply supported elastic plate on Maxwell and Kelvin foundations.

1. INTRODUCTION

The use of plates on foundations arise in mat foundations. A mat foundation is a large concrete slab which supports a number of columns or an entire structure. Due to the time dependent behavior of soil foundation systems, viscoelastic models have been employed to represent soil characteristics as discussed in references [1-2]. The force-deformation relationship of a viscoelastic foundation can be represented by a differential equation or an integral equation of hereditary type. For simple problems, such as an infinite beam supported by a two element viscoelastic model (e.g. Maxwell and Kelvin models), solutions have been obtained by using the Laplace transformation method [3]. For complicated geometries, general material models, and complex loading and boundary conditions, a numerical approach for elastic beams on viscoelastic foundations was developed by Ting and Yang [4].

Based on the concepts of Ting and Yang [4] for elastic beams on viscoelastic foundations, a formulation for elastic plates on viscoelastic foundations is developed. In Section 2, the governing equations for elastic plates on Maxwell and Kelvin foundations are discussed. Section 3 provides an overview of the finite element formulation while Section 4 looks at numerical examples. Finally, the conclusion are stated in Section 5.

2. GOVERNING EQUATION AND VISCOELASTIC MODELS

2.1. Governing Partial Differential Equation

The partial differential equation that governs the behavior of an elastic plate supported by a viscoelastic foundation subjected to a time-independent transverse load $p(x,y)$ as shown in the Fig. 1 is given by

$$D\nabla^4 w(x,y,t) = p(x,y) - q(x,y,t) \qquad (1)$$

where $D = Et^3/12(1-\nu^2)$ is the flexural rigidity, ∇^4 is the biharmonic, t is the plate thickness, E is Young's modulus and ν is the poisson's ratio. Also, $q(x,y,t)$ is the bearing pressure which depends on the viscoelastic foundation.

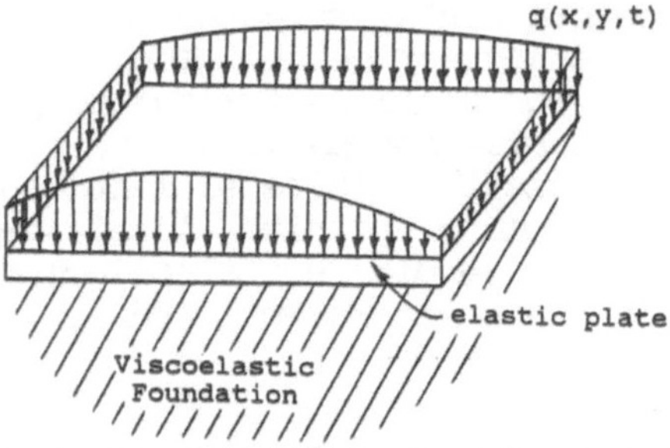

a. Plate on general viscoelastic foundation.

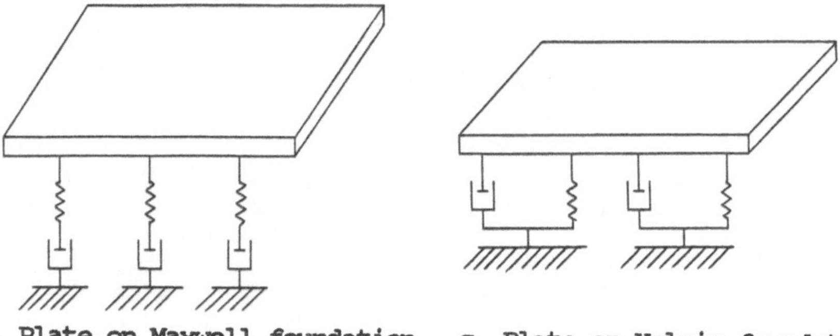

b. Plate on Maxwell foundation   c. Plate on Kelvin foundation

Figure 1. Plate on various viscoelastic foundations.

2.2. Viscoelastic Foundation Models

2.2.1. Maxwell Model

A Maxwell foundation as shown in Fig. 1b can be introduced into Equ. (1) by defining the bearing pressure q(x,y,t) as the integral of the hereditary type [5]

$$q(x,y,t) = \int_{0^-}^{t} G(t-\tau)\ \partial w(x,y,\tau)/\partial\tau \ d\tau \tag{2}$$

where w is the transverse displacement. For a Maxwell foundation, the relaxation function G(t) has the form

$$G(t) = k\ e^{-t/t_r} \tag{3}$$

where k is the elastic foundation modulus of the soil and t_r is the relaxation time. To remove the discontinuity at t=0, Equ. (2) is integrated by parts which yields

$$q(x,y,t)=kw(x,y,t)+\int_{0^+}^{t} \partial G(t-\tau)/\partial\tau\ w(x,y,\tau)d\tau \tag{4}$$

A standard technique which is employed to solve Equ. (4) is the Laplace transformation method. However, for general viscoelastic models, this technique is too complicated. In this paper a trapezoidal integration rule is utilized to evaluate the second term of Equ. (4). Assuming $w(x,y,t_{i+1}) \approx w(x,y,t_i)$ and introducing $T = t/t_r$ into Equ. (4) yields

$$k\int_{T_i}^{T_{i+1}} (T_{n+1}-\tau)w(x,y,\tau)d\tau \approx 0.5k\Delta T[(T_{n+1}-T_{i-1}) + (T_{n+1}-T_i)]w(x,y,T_i) \tag{5}$$

where the time step $\Delta T = (T_{i+1}-T_i)$ and the subsripts indicate the time steps.

Substituting Equs. (3) and (5) into Equs. (1) and (4) yields the governing partial differential equation of an elastic plate on Maxwell foundation

$$D\nabla^4 w(x,y,T_{n+1}) = p(x,y) - k\ w(x,y,T_{n+1}) - \emptyset(x,y,T_{n+1}) \tag{6a}$$

with

$$\emptyset(x,y,T_{n+1}) = 0.5\ k \sum_{i=1}^{n} w(x,y,T_i)[e^{-T_{n+1}+T_{i+1}} + e^{-T_{n+1}+T_i}]\Delta T \tag{6b}$$

Equ. (6) states that the value of the transverse displacement w at the next time step (T_{n+1}) depends on the history of the w's already obtained. The initial value of transverse displacement w can be found from Equ. (6) as

$$D\nabla^4 w(x,y,T_1) = p(x,y) - k\ w(x,y,T_1) \tag{7}$$

Equ. (7) corresponds to the response of an elastic plate on the elastic foundation.

2.2.2. Kelvin Model

A Kelvin foundation as shown in Fig. 1c can be introduced into Equ. (1) by defining the bearing pressure $q(x,y,t)$ as the differential form [5]

$$q(x,y,t) = k\,[w(x,y,t) + t_r\,\partial w(x,y,t)/\partial t] \qquad (8)$$

where k is the foundation modulus of the soil and t_r is the relaxation time.

To evaluate the time derivative of the transverse displacement w in Equ. (8), a forward difference equation is utilized which yields

$$\partial w(x,y,t_{n+1})/\partial t \approx [(w(x,y,t_{n+1})-w(x,y,t_n)]/(t_{n+1}-t_n) \qquad (9)$$

where the subscript indicate the time step. Substituting Equs. (8) and (9) into Equ. (1) with $T = t/t_r$, yields the governing partial differential equation of an elastic plate on a Kelvin foundation

$$D\nabla^4 w(x,y,T_{n+1}) = p(x,y) - k(1+1/\Delta T)\,w(x,y,T_{n+1}) + k\,w(x,y,T_n)/\Delta T \qquad (10)$$

where the time step $\Delta T = (T_{n+1}-T_n)$. Equ. (10) states that the current value of the transverse displacement at time T_{n+1} can be found by knowing the displacement at the previous time T_n. The initial value of the transverse displacement for a elastic plate on a Kelvin foundation is

$$w(x,y,0) = 0 \qquad (11)$$

Equ. (11) states that the initial response of an elastic plate is on a perfectly rigid foundation.

3. FINITE ELEMENT FORMULATION

3.1. Global Finite Element Equations

The Galerkin weighted residual method [6,7] can be utilized to transform the governing partial differential equation of an elastic plate on a viscoelastic foundation (Equ. (1)) into an equivalent integral equation. After introducing the finite element approximation into the integral equation, the general global finite element equation for an elastic plate on a viscoelastic foundation is in the form

$$(\underline{K}_P + \underline{K}_F)\,\underline{w} = \underline{f}_L + \underline{f}_V \qquad (12a)$$

with

$$\underline{K}_P = \sum_e \underline{K}_P^e \qquad \underline{K}_F = \sum_e \underline{K}_F^e$$

$$\underline{f}_L = \sum_e \underline{f}_L^e \qquad \underline{f}_V = \sum_e \underline{f}_V^e \qquad (12b)$$

where superscripts signify the element, the summation denotes that all element matrices are assembled properly according to the direct stiffness method, $\underline{w}$ denotes a global vector composed of nodal displacements and rotations, $\underline{K}_P$ is the global stiffness matrix of the plate, and $\underline{f}_L$ is the global force vector due to external loading on the plate. Also, $\underline{K}_F$ is the global foundation stiffness matrix and $\underline{f}_V$ is the global visco-force vector. Both $\underline{K}_F$ and $\underline{f}_V$ depend on the viscoelastic foundation employed. A discussion on these element matrices and vectors is given in the next two subsections.

3.2. $\underline{K}_P^e$ and $\underline{f}_L^e$

The elastic plate element stiffness matrix $\underline{K}_P^e$ and the element force vector $\underline{f}_L^e$ depend on the elastic plate element chosen. A variety of plate elements are available in [6,7]. The plate element utilized in this study is based on Mindlin plate theory and is called a "Heterosis" plate bending element [8]. The "Heterosis" plate element assumes that the transverse displacement w shape functions to be of the 8 node Serendipity form while the shape functions of the θ_x and θ_y rotations are of the 9 node Lagrange form.

The element stiffness matrix $\underline{K}_P^e$ of the plate is of order 26 x 26, while the load vector $\underline{f}_L^e$ is of order 26 x 1. This plate element has been shown to have high-accuracy for thin and thick plate bending applications [9].

3.3. $\underline{K}_F^e$ and $\underline{f}_V^e$

3.3.1. Maxwell Foundation

For a Maxwell foundation, the element foundation stiffness matrix $\underline{K}_F^e$ and the element visco-force vector $\underline{f}_V^e$ of Equ. (12b) have the form

$$\underline{K}_F^e = k \int_A \underline{N}^T \underline{N} \, dA \tag{13a}$$

and

$$\underline{f}_V^e = \beta \int_A \underline{N}^T \underline{N} \, dA \; \underline{w}^e(x,y,T_i) \tag{13b}$$

with

$$\beta = 0.5 \; k \sum_{i=1}^{n} [\, e^{-T_{n+1}+T_{i+1}} + e^{-T_{n+1}+T_i}]\Delta T \tag{13c}$$

$$\Delta T = (T_{i+1} - T_i) \tag{13d}$$

where $\underline{N}$ is the vector of Serendipity shape functions [9] which defines the transverse displacement w. Te order of $\underline{K}_F^e$ is a 26 x 26 while $\underline{f}_V$ is a 26 x 1. One should note that the form of the element foundation stiffness matrix $\underline{K}_F^e$ of Equ. (13a) is similar to the consistent element mass matrix for mechanical plate vibration if k is replaced by the density.

3.3.2. Kelvin Foundation

For a Kelvin foundation, the element foundation stiffness matrix $\underline{K}_F^e$ and the element visco-force vector $\underline{f}_V^e$ of Equ. (12b) have the form

$$\underline{K}_F^e = k\ (1 + 1/\Delta T) \int_A \underline{N}^T\ \underline{N}\ dA \tag{14a}$$

and

$$\underline{f}_V^e = \beta \int_A \underline{N}^T\ \underline{N}\ dA\ \underline{w}^e(x,y,T_n) \tag{14b}$$

with

$$\beta = k/\Delta T \tag{14c}$$

$$\Delta T = (T_{i+1} - T_i) \tag{14d}$$

where $\underline{N}$ is the vector of Serendipity shape functions [12] which defines the transverse displacement w. The order of $\underline{K}_F^e$ is a 26 x 26 while $\underline{f}_V^e$ is a 26 x 1. Once again the analogy between the element foundation stiffness matrix $\underline{K}_F^e$ and the element consistent mass matrix exists.

4. NUMERICAL EXAMPLES

4.1. Example 1 - Beam

To evaluate the accuracy of the proposed finite element model, an elastic continuous beam on a Kelvin foundation and loaded by equal column loads is considered [3] as shown in Fig. 2a. The elastic reinforced concrete beam is 0.762 m wide and 0.508 m deep with a modulus of elasticity of E = 17237.5 mPa. A Poisson's ratio of $\nu = 0$ was assumed. The beam is loaded by 0.762 x 0.762 m^2 columns every 6.096 meters. Each column load of 1334.4 kN is considered uniformly distributed over the contact area. The foundation modulus of the soil k is 203,592.52 kPa/m.

Due to the symmetry of the problem, only a 3.048 m long finite element model is used as shown in Fig. 2b. The boundary

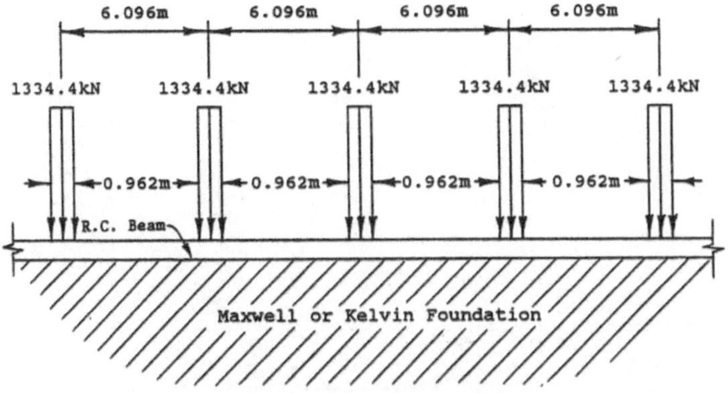

a. Infinite beam on viscoelastic foundation.

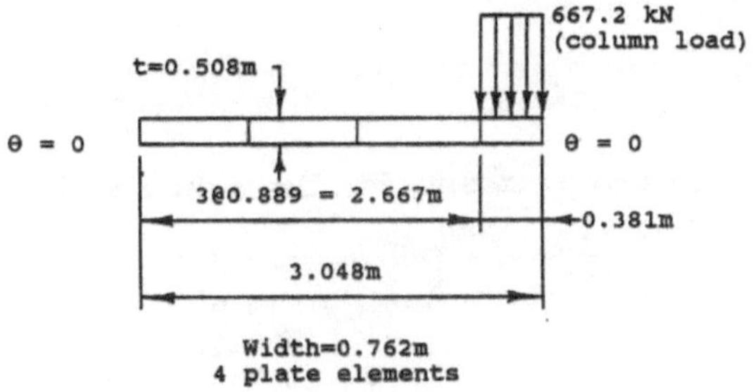

b. Finite element model.
Figure 2. Example 1 - Beam.

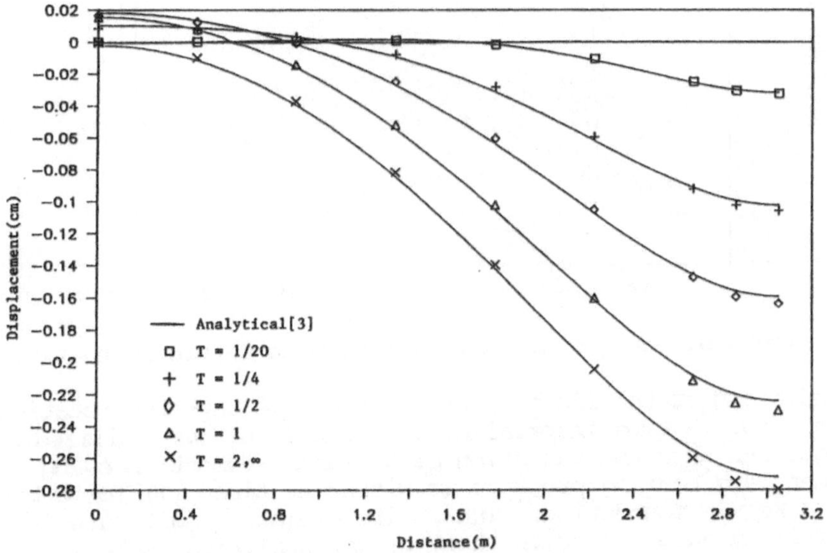

Figure 3. Deflection for beam on Kelvin foundation.

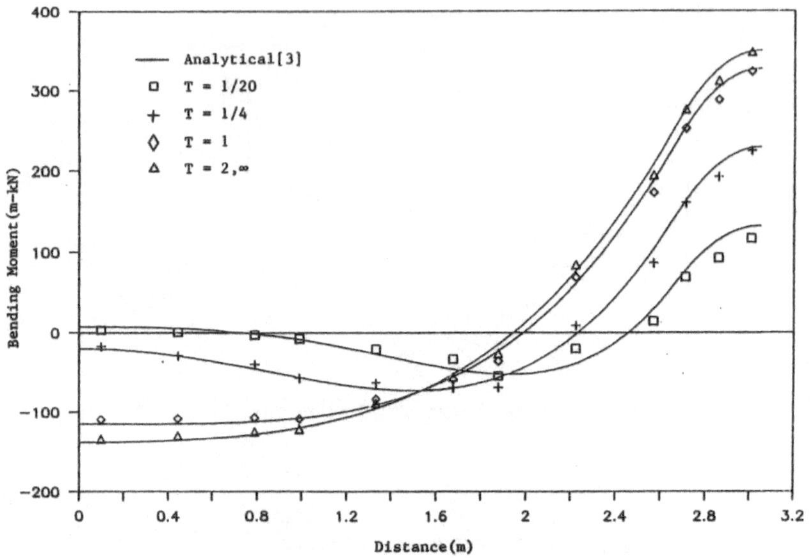

Figure 4. Bending moment for beam on Kelvin foundation.

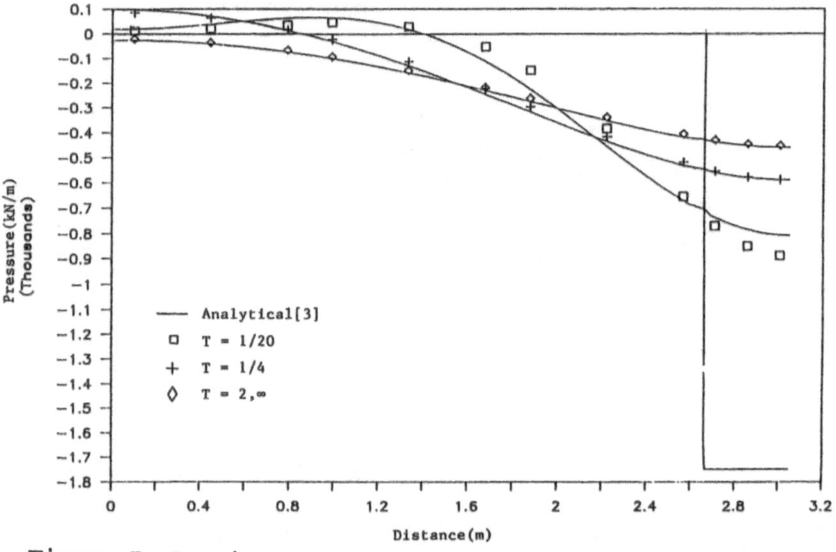

Figure 5. Bearing pressure for beam on Kelvin foundation.

conditions at the lines of symmetry assume that the rotations are zero. A time interval of $\Delta T=\Delta t/t_r=0.025$ was utilized. Figs. 3-5 show the variation of the displacement, bending moment and bearing pressure at different times (T) for a beam on a Kelvin foundation. The finite element results are in excellent agreement with those of the analytical solutions of reference [3].

<u>4.2 Example 2 - Simply Supported Slab</u>

A simply supported concrete slab (plate) on Maxwell and Kelvin foundations is considered as shown in Fig. 6. The concrete slab is 4 x 4 m^2 in dimension and 12 cm thick. The concrete slab has a modulus of elasticity of E = 17237.5 mPa, poisson's ratio ν = 0.3, and the foundation modulus of the soil is k = 203592.52 kPa/m. A uniformly distributed load of 10000 kPa is applied.

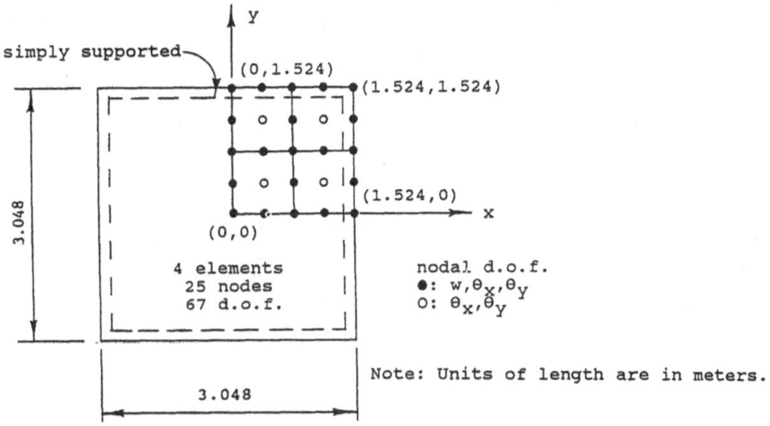

Figure 6. Example 2-Simply supported square plate on a viscoelastic foundation

Due to symmetry, only one quadrant of the problem is discretized as shown in Fig. 6. The variation of the displacement for an elastic plate on a Maxwell foundation is shown in Fig. 7. A time interval of $\Delta T = \Delta t/t_r = 0.1$ was used. At T = 0, the Maxwell foundations response is an elastic plate on an elastic foundation. This response is in agreement with the analytical solution of Timoshenko and Woinowsky-Krieger [9]. The displacement responses of a plate on a Kelvin foundation shown in Fig. 8. A time interval of $\Delta T = \Delta t/t_r = 0.025$ was employed. The initial response (T=0) of the Kelvin foundation corresponds to a elastic plate on a perfectly rigid foundation [9].

5. CONCLUSIONS

A numerical technique for the analysis of elastic plates on viscoelastic foundations has been developed. The viscoelastic foundations considered are of the Maxwell and Kelvin type. The results were found to be in agreement with the analytical solutions when available.

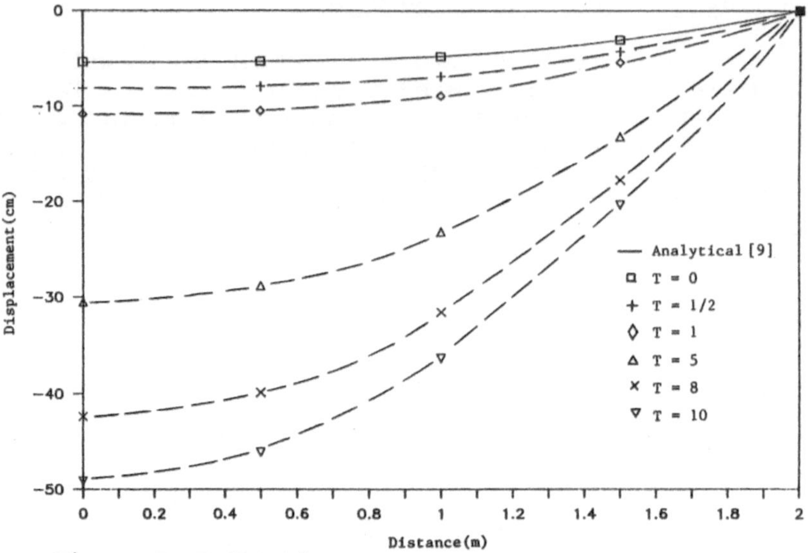

Figure 7. Deflection for plate on Maxwell foundation.

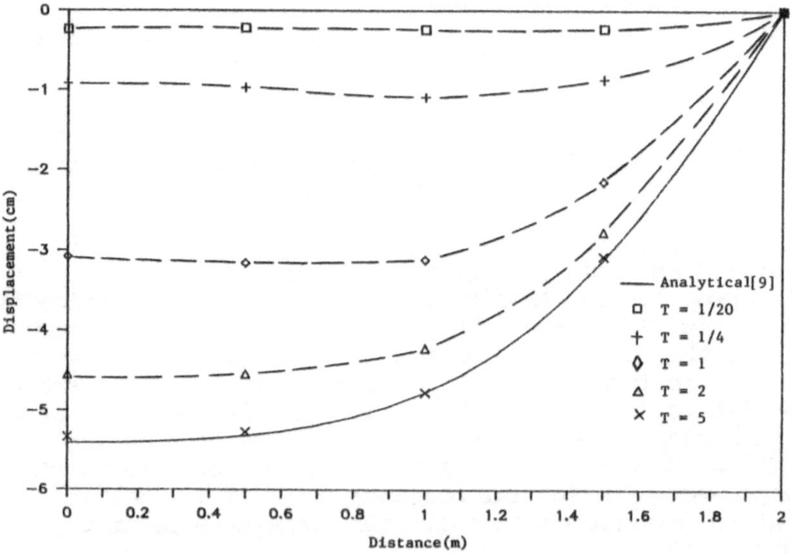

Figure 8. Deflection for plate on Kelvin foundation.

ACKNOWLEDGEMENTS

JJR would like to thank Professor D.N. Zwiep, Head of Mechanical Engineering Department at Worcester Polytechnic Institute, for his providing the funds to attend this conference.

REFERENCES

1. Bishop, A.W. and Levenbury, H.V., 'Creep Characterization of Unsaturated Clays', Intl. Conf. on Soil Mech. and Foundation Eng., Mexico, Vol. 1, 1969.

2. Pagen, C.A. and Jagannath, B.N., 'Evaluation of Soil Compaction by Rheological Techniques', Highway Research Record 117, Highway Research Board, 1967.

3. Freudenthal, A.M. and Lorsch, H.G., 'The Infinite Elastic Beam on a Viscoelastic Foundation', J. Eng. Mech., ASCE, 83, Jan., 1957.

4. Ting, E.C. and Yang, H.T., 'A Numerical Procedure for the Analysis of Structures Supported by Viscoelastic Foundations', Developments in Mechanics, Proceedings of 13th Midwestern Mechanics Conference, J.I. Abrams, T.C. Woo, C.P. Mangelsdorf, and J.L.S. Chen (eds.), University of Pittsburgh, Pittsburgh, PA, 1973, 959-975.

5. Flügge, W. - Viscoelasticity, Springer-Verlag, New York, 2nd ed., 1975, p. 42 & p. 22.

6. Zienkiewicz, O.C. - The Finite Element Method, McGraw-Hill, New York, 3rd ed., 1977, pp. 42-58 & p. 529.

7. Gallagher, R.H., - Finite Element Analysis, Prentice-Hall, New Jersey, 1975, pp. 126-132 & pp. 327-369.

8. Hughes, T.J.R. and Cohen, M., 'The "Heterosis" Finite Element for Plate Bending', Comp. Struc., 9, 1978, 445-450.

9. Timoshenko, S. and Woinowsky-Krieger, S., - Theory of Plates and Shells, McGraw-Hill, 2nd ed., 1959, pp. 269-273.

COMPUTER AIDED SIMULATION OF TOOTH CONTACT ANALYSIS FOR HELICAL GEARS WITH INVOLUTE SHAPE TEETH

C.B. TSAY

Dept. of Mechanical Engineeing, National Chiao Tung Univ.
Hsinchu, Taiwan 30049, R.O.C.

ABSTRACT

The level of kinematic errors is one of the main source of the gear noise and vibration. Thus, one of the goal of mechanical engineers is to solve the noise and vibration problems of the mating gears. In this paper, we assume that the equations of mating pinion and gear are given.

The paper covered: (1) Computer simulation of the conditions of meshing and bearing contact and (2) Investigation of the sensitivity of gears to the errors of manufacturing and assembly (due to the change of center distance and axes misalignment). A method of compensation for the dislocation of the bearing contact induced by errors of manufacturing and assembly has been proposed. Three numerical examples have also been presented to illustrate the influence of the abovementioned errors and the method of compensation of the dislocation of bearing contact.

1. INTRODUCTION

There are two typical purposes when gearing analysis is applied: (1) to determine the kinematic errors in a gear mechanism which are induced by errors of manufacturing and assembly; and (2) for the optimal synthesis of gears. The optimal synthesis of gearings is usually an interative computational procedure which requires intermediate analysis between iterations.

Consider that the equations of pinion and gear tooth surfaces of helical gears are given in coordinate systems $S_1(X_1, Y_1, Z_1)$ and $S_2(X_2, Y_2, Z_2)$, respectively, as follows:

$$x_1=(\ell_F\cos\Psi_n^{(F)}-a_F+r_1)\cos\phi_1+(\ell_F\cos\Psi_n^{(F)}-a_F)\cot\Psi_n^{(F)}\sin\lambda_F\sin\phi_1$$
$$y_1=(\ell_F\cos\Psi_n^{(F)}-a_F+r_1)\sin\phi_1-(\ell_F\cos\Psi_n^{(F)}-a_F)\cot\Psi_n^{(F)}\sin\lambda_F\cos\phi_1$$
$$z_1=(a_F\tan\Psi_n^{(F)}-\ell_F\sin\Psi_n^{(F)})\cos\lambda_F+\left(\frac{a_F}{\cos\Psi_n^{(F)}\sin\Psi_n^{(F)}}-\frac{\ell_F}{\sin\Psi_n^{(F)}}\right)\times\tan\lambda_F\sin\lambda_F+\frac{b_F}{\cos\lambda_F}+r_1\phi_1\tan\lambda_F \quad (1)$$

$$x_2=(\ell_P\cos\Psi_n^{(P)}-a_P-r_2)\cos\phi_2-(\ell_P\cos\Psi_n^{(P)}-a_P)\cot\Psi_n^{(P)}\sin\lambda_P\sin\phi_2$$
$$y_2=-(\ell_P\cos\Psi_n^{(P)}-a_P-r_2)\sin\phi_2-(\ell_P\cos\Psi_n^{(P)}-a_P)\cot\Psi_n^{(P)}\sin\lambda_P\cos\phi_2 \quad (2)$$
$$z_2=(a_P\tan\Psi_n^{(P)}-\ell_P\sin\Psi_n^{(P)})\cos\lambda_P+\left(\frac{a_P}{\cos\Psi_n^{(P)}\sin\Psi_n^{(P)}}-\frac{\ell_P}{\sin\Psi_n^{(P)}}\right)\times\tan\lambda_P\sin\lambda_P+\frac{b_P}{\cos\lambda_P}+r_2\phi_2\tan\lambda_P$$

The corresponding tooth surfaces' unit normals can be obtained as follows:

$$n_{1x}=\sin\Psi_n^{(F)}\cos\phi_1+\cos\Psi_n^{(F)}\sin\lambda_F\sin\phi_1$$
$$n_{1y}=\sin\Psi_n^{(F)}\sin\phi_1-\cos\Psi_n^{(F)}\sin\lambda_F\cos\phi_1 \quad (3)$$
$$n_{1z}=\cos\Psi_n^{(F)}\cos\lambda_F$$

$$n_{2x}=\sin\Psi_n^{(P)}\cos\phi_2-\cos\Psi_n^{(P)}\sin\lambda_P\sin\phi_2$$
$$n_{2y}=-\sin\Psi_n^{(P)}\sin\phi_2-\cos\Psi_n^{(P)}\sin\lambda_P\cos\phi_2 \quad (4)$$
$$n_{2z}=\cos\Psi_n^{(P)}\cos\lambda_P$$

2. SIMULATION OF MESHING AND TOOTH CONTACT ANALYSIS

We may simulate the conditions of meshing by changing the settings and orientation of the coordinate system $S_h(X_h, Y_h, X_h)$ with respect to $S_f(X_f, Y_f, Z_f)$ as shown in Fig.1. For instance,

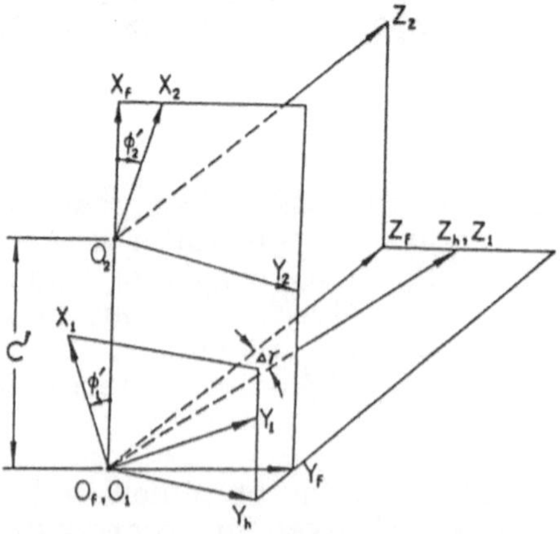

Fig.1 Relations Between Coordinate Systems S_1, S_2, S_h and S_f

simulating the change of center distance Δc, we may displace the origin O_h of the coordinate system S_h by Δc with respect to origin O_f of the coordinate system S_f. Similarly, when we simulate the axes of misalignment $\Delta\gamma$, we may rotate the coordinate system S_h about axis X_h an angle $\Delta\gamma$ with respect to the coordinate system S_f. Then, using the coordinate transformation from system S_h to S_f, we can represent the equations of pinion tooth surface Σ_1 and its surface normal in coordinate system S_f. If we simulate the change of center distance Δc and the axes of misalignment Δr simultaneously, then the following coordinate transformation matrix equations can be applied [1]:

$$[r_f^{(1)}]=[M_{fh}][M_{h1}][r_1] \tag{5}$$

$$[r_f^{(2)}]=[M_{f2}][r_2] \tag{6}$$

$$[n_f^{(1)}]=[L_{fh}][L_{h1}][n_1] \tag{7}$$

$$[n_f^{(2)}]=[L_{f2}][n_2] \tag{8}$$

where

$$[M_{fh}]=\begin{bmatrix} 1 & 0 & 0 & 0 \\ 0 & \cos\Delta\gamma & \sin\Delta\gamma & 0 \\ 0 & -\sin\Delta\gamma & \cos\Delta\gamma & 0 \\ 0 & 0 & 0 & 1 \end{bmatrix}$$

$$[M_{h1}]=\begin{bmatrix} \cos\phi_1' & \sin\phi_1' & 0 & 0 \\ -\sin\phi_1' & \cos\phi_1' & 0 & 0 \\ 0 & 0 & 1 & 0 \\ 0 & 0 & 0 & 1 \end{bmatrix}$$

$$[M_{f2}]=\begin{bmatrix} \cos\phi_2' & -\sin\phi_2' & 0 & C' \\ \sin\phi_2' & \cos\phi_2 & 0 & 0 \\ 0 & 0 & 1 & 0 \\ 0 & 0 & 0 & 1 \end{bmatrix}$$

$$[L_{fh}]=\begin{bmatrix} 1 & 0 & 0 \\ 0 & \cos\Delta\gamma & \sin\Delta\gamma \\ 0 & -\sin\Delta\gamma & \cos\Delta\gamma \end{bmatrix}$$

$$[L_{h1}] = \begin{bmatrix} \cos\phi_1' & \sin\phi_1' & 0 \\ -\sin\phi_1' & \cos\phi_1' & 0 \\ 0 & 0 & 1 \end{bmatrix}$$

$$[L_{f2}] = \begin{bmatrix} \cos\phi_2' & -\sin\phi_2' & 0 \\ \sin\phi_2' & \cos\phi_2' & 0 \\ 0 & 0 & 1 \end{bmatrix}$$

The conditions of continuous tangency of gear tooth surfaces Σ_1 and Σ_2 are represented by the following equations [1-4]:

$$\underset{\sim}{r}_f^{(1)} = \underset{\sim}{r}_f^{(2)} \tag{9}$$

and

$$\underset{\sim}{n}_f^{(1)} = \underset{\sim}{n}_f^{(2)} \tag{10}$$

Equation (9) expresses that pinion tooth surface Σ_1 and gear tooth surface Σ_2 have a common contact point determined with the position vectors $\underset{\sim}{r}_f^{(1)}$ and $\underset{\sim}{r}_f^{(2)}$. Equation (10) indicates that surfaces Σ_1 and Σ_2 have a common unit normal at their contact point. Equations (9) and (10), if considered simultaneously, yield a system of five independent equations with six unknowns: $\ell_F, \ell_p, \phi_1, \phi_2, \phi_1'$ and ϕ_2', since $|\underset{\sim}{n}_f^{(1)}| = |\underset{\sim}{n}_f^{(2)}| = 1$. Here ℓ_F and ℓ_p are surface coordinates of the rack cutter Σ_F and Σ_p ; ϕ_1 and ϕ_2 are the angle of rotation of pinion and gear, respectively, when generated by the rack cutter; ϕ_1' and ϕ_2' are the real angle of rotation of pinion and gear, respectively, when in meshing with each other (Fig.1). Thus one of these unknowns, say ϕ_1', may be considered as an input variable, then solve five independent equations with five unknowns.

Equations (1) to (10) yield the following six equations:

$$A_1\cos\mu_1 + B_1\sin\mu_1 = A_2\cos\mu_2 - B_2\sin\mu_2 + C' \tag{11}$$

$$(A_1\sin\mu_1 - B_1\cos\mu_1)\cos\triangle\gamma + \{(a_0\tan\Psi_n^{(F)} - \ell_F\sin\Psi_n^{(F)})\cos\lambda_F + (\frac{a_F}{\cos\Psi_n^{(F)}\sin\Psi_n^{(F)}} - \frac{\ell_F}{\sin\Psi_n^{(F)}})\tan\lambda_F\sin\lambda_F + \frac{b_F}{\cos\lambda_F} + r_1\phi_1\tan\lambda_F\}\sin\triangle\gamma = -A_2\sin\mu_2 - B_2\cos\mu_2 \tag{12}$$

$$-(A_1\sin\mu_1 - B_1\cos\mu_1)\sin\triangle\gamma + \{(a_F\tan\Psi_n^{(F)} - \ell_F\sin\Psi_n^{(F)})\cos\lambda_F + (\frac{a_F}{\cos\Psi_n^{(F)}\sin\Psi_n^{(F)}} - \frac{\ell_F}{\sin\Psi_n^{(F)}})\tan\lambda_F\sin\lambda_F + \frac{b_F}{\cos\lambda_F} + r_1\phi_1\tan\lambda_F\}\cos\triangle\gamma$$
$$= (a_P\tan\phi_n^{(P)} - \ell_P\sin\Psi_n^{(P)})\cos\lambda_P + (\frac{a_P}{\cos\Psi_n^{(P)}\sin\Psi_n^{(P)}} - \frac{\ell_P}{\sin\Psi_n^{(P)}})\tan\lambda_P\sin\lambda_P + \frac{b_P}{\cos\lambda_P} + r_2\phi_2\tan\lambda_P \tag{13}$$

$$\sin\Psi_n^{(F)}\cos\mu_1+\cos\Psi_n^{(F)}\sin\lambda_F\sin\mu_1=\sin\Psi_n^{(P)}\cos\mu_2-\cos\Psi_n^{(P)}\sin\lambda_P\sin\mu_2 \tag{14}$$

$$(\sin\Psi_n^{(F)}\sin\mu_1-\cos\Psi_n^{(F)}\sin\lambda_F\cos\mu_1)\cos\triangle\gamma+\cos\Psi_n^{(F)}\cos\lambda_F\sin\triangle\gamma$$
$$=-\sin\Psi_n^{(P)}\sin\mu_2-\cos\Psi_n^{(P)}\sin\lambda_P\cos\mu_2 \tag{15}$$

$$-(\sin\Psi_n^{(F)}\sin\mu_1-\cos\Psi_n^{(F)}\sin\lambda_F\cos\mu_1)\sin\triangle\gamma+\cos\Psi_n^{(F)}\cos\lambda_F\cos\triangle\gamma$$
$$=\cos\Psi_n^{(P)}\cos\lambda_P \tag{16}$$

where

$\mu_1=\phi_1-\phi_1'$ $\qquad$ $\mu_2=\phi_2-\phi_2'$

$A_1=\ell_F\cos\Psi_n^{(F)}-a_F+r_1$ $\qquad$ $A_2=\ell_P\cos\Psi_n^{(P)}-a_P-r_2$

$B_1=(\ell_F\cos\Psi_n^{(F)}-a_F)\cot\Psi_n^{(F)}\sin\lambda_F$ $\qquad$ $B_2=(\ell_P\cos\Psi_n^{(P)}-a_P)\cot\Psi_n^{(P)}\sin\lambda_P$

$C'=r_1+r_2+\triangle C$

We can obtain the relations between the angles ϕ_2' and ϕ_1' of gear rotation. Function $\phi_2'(\phi_1')$ is a nonlinear function and its deviation from the linear function is given by

$$\Delta\phi_2'(\phi_1') = \phi_2'(\phi_1') - \frac{N_1}{N_2}\phi_1' \tag{17}$$

Here N_1 and N_2 represent the tooth numbers of pinion and gear, respectively; $\Delta\phi_2'(\phi_1')$ represent the kinematic errors of the gear train; $\ell_F(\phi_1')$ and $\ell_P(\phi_1')$ represent the change of location of the bearing contact induced by the misalignment of gear axes.

3. IDEAL CONDITION OF MESHING

Example 1: The rack cutter parameters as shown in Fig.2 and Fig.3 with normal pressure angle $\psi_n^{(F)}=\psi_n^{(P)}=25°$

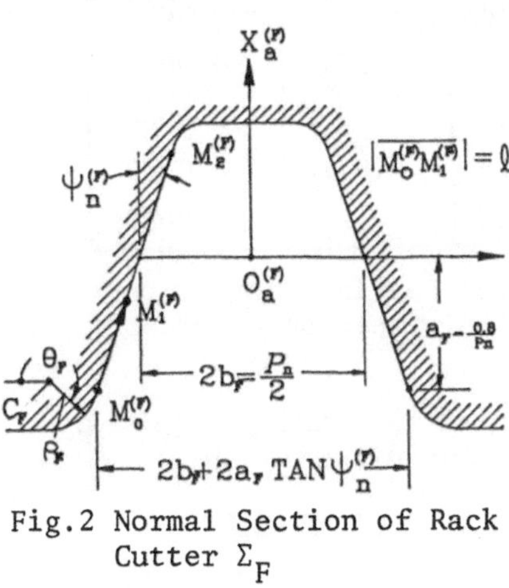

Fig.2 Normal Section of Rack Cutter Σ_F

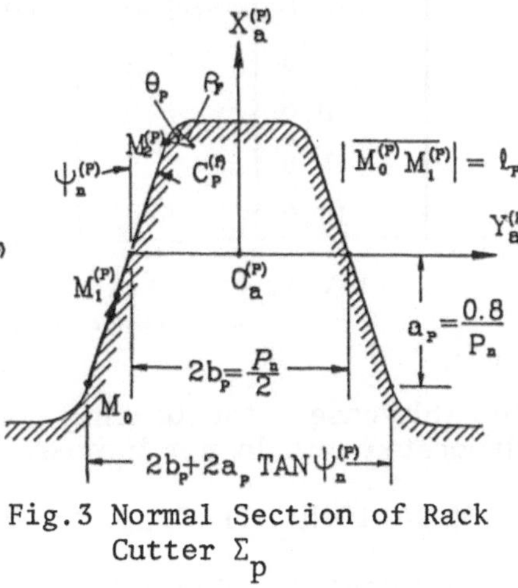

Fig.3 Normal Section of Rack Cutter Σ_P

The gear parameters are as follows: tooth numbers $N_1 = 27$, $N_2 = 54$; the lead angles $\lambda_F = \lambda_p = 60^o$; the module $m = 4.888^{mm}$, the radii of pitch circle $r_1 = 66^{mm}$, $r_2 = 132^{mm}$, thus ideal center distnace $c = 198^{mm}$.

If two mating helical gears are in meshing neither axes misalignment nor change of axes' distance, that is $\Delta r = 0^o$ and $\Delta c = 0$, the contact of two gear surfaces is a straight line at every instant as shown in Figure 4. Also, the gear train does not induce kinematic errors for this case.

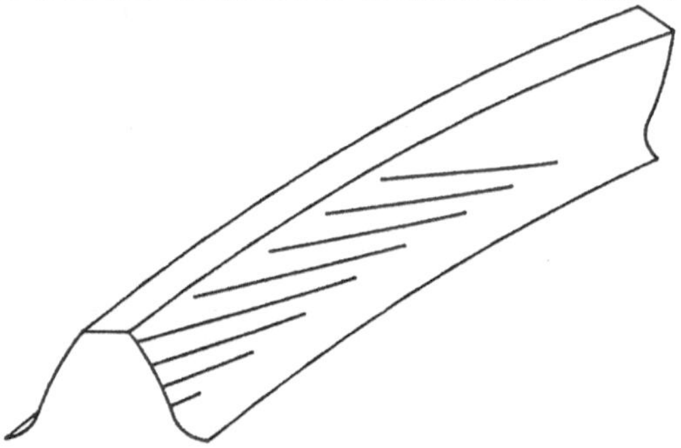

Fig.4 Contacting Lines of Two Mating Gears

4. CHANGE OF AXES' DISTANCE

Example 2: the rack cutter and gear parameters are the same as given in example 1 with $\Delta r = 0^o$ and $\Delta c = 2^{mm}$ (in Fig. 1, $c' = c + \Delta c$). The bearing contact and kinematic errors are shown in Table 1.

ϕ_1' (deg.)	ϕ_2' (deg.)	ℓ_F (mm)	ℓ_p (mm)	$\Delta\phi_2'$ (sec.)
-10.0	-5.0	0.7554	-1.4145	0.0000
- 5.0	-2.5	3.4078	1.2379	0.0000
0.0	0.0	6.0602	3.8902	0.0000
5.0	2.5	8.7126	6.5426	0.0000
10.0	5.0	11.3649	9.1950	0.0000

Table 1. Kinematic Errors and Bearing Contact Due to Change of Axes' Distance $\Delta c = 2^{mm}$.

In this case, the bearing contact of two mating gears are dislocated but do not induce kinematic errors at all.

5. MISALIGNMENT OF AXES OF GEAR ROTATION

Example 3. The axis of pinion rotation is not parallel to

the axis of gear rotation and forms a crossed angle $\Delta r = 0.1^{o}$ (Fig.1). The bearing contact of gear surfaces becomes a point contact at the cross section of $z_f = 0$ instead of a line contact. The kinematic errors and bearing contact are shown in Table 2.

ϕ_1'(deg.)	ϕ_2'(deg.)	ℓ_F(mm)	ℓ_p(mm)	$\Delta\phi_2'$(sec.)
-10.0	-5.0039	0.1880	0.1908	-14.094
- 8.0	-4.0031	1.2501	1.2520	-11.275
- 6.0	-3.0024	2.3122	2.3133	- 8.456
- 4.0	-2.0016	3.3742	3.3745	- 5.638
- 2.0	-2.0008	4.4363	4.4375	- 2.819
0.0	0.0000	5.4983	5.4970	0.000
2.0	1.0008	6.5604	6.5582	2.819
4.0	2.0016	7.6225	7.6194	5.638
6.0	3.0024	8.6845	8.6806	8.456
8.0	4.0031	9.7466	9.7419	11.275
10.0	5.0039	10.8086	10.8031	14.094

Table 2. Kinematic Errors and Bearing Contact Due to Crossed Angle $\Delta r = 0.1^{o}$

6. COMPENSATION FOR THE INFLUENCE OF MISALIGNMENT OF GEAR AXES

The compensation of kinematic errors and dislocations of bearing contact is achieved by regrinding the pinion with a modified lead angle $\lambda_F = 60.1^{o}$ ($\Delta\lambda_F = 0.1^{o}$). The kinematic errors and bearing contact after compensation are given in Table 3. Using the proposed method of compensation we can reduce the kinematic errors substantially.

7. CONCLUSION

A method for the simulation of the conditions of meshing and bearing contact has been proposed. The sensitivity of the mating gears to the change of axes' distance and to the misalignment of gears has also been investigated. A compensate method for the improvement of the dislocations of bearing contact and kinematic errors for misaligned gears has been proposed.

ϕ_1'(deg.)	ϕ_2'(deg.)	ℓ_F(mm)	ℓ_p(mm)	$\Delta\phi_2'$($x10^{-6}$sec.)
-10	-5.0000	0.2021	0.2021	- 60
- 8	-4.0000	1.2625	1.2625	- 20
- 6	-3.0000	2.3229	2.3229	10
- 4	-2.0000	3.3833	3.3833	20
- 2	-1.0000	4.4437	4.4437	20
0	0.0000	5.5041	5.5041	0
2	1.0000	6.5645	6.5645	- 30
4	2.0000	7.6249	7.6249	- 70
6	3.0000	8.6853	8.6853	-130
8	4.0000	9.7457	9.7457	-200
10	5.0000	10.8061	10.8061	-290

Table 3. Compensated Kinematic Errors and Bearing Contact.

REFERENCES

1. Litvin, F.L. and Tsay, C.B., 'Helical Gears with Circular Arc Teeth: Simulation of Conditions of Meshing and Bearing Contact', ASME 1984 Des. Eng. Tech. Conf.; Also Trans. ASME, J. Mech. Transm. Autom. Des., Vol. 107, No. 4, (1985) pp. 556-546.

2. Litvin, F.L., - Theory of Gearing, 2nd ed., Nauka, 1968 (in Russian). The new edition (in English), revised, completed and sponsored by NASA, is in press.

3. Litvin, F.L., 'Methods for Generation of Gear Tooth Surfaces and Basic Principles of Computer Aided Tooth Contact Analysis', Proc. of Computers in Eng. Vol. 1, pp. 329-333, 1985.

4. Litvin, F.L., Rahman, P. and Goldrich, R.N., 'Mathematical Models for the Synthesis and Optimization of Spiral Bevel Gear Tooth Surfaces', NASA Contractor Report 3553 (1982).

STRESS AND VELOCITY FIELDS AT DISCHARGING OF SILOS

J. Eibl, G. Rombach

University of Karlsruhe, IfMB
Postfach 6980, D-7500 Karlsruhe 1, Germany

1. SUMMARY

An Eulerian finite element method for the calculation of transient velocity and stress fields of path dependant materials in mass- and core flow silos based upon a consistent continuum mechanics approach is presented. Geometric nonlinearities as well as inertia effects of the solid are included. An incremental viscoplastic law for the granular bulk material is formulated that covers the solid-like behavior of the material during filling conditions as well as the fluid-like behavior during discharging. A compatible contact element is used for the simulation of the interface zone between the granular bulk material and the structure. Computations are done from static conditions up to discharging velocities of more than four meter per second.

2. GOVERNING EQUATIONS

In solid mechanics the deformation of nonlinear path dependant materials is mostly described by the Lagrangian formulation where the finite element mesh coincide with the quadrature resp. material points throughout the computation. This method encounters difficulties for the calculation of velocity and stress fields in silos as during the discharge period large distortions of the material occur in the outlet region of the bin.

To avoid these difficulties the Eulerian formulation is used [1], [2]. Within this description the element mesh does not coincide with material points during the deformation process. Therefore when nonlinear materials are used, the history of all path dependant quantities like e.g. yield stresses or several hardening parameters need to be traced throughout the mesh. Details of relevant update procedures are given later in this text.

On the basis of the well-established theory of continuum mechanics the motion of a solid is characterized by the following set of 18 nonlinear differential equations:

momentum equation: $\nabla \underline{T} + \rho\,(\underline{b}_V - \dot{\underline{v}}) = 0$ (1)

kinematic relations: $\underline{d} = 0.5\,(\nabla \underline{v} + \nabla \underline{v}^T)$ (2a)

$\underline{w} = 0.5\,(\nabla \underline{v} - \nabla \underline{v}^T)$ (2b)

constitutive relations: $\underline{T} = \underline{T}\,(\underline{d}, \overset{\circ}{\underline{d}}, C_i \ldots)$ (3)

In the above equations, is the bulk density, $\underline{b}_V$ is the body force per unit volume, $\underline{d}$ is the velocity strain tensor, $\underline{w}$ is the spin tensor. $\underline{T}$ are the Cauchy stresses and $\dot{\underline{v}}$ is the acceleration which is given by $\dot{\underline{v}} = \frac{\partial \underline{v}}{\partial t} + \nabla \underline{v} \cdot \underline{v}$.

For reasons of simplicity, ρ is assumed to be constant in space and time. Both geometric and material nonlinearities are included in the setting of equations (1) - (3).

3. NUMERICAL SOLUTION METHOD

By means of the principle of virtual velocities the equations of dynamic equilibrium in an Eulerian frame of reference (1) can be formulated as follows:

$$\int_V \delta \underline{d}^T\, T\, dV + \int_V \delta \underline{v}^T \left(\frac{\delta \underline{v}}{\delta t} + \nabla \underline{v} \cdot \underline{v}\right) \rho\, dV =$$

$$= \int_V \delta\, \underline{v}^T\, \underline{b}_V \rho\, dV + \int_S \delta \underline{v}^T\, \underline{b}_t\, dS \tag{4}$$

In the solution process the velocity field is approximated by $\underline{v}(x,t) = \underline{N}(x)\underline{a}(t)$ resp. the strain rate by $\underline{d}(x,t) = \underline{B}(x)a(t)$ where the matrix of shape functions $\underline{N}$ and the matrix of its spatial derivatives $\underline{B}$ depend only on the fixed spatial coordinate $\underline{x}$ and the vector of nodal velocities $\underline{v}$ only on time.
In the above equation the Cauchy stress $\underline{T}$ is computed according to equation[10], which is discussed in greater detail in chapter 4. The velocity gradient $\nabla \underline{v}$ is computed within each element by finite difference techniques. Therefore no further shape functions have to be introduced. Computations have been done with plane three-node triangle elements as well as with isoparametric four to nine node elements.

With the following abbreviations:

$$\underline{M} = \int_V \underline{N}^T \underline{N} \rho dV \qquad \text{constant mass matrix} \qquad (5a)$$

$$\underline{M}_c = \int_V \underline{N}^T \nabla \underline{v}\, \underline{N} \rho dV \qquad \text{convective mass matrix} \qquad (5b)$$

$$\underline{k} = \int_V \underline{B}^T \cdot \underline{T}\, dV \qquad \text{vector of inner forces} \qquad (5c)$$

$$\underline{p} = \int_V \underline{N}^T \underline{b}_v\, \rho dV + \int_S \underline{N}^T \underline{b}_t\, dS \qquad \text{vector of outer forces} \qquad (5d)$$

equation (4) leads to a set of first order differential equations in the unknown nodal velocities which are solved by the Euler forward method and the modified Newton Raphson scheme.

$$\underline{M} \cdot \frac{\partial \underline{a}}{\partial t} + \underline{M}_c \cdot \underline{a} + \underline{k} = \underline{p} \qquad (6)$$

The evaluation of the stress rate is to be described in close connection with the constitutiv laws as done in the following chapter.

4. VISCOPLASTIC CONSTITUTIVE LAW - CALCULATION OF STRESSES

Besides numerical difficulties the essential problem are the constitutive relations for the bulk material. With regard to the filling and discharging process one needs a complete description ranging from the static up to free flowing conditions of the solid. To cover both the Cauchy stress $\underline{T}$ is divided into a rate independant part $\underline{T}_s$ and a rate dependant part $\underline{T}_v$.

$$\underline{T}\,(\underline{d},\overset{\circ}{\underline{d}}) = \underline{T}_s(\underline{d}) + \underline{T}_v(\overset{\circ}{\underline{d}}) = \underline{H} \cdot \underline{d} + \underline{G} \cdot \overset{\circ}{\underline{d}} \qquad (7)$$

For reasons of objectivity the co-rotational stress rate due to Jaumann and a similar approach for the strain rate is used:

$$\overset{\circ}{\underline{T}} = \dot{\underline{T}} + \underline{T}\,\underline{w} - \underline{w}\,\underline{T} \qquad (8)$$

$$\overset{\circ}{\underline{d}} = \dot{\underline{d}} + \underline{d}\,\underline{w} - \underline{w}\,\underline{d} \qquad (9)$$

Regarding eq. (7) - (9) the local stress rate in the Eulerian frame reads:

$$\frac{\partial \underline{T}}{\partial t} = \underline{H} \cdot \underline{d} + \underline{G} \cdot \frac{\partial \underline{d}}{\partial t} - (\underline{T}\,\underline{w} - \underline{w}\,\underline{T} + \nabla\underline{T}\,\underline{v}) + \underline{G} \cdot (\underline{d}\,\underline{w} - \underline{w}\,\underline{d} + \nabla\underline{d}\,\underline{v}) \qquad (10)$$

The physical properties of the material are represented by the first and second part on the right, the geometric nonlinearities show up in the 3rd and 4th part.

The time derivative $\frac{\partial \underline{d}}{\partial t}$ and spatial gradients $\nabla \underline{T}$, $\nabla \underline{d}$ are approximated by finite difference techniques.

The rate independent material tensor $\underline{\underline{H}}$ is represented by an incremental elasto-plastic law first proposed by Lade [3] and meanwhile extended to large strains and displacements. Within this formulation a strain increment consists of three different parts:

$$\underline{e} = \underline{e}_e + \underline{e}_c + \underline{e}_p \tag{11}$$

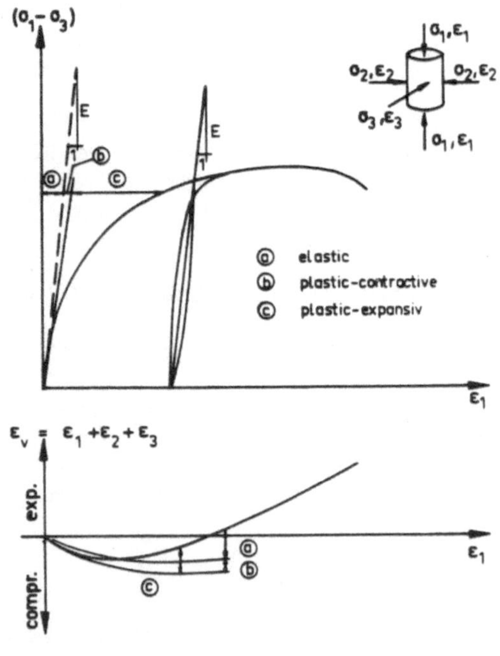

fig. 1: Stress-strain relations in triaxial compression test

The elastic strains $\underline{e}_e$ are calculated by means of a modulus E, depending on the minimal principal stress T_3. The so-called plastic contractive component $\underline{e}_c$ due to particle collapse of the grains is gained from a plastic potential with an associated flow rule, whereas the plastic expansiv component $\underline{e}_p$ is calculated with a non-associated flow rule from another potential function.

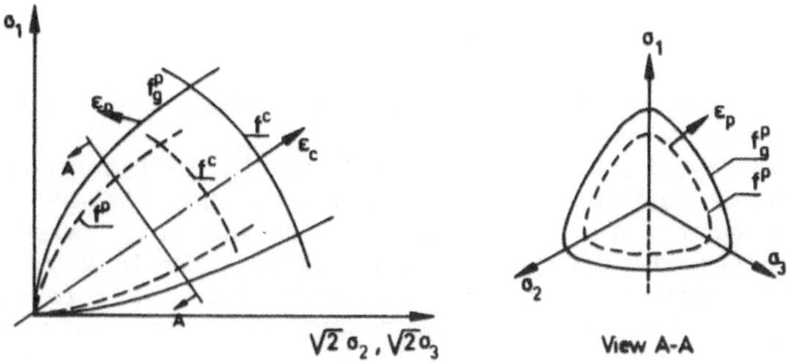

fig. 2: Yield surfaces

The mayor disadvantage of such a law based on associated resp. non-associated flow rules in an Eulerian frame is that all quantities like yield stresses or the different hardening components have to be traced throughout the whole computation, needing much storage and computing time.

Therefore a different constitutive formulation for the calculation of $\underline{T}_s$ (eq. 12), presented by Gudehus and Kolymbas [4], [5] in 1986 has been studied.

$$\overset{\circ}{\underline{T}} = C_1\ (\underline{Td} + \underline{dT})/2 + C_2\ \mathrm{tr}\ (\underline{Td})\underline{1} + C_3\underline{T} \cdot \sqrt{\mathrm{tr}\ \underline{d}^2} + C_4 \frac{\underline{T}^2}{\mathrm{tr}\underline{T}} \sqrt{\mathrm{tr}\ \underline{d}^2} \tag{12}$$

Only four material parameters C_1 - C_4 are used which can be determined by a single triaxial test. There is no destinction between plastic and elastic strains. No yield surfaces have to be calculated. The constitutive equation is well-applicable to problems of small strains as well as to large deformations. After comparision calculations with the aforementioned laws given by Lade the authors prefer the latter formulation as it is much easer to handle.

The behavior of the free flowing material is mainly covered by the tensor $\underline{G}$ in equation (10). Several laws for steady state conditions have been proposed [6], [7] which for reasons of complexity can hardly be used. The significant phenomenas can also be described on the basis of the kinematic gas theory as it was proposed by Buggisch [8]. This formulation leads to a stress strain relation $\underline{G}$ which is similar to the deviatoric part of an incompressible fluid. The viscosity depends on the trace ot the strain rates:

$$\underline{T}_v = \underline{G} \cdot \underline{d} \qquad \underline{G}_{ijrs} = 2\ \nu\ \mathrm{tr}\ \underline{d} \cdot (\delta_{ir}\delta_{js} - \tfrac{1}{3}\ \delta_{ij}\delta_{rs}) \tag{13}$$

This viscosity parameter ν is assumed to be 1 KN sec/m^2 according to a comparision of calculated and experimentally gained results.

5. BOUNDARY CONDITIONS - CONTACT ELEMENT

Both physical and kinematic boundary conditions have to be considered in computation.
As stress states in silos are significantly influenced by the soil structure interaction, a compatible thin layer contact element has been evaluated, which allows the bulk material to move only perpendicular to the wall normal. Besides debonding displacements normal to the wall are excluded. The material behavior of this interface zone is described by the Mohr-Coulomb friction law.

For the discharging process the boundary conditions of the outlet region have to be modified. While the filling process all nodes at the bottom of the bin are fixed in space, at discharging the nodes at the opening have to be free to move.
Furthermore during discharging the changing surface of the material at top has to be traced throughout the whole computation as integration is done over an Eulerian domain.

6. NUMERICAL EXAMPLES

By means of such a programm velocity and stress fields in silos can be calculated for different silo geometries and material parameters. For reasons of computer capacity until now the calculations are restricted to symmetry and plane strain conditions.
The geometry of an excentric discharged silobin and parameters for the bulk material dry sand are given in fig. (3).

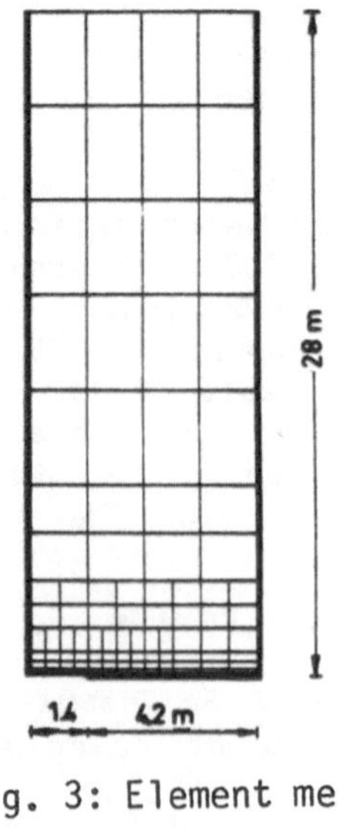

Discredisation parameters
470 Nodes
718 DoF

Bulk material: sand
Unit weight $\gamma = 16$ kN/m
Coefficient of wall friction $\mu = 0.5$
Time step $\Delta t = 0.003$ sec

fig. 3: Element mesh

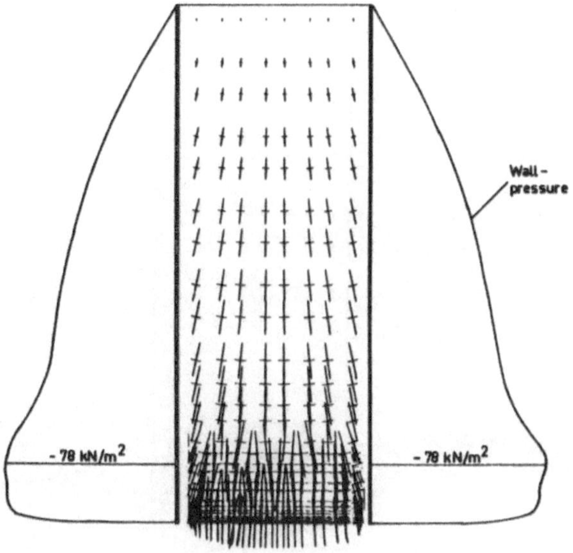

fig. 4: Principal stress field and normal wall pressure after filling

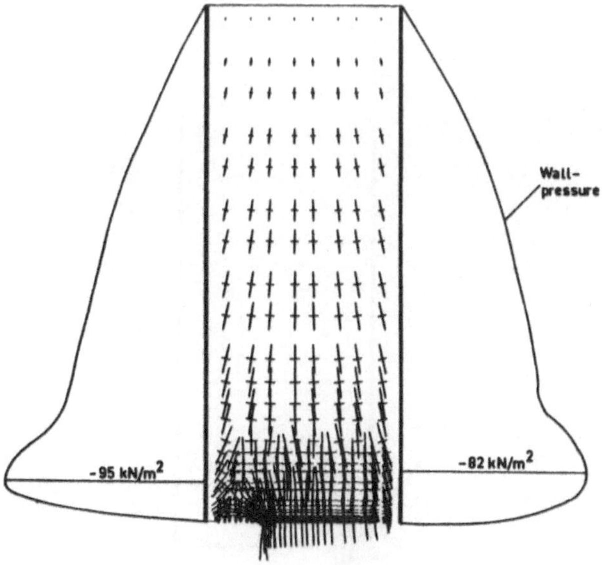

fig. 5: Principal stress field and normal wall pressure after opending phase

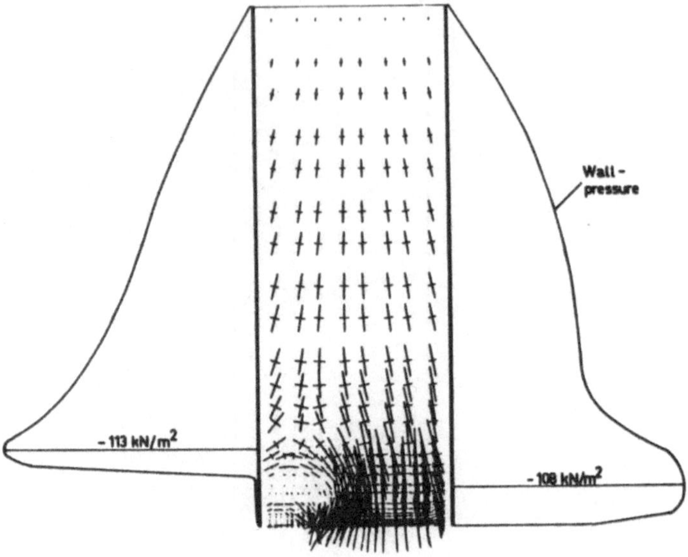

fig. 6: Principal stress field and normal wall pressure after a discharging period of t = 0.60 sec

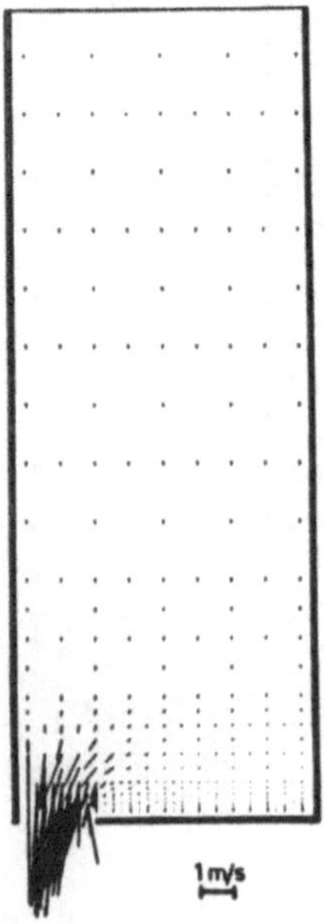

fig. 7: Velocity field after a discharging period of t = 0.6 sec

To simulate the filling phase, the silo is incrementally loaded. Viscous and mass terms have to be omitted. All bottom nodes are fixed in vertical direction (see fig. 4).

The discharging process starts with an opening phase, where during a fixed time all nodes at the outlet are free to move and therefore the bulk material is accelerated towards the outlet. The vertical pressure directly above the outlet decreases to zero. The orientation of the larger principal stresses changes from vertical to horizontal leading to a significant increase of wall pressure at the transition zone (see fig. 5).

Afterwards the discharging phase follows, where the zone of low stresses expands into the silo until a constant state of velocity and stress is reached. The wall pressure peaks increase (see fig. 6). In the upper part of the bin the flow velocity is approximately constant, what means that the bulk material moves more or less like a rigid body (see fig. 7).

Results of this type are of great importance for engineering practice as such unsymmetric pressures and load peaks lead to bending moments in silowalls. In silo design they have to be regarded otherwise as experimence teaches great damages, e.g. buckling of steel bins, occur.

References

[1] Häußler, U., Eibl, J.: 'Numerical Investigations on Discharging Silos', ASCE, Vol. 110, No. 6, June 1984

[2] Gladen, W.: 'Numerische Untersuchung der Lasten in Silozellen beim exzentrischen Entleeren' D Thesis 1985, Karlsruhe

[3] Lade, P.V.: 'Elastic-Plastic Stress Strain Theory for Cohesionsless Soils with Curved Yield Surfaces', Int. J. of Solids and Structures, Vol. 13, 1977

[4] Gudehus, G., Kolymbas, D.: 'An approach to cohesionsless materials and cylindrical silos', 1st World Congress Particle Technologie, 1986

[5] Kolymbas, D.: 'A Constitutive Law of the Rate Type for Soils and other Granular Materials' Numerical Methods in Geomech., Kosice, 1987

[6] Savage, S.B.: 'Gravity flow of Cohesionless Granular Materials in Chutes and Channels' J. of Fluid Mech., 92, 1979

[7] Haff, P.K.: 'Grain Flow as a Fluid-mechanical Phenomenon', J. of Fluid Mech., 134, 1983

[8] Buggisch, W.: 'Rheologie of Rapid Flowing Granular Powders', Int. Fine Particle Research Inst., Brüssel, 1986

THE CONVERGENCE PROPERTIES OF A SERIES OF R-FUNCTIONS FOR SIMPLE POLYGONAL SHAPES

D.V.Altiparmakov[1] and M.S.Milgram[2]

[1]The Boris Kidrič Institute of Nuclear Sciences, Vinča
P.O.Box 522, 11001 Belgrade, Yugoslavia

[2]Atomic Energy of Canada Ltd., Chalk River Nuclear Laboratories
Chalk River, Ontario, Canada KOJ 1JO

SUMMARY

A numerical study of the convergence of the R-function solution to the Helmholtz equation is presented in this paper. Several two-dimensional domains of simple polygonal shape have been considered. Calculations have been carried out by four types of trial functions derived from two different solution structures. In addition, a singular function series is applied for the purpose of comparison. In the case of convex domains, one of the presented approximations yields an accurate solution with a very low number of degrees of freedom. However, the accuracy is very poor for the reentrant region and a separate treatment of singularity seems to be necessary.

1. INTRODUCTION

A general approach to the problem of geometry in the solution of boundary value problems has been offered by Rvachev's R-function theory [1]. Using this theory, a formal solution, that a priori satisfies all boundary conditions, can be easily constructed for various engineering problems dealing with a complex geometry. This paper presents a numerical study of the convergence of R-function approximation. We are concerned with the solution of the Helmholtz equation

$$\nabla^2\phi + \lambda\phi = 0 \,, \tag{1}$$

in a two-dimensional domain Ω, subject to the Dirichlet condition on the boundary $\delta\Omega$

$$\phi = 0 \quad \text{on} \quad \delta\Omega. \tag{2}$$

For the sake of simplicity, the attention is concentrated on the determination of the smallest eigenvalue λ_1 and the related eigenfunction ϕ_1. The following domains have been considered:

i) Regular triangle ($a = 2\sqrt{3}$),

ii) Square ($a = 2$),

iii) Regular hexagon ($a = 2/\sqrt{3}$),

iv) Rhombus ($a = 1$, $\Theta = \pi/4$) and

v) L-shaped domain consisting of three unit squares,

where a denotes the side length and Θ is the skew angle.

2. APPROXIMATE SOLUTION

The smallest eigenvalue λ_1 is equal to the minimal value of the Rayleigh quotient over the space of continuously differentiable functions on Ω which vanish on $\delta\Omega$. As suggested by Kantorovich and Krylov [2], an approximate value $\lambda_1^{(n)}$ can be determined by looking for the stationary point of the Rayleigh quotient over an n-dimensional space with the following basis:

$$\left\{ \omega\xi_i \;:\; \omega, \xi_i \in C^1(\Omega),\;\; \omega > 0 \text{ within } \Omega, \omega = 0 \text{ on } \delta\Omega \right\}_{i=1,2,..n} \tag{3}$$

where $\{\xi_i\}$ is a complete set of functions. Owing to the completeness of $\{\xi_i\}$, the approximate solution would converge to the exact one, $\lambda_1^{(n)} \to \lambda_1$ and $\phi_1^{(n)} \to \phi_1$, when $n \to \infty$.

2.1 The function ω

In some cases the function ω can be very simply defined. For example, if the domain Ω is an analytic figure which boundary $\delta\Omega$ is defined by an equation of the form $f(x,y) = 0$, then one can adopt $\omega = \pm f$. Similarly, for the case of a convex polygon, the equations of whose sides are $\omega_k(x,y) = 0$, one can adopt $\omega = \pm\prod_k \omega_k$. However, the matter can be considerably more complicated in a general case. The R-function theory yields a simple technique of constructing the function ω for arbitrary semi-analytic region. Among various systems of R-functions, which are related to the complete set of logical functions, our choice has been made on $\overset{0}{R}_m$-system. The corresponding functions belong to

$C^{m-1}(\Omega)$ and conserve the normalization, i.e.

$$\frac{\partial \omega_k}{\partial \nu}\bigg|_{\delta\Omega_k} = 1 \quad \Rightarrow \quad \frac{\partial \omega}{\partial \nu}\bigg|_{\delta\Omega_k} = 1 \;, \qquad (\delta\Omega = \bigcup_k \delta\Omega_k), \tag{4}$$

where ν denotes the inward normal to $\delta\Omega$, ω is $\overset{0}{R}_m$-function of the domain Ω and $\delta\Omega$ consists of several parts $\delta\Omega_k$ whose equations are $\omega_k = 0$. Another important property of $\overset{0}{R}_m$-system is that the geometric symmetry of Ω is reflected on the function ω as well. As an illustration, here is given the explicit form of $\overset{0}{R}_m$-conjunction,

$$\Omega_1 \cap \Omega_2 \quad \rightarrow \quad \omega_1 \wedge \omega_2 \;\equiv\; \begin{cases} \omega_1\omega_2(\omega_1^m + \omega_2^m)^{-1/m}, & \forall\, \omega_1 > 0,\ \omega_2 > 0 \\ \omega_1, & \forall\, \omega_1 \leqslant 0,\ \omega_2 \geqslant 0 \\ \omega_2, & \forall\, \omega_1 \geqslant 0,\ \omega_2 \leqslant 0 \\ (-1)^{m+1}(\omega_1^m + \omega_2^m)^{1/m}, & \forall\, \omega_1 < 0,\ \omega_2 < 0 \end{cases} \tag{5}$$

In the case of a domain which logical expression is an intersection of a number of component domains Ω_k $(\Omega = \bigcap_k \Omega_k)$, after a certain amount of algebra, the resulting function can be written as follows:

$$\omega = \left(\sum_k \omega_k^{-m} \right)^{-1/m} . \tag{6}$$

2.2. Solution structures

Assuming that a function $\omega \in C^1(\Omega)$ is constructed, the solution of Eq.(1) under the boundary condition (2) can be expressed in the following form:

$$\phi = \omega \Psi, \tag{7}$$

where Ψ is an inner product of the approximation skeleton $\{a_i\}$ and a basis $\{\xi_i\}$.

In the case of geometric symmetry, the original problem, Eqs.(1) and (2), can be modified and instead of looking for a solution over the entier domain Ω, one can consider a symmetric part Ω_s only. Thus, Eq.(1) has to be solved under the following boundary conditions:

$$\phi = 0 \qquad \text{on } \delta\Omega_0, \tag{8}$$

$$\frac{\partial \phi}{\partial n} = 0 \qquad \text{on } \delta\Omega_1, \tag{9}$$

where Ω_s is bounded by $\delta\Omega_s = \delta\Omega_0 \cup \delta\Omega_1$, $\delta\Omega_0 \subset \delta\Omega$, $\delta\Omega_1$ is formed by the axes of symmetry and n denotes the outward normal to $\delta\Omega$. Let us denote by ω_0, ω_1 and ω_s the normalized R-functions of the domains Ω_0, Ω_1 and Ω_s ($\Omega_s \subset \Omega_0, \Omega_1$), respectively. A solution of the problem defined by Eqs.(1),(8) and (9) can be constructed as follows:

$$\phi = \omega_1 \Psi_1 - \omega_s (\nabla\omega_2, \nabla(\omega_1 \Psi_1)) + \omega_s^2 \Psi_2 \ , \tag{10}$$

where, similarly to (7), the components Ψ_1 and Ψ_2 are expressed by the unknown coefficients $a_{1,i}$, $a_{2,i}$ and some basis functions. For a normalized function ω, the following relationship holds:

$$\frac{\partial f}{\partial n} = -(\nabla\omega, \nabla f) \qquad \text{on} \quad \delta\Omega. \tag{11}$$

Hence, it is easy to show that the boundary conditions (8) and (9) are a priori satisfied by the structure (10).

2.3. Corner singularity

It has been shown [3] that in a domain with corners, the solution of elliptic problems has a singular behaviour. In particular, near to a corner, the solution of the Helmholtz equation can be represented by the following asymptotic expansion:

$$\phi(r,\theta) \sim \sum_{\ell} c_\ell \sin(\ell\theta/\alpha)\, J_{\ell/\alpha}(\sqrt{\lambda}\, r) \ , \tag{12}$$

where (r,θ) are polar coordinates, $\alpha\pi$ is the angle and $J_{\ell/\alpha}$ is the Bessel function of order ℓ/α. In order to increase the accuracy, the leading term $r^{1/\alpha}$ of the Bessel function expansion can be incorporated into the approximate solution. For that purpose, the functions ω and ω_1, in the structures (7) and (10), have to be multiplied by

$$g_c = \prod_k |r - r_k|^{1/\alpha_k - 1}. \tag{13}$$

3. NUMERICAL RESULTS

The results of calculations of $\lambda_1^{(n)}$, obtained by varying number of degrees of freedom $(n = 1,2,\ldots 10)$, are presented in Table 1. (regular polygons) and Table 2. (rhombus and L-shaped domain). In addition, for the case of triangle and square, the global and local errors (L_2 and L_∞ norms) of the first eigenfunction are given in Fig.1. The calculations have been carried out by four types of trial functions, which are denoted by A,B, C and D and defined in the sequel.

A : Structure (7) and $\Psi = \sum_i a_i P_i(x,y)$

Domain	n	$\lambda_1^{(n)}$ Approximation A	B	C	D	E
Triangle	1	5.241	4.389	6.082	5.461	4.400
	2	4.495	4.3882	6.081	5.334	4.3945
	3	4.460	4.3881	5.001	4.701	4.3871
	4	4.3889	4.3869	4.929	4.679	4.386503
	5	4.3881	4.38692	4.597	4.403	4.3864929
	6	4.38684	4.38685	4.596	4.4027	4.3864923
	7	4.38653	4.38659	4.468	4.4026	4.3864915
	8	4.386502	4.386545	4.461	4.4022	4.3864908
	9	4.386502	4.386543	4.432	4.3953	4.3864908
	10	4.386502	4.386543	4.431	4.3951	
	Exact value $\lambda_\Gamma = 4.3864909$					
Square	1	5.531	4.9631	6.149	5.955	5.000
	2	4.997	4.9386	6.149	5.876	4.9369
	3	4.9687	4.9362	6.136	5.078	4.93543
	4	4.9357	4.9351	6.108	5.067	4.9348023
	5	4.9354	4.93486	5.286	4.950	4.9348022
	6	4.93511	4.93484	5.230	4.948	4.9348022
	7	4.93509	4.934813	5.116	4.9438	
	8	4.934844	4.934809	5.116	4.9433	
	9	4.934844	4.934808	4.942	4.9425	
	10	4.934834	4.934807	4.9417	4.9425	
	Exact value $\lambda_1 = 4.9348022$					
Hexagon	1	5.970	5.525	6.109	6.229	5.3836
	2	5.385	5.376	6.109	6.187	5.3670
	3	5.384	5.372	6.041	5.392	5.36689
	4	5.3823	5.3718	6.023	5.388	5.36668
	5	5.3675	5.3683	5.664	5.383	5.366512
	6	5.36749	5.36817	5.619	5.381	5.3665056
	7	5.36696	5.36718	5.504	5.3767	5.3665042
	8	5.36693	5.36718	5.504	5.3761	5.3665042
	9	5.36683	5.36718	5.383	5.3719	5.3665042
	10	5.36682	5.36718	5.379	5.3718	5.3665040

Table 1: Approximate values $\lambda_1^{(n)}$ for the regular polygons

Domain	n	$\lambda_1^{(n)}$ Approximation A	B	C	D	E
Rhombus	1	44.20	34.87	44.87	42.20	40.49
	2	44.20	34.82	44.83	40.79	35.239
	3	35.89	34.82	44.06	40.64	34.919
	4	35.78	34.80	44.02	40.54	34.846
	5	35.78	34.79	36.41	35.15	34.787
	6	34.82	34.786	36.12	35.10	34.784
	7	34.81	34.786	36.01	34.95	34.766
	8	34.802	34.784	36.00	34.89	34.7535
	9	34.802	34.772	36.00	34.83	34.7468
	10	34.751	34.771	35.98	34.77	34.7468
	Bounds [4]	$\lambda_1^-=34.7114$	$< \lambda_1$	$<$	$35.028 = \lambda_1^+$	
L-shaped domain	1	11.67	9.972	14.99	10.37	9.868
	2	11.43	9.767	14.64	10.21	9.6628
	3	11.34	9.747	13.62	10.21	9.6573
	4	10.15	9.740	13.57	10.12	9.6463
	5	10.15	9.735	11.92	10.04	9.6463
	6	10.15	9.733	11.90	10.03	9.6409
	7	10.11	9.718	10.79	10.02	9.64048
	8	10.07	9.718	10.79	10.02	9.64040
	9	10.07	9.715	10.46	10.02	9.63967
	10	10.04	9.715	10.44	9.978	9.63967
	Reference value [5]	$\lambda_1 = 9.6397238$				

Table 2: Approximate values $\lambda_1^{(n)}$ for the rhombus and L-shaped domain

B : Structure (7), $\Psi = g_c \sum_i a_i P_i(x,y)$, $g_c = \prod_\ell \omega_{c\ell}^{1/\alpha_\ell}$ and $\omega_{c\ell} = \bigwedge_{s=1,2,.} |r-r_{\ell s}|$

where the index s denotes the symmetric counterparts of the corner ℓ and $P_i(x,y)$ are polynomials of x and y, which satisfy the geometric symmetry of Ω.

C : Structure (10), $\Psi_t = \sum_{i,j}^{i+j=n} a_{t,ij}\, x^i y^j$, t=1,2.

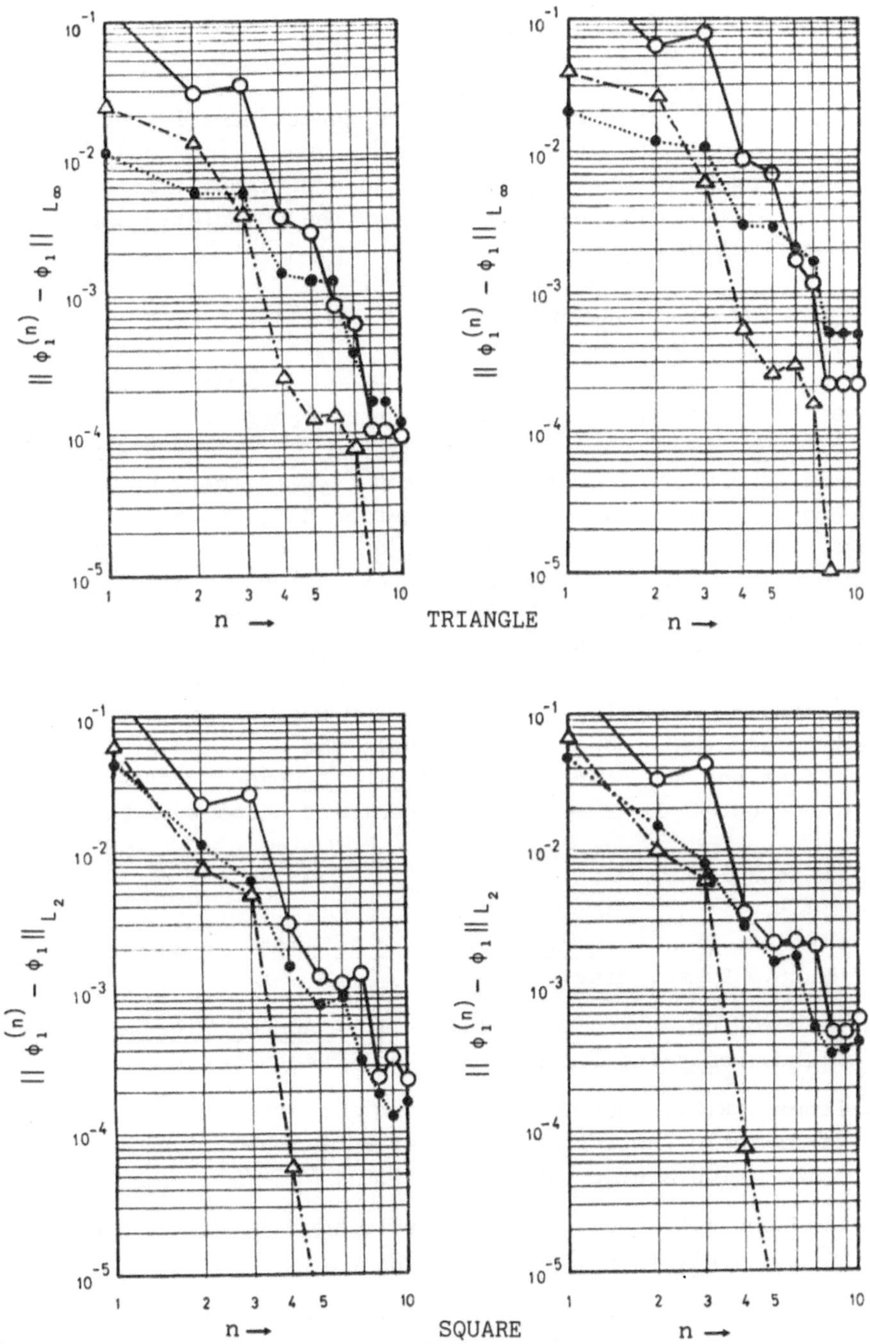

Fig.1 : Global and local errors of $\phi_1^{(n)}$ for triangle and square

Approximations: A O—O—O , B •·····•·····• , E Δ-·-Δ-·-Δ

$$\text{D : Structure (10),} \quad \Psi_t = \prod_{\ell} |r - r_\ell|^{1/\alpha_\ell - t} \sum_{i,j}^{i+j=n} a_{tij}\, x^i y^j, \quad t=1,2$$

where the index ℓ is related only to the corners lying on $\delta\Omega_1$.

For the purpose of comparison, an additional trial function, denoted by E, is defined according to the asymptotic expansion (12) as follows

$$\text{E :} \quad \phi \quad \sum_{i,j,k} c_{ijk} \; w_k^i \, |r - r_k|^{\frac{2j+1}{\alpha_k}} \; \sin \frac{2j+1}{\alpha_k}(\theta - \theta_k)$$

where w is the product of all ω_i, exept the two which are related to the polygon sides forming the corner k.

It is obvious from the presented results that, concerning the convergence rate, the approximation E (singular function series) is superior to the others. On the other hand, the convergence rate of the R-function aproximations is practically the same in all of the considered cases, but a great difference in the accuracy can be observed. In the case of convex domains, the approximation B yields a sufficiently accurate solution by only one unknown. Comparing the results for regular polygons, it is easy to see that increasing the angle the solution is aggravated. That fact and the poor results for the L-shaped domain indicate that a special treatment of the reentrant corners is necessary.

REFERENCES

1. Rvachev, V.L. - Teorya R-Funktsij i Nekotorye ee Prilozhenya, Naukova Dumka, Kiev, 1982.

2. Kantorovich, L.V. and Krylov, V.I. - Approximate Methods of Higher Analysis, Interscience Publishers, New York.

3. Lehman, R.S., 'Developments at an Analytic Corner of Solutions of Elliptic Partial Differential Equations', Journal of Mathematics and Mechanics, 8 (1959), 727-760.

4. Stadter, J.R.,'Bounds to Eigenvalues of Rhombical Membranes', J.SIAM Appl. Math.,14 (1966), 324-341.

5. Fix, G.J., Gulati, S. and Wakoff, G.I., 'On the Use of Singular Functions with Finite Element Approximations', J.Comp. Physics, 13 (1973), 209-228.

CREATION OF SHIP BODY FORM WITH MINIMUM THEORETICAL RESISTANCE USING FINITE ELEMENT METHOD

Jan P. Michalski*, Antti Pramila, Simo Virtanen
Tampere University of Technology, 33101 TAMPERE, FINLAND
*On leave from Technical University of Gdansk, GDANSK, POLAND

SUMMARY

The paper presents an approximate solution of a nonlinear, isoperimetric and stationary variational problem. This is a mathematical model of an interesting engineering task in the field of Naval Architecture: creation process of such a surface $S(\Omega)\subset R^3$ which describes a ship body form with minimum theoretical total resistance for a given velocity of the ship. The surface S is a graph of an extremal which minimizes a functional representing mathematical model of ship resistance.

An analytical model of the problem is formulated, and next, it is discretised using FEM approximation procedure with such basis and transformation functions that local domains are convex and global solution is of class $C^1(\Omega)$ on a given domain $\Omega\subset R^2$.

Using Ritz-finite element method the conditions for a stationary point are formulated and the solution is sought by mathematical programming technique. An illustrative example on an engineering task and a graphical representation of the solution is shown.

1. INTRODUCTION

A successful process of designing a ship body form with possible low resistance and with a given set of design conditions, as a rule, is the result of labour-consuming drafting techniques, and next, expensive experiments in a towing tank or, can be a consequence of the designers genius, but success by a chance can also be a case. However, by such approach we never know whether the result is the real optimum.

The attempts to find an algorithmic method for creation

minimum resistance ship form, based on mathematical formulations resulting from physical theory, has a long tradition since Newton's considerations in "Principles". The searches result mainly from an economical aspect of the following engineering problem: minimization of ship fuel consumption and ship machinery investment costs.

The analytical solution of the problem is not known but some interesting results for an approximate model were obtained by Wehausen et al. [1], [2] and later by Hsiung [3]. However, from engineering applications point of view the quoted solutions have same limitations: simple boundaries (rectangular domain) and small number of degrees of freedom in the former case (global approximation by finite Fourier series), and the same simple boundaries and only $C^0(\Omega)$ continuous global solution in the latter case (local approximation method). We try to overcome these limitations by a method with high degree of freedom, with a large class of admissible boundaries (it is an important problem especially at the stern and stem of a ship), and with trial functions of class $C^1(\Omega)\subset C^0\cap C^1$.

2. CONTINUOUS MATHEMATICAL MODEL

The total theoretical resistance R_T of a ship moving on the water surface with uniform velocity v can be expressed, with some simplification, as a sum of frictional resistance component R_F and wave-making resistance component R_W. The nature of these two forces reflects, in the first case, viscous friction between water and hull surface and, in the second case, gravity forces and pressure disturbances in water, caused by moving ship.

According to Froude hypothesis R_F is proportional to a friction coefficient C_F and to the wetted area A of the hull surface S. So, minimizing the area of the wetted surface the fricitonal resistance is simultaneously minimized. Theoretical wave resistance can be expressed as an integral which results from variational formulation of a Neumann type boundary value problem. In the scope of present work we express the wave resistance by Michell integral [4] which results from the thin ship wave resistance theory. According to our assumption, the total theoretical resistance R_T can be expressed as

$$R_T = R_F + R_W \equiv \Phi_T(u) = \Phi_F(u) + \Phi_W(u) \tag{2.1}$$

where frictional resistance

$$R_F = \Phi_F(u) = \tfrac{1}{2}\rho v^2 C_F A = C_f \iint [1 + (\frac{\partial u}{\partial x})^2 + (\frac{\partial u}{\partial y})^2]^{\frac{1}{2}} d\Omega \tag{2.2}$$

and Michell integral

$$R_W = \Phi_W(u) = \frac{4\rho g^2}{\pi v^2} \int_0^{\pi/2} (H_S^2 + H_C^2) \frac{d\Theta}{\cos^3\Theta} \tag{2.3}$$

were

$$\left.\begin{matrix} H_S \\ H_C \end{matrix}\right\} = \iint\limits_{\Omega_u} \frac{\partial u}{\partial x} \exp\left(- \frac{g \cdot y}{v^2 \cdot \cos^2\Theta}\right) \cdot \left\{\begin{matrix} \sin \\ \cos \end{matrix}\right. \left(\frac{g \cdot x}{v^2 \cdot \cos\Theta}\right) d\Omega_u \qquad (2.4)$$

and ρ denotes density of water, g is gravitational constant, v is ship velocity and Θ is the direction of wave propagation. Both functionals depend on a function u, $u:\Omega_u \subset R^2 \rightarrow u(x,y) \in R^1$, and $\Phi_T: u \in D_\Phi \rightarrow \Phi(u) = \int_\Delta F(\cdot) d\Delta \in R^1$ where D_Φ is a space of admissible functions and Ω_u is a given open domain defined by its boundary $\partial\Omega_u$. In ϕ_F the integrand $F(x,y,u,u_x,u_y):\Omega_u \subset R^2 \rightarrow R^1$ and in ϕ_W, $F(x,y,\Theta,u,u_x):R^5 \supset \Omega_u \times \Omega_u \times \{\Theta:\Theta \in (0;\pi/2)\} \rightarrow R^1$. As isoperimetric constraints we have a set of linear geometrical conditions

$$K: \iint\limits_{\Omega_u} K(x,y,u,u')d\Omega_u = C \quad , \; C \in R^m \quad . \qquad (2.5)$$

On the boundary $\partial\Omega_u$ are given boundary conditions $u = \bar{f}(x,y)$, $u_{,n} = \bar{g}(x,y)$ or $u(s) = \bar{u}(s)$, $u_{,n} = \bar{g}(s)$ where a bar denotes given function and s is a co-ordinate along $\partial\Omega_u$. The graph of the function u is the surface S which defines the ship body form: $S : S \equiv \{(x,y,u(x,y)) : (x,y) \in \Omega_u\} \subset R^3$. Because of the above formulation and application aspects it is demanded that $u \in D_\Phi \subset C^1$, at least. The final formulation of the problem under consideration is: find such an element u_0 from the admissible function space that

$$\Phi_T(u_0) \leq \Phi_T(u) \wedge u \in D_\Phi \qquad (2.6)$$

A solution, in a classical sense, of the variational problem formulated above is unknown. For the functional Φ_F a classical solution is known but only for very simple cases (e.g., a sphere) so that it is out of interest from point of view of present work.

3. APPROXIMATE MODEL

3.1 Preliminaries

Based on the theory of non-linear operators in a Banach space it is possible to show [5] that the functional $\Phi_T(u)$ fulfils the following condition:

$$\Phi_T(u) + \Phi_T(v) - 2\,\Phi_T\left(\frac{u+v}{2}\right) \geq 0 \;\; ; \;\; u,v \in D_\Phi$$

The equality sign only holds for u = v. It means, that if a solution exists, it is unique. In the practical case it is sufficient since the solution is based on physical grounds and the question of interest is whether or not it will be a unique solution.

Explicit form of a function, in which the above continuous model has been expressed, has well known limitations when used as a graph of a surface, e.g., it is not possible to express a surface with an infinite gradient. Since in our problem it is a fundamental matter to avoid such limitations we express the surface S in a parametric form which, in the discussed type of problems, has obvious advantages [6]. A parametrized surface can be expressed by a vector function r, $r:\Omega_r \subset R^2 \to R^3$. The surface S corresponding to the function r is its image: $S \equiv r(\Omega_r)$. We can write $r = (x=x,\ y=y,\ z=z(x,y))$ and if $x,y,z(x,y)$ are functions globally of class C^1 then S is also a surface globally of class C^1.

The expression for Michell integral, formulas (2,3) and (2,4), are modified in such a way that the outer integral (with respect to Θ) is replaced by Gauss-Tshebyshev quadrature. Putting $t = \cos\Theta$ which yields

$$\Phi_W(r) = C_W \int_0^1 t^{-3}(1-t^2)^{-\frac{1}{2}}[H_S^2(t) + H_C^2(t)]\,dt \tag{3.1}$$

The orthogonal system of polynomials which correspond to the segment [0;1] and the weight function $(1-t^2)^{-\frac{1}{2}}$ is the system of Chebyshev polynomials of the first kind

$$T_n = \cos(n\cdot\arccos(t)) \quad , \ t \in R^1 \tag{3.2}$$

The roots of T_n are the nodes to be used in the quadrature formula; these are

$$t_K = \cos\left(\frac{2\cdot K-1}{4\cdot n}\pi\right) \quad , \ K = 1,2,\ldots,n \tag{3.2}$$

and hence we obtain

$$\Phi_W(r) \cong \tilde{\Phi}_W(r) = C_W \frac{\pi}{n} \sum_{K=1}^{n} \left\{ t_K^{-3}[H_S^2(t_K) + H_C^2(t_K)] \right\} \tag{3.4}$$

The above quadrature with n nodes is exact for polynomials of degree $2\cdot n-1$ [7].

3.2 Discretization by FEM

The original problem with infinite degrees of freedom is transformed into an approximate mathematical model with finite degrees of freedom by Ritz-finite element method procedure.

The given open and bounded domain Ω_r is closed by its boundary $\partial\Omega_r$ where $\partial\Omega_r = \bigcup_{i=1}^{I} \Gamma_i$, $\Gamma_i \in C^1(s)$. The closure $\bar{\Omega}_r$ of Ω_r is partioned into E subdomains $\bar{\Omega}_r^e$ in such a way that every element $\bar{\Omega}_r^e$ is closed and consists of a nonempty interior Ω_r^e,

$$\bar{\Omega}_r = \bigcup_{e=1}^{E} \bar{\Omega}_r^e \ , \text{ and } \Omega_r^e \cap \Omega_r^f = 0 \text{ for } e \neq f.$$

Because of specific features of ship surfaces and kinds of geometrical conditions imposed on the surface we choose a topologically quadrilateral shape of the element. On each element the sought function r is approximated by a linear combination of the form

$$r^e(x,y) = \sum_{i=1}^{16} a_i^e \Psi_i^e(x,y) \quad , \ a_i^e \in R^3 \ , \ r^e:\Omega^r \to R^3 \quad , \tag{3.5}$$

where vectors $\underline{a}_i^e$ are treated as independent variables of the problem. They are taken as the values of r and its derivatives r_x, r_y and r_{xy} in the nodal points on each element $\bar{\Omega}_r^e$. The set of basis functions $\{\Psi_i^e(x,y)\}$ spans a finite-dimensional space of polynomials which are linearly independent and have locally compact support, $\bar{\Omega}_r^e = \text{supp}\{\Psi_i^e(x,y)\}_{e=1}^E$.

To unify the expressions for numerical integrations within all elements, which can be of different shapes in global system, an element of reference V_r is introduced, which is defined in an abstract nondimensional space. The V_r element is a rectangle $[-1;1] \times [-1;1]$. The geometry of the reference element is then mapped into the geometry of the real element by an one to one transformations

$$\tau_x : x \to x(\xi,\eta) = [\Psi(\xi,\eta)]\{\hat{x}^e\}, \ \tau_y : y \to y(\xi,\eta) = [\Psi(\xi,\eta)]\{\hat{y}^e\}$$
$$z \to x(\xi,\eta) = [\Psi(\xi,\eta)]\{\hat{x}^e\} \tag{3.6}$$

The basis functions are chosen to fulfill the continuity conditions imposed on the surface S. The same basis functions are used for construction of the geometrical transformation functions (isoparametric method). With basis of Hermitian polynomials an elementary patch of the surface can be expressed in full as

$$r^e(\xi,\eta) = [\Psi_1^e(\xi)\,\Psi_2^e(\xi)\,\Psi_3^e(\xi)\,\Psi_4^e(\xi)] \cdot \begin{bmatrix} r(-1,-1) & r(-1,1) & r_\eta(-1,-1) & r_\eta(-1,1) \\ r(1,-1) & r(1,1) & r_\eta(1,-1) & r_\eta(1,1) \\ r_\xi(-1,-1) & r_\xi(-1,1) & r_{\xi\eta}(-1,-1) & r_{\xi\eta}(-1,1) \\ r_\xi(1,-1) & r_\xi(1,1) & r_{\xi\eta}(1,-1) & r_{\xi\eta}(1,1) \end{bmatrix}^e \begin{bmatrix} \Psi_1^e(\eta) \\ \Psi_2^e(\eta) \\ \Psi_3^e(\eta) \\ \Psi_4^e(\eta) \end{bmatrix} \tag{3.6}$$

or shortly

$$r^e(\xi,\eta) = [\Psi_i^e(\xi)][Q^e][\Psi_i^e(\eta)]^T \tag{3.7}$$

The elementary patch is completely defined in terms of the vectors $r^e, r^e_\xi, r^e_\eta, r^e_{\xi\eta}$ at its four corners and we get a tensor-product patch [6], which is acceptable for this type of surfaces and further design process. It is only necessary to match these four vector quantities at contiguos corners of adjacent patches to achieve ordinate and gradient continuity across all boundaries [6]. All derivatives in $\Phi_T(r)$ are mapped by Jacobian matrix of the geometrical transformations

$$\begin{bmatrix} \frac{\partial}{\partial x} \\ \frac{\partial}{\partial x} \end{bmatrix} = \begin{bmatrix} \frac{\partial \xi}{\partial x} & \frac{\partial \eta}{\partial x} \\ \frac{\partial \xi}{\partial y} & \frac{\partial \eta}{\partial y} \end{bmatrix} \begin{bmatrix} \frac{\partial}{\partial \xi} \\ \frac{\partial}{\partial \eta} \end{bmatrix} \tag{3.8}$$

Now, the original problem can be expressed in the following form

$$\Phi_F(r) \cong \tilde{\Phi}_F(r) = \sum_{e=1}^{E} \tilde{\Phi}_F^e(r) = C_F \bigcup_{e=1}^{E} \iint_{V_r} \left\{1 + [\frac{\partial z(\xi,\eta;\{\hat{z}\}^e)}{\partial \xi}]^2 + [\frac{\partial x(\xi,\eta;\{\hat{z}\}^e)}{\partial \eta}]^2\right\}^{\frac{1}{2}} \det{}^e(J) dV_r \tag{3.9}$$

and

$$\tilde{\Phi}_w(r) = \sum_{e=1}^{E} \tilde{\Phi}_w^e(r) = C_w \bigcup_{e=1}^{E} \left\{\frac{\pi}{n} \sum_{K=1}^{n} [\iint_{V_r} \frac{\partial z(\xi,\eta;t_K;\{\hat{z}\}^e)}{\partial \xi} \exp(\omega \cdot y(\xi,\eta; t_K;\{\hat{y}\}^e)) \cdot {\cos \atop \sin}(\mu \cdot x(\xi,\eta;t_K;\{\hat{x}\}^e)) \det{}^e(J)\, dV_r]^2 \cdot t_K^{-3}\right\} \tag{3.10}$$

The $\tilde{\Phi}_w^e$ functionals are not easy integrable because of multioscillation character of the integrand which, for practical data, can oscillate hundred times on the range of integration. Numerical integration technique generally used in FEM, e.g., a Gauss quadrature, can not be used for these calculations because of very poor convergence. In order to get possibly exact quadrature the reference element is partitioned into subelements and on each subelement the following approximations are used by bilinear form

$$z(\xi,\eta) \cong \tilde{z}(\xi,\eta) = a_z(\xi)\cdot b_z(\eta) + c_z, \quad z: V_r^s \to R^1 \tag{3.11}$$

and by linear forms

$$\begin{aligned} x(\xi,\eta) &\cong \tilde{x}(\xi,\eta) = a_x(\xi) + b_x(\eta) + c_x \\ y(\xi,\eta) &\cong \tilde{y}(\xi,\eta) = a_y(\xi) + b_y(\eta) + c_y \end{aligned} \quad , \; x,y: V_r^s \to R^1 \tag{3.12}$$

With these approximations the integrals are calculated analytically and the tests performed have given very good results.

As the next step a routine assembly process is performed.

Following the Ritz procedure the conditions for the stationary point are formulated as

$$\frac{\partial\tilde{\Phi}_T(r)}{\partial\hat{z}_i} = 0 \quad , \hat{z}_i \in \{\hat{z}_j^l\} \begin{smallmatrix} l=1,2,\dots,N \\ j=1,2,3,4 \end{smallmatrix} \tag{3.13}$$

where N is the number of nodes. Using Lagrangian method the linear constrains of the problem can be taken into account by reformulation of the objective functional into the form

$$\tilde{\Phi}_T(r) = \tilde{\Phi}_T(r) + \lambda(\tilde{K}(r) - C_K) \ , \ \tilde{K}:R^n \to R^{n-m}, \ \lambda, \ C_K \in R^m \tag{3.14}$$

Now, if a solution exists, the following system of algebraic equations has to be solved

$$\begin{aligned} \frac{\partial\tilde{\Phi}_T(r)}{\partial\hat{z}_i} &\doteq 0 \qquad i = 1,2,\dots,n = 4 \cdot N \\ \frac{\partial\tilde{\Phi}_T(r)}{\partial\lambda_p} &= 0 \qquad p = 1,2,\dots,m \end{aligned} \tag{3.15}$$

The $\tilde{\Phi}_W(r)$ functional is quadratic so after differentation we get a system of linear equations. The functional $\tilde{\Phi}_F(r)$ is strongly nonlinear so the system of algebraic equations is also nonlinear because the derivatives of z squared appear under a square root sign

$$\frac{\partial\tilde{\Phi}^e(r)}{\partial\hat{z}_i} = \iint_{V_r} [1 + (\frac{\partial z}{\partial x})^2 + (\frac{\partial z}{\partial y})^2]^{-\frac{1}{2}} \cdot [\frac{\partial z}{\partial x}\frac{\partial^2 z}{\partial x\partial\hat{z}_i} + \frac{\partial z}{\partial y}\frac{\partial^2 z}{\partial y\partial\hat{z}_i}] \, dV_r \tag{3.16}$$

To overcome this problem an iterative process, based on successive solutions of linear equations system is used by considering the square root as a known function from the previous iteration cycle. In this way we get a system of linear equations which approximates the set of nonlinear equations. Because $\tilde{\Phi}_T(r)$ is a linear form of $\tilde{\Phi}_F(r)$ and $\tilde{\Phi}_W(r)$ so we can add the coefficients at the same variables. The sets of values $\{\hat{x}_i\}$, $\{\hat{y}_i\}$, $\{\hat{z}_i\}$ together with the $\{\Psi_i^e(\xi,\eta)\}$ and the transformation functions (3.6) uniquely describe the surface S.

4. ILLUSTRATIVE EXAMPLE AND CONCLUSIONS

For a given displacement D and breath B (length and draft of the ship are defined by the contour line Γ_B) our task is to find such a shape of the ship that its resistance approaches the possible minimum for a given velocity v. Preparation of such set of input data is a trivial task for a naval architect. To find an answer for the question it is an extremely difficult task. The result obtained by the present method is shown on Figure 1. Thus we have got an answer to a very interesting question: how looks like a ship body with minimum resistance under certain constraints (e.g. displacement) and in the scope of chosen theory when assumed approximations are used. The results obtained thus far are promising. The industrial value of

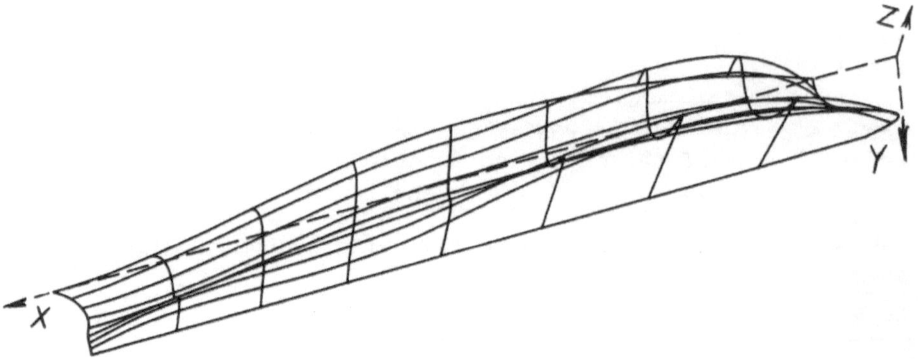

Figure 1. A ship of minimum theoretical resistance

the method can, however, only be verified by a proper series of experiments with models in a towing tank. The computer system is still developed. The graphical representation of the surface should be improved in order to obtain quality of the drawings on the level acceptable by industry standard.

ACKNOWLEDGEMENT

This work was completed during the course of a research project in the Institution for Applied Mechanics at Tampere University of Technology. The authors, and particularly J.P.M., gratefully like to acknowledge the support by the Academy of Finland for this work.

REFERENCES

1 Webster, W.C. and Wehausen, J.V., 'Schiffe Geringsten Wellenwiderstandes mit Vorgegebenem Hinterschiff', Schiffstechnik, Vol. 9 (1962), 62-68.
2 Lin, W.C., Webster, W.C., and Wehausen, J.V., 'Ships of Minimum Total Resistance', Report No. NA-63-7, Institute of Engineering Research, University of California at Berkeley, Aug. (1963).
3 Hsiung, C.C. and Shenyan, D., 'Optimal Ship Forms for Minimum Total Resistance', Journal of Ship Research', 3 (1984), 163-172.
4 Michell, J.H., 'The Wave Resistance of a Ship', Philosophical Magazine, 45, (1898) 106-123.
5 Mikhlin, S.G. - The Numerical Performance of Variational Methods, Wolters-Nordhoff, The Netherlands, 1971.
6 Faux, I.D. and Pratt, M.A. - Computational Geometry for Design and Manufacture, Ellis Horwood Limited, Chichester, 1985.
7 Krylov, I.V. - Approximate Calculation of Integrals, The Macmillan Company, New York, 1962.

TRANSITION PLATE BENDING ELEMENTS WITH VARIABLE NODES

Chang-Koon Choi and Yong-Myung Park

Department of Civil Engineering, Korea Advanced Institute of Science and Technology, Seoul 131, Korea

ABSTRACT

In this study, the developement of 5- and 6-node transition plate bending elements was presented. The mixed use of these elements and regular 4-node elements enable us to refine the mesh of plate locally where the steep displacement or stress gradient exists. The overestimation of the stiffness pertinent to shear in isoparametric plate bending elements can be cured efficiently by addition of nonconforming displacement modes.

INTRODUCTION

The degenerated plate/shell elements are more frequently used in the practical problem than the original form of three dimensional isoparametric elements. Some of the 4-node or 8-node plate bending elements which have symmetric axes in x,y direction, give good results in the analysis of plates[2,3,7]. It is, however, inconvenient to refine the mesh locally by using these elements. If a mesh can be refined locally, i.e., a part of structure where the steepest stress gradient exists such as stress concentration is modelled with finer mesh while the rest of the structure is modelled with rather coarse mesh, it will be more economical and effective in plate bending problem.

In this study, some transition plate bending elements which have one or two additional mid-side nodes and can be efficiently used together with 4-node elements to refine the given mesh locally as shown in Fig.1 (a) were developed. The mixed use of different element types in mesh generating has been made previously in inplane problems. This type of use in plate bending problems, however, can be seldom found in the published literature. There are two types of transition element in the past studies. One is such an element that has mid-side nodes in certain edges of the element and has been used in transition zone to connect lower order and higher order elements as shown in Fig.1 (b) [1,4,14]. The other is such an element that has transition node, which enables to refine the mesh locally[5]. These transition elements are used in inplane problems and compatibility is achieved by enforced compatibility formulation, if necessary.

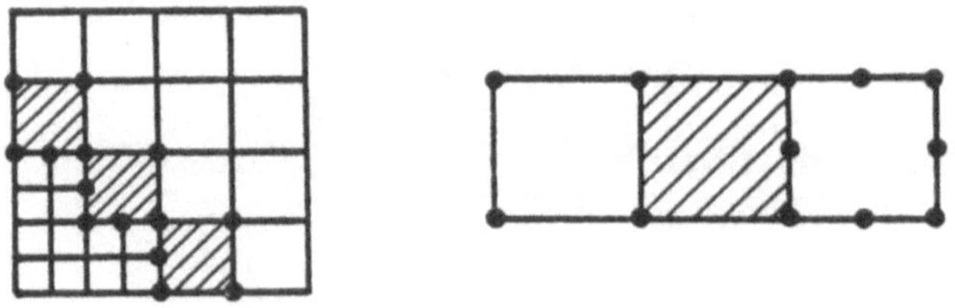

(a) (b)

Fig. 1 Examples using Transition Elements

The Mindlin plate bending elements generated from isoparametric element tend to overestimate its shear stiffness. Therefore, in order that the 5- and 6-node plate bending elements are used successfully as transition elements in structural modelling, their behavior should be improved. Such an improvement of the elements are achieved by addition of nonconforming displacement modes in this study.

STRAIN ENERGY OF PLATE BENDING ELEMENT

When a plate bending element has three degrees of freedom at each node, i.e., a transverse displacement w and two rotations about x, y axis α, β, respectively, the strain energy of a plate bending element based on Mindlin plate theory is given by the following equation.

$$U = 1/2 \int_A \kappa^T [C_b] \kappa \, dA + 1/2 \int_A \gamma^T [C_s] \gamma \, dA \tag{1}$$

where

$$\kappa = \begin{bmatrix} \dfrac{\partial \alpha}{\partial x} \\ \dfrac{\partial \beta}{\partial y} \\ \dfrac{\partial \beta}{\partial x} + \dfrac{\partial \alpha}{\partial y} \end{bmatrix} \qquad \gamma = \begin{bmatrix} \dfrac{\partial w}{\partial x} - \alpha \\ \dfrac{\partial w}{\partial y} - \beta \end{bmatrix} \tag{2}$$

Here, $[C_b]$ and $[C_s]$ denote plate bending and shear material matrix, respectively. The first integral in Eq.(1) is plate bending strain energy and the second integral shear strain energy. In the isoparametric plate element, the lower order element in particular, overestimation of shear stiffness or shear locking phenomena in very thin plate is resulted due to the overestimation of shear strain energy.

The displacement fields w^n, α^n and β^n of the degenerated plate elements can be defined by following equations. Here, n indicates the number of nodes in an element.

$$w^n = \sum_{i=1}^{n} N_i w_i \ , \quad \alpha^n = \sum_{i=1}^{n} N_i \alpha_i \ , \quad \beta^n = \sum_{i=1}^{n} N_i \beta_i \qquad (3)$$

In general, bending behavior of an isoparametric plate element of which displacement fields are assumed by Eq.(3) is represented by shear type deformation[2]. Thus, the shear strain energy (and/or shear stiffness) in Eq.(1) is overestimated. To eliminate the excessive shear effect and evaluate shear stiffness properly, reduced (selective) integration techniques[6,7, 8,10] or addition of nonconforming displacement modes [2,3,11, 13] was successfully used previously.

ADDITION OF NONCONFORMING DISPLACEMENT MODES

Addition of nonconforming modes may cause a defect that violates the interelement displacement compatibility. The element of this type, however, give good results for plane stress problems or plate and shell analysis problems restoring the real deformation to the element flexure.

The nonconforming displacement modes used in this study are given in Eq.(4) and shown in Fig.2. These modes have the same shape functions of mid-side nodes of an 8-node element which change the linear variation of displacements to quadratic variation along the edges without mid-side node in the transition element. Thus, these are formulated in a different way from the previous study[3,13]. These modes are considered as internal nodes in a sense.

$$\begin{aligned} \bar{N}_5 &= (1/2)(1-\xi^2)(1-\eta) \\ \bar{N}_6 &= (1/2)(1+\xi)(1-\eta^2) \\ \bar{N}_7 &= (1/2)(1-\xi^2)(1+\eta) \\ \bar{N}_8 &= (1/2)(1-\xi)(1-\eta^2) \end{aligned} \qquad (4)$$

The nonconforming modes in Fig.2 are added selectively to the edges of 5- and 6-node elements that do not have any mid-side nodes. To correct only shear stiffness of 5- and 6-node plate elements, nonconforming modes will be added only to transverse displacement w. The nonconforming modes $\bar{N}_6$,$\bar{N}_7$ and $\bar{N}_8$ are added to 5-node element, and $\bar{N}_7$ and $\bar{N}_8$ to 6-node element to form a more flexible plate element.

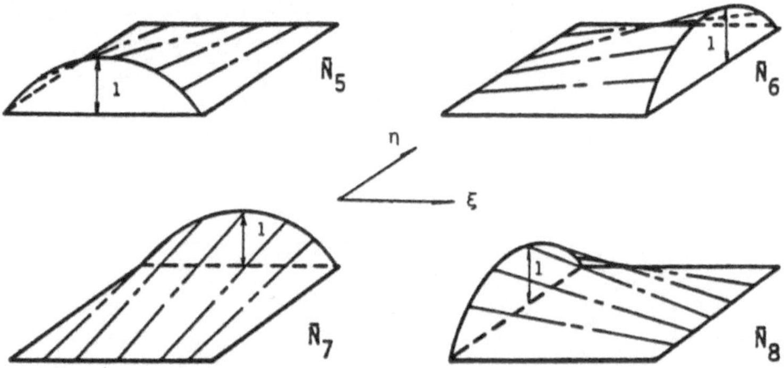

Fig. 2 Nonconforming Modes introduced in this study

CONSTRUCTION OF ELEMENT STIFFNESS MATRIX K^e

The force-displacement equation for an element which has nonconforming modes can be obtained by the means of minimum potential theory and given by

$$\begin{bmatrix} K_{cc} & K_{cn} \\ K_{cn}^T & K_{nn} \end{bmatrix} \begin{Bmatrix} u_c \\ u_n \end{Bmatrix} = \begin{Bmatrix} F \\ 0 \end{Bmatrix} \tag{5}$$

where

$$K_{cc} = \int_{-1}^{+1} \int_{-1}^{+1} [B_c]^T [D] [B_c] \; |J| \; d\xi d\eta$$

$$K_{cn} = \int_{-1}^{+1} \int_{-1}^{+1} [B_c]^T [D] [B_n] \; |J| \; d\xi d\eta \tag{6}$$

$$K_{nn} = \int_{-1}^{+1} \int_{-1}^{+1} [B_n]^T [D] [B_n] \; |J| \; d\xi d\eta$$

In Eq.(6), $[B_c]$ is strain matrix constructed from conforming shape functions and $[B_n]$ from nonconforming modes. Also, [D] is material property matrix composed of bending and shear rigidity in Eq.(1). And $\{u_c\}$ is unknown nodal vector pertaining to conforming part, and $\{u_n\}$ to nonconforming part. Then, by static condensation of Eq.(5), we can get element stiffness matrix K^e given by following equation.

$$K^e = K_{cc} - K_{cn}.K_{nn}^{-1}.K_{cn}^T \tag{7}$$

It is possible to form various elements by using different quadrature rules to calculate the submatrices in Eq.(5). The element designations and quadrature rules used in this study are summarized in Table 1.

Table 1 Plate Element designations

No. of Nodes	Element Designations	N.C. Modes		Quadrature rule	
		W	α, β	Bending	shear
4	C4	none	none	2 X 2	2 X 2
	C4 - R	none	none	2 X 2	1 X 1
	* NC4 - 2.2	$\bar{N}_5, \bar{N}_6, \bar{N}_7, \bar{N}_8$	none	2 X 2	2 X 2
5	C5	none	none	2 X 2	2 X 2
	* NC5 - 2.2	$\bar{N}_6, \bar{N}_7, \bar{N}_8$	none	2 X 2	2 X 2
	* NC5 - 3.3	$\bar{N}_6, \bar{N}_7, \bar{N}_8$	none	3 X 3	3 X 3
6	C6	none	none	2 X 2	2 X 2
	* NC6 - 2.2	$\bar{N}_7, \bar{N}_8$	none	2 X 2	2 X 2
	* NC6 - 3.3	$\bar{N}_7, \bar{N}_8$	none	3 X 3	3 X 3

* Element introduced in this study

NUMERICAL TEST

Cantilever plate - The test meshes for 4-node, 5-node and 6-node plate elements are shown in Fig.3 and results are collected in Table 2. Here, it is shown that all the nonconforming elements developed in this study give good results for both regular and irregular mesh shapes. The test results in Table 3 show good convergence of these elements to exact value, with increasing number of nodes.

Square plate - The test meshes (1/4 modelling) are shown in Fig.4 and results are given in Table 4. Analysis results by mesh A using nonconforming 5- and 6-node transition elements are nearly equal to the results obtained from mesh B, with respect to maximum displacement.

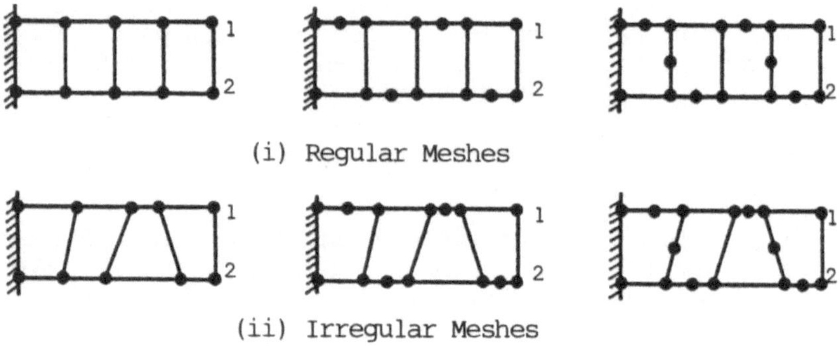

Fig. 3 Test Meshes for 4-, 5- and 6-node element

Table 2 Results of 4-, 5- and 6-node element test(Fig.3)
($E = 10.0$, $\nu = 0.3$, $t = 0.5$, $q = 0.1$, $L = 9.0$, $H=3.0$)

(i) Regular meshes

Element	D.O.F.	w_1 : w_2	M_{max}	$w_{normalized}$
C4	24	106.35	-0.457	13.90
C4-R	24	767.34	-3.246	100.31
NC4-2.2	24	764.99	-3.238	100.00
C5	36	441.76 : 400.18	-2.212	55.03
NC5-2.2	36	769.66 : 764.46	-3.649	100.27
NC5-3.3	36	755.15 : 751.35	-3.567	98.46
C6	42	457.76 : 415.77	-2.193	57.09
NC6-2.2	42	773.86 : 768.73	-3.647	100.82
NC6-3.3	42	758.03 : 753.38	-3.568	98.80

(ii) irregular meshes

Element	D.O.F.	w_1 : w_2	M_{max}	$w_{normalized}$
C4	24	91.99 : 95.64	-0.400	24.53
C4-R	24	680.31 : 854.81	-3.215	-
NC4-2.2	24	763.57 : 764.04	-3.180	99.84
C5	32	494.64 : 448.99	-2.388	61.68
NC5-2.2	32	769.60 : 762.70	-3.642	100.15
NC5-3.3	32	752.06 : 747.67	-3.472	98.02
C6	42	514.93 : 471.55	-2.416	64.48
NC6-2.2	42	773.17 : 765.83	-3.609	100.59
NC6-3.3	42	754.51 : 748.25	-3.437	98.22

Table 3 Test Results of Convergence check

Element	D.O.F.	w_{max}	M_{max}	$w_{normalized}$
C4-R	180	766.21	-4.052	100.03
NC4-2.2	180	766.54	-4.270	100.07
NC5-2.2	288	769.31	-3.943	100.25
NC6-2.2	360	768.03	-4.240	100.27

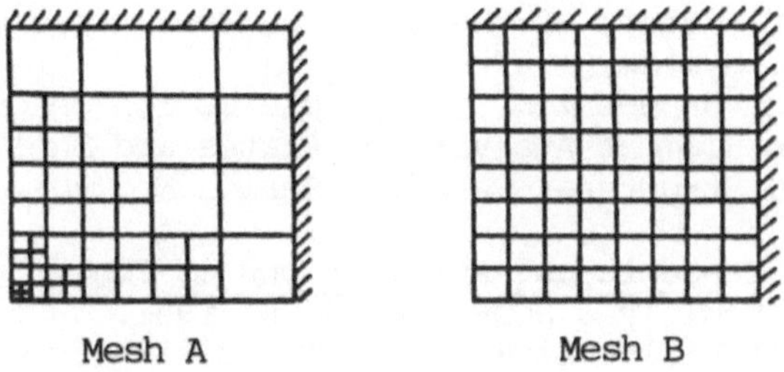

Fig. 4 Test Meshes for Square Plate Test

Table 4 Results of Square Plate Test (Fig.4)
(E=3600.0, ν=0.3, t=0.2, L=20, P=400.0, L/t=100.0)

Mesh	Element	D.O.F.	w_{max}	M_{max}	$W_{normalized}$
A	C4-R + NC5-2.2 + NC6-2.2	174	338.81	147.4	99.73
	C4-R + C5 + C6	174	176.30	138.7	51.89
B	C4 - R	176	338.95	91.1	99.77
* Exact Value (Thin Theory)			339.73	∞	

* ref. 12

CONCLUSION

The 5- and 6-node nonconforming plate bending element can be used effectively as transition elements. There are no zero energy mechanisms identified by eigenvalue analysis. Although the addition of nonconforming displacement modes to the original element violates the compatibility along the interelement boundaries, the improvement achieved by such addition was significant making the use of this type of element in modelling transition zone very effective.

It is also noted that the elements integrated by (2 X 2) quadrature (NC5-2.2 and NC6-2.2) produced better results than the elements integrated by (3 X 3) quadrature rule (NC5-3.3 and NC6-3.3).

REFERENCES

1 Bathe, K.J. - Finite Element Procedures in Engineering Analysis, Prentice Hall, New Jersey, 1982.

2 Choi, C.K. and Kim, S.H., 'Improvement of Degenerated Plate/ Shell Element', Proc. of the US-Korea Seminar/Workshop on Critical Eng. system, 1987

3 Choi, C.K. and Schnobrich, W.C., 'Use of Nonconforming Modes in the Finite Element Analysis of Plates and Shell', Civil Eng. Studies, Structural Research Series No. 401, Univ. of Illinois, 1973

4 Cook, R.D. - Concepts and Applications of Finite Element Analysis, John Willy & Sons, New York, 1981.

5 Gupta, A.K. 'A Finite Element for Transition from a fine to a Coarse Grid', Int. J' for Numer. Meth. in Eng. 12 (1978), 35-45

6 Hughes, J.R., Cohen, M. and Haroun, M., 'Reduced and Selective Integration Techniques in the Finite Element Analysis', Nuclear Eng. 46(1977), 203-222

7 Hughes, J.R., Taylor, R.L. and Kanoknukulchai, W., 'A Simple and Efficient Element for Plate Bending', Int. J' for Numer. Meth. in Eng. 11(1977), 1529-1543

8 Macneal, R.H., 'A Simple Quadrilateral Shell Element', Comp. & struct. 46(1976), 175-183

9 Prathap,G. and Viswanth, S., 'An Optimally Integrated Four-node Quadrilateral Plate Bending Element', Int.J' for Numer. Meth. in Eng. 19(1983), 831-840

10 Pugh, E.D.L, Hinton, E. and Zienkiewicz, O.C., 'A Study of Quadrilateral Plate Bending Elements with Reduced Integration', Int. J' for Numer. Meth. in Eng. 12(1978), 1059-1079

11 Taylor, R.L. and Beresford, P.J., 'A Nonconforming Element for Stress Analysis', Int. J' for Numer. Meth. in Eng. 10 (1976), 1211-1210

12 Timoshenko, S.P. and Kriger, S.W., - Theory of Plates and Shells, McGraw-Hill, New-York, 1959.

13 Wilson, E.L., Taylor, R.L., Doherty, W.P. and Ghaboussi,J., 'Incompatible Displacement Modes', Int. Sym. on Numer. and Comp. Meth. in Struct. Mech., Univ. of Illinois, 1971

14 Zienkiewicz, O.C., - The Finite Element Method, McGraw-Hill, New-York, 1977

A SIMPLE ADAPTIVE SCHEME BASED ON A NEW HYBRID FE MODEL

J. Jirousek
IREM - Swiss Federal Institute of Technology, Lausanne
CH-1015 Lausanne, Switzerland

SUMMARY

The paper presents the p-version of the so-called hybrid-Trefftz (HT) finite element model and shows its excellent suitability for adaptive solutions. The HT model uses assumed displacement fields, chosen so as to satisfy a priori the governing differential equations of the problem, and enforces the interelement continuity and the boundary conditions. Optional classes of functions are available for various singularities and stress concentrations and make unnecessary troublesome local refinement of FE meshes. A simple local a posteriori error estimator is provided for controling the predicted stresses. The new approach may be implemented in existing FE codes. The high practical efficiency is demonstrated on analysis of an involved plate bending problem.

1. INTRODUCTION

It is now well established that the p-method is superior to the h-method. However, when the conventional p-method is used, singular points have to be isolated by one or two layers of small elements graded in a geometrical progression towards the singularity [1]. Also in presence of movable concentrated loads, suitable refinement may be needed nearly everywhere.

The new approach presented in this paper attempts to circumvent this drawback by making use of the possibilities offered by the new model.

2. THEORY

Although the basic theory has already been presented elsewhere [2-4], the HT model is not largely known and its principle will first be shortly summarized for the reader's convenience.

The approach is based on application over the element of an assumed displacement field that fulfils a priori the governing differential problem equations (Trefftz-type approach) and, consequently, satisfies simultaneously both the equilibrium and the compatibility. Let the internal equilibrium of an elastic continuum occupying the region Ω bounded by $\Gamma = \partial\Omega$ be expressed in terms of unknown displacements $\mathbf{u}$ by

$$\mathbf{L}\mathbf{u} = \bar{\mathbf{f}} \ , \tag{1}$$

where $\mathbf{L}$ is a differential operator matrix and $\bar{\mathbf{f}}$ are generalized body forces. Then over a particular finite element $\Omega_e \in \Omega$ the displacement field is approached by

$$\mathbf{u}_e = \overset{o}{\mathbf{u}}_e = \sum_{j=1}^{m} \boldsymbol{\phi}_j \mathbf{c}_j = \overset{o}{\mathbf{u}}_e + \boldsymbol{\phi}_e \mathbf{c}_e \ , \tag{2}$$

where $\mathbf{c}_e$ is a vector of undetermined coefficients and $\overset{o}{\mathbf{u}}_e$ and $\boldsymbol{\phi}_e$ are known trial functions such that

$$\mathbf{L}\overset{o}{\mathbf{u}}_e = \bar{\mathbf{f}} \quad \text{and} \quad \mathbf{L}\boldsymbol{\phi}_e = \mathbf{0} \text{ on } \Omega_e \ . \tag{2a,b}$$

In general, the boundary conditions and the interelement continuity will be specified in terms of suitably defined generalized boundary displacements $\mathbf{v} = \mathbf{v}(\mathbf{u})$ and the conjugated boundary tractions $\mathbf{T} = \mathbf{T}(\mathbf{u})$. As a means of enforcing the interelement continuity of $\mathbf{v}$ and the kinematical boundary conditions on the field (2) at Γ_v, assume an auxiliary conforming field $\tilde{\mathbf{u}}_e$ defined in the usual way in terms of nodal parameters $\mathbf{d}_e$ and a matrix of conforming shape functions $\tilde{\mathbf{N}}_e$:

$$\tilde{\mathbf{u}}_e = \tilde{\mathbf{N}}_e \mathbf{d}_e \quad \text{on} \quad \Omega_e \ . \tag{3}$$

Then minimizing with respect to $\mathbf{c}_e$ the strain energy of the difference $\mathbf{u}_e - \tilde{\mathbf{u}}_e$,

$$U(\mathbf{u}_e - \tilde{\mathbf{u}}_e) = \frac{1}{2} \int_{\Omega_e} (\boldsymbol{\varepsilon}_e - \tilde{\boldsymbol{\varepsilon}}_e)^t \mathbf{D} (\boldsymbol{\varepsilon}_e - \tilde{\boldsymbol{\varepsilon}}_e) d\Omega = \min \ , \tag{4}$$

makes it possible to express the undetermined coefficients $\mathbf{c}_e$ in terms of nodal parameters $\mathbf{d}_e$ and body forces $\bar{\mathbf{f}}_e$. As a result, equation (2) with $\mathbf{c}_e = \mathbf{c}_e(\mathbf{d}_e, \overline{\mathbf{f}}_e)$ yields a Trefftz's field with enforced conformity :

$$\mathbf{u}_e = \overset{o}{\mathbf{u}}'_e + \mathbf{N}'_e \mathbf{d}_e \quad . \tag{4a}$$

Next we simply use the virtual work principle (with

Parameters :

○ $\cdots$ 3 DOF (w , w_x , w_y)

△ $\cdots$ variable number of DOF

$(a_1 , b_1 , a_2 , b_2 , \ldots)$

Frame functions :

$\widetilde{W}_n$ (normal slope) $\widetilde{W}$ (displacement)

a) Modes associated with corner nodes o

b) Modes associated with mid side nodes △

$$L_i = \xi^{i-1} (1 - \xi^2) \qquad\qquad M_i = \xi^{i-1} (1 - \xi^2)^2$$

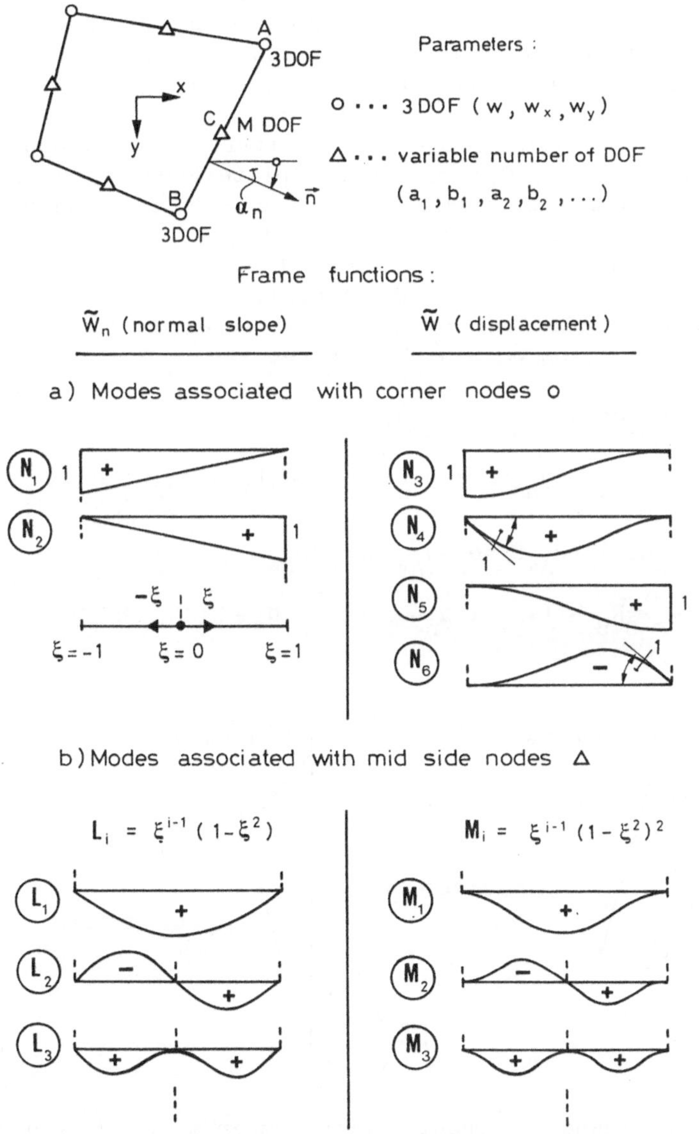

Fig.1. Hybrid-Trefftz p-version plate element.

$\delta\tilde{\mathbf{u}}_e = \tilde{\mathbf{N}}_e \delta \tilde{\mathbf{d}}_e$ as the virtual displacement field) to enforce the continuity of boundary tractions across interelement boundaries and the statical boundary conditions at Γ_T . This results in the customary force-displacement relationship with a symmetric positive definite element stiffness matrix and a load dependent vector (effect of $\bar{\mathbf{f}}_e$) of equivalent nodal forces.

An important point arising in evaluation of the element matrices is the fact that by a systematic use of the divergence theorem all integrals over the element area Ω_e may be replaced by equivalent expressions involving only integration over the element boundary Γ_e . As a result, the explicit definition of the conforming field $\tilde{\mathbf{u}}_e$ over Ω_e , difficult to obtain for applications requiring C_1 continuity (plate bending) becomes unnecessary and may be replaced by a direct definition of the corresponding generalized boundary displacement - the frame function $\tilde{\mathbf{v}}_e$. As an example, figure 1 shows frame functions of a p-version HT plate bending element, case where $\tilde{\mathbf{v}}_e$ includes suitable interpolations of the transverse displacement $\tilde{w}$ and the normal rotation $\tilde{w}_n = \partial\tilde{w}/\partial n$ defined as

$$\tilde{w}_n = (w_{Ax}\cos\alpha_n + w_{Ay}\sin\alpha_n)N_1 + \\ + (w_{Bx}\cos\alpha_n + w_{By}\sin\alpha_n)N_2 + a_{C1}L_1 + a_{C2}L_2 + \ldots , \quad (5a)$$

$$\tilde{w} = w_A N_3 - (w_{Ax}\sin\alpha_n - w_{Ay}\cos\alpha_n)N_4 + \\ + w_B N_5 - (w_{Bx}\sin\alpha_n - w_{By}\cos\alpha_n)N_6 + b_{C1}M_1 + b_{C2}M_2 + \ldots \quad (5b)$$

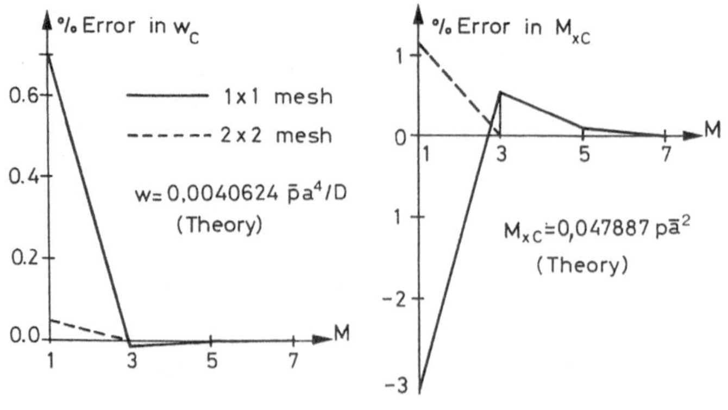

Fig.2. Convergence of results with number M of DOF at mid-side nodes. Simply supported uniformly loaded square plate (ν=0.3) with fixe FE mesh over a symmetric quadrant.

In addition to the simplest interpolation functions associated with the corner nodes, $\tilde{w}_n$ and $\tilde{w}$ include an optional number of "bubble" modes, the additional DOF of which (amplitudes a_1, a_2 ... for normal slope $\tilde{w}_n$ and b_1, b_2 ... for transverse displacement $\tilde{w}$) may be formally associated with mid-side nodes. The solution accuracy is then modified (figure 2) by changing simply the number M of DOF specified at mid-side nodes; the number of internal functions ϕ_j (here biharmonic polynomials) is at the same time automatically suitably adjusted by the program (see [2] to [4] for further details).

3. CONTROL OF THE ADAPTIVE PROCESS

Since our concern is usually with stresses we will control the error generated in suitably defined equivalent stresses. For plate bending, in particular, we introduce equivalent moments $M_{eqv} = \sqrt{DU_o}$ (U_o = strain energy density, $D = Et^3/12(1-\nu^2)$) and define the error measure as

$$\eta = M_{eqv}(e) = \sqrt{DU_o(e)} \ , \qquad \text{(6a)}$$

where $e = w - w_{FE}$ stands for the displacement error of the FE approximation.

In the present FE model the governing differential problem

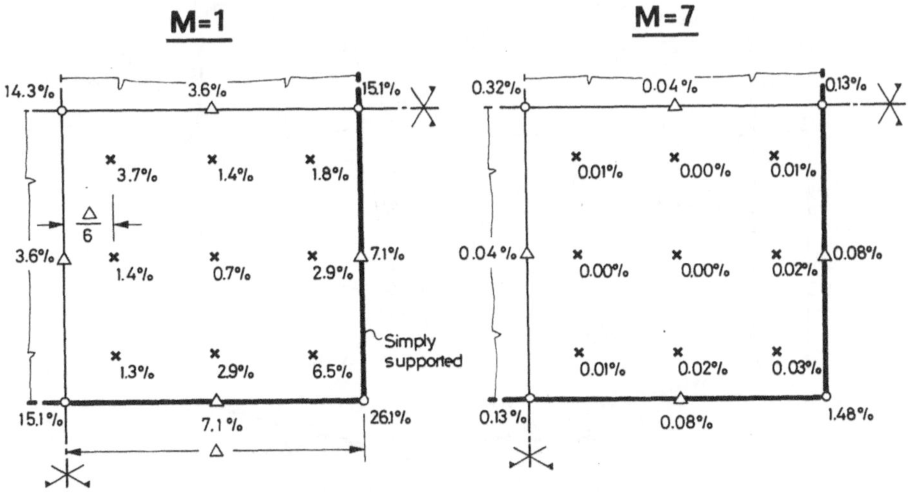

Fig.3. - Distribution of percentage error η% in M_{eqv}. Single element over a quadrant of uniformly loaded square plate (ν=0.3). 100% = average $M_{eqv} = \sqrt{DU/a^2}$ ($a = 2\Delta$ = plate side).

equations are verified over each element rigorously and the only source of solution errors arises from imperfect satisfaction of the interelement and boundary condition requirements. The study of the undue displacement and traction jumps at element boundaries [4] shows that :

- the jump distributions are of oscillatory nature;
- their period and amplitude decrease rapidly when the p-refinement level (number M of DOF at mid-side nodes) is increased;
- the displacement jumps are very small with respect to the traction jumps.

Thus in the first approximation the solution errors may be assimilated to the effect of parasitic forces represented by the undue traction jumps. Moreover, these forces are self-equilibrating and as a consequence (figure 3) the error decays rapidly with the distance from the element boundary and leaves in each element a large interior high precision zone in which the error is one order of magnitude, or more, lower than in the narrow perturbed low precision zone adjacent to element boundary. Thus when a suitable smoothing technique, the

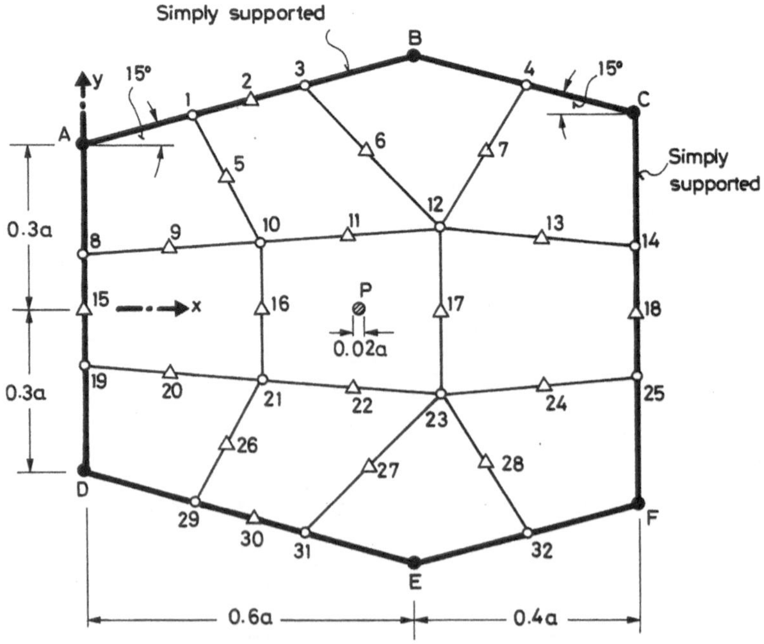

Fig.4. Example of a plate solved by using p-version hybrid-Trefftz elements.

so-called "Krigeing" [5] is used, the results in the low precision zone may be considerably improved and their difference with respect to the unsmoothed values used to estimate the error η as

$$\eta = \sqrt{DU_o(e)} \simeq \sqrt{[\Delta M_x^2 - 2\nu\Delta M_x \Delta M_y + \Delta M_y^2 + 2(1+\nu)\Delta M_{xy}^2]/(1-\nu^2)}, \quad (6b)$$

where ΔM_x , ΔM_y and ΔM_{xy} stand for differences between the smoothed and unsmoothed values. The idea of using the smoothed values as reference solution for error estimation has first been applied successfully by Zienkiewicz and Zhu [6] with integral error norms. In the present FE model where the largest errors appear always at element corners it was found preferable to replace the use of such norms (liable to smooth out local error concentrations) by pointwise evaluation of the error measure (6b) at element corner nodes. When using the "Krigeing", the smoothing consists in alloting to a limited number of high precision internal sampling points i in the neighbourhood of a given node N appropriate weights λ_i (obtained from a small system of simultaneous equations independent of the interpolated field F) and in setting (see [5]) :

Load	Node	M = 1		M = 3		M = 5		M = 7	
		η_1%	η_2%	η_1%	η_2%	η_1%	η_2%	η_1%	η_2%
Uniform	1	3.24	3.55	0.22	0.41	0.45	0.46	0.07	0.13
	3	6.06	6.18	0.69	1.20	0.35	0.45	0.24	0.27
	4	0.00	1.12	0.00	0.29	0.00	0.02	0.00	0.06
	8	1.95	3.03	0.43	0.44	0.11	0.14	0.06	0.07
	10	2.25	3.79	0.56	0.62	0.09	0.06	0.04	0.06
	12	1.29	2.10	0.78	0.53	0.70	0.60	0.19	0.15
	14	1.47	2.11	0.40	0.54	0.07	0.13	0.07	0.07
Concentrated	1	1.97	1.97	0.37	0.37	0.24	0.61	0.08	0.08
	3	0.99	1.63	0.57	0.57	0.30	0.41	0.05	0.05
	4	0.00	0.37	0.00	0.09	0.00	0.07	0.00	0.03
	8	0.28	0.62	0.54	0.55	0.11	0.33	0.09	0.09
	10	0.47	0.29	0.42	0.39	0.53	0.66	0.20	0.31
	12	0.99	1.85	0.70	0.61	0.81	0.90	0.16	0.37
	14	0.78	0.79	0.66	0.70	0.14	0.17	0.10	0.10

Table 1. Plate of figure 4 (ν=0.3). Variation of % error with number M of DOF at mid-side nodes (100% = M_{eqv} at plate center, x=a/2). η_1,η_2 estimated and true errors in nodal averages.

$$F_N = \sum_i F_i \lambda_i \ . \qquad (7)$$

This procedure is applied to M_x, M_y and M_{xy} to check the error at all internal corner nodes of the FE mesh. To avoid extrapolation, which would result in less reliable values, the error at nodes situated at the plate boundary may be estimated by using simply as reference the known accurate boundary values (for example $M_n = M_t = 0$ at the simply supported, $M_{nt} = 0$, $M_t = \nu M_n$ at the clamped edge, etc.).

When using the hybrid-Trefftz model, it is usual [2,4] that the singularities are properly taken into account through the application of suitable optional special purpose functions, $\overset{o}{\mathbf{u}}$ and $\boldsymbol{\phi}_j$, for the element containing a singular point. As an example the displacement field of the six singular corner elements in figure 4 accurately fulfils the boundary conditions along the two sides forming the singular corner. Since such elements exhibit the highest accuracy in the neighbourhood of the singular points, the singular corners A to F may be discarded from the checking process.

The comparison of the estimated and the true errors in Table 1 seems to indicate that augmenting the estimated errors by 3/4 of the largest error will result in a conservative estimate of the upper bound on the error.

REFERENCES

1. Szabo, B.A., 'Mesh design for the p-version of the finite element method', Comp. Meth. Appl. Mech. Engng., 55, 181-197 (1986).
2. Jirousek, J., 'The hybrid-Trefftz finite element model and its application to plate bending', Int. J. Num. Meth. Eng., 23, 651-693 (1986).
3. Jirousek, J., 'Hybrid-Trefftz plate bending elements with p-method capabilities', Int. J. Num. Meth. Engng.; in press.
4. Jirousek, J., 'The hybrid-Trefftz model - A finite element model with special suitability to adaptive solutions and local effect calculations', Finite Elements Methods for Plate and Shell Structures (edited by T.J.R. Hughes and E. Hinton), Pineridge Press, 1986, Vol. 1. Chapt. 9.
5. Jirousek, J., Bouberguig, A. and Frey, F., 'An efficient unified post-processing approach based on a probabilistic concept', Proc. Int. Conf. NUMETA 85, University of Swansea, U.K., 723-732 (1985).
6. Zienkiewicz, O.C. and Zhu, J.Z., 'A simple error estimator and adaptive procedure for practical engineering analysis', Int. J. Num. Meth. Engng., 24, 337-357 (1987).

TWO-DIMENSIONAL SOLIDIFICATION ANALYSIS FOR TWIN-ROLL CONTINUOUS CASTING

by C.G.Kang, H.Hojo, T.Saitoh, and H.Yaguchi
Dept. of Mechanical Engineering II
Tohoku University
Sendai 980, Japan.

ABSTRACT:

A numerical algorithm for the two-dimensional solidification problem in the twin-roll continuous casting(C.C.) system is presented in this paper. Attention is focused on the elucidation of heat transfer and flow characteristics in both the liquid and solid phases. The present mathematical model can be applied to general full Navier-Stokes and energy equations, thereby covering the wide range of twin-roll casting conditions. The boundary fixing method(BFM) was adopted to handle the moving boundary.

In this paper, a general numerical methodology is presented for the C.C. problem and the quantitative relationship between the important control parameters in continuous casting of twin-roll type.

1. INTRODUCTION

In recent years a growing interest has been concentrated on the twin-roll continuous casting(C.C.) processes which have many advantages over other existing methods. The rapid solidification process using twin-roll is promising because it: (i) permits greater control over the material structure such as the formation of a quasi-steady stable phase, improves the solubility limit, establishes microcrystalline structure and mechanical properties, prevents segregation, and (ii) saves energy through a reduction of processes.

Although the twin-roll C.C. process is similar to the one as mentioned above, it is more difficult to control because of the existence of the deformation of the roll itself due to thermal expansion or thermal stress owing to the narrow roll gap. Further, if solidification is completed before the liquid reaches the minimum clearance point between the rolls, then deformation of the solid will occur. The plastic deformation of the material then has to be considered in this case.

For the above mentioned reasons, it is necessary to develop efficient numerical tools to elucidate the complicated flow and heat transfer mechanism, which will be especially useful for designing optimum twin-roll C.C. systems.

Solidification analyses for the C.C. of a slab have been done by Siegel[2,3,4], Koikkalainen et al.[5], Wang and Inoue[6], Lu and Zhi[7], Ohnaka[8], Ohnaka and Kobayashi[9], and others. A complete solidification analysis for the twin-roll C.C. including flow and heat transfer in the liquid and solid regions is limited, to the best of our knowledge, to the report by Miyazawa and Szekely[1]. The C.C. problems including the twin-roll technique have been reviewed by Szekely[11], Ohnaka[12], and recently by Ohashi[13].

This paper presents a numerical methodology for the two-dimensional heat transfer and flow phenomena in the liquid and solid regions in a twin-roll continuous casting system. The mathematical model presented covers the wide range of casting parameters since the two-dimensional transport in both phases was taken into account. Another objective is to provide a quantitative relationship between the principal casting parameters.

2. MATHEMATICAL FORMULATION

2.1 Numerical Model and the Governing Equations

A numerical model and the coordinate system are schematically shown in Fig.1. Molten metal is being fed from upstream into the nip of the two rolls rotating in opposite direction with an angular velocity ω. The surface of the molten metal $O'P$ is always kept constant by overflowing the excess molten metal from the small leveling mouth. As soon as the molten material is poured into the nip, solidification takes place on the roll surface which is cooled by recirculating water inside the roll. Now, we define two coordinate systems; (r,ϕ) is the coordinate for the solid phase with the origin at the center of the roll, (r',ϕ') is the coordinate for the liquid phase with the origin at point O'. The solidified shell which grew by the time $t=t$ is represented by the functions $F(\phi,t)$ and $G(\phi',t)$ in the solid and the liquid phase coordinates, respectively.

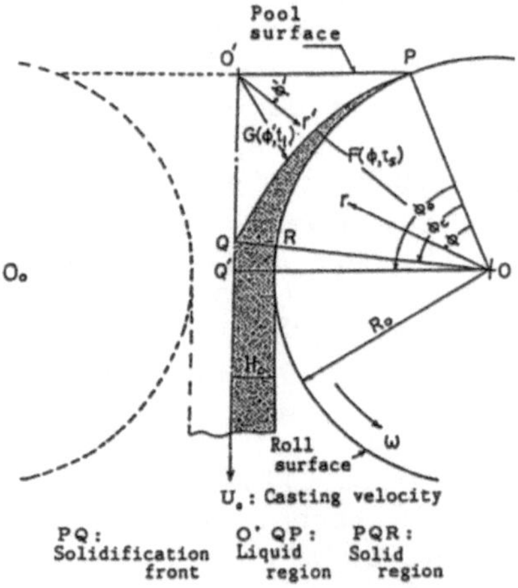

Fig.1 Mathematical model and coordinate system for twin-roll continuous casting.

The problem is then to find the steady solidification profile and the distributions of temperature and fluid velocities in both phases. Special attention is focused on the end point of solidification since the casting phenomena will change drastically depending on whether or not this point is located before the end point.

The following principal assumptions and restrictions may be made for the analysis in the development of the basic equations: 1) The hydrostatic pressures in the liquid and the solid regions remain constant in the casting direction. 2) The heat transfer is two-dimensional. 3) The relation between the solidification rate and temperature in the solid-liquid coexisting region is linear. 4) Thermophysical properties are constant. 5) The principal momentum transfer is in the center line direction.

2.1.1 Boundary Fixing Formulation

The problem of finding the solidification front in twin-roll continuous casting belongs mathematically to the so-called moving(or free) boundary problem(MBP) which is characterized by having a moving interface dividing the relevant field into two regions. Here, the Boundary Fixing Method(BFM)[14,15] was adopted. The BFM considers arbitrary geometry of both the moving interface and the domain boundary via the change of an independent variable. The next two independent variables are introduced for the present case:

Solid region: $$\xi=\frac{r-B(\phi)}{F(\phi,t)-B(\phi)} \quad (1)$$ Liquid region: $$\eta=\frac{r'}{G(\phi',t)} \quad (2)$$

The correspondence between the physical plane and the transformed plane is schematically shown in Fig.2.

The following dimensionless variables are used for descreption of the governing equations.

$$F^*=\frac{F}{R_0} \quad G^*=\frac{G}{R_0} \quad B^*=\frac{B}{R_0} \quad 2H_0^*=\frac{2H_0}{R_0}$$

$$T_l^*=\frac{T_l}{T_0-T_f} \quad T_s^*=\frac{T_s}{T_0-T_f} \quad T_f^*=\frac{T_f}{T_0-T_f}$$

$$T_0^*=\frac{T_0}{T_0-T_f} \quad U(\phi)^*=\frac{U(\phi)}{U_0} \quad \phi'^*=\frac{\phi'}{\phi_0'} \quad (3)$$

$$r^*=\frac{r}{R_0} \quad \phi^*=\frac{\phi}{\phi_0} \quad r'^*=\frac{r'}{R_0} \quad \delta^*(\phi)=\frac{\delta(\phi)}{R_0}$$

$$\gamma'^*=\frac{\beta'}{\phi_0'} \quad \gamma^*=\frac{\beta}{\phi_0} \quad t^*=\frac{\alpha t}{R_0} \quad h^*=\frac{h_s}{h}$$

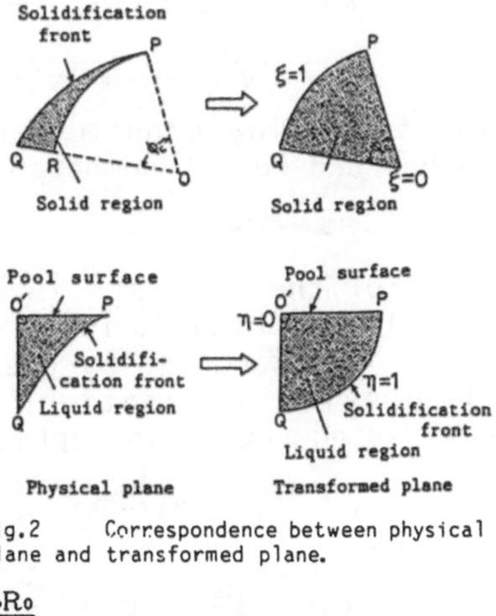

Fig.2 Correspondence between physical plane and transformed plane.

$$U_r^*=\frac{U_r R_0}{\alpha_s} \quad U_\phi^*=\frac{U_\phi R_0}{\alpha_s} \quad U_r'^*=\frac{U_r' R_0}{\alpha_s} \quad U_\phi'^*=\frac{U_\phi' R_0}{\alpha_s}$$

Symbols * will be omitted hereafter for simplicity.

The resultant governing equations for the solid region are written together with the boundary conditions as,

$$\frac{\partial T_s}{\partial t}=\left[\frac{1}{(F-1)^2}+\frac{1}{R_s}\left(\frac{\partial \xi}{\partial \phi}\right)^2\right]\frac{\partial^2 T_s}{\partial \xi^2}+\left[\frac{1}{r}\frac{1}{F-1}+\frac{1}{R_s}\frac{\partial^2 \xi}{\partial \phi^2}-\frac{U_\phi}{\partial \phi_0}\frac{\partial \xi}{\partial \phi}\right.$$

$$-U_r\frac{1}{F-1}-\frac{\partial\xi}{\partial t}\Big]\frac{\partial T_s}{\partial\xi}+\frac{2}{R_s}\frac{\partial\xi}{\partial\phi}\frac{\partial^2 T_s}{\partial\xi\,\partial}+\frac{1}{R_s}\frac{\partial^2 T_s}{\partial\phi^2}-\frac{U'_\phi}{r'\phi'_0}\frac{\partial T_\ell}{\partial\phi'} \qquad (4)$$

$$\xi=0 \quad T=T_w \qquad (5) \qquad\qquad \xi=1 \quad T=T_f \qquad (6)$$

Further, the transformed equations for the liquid region are also written as,

$$\frac{\partial T_\ell}{\partial t}=\Big[\frac{1}{G(\phi)^2}+\frac{1}{R_\ell}\Big(\frac{\partial\eta}{\partial\phi}\Big)^2\Big]\frac{\partial^2 T_\ell}{\partial\eta^2}$$
$$+\Big[\frac{1}{r'}\frac{1}{G(\phi)}+\frac{1}{R_\ell}\frac{\partial^2\eta}{\partial\phi'^2}-\frac{U'_\phi}{r'}\frac{1}{\phi'_0}\frac{\partial\eta}{\partial\phi}-\frac{U r'}{G(\phi)}-\frac{\partial\eta}{\partial t}\Big]\frac{\partial T_\ell}{\partial\eta} \qquad (7)$$
$$+2\frac{1}{R_\ell}\frac{\partial\eta}{\partial\phi'}\frac{\partial^2 T_\ell}{\partial\eta\,\partial\phi'}+\frac{1}{R_\ell}\frac{\partial^2\eta}{\partial\phi'^2}-\frac{U'_\phi}{r'\phi_0'}\frac{\partial T_\ell}{\partial\phi'}$$

$$\eta=1 \quad T=T_f \qquad (8) \qquad\qquad \phi'=1 \quad \frac{\partial T_\ell}{\partial\eta}=0 \qquad (9)$$

Here, $\partial\xi/\partial\phi$, $\partial^2\xi/\partial\phi^2$, $\partial\eta/\partial\phi$, ... in the above equations are obtained by

$$\frac{\partial\xi}{\partial\phi}=-\xi\frac{1}{F-1}\frac{\partial F}{\partial\phi} \qquad \frac{\partial^2\xi}{\partial\phi^2}=-\frac{1}{F-1}\Big[2\frac{\partial F}{\partial\phi}\frac{\partial\xi}{\partial\phi}+\xi\frac{\partial^2 F}{\partial\phi^2}\Big]$$
$$\frac{\partial\eta}{\partial\phi'}=-\eta\frac{1}{G(\phi',t)}\frac{\partial G}{\partial\phi'} \qquad \frac{\partial^2\eta}{\partial\phi'^2}=-\frac{1}{G(\phi',t)}\Big[2\frac{\partial G}{\partial\phi'}\frac{\partial\eta}{\partial\phi'}+\eta\frac{\partial^2 G}{\partial\phi'^2}\Big] \qquad (10)$$

$$R_s=(r\phi_0)^2 \qquad R_\ell=(r'\phi_0')^2 \qquad \Phi=\phi_0'\phi'^*$$

The heat balance equation at the solid-liquid interface is transformed to

$$\frac{\partial F}{\partial\phi}=\frac{Ste}{U_\phi^*}\Big(\frac{\phi_0}{R_0}\Big)\Big[\frac{1}{F(\phi,t)-1}\Big\{1+\Big(\frac{R_0}{\phi_0}\frac{\partial F}{\partial\phi}\Big)^2\Big\}\frac{\partial T_s}{\partial\xi}$$
$$-\frac{\sigma}{G(\phi',t)}\frac{1}{\cos(\phi_0\gamma)\cos(\phi_0'\gamma')}\frac{\partial T_\ell}{\partial\eta}\Big]. \qquad (11)$$

The following nondimensional parametyers including Stefan number and Peclet number have been defined.

$$Ste=\frac{T_0-T_f}{L}C_s \qquad Pe=\frac{U_0 R_0}{\alpha} \qquad (12)$$

For the analysis in the liquid region, it is postulated that latent heat is rejected according to the solid fraction which is determined by the liquid phase temperature. By use of the previous assumption, the equivalent specific heat in the coexisting region is expressed as

$$C_m^*=C_\ell^*+\frac{L^*}{T_L^*-T_s^*}\ . \qquad (13)$$

Here,

$$C_m^*=\frac{C_m}{C_s},\quad C_\ell^*=\frac{C_\ell}{C_s},\quad L^*=\frac{L}{C_s(T_0-T_f)},\quad T_L^*=\frac{T_L}{T_0-T_f} \qquad (14)$$

The computation of the energy equation (7) in the solid-liquid coexisting region was performed by using the specific heat shown by the above equation.

2.3 Numerical Procedure

The solution of the steady continuous casting problem in twin-roll geometry was obtained as a steady ultimate solution of a false transient problem. First, the solidification profile and the initial temperature distributions in both the liquid and solid phases were assumed for given casting conditions. Then, the temperature distributions in both phases were obtained by use of equations (4) and (7) with boundary conditions (5),(6),(8), and (9). Next, the new solidification front $F(\phi,t)$ was calculated by virtue of equation (11) by using newly obtained temperature information in the vicinity of the interface. This procedure is repeated until the final steady solidification front is attained.

The usual 3-point explicit finite difference scheme was used for the computation of energy equations. The computer running time for a typical case was approximately 10 seconds on Tohoku University's SX-1. The principal data used for this computation are listed in Table 1. The materials selected were Sn-15Pb and steel.

3. Numerical Results and Discussion

The results of computation including heat transfer in both phases and fluid flow in the liquid phase is indicated in Fig.3(a) in the case of $2H_0^*=0.073$, Pe,s=47.2, $\sigma=0.417$, and $T_p^*=7.1$. The isotherms are shown by the solid lines including the solidus line. The experimental solidus and liquidus lines are also shown by the broken lines. The experiment[17] was done separately with this study and the detailed description is omitted. The comparison of the present calculation results and the experimental data reveals a moderate coincidence. In Fig.3(b), the computed result without consideration of heat transfer and fluid flow in the liquid phase is depicted for the purpose of comparison. It is shown that results of full heat and fluid flow analysis are close to the experimental data. However, its difference is not large. This consequence is in line with that of Kroeger and Ostrach[1] who analyzed a plane continuous casting problem via a conformal mapping method and showed that the natural convection effect has a negligible effect on the solid-liquid interface position, at least in the range of the parameters they investigated.

Figure 4 shows the velocity

Thermal conductivity (Solid)	K_s	(w/m K)	50.2
Thermal conductivity (Liquid)	K_ℓ	(W/m K)	21.0
Specific heat (Liquid)	C_ℓ	(kJ/kg K)	0.23
Density (Solid)	ρ	(kg/m³)	7200
Latent heat	L	(kJ/kg)	38.0
Liquidus temp.	T_L	(℃)	208
Solidus temp.	T_s	(℃)	183

Table 1 Principal physical properties of Sn-15Pb.

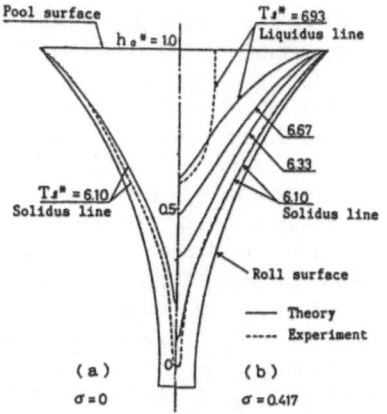

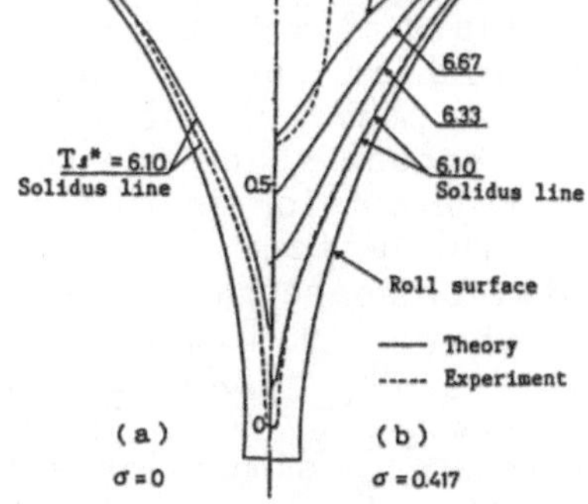

Fig.3 Solidified shell profile and isotherms in the liquid region.

in the liquid and solid regions. Figure 5 plots the temperature distribution in the vertical center plane(O'Q' plane) of the liquid domain under the same conditions as in Fig.3(a). The squares and the solid line designate experimental data and theoretical results, respectively. The entire coincidence is fairly good except in the solid-liquid coexisting region($T_l^* = 6.1 - 6.9$) where a slight difference is seen.

Figure 6 shows the solidified shell thickness versus angle measured from the initiation point P(see Fig.1). The figure compares the experimental results with the theoretical ones obtained under assumptions of (i) a 1-D heat flow, (ii) a 2-D heat flow without consideration of liquid heat transfer and fluid flow, and (iii) a 2-D heat flow under constant mass flow, (iv) a 2-D solution considering full heat and fluid flow. It is seen from the figure that the full solution gives the most accurate result since the experimental data can be considered to be reliable at present. It is noted here that a 1-D analysis does not provide a good result. This point is especially important since most prior studies have been based on this assumption[10]. The heat and the fluid flow in the vicinity of the end of the solidification point(point Q) becomes two-dimensional since the solidified layer gets thick in this region.

The effect of the roll's rotation speed on solidification is shown in Fig.7 for roll spacing $2H_0^*=0.068$ and initial temperature $T_0^*=7.1$. The ordinate designates a dimensionless angle and the abscissa indicates the solidified layer thickness. The rotation speed of the roll is

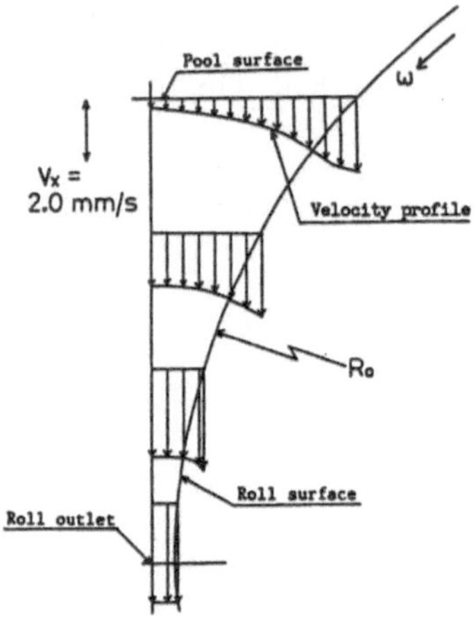

Fig.4 Simplified block flow chart of calculations.

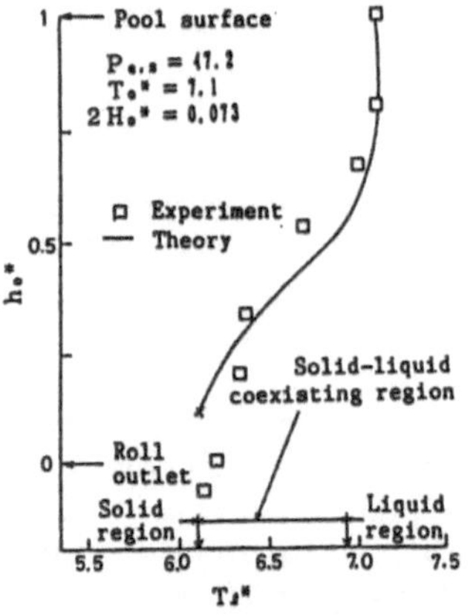

Fig.5 Temperature distribution along the center line of molten material.

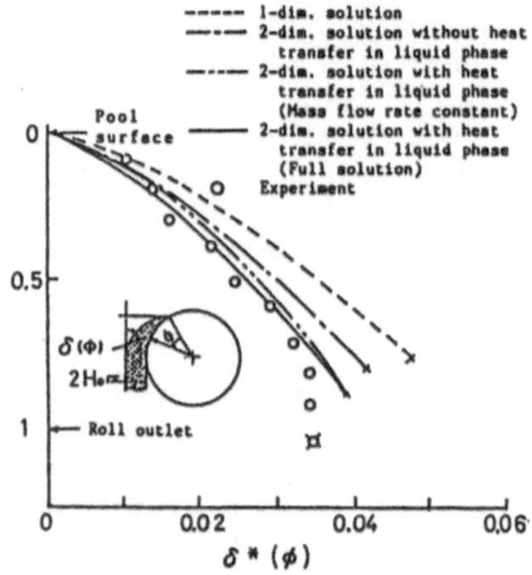

Fig.6 Comparison between calculated and experimented values for solidified shell thickness.

included in the Peclet number, Pe,s. The value of Pe,s when the end point of solidification just coincides with the minimum gap position Q (i.e. roll outlet, shown by an arrow in the figure) which is around 102. The complete solidification point moves downward from the minimum gap point if the Pe,s exceeds this critical value. Whether the end point of solidification is located before the minimum gap point or not may be an important criterion since deformation of the solid, i.e. plastic flow of the solid, will occur if solidification is completed before the material reaches the minimum clearance position. This point will also be important from the view point of quality control of the cast materials, for example, in the prevention of cracks in the center portion of the cast material, which has been clarified by the experiment[17].

Figure 8 shows a relationship between the Peclet number, Pe,s and the roll spacing $2H_0^*$ when the end point of solidification just coincides with the minimum clearance point Q under a condition of $T_0^*=8.04$ and $\sigma=0.417$. As mentioned earlier, this diagram may be helpful in designing a real twin-roll casting machine. The relation can be expressed by the following simple equation:

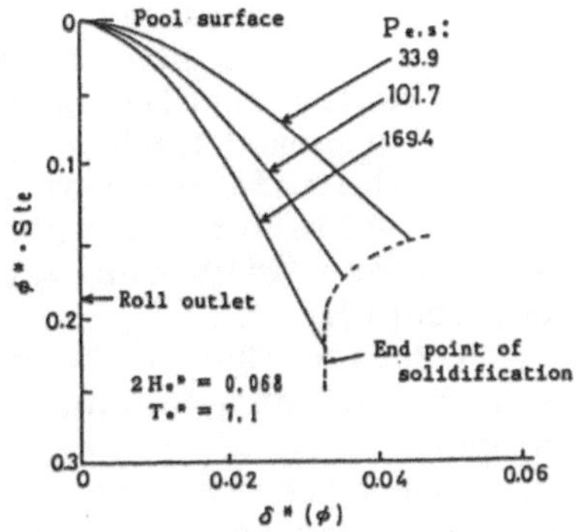

Fig.7 Variation of solidified shell thickness with the roll speed.

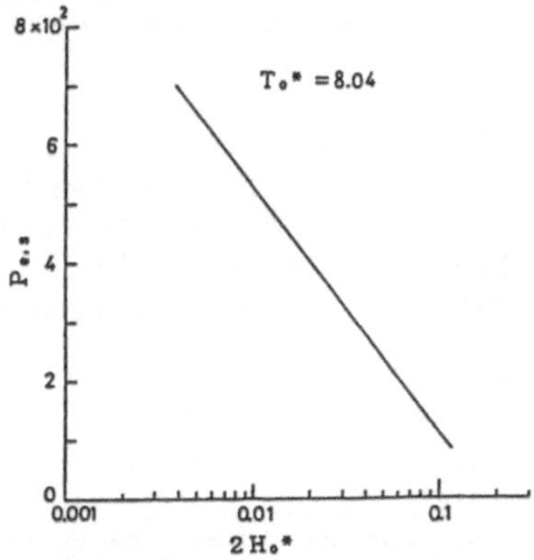

Fig.8 Showing the relationship between the Peclet number and the roll spacing when end point of solidification just coincides with the minimum clearance point.

$$Pe,s=-4.11\ln(2H_0)+5.26 \qquad (15)$$

4. CONCLUDING REMARKS

A numerical algorithm for the two-dimensional solidification problem in the twin-roll continuous casting system has been presented in this paper. Attention was focused on the elucidation of flow and heat transfer characteristics in both the liquid and solid phases. The following conclusions may be drawn from the present study:

(i) A general numerical methodology was presented and the quantitative relationship between the important control parameters in continuous casting of twin-roll types, such as roll speed, the roll gap, the initial temperature of molten materials, the material properties, the solidification profile, and the end point of solidification and the like was clarified in detail. The present numerical results were compared with experimental results obtained separately to check the validity of the proposed method.

(ii) The influence of the liquid phase heat transfer on the entire solidification profile is not significant, which is in accordance with the consequence presented by Kroeger and Ostrach[1].

(iii) Even though the roll speed has a great influence on the solidification profile, the influence of the roll spacing is minor.

In closing, it is noted here that a thorough two-dimensional analysis including full heat and flow equations will be necessary to get more insight into the natural convection effect in the liquid phase. Such tasks will be important in reducing segregation and also in improving material quality.

5. ACKNOWLEDGMENTS

The authors extend their sincere thanks to the Computer Center of Tohoku University for the use of its SX-1 Time Sharing Systems. They also extend thanks to Mr.S.Sugisaki for his assistance to prepare the present paper.

NOMENCLATURE:

$B(\phi)$: shape function of roll
C : specific heat
C_m^* : equivalent specific heat
F,G : solidification fronts
h : vertical distance along the center line
h_0 : distance between the molten surface and the roll outlet
H_0 : roll spacing
k : thermal conductivity
L : latent heat of solidification
Pe , : Peclet number($=U_0R_0/\alpha$)
r,r : radial coordinates
R_0 : roll radius
t : time or Fourier number
T : temperature
T_0 : initial temperature of molten metal
T_f : solidus temperature
T_l : liquidus temperature
T_w : roll surface temperature
U_0, : casting speed
U_r , U_ϕ : velocities in the r and ϕ directions, respectively

Greek Symbols

α : thermal diffusivity
β , β' : angles between radial directions and normal to solidification front
$\delta(\phi)$: thickness of solidified layer
ρ : density
σ : $=k_l/k_s$
ϕ : angle
ϕ_0 : angle between the molten metal surface and the roll outlet

Subscripts and Superscripts:

l : liquid region
m : two-phase region
s : solid region
* : dimensionless quantity
' : relevant to variables in the liquid region

REFERENCES

[1] P.G.Kroeger and S.Ostrach, Int.J.Heat Mass Transfer, 17(1974),pp.1191-1207.
[2] R.Siegel, Int.J.Heat Mass Transfer, 21 (1978), pp.1421-1430.
[3] R.Siegel, J.Heat Transfer, Trans. ASME, 106(1984),pp.237-240.
[4] R.Siegel, J.Heat Transfer, Trans. ASME, 106 (1984),pp.506-511.
[5] P.Koikkalainen, E.Laitinen, S.Louhenkilpi, P.Neittaanmaki, and L.Holappa, Proc. Int.Conf. on Comp.Mech., Vol.III (1986),pp.29-35.
[6] Z.G.Wang and T.Inoue, Proc. Int.Conf. on Comp.Mech.,Vol.VIII (1986),pp.103-108.
[7] Z.F.Lu and J.Zhi, Proc. Int. Conf. on Comp. Mech., Vol.VII (1986), pp.23-28.
[8] I.Ohnaka, Proc. MRS-Europe Symp., Strasbourg (1986),pp.1-13.
[9] I.Ohnaka and K.Kobayashi, Trans. of the Iron and Steel Inst. of Japan,Vol.26(1986),pp.781-789.
[10] K.Miyazawa and J.Szelkely, Metallurgical Trans., Vol.12A(1981),pp.1047-1057.
[11] J.Szekely, in "Free Boundary Problems: Theory and Applications", vol.II(1981),pp.283-292.
[12] I.Ohnaka, "Production of Amorphous Alloy Materials and Rapid Quenching Techniques", J.JSME,88-802(1985),pp.1060-1084.
[13] T.Ohashi, "Present Status of Continuous Casting of Steel and Future Developments", Bull.Japan Inst.Metal 25(1986),pp.505-513.
[14] T.Saitoh, Proc. National Heat Transfer Symp.Japan(1974), pp.333-336.
[15] T.Saitoh, J.Heat Transfer, Trans. ASME,100 (1978), pp.294-299.
[16] H.Hojo,C.G.Kang,K.Kato,N.Tamagawa, and H.Yaguchi, Proc. of the 37th Joint Conf. of the Japan Soc. for Technology of Plasticity, pp.517-520.

VARIATIONAL FORMULATION BY INTEGRAL EQUATIONS FOR THE SOUND RADIATION IN A NON UNIFORM FLOW

M. Ben Tahar and M.A. Hamdi
Division Acoustique et Vibrations Industrielles
Université de Technologie de Compiègne
BP 233, 60206 Compiègne Cédex, France

Abstract

We present in this work a new variational formulation by integral equations which enables the calculation of acoustic fields radiated in non uniform compressible flows. Indeed, this formulation reduces the integral domain without modifying the Sommerfeld's radiation condition. Moreover it avoids computation of the "finite part" of singular integrals and leads to a compact symetrical linear system after discretisation by finite elements. This system is then solved by an iterative method.

1. Introduction

In many industrial fields, the calculation of sound radiation and it's introductions with non-uniform subsonic flows is necessary. Examples of such applications include various components of air-breathing engines, winds tunnels, turbo-fan aircraft and so on. In fact, with the imposition of strict regulations pertaining to noise in order to protect cities, the acoustical aspect is becoming an important feature of quality. For these reasons, we present in this paper a numerical method for the study of sound radiation from ducts with non-uniform flow. Except the academical applications for which exist analytical solutions, only the numerical methods are able to predict the field radiation for finite length ducts with arbitrary shapes.

Contributions to the recent literature have included proposals for a number of different techniques of mathematical modelling and computation for design and analysis purposes. We found the work done by Eversman and Astley [1], [2] on the problem of transmission of sound through non uniform ducts carrying a high speed subsonic compressible flow. They use the method of weighted residuals and the finite element method. The mean flow is not restricted to be irrotational and in the model the non-uniformity joins two infinite uniform ducts and the radiation conditions takes the form of propagation in a known form in these uniform sections. A similar approach with some what different considerations, has been discussed by Nayfeh et al [3], [4] and Uenishi and Myers [5]. Sigman, Najjigi and Zinn [6] used the finite element method in the determination of the acoustic properties of turbo inlets containing high subsonic Mach number steady flow. Due to the complex nature of the reflection process at the inlet entrance plane they consider the Rice hypothesis. In this model except for modes near cutoff frequencies, the assumption of no reflection of "internal" duct waves at the inlet entrances plane is resonable. Baumeister and Majjigi [7] extended previous formulation and incorporated sheared flow and considered the "partitioning approach" that comes from the marching technique developed in [8]. For high frequency sound where reflexions are small, the partition could be taken at the exit. For low frequency sound or for an arbitrary in put with multiple nodes, the partition should be moved from the exit to the far field where the no reflexion hypothesis is valid. Horowitz, Sigman and Zinn [9] propose a new iterative solution where the sound field is divided into two regions : the sound field within and near the inlet which is computed using the finite element method and the radiation field beyoud the inlet which is calculated using an integral solution technique. A continuous solution is obtained by matching the finite element and integral solutions at the interface between the two regions.

Two important remarks can be made from this brief bibliography. The first one is the difficulty to satisfy the radiation condition using the classical finite elements technique. The second one is due to the complexity of the convect equation which excludes the utilisation of the integral equations alone to avoid these difficulties. We propose in this paper to use a new variational formulation by integral equations which combine the two methods : integral equations and finite elements for the resolution of the problem of sound radiation from finite length ducts with non uniform flow. In fact, an integral representation can satisfy the Sommerfeld's radiation condition exactly and limit the domain of integration only where the flow is not zero. With the variational formulation, the explicit evaluation of the "finite part" of the singular integral is avoided and the final system is symmetrical. We solved the implicit system obtained after discretization using an iterative process. For simplicity, the current presentation is confined to the case of a plane formulation.

2. Formulation of the problem

2.1. Acoustic equations

Consider the inlet shown in figure 1, the flow is assumed to be inviscid, perfect isentropic, irrotationnal and stationnary. We assumed that the acoustical sources are a harmonic function of time. The linearised equation for the complex acoustical potential ϕ and the boundary conditions are then :

$$\Delta\phi + k^2 \phi + 2ik \langle \vec{M}, \overrightarrow{\text{grad}}\, \phi\rangle - \langle \vec{M}, \overrightarrow{\text{grad}} \langle \vec{M}, \overrightarrow{\text{grad}}\, \phi\rangle\rangle = 0 \text{ in } \Omega \quad (1)$$

$$\phi - f\Big|_{S_1} = 0 \quad (2)$$

$$\frac{\partial\phi}{\partial n} - g\Big|_{S_2} = 0 \quad (3)$$

$$\lim_{r \to \infty} \sqrt{r} \left|\frac{\partial\phi}{\partial r} - ik\, \phi\right| = 0 \quad (4)$$

where $k = \omega/c$ (c is the local sound speed), M is the local vector Mach number (2), (3) are the boundary conditions and (4) is the Sommerfeld's radiation condition.

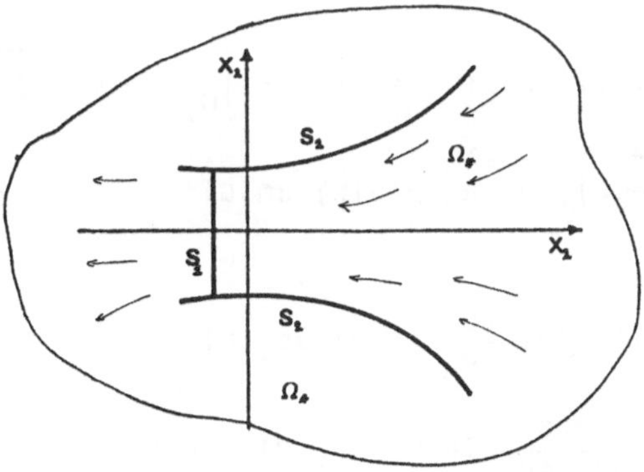

Figure 1 - Geometry of an inlet model

2.2. Integral representation

To find an integral representation for the complex potential $\phi(x,y)$ we use an "indirect method" based on singular and double-layer potentials. After writing the equation (1) from the distribution sense and it's convolution by Green function in the free space $\mathbb{R}^2$ which satisfy Sommerfeld's radiation condition, the potential solution of equation (1) have the following mixed implicit integral representation :

$$\phi(P) = \int_{S_1} (1 - M_n^2(Q))\, \sigma(Q)\, G(P,Q)\, ds_1(Q)$$
$$- \int_{S_2} \mu(Q) \frac{\partial G(P,Q)}{\partial n_Q} ds_2(Q) \tag{5}$$
$$+ \int_{\Omega_*} \langle \vec{M}(Q), \overrightarrow{\mathrm{grad}_Q}\, G(P,Q)\rangle \left\{2ik\, \phi(Q) - \langle \vec{M}(Q), \overrightarrow{\mathrm{grad}_Q}\, \phi(Q)\rangle\right\} d\Omega(Q)$$

where $G(P,Q) = - i/4 \; H_0{}^1(kR)$, ($R = |PQ|$) is the free space Green function which is the elementary solution of Helmholtz's equation satisfying the radiation condition (4) σ and μ are the complex normal speed jump through S_1 (where the potential is known) and the complex potential jump through S_2 (where the normal speed is known) respectively. Ω_* is the part domain of Ω ($\Omega = \mathbb{R}^2 - \{S_1 \cup S_2\}$ where the Mach vector number is not equal to zero.

Apply the boundary conditions (2) and (3) where point P tends to S_1 and S_2 respectively, the following integral system is obtained :

$$\left\{\begin{array}{l}
\int_{S_1} (1 - M_n^2(Q))\, \sigma(Q)\, G(P,Q)\, ds_1(Q) - \\
\int_{S_2} \mu(Q) \frac{\partial G}{\partial n_Q}(P,Q)\, ds_2(Q) = f(P) - \int_{\Omega_*} \langle \vec{M}(Q), \overrightarrow{\mathrm{grad}_Q}\, G(P,Q)\rangle \times \\
\{2ik\, \phi(Q) - \langle \vec{M}(Q), \overrightarrow{\mathrm{grad}_Q}\, \phi(Q)\rangle\}\, d\Omega(Q) \qquad P \text{ in } S_1 \\
\\
\int_{S_1} (1 - M_n^2(Q))\, \sigma(Q) \frac{\partial G}{\partial n_p}(P,Q)\, ds_1(Q) - \\
\mathrm{F.P.} \int_{S_2} \mu(Q) \frac{\partial^2 G}{\partial n_p\, \partial n_Q}(P,Q)\, ds_2(Q) = g(P) \\
- \frac{\partial}{\partial n_p} \int_{\Omega_*} \langle \vec{M}(Q), \overrightarrow{\mathrm{grad}_Q}\, G(P,Q)\, \{2ik\, \phi(Q) - \\
\langle \vec{M}(Q), \overrightarrow{\mathrm{grad}_Q}\, \phi(Q)\}\, d\Omega(Q) \qquad P \text{ in } S_2
\end{array}\right. \tag{6}$$

Usually the solution is calculated by the collocation technique, but this method have two disadvantages : firstly, numerical computation of the double layer term is difficult and secondly, the final algebraic system is non-symmetrical.

2.3. Associated variational formulation

To overcome these difficulties, we suggest to associate a variational formulation to this system. This formulation helps us avoid the explicit calculation of the "finite part" of the singular integrals. In this new formulation we consider the following bilinear form :

$$A((\sigma,\mu),(\sigma',\mu')) = \int_{S_1}\int_{S_1} \sigma'(Q)\,(1-M_n^2(Q))\,(1-M_n^2(P))\,\sigma(P)\, G(P,Q)\,ds_1(P)ds_1(Q) + \int_{S_2}\int_{S_2} \mu'(Q)\,\frac{\partial^2 G(P,Q)}{\partial n_p\,\partial n_Q}\,\mu(P)\,ds_2(P)\,ds_2(Q)$$
$$- \int_{S_1}\int_{S_2} \sigma'(Q)\,(1-M_n^2(Q))\,\frac{\partial G(P,Q)}{\partial n_p}\,\mu(P)\,ds_2(P)\,ds_1(Q) \qquad (7)$$
$$+ \int_{S_2}\int_{S_1} \mu'(Q)\,(1-M_n^2(P))\,\frac{\partial G(P,Q)}{\partial n_Q}\,\sigma(P)\,ds_1(P)\,ds_2(Q)$$

Since A is the symmetrical bilinear form. $A((\sigma,\mu),(\sigma',\mu')) = A((\sigma',\mu'),(\sigma,\mu))$, the solution of the system (6) makes stationary the following functional $L((\sigma,\mu),(\sigma,\mu))$ given by :

$$L((\sigma,\mu),(\sigma,\mu)) = \frac{1}{2}\,A((\sigma,\mu),(\sigma,\mu)) -$$
$$\int_{S_1}\int_{\Omega_*} (1-M_n^2(P))\,\sigma(P)\,\{f(P) - \langle \vec{M}(Q),\,\overrightarrow{\mathrm{grad}}_Q\,G(P,Q)\rangle \times$$
$$\{2ik\,\phi(Q) - \langle \vec{M}(Q),\,\overrightarrow{\mathrm{grad}}_Q\,\phi(Q)\rangle\}\}d\Omega(Q)\,ds_1(P) + \int_{S_2}\int_{\Omega_*} \mu(P)$$
$$\{g(P) - \frac{\partial}{\partial n_p}\,[\langle \vec{M}(Q),\,\overrightarrow{\mathrm{grad}}_Q\,G(P,Q)\rangle\,.\,(2ik\,\phi(Q) - \qquad (8)$$
$$\langle \vec{M}(Q),\,\overrightarrow{\mathrm{grad}}_Q\,\phi(Q)\rangle)]\}d\Omega(Q)\,dS_2(P)$$

The imposition of the condition $\mu = 0$ on the boundary of S_2 and the simultaneous transformation of :

$$\partial^2 G(P,Q)/\partial n_p \partial n_Q \quad \text{and} \quad \langle \vec{M},\,\partial/\partial n_p\,\overrightarrow{\mathrm{grad}}_Q\,G(P,Q)\rangle$$

permit to write the integrals of $S_2 \times S_2$ and $S_2 \times \Omega_*$ as following [10], [11].

$$\int_{S_2}\int_{S_2} [k^2 \langle \vec{n}_p, \vec{n}_Q \rangle \mu(P)\, \mu(Q) - \langle \vec{n}_Q \wedge \overrightarrow{grad_Q}\, \mu(Q),\ \vec{n}_p \wedge \overrightarrow{grad_p}\, \mu(P) \rangle] \times$$

$$G(P,Q)\, ds_2(P)\, ds(Q)$$

$$\int_{S_2}\int_{\Omega_*} \langle \vec{n}_p, \vec{M}(Q) \rangle\, k^2\, G(P,Q) +$$

$$\langle \vec{n}_p \wedge \overrightarrow{grad_p}\, \mu(P),\ \vec{M}(Q) \wedge \overrightarrow{grad_Q}\, G(P,Q) \rangle \times$$

$$\{2ik\, \phi(Q) - \langle \vec{M}(Q), \overrightarrow{grad_Q}\, \phi(Q) \rangle\}\, d\Omega(Q)\, ds_2(P)$$

written as above, all integrals are well-defined and their numerical computational is obvious.

To find the solution which makes the functional $L((\sigma,\mu),(\sigma,\mu))$ stationary, we use the finite elements method. S_1 and S_2 are discretized by straight linear elements, whereas the surface Ω_* is discretized by triangular linear elements. The discretized stationary functional leads to the following matrix system :

$$\begin{bmatrix} B & C \\ C^T & D \end{bmatrix} \begin{Bmatrix} \sigma \\ \mu \end{Bmatrix} + \begin{bmatrix} S \end{bmatrix} \begin{Bmatrix} \phi \end{Bmatrix} = \begin{Bmatrix} F \\ G \end{Bmatrix} \qquad (9)$$

where B, C, D are the matrix of single layer, the matrix of coupled potential between single and double layer and the matrix of double layer respectively. The matrix [S] represents the effect of the convertive term. The vectors F and G are the global discretized acoustical sources.

In order to resolve the matrix system (9) we choose the following iterative method :

At the $(i)^{th}$ iteration we have :

$$\begin{Bmatrix} \sigma \\ \mu \end{Bmatrix}^{i} = \begin{bmatrix} B & C \\ C^T & D \end{bmatrix}^{-1} \begin{Bmatrix} F \\ G \end{Bmatrix} + \begin{bmatrix} B & C \\ C^T & D \end{bmatrix}^{-1} \begin{bmatrix} S \end{bmatrix} \begin{Bmatrix} \phi \end{Bmatrix}^{i-1}$$

and the following integral representation (5) :

$$\{\phi\}^{i} = [H_1]\ \{\sigma\}^{i} - [H_2]\ \{\mu\}^{i} + [H_3]\ \{\phi\}^{i-1}$$

where the vector $\{\phi\}^{i}$ is the discretized complex potential in Ω_*.

This iterative aspect does not constitute any disavantage for the method because all the matrices are calculated and inverted only once.

3. Results

We consider the inlet shown in figure 2. The potential ϕ imposed is equal to 1 in S_1 and $\partial\phi/\partial n = 1$ in S_2. The wave number k = 1 (k = ω/c). The steady flow computation consists of a potential flow solution governed by Laplace's equation (k = 0). In this case we imposed $\partial\phi/\partial n = 0$ in S_2 and M_x = 0.3, 0.4 in the section $x_1 = 0$ (eq. S_1)

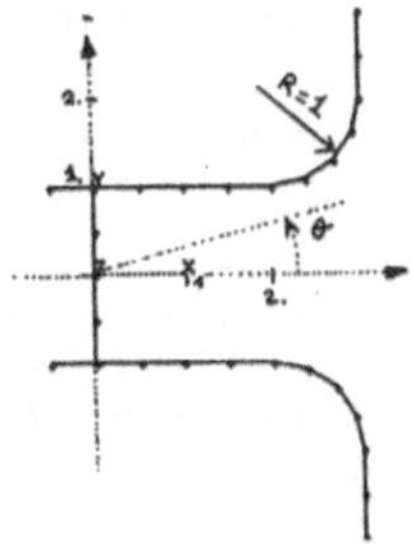

Figure 2 - Geometry of an inlet

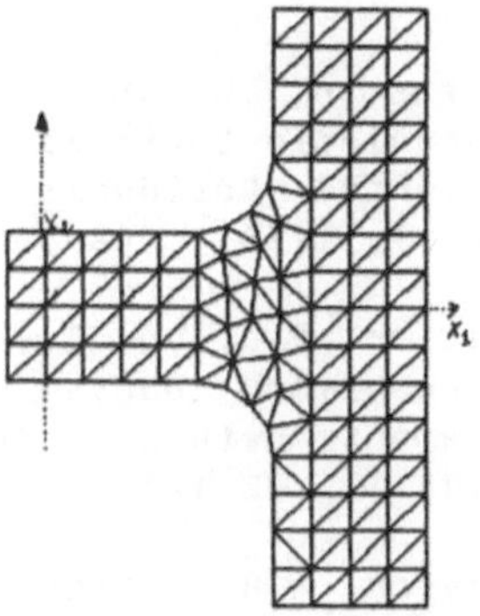

Figure 3 - Mech of domain Ω_*

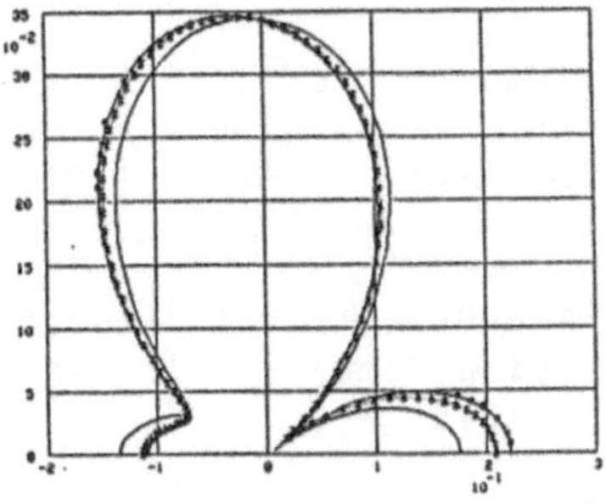

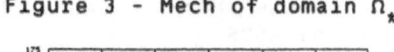

Figure 4 - Directivity with in flow Mach number (—— M = 0, —*— M = 0.3, —•— M = 0.4)

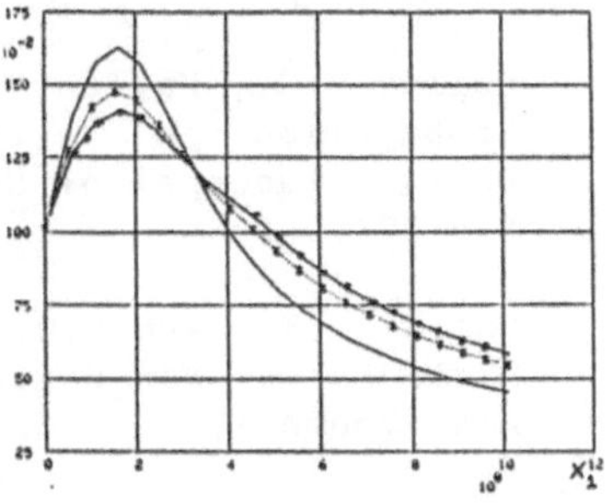

Figure 5 - Amplitude potential along x axis (—— M = 0, —*— M = 0.3, —•— M = 0.4)

Figure 4 and 5 show the convertive effects ($M = 0$, $M = 0.3$ and $M = 0.4$) on the directivity (between polar angle $\theta = 0^\circ$ and $\theta = 180^\circ$) and the variation of the amplitude potential along x_1 axis ($0 \leq x_1 \leq 10$).

4. Conclusion

To study the sound radiation in non uniform subsonic flows, we have suggested to use an integral representation of the complex potential which takes into consideration the condition of the radiation and reduces the integral domain to a single region where the flow is not trivial. The association of the variational formulation avoids the problem of singular integrals and defines a symmetrical system. Moreover the iterative process doesn't have an effect on the computer time. The comparison between the analytical results from the simple cases and this formulations has proved it's efficacy.

5. References

1. Eversman, W. and Astley, R.J., 'Acoustic Transmission in NonUniform Ducts with Nean Flow , Part I : The Method of Weighted Residuals', Journal of Sound and Vibration, 74(1), (1981), 89-101.

2. Astley, R.J. and Eversman, W., 'Acousticf Transmission in Non-Uniform Ducts with Nean Flow, Part II : The Finite Element Method', Journal of Sound and Vibration, 74(1), (1981), 103-121.

3. Nayfeh, A.H., Shaker, B.S. and Kaiser, J.E., 'Transmission of Sound Through Non-Uniform Circular Ducts with Compressible Mean Flows', AIAA Journal, Vol. 18(5), (1980), 515-525.

4. Kelly, J.J., Nayfeh, A.H., and Watson, L.T., 'Acoustic Propagation in Partially Choked Converging-Diverging Ducts', Journal of Sound and Vibration, 81(4), (1982), 519-534.

5. Unishi, K., and Myers, M.K., 'Two-Dimentional Acoustic Field in a Non-Uniform Duct Carrying Compressible Flow', AIAA Journal, Vol. 22(9)j, (1984), 1242-1248.

6. Sigman, R.K., Majjigi, R.K. and Zinn, B.T., 'Determination of Turbofan Inlet Acoustics Using Finite Elements', AIAA Journal, Vol. 16(11), (1978), 1139-1145.

7. Baumeister, K.J. and Majjigi, R.K., 'Application of Velocity Potential Function to Acoustic Duct Propagation Using Finite Elements', AIAA Journal, Vol. 18(5), (1980), 509-514.

8. Baumeister, K.J., 'Numerical Spatial Marching Techniques in Ducts Acoustics', Journal of Acoustical Society of America, Vol. 65, (1979), 297-306.

9. Horowitz, S.J., Sigman, R.K., and Zinn, B.T., 'An Iterative Finite Element - Integral Technique for Predicting Sound Radiation from Turbofan Inlets in Steady Flight', AIAA Journal, Vol. 24(8), (1986), 1256-1262.

10. Hamdi, M.A., 'Formulation variationnelle par équations intégrales pour le calcul de champs acoustiques linéaires proches et lointains', Thèse d'Etat, Université de Technologie de Compiègne, (1982).

11. Ben Tahar, M., Hamdi, M.A., 'Formulation variationnelle par équations intégrales pour le rayonnement acoustique dans un écoulement non uniforme', 7ème JESPA, Lyon (France), 1986).

DESIGN AND STUDY OF BEHAVIOUR OF BRUSH BEARING PLATENS BY THE FINITE ELEMENT METHOD

J.M. Torrenti and P. Royis
L.C.P.C. Paris, France ; E.N.T.P.E. Vaulx-en-Velin, France

ABSTRACT

The finite element method is very valuable when analytical solutions are not feasible . In our case we want to know stresses , strains and displacements in a concrete sample loaded with brush bearing platens . The method is also used for platens design . These platens are used to minimize edge effects in triaxial tests performed at the ENTPE . The software, also developed at the ENTPE, generates meshes using a series of conformal transformations . The solver is a non symetrical front-end one . The analysis of the results obtained with different bearing devices permitted a correct design of our platens, of the boundary conditions it imposes and of the resulting stress and strain fields in the specimen .

1. INTRODUCTION

As the number of structures used for nuclear enngineering and offshore applications has grown , there has arisen a need to understand the behaviour of concrete under multiaxial compression . The easiest test to apply this purpose uses a cubical specimen. Unfortunetaly , since the slenderness ratio of the specimen is then unity , edge effects assume an importance that must be determined and held to a minimum . The principal effect involved is lateral confinement . It results from the fact that the bearing device , generally massive and made of steel , is not deformed in the same manner as the specimen in contact with it .

Because of friction at the interface between the steel and concrete , the concrete cannot be deformed freely . Locally , the stress field is no longer uniaxial but triaxial , leading to an overestimate of the strength in a compression test. What is true of uniaxial loading applies even more to multiaxial loadings . It is therefore vital to eliminate restraint . Quite simply , we are going to see that the approach chosen , the brush bearing platen , has its own drawbacks , which we are going to analyse and quantify using the finite element method .

2. MINIMIZING EDGE EFFECTS : BRUSH BEARING PLATENS

One way of eliminating lateral confinement consists of making the bearing device capable of lateral deformation so that it can "follow" the deformations of the concrete . The first bearing plate designed in accordance with this principle was developed by Hilsdorf [1] for biaxial tests . The operation of this device is such as to eliminate a large part of restraint (see Winkler's comparative study [2]) , but it introduces other effects .It can be seen that this type of bearing platen makes it impossible to control either stress or strain , since the displacement and stress are largest at the center .

The E.N.T.P.E. , when acquired a triaxial press , choose , despite their defects , but encouraged by the results they make it possible to obtain , to fit it with brush bearing platens . Each of these bearing platens consists of a ball joint (A) , a steel base (B) , and steel rods (C) having a cross-section 5 mm square , 80 mm long (see figure 2.1) . The ball joint is necessary with triaxial loading to palliate the parallelism and orthogonality defects of the surfaces of the specimens . It can be seen , however , that the displacement of the bearing platen is transmitted by the ball joint , and so thickness e must be great enough to make bending of the base negligible .

It was accordingly to design the base , but also to determine the boundary conditions on our specimen and their influence on the stress and strain fields , that we undertook a finite element study . We shall also take advantage of it to verify the improvements introduced by the brush bearing platen .

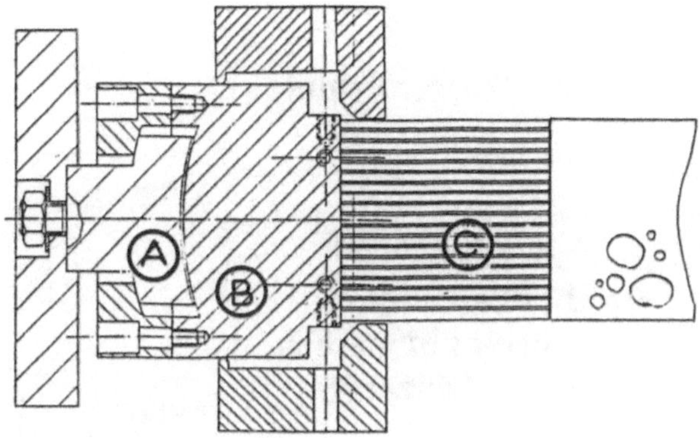

Fig.2.1: The brush bearing platen.

3. THE FINITE ELEMENT SOFTWARE

The software used , ELFIM , conceived and developed at the E.N.T.P.E. by P. Royis [3] , was designed to satisfy the following requirements as well as possible : to create flexible software , easy to use , allowing for rapid analysis of results ; to allow for future development (the introduction of new elements , allowance for new laws of behaviour , ...) ; to develop software that is optimized in terms of computing time .

3.1 The preprocessor MATC

Using a series of conformal transformations , the MATC processor generates the meshes of simply-related plane domains . Exept for a few details , the method used is the one recommended by Chambon [4] . The contour of the domain D to be meshed is transformed into a rectangle . The transformation is conformal for all points inside D . The rectangle is then put into discrete form by meshes that are as close to square as possible , which has repercussions on the initial domain . The reverse transformation delivers the nodes of the mesh inside D .

Figure 3.1 shows the various stages of the meshing of domain D . The nodes and elements are numbered on the rectangle , along each parallel to the width , always in the same direction . Figures 3.2a and 3.2b illustrate this numbering mode in the case of 4 and 8 nodes quadrilateral elements of the Serendip family.

o " corner " points.

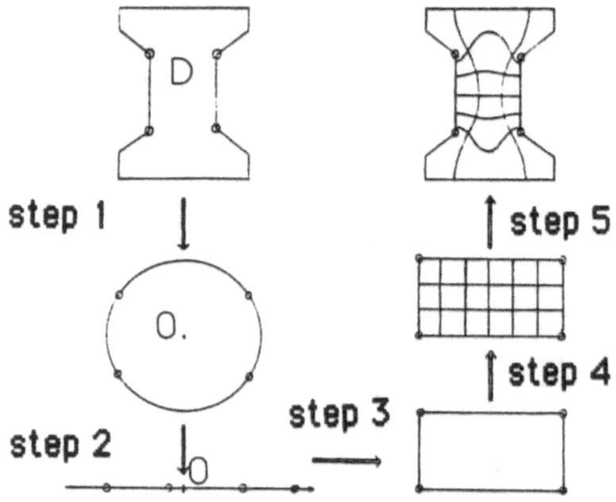

Fig. 3.1: Description of stages of meshing of domain D.

o nodes

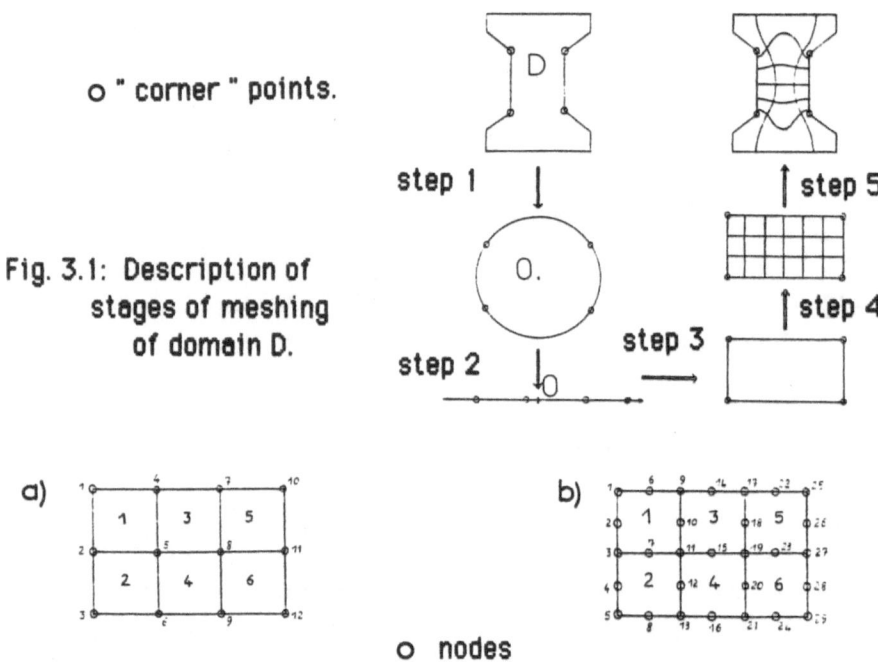

Fig. 3.2: Numbering of nodes and elements.

3.2 The EFEVEL finite element computing processor

The linear systems are solved by a non-symmetrical front-end solver . The originality of this solver lies in the fact that the unknowns are eliminated not after the assembly of each element , as with a conventional front-end solver (Irons [5]) , but after the assembly of the last element of each row (figure 3.2b) . Let us consider , for example , the meshwork shown in figure 3.2b . Elements 1 and 2 are assembled , then the unknowns relative to nodes 1 to 8 are eliminated . Then elements 3 and 4 are assembled and the unknowns relative to nodes 9 to 16 eliminated , and so on . For more details , refer to Royis [3] .

The calculations were carried out using an isoparametric quadrilateral element (substructure) made up of four linear triangles in which the central node was eliminated by imposing on its displacement values equality with the mean of the displacement values at the four corners of the quadrilateral . This element is especially well suited to problems involving several materials having very different mechanical properties .

4. NUMERICAL SIMULATION OF BEARING DEVICES

We tested two types of bearing devices , massive bearings and brush bearing platens , varying the thickness of the base for each : 2 cm , 5 cm , and ∞ (i.e. not bending) . For reasons of symmetry , only a quarter of the specimen and half of the bearing device were meshed . All six cases are shown in fig. 4.1 .

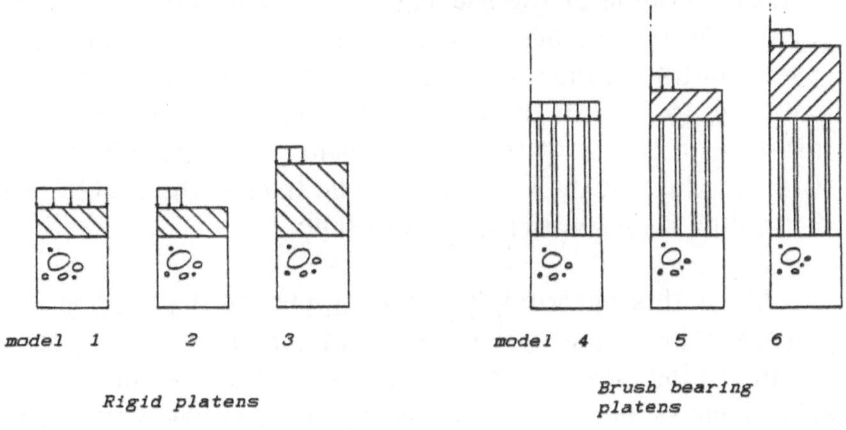

Fig. 4.1: The various cases tested.

Since the MATC preprocessor operates on only a single material , we created the meshwork manually , using a node and element numbering order such that the front-end solver could still be used .We worked in plane stresses , with linear elastic behaviour laws having the following characteristics :

-concrete : E=20,000 MPa ; $\nu = 0,2$
-steel : E=200,000 MPa , $\nu = 0,3$
-void (gap between brush elements) : E=10 MPa

5. RESULTS OF SIMULATIONS

We first analyzed the incidence of bending of the base . From the displacements of the nodes of the meshwork we then obtain :

$$\varepsilon_1 = \frac{1}{\ell_1}\int_0^{\ell_1} du_1 \quad ; \quad \varepsilon_2 = \frac{1}{\ell_2}\int_0^{\ell_2} du_2$$

We can in this way deduce the Poisson's ratio measured for each meshwork : $\nu = -\varepsilon_2/\varepsilon_1$

Meshwork	1	2	3	4	5	6
Poisson's ratio	0.194	0.290	0.205	0.196	0.250	0.200

Its should be recalled that the Poisson's ratio used in the calculations was 0.200 . This table therefore shows the harmful effect of bending of the bearing device . This influence is again significant for the boundary conditions . For example , with a base 2 cm thick , the stress at the edges of the cube is 60 % less than that applied to the centre . The loss is only 15 % with a thickness of 5 cm . As for displacement , with mesh 5 (e = 2 cm) it is 30 % less at the edge of the cube than at the center , while it is only 7 % less with mesh 6 (e= 5 cm).

All of this accordingly led us to choose the bearing device corresponding to mesh 6 . Finally , we considered the stress and strain fields inside the specimen , and to measure the effectiveness of the brush bearing platens we compared the results of meshes 3 and 6 . Consider for example the displacement isovalue curves in the direction perpendicular to loading (figure 5.1) . The improvement introduced by the brush bearing platen , for which the displacement fields is close to that obtained without lateral confinement , is clear . If we now consider the stress isovalue curves in the direction perpendicular to loading , we find here again that with the brush bearing platen the stress field is much more homogeneous and close to zero , the value that would obtain without restraint (which confirms , in this case only , the assumption of plane strains) (figure 5.2) .

To complete this study , we present the displacement and stress isovalue curves in the direction of loading for our device (figure 5.3 and 5.4) . There are of course again effects on the boundary conditions (a larger displacement and stress at the centre of the specimen) . The stress decreases towards the free edge , but the loss is limited to 5 % if we eliminate a zone of 1 cm located under the bearing device and highly perturbed by its discontinuities .

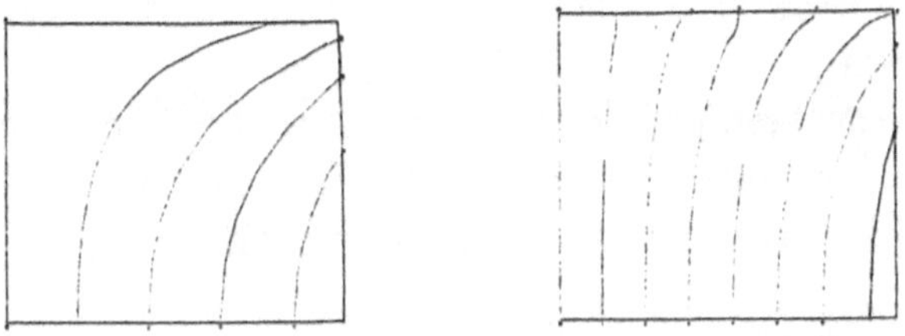

meshwork 3 meshwork 6

Fig. 5.1: Displacement isovalue curves in the direction perpendicular to loading.

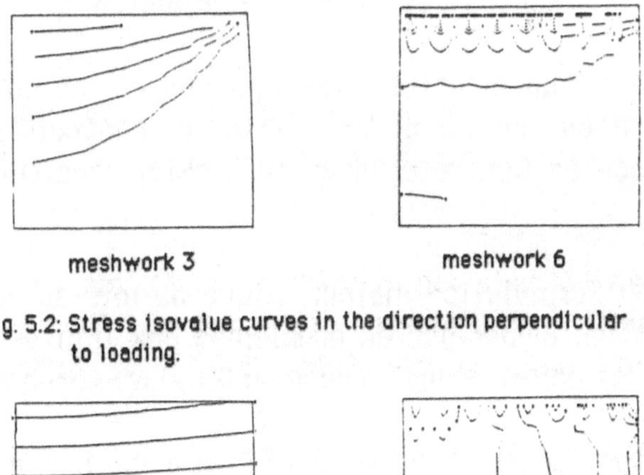

meshwork 3 meshwork 6

fig. 5.2: Stress isovalue curves in the direction perpendicular to loading.

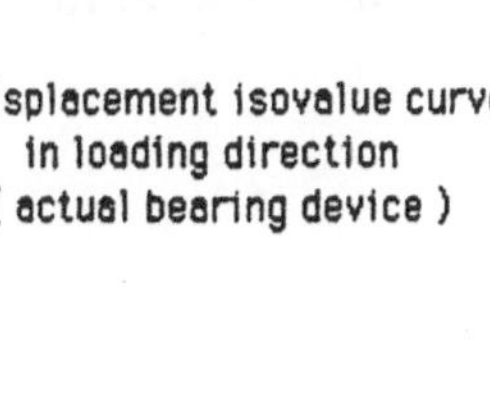

Fig. 5.3: Displacement isovalue curves in loading direction (actual bearing device)

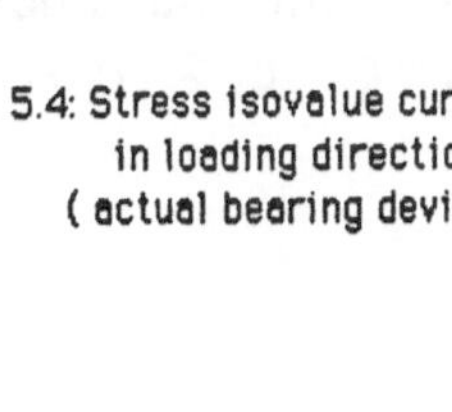

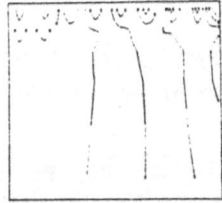

Fig. 5.4: Stress isovalue curves in loading direction. (actual bearing device)

6. CONCLUSIONS

We used a finite element program with a non-symmetrical front-end solver . The calculations carried out using an isoparametric quadrilateral element enabled us :

-to quantify the edge effects result from lateral confinement and the bending of the bearing devices ;
-to design our bearing device in the light of these effects ;
-to check that the edge effects are then to a large extent eliminated ;
-to determine more precisely how our bearing device works , the boundary conditions it imposes , and the resulting stress and strain fields in the specimen .

7. REFERENCES

1. Hilsdorf, "Die Bestimmung der Zweiachsigen Festigkeit des Betons", Deutscher Ausschuss für Stahlbeton, n^{0} 173, 1965 .

2. Winkler, "Fundamental investigations on the influence of test equipment on multiaxial test result of concrete", International Conference on Concrete under multiaxial conditions, Toulouse . 1984

3. Royis, "Formulation mathématique de lois de comportement . Modelisation numérique de problèmes aux limites en mécanique des solides déformables", Thèse de D.I., I.N.P. Grenoble, 1896 .

4. Chambon, "application de la M.E.F. et d'une loi incrémentale aux problèmes de mécanique des sols", Grenoble, 1975.

5. Irons, "A frontal solution program for finite element analysis", International Journal of Numerical Methods in Engineering, Vol 2 (1970), 5-32 .

WATER POLLUTION CONTROL USING FINITE ELEMENT MODEL

T.Oikawa and M.Kawahara

Chuo University
Kasuga,Bunkyo-ku,Tokyo,112,Japan

SUMMARY

A finite element model for the water pollution control of a nearly closed water area is described. The principal constraints are to attain the chemical oxygen demand (COD) standards. The finite element method is used for the derivation of linear constraints of the constrained optimization problem. The objective function is employed the total value of removal COD concentration discharging into the water. This model is applied to Tokyo Bay which is a nearly closed water area in Japan.

INTRODUCTION

The combination method of linear programming and finite element methods has proven to be an effective tool for constrained optimization problems. Several investigators have solved the water pollution control or the thermal diffusion control by this method. The finite element method is not only suitable for solving the problems with irregular boundaries but also a useful technique for the derivation of linear constraints of the optimization problem.

In this paper the numerical model is shown to apply to the water pollution control of a nearly closed water area. This model requires the permanent current and substance dispersion pattarns. Although tidal residual current is generally used as permanent current, it is difficult to simulated the actual current. The required one is complex. For simplicity, the present study employs the current due to the influence of river discharge. Regarding substance dispersion, the model based on two-dimensional dispersion is used for the calculation of COD concentration. For the numerical simulations of them, different kinds of finite element idealization are employed. Namely, the finite element idealization for the current is smaller than that for the substance dispersion. Because the smaller the size of finite elements, the better the accuracy, as far as the current is concerned. Moreovere, the current is one of the most important factors to detarmine the dispersion phenomenon.

The model presented in this paper makes it possible to apply to the water pollution control of a lake or a nearly closed water area.

MODEL FORMULATION

In this paper, to get constraints of the linear programming the formulation of the substance dispersion model using the finite element method is applied.

The diffusion-convection equation on COD in an arbitrary domain V is written as

$$u\, c,_x + v\, c,_y - D_x\, c,_{xx} - D_y\, c,_{yy} + R\, c - Q = 0 \qquad (1)$$

where c is COD concentration, u and v are flow velocities in x- and y- directions, respectively, D_x and D_y are dispersion coefficients in x- and y- directions, R is substance decaying rate, and Q is the increasing rate of substance concentration due to a source.

Fig.1 shows a typical domain V for the analysis. The domain boundary can be classified into two types. They are S_1 with specified COD concentration and S_2 with specified flux of concentration.

Boundary conditions on S_1 and S_2 are

$$c = \hat{c} \qquad \text{on } S_1 \qquad (2)$$

$$t = q_x\, l + q_y\, m = \hat{t} \qquad \text{on } S_2 \qquad (3)$$

where ^ denotes the specified value on the boundary S_1 and S_2, q_x and q_y are flux in x- and y- directions, l and m are directions cosines of the unite outward normal to S_2 and t represents flux of concentration.

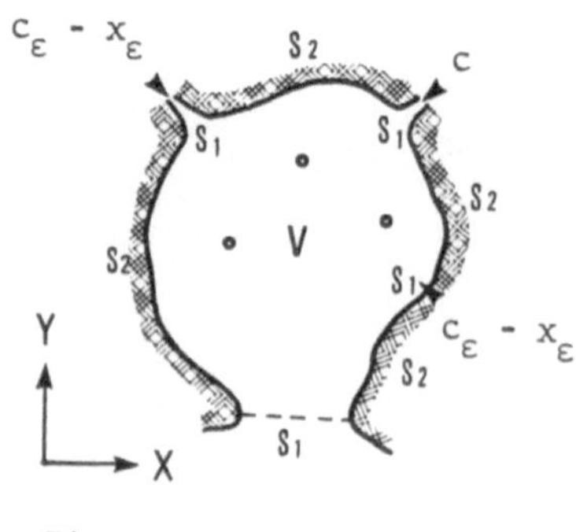

Fig.1 Problem Definition

The Garelkin method is used for the formulation of the finite element method. Using a linear interpolation based on a three-node triangular element, it is obtained that

$$K_{\alpha\beta\gamma}{}^{x}\, u_\beta\, c_\gamma + K_{\alpha\beta\gamma}{}^{y}\, v_\beta\, c_\gamma + S_{\alpha\beta}\, c_\beta + H_{\alpha\beta}\, c_\beta = \hat{\Omega}_\alpha \qquad (4)$$

Let $(c_\varepsilon - x_\varepsilon)$ be concentration of the control inflow, provided that x_ε is removal COD concentration at nodal point ε, as shown in Fig.1. Then, the usual superposition procedure leads to;

$$E_{\alpha\beta}\, c_\beta = E_{\alpha\lambda}\, c_\lambda + E_{\alpha\varepsilon}\,(c_\varepsilon - x_\varepsilon) + \hat{\Omega}_\alpha \tag{5}$$

where

$$E_{\alpha\beta} = K_{\alpha\gamma\beta}{}^{x}\, u_\gamma + K_{\alpha\gamma\beta}{}^{y}\, v_\gamma + S_{\alpha\beta} + H_{\alpha\beta} \tag{6}$$

Multiplying both sides of Eq.(5) by the inverse matrix of the left-hand side coefficient matrix, equations for c_β can be obtained.

$$\begin{aligned} c_\beta &= E_{\alpha\beta}{}^{-1}\,\{ E_{\alpha\lambda}\, c_\lambda + E_{\alpha\varepsilon}\,(c_\varepsilon - x_\varepsilon) + \hat{\Omega}_\alpha \} \\ &= j_\beta + t_\beta - A_{\beta\varepsilon}\, x_\varepsilon + h_\beta \end{aligned} \tag{7}$$

The principal constraints of the problem are to be satisfied COD standerds. Therefore, water quality constraints at observation points are

$$c_\beta \leqq c_\beta{}^{a} \tag{8}$$

where $c_\beta{}^{a}$ is the allowable COD concentration.

Substituting Eq.(7) into Eq.(8), the constraints of this problem become

$$A_{\beta\varepsilon}\, x_\varepsilon \geqq j_\beta + t_\beta + h_\beta - c_\beta{}^{a} \tag{9}$$

In case that the water pollution control in a river basin is studied, the objective function, which can be expressed by the sum of the operation cost of treatment plants in the region, is usually used. But in caes that the water pollution control in an estuary and a coastal zone is studied, it may be difficult how to choose the most suitable objective function.

In this paper, for the sake of simplicity, the objective function is chosen to be a linear function of the form

$$J = \boldsymbol{w}^{\mathsf{T}}\, \boldsymbol{x} \tag{10}$$

where $\boldsymbol{w}$ is an n-vector representing wight to estimate the rate of each municipality discharging wastes and J is the total value of COD concentration to reduce at control inflows.

Hence, the objective function and constraints are presented as follows:

$$\min \sum_{j=1}^{n} w_j x_j \tag{11}$$

$$\text{subj. to} \sum_{j=1}^{n} a_{ij} x_j \geqq b_j \qquad i = 1,2,\cdots,m \tag{12}$$

$$l_j \leqq x_j \leqq u_j \qquad j = 1,2,\cdots,n \tag{13}$$

where Eq.(12) denotes Eq.(9), l_j and u_j are the lower bound and the upper bound of control valuables which are treatment concentrations of COD, m is the number of observation points of water quality and n is the number of points of inflow.

Solving Eqs. (11)-(13) by simplex method, the optimal removal concentration x_j of the control inflow j are obtained.

APPLICATION TO TOKYO BAY

As an example a brief water quality assesment of Tokyo Bay is treated, of which configuration is shown in Fig.2. Tokyo Bay is the tipical nearly closed water area with the distance of about 50 km in the direction of south and north, with about 20 km in the direction of east and west, and with the mean water depth of about 18 m. There are not many points achieved the COD standards.

The finite element idealization used for the analysis of COD is shown in Fig.3. The numbers of the finite elements and nodal points are 331 and 202, respectively.

Table 1 COD Concentration, Wight and Computational Result at Twelve Points (E1-E12)

Inflow	E1	E2	E3	E4	E5	E6	E7	E8	E9	E10	E11	E12
COD (mg/l)	1.0	3.0	2.0	3.5	4.0	7.0	7.8	4.3	9.0	5.0	4.1	3.0
Weight	-	1.0	-	1.1	1.2	1.4	1.4	1.25	1.2	1.0	1.0	0.75
Optimal Removal COD (mg/l)	-	0	-	0	3.0	1.45	0	2.44	0	0	0	0
Desirable COD(mg/l)	1.0	3.0	2.0	3.5	1.0	5.55	7.8	1.86	9.0	5.0	4.1	3.0

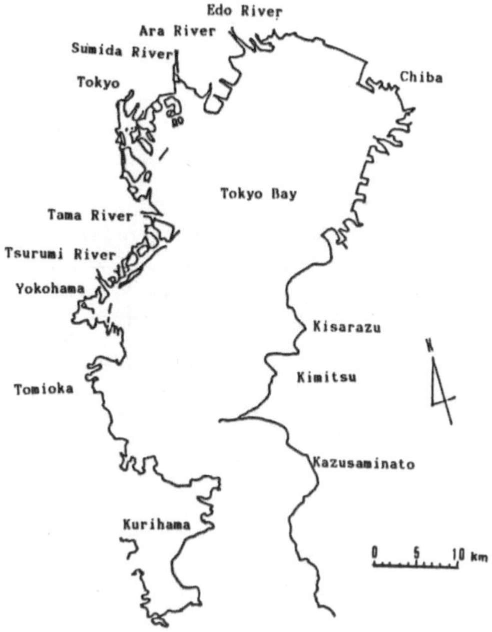

Fig.2 Location

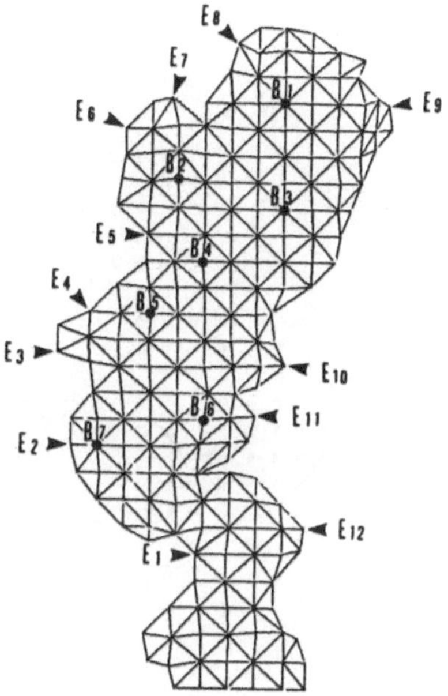

Fig.3 Finite Element Idealization

In this example, twelve points along the coastal line discharge effluents into Tokyo Bay. The data of COD concentration including the effluent discharges of these points are summarized in Table 1. Flow velocities (see Fig.4) are known functions which are linearly interpolated the result of the another finite element idealization divided into smaller sections. First, making use of these data, COD concentration in the domain at existing stage is simulated. As the boundary conditions, COD concentration is specified zero on the ocean boundary and flux of concentration t is specified zero on the coastline boundary. The value of dispersion coefficients D_x and D_y is determined in terms of the the numerical simulations. Table 2 summarizes the constants used in this computation. B1 - B 7 in Fig.3 denote observation points of water quallity. The COD standards at the observation points are all 3.0 mg/l. The upper bound of the removal COD concentration at each inflow is 75 % of the existing COD. The lower bound at each inflow is equal to zero. Fig.5 shows the concentration distribution of COD at the existing stage, which violate the standards at several observation points. Using the above results, it is possible to formulate the optimal problem that makes the standards satisfy and makes the appropriate objective function minimize. The weight w_j of the control inflows summarised in Table 1. Since COD concentrations at E1 and E3 are less than 3 mg/l, E1 and E3 are not considered control inflows.

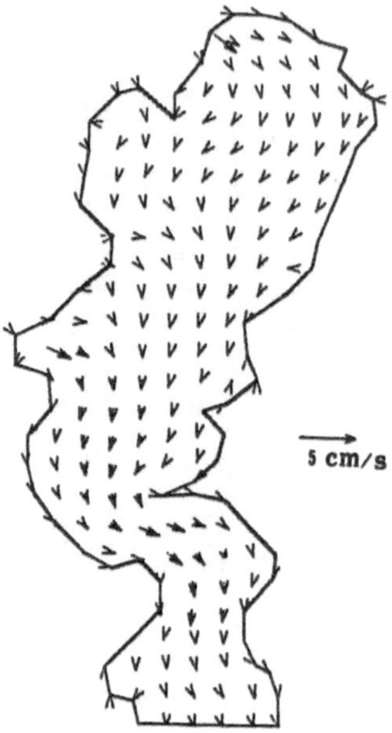

Fig.4 Velocity

Table 2 Computational Data

Dx	m^2/s	2,500
Dy	m^2/s	2,500
R	day^{-1}	1.0
Q	mg/l·day	0.0

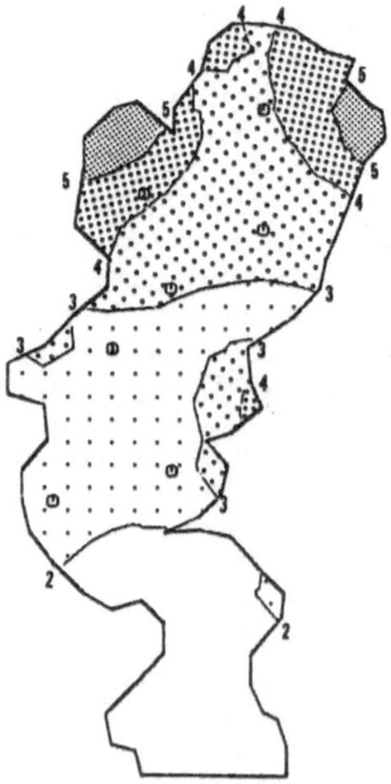

Fig.5 Concentration Distribution of COD at Existing Stage

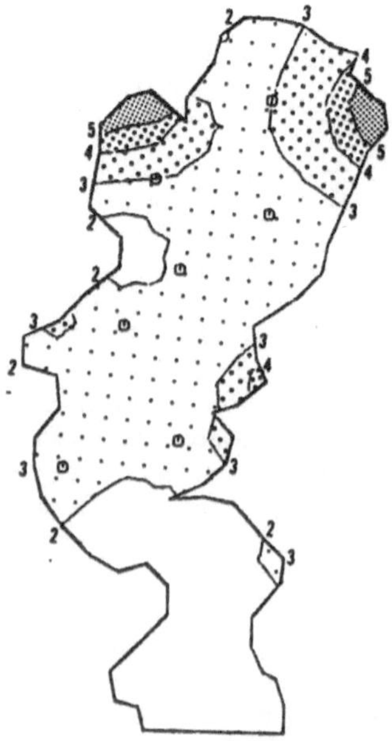

Fig.6 Concentration Distribution of COD after the Control of the Inflow

Fig.6 shows the concentration distribution of COD after the control of the inflow computed by the finite element model. The optimal removal COD concentration at each control inflow is shown in Table 1. According to the result, the COD standards are satisfied at all observation points.

CONCLUSIONS

A method for the water pollution control is presented in this paper. The formulation is based on the combination of linear programming and the finite element method. This model is adaptable for the region in two and three dimensional space. The model is assumed as the steady state formulation. The model can be extended to unsteady state analysis in principle, but the cost would be much more expensive. Further improvements and applications to practical problems can be anticipated in the near future.

ACKNOWLEDGEMENTS

The authors wish to express their appreciation to Prof. Y. Matsuo, Chuo University, for his valuable advice and Messers. Y. Tanaka and N. Tsurumi for supports during the preparation of this paper.

REFFERENCES

[1] T. Futagami, "The FE (Finite Element) and DP (Dynamic Programming) Method in Optimization of Field Problems," Finite Elemente Flow Analysis, University of Tokyo Press, pp. 717-724, 1982.
[2] Revelle,C.S., Loucks,D.P., and Lynn,W.R., "Linear Programming Applied to Water Quality Management," Water Resources Research, Vol. 4, No. 1, 1968, pp. 1-9.
[3] O. Fujiwara, Selvanayagam,K.G., and S. Ohgaki, "River Quality Management under Stochastic Streamflow," Journal of the Environmental Engineering, ASCE, Vol. 112, No. 2, 1986, pp. 185-198.
[4] G.B. Dantzing, "Linear Programming and Extensions," Princeton Univ. Press, 1963.
[5] T.Horie, "Hydraulic Investestigation on Seawater Flow and Substance Dispersion in Estuarine and Coastal Regions," Technical Note of The Port & Harbour Research Institute, No. 360, pp. 1-222, 1980, (in Japanese).

THREE-DIMENSIONAL FINITE ELEMENT ANALYSES FOR A MAXWELL FLUID USING THE PENALTY FUNCTION METHOD

Takeo Shiojima* and Yoji Shimazaki**

* Idemitsu Petrochemical Co.,Ltd., Chiba,Japan
** Tokai University,Kanagawa,Japan

SUMMARY

The penalty function formulation of the three-dimensional finite element method is applied for analyzing the extrudate swells of a Maxwell fluid. The momentum and the constitutive equations are solved separately until the convergence is achieved, in which the standard Galerkin's method is applied to solve the velocity and the tangential extra-stresses, and the least square finite element method is applied to the normal extra-stresses.

1. INTRODUCTION

In operating polymer processes in the steady-state flows, such as those found in the extrudate molding or the fiber spinning, we often observe a most interesting fact that the shape of a final product can be quite different from that at the die exit[1]. In order to estimate this phenomenon by the numerical simulation, elastic effects should be considered in the equations of governing the flow of viscous fluid. A primary difficulty to analyze such a model is the inability to eliminate stresses from the governing equations. Several approaches are available to overcome this difficulty. Among those, the mixed finite element method which solves velocity, pressure and extra-stresses simultaneously is frequently used[2-5]. Because the method requires a large storage area for the stiffness matrix, the amount of computer time is increased for reducing the matrix. On the other hand, the method which decouples the momentum and continuity equations from the constitutive equation requires a less storage area for the stiffness matrix[6-8]. This method, however, is not only unstable for highly non-linear problems, but also needs a quadratic element for the velocity field. These disadvantages can be improved if we combine the Petrov-Galerkin's method for the extra-stress field with the penalty function formulation of the standard Galerkin's method for the velocity field[9-10].

In this study, we describe a compact three-dimensional finite element method for computing the extrudate swells of a Maxwell (upper convected) fluid. The standard Galerkin's method is applied to solve both the velocity and tangential extra-stresses, and the least square finite element method is applied to the normal extra-stresses. An eight-node isoparametric hexahedral element is used for the approximation of both velocity and extra-stress fields. The pressure replaced by a penalty function is determined with the least square finite element method using the calculated extra-stress tensor and velocity vector[11]. The swells of a planar die are first solved to verify the scheme and check the stability. Both the calculated swells and the numerical stability agree well with each of those obtained by other authors. In the next example, we calculate the swells of the fluid being extruded through a rectangular die. In order to illustrate the swell behavior, the die with several different aspect ratios is considered.

2. GOVERNING EQUATIONS

The governing equations for the slow steady-state flow of an incompressible Maxwell fluid without body forces in rectangular Cartesian coordinates are

Equilibrium: $$\sigma_{ij,j} = 0 \tag{1}$$

Continuity : $$u_{i,i} = \varepsilon_{ii} = 0 \tag{2}$$

Constitutive relationship:

$$\sigma'_{ij} = \sigma_{ij} + P\delta_{ij} \tag{3}$$

$$\varepsilon_{ij} = \frac{1}{2\mu}\sigma'_{ij} + \frac{1}{2G}\overset{\nabla}{\sigma}'_{ij} \tag{4}$$

$$\overset{\nabla}{\sigma}'_{ij} = u_k\sigma'_{ij,k} - u_{j,k}\sigma'_{ik} - u_{i,k}\sigma'_{kj} \tag{5}$$

$$\varepsilon_{ij} = \frac{1}{2}(u_{i,j} + u_{j,i}) \tag{6}$$

$$P = -\frac{1}{3}\sigma_{ii} \tag{7}$$

Boundary condition :

$$u_i = \bar{u}_i \quad \text{on } S_u \tag{8}$$

$$\nu_j\sigma_{ij} = \bar{T}_i \quad \text{on } S_t \tag{9}$$

$$\sigma'_{ij} = \bar{\sigma}'_{ij} \quad \text{on } S_\sigma \tag{10}$$

where σ_{ij} is the total stress, σ'_{ij} is the extra-stress, u_i is the velocity, ε_{ij} is the strain-rate, μ is the viscosity, G is the shear modulus of elasticity, p is the pressure, $\bar{u}_i$ is the specified velocity on S_u, $\bar{T}_i$ is the specified traction on S_t with unit outward normal vector ν_j, and $\bar{\sigma}_{ij}$ is the specified stress on S_σ.

3. FINITE ELEMENT FORMULATION

In order to obtain the solutions for equations (1), (2) and (3), we use the method decoupling the equilibrium and continuity equations from the constitutive equation. The algorithm is shown in Figure 1.

3.1 Equilibrium and continuity equations

Consider the following approximation to the velocity

$$u_i \cong N_\alpha u_{\alpha i} \tag{11}$$

Galerkin's method with the use of the finite element equation (11) gives us

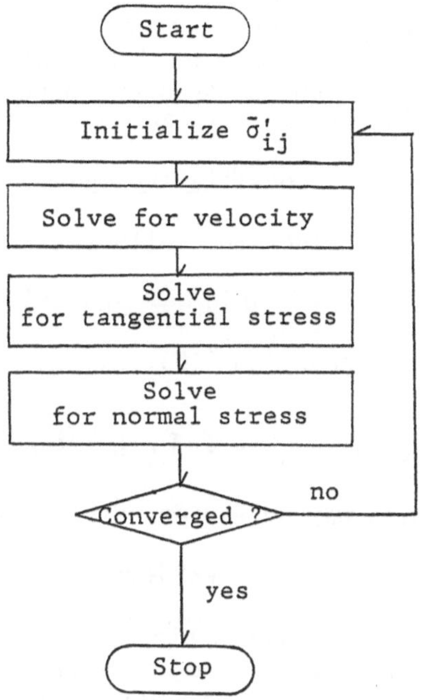

Figure 1. Solution algorithm

$$\int_V N_{\alpha,j}(\sigma'_{ij})\,dv - \int_V N_{\alpha,j}\delta_{ij}P\,dv - \int_{S_t} N_\alpha \bar{T}_i\,ds = 0 \tag{12}$$

in which the first and third terms of the integrals are obtained by the partial integration. We now approximate the pressure using a large positive number λ [12], as

$$P = -\lambda\,\varepsilon_{ii} \tag{13}$$

We note that the incompressibility condition, equation (2), is imposed in equation (13). Substituting equation (4) and (13) into equation (12), we obtain

$$\int_V N_{\alpha,j}(2\mu\,\varepsilon_{ij})\,dv - \int_V N_{\alpha,j}\frac{\mu}{G}\overset{\nabla}{\sigma}{}'_{ij}\,dv + \int_V N_{\alpha,j}\delta_{ij}\lambda\,\varepsilon_{kk}\,dv - \int_{S_t} N_\alpha \bar{T}_i\,ds = 0 \tag{14}$$

In matrix notation, we have

$$[K1]\{u\} + \lambda [K2]\{u\} = \{F_u\} \tag{15}$$

where K1 represents the coefficients obtained from the left hand side of the first and second integrals in equation (14), and K2 from the third integral, and F_u represents the coefficients obtained from the last integral in equation (14). The extra-stress tensor appeared in equation (14) is considered known and held constant at each nodal point.

3.2 Constitutive equation for the tangential extra-stresses

Consider the following finite element approximation

$$\sigma'_{ij} \cong L_\alpha \sigma'_{\alpha ij} \tag{16}$$

Using equation (16), Galerkin's method gives us the equation

$$\int_V L_\alpha \sigma'_{ij}\, dv + \int_V L_\alpha \frac{\mu}{G} \overset{\nabla}{\sigma}{}'_{ij}\, dv - 2\mu \int_V L_\alpha \varepsilon_{ij}\, dv = 0 \tag{17}$$

which may be expressed more compactly, in matrix notation, as

$$[K3]\{\sigma'_t\} = \{F_t\} \tag{18}$$

where K3 represents the coefficients obtained from the integrals with tangential extra-stresses in equation (17) and σ'_t terms are the unknown components of the nodal tangential extra-stresses. In equation (18), we consider the velocity and the normal extra-stresses to be known and held constant at each nodal point. Thus, F_t represents the coefficients obtained from the third integral in equation (17) as well as those obtained from the second integral with the normal extra-stresses.

3.3 Constitutive equation for the normal-stresses

Using equation (16), the least square finite element method gives us the equation

$$\int_V \{L_\alpha + \frac{\mu}{G}(u_k L_{\alpha,k} - 2u_{i,j} L_\alpha)\}$$

$$(\sigma'_{ij} + \frac{\mu}{G} \overset{\nabla}{\sigma}{}'_{ij} - 2\mu\, \varepsilon_{ij})\, dv = 0 \tag{19}$$

which may be written in matrix notation as

$$[K4]\{\sigma'_n\} = \{F_n\} \tag{20}$$

where K4 represents the coefficients obtained from the integrals with the normal extra-stresses in equation (19) and σ_n' terms are the unknown components of the nodal normal extra-stresses. In this case we consider the velocity and the tangential extra-stresses to be known. Thus, F_n coefficients represent the integral with the known terms in equation (19).

4. EXAMPLES

4.1 Extrusion from a planar die

In order to verify the scheme and check the numerical stability, we first solve the problem of the extrusion from a planar die. In this example, $W=(\mu/G)\gamma_w$ is considered a Weissenberg number in which γ_w denotes the shear-rate on the channel wall.

Figure 2 shows the deformed element patterns. In Figure 3, swell ratios obtained by the present method are compared with those obtained by two-dimensional analyses.

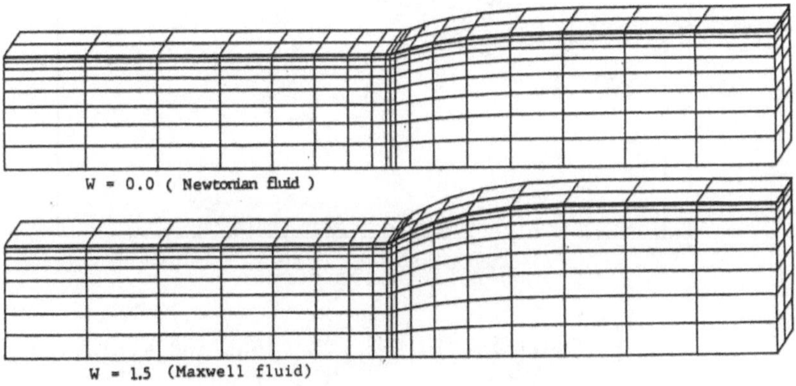

Figure 2. Deformed element patterns for the planar die

4.2 Extrusion from a rectangular die

As a practical example, the present method is applied for analyzing the flow through the rectangular die whose aspect ratio (AR: width/thickness) is 1.0, 3.0 or 6.0. In order to determine the boundary condition of the velocity entered into the control volume, the velocity of the steady-state Newtonian fluid flow through the rectangular duct is specified.

Figure 4 shows the deformed shapes with AR=1.0. The Weissenberg number shown in the figure is defined as $W=\mu V/Ga$, where V is the average velocity, and a is the half thickness of the die. The upper limits of the convergence for AR=1.0, 3.0 and 6.0 are 0.3, 0.5 and 0.4 respectively. Figure 5 shows the final shapes of the cross sections with the three different aspect ratios. The difference in the final shapes according to the Weissenberg number is also shown in the figure.

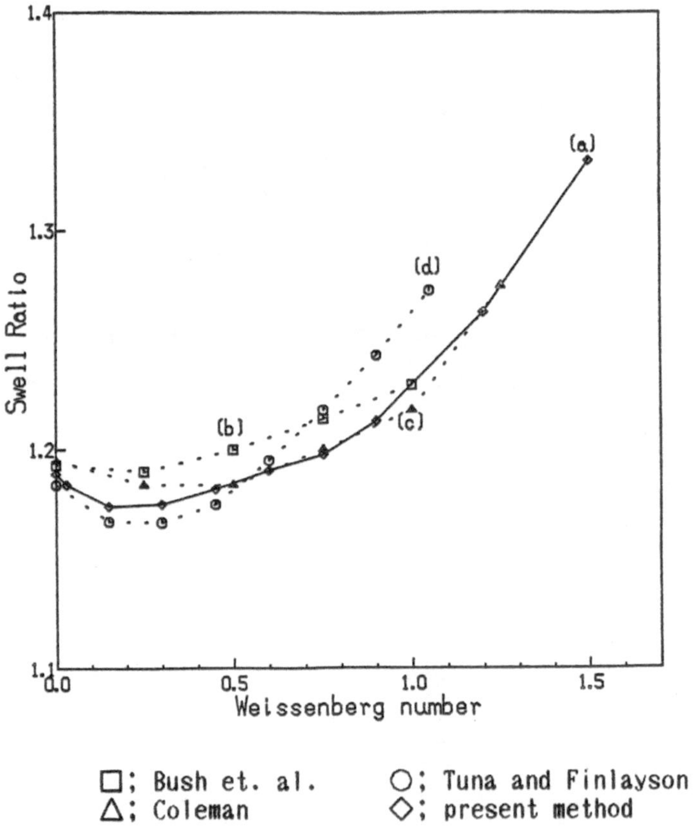

Figure 3. Swell ratios for the planar die problem

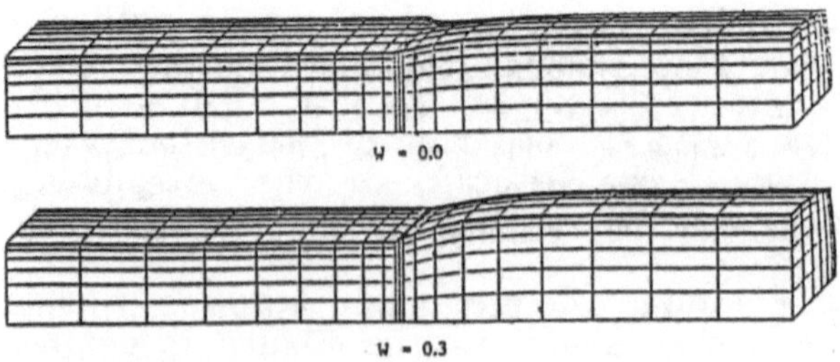

Figure 4. Deformed element patterns for the rectangular die problem (AR=1.0)

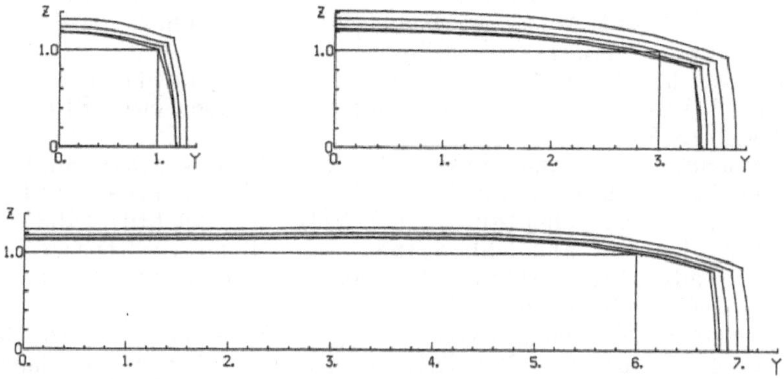

Figure 5. Shapes of the cross section for the rectangular die

5. CONCLUSIONS

By decoupling the momentum equation from the constitutive equation, a compact three dimensional finite element method with the penalty function formulation was able to be applied for analyzing the extrudate swells of the upper convected Maxwell fluid. Convergence for a higher elastic response problem was attained by applying the standard Galerkin's method for solving both the velocity and the tangential extra-stresses, and the least square finite element method for the normal extra stresses.

ACKNOWLEDGMENT

The authors wish to thank Professor H. Daiguji, Tohoku University, for his helpful discussions.

REFERENCES

1. Han, C. D,'Rheology in Polymer Processing', Academic Press(1976).
2. Kawahara, M. and Takeuchi, N.,'Mixed Finite Element Method for Analysis of Viscoelastic Fluid Flow', Computers and Fluids, 5(1977), 33- 45.
3. Crochet, M. J. and Bezy, M., 'Numerical Solution for the Flow of Viscoelastic Fluids', J. Non-Newt. Fluid Mech., 5(1979), 201-218.
4. Crochet, M. J. and Keunings, R., 'Finite Element Analysis of Die Swell of a Highly Elastic Fluid', J. Non-Newt. Fluid Mech., 10(1982),339-356.
5. Bush, M. B., Milthorpe, J. F., and Tanner, R. I., 'Finite Element and Boundary Element Methods for Extrusion Computations',

J. Non-Newt. Fluid Mech., 16(1984), 37-51.
6. Shimazaki, Y. and Thompson, E. G., 'Elasto Visco-plastic Flow with Special Attention to Boundary Conditions', Int. J. Num. Meth. Engng., 17(1981), 97-112.
7. Josse, S. L. and Finlayson, B. A., 'Reflection on the Numerical Viscoelastic Flow Problem', J. Non-Newt. Fluid Mech., 16(1984), 13-36.
8. Thompson, E. G. and Berman, H. M., 'Steady-state Analysis of Elasto-viscoplastic Flow during Rolling', Numerical Analysis of Forming Processes, Chapter 9, John Wiley & Sons Ltd.,(1984).
9. Shimazaki, Y. and Shiojima, T., 'Elastic Strain Effects during Steady-state Rolling of Visco-elastic Slabs', To appear in Int. J. Num. Meth. Engng.(1987).
10. Shimazaki, Y. and Shiojima, T., 'A Penalty Function Method for Maxwell Fluids', 2nd International Conference on Numerical Methods in Industrial Forming Processes, Sweden,(1986).
11. Shiojima, T. and Shimazaki, Y., 'A Pressure Smoothing Scheme for Incompressible Flow Problems using Finite Element Method', 2nd International Conference on Numerical Methods in Industrial Forming Processes, Sweden,(1986).
12. Hughes, T. J. R., Liu, W. K. and Brooks, A., ' Finite Element Analysis of Incompressible Viscous Flows by the Penalty Function Formulation', J. Comp. Physics, 30(1979),1-60.
13. Tuna, N. Y. and Finlayson, B. A., 'Exit Pressure Calculations from Numerical Extrudate Swell Results', J. Rheology, 28(1984), 79-93.
14. Coleman, C. J.,'A Finite Element Routine for Analysing Non-Newtonian Flows', J. Non-Newt. Fluid Mech., 8(1981), 261-270.

OPTIMIZATION OF CONTINUOUS PRESTRESSED BEAM FOR DIFFERENT CONSTRUCTION STAGES

Dragan Radić, Josip Dvornik, Vinko Čandrlić, Đuro Dekanović, Joško Ožbolt
GRAĐEVINSKI INSTITUT
Janka Rakuše 1, 41000 Zagreb, Yugoslavia

SUMMARY

This report presents a method of linearized problem optimization for full, limited and partial prestressing. A structure of known dimensions, material properties and construction stages has been analyzed. The structural model is approximated by finite beam elements. The minimization of the prestressing force, reduced for losses, has been achieved iteratively. In each iteration step, one of the possible solutions is determined by linear programming, wich provides for procedure stability The method for the determination of influence coefficients tendon axis and friction losses. Optimization yields the minimum presstresing force and optimal cable curve for each construction stage, prestressing stage and/or final exploitation stage.

1.0 INTRODUCTION

The idea of prestressed girder optimization is as old as prestressing technique themselves [1]. Significant advances in this field are a result of developments in the area of structural analysis, computers and optimization methods.

The trend in the development of optimization algorithms is toward the solution of specific sub-problems, which are the necessary elements of the design and construction of prestressed concrete structures.

Optimization of continuous girders in cases of full or limited prestressing is treated in [4,5 and 13]. In [6,7,15 and 17] statically determined girders are treated by considering the effect of an arbitrary degree of prestressing during various stages of prestressing, construction and failure. A relation between cross-section dimensions and the general girder behaviour is established.

The aim of the work presented here is to improve the girder optimization process during all construction stages as well as in the final exploitation stage. Particular attention is payed to the determination of the degree of prestressing and of the cable curve.

2.0 BASIC CONCEPT OF THE OPTIMIZATION PROCEDURE

Optimization procedure is used for the determination of the necessary degree of prestressing and of the cable curve for a girder with known cross-sections and loading. The object function is the optimal prestressing state in all cross-sections and during various construction stages. The optimal prestressing state is expressed numerically using the value of the prestressing force at the location of anchorage. Structural constraints are the normal stresses during prestressing stages and during exploitation. Constructive constraints are the cable position and the radius of cable curvature. Other constraints such as deflections, transversal forces and failure moments ..., can be included in the analysis as shown in [3,4,5,6,7,15,16,17 and 18]. Prestressing acts on the concrete girder at anchorage locations and at cable deviation locations. In the general case, it is necessary to consider the deviation from the principal direction in two perpendicular planes. The effect of prestressing can be accounted for by employing the equivalent load method. Since in the majority of cases the angle between the cable and the girder axis is very small, it is assumed that the direction of equivalent forces is perpendicular to the girder axis.

It is necessary to express the bending moments due to prestressing as a function of the prestressing force at the anchorage location and of the sum of linear functions and cable ordinates [4,13,14]. Expression (1) shows the moment due to prestressing, containing the effect of the cable shape (force losess due to friction), of the degree of static indeterminacy and of force losses determined using an iterative procedure [13].

$$M_i = N_0 \sum_j^n \bar{\beta}_{ij} Y_j \tag{1}$$

By substituting the expression (1) into stress inequalities for the top and bottom girder edge and by substituting expressions from [4,13], $N = X_0$ and $Y_i X_0 = X_i$, expressions (2), to be utilized for each cross-section for which the equivalent forces have been calculated, are derived.

$$\left(\frac{-1}{A_i} + \frac{\alpha_{iE}\, y_E - \alpha_{iS}\, y_S}{W_{it}}\right) X_0 + \frac{1}{W_{it}} \left(\bar{\beta}_{i1} X_1 + \bar{\beta}_{i2} X_2 + \ldots\ldots + \bar{\beta}_{in} X_n\right) - \frac{M_{ig1}}{W_{it}} \leq \left|\sigma_t^P\right|$$

$$\left(\frac{1}{A_i} - \frac{\alpha_{iE}\, y_E - \alpha_{iS}\, y_S}{W_{it}}\right) X_0 + \frac{1}{W_{it}} \left(\bar{\beta}_{i1} X_1 + \bar{\beta}_{i2} X_2 + \ldots\ldots + \bar{\beta}_{in} X_n\right) + \frac{M_{ig1}}{W_{it}} + \frac{\Sigma M_{ig2}}{W_{it}} + \frac{\Sigma M_{ip}}{W_{it}} \leq \left|\sigma_c^E\right|$$

$$\left(\frac{1}{A_i} + \frac{\alpha_{iS}\, y_S - \alpha_{iE}\, y_E}{W_{ib}}\right) X_0 + \frac{1}{W_{ib}} \left(\bar{\beta}_{i1} X_1 + \bar{\beta}_{i2} X_2 + \ldots\ldots + \bar{\beta}_{in} X_n\right) - \frac{M_{ig1}}{W_{ib}} \leq \left|\sigma_t^P\right|$$

$$\left(\frac{-1}{A_i} - \frac{\alpha_{iS}\, y_S + \alpha_{iE}\, y_E}{W_{ib}}\right) X_0 + \frac{1}{W_{ib}} \left(\bar{\beta}_{i1} X_1 + \bar{\beta}_{i2} X_2 + \ldots\ldots + \bar{\beta}_{in} X_n\right) + \frac{M_{ig1}}{W_{ib}} + \frac{\Sigma M_{ig2}}{W_{ib}} + \frac{\Sigma M_{ip}}{W_{ib}} \leq \left|\sigma_c^E\right|$$

To this, constructive constraints on the cable position and the cable curvature radius should be added.

$$Y_i^b X_0 \leq X_i \leq Y_i^t X_0 \tag{3}$$

$$R^{min} X_0 \leq X_{i-1} - 2X_{ii} + X_{i+1} \leq R^{max} X_0 \tag{4}$$

The prestressing force X_0 is determined by linear programming, in a single iteration step. The final, minimal force is determined by employing an iterative procedure. Iteration is necessary since time dependant deformations of concrete are directly related to the degree of prestressing, which is not known in advance.

Example of a girder from Fig. 1 will be used to show how various constraints define the space whose principal axes are the prestressing force and cable ordinates (Figs. 2a, b, c, d, e). Due to symmetry, the space considered is 3-dimensional with unknowns $X_0 = N$, $X_1 = Y_1X_0$ and $X_2 = Y_2X_0$.

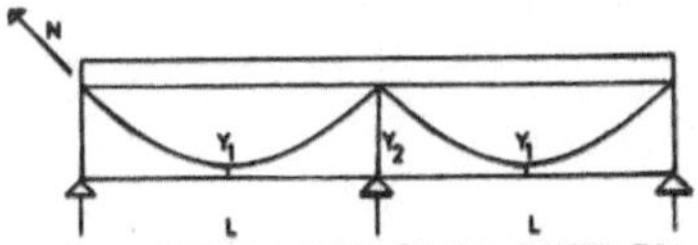

FIG.1. GENERAL VIEW OF 3-D PROBLEM

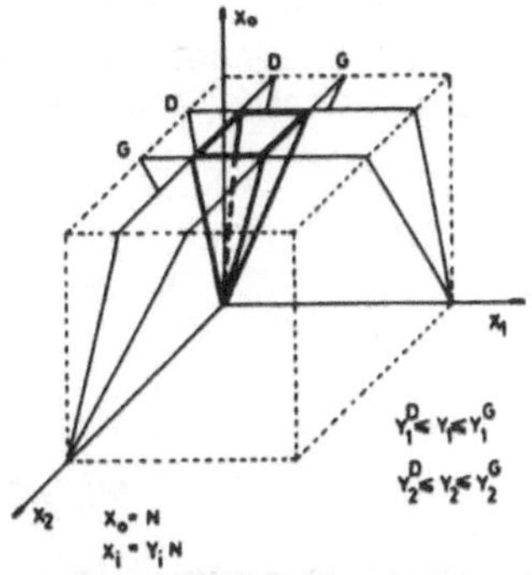

FIG. 2A. TENDON EXTREMAL POSITIONS CONSTRAINTS

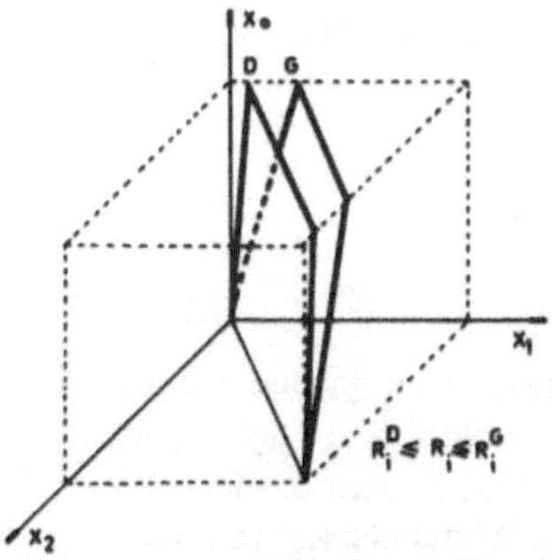

FIG. 2B. TENDON RADIUS OF CURVATURE CONSTRAINTS

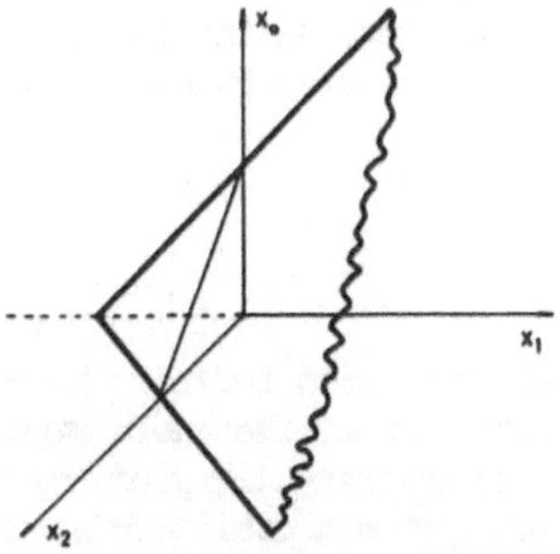

FIG. 2C. STRUCTURAL BEHAVIOUR CONSTRAINT

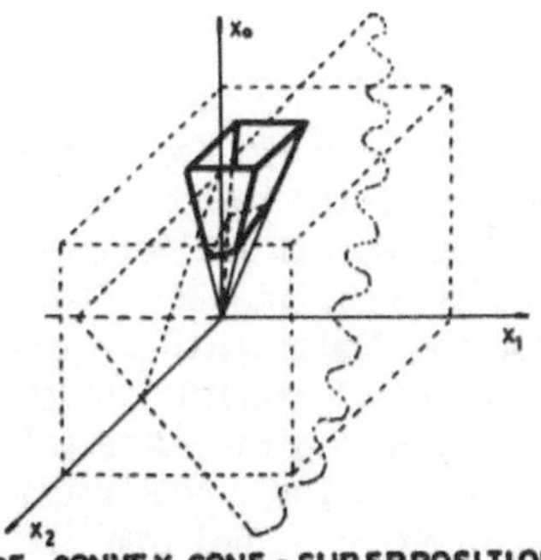

FIG. 2E. CONVEX CONE - SUPERPOSITION FROM FIG. 2A. AND FIG. 2C.

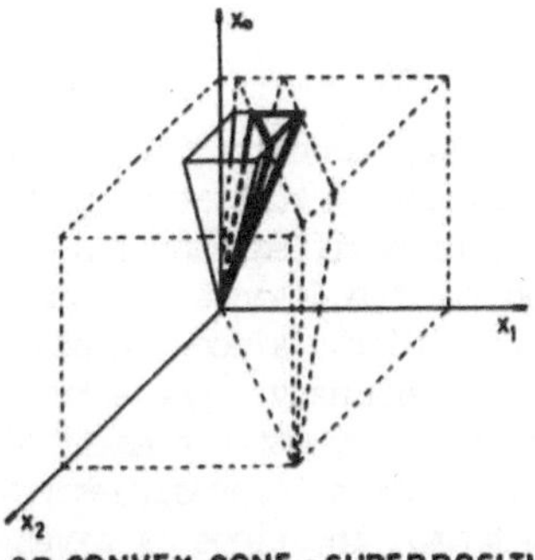

FIG. 2D. CONVEX CONE - SUPERPOSITION FROM FIG. 2A. AND FIG. 2B.

2.1 ALGORITHM FOR THE DETERMINATION OF INFLUENCE COEFFICIENTS

The moment M_{ij} at location *'i'* due to the force at location *'j'* can be expresse by using coefficients β_{ij}, expression (5), where K_{tj} is the friction loss coefficient. W_j is the equivalent force due to the equivalent load at location *'j'*, evalued by employing the expression (6), where D is the distance between cable ordinates. It is assumed that D = constant. Expression (8) is obtained by substituting (7) into (6) and (5).

$$M_{ij}^{w} = \beta_{ij}\, k_j^{T}\, W_j \tag{5}$$

$$W_j = \frac{N}{D}\left(Y_{ij-1} - 2Y_{ij} + Y_{ij+1}\right) \tag{6}$$

$$\beta_{ij}^{\infty} = \beta_{ij}\, k_j^{T} \tag{7}$$

$$M_{ij}^{w} = \beta_{ij}^{\infty}\frac{N}{D}\left(Y_{ij-1} - 2Y_{ij} + Y_{ij+1}\right) \tag{8}$$

The influence coefficient β_{ij}^{∞} is determined by using the expression (9), where M_{ij} *is a moment at 'i'* due to the equivalent force at *'j'*. The equivalent load usec in the evaluation of M_{ij} has to be reduced to account for friction losses of the cable curve for which the prestressing effects are being computed. It is important to note that the magnitude of the equivalent force W_j must not be reduced. By superimposing the bending moments it is easy to show that the influence coefficients β_{ij} can be determined by using expressions (10).

$$\beta_{ij}^{\infty} = \frac{M_{ij}^{w}}{W_j} \tag{9}$$

$$\bar{\beta}_{ij} = \beta_{ij-1}^{\infty} - 2\beta_{ij}^{\infty} + \beta_{ij+1}^{\infty} \qquad \bar{\beta}_{i1} = \beta_{i2}^{\infty} \qquad \bar{\beta}_{in} = \beta_{in-1}^{\infty} \tag{10}$$

The final expression for the moments in a continuous girder due to prestressing , taking into account the cable curve shape, force losses and the effect of the girdeı static indeterminacy, is given by the expression (11). Expression (12) is obtained after, for a chosen cable curve, the multiplication and summation for $j = 1, \ldots ,n$ is carried out. Here, e_r is the equivalent excentricity containing the cable curve shape with force losses and the effect of the continuous girder static indeterminacy

$$M_i = N_0 \sum_j^{n} \bar{\beta}_{ij}\, Y_j \tag{11}$$

$$M_i = N e_r \tag{12}$$

Equivalent excentricity determined in this way extends the application of Mangel's lines [1] to continuous girders. Advantages of the method for the derivation of modified Mangel's lines are explained in detail in [20]. Also explained are the disadvantages of the application of unmodified Mangel's lines for continuous girders, since the basic assumption for the application of Mangel's lines is static determinacy. Concentrated moments due to the excentricity of the cable anchorage are very often present at the ends of continuous girders. Final expressic for the moment due to prestressing effects and accounting for concentrated moments ε cable anchorage locations is given by (13). Coefficients α_{ios} and α_{ioE} are determine in the same way as the coefficients β_{ij}.

$$M_i = N_o\left(-\alpha_{i\overset{\circ}{S}}\, y_S + (\alpha_{i\overset{\circ}{S}} - \beta_{i1}^{\infty} + \beta_{i2}^{\infty})Y_1 + \sum_{j=2}^{n-1}\bar{\beta}_{ij}Y_j + (-\alpha_{i\overset{\circ}{E}} - \beta_{in}^{\infty} + \beta_{in-1}^{\infty})Y_n + \alpha_{i\overset{\circ}{E}}\, y_E\right) \tag{13}$$

2.2 PURPOSEFULNESS OF THE USAGE OF INFLUENCE COEFFICIENTS β_{ij}

Girders over two equal spans with two different cable curves are shown in Figs. 3 and 4. Cable curve in Fig. 3 consists of successive parabola segments with different lengths. Such a cable curve is very frequent in practical applications. Cable curve in Fig. 4 is totally useless for practical application, but it illustrates and explains the usage of influence coefficients β_{ij} in the optimization process. Loading schemes for which optimization, i.e. force determination without changing the cable curve shape will be performed, are shown in Fig. 5. Prestressing is carried out from both girder ends.

Comparative computations for cabel curves on fig.3 and 4. have been carried out:
A) No force losses due to friction on cable curves ($\mu = 0.0$ and $K = 0.0$ *rad/m*).
B) Force losses due to friction on cable curves ($\mu = 0.25$ and $K = 0.016$ *rad/m*).

Analysis results are shown in Table *I*. It can be seen that in case *A* forces are the same regardless of the cable curve shape, whereas in case *B* the forces are different. Since the purpose of the chosen example is only to illustrate theoretical considerations, relative force difference is of lesser importance than the fact that the difference exists. Hence, it can be concluded that the employment of the coefficients β_{ij} is justified. The procedure can be generalized in such a way that force losses due to time effects can also be taken into account with influence coefficients β_{ij}. Cable coordinates for both curves are shown in Table *II*.

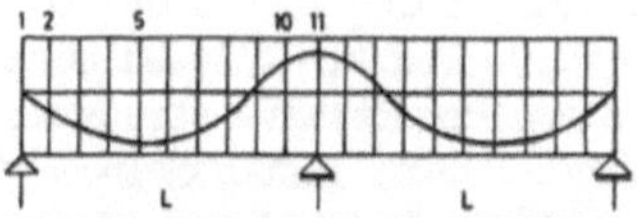

FIG. 3. EXAMPLE 1: PARABOLIC TENDON

TABLE I. PRESTRESSING FORCE - EXAMPLE 1

	A	B
TENDON POSITION FIG.3	1307	1397
TENDON POSITION FIG.4	1307	1824

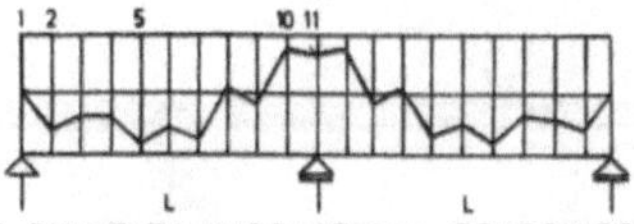

FIG. 4. EXAMPLE 1: IMPRACTICAL TENDON POSITION

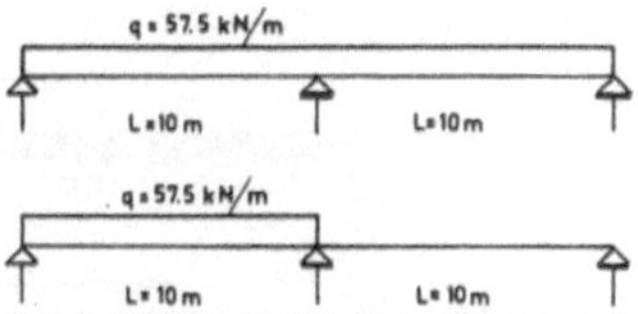

FIG. 5. EXAMPLE 1: LOADING CONDITIONS

3.0 APPLICATION OF THE OPTIMIZATION PROCESS FOR VARIOUS CONTINUOUS GIRDER CONSTRUCTION METHODS

The general problem of the optimization of prestressed girders and possible numerical difficulties in practical applications are explained in detail in [16]. Various formulation types of optimization are systematically presented. In [21], a detailed description of the optimization process for a case of the span-by-span construction method is given. Here, only basic principles will be presented while a numerical example will be used for the practical explanation of the procedure.

It is assumed that continuous girder construction method, cross-section shape during all construction stages, static systems and loadings during all construction stages, sequence of cable prestressing are known.

Degree of prestressing and cable position during various construction stages will be determined by optimization. The procedure itself is iterative. In each iteration, cable position and the force, including losses, are determined by linear programming. The final, minimal prestressing force is determined using an iterative procedure which is necessary because of the time effects.

Various construction methods, such as cast-in-place on the falsework method, span-by-span method, incremental launching method and the cantilever method have bee used in the civil engineering practice. Common characteristics of these methods are

- The existence of the basic system during some of the construction stages
- The effect of the final, exploitation system on the basic system
- The effect of the subsequent construction stage system on the basic system
- Choice of the construction method related to the ratio of the neighbouring spans
- Choice of the degree of prestressing that is desired or that can be achieved durin particular construction stages and in the final, exploitation stage. Here, considerations can be extended to the number of segments participating in the load transfer during the prestressing stage and in the current construction stage.

3.1 DETERMINATION OF THE NECESSARY DEGREE OF PRESTRESSING FOR VARIOUS CONSTRUCTION S

Optimization procedure will be explained using the example from Fig. 6. It is assumed that the girder with a smaller span of 30.0 m is constructed first, on falsework, and than the construction of the second girder with a larger span of 45.(will start. Since the spans are large, it is justified to construct a continuous girder over two spans. For the sake of briefness, the change of the prestressing force due to time effects will not be considered. Only force losses due to frictior will be taken into account, with $\mu = 0.16$ and $K = 0.016$ *rad/m*. Allowed normal stresses are: allowed prestressing tensile stress and compressive stress and workir tensile stress and compressive stress are *3000, 19500, 1800, 16000 kN/m²* respectively. Prestressing sequence during construction stages *I* and *II* is shown ir Fig. 7. Moment above the middle support will be carried by overlapping cables fron both stages.

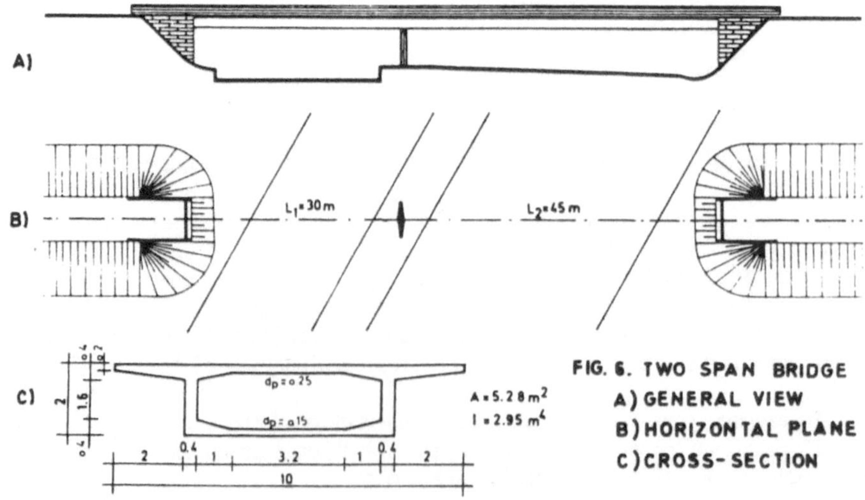

FIG. 6. TWO SPAN BRIDGE
A) GENERAL VIEW
B) HORIZONTAL PLANE
C) CROSS-SECTION

Basic system for prestressing during the first construction stage is shown in Fig. 8a. Static system and loading activated at the moment of prestressing are shown in Fig. 8b. The effect of the final exploitation system on the basic system is shown in Fig. 8c. In order to carry out the optimization procedure for the first construction stage properly, it is necessary to add the system shown in Fig. 8d. It is the same system as in Fig. 8b, but in the optimization process it is associated with groups of stresses allowable in the exploitation stage. It is also possible to add systems which would make it possible to make a detailed analysis of time effects. Loading and static systems comprising final exploitation stage effects and accounting for some intermediate construction stages , are shown on Fig. 8e. Effect of the *II* construction stage at the moment of prestressing on the basic system is shown in Fig. 8f. In the optimization process for the first construction stage, the relevant system was the one shown in Fig. 8e. Prestressing force for the first construction stage is $N_1=9281$ *kN* at anchorage. With the prestressing force determined in this way, the basic system will be fully prestressed during the first exploitation stage, while in the final exploitation stage, prestressing will be limited. Since behaviour constraints have been given by normal stresses on the top and bottom edge, degree of prestressing could only vary between full and limited. By applying additional bahaviour constraints, it is also possible to achieve the degree of partial prestressing. Basic system for the optimization of the second construction stage and the loading activated at that moment are shown in Figs. 9a and 9b. The effect of the final exploitation system on the basic system is shown in Figs. 9c and 9d. Here, the effect of the previous on the subsequent construction stage has to be added, Fig. 9e. In this case, since there are only two stages, the effect of the first on the final stage has to be added. The effect of the final exploitation stage, taking into account the intermediate construction stage is shown in Fig. 9f. Prestressing force in the second construction stage, at the anchorage, is $N_2=30400kN$. With this force, limted prestressing will be achieved in the larger span. Final optimization results are shown in Fig. 10.

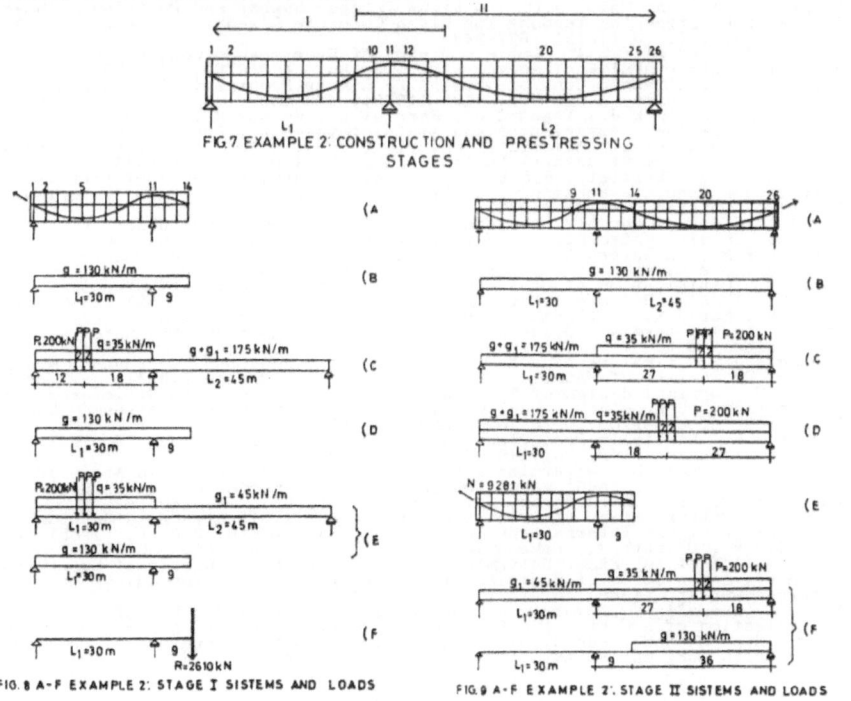

FIG.7 EXAMPLE 2: CONSTRUCTION AND PRESTRESSING STAGES

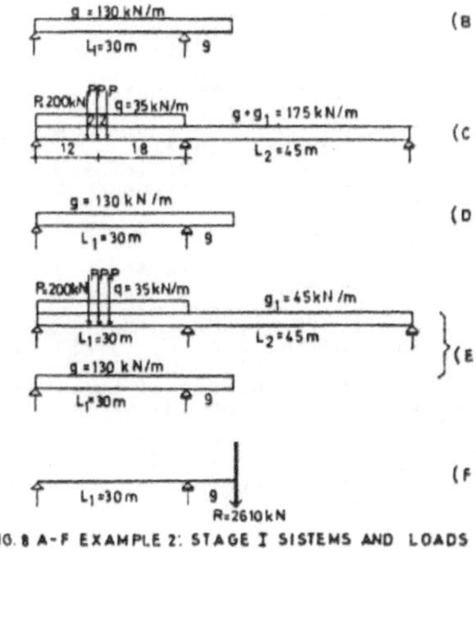

FIG. 8 A-F EXAMPLE 2: STAGE I SISTEMS AND LOADS

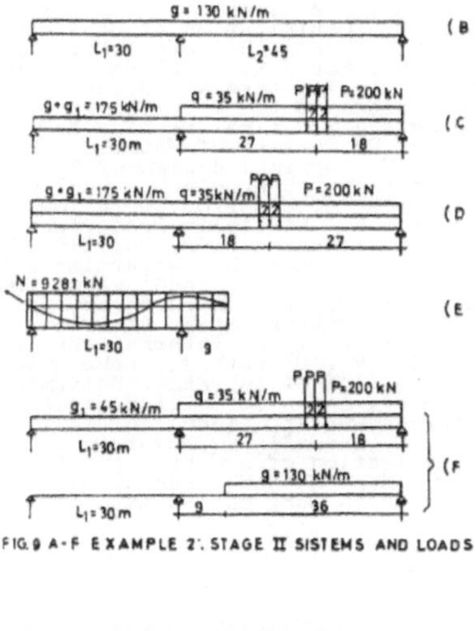

FIG.9 A-F EXAMPLE 2: STAGE II SISTEMS AND LOADS

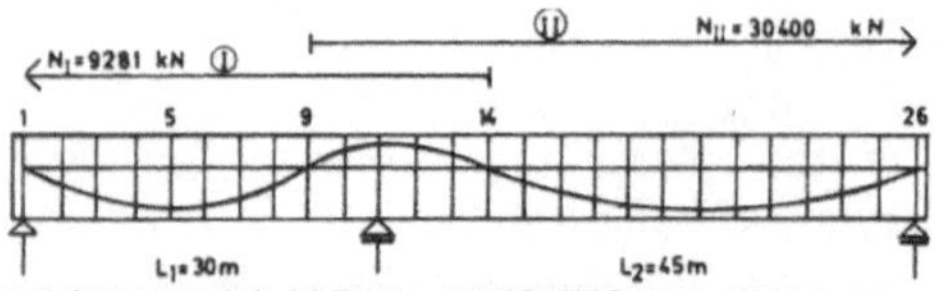

FIG.10. EXAMPLE 2: OPTIMAL PRESTRESSING FORCE FOR STAGES I AND II

TABLE II. TENDON POSITION

SECTION	FIG. 3	FIG.4
1	0.4333	0.4333
2	0.2875	0.1875
3	0.1833	0.2833
4	0.1208	0.3208
5	0.1000	0.0500
6	0.1278	0.2278
7	0.2111	0.1111
8	0.3500	0.4500
9	0.5444	0.3444
10	0.7111	0.8111
11	0.7666	0.7666

4.0 CONCLUSION

Optimization procedure for the determination of the necessary degree of prestressing and of the optimal prestressing state in all girder cross-sections and during all construction stages is applicable to very complex continuous girder construction methods. It is essential to note that the girder is considered as a whole and not at the most-stressed cross-section level. The understanding of this will be helpful in further research projects where safety and load bearing capacity of prestressed continuous girders will be considered. Advantages of the employment of influence coefficients β_{ij} for the correct accounting of prestressing effects during all construction stages is emphasized. Also, Magnel's lines have to be modified in order to make them applicable in the analysis of continuous girders. It is possible to extend the procedure to the determination of necessary cross-section dimensions and of the necessary degree of prestressing of continuous girders.

REFERENCES

1. Magnel G.,Prestressed Concrete, McGraw Hill Book Co., (1954)
2. Dvornik J.,'Optimal Dimensioning of Reinforced Cocncrete Cross-Sections' Ph.D. Thesis, University of Zagreb, (1972) (in Croatian)
3. Kirsch U.,'Optimum Design of Prestressed Beams', Computers and Structures, Vol. 2 (1972), 573-583.
4. Kirsch U.,'Optimized Prestressing by Linear Programming', International Journal for Numerical Methods in Engineering, Vol. 7, No. 2 (1973), 125-136
5. Bengtsson, A. and Wolf, J.P., 'Optimum Integer Number and Position of Several Groups of Prestressing Tendons for Given Concrete Dimensions', Computers & Structures, Vol. 3 (1973), 827-848
6. Desayi, P. and Ali, S.A, 'Optimum Design of Prestressed Concrete Girders', Journal of Structural Engineering, Vol. 3, No. 4 (1976), 192-200
7. Morris D.,'Prestressed Concrete Design by Linear Programming', Journal of the Structural Division, ASCE, Vol. 104 (1978), 439-452
8. Bičanić N., Dvornik J., Ivančić N., Nardini D., Werner H., 'Automatic Structural Design - experiences and state-of-the-art', Proceedings of the Simposium Computer at University, Cavtat, (1980), 3.13 - 3.21 (in Croatian)
9. Kirsch U., Optimum Structural Design, McGraw Hill Book Co. (1981)
10. Cohn, M.Z. and Bartlett, F.M.P.,'Nonlinear Flexural Response of Partially Prestressed Concrete Sections', SM Study No. 168, University of Waterloo, Waterloo, Ontario, Canada (1981)
11. Bartlett F.M.P.,'Computer Analysis of Partially Prestressed Concrete', M.A.Sc. Thesis, University of Waterloo, Waterloo, Ontario, Canada (1982)
12. Johnson F.R.,'An Interactive Design Algorithm for Prestressed Concrete Girders', Computers & Structures, Vol. 2 (1982), 1075-1088
13. Radić D.,'Optimization of Prestressed Beam System in Concrete Structures', M.A.Sc. Thesis, University of Zagreb, Zagreb, Yugoslavia (1982) (in Croatian)
14. Radić D., Marić Z.,'Influence of Prestressing (With Comprising of Losses) in Statically Indeterminte Griders', Association of Yugoslav Structural Engineers, Proceedings of the 7.th Congress, Cavtat, (1983), 325-332 (in Croatian)
15. MacRae A.J.,'Optimal Design of Partially Prestressed Concrete Beams', M.A.Sc. Thesis, University of Waterloo, Waterloo, Ontario, Canada (1983)
16. Kirsch U.,'Optimal design of Continuous Prestressed Concrete Srtuctures', International Symposium on Nonlinearity and Continuity in Prestressed Concrete, University of Waterloo, Ontario, Canada, (1983) 175-204
17. Saouma V.E.,'Partially Prestressed Concrete Beam Optimization', Journal of Structural Engineering, Vol.110, 3 (1984), 589-604
18. Dvornik J., Radic D.,'Determination of Optimal Reinforcement in Arbitrary Cross-Sections', Proceedings of the International Conference on Computer Aided Design of Concrete Structures, Split, part II (1984), 747-760
19. Radić D.,'Determination of Prestressed Force for Arbitrary Cross-Sections', Proceedings of the International Conference for CAD/CAM, Zagreb, (1985), 231-236
20. Čandrlić V., Dvornik J., Dekanović Đ., Ožbolt J., Radić D., 'Graphic Presentation of the Exact Determination of the Force in Cabels of Prestressed Girders', Association of Croatian Structural Engineers, Proceedings of the 2.nd Congress, Split, (1986), 61-73 (in Croatian)
21. Radić D.,Čandrlić V., Dvornik J., Dekanović Đ., Ožbolt J.,'Cavtat 'Optimisation of a Continuous Prestressed Concrete Grider Constructed by the Spa - by - Span Method', Association of Yugoslav Structural Engineers, Proceedings of the 8.th Congres, Cavtat, (1987), 73-79 (in Croatian)

SLOPE STABILITY COMPUTATIONS WITH NONLINEAR FAILURE ENVELOPE USING GENERALIZED PROCEDURE OF SLICES AND OPTIMIZATION TECHNIQUES

by

Yudhbir, P.K. Basudhar S.K.Bhowmik

Civil Engineering Department
Indian Institute of Technology
Kanpur-208016, India

SUMMARY

The study pertains to the development of an autosearch technique for locating the critical shear surface and the corresponding minimum factor of safety without any prior assumption regarding its shape, for slopes in soils having nonlinear strength envelopes. Sequential unconstrained optimization in conjunction with Janbu´s generalized procedure of slices has been used in the analysis. Results have been obtained for both linear and nonlinear failure envelopes and compared. It has been observed that in case of nonlinear strength idealisation, the critical slip surface is sensitive to the magnitude of pore water pressure parameter (r_u) and the factor of safety versus r_u relationship is nonlinear. The developed method has been found to be effective for the problems studied.

1 INTRODUCTION

Slope stability is essentially a problem of optimization namely the determination of the slip surface that yields the minimum factor of safety. In most of the analyses the slip surface has been assumed to be of particular geometry. The assumptions regarding the shape of the slip surface greatly simplify the computations involved in the analysis but the restriction imposed on the slip surface geometry may lead to the bypassing of the actual critical surface. Most of the slope stability software using the limit equilibrium analysis, described in the literature, provided an automated version of the existing methods of slope stability analysis. The need for autosearch led to the use of sophisticated optimization algorithms. However most of these programs uses the linear Mohr-Coulomb relationship in the analysis. There is considerable experimental evidence to show that the Mohr-Coulomb failure envelope exhibits significant curvature for many different types of soils e.g. stiff clays, dense sands and compacted rock fill.

Maksimovic [1979], Charlse and Soares [1984], have presented useful contribution in the field of stability analysis of slopes in soils with nonlinear failure envelope. However, these analyses used the circular slip surfaces only.

2 ANALYSIS

The generalized procedure of slices developed by Janbu [1973] for homogenous soils with linear Mohr-coulomb relationship has been adopted in analysing the slope with the proper choice of the shear strength parameters corresponding to the normal effective stress at the base of each slice depending on its location.

The following relationship suggested by De Mello [1977] has been introduced to represent the curved failure envelope typically exhibited by many soils over a wide range of normal effective stress:

$$\tau_f = \alpha(\sigma')^{\beta} \tag{1}$$

τ_f is the shear strength and the dimensionless parameter β defines the degree of curveture of the envelope. The parameter α has dimension $(\sigma')^{1-\beta}$ and for any value of σ' the magnitude of τ_f is proportional to this parameter.

The equilibrim shear stress along the shear surface is given by the equation.

$$\tau = \frac{\tau_f}{F} = \frac{\alpha}{F}(\sigma')^{\beta} \tag{2}$$

F is the factor of safety.

Fig 1 shows the geometry of the slope with a general potential slip surface and with the potential mass divided into N number of slices.

For the given geometry of the dam section and soil properties, the factor of safety is a function of the shape and location of the potential slip surface.

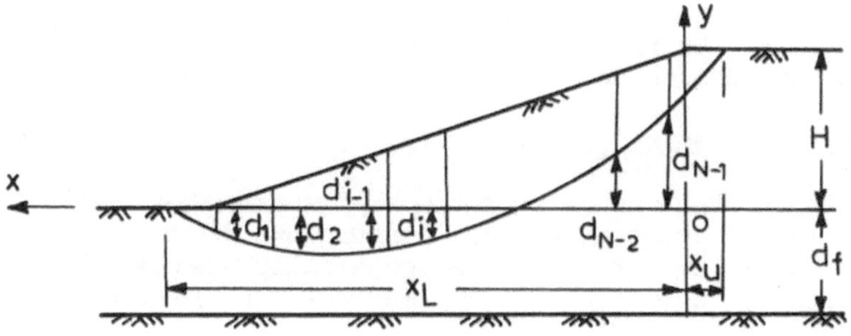

Fig.1 Sketch defining coordinate axes & decision variables.

To evaluate the factor of safety for a nonlinear failure envelope (Eq.1) using the expression developed by Janbu [1973] for linear Mohr-Coulomb strength criterion, the point to point variation in the effective strength parameters C_i' and ϕ_i' should be used. The effective strength parameters c_i' and ϕ_i' corresponding to any effective normal stress σ_i' are as follows:

$$c_i' = (1-\beta)\ \alpha\ \sigma_i'^{\beta} \tag{3a}$$

$$\phi_i' = \tan^{-1}[\alpha\beta(\sigma_i')^{1-\beta}] \tag{3b}$$

c_i' represent the intercept of the tangent to the curved failure envelope at the effective normal stress, σ_i' and ϕ_i' is the inclination of the aforementioned tangent with the horizontal.

The factor of safety is a function of slip surfce coordinates (Fig.1) given the geometry of the slope, soil properties, field conditions and the loads. Hence the function is minimized with respect to the coordinates and the design vector is as follows:

$$\bar{D}^T = (x_u, x_L, d_1, d_2, d_3, \ldots..) \tag{4}$$

Writing $d_N = x_u$ and $d_{N+1} = x_L$ one obtains

$$\bar{D}^T = (d_1, d_2, d_3, \ldots\ldots d_N, d_{N+1}) \tag{5}$$

So the problem involves (N+1) design variables where N is the number of slices ito which the sliding mass is divided.

Since the objective is to minimize the factor of safety with respect to the design vector D , the objective function, is the factor of safety and can be expressed in terms of the design vector D as

$$F = f(\bar{D}) = f(d_1, d_2, d_3, \ldots\ldots, d_{N+1}) \qquad (6)$$

In order to ensure that the slip surface is physically reasonable and acceptable some constraints are to be placed on the decision variables. A physically reasonable and geometrically compatible slip surface should stisfy the following constraints:

a) Unless there is a discontinuity in the properties of the soil mass comprising the slope, the slip surface should not have downward concavity at any point. This requires

$$-(d_{i-1} - 2d_i + d_{i+1}) \leqslant 0 \ldots. \qquad (7)$$

where i varies from 1 to N-1

b) There is no existance of slip surface above the crest line and the slip surfce would not cut across the intact hard rock bed. This requires

$$d_i - H \leqslant 0 \; ; \text{ for } d_i > 0 \ldots \qquad (8)$$

$$d_f - d_i \leqslant 0 \; ; \text{ for } d_i < 0 \ldots \qquad (9)$$

where H is the height of the slope and d is the depth to hard stratum and i varies from 1 to N-1.

c) The slope of slip surface at its intersection with the top surface of the slope should satisfy active earth pressure condition, which requires

$$\alpha = 45^\circ + \frac{\phi'_m}{2} \quad \ldots \qquad (10a)$$

If the shear surface has negative slope at the intersection with the base, then it should satisfy passive earth pressure condition, which requires

$$|\alpha_1| = 45^\circ - \frac{\phi'_m}{2} ; \quad \text{for} \quad < 0 \ldots \qquad (10b)$$

The constraints denoted by equation (10a) and (10b) imply that at the first and the last points of the slip surface the vertical and horizontal directions are the principal stress directions respectively. Equation (10b) is not valid if the exit point of the slip surface is at the toe or above the toe of the slope.

d) The factor of safety is a positive number. This requires

$$- F < 0 \quad \ldots \qquad (11)$$

To locate the critical slip surface, the factor of safety is to be minimized. The problem is stated as an optimization problem as follows:

Find the decision vector $\bar{D}_m$ such that,

$F = f(\bar{D}_m)$ is the minimum of $f(\bar{D})$, subject to
$g_j(\bar{D}_m) \leqslant 0$; $j = 1, 2, \ldots\ldots M$

where M is the total number of inequality constraints.

There are number of approches for solving costrained optimization problem [Fox, 1971] . The extended penalty function method enunciated by Kavlie [1971] has been used in the present study. In this method the constrained problem is blended into a unconstrained minimization by blending the contraints into a composite function $\phi(\bar{D}, r_k)$ and making it possible to ignore them at the minimization stage. The problem is stated as follows:

$$\min_{\bar{D}} \ \phi(D, r_k) = f(\bar{D}) - r_k \sum_{j=1}^{M} G\left[g_j(\bar{D})\right] \qquad (12)$$

The function $G[g_j(\bar{D})]$ is chosen as follows:

$$G\left[g_j(\bar{D})\right] = 1/g_j(\bar{D}) \quad ; \quad g_j(\bar{D}) \leqslant \epsilon$$
$$= \left[2\epsilon - g_j(D)\right]/ \ ; \ g_j(\bar{D}) > \epsilon$$

where $\epsilon = \ - r_k/\delta_t$; r_k is a problem dependent parameter, called penalty parameter, whose value is made successively smaller in order to obtain the constrained minimum of $f(\bar{D})$; δ_t is a constant that define the transition between the two types of penalty terms.

In this study the sequential unconstrained minimization has been carried out using the Extended Penalty Function Method in combination with Powell´s method for multidimensional search and quadratic fit [Fox, 1971] for unidimensional search.

3 RESULTS AND DISCUSSION

In order to investigate the effect of non-linear failure envelope on the stability of a slope, strength parameters were chosen for the highly fissured London Clay from Wraysbury[Marsland, 1971] . A 4:1 slope in London clay as shown in Fig 2 was analysed. This problem has been studied earlier by Charles and Soares[1984].

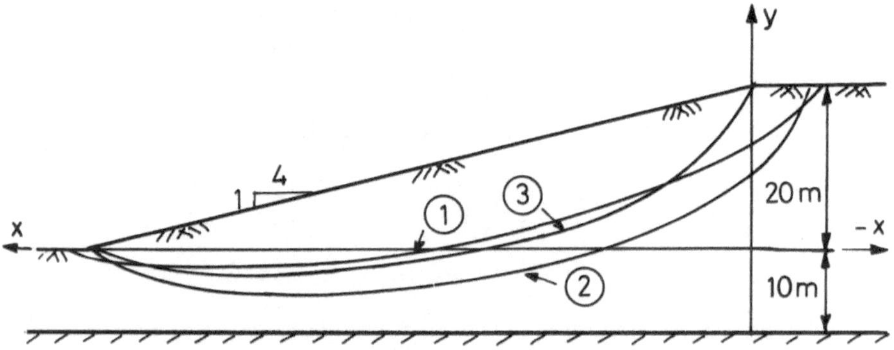

Fig.2 Comparison of the results obtained with linear and nonlinear strength envelope.

The α and β parameters in the expression $\tau_f = \alpha\,(\sigma')^{\beta}$ were evaluated as 1.4 & 0.8 respectively. This expression fits the experimental data reported by Marsland.

Fig. 2 shows the results of the analysis for both the cases. An initial general slip surface indicated as (1) was chosen and the critical slip surface was searched for the minimum value of factor of safety for both linear and non-linear failure envelopes. It is important to note that a deeper critical slip surfce (2) is obtained in case of linear failure envelope whereas the critical slip surface (3) corresponding to the nonlinear falure envelope is relatively a shallower one. The values of minimum factor of safety are 1.96 and 1.70 respectively for linear and nonlinear idealisations of failure envelope. In these computations r_u = 0.3 was used. Charles and Soares reported a factor of safety of 1.66 for a circular slip surface; the critical circle obtained is shallower than the general slip surface reported in this study.

In order to evaluate the effect of pore water pressures, computations were made for both linear and nonlinear envelopes. Fig.3 depicts the influence of r_u on the location of the critical slip surfce and the factor of safety. For nonlinear shear strength envelope (Fig.3) it is seen that, as expected, slip surfce becomes shallow as r_u increases from 0 to 0.5 and the factor of safety vs r_u

relationship is non-linear. However in case of linear failure envelope, the critical slip surface did not show any significant shift in position for values of r_u varying from 0 to 0.5. For toe circles using Lowe and Karafiath's procedure, Wright [1969] has, however, shown the dependence of the centres of critical circles on the value of r_u.

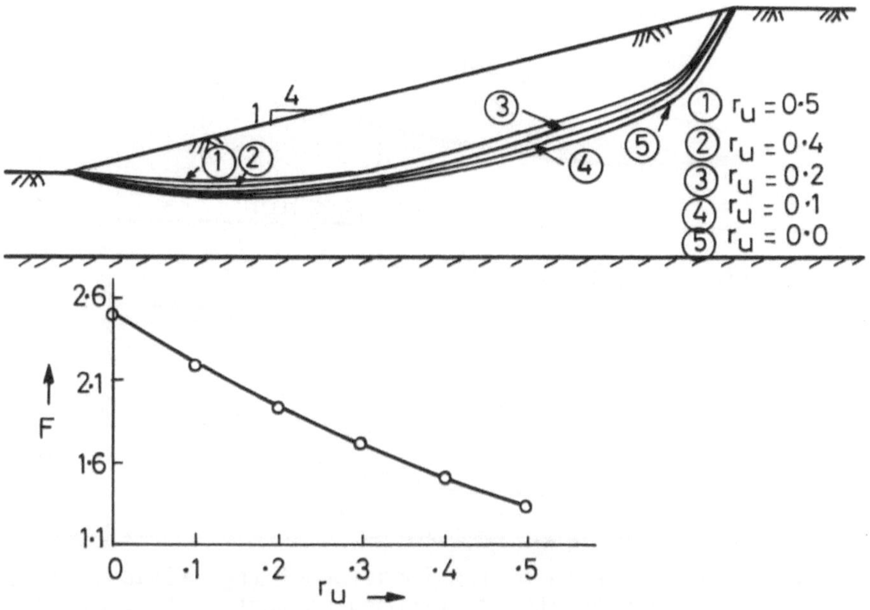

Fig.3 Influence of ru on the location of the critical slip surface and the corresponding factor of safety (Non linear strength envelope)

Typical results for $\phi' = 19°$ and $C' = 4.125$ & 33 KN/m^2 are shown in Fig 4(a) and (b) respectively.

It would be instructive to contrast these results with the coventional use of linear failure envelope. Fig.5 depicts the location of critical slip surface for a wide range of C' values.

It would be useful to note that for every C' there exists a unique slip surface for a given slope. For higher values of C' the critical slip surface is deeper and with the reduction in C' values the critical slip surface rises towards the slope. Influence of ϕ', for a fixed C', on the location of critical slip surface was also investigated. It was observed that for ϕ' varying from 10° to 19°, the location of the critical slip surface remained essentially unaltered though the factor of safety increased linearty with

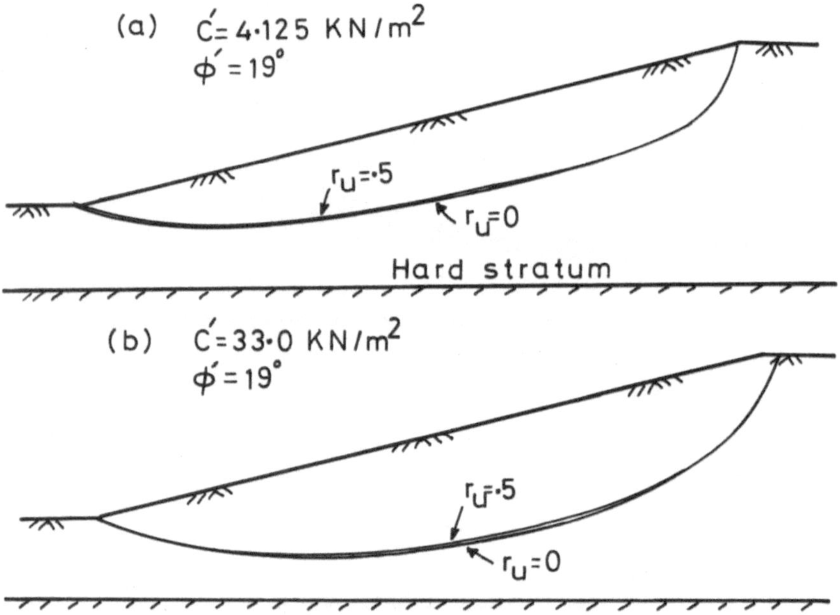

Fig.4 Influence of r_u on the location of the critical slip surface and the corresponding factor of safety (Linear strength envelope).

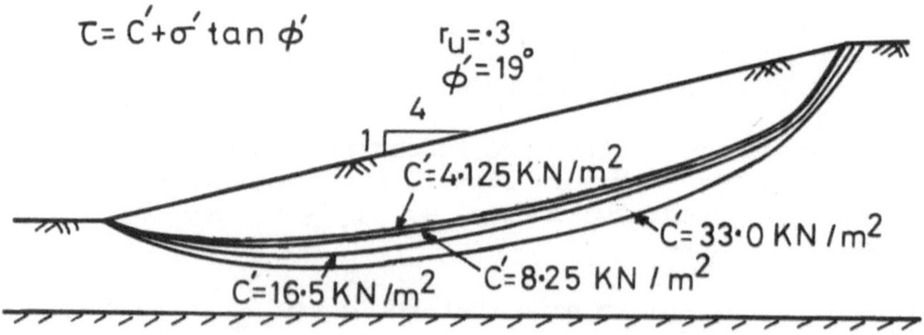

Fig.5 Influence of C′ on the location of the critical slip surface and the corresponding factor of safety.

4 CONCLUSIONS

The method of analysis discussed here allows one to incorporate the non-linear strength envelope of soils in the standard slope stability computational methods. In this procedure no prior assumption is needed as to the shape of the slip surface. This is an autosearch method and for the cases investigated the critical slip surfaces are non-circular. It has been shown quantitatively (Fig.2) that non-linear failure envelope gives shallower slip surface with a lower factor of safety as compared to the conventional approach using constant values of C' & ϕ'.

It is also shown that in case of non-linear strength idealisation, the critical slip surface is sensitive to the magnitude of pore water pressure. This is in contrast with the results for constant C', where the general critical slip surface essentially remains unaltered for r_u values ranging from 0 to 0.5. Variations in ϕ' (for a given C') also do not significantly alter the location of the critical slip surface.

It has been demonstrated that the position of the critical slip surface is very sensitive to the magnitude of C' adopted in calculations. Considering the uncertainities in evaluating C' and also the marked influence it has on the position of the critical slip surface, the procedure using non-linear strength envelope in stability computations appears a much better choice.

REFERENCES

1 Charles, J.A. and Soares, M.M., "The Stability of slopes in soils with non-linear failure envelopes", Can. Geotech. J. Vol.21, (1984), pp. 397-406.

2 De Mello, V.F.B., "Reflections on design decisions of practical significance to embankment dams", 17th Rankine lecture, Geotechnique 27 (1977), No.3, pp 281-354.

3 Fox, R.L., "Optimization methods for engineering design", Addison-Wesley, Reading, Mass, 1971.

4 Janbu, N. "Slope stability computations", Embankment Dam Engineering, Casgrande Vol. , Ed. Hirschfeld and S.J. Paulos, John Wiley & Sons, NY, pp 47-86, 1973.

5 Kavlie, D., "Optimum design of statically indeterminate structures", PhD thesis, Univ. of California, Berkeley, 1971.

6 Maksimovik, M., "Limit Equilibrium for nonlinear failure envelope and arbitrary slip surface", Proc. 3rd Int. Conf. Numer. Meth. Geomech., Aachen 2, pp 769-777, 1979.

7 Marsland, A., "The shear strength of stiff fissured clays", Building Research Station Current Paper, 21/71, 1971.

8 Wright, S.G., "A Study of stability and the undrained shear strength of clay shales, PhD Thesis, Univ of California, Berkeley, 1969.

A COMPUTER CODE FOR THE STRESS ANALYSIS OF CYCLICALLY SYMMETRIC COMPONENTS

Y.V.L.N.MURTHY C.P.AGRAWAL

BHARAT HEAVY ELECTRICALS LIMITED, CORPORATE R&D
VIKASNAGAR, HYDERABAD -500 593, INDIA

SUMMARY

A computer program, using three dimensional finite element method, has been developed for the analysis of centrifugal impellers based on sector symmetry considerations .The program is applied to the analysis of impeller with curved vanes subjected to centrifugal loads and results are presented .A simplified analysis model of the impeller with axisymmetric idealisation is included after a comparative study of its results with that of three dimensional analysis. Region of maximum stress values occurring in the component has been identified by the axisymmetric model suggesting its application at the preliminary design stages.

1. INTRODUCTON

A rigorous analysis of curved vane impeller is possible only by means of a three dimensional finite element method.The loads on the impeller include surface forces due to pressure and centrifugal forces due to rotation.Sectorial symmetry of the impeller with reference to geometry, loading and boundary conditions enable the analysis of the whole structre by considering a representative sector segment.Basic principles on these consideration are given in [1] and some related three dimensional finite element formulations can be found in [2].

A computer program has been developed using three dimensional quadratic isoparametric elements including routines for explicit evaluation of stiffness coefficients and for the generation of consistent nodal load vectors of surface forces, body forces and thermal gradients. Frontal solution technique is used incorporating steps for the application of necessary conditions at the sector boundaries after carrying out the required transformations of stiffness coefficients and load vectors at the nodes along the boundaries. The program is coupled with routines for the generation of contour maps based on search algorithm by node-wise elimination process.

The impeller considered for the analysis consists of six large vanes and six small vanes which are equispaced and placed alternately between hub and cover disks.A 60 degree sector sub-model comprising one large vane and one small vane with boundaries following the contour of the vane is considered for the analysis. The von Mises stress distribution at the vane sections and the deformation patterns are presented.

Although the three dimensional analysis enables accurate determination of stresses, the cost of analysis and the time taken in obtaining the results are considerable.A simplified approach of straight radial vaned components using axisymmetric idealisation has been proposed in the literature elsewhere [3].As per the philosophy, the hub and cover disks can be idealised with ring elements and the vanes with plane stress elements.A further modification considering the vanes as an equivalent fictitious ring of orthotropic material has been given in [4]. The approach is used in the present work for the analysis of curved vane impeller with a view to evolve a model for use in the initial stages of design. The results obtained by the axisymmetric model are compared with that of three dimensional analysis to check the extent of deviations.

2.THREE DIMENSIONAL ANALYSIS

The impeller has been analysed considering a sector of one pitch covering a large vane and a

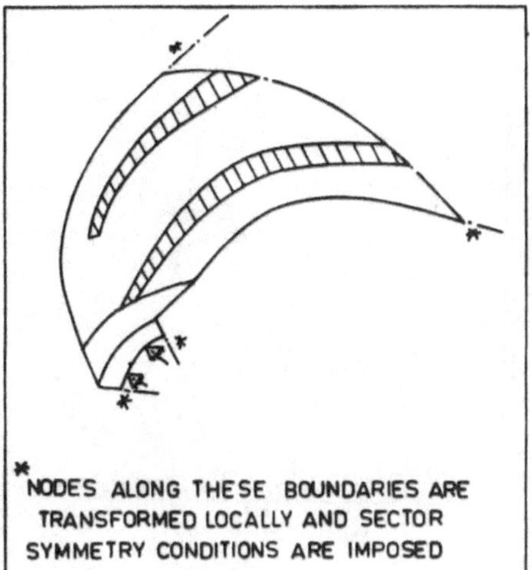

SECTOR OF IMPELLER FOR 3D ANALYSIS

FIG. 1

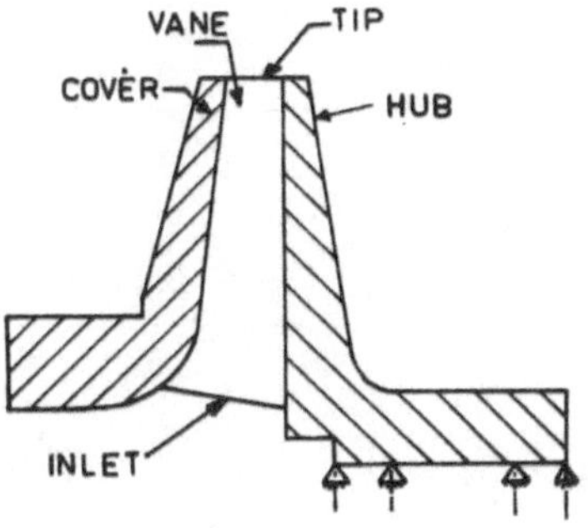

MODEL FOR THE AXISYMMETRIC ANALYSIS

FIG. 2

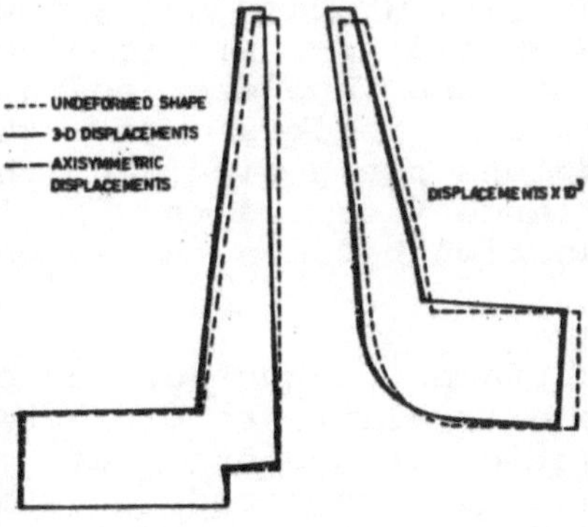

DEFORMATION PATTERN BETWEEN THE VANES

FIG. 3

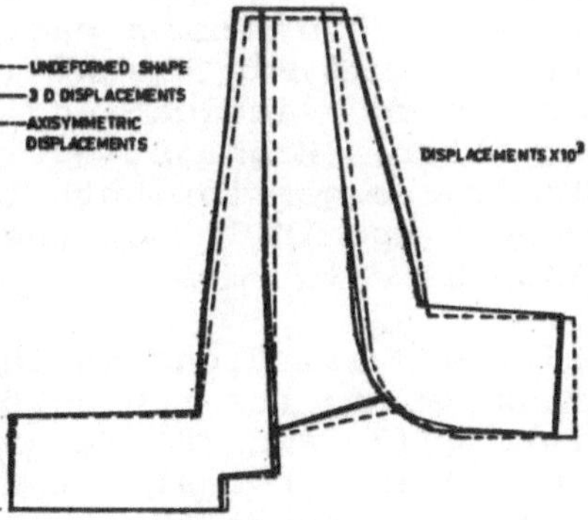

DEFORMATION PATTERN AT THE VANES

FIG. 4

small vane as shown in Fig (1). Loads due to the centrifugal forces are only considered owing to the large extent of their contribution. The boundary conditions include identical displacement conditions at the sector edges consistent with the coordinate system and fixity conditions along the inner

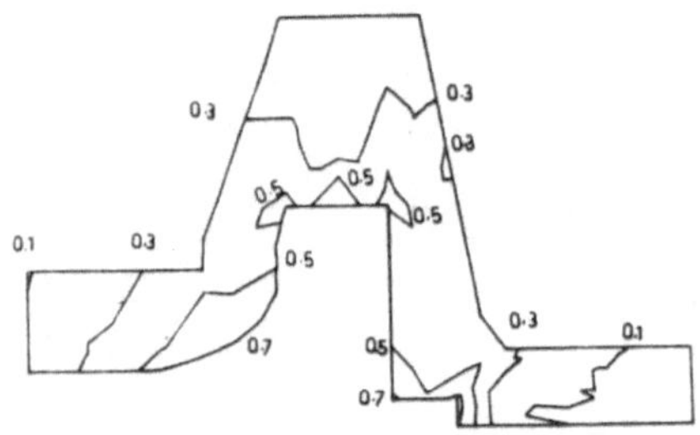

VON-MISES STRESS PATTERN AT SMALL VANE

FIG. 5

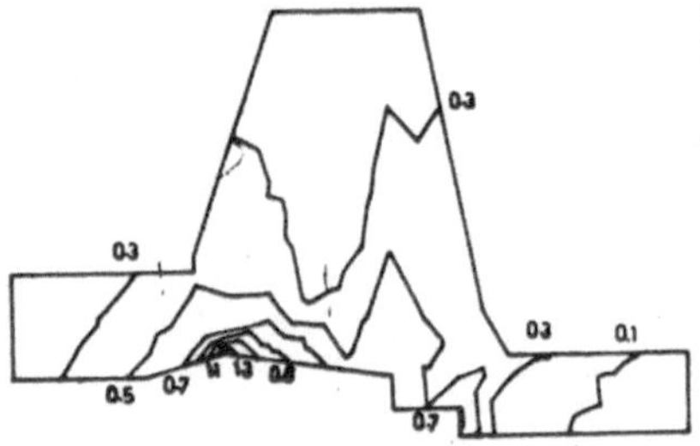

VON-MISES STRESS PATTERN AT LARGE VANE

FIG. 6

periphery of impeller hub. The sector sub-model has been discretized using 20-node quadratic elements with a total of 3300 nodal variables. A thorough check of the geometry is carried out using interactive graphics by screening through various views. Deformation patterns, Fig(3) and (4), iso-von Mises plots, Fig(5) and (6), at the vane sections are presented indicating critical regions of the impeller. High stress regions are noticed at the leading edges for both the large vane and the small vane, specially at the junction between the vanes and the cover disk.

Stress variations on the surfaces of hub and cover disk along the radius of the impeller are presented in Fig (7) and (8). The reduction in the stresses in the disks beneath the vanes as compared to the region between the vanes indicates the stiffening effect of the vanes on the disk.

Stresses are non-dimensionalised using the relation

$$\sigma = \sigma/(\rho/g)\,\omega^2 r_t^2$$

and are plotted against the non-dimensionalised radius $r = r/r_t$

where ρ : Mass density of elemental material
ω : Angular velocity
r_t : Tip radius of the impeller
g : Acceleration due to Gravity

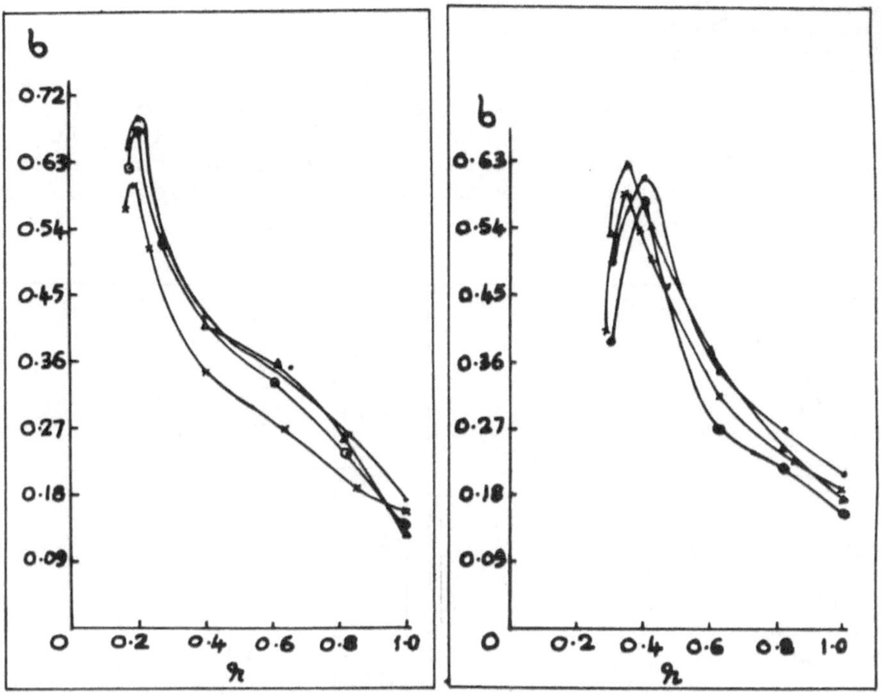

(a) Hub (b) Cover

FIG. 7 : VON MISES STRESSES ALONG RADIUS (VANED FACE)

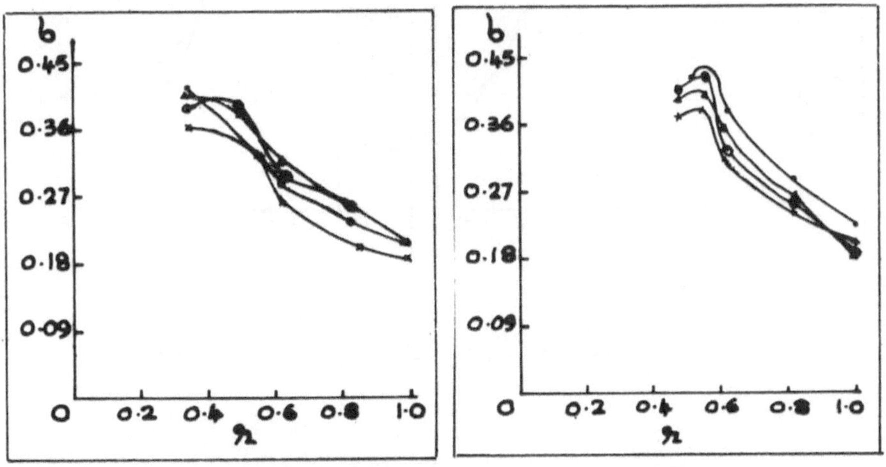

(a) Hub (b) Cover

FIG. 8 : VON MISES STRESSES ALONG RADIUS (REAR FACE)

—·— Between the vanes (3D) —▲— Beneath the small vanes (3D)

—⊙— Beneath the large vanes (3D) —×— Axisymmetric analysis

Vane root stress distributions, Fig (9), are plotted to check the order of stress rise at the junction between the vanes and the disks. It is noticed that the stresses increase as the leading edge is approached with a maximum value occurring at the leading edge.

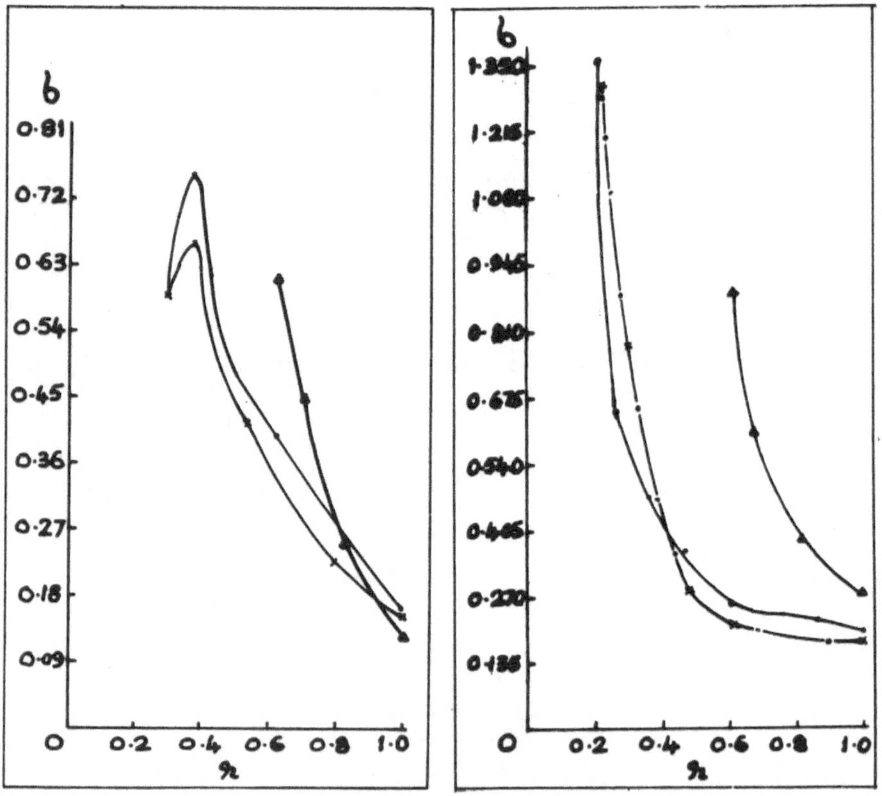

(a) Hub (b) Cover

FIG. 9 : STRESSES AT THE VANE ROOT

—·— Near large vane (3 D)

—▲— Near small vane (3 D)

—x— Axisymmetric analysis

3. AXISYMMETRIC ANALYSIS:

The analysis is carried out assuming the whole impeller including the vanes as a body of revolution, Fig(2). The vanes are assumed to be made of transversely orthotropic material with negligible elastic modulus associated with the tangential direction. The solidity of such a fictitious ring increases from a low value at the tip of impeller to a high value at the inlet. Accordingly, the effect of centrifugal loads exerted by the vanes on the disks is simulated by reducing the mass density of the vane elements by solidity factor varying from tip to inlet. To obtain possible stress concentrations at the root of the vane, the size of the elements at the interface is reduced.

A high precision triangular element with continuity of strains at the nodes has been used for the idealisation of the axisymmetric model. The numerical details of the element are obtained from [3].

Deformations computed from the axisymmetric analysis are indicated in Figures (3) and (4) along with the results of three dimensional analysis. The deformation patterns are almost identical excepting near the inlet regions of the impeller. The deformations noticed at the leading edge, Fig (4), of the vane are similar to the three dimensional analysis results indicating that the maximum stress region is identified by the simplified analysis.

von Mises stress variations along radii obtained from the axisymmetric analysis are also plotted in Fig(7) & (8). The stress variations predicted by axisymmetric model are similar, although on the lower side in comparison with the results of the three dimensional analysis.

Stresses at the vane root computed by axisymmetric model compared well with those obtained by three dimensional analysis near the large vane as shown in Fig(9). Close correlation is noticed in case of maximum stress value at the leading edge of the vane.

4. CONCLUSIONS:

Three dimensional analysis of a curved vane impeller subjected to centrifugal loads has been carried out considering a repetitive sector segment. The results of analysis are used for evaluating a simplified analysis using axisymmetric model.

The stresses obtained by axisymmetric model are lower compared to that of three dimensional analysis. Close correlations are noticed on the cover disk and in particular near the large vane region. Maximum stress values occurring at the junction of large vane leading edge with the cover disk are predicted to sufficient accuracy by the axisymmetric model indicating that the model can serve as an useful tool at the initial stages of design and for preliminary investigations of a failure case.

5.REFERENCES:

1. Zienkiewicz, O.C., 'On the Principle of Repeatability and its Application in Analysis of Turbine and Pump Impellers.' Int. J. of Numerical Methods in Engg., Vol4, 445-452 (1972)

2. Jurno Imamasa et al, 'A Stress Analysis Program by FEM for radial Flow Impellers', Technical Review , Mitsubishi Heavy Industries Limited, Japan, October 1974, pp 191-202

3. Chacour, S., 'A High Precision Triangular Element Used in the Analysis of Hydraulic Turbine Components ', Journal of Basic Engineering, December 1970.

4. Grasso, A., et al, 'Some Theoretical and Experimental Investigations of Stresses and Vibrations in a Radial Flow Rotor ', AGARD Conference Proceedings N248, 1978 , pp 11-1 to 11-19.

A NUMERICAL METHOD FOR THE DETERMINATION OF THE MOMENT-ROTATION-CAPACITY OF THIN WALLED MEMBERS

Dr.-Ing. Ch. Stutzki and Prof.Dr.-Ing. G. Sedlacek
Lehrstuhl für Stahlbau, RWTH Aachen, Fed. Rep. of Germany

SUMMARY

For the application of elastic-plastic design of steel frames the rotational capacity of the plastic hinges is of great importance. This rotational capacity is limited by local buckling of the beam in the plastic zone. The sensivity of buckling is mainly governed by the relation of width/ thickness of the thin walled parts of the profile. The behaviour in a plastic hinge is determined by the moment- rotation- behaviour in the post buckling range. A calculation model is presented which reduces the problem of local buckling in the plastic range with the help of a static condensation. The three-dimensional structure is transformed into a linear beam structure with additionally defined degrees of freedom for local buckling effects.

STRUCTURAL MODEL

The spatial form of a steel girder (fig.1a) is modelled with the help of beam elements (fig.1b) in which each beam element consists of a substructure of plate elements.

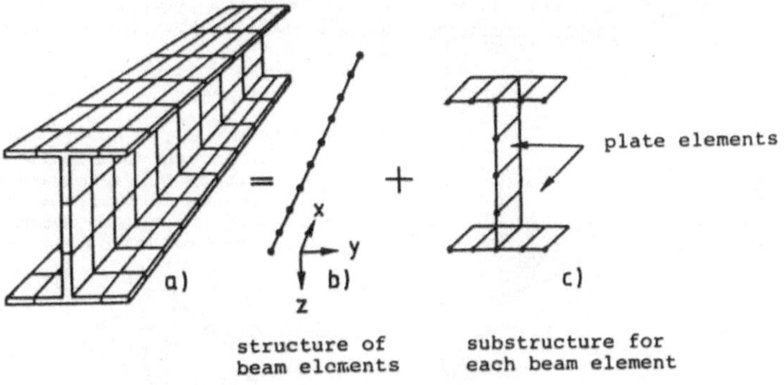

fig.1 Modelling of an I-shaped beam as a structure of beam elements,each consisting of an assemblage of plate elements

The beam elements are generated by condensation of the inner degrees of freedom of the substructure.

In a preparatory step the cross section is regarded as a plane frame whose first natural frequencies are calculated.

The eigenvectors $\mathbf{q}_i$ give quite useful descriptions of possible distorsions of the profile (fig.2a). Also the usual deflections of the beam are written as vectors (fig.2b) whereby each cross- sectional node posesses three degrees of freedom in the y-z- plane.

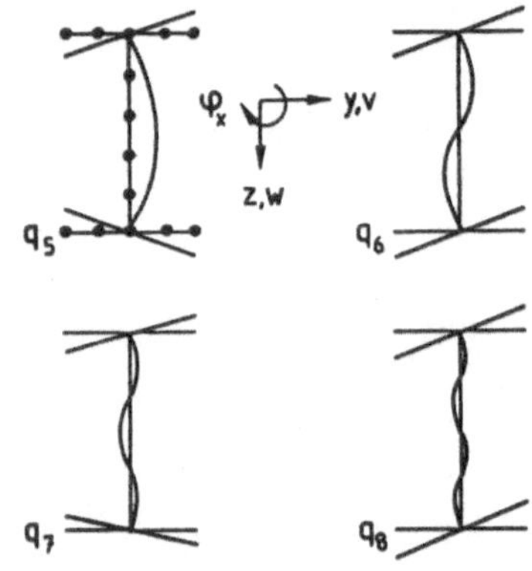

fig.2a **Regarding the cross section as a plane frame, possible distortions of the section can be determined by an eigenvalue calculation. The eigenvectors q_i describe the shapes of distorsion.**

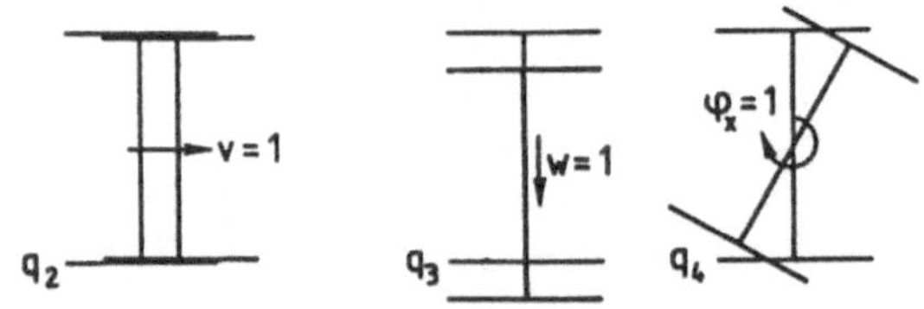

fig.2b **Movements of the profile in bending and torsion**

The condition for the determination of displacements in the longitudinal (x-) direction, according to the transversal (y-z-) displacements, is the Bernoulli- resp. Kirchoff- hypothesis which postulates the shear strains to be zero.(fig.3). This assumption leads to a relation between transversal and longitudinal displacements (eq.1.2, 1.3).

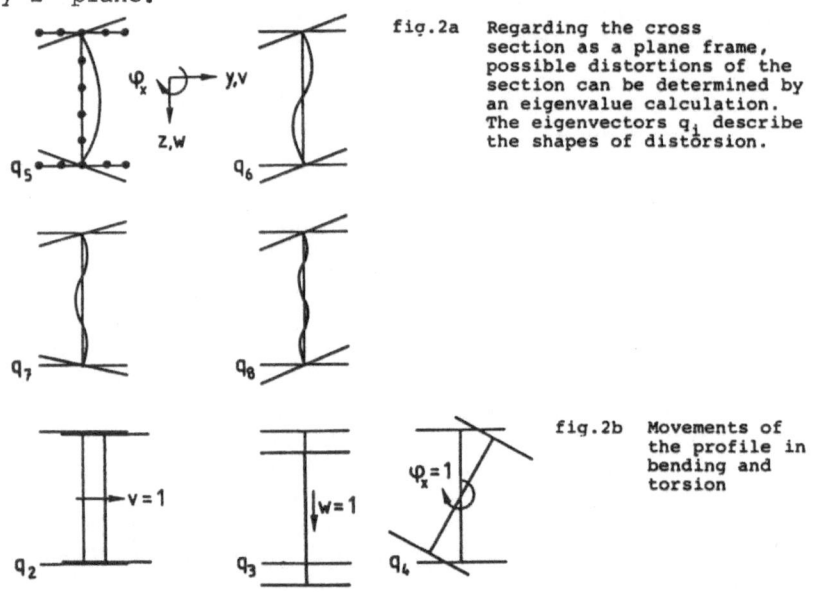

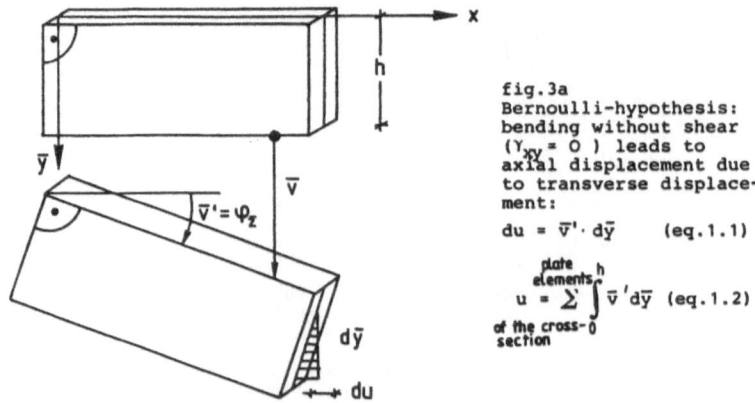

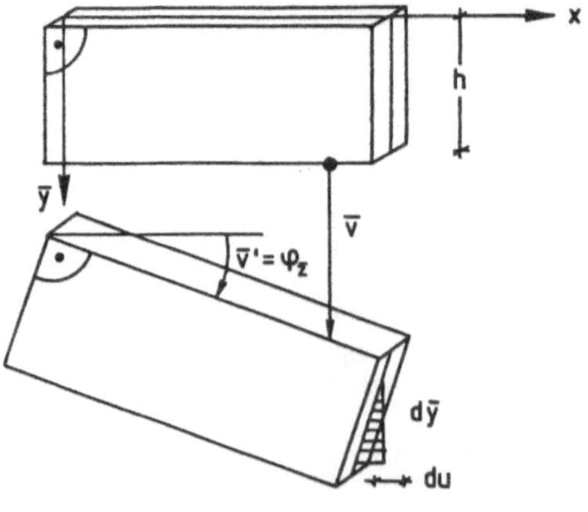

fig.3a
Bernoulli-hypothesis: bending without shear ($\gamma_{xy} = 0$) leads to axial displacement due to transverse displacement:

$$du = \bar{v}' \cdot d\bar{y} \qquad \text{(eq.1.1)}$$

$$u = \sum_{\substack{\text{plate elements} \\ \text{of the cross-section}}} \int_0^h \bar{v}' d\bar{y} \qquad \text{(eq.1.2)}$$

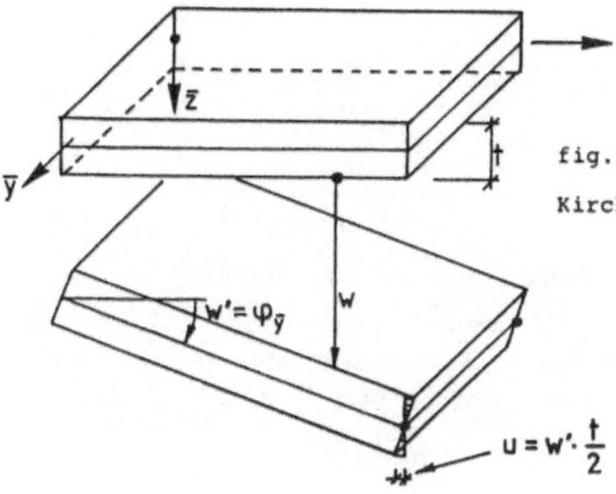

fig.3b:

Kirchhoff-hypothesis ($\gamma_{xz} = 0$)
(eq.1.3)

fig.3 Derivation of longitudinal displacements

Each vector $\mathbf{q}_i$ which describes a transversal movement thus corresponds to a vector $\boldsymbol{\omega}_i$ describing the movement out of the y-z- plane. These vectors are collected as columns of the matrices $[\psi]$ and $[\Omega]$ (fig.4a). Examples for the contents of the vectors are given in fig.4b.

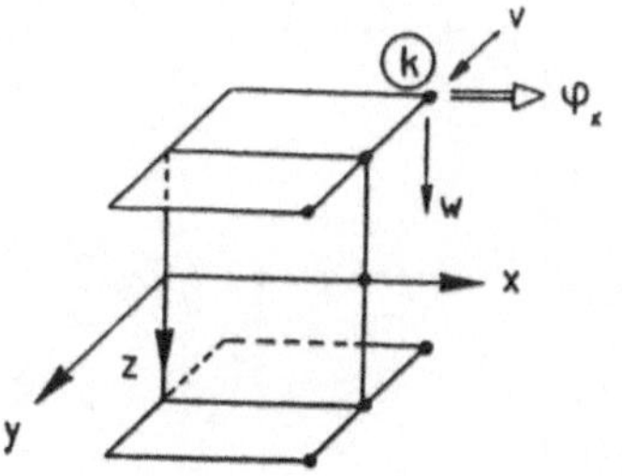

Description of movements in the y-z-plane:

$$[\psi] = \left[\begin{array}{c|c|c|c|c|c|c} q_1 & q_2 & q_3 & q_4 & q_5 & .. & q_8 \\ & \vdots & & & & & \\ & \varphi_{xk} & & & & & \\ & v_k & & & & & \\ & w_k & & & & & \end{array} \right] \qquad \text{(eq.2.1)}$$

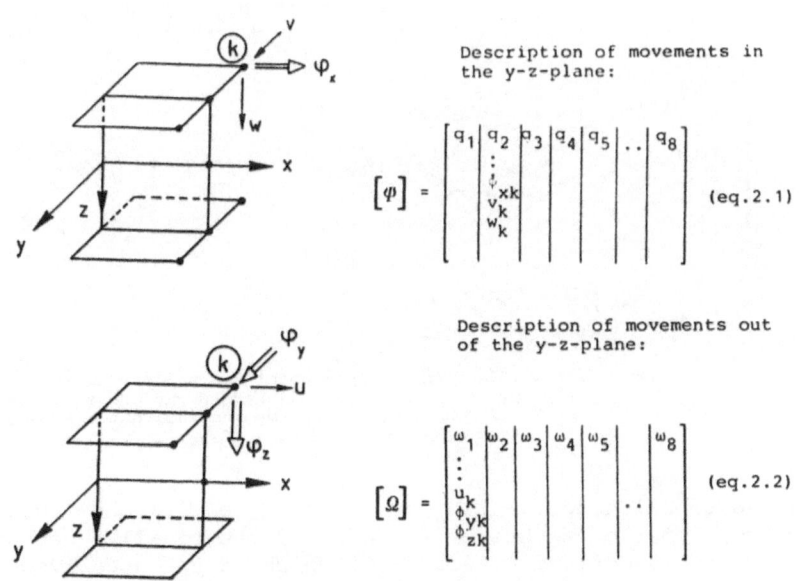

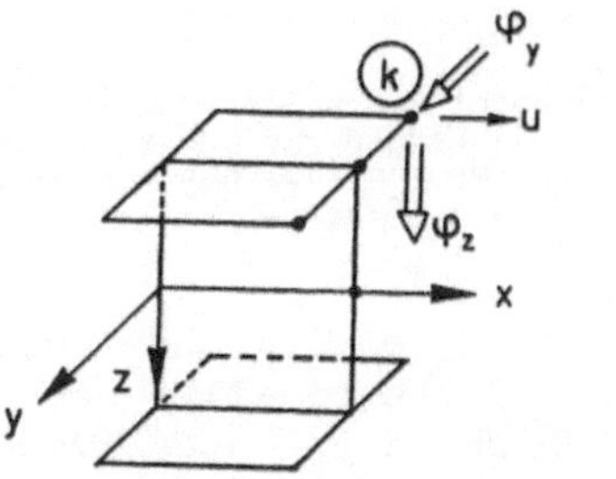

Description of movements out of the y-z-plane:

$$[\Omega] = \left[\begin{array}{c|c|c|c|c|c|c} \omega_1 & \omega_2 & \omega_3 & \omega_4 & \omega_5 & & \omega_8 \\ \vdots & & & & & & \\ u_k & & & & & .. & \\ \phi_{yk} & & & & & & \\ \phi_{zk} & & & & & & \end{array} \right] \qquad \text{(eq.2.2)}$$

fig.4a Definition of the Matrices $[\psi]$ and $[\Omega]$ describing the movements of the cross section in and out of the y-z-plane

q_1	:	zero-vector	
ω_1	:	(1,0,0, 1,0,0 ...)	:unit-displacements due to axial force
q_2	:	(0,1,0, 0,1,0 ...)	:unit-displacement in y-direction
ω_2	:	(y_1,0,1, y_k,0,1 ...)	:y-coordinates of the cross-sectional nodes

fig.4b examples for the description of the movements of the cross sectional nodes

Each unit- displacement q_i resp. ω_i is multiplied by a function f_i, whose meanings are the usual beam deformations plus additional functions for the distorsions of the profile.(fig.5, eq.3.3, 3.4). The total displacements of the

$$\{q\} = [\Psi]\cdot\{f\} \quad \text{(eq.3.1)}$$
$$\{\omega\} = [\Omega]\cdot\{f\}, \quad \{v_{cross}\} = \begin{Bmatrix} q \\ \omega \end{Bmatrix} \quad \text{(eq.3.2)}$$
$$\{f\}^T = \{\ -\ ,\ v\ ,\ w\ ,\ \phi_x, w_5\ ,\ w_8\ \} \quad \text{(eq.3.3)}$$
$$\{f\}^T = \{\ u\ ,\ v',\ w',\ \phi_x', w_5'\ ,\ w_8'\ \} \quad \text{(eq.3.4)}$$
$$\{v_{beam}\} = \{\{f'\}_{x=0}\{f\}_{x=0}, \{f'\}_{x=1}, \{f\}_{x=1}\} \quad \text{(eq.3.5)}$$

= vector of the unknown for the whole beam element

fig. 5 Definition of longitudinal functions f_i, connecting the displacements of the crosssectional modes with the degrees of freedom of the beam element $\{v_{beam}\}$.

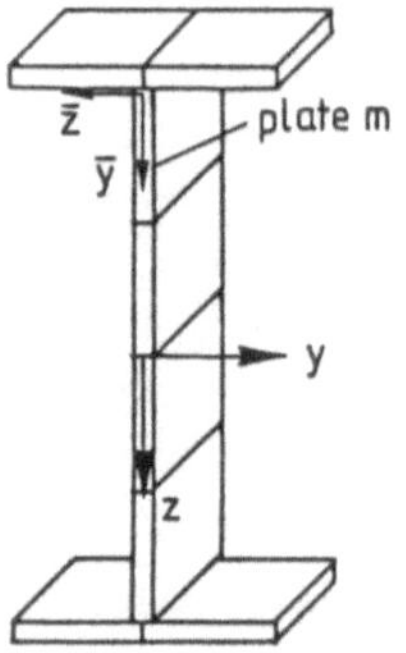

$$\{v_{plate}\} = [T_m]\cdot\{v_{cross}\} \quad \text{(eq.4)}$$

fig.6 The local displacement at the four nodes of each plate element are linked to the cross-sectional displacement vector by the incidente- and transformation matrix $[T_m]$

cross- sectional nodes, $\{v_{cross}\}$, are given by eq.3.1 and eq.3.2. The vector of the degrees of f reedom of the finally generated beam element is defined by eq. 3.5, which is the usual vector, extended by the additional functions for the distorsions of the profile.

The functions for the displacements in the middle plane of a single plate element are given in fig.7. The displacements are at first defined in the transversal direction (eq.5.1, 5.2), depending on parameters at the edges i and k, which are functions of x (eq. 5.3 - 5.8).

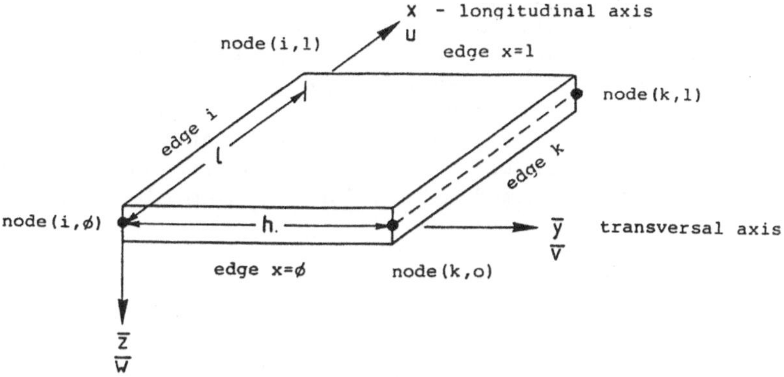

$$\bar{w}(x,\bar{y}) = \left[B_1(\bar{y})\right]\cdot\begin{Bmatrix}\dot{w}_i(x)\\ w_i(x)\\ \dot{w}_k(x)\\ w_k(x)\end{Bmatrix} \quad \text{(eq.5.1)}; \quad \dot{w}_i(x) = \left[B_2(x)\right]\cdot\begin{Bmatrix}\dot{w}'_{io}\\ \dot{w}_{io}\\ \dot{w}'_{il}\\ \dot{w}_{il}\end{Bmatrix} \quad \text{(eq.5.5)}$$

$$\bar{u}(x,\bar{y}) = \left[B_3(\bar{y})\right]\cdot\begin{Bmatrix}u_i(x)\\ u_k(x)\end{Bmatrix} \quad \text{(eq.5.2)}; \quad w_i(x) = \left[B_2(x)\right]\cdot\begin{Bmatrix}w'_{io}\\ w_{io}\\ w'_{il}\\ w_{il}\end{Bmatrix} \quad \text{(eq.5.6)}$$

$$u_i(\bar{x}) = \left[B_2(x)\right]\cdot\begin{Bmatrix}u'_{io}\\ u_{io}\\ u'_{il}\\ u_{il}\end{Bmatrix} \quad \text{(eq.5.3)}; \quad \dot{w}_k(x) = \left[B_2(x)\right]\begin{Bmatrix}\dot{w}'_{ko}\\ \dot{w}_{ko}\\ \dot{w}'_{kl}\\ \dot{w}_{kl}\end{Bmatrix} \quad \text{(eq.5.7)}$$

$$u_k(x) = \left[B_2(x)\right]\begin{Bmatrix}u'_{ko}\\ u_{ko}\\ u'_{kl}\\ u_{kl}\end{Bmatrix} \quad \text{(eq.5.4)}; \quad w_k(x) = \left[B_2(x)\right]\begin{Bmatrix}w'_{ko}\\ w_{ko}\\ w'_{kl}\\ w_{kl}\end{Bmatrix} \quad \text{(eq.5.8)}$$

$$\left[B_1(\bar{y})\right] = \left[\frac{y^3}{h^2} - \frac{2y^2}{h} + y\ ,\ 2\frac{y^3}{h^3} - 3\frac{y^2}{h^2} + 1\ ,\ \frac{y^3}{h^2} - \frac{y^2}{h}\ ,\ -\frac{2y^3}{h^3} + 3\frac{y^2}{h^2}\right] \quad \text{(eq.5.9)}$$

$$\left[B_2(x)\right] = \left[\frac{x^3}{l^2} - \frac{2x^2}{l} + x\ ,\ 2\frac{x^3}{l^3} - 3\frac{x^2}{l^2} + 1\ ,\ \frac{x^3}{l^2} - \frac{x^2}{l}\ ,\ -\frac{2x^3}{l^3} + 3\frac{x^2}{l^2}\right] \quad \text{(eq.5.10)}$$

$$\left[B_3(\bar{y})\right] = \left[\frac{h-y}{h}\ ,\ \frac{y}{h}\right] \quad \text{(eq.5.11)}$$

in general : $$\begin{Bmatrix}\bar{u}\\ \bar{w}\end{Bmatrix} = \left[B\right]\left\{\bar{v}_{plate}\right\} \quad \text{(eq.5.12)}$$

fig.7 Displacement field in a plate element

With the help of the strain- displacement- relations (fig.8, eq.6.1- 6.3), where only the linear term is used so far, with the material law for elastic thin plates, and with the principle of virtual displacements, the stiffness matrix for the substructure can be derived.
Using eq.3.1- 3.5 and eq.4 this stiffness matrix is condensed by pre- and post- multiplication with the matrices $[\Psi]$ and $[\Omega]$. The result is the condensed matrix for the beam element.

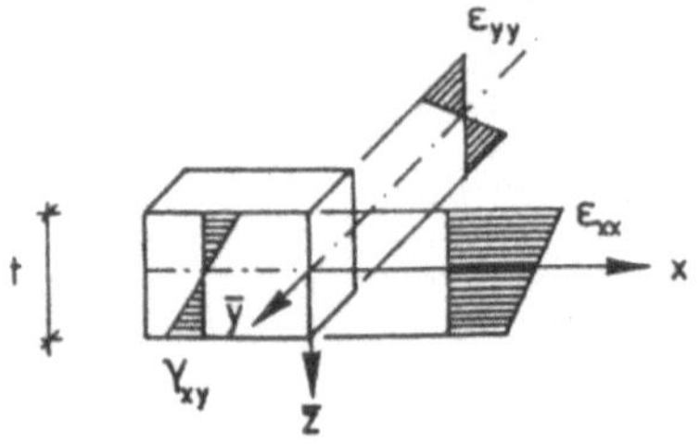

$$\varepsilon_{yy} = z \cdot \ddot{w} \qquad \text{(eq.6.1)}$$

$$\varepsilon_{xx} = \underbrace{z \cdot w'' + u'}_{\text{linear}} + \underbrace{\tfrac{1}{2}\,(v'^2 + w'^2)}_{\text{geom.nonlinear term}} \qquad \text{(eq.6.2)}$$

$$\gamma_{xy} = \underbrace{\dot{u} + v'}_{\text{closed}} - \underbrace{2 \cdot z \cdot \dot{w}'}_{\text{open}} \qquad \text{(eq.6.3)}$$

sections St.-Venant - Torsion

fig.8 Definition of strains

NONLINEAR PROCEDURE

The load- delection path is computed incrementally, by incrementing the length of the deformation vector, that is kept constant during the equilibrium iteration. The condition $\gamma \cdot$load + reaction $< \varepsilon$ is satisfied by an iteration, adjusting the load- factor γ and the direction of the incremental deformation- vector. The iteration method employed here, developed by Lopetegui /1/, uses load- deflection states which are adjusted continously to the changing stiffness, whereas the elastic, linear stiffness matrix of the global system remains unchanged throughout the whole calculation. The reaction forces are computed by a numerical integration of stresses (fig.9, with eq. 6 from fig.8 including the nonlinear term). The forces and moments at the nodes of the plate elements are condensed with the matrices $[\Psi]$ and $[\Omega]$, which leads to beam-end- forces and -moments that corrspond to the degrees of freedom defined in eq.3.5.
The yield criterion and the formulation of the elastic- plastic material law is used as given by Owen and Hinton in /2/ for shells and plates.

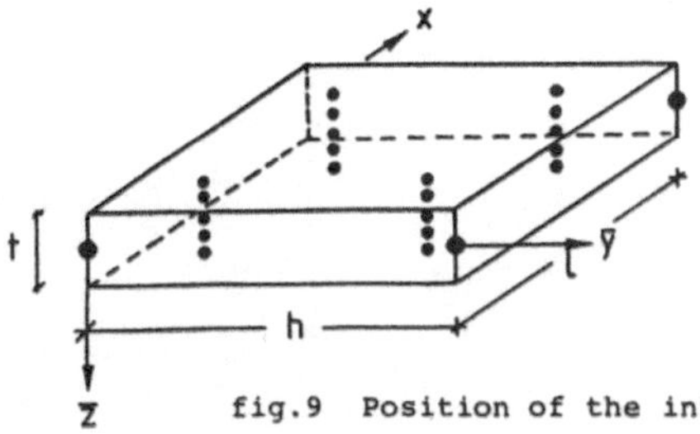

EXAMPLE

fig.9 Position of the integration points

As an example the numerical simulation of an experimental result by Perlynn and Kulak / 3 / is presented.
Compared with a finite element analysis with shell elements the method presented above turned out to be very effective, without loss of accuracy, as far as the range of application is limited to thin walled profiles under bending and compression.

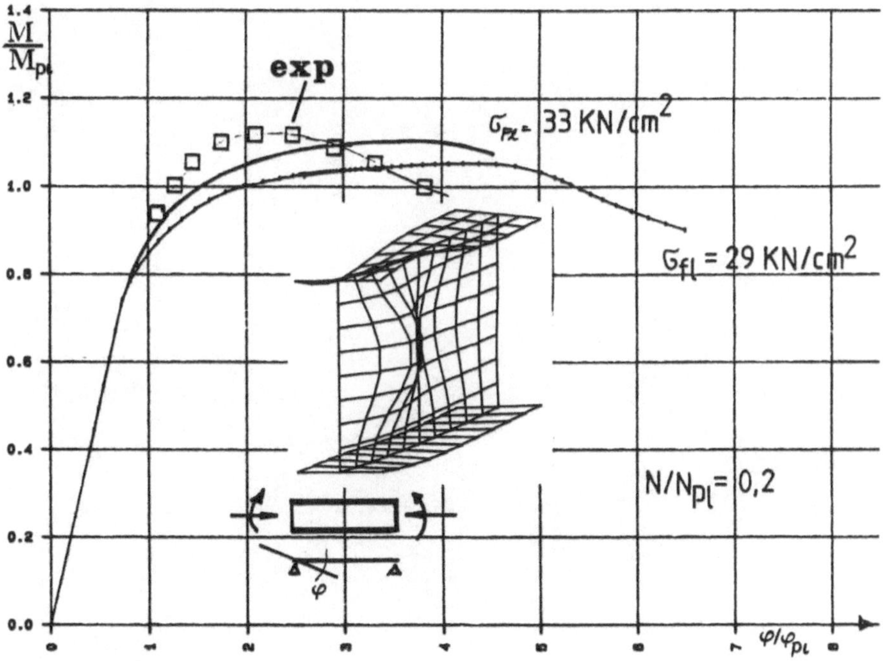

REFERENCES

/1/ Lopetegui,J., Procedure for the solution of nonlinear problems using load - deflection - states. Doctoral thesis, University of Aachen, 1983

/2/ Hinton,E., Owen,D.R.J.: Finite Element Software for plates and shells. Pineridge press, Swansea, U.K. 1984

/3/ Perlynn,M.J., Kulak,G.L.: Web Slenderness Limits for Compact Beam-Columns. Struct. Eng. Report No.50, Univ. of Alberta, Edmonton, Canada 1974

SOME APPLICATIONS OF LOAD-DEFORMATION STATES

Dr.Ing. J. Lopetegui and Prof. Dr. Ing. G. Sedlacek

Lehrstuhl fuer Stahlbau,RWTH-Aachen, Fed. Rep. of Germany

The method of load deformation states has proved to be useful, /1/, /2/, /3/, /4/, /5/, /6/, /7/, in solving statical and dynamical problems. It has furthermore the advantage that a clear physical interpretation can be given to the mathematical steps of the method.
An attempt will be made in the following pages to define the idea of the load deformation states and to present the possibilities of such a method together with various applications such as :
1) Solution of large linear systems of equations
2) Solution of non linear problems
3) Solution of eigenvalue problems

LOAD DEFORMATION STATES

The stiffness method of structural analysis leads to an equation which relates the applied loads, structure stiffness, and resultant joint deflections .

$$\mathbf{K} v = P \qquad (1)$$

where:
K: is the stiffness matrix of the system
v: the unknown deformation vector
P: the loading vector

A load deformation state is defined as a pair of vectors, P and v, that satisfy the matrix equation (1). This equation

expresses a unique correspondence in the case of linear systems.
We then write: $P_1 \leftrightarrow v_1$. (2)

We define orthogonality of two load deformation states $P_1 \leftrightarrow v_1$, $P_2 \leftrightarrow v_2$, when the load of one configuration has no work contribution in the other configuration, that means

$$P_1 \cdot v_2 = P_2 \cdot v_1 = 0 .$$

Two load vectors or two deformation vectors are parallel when there exists a scalar x such that, $P_1 = x \cdot P_2$ or $v_1 = x \cdot v_2$.

Orthogonalisation of two load deformation states can always be achieved. One expresses, for example, the load vector P_2 in two components; one component parallel to the load vector P_1 and the other component orthogonal to the deformation vector v_1 ,

$$\begin{aligned} P_2 &= P_2^{\parallel P_1} + P_2^{\perp v_1} \\ &= x_1 P_1 + P_2^* . \end{aligned} \tag{3}$$

P_2^* is the component of P_2 which is orthogonal to v_1

$$P_2^* = P_2^{\perp v_1} = P_2 - x_1 P_1 . \tag{4}$$

The scalar x_1 is calculated from the condition of orthogonality between the vectors P_2^* and v_1

$$(P_2 - x_1 P_1) \cdot v_1 = 0 ; \quad x_1 = \frac{P_2 \cdot v_1}{P_1 \cdot v_1} . \tag{5}$$

The displacement vector v_2 can also be decomposed in two components; one component parallel to the displacement vector v_1 , the other component orthogonal to the load vector P_1 ,

$$\begin{aligned} v_2 &= v_2^{\parallel v_1} + v_2^{\perp P_1} \\ &= y_1 v_1 + v_2^* \end{aligned} \tag{6}$$

v_2^* is the component of v_2 which is orthogonal to P_1 . The scalar y_1 can also be calculated using the dual orthogonality condition $P_1 \cdot v_2^* = 0$

$$(v_2 - y_1 v_1) \cdot P_1 = 0 ; \quad y_1 = \frac{P_1 \cdot v_2}{P_1 \cdot v_1} \tag{7}$$

The general principle ascribed to Betti leads to the equation

$$P_1 \cdot v_2 = P_2 \cdot v_1$$

It follows then that $x_1 = y_1$

The orthogonal state 2 can then be expressed as $P_2^* \leftrightarrow v_2^*$

with $P_2^* = P_2^{\perp v_1} = P_2 - x_1 P_1$ and $v_2^* = v_2^{\perp P_1} = v_2 - y_1 \cdot v_1$. (8)

WORK AND STRAIN ENERGY

The work equation for a load deformation state of a structural system is:

$$W = \frac{1}{2} P_1 \cdot v_1 \tag{9}$$

where P_1 is the generalized load vector and v_1 is the corresponding generalized displacement vector.
Equation (9) expresses the work done by the load vector P_1 on the deformation state represented by the displacement v_1. This work is stored as strain energy. The strain energy for structures that primarily store energy in the form of bending can be written as

$$U = \frac{1}{2} \sum \int_0^l EI \cdot v''(x)^2 \cdot \mathrm{d}x \tag{10}$$

where v(x) is the displacement function.
According to the engineering bending theory of beams,

$$EI \cdot v''(x) = -M(x)$$

It follows that

$$U = \frac{1}{2} \sum \int_0^l \frac{M(x)^2}{EI} \cdot \mathrm{d}x. \tag{11}$$

WORK OF A LOAD ON THE DISPLACEMENT CAUSED BY ANOTHER LOAD.

Considering two load deformation states denoted by $P_1 \leftrightarrow v_1$ and $P_2 \leftrightarrow v_2$.

It follows then: $P_1 \cdot v_2 = P_2 \cdot v_1 = \sum \int_0^l EI \cdot v_1''(x) \cdot v_2''(x) \cdot \mathrm{d}x$

$$= \sum \int_0^l M_1(x) \cdot M_2(x)/EI \cdot \mathrm{d}x. \tag{12}$$

$M(x)$ represents the bending moment diagram of the structure due to P_i.

It is possible to work on the level of the whole structure with load deformation states in the same way as with Ritz-functions for a continuum.If load deformation states are available , or if they can be generated using an existing stiffness matrix then all methods of statical analysis can be applied replacing functions and integrals by load deformation states. For the analysis of non linear systems, the stiffness matrix for generating the load deformation states could be that of the linear or simplified system.

As an example , the determination of an approximate solution with a subspace that is defined by some load deformation states will be shown below.

APPROXIMATE SOLUTIONS

Let $P_1 \leftrightarrow v_1$, $P_2 \leftrightarrow v_2$,...., $P_n \leftrightarrow v_n$ be load displacement states for a structural system under an external load vector F_0 . An approximate solution for a displacement field can be written as:

$$v = x_1 \cdot v_1 + x_2 \cdot v_2 + \cdots + x_n \cdot v_n. \tag{13}$$

The reaction forces corresponding to the deformed state are :

$$R = -x_1 \cdot P_1 - x_2 \cdot P_2 - \cdots - x_n \cdot P_n, \tag{14}$$

The out of balance forces can be written as:

$$\begin{aligned} F &= F_0 + R \\ &= F_0 - x_1 \cdot P_1 - x_2 \cdot P_2 - \cdots - x_n \cdot P_n. \end{aligned} \tag{15}$$

The deformed state is successively varied in such a way, that each deformation component is increased from $x_i \cdot v_i$ to $(x_1 + \mathrm{d}x_i) v_i$

The new deformation is given by : $v + \Delta v_i$, with $\Delta v_i = \mathrm{d}x_i \cdot v_i$,

The differential $\mathrm{d}x_i$ is a scalar. The work corresponding to the variation $\mathrm{d}v_i$ can be expressed as:

$$\begin{aligned} \mathrm{d}U_i^* &= F \cdot \mathrm{d}v_i \\ &= (F_0 - x_1 \cdot P_1 - x_2 \cdot P_2 - \cdots - x_n \cdot P_n) \cdot v_i \cdot \mathrm{d}x_i. \end{aligned} \tag{16}$$

The virtual work is then zero when the scalar product, $F \cdot \mathrm{d}v_i$ is zero. Setting the individual work variations $\mathrm{d}U_i^*$ to zero, a system of equations is then obtained

$$\begin{bmatrix} P_1 \cdot v_1 & P_2 \cdot v_1 & \dots & \dots & \dots & P_n \cdot v_1 \\ P_1 \cdot v_2 & P_2 \cdot v_2 & \dots & \dots & \dots & P_n \cdot v_2 \\ \dots & \dots & \dots & \dots & \dots & \dots \\ P_1 \cdot v_n & P_2 \cdot v_n & \dots & \dots & \dots & P_n \cdot v_n \end{bmatrix} \begin{Bmatrix} x_1 \\ x_2 \\ . \\ x_n \end{Bmatrix} = \begin{Bmatrix} F_0 \cdot v_1 \\ F_0 \cdot v_2 \\ \dots \\ F_0 \cdot v_n \end{Bmatrix}. \tag{17}$$

The system of equations has a solution if the coefficient matrix is not singular and this is the case when the load deformation states are linearly independent .

The best possible approximation of the deformation vector of the structure is v (eq. 13) and the corresponding out of balance forces to this deformation state is F (eq.15). The solution provided is then an exact solution when the given load vector is a linear combination of the known load vectors P_1 to P_n. The coefficient matrix is a diagonal matrix in the case that the load deformation states are orthogonal, i.e the products $P_i \cdot v_k = 0$ for $i \neq k$, the scalars xi are then :

$$x_i = \frac{F_0 \cdot v_i}{P_i \cdot v_i} \tag{18}$$

If the load deformation state are not orthogonal, an orthogonalisation procedure can be carried out.

Another way at arriving at the same result is to decompose the load vector Fo in components parallel to the individual loads P_i and in a component perpendicular to all displacement vectors of the load deformation states.

$$\begin{aligned} F_0 &= F_0^{\parallel P_1} + F_0^{\parallel P_2} + \cdots + F_0^{\parallel P_n} + F_0^{\perp v_1, v_2, \ldots, v_n} \\ &= x_1 \cdot P_1 + x_2 \cdot P_2 + \cdots + x_n \cdot P_n + F^*, \end{aligned} \tag{19}$$

i.e

$$F^* = F_0 - x_1 \cdot P_1 - x_2 \cdot P_2 - \cdots - x_n \cdot P_n. \tag{20}$$

The condition of orthogonality $F^* \cdot v_i = 0$ leads also to the system of equations(17).

SOLVING LARGE SETS OF LINEAR EQUATIONS

In order to solve large sets of linear equation with the load deformation state method, load deformation states must be found. An algorithm is given below in order to find such states in the case that the system of equation is partitionned after equation (21).

$$\left[\begin{array}{c|c} A_{11} & A_{12} \\ \hline A_{21} & A_{22} \end{array}\right] \left\{\begin{array}{c} V_a \\ \hline V_b \end{array}\right\} = \left\{\begin{array}{c} P_a \\ \hline P_b \end{array}\right\} \tag{21}$$

that is

$$\begin{aligned} A_{11} V_a + A_{12} V_b &= P_a \\ A_{22} V_a + A_{22} V_b &= P_b \end{aligned} \tag{22}$$

We first attempt to solve the following system of equations

$$A_{11}\cdot V_a + A_{12}\cdot V_b = P_a - P_{00} \qquad (23)$$
$$A_{21}\cdot V_a + A_{22}\cdot V_b = P_b$$

whereby the matrices A_{ik} with $i,k=1,2$, the vectors P_a and P_b are given and the solution is sought for V_a, V_b and P_{00}. A possible solution strategy would be first to solve for v_a in the following equation

$$A_{11}\cdot V_a = P_a \qquad (24)$$

With a known vector v_a, the following system of equations is then solved

$$A_{22}\cdot v_b = P_b - A_{21}\cdot v_a \qquad (25)$$

A solution is then obtained for the displacement vector v_b. The vector P_{00} which is unknown at this stage, is determined from:

$$P_{00} = -A_{12}\cdot v_b \qquad (26)$$

The vector P_{00} is in structural analysis the vector of the out of balance forces which are present when the deformation vector is v. The vector P_{00} has components not equal to zero only in its upper part. When every load deformation state has no component in the lower part of the load vector then the orthogonalisation procedure proves to be very advantageous in as much as the scalar products $P_i \cdot v_k$ need only be evaluated for the upper part of the vectors.
For the solution of the system of equations it is better, in the first step, to determine a deformation vector v_0 and a residual force $P_1 = P_{00}$. The solution is then postulated in the form,

$$v = v_0 + x_1 v_1 + \cdots + x_n v_n . \qquad (27)$$

The calculation procedure for the following steps is then :

1) Determination of v_i and $P_{i,i}$ using P_i (i=1,2,.....,n). The load deformation state is : $P_i - P_{i,i} \leftrightarrow v_i$

2) The load deformation state is orthogonalised for i>1. Every load deformation state, which is obtained with this algorithm is orthogonal to all foregoing states but the last one, therefore it needs only be orthogonalised from the components of the last load deformation state. The orthogonalised load deformation state is denoted by

$$P_i^* \leftrightarrow v_i^* . \qquad (28)$$

3) The approximation for the solution is then improved,

$$\nu_{new} = \nu_{old} + x_i \, \nu_i^*. \qquad (29)$$

whereby ν_i^* is the deformation vector of the orthogonalised load deformation state and

$$x_i = \frac{P_i \cdot \nu_i^*}{P_i^* \cdot \nu_i^*} \qquad (30)$$

4)The out of balance forces corresponding to ν_{new} is : (31)

$$P_{i+1} = P_i - x_i \cdot P_i^*$$

5) If the required accuracy is not reached, the process is restarted from 1) with i = i+1

There are many advantages to be gained in solving large set of equations when the above explained algorithm is used together with a substructure technique. One defines a set of boundary nodes which would divide the structure into a number of independent substructures if the boundary nodes were fixed.The numbering of the nodes is such that the degrees of freedom of the chosen boundary nodes appear in the upper part of the system of equations; the rest of the nodes are numbered so that the bandwidths of the substructures are kept to a minimum. In the case of clearly partitioned structures, the saving of computing time can be considerable, furthermore various strategies could be sought for special problems. For the structure given in Fig.1 with 172 nodes, 984 d.o.f. and an optimal bandwidth of 132, the present algorithm needs 4,3 10**6 operations, about half the amount of operations usually needed by the standard CHOLESKY technique. Further reduction of computer time is also possible when the structure to be analysed is composed of several equal substructures. The stiffness matrix of the repeated substructure is set up only once and solved under various right hand sides i.e. load vectors. In the case that the entire structure is to be solved with different right hand sides, which are themselves the result of further calculations, the load deformation states must be stored, this happens e.g. in the case of numerical calculations involving non linear structures.

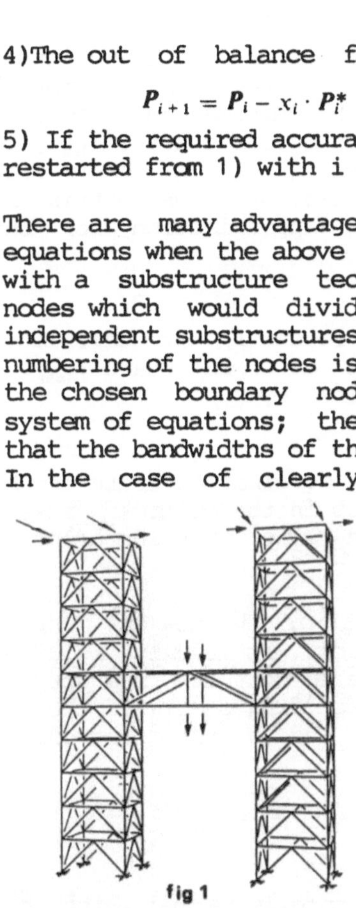

fig 1

ANALYSIS OF NON LINEAR SYSTEMS

In the case of non linear systems it is also possible to calculate the deformation vector if several load deformation states $P_i^* \leftrightarrow v_i^*$ are known. The asterisk indicates that the load deformation states are semi-orthogonal, that is

$$P_i^* \cdot v_k^* = 0 \qquad \text{if } i > k \,.$$

The approach for the deformation that defines the wanted state of equilibrium is

$$v = x_1 \cdot v_1^* + x_2 \cdot v_2^* + \cdots + x_n \cdot v_n^*$$

The following procedure is performed in order to build up a system of orthogonalized load deformation states and to improve simultaneously the approximation for the deformation vector.
Let a non linear structure in the deformed state $v_{\mathbf{old}}$, be loaded with the out of balance forces F_i. A deformation vector v_i is calculated with the initial stiffness matrix and with F_i as a load vector. The new approximation for the deformation vector is

$$v_{\mathbf{new}} = v_{\mathbf{old}} + v_i$$

In the next step the out of balance forces F_{ii} are calculated with the real non linear system in the deformed state $v_{\mathbf{new}}$. The new load deformation state is defined by

$$F_i - F_{ii} \leftrightarrow v_i$$

On the left hand side of this load deformation state is the difference of the out of balance forces in the deformed states $v_{\mathbf{old}}$ and $v_{\mathbf{new}}$, and on the right side the increment of deformation v_i.

If other load deformation states are known, the new one must be orthogonalized from all foregoing states. It is only required that the load vector of this new state do not perform a work in the deformation vectors of all the preceeding ones. If the new load deformation state is $P_i \leftrightarrow v_i$ and a foregoing one is $P_k^* \leftrightarrow v_k^*$ (k=1,2,..,i-1), then

$$P_i^* = P_i - z_k \cdot P_k^* \qquad\qquad \text{with} \qquad z_i = \frac{P_i \cdot v_k^*}{P_k^* \cdot v_k^*}\,.$$
$$v_i^* = v_i - z_k \cdot v_k^*$$

This orthogonalized load-deformation state is varied with the assumption of linearity in the surroundings of $v_{\mathbf{new}}$, which means that

$$x_i\, P_i^* \leftrightarrow x_i\; v_i^*$$

is also correct. The new approximation for the deformation vector is

$$\nu_{new} = \nu_{old} + x_i \, \nu_i^* .$$

The factor x_i is determined in such a way that the scalar product of the so calculated out of balance force with ν_i^* is zero.

$$x_i = \frac{F_i \cdot \nu_i^*}{P_i^* \cdot \nu_i^*} \qquad\qquad F_{i+1} = F_i - x_i \cdot P_i^* .$$

The out of balance force for the next step is F_{i+1} .

DISPLACEMENT CONTROL

For the displacement controled calculation, a starting deformation line ν_1 is determined and with a variation on ν_1 , in such a way that the amplitude of the deformation vector is kept near enough constant, the out of balance forces are reduced to a minimum. For the deflected line ν_1 , the force R required to keep the system in equilibrium is calculated. It is assumed that

$R = \lambda F_0 + F$; λ denoting a load factor and
F_0 = load to be taken by the system
F = out of balance forces

The load factor is so determined that the out of balance forces do not perform any work in the actual deflection ν_1

$$(R - \lambda \cdot F_0) \, \nu_1 = 0 \text{ also } \qquad \lambda = \frac{R \cdot \nu_1}{F_0 \cdot \nu_1}$$

A deformation increment ν_2 is calculated with the help of the residual force F and the resultant deformation ($\nu_1 + \nu_2$) is normalised in such a way that it has nearly the same amplitude of ν_1 . Two possibilities are given below

1) The new vector ($\nu_1 + \nu_2$) is multiplied with a reduction factor a, so that the total deformation vector a($\nu_1 + \nu_2$) and the vector ν_1 are of equal length :

$$\sum_{i=1}^{n} \nu_{i1}^2 = a^2 \sum_{i=1}^{n} (\nu_{i1} + \nu_{i2})^2 \quad ; \qquad a = \sqrt{\frac{\Sigma \, \nu_{i1}}{\Sigma \, (\nu_{i1} + \nu_{i2})}}$$

2) The deformation ν_2 is the only one which is multiplied with the factor a, so that the lengths of the vectors ν_1 and ($\nu_1 + a \, \nu_2$) are equal.

$$\Sigma \, \nu_{i1}^2 = \Sigma \, \nu_{i1}^2 + 2 \, a \cdot \Sigma \, \nu_{i1} \cdot \nu_{i2} + a^2 \cdot \Sigma \, \nu_{i2}^2 \quad ; \quad a = - \frac{2 \cdot \Sigma \, \nu_{i1} \cdot \nu_{i2}}{\Sigma \, \nu_{i2}^2}$$

The force R_1 which holds the system in the deformed position ν_{new} is calculated. The improved deformation increment is : $\nu_{new} - \nu_1$. The newly obtained load deformation state is :

$$R - R_1 \longleftrightarrow \nu_{new} - \nu_1$$

The iteration is performed in the same way , as it is performed for the load control. The length of the deformation vector is kept constant only for the determination of the load deformation states.

EXAMPLES.

The following three examples demonstrate the efficency of the iteration.

Example 1

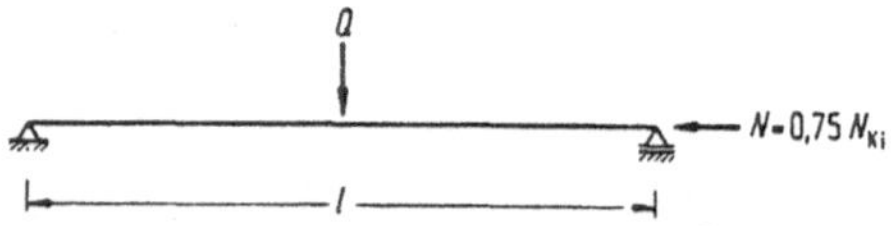

Mode of calculation:
- elastic with second order effects
- load control, with initial stiffness matrix.

Starting load factor $\gamma = 1.0$
Results of the iteration are shown in table 1, with
$|F| = \sqrt{\sum F_i^2}$; F = vector of the residual forces
(moment in KNm, forces in KN)
$|\nu| = \sqrt{\sum \nu_i^2}$; v = displacement vector
(angles in radians, deflection in mm)

Example 2

System and loading see example 1.
Mode of calculation
- elastic, large deflection theory, displacement control.

Starting load factor $\gamma = 1.0$
The stiffness matrix is only recalculated after the first step. The length of the displacement vector is kept constant (Table 2) during the iteration. The load factor is recalculated in every step and is equal to 0.57510 after the last step. As a control the Newton-Raphson method was used with recalculation of the stiffness matrix (for the geometrical non linearity) after every step. Here, it took 35 steps to reach a comparable accuracy which was already reached after 8 steps by the first method.

Example 3

Mode of calculation - First order theory, elastic-plastic calculation taking into account the spread of plasticity zones.
- Load control with initial stiffness method of the elastic system. Results for $\gamma = 1.3$ see table 3. Plasticity sets in middle of beam with a depth of 5,53 cm

Fig.2: Examples 1 and 2. N = 906 kN (Euler load)
$F = 149\ cm^2$; $I = 25170\ cm^4$ (IPB 300);
$E = 21\ 000\ kN/cm^2$; $l = 2400$ cm; $Q = 50$ kN;

Table 1. Results of the iteration for example 1

step	out of balance force	deflection
1	12.075088	0.4280
2	6.445579	1.1823
3	2.594143	1.6483
4	0.192282	1.6225
5	0.040878	1.6129
6	0.001178	1.6126
7	0.000393	1.6126

Table 2. Iteration for example 2

Results with orthogonalised load deflection states			Results of the iteration with the tangent stiffness matrix (Newton-Raphson method)		
step	out of balance forces	load-factor	step	out of balance forces	load-factor
1	2166.34	1.29	1	2166.34	1.29
2	1130.0	0.53496	2	1130.0	0.53
3	577.99	0.51869	3	171.06	0.61
4	255.16	0.54120	4	111.63	0.56
5	33.2458	0.58054	5	66.63	0.59
6	4.3549	0.57580	6	42.35	0.57
7	0.01838	0.57510	7	25.98	0.58
8	0.0000319	0.57510	8	16.31	0.57

Fig.3: Example 3. Table 3. Results for $\gamma = 1.3$. The elastic limit is just reached for $\gamma = 1.0$. Ultimate load $Q_{cr} = 1.5\ Q$.
$F = 120\ cm^2$; $l = 900$ cm ; $E = 21\ 000\ kN/cm^2$; $\beta_s = 24\ kN/cm^2$

step	out of balance force	deflection
1	42.957565	2.063
2	31.342958	2.102
3	18.423325	2.143
4	6.241236	2.159
5	0.610650	2.150
6	0.259967	2.151
7	0.051931	2.151
8	0.000929	2.151

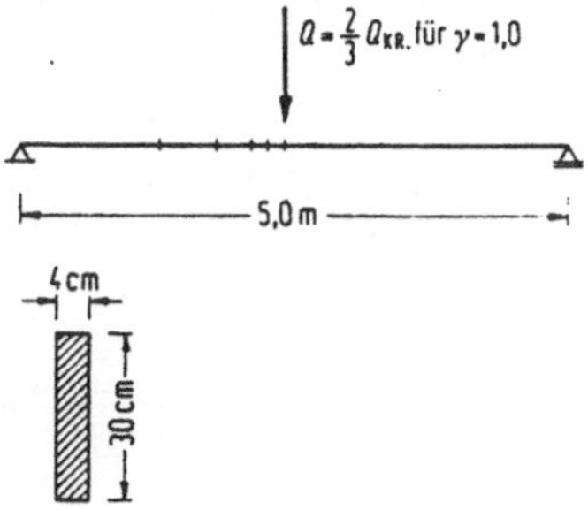

SOLUTION OF EIGENVALUE PROBLEMS

Approximations to eigenvalues can easily be postulated with the help of load deformation states, when a number of such states are known. In the case that two load deformation states $P_1 \longleftrightarrow V_1$ and $P_2 \longleftrightarrow V_2$ are known, it is possible to combine them linearly with the scalar factors x and y

$$P_a = P_1 + x\,P_2 \longleftrightarrow v_a = v_1 + x\,v_2$$
$$P_b = P_1 + y\,P_2 \longleftrightarrow v_b = v_1 + y\,v_2$$

The factors x and y could be choosen in such a way that the new load deformation states are good approximations to the eigen-vectors. If the states a and b are different eigenvectors, the following orthogonality conditions are fulfilled.

1) $P_a\ P_b = 0$; 2) $P_a\ v_b = 0$; 3) $v_a\ v_b = 0$.

In order to force the states a and b to be good approximations to two different eigen-vectors, the factors x and y are so determined that two of the three conditions given above are fulfilled. If one takes conditions 2) and 3), it follows :

$$P_1 v_1 + (x+y)\ P_1 v_2 + x\,y\ P_2 v_2 = 0$$
$$v_1 v_1 + (x+y)\ v_1 v_2 + x\,y\ v_2 v_2 = 0$$

with

$$A = P_1 v_1 \ ,\quad B = P_1 v_2 \ ,\quad C = P_2 v_2$$
$$D = v_1 v_1 \ ,\quad E = v_1 v_2 \ ,\quad F = v_2 v_2$$

using the variables $p = x + y$ and $q = x\,y$, one finds that :

$$p = -\frac{A\,F - C\,D}{B\,F - C\,E} \quad ; \quad q = -\frac{B\,D - A\,E}{B\,F - C\,E}$$

$$x = 1/2\ (\ p + \sqrt{(\ p^2 - 4\,q\ )})$$

$$y = 1/2\ (\ p - \sqrt{(\ p^2 - 4\,q\ )})$$

If one requires also that the vectors v_a and v_b be orthonormal then

$$P_a = a\ (\ P_1 + x\,P_2\ )\ ;\ v_a = a\ (\ v_1 + x\,v_2\ )$$
$$P_b = b\ (\ P_1 + y\,P_2\ )\ ;\ v_b = b\ (\ v_1 + y\,v_2\ )$$

with $|v_a| = |v_b| = 1$

$$a = \frac{1}{\sqrt{(\ D + 2\,x\,E + x^2 F)}} \quad ; \quad b = \frac{1}{\sqrt{(\ D + 2\,y\,E + y^2 F)}}$$

The eigen-values can be obtained from :

$$\lambda_a = \frac{P_a\ v_a}{v_a\ v_a} \quad \text{and} \quad \lambda_b = \frac{P_b\ v_b}{v_b\ v_b}$$

The residual forces vector for a load deflection state $P_i \longleftrightarrow v_i$ with the eigen-value λ_i is given by

$$W_i = 1/\lambda_i\, P_i - v_i$$

If the components of the residual forces are still too large, a new load deflection state can be determined using the deflection vector as a load vector in order to determine a new deflection vector. The load deflection state which is finally obtained can be in turn combined with the earlier ones.

The authors are grateful to the Deutsche Forschungsgemeinschaft for its financial support of this research.

REFERENCES

(1) Lopetegui,J.: Verfahren der orthogonalisierten Last-Verformungszustaende zur Loesung nichtlineare Probleme der Stabstatik.Diss. RWTH Aachen 1983.

(2) Sedlacek, G.; Lopetegui, J.; Saleh, A.; Stutzki, Ch.: Ein computerorientiertes Verfahren zur statischen Berechnung raeumlicher Stabwerke unter Beruecksichtigung nichtlinearer Effekte. Bauingenieur 60 (1985) 297-305.

(3) Saleh,A.; Traglastberechnung von raeumlichen Stabwerken mit grossen Verformungen und Plastizierung. Diss. RWTH Aachen 1982.

(4) Stutzki, Ch.:Traglastberechnung raeumlicher Stabwerke unter Beruecksichtigung verformter Anschluesse. Diss. RWTH Aachen 1982.

(5) Kaufels, G.: Beitrag zur nichtlinearen Berechnung ebener Flaechentragwerke unter besonderer Beruecksichtigung von Stahlbetonplatten. Diss. RWTH Aachen 1985.

(6) Stoverink, H.: Beitrag zur Ermittlung der Gesamtstabilitaet von Hallenrahmen unter Beruecksichtigung von Vouten und Steifen. Diss. RWTH Aachen 1985.

(7) Lopetegui, J.; Sedlacek, G.: Zur Loesung von Gleichungssystemen der Baumechanik mittels Last-Verformungszustaende. Bauingenieur 61 (1986)

A SECOND GENERATION STRUCTURAL SHAPE OPTIMIZATION CAPABILITY EMPLOYING A BOUNDARY ELEMENT FORMULATION

J. H. Kane

Worcester Polytechnic Institute
Worcester, Massachusetts 01609

Summary

References [1-3] describe a novel analytical formulation of numerical techniques and their implementation in a computer program to perform shape optimization research. This work involved coupling a boundary element fomulation for structural analysis and design sensitivity analysis with geometry, meshing, numerical optimization, and computer graphics. This paper describes research efforts to greatly enhance this technique including boundary element substructuring (matrix condensation) re-analysis techniques, and matrix sparsity exploitation. Implicit differentiation of the discretized boundary integral equations has been shown to produce a computationally efficient, accurate, and general form of design sensitivity analysis. In this paper it is shown that a form of boundary element substructuring (matrix condensation) can be utilized to form a dramatically economized "reduced problem." Expressions for the efficient calculation of required reduced matrix sensitivities are also given. Re-analysis strategies are presented that re-use unchanging portions of matrix triangular factorizations during the optimization process.

1.0 MULTI-ZONE BOUNDARY ELEMENT ANALYSIS

A boundary integral equation [4-13] generally called Somigliana's Identity can be formed from the elastostatic reciprocal theorem. This is accomplished by specializing one system of loading and response involved in the relationship to be that associated with the fundamental solution (Kelvin's Solution). Somigliana's Identity can then be discretized using boundary elements to form an algebraic relationship for each location and direction of the load point used in the fundamental solution. Such a technique is called a boundary element formulation. Strategic placement of the load point on or near each node in the model can be employed to produce a system of simultaneous algebraic equations. The boundary element formulation of the structural analysis problem can be used to develop the system of equations shown below [4-13].

$$A \; U = B \; T \tag{1}$$

In this equation U and T are respectively the vector of node point displacements and tractions. Renumbering and partitioning of the items is done so that all unknown displacements occur first, followed by all specified tractions

$$\left[\begin{array}{c:c} A_{11} & A_{12} \\ \hdashline A_{21} & A_{22} \end{array}\right] \left[\begin{array}{c} \bar{U}_1 \\ \hdashline U_2 \end{array}\right] = \left[\begin{array}{c:c} B_{11} & B_{12} \\ \hdashline B_{21} & B_{22} \end{array}\right] \left[\begin{array}{c} T_1 \\ \hdashline T_2 \end{array}\right] \tag{2}$$

Note that the bar symbol is used to indicate vectors containing items that are specified in advance. Manipulation of equation (2) is easily performed to put all unknown quantities on the right hand side.

$$\left[\begin{array}{c:c} -B_{11} & A_{12} \\ \hdashline -B_{21} & A_{22} \end{array}\right] \left[\begin{array}{c} T_1 \\ \hdashline U_2 \end{array}\right] = \left[\begin{array}{c:c} -A_{11} & B_{12} \\ \hdashline -A_{21} & B_{22} \end{array}\right] \left[\begin{array}{c} \bar{U}_1 \\ \hdashline T_2 \end{array}\right] \tag{3}$$

or

$$A \; x = b \tag{4}$$

For a multizone boundary element model one can write discretized boundary integral equations for each zone, along with relationships involving continuity of displacements and equilibrium of surface tractions at inter-zone boundaries. Expanding the accounting scheme for nodal point degrees of freedom, and strategically renumbering and partitioning information according to whether the nodes are all in zone one ($U_1^{(1)}$, or on the interface between zone one and zone two ($U_{12}^{(1)}$, $U_{12}^{(2)}$ and so on, we obtain matrices that are sparse and blocked. These equations can still be written symbolically as equation (4). Figure 2 illustrates the population of the left hand side of the boundary element system equations associated with the three zone boundary mesh shown in Figure 1. These are the types of matrices normally encountered in the equation solving step of conventional boundary element analysis. Sparse blocked equation solvers are normally employed to effectively handle the equation solving task. Block triangular decomposition of the left hand side matrix is performed, followed by forward reduction and back substitution of all required right hand side vectors.

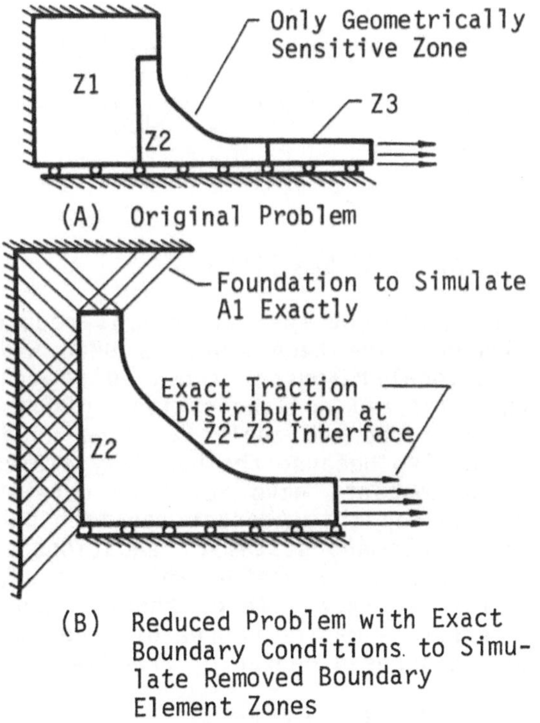

Figure 1: 3 Zone Model. Physical Interpretation of the Condensation of Geometrically Insensitive Zones on a Multi-Zone Boundary Element Analysis.

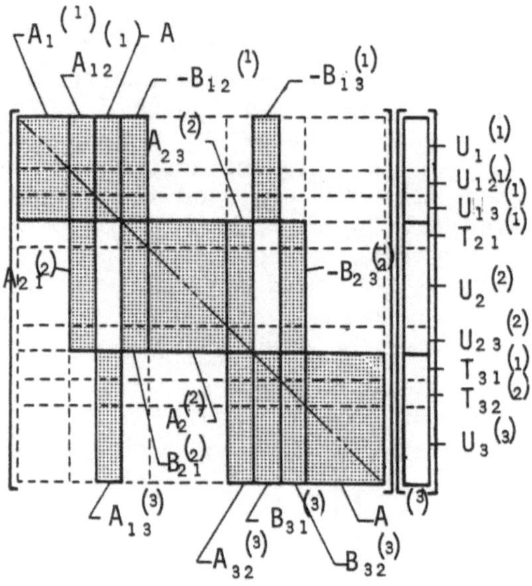

2.0 CONDENSATION OF BEA ZONES

In finite element analysis matrix reduction (condensation) is well known. The technique has been widely used to reduce the computational resources required to solve nonlinear problems with localized nonlinearities. Probably because BEA is perceived as a reduced analysis (i.e. no interior degrees of freedom) and also because the boundary element method is newer, condensation concepts have not been developed to such an extent. To develop this concept consider partitioning a single zone boundary element system of equations as shown below. In this equation F is a vector which accounts for either body force or initial stress (e.g. thermal expansion), the M subscript refers to a master degree of freedom and the C subscript refers to a condensed degree of freedom.

$$A \quad U = B \quad T + F \tag{5}$$

$$\left[\begin{array}{c:c} A_{MM} & A_{MC} \\ \hdashline A_{CM} & A_{CC} \end{array}\right] \left[\begin{array}{c} U_M \\ \hdashline U_C \end{array}\right] = \left[\begin{array}{c:c} B_{MM} & B_{MC} \\ \hdashline B_{CM} & B_{CC} \end{array}\right] \left[\begin{array}{c} T_M \\ \hdashline T_C \end{array}\right] + \left[\begin{array}{c} F_M \\ \hdashline F_C \end{array}\right] \tag{6}$$

The matrix equation shown in the lower partition of equation can be solved symbolically for U_c. This relationship can then

be substituted into upper partition of equation (6) to produce equation (8) shown below.

$$U_C = A_{CC}^{-1} A_{CM} U_M = A_{CC}^{-1} B_{CM} T_M = A_{CC} T_C = A_{CC}^{-1} F_C \qquad (7)$$

$$M_1 \; U_M = M_2 \; T_M = F_M = M_3 \; T_C = M_4 \; F_C \qquad (8)$$

where

$$M_1 = A_{MM} - A_{MC} \; A_{CC}^{-1} \; A_{CM} \qquad (9)$$

$$M_2 = B_{MM} - A_{MC} \; A_{CC}^{-1} \; B_{CM} \qquad (10)$$

$$M_3 = B_{MC} - A_{MC} \; A_{CC}^{-1} \; B_{CC} \qquad (11)$$

$$M_4 = - \; A_{MC} \; A_{CC}^{-1} \qquad (12)$$

Equation (8) can be referred to as the reduced matrix equation and the matrices M_1 through M_4 referred to as reduced matrices.

Equation (14) can then be called the expansion equation used to obtain deflections at the condensed degrees of freedom given the other information shown on the right hand side. Notice that this manipulation requires only the tringular factorization of the partition A_{cc} and not its actual inversion. Note further that the use of U and T in these equations is symbolic for the unknown and specified node point quantities respectively.

Now considering the three zone problem associated with the mesh depicted in Figure 1 there are various ways that one can pick master degrees of freedom. One possibility is to pick master modes at each zone interface, and write three systems of reduced equations.

$$M_1^{(1)} \; U_M^{(1)} = M_2^{(1)} \; T_M^{(1)} + F_M^{(1)} + \; M_S^{(1)} \; T_C^{(1)} + M_4^{(1)} \; F_C^{(1)}$$

$$M_1^{(2)} \; U_M^{(2)} = M_2^{(2)} \; T_M^{(2)} + F_M^{(2)} + \; M_3^{(2)} \; T_C^{(2)} + M_4^{(2)} \; F_C^{(2)}$$

$$M_1^{(3)} \; U_M^{(3)} = M_2^{(3)} \; T_M^{(3)} + F_M^{(3)} + \; M_3^{(3)} \; T_C^{(3)} + M_4^{(3)} \; F_C^{(3)}$$

or

$$M_1^{(1)}\ U_M^{(1)} \ = \ M_2^{(1)}\ T_M^{(1)} \ + \ F_M^{(1)} \ + \ V^{(1)} \qquad (13)$$

$$M_1^{(2)}\ U_M^{(2)} \ = \ M_2^{(2)}\ T_M^{(2)} \ + \ F_M^{(2)} \ + \ V^{(2)} \qquad (14)$$

$$M_1^{(3)}\ U_M^{(3)} \ = \ M_2^{(3)}\ T_M^{(3)} \ + \ F_M^{(3)} \ + \ V^{(3)} \qquad (15)$$

Expanding the size of these equations and utilizing the interzone compatibility and equilibrium relations, one obtains the assembled reduced system of equations for the three zone problem.

$$\begin{bmatrix} M_{1'12}^{(1)} & M_{1'13}^{(1)} & -M_{2'12}^{(1)} & 0 & -M_{2'13}^{(1)} & 0 \\ M_{1'21}^{(2)} & 0 & M_{2'21}^{(2)} & M_{1'23}^{(2)} & 0 & -M_{2'23}(2) \\ 0 & M_{1'31}^{(3)} & 0 & M_{1'32}^{(3)} & M_{2'31}(3) & M_{2'32}^{(3)} \end{bmatrix} \begin{bmatrix} U_{M'12}^{(1)} \\ U_{M'13}^{(1)} \\ T_{M,21}^{(1)} \\ U_{M'23}^{(2)} \\ T_{M,31}^{(1)} \\ T_{M'32}^{(2)} \end{bmatrix}$$

$$= \begin{bmatrix} F_{M'12}^{(1)} \\ F_{M'13}^{(1)} \\ F_{M'21}^{(2)} \\ F_{M'23}^{(2)} \\ F_{M'31}^{(3)} \\ F_{M'32}^{(3)} \end{bmatrix} + \begin{bmatrix} V_{M'12}^{(1)} \\ V_{M'13}^{(1)} \\ V_{M'21}^{(2)} \\ V_{M'23}^{(2)} \\ V_{M'31}^{(3)} \\ V_{M'32}^{(3)} \end{bmatrix} \qquad (16)$$

Figure 3 illustrates the sparse blocked nature of this system of equations. Comparison of the figure with Figure 2 reveals the absence of all blocks associated with degrees of freedom which are not on zone interfaces. Also in this illustration is the dramatic reduction in the size of the system matrix which must now be factored. For evolving shapes, matrix blocks associated with unchanging portions can be condensed once in the beginning of the shape optimization, saved on disk, and simply reused in subsequent analyses.

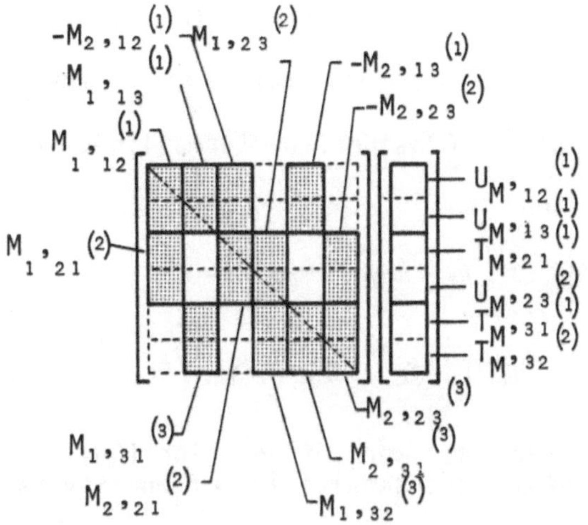

Figure 3: Sparse Blocked Condensed Left Hand Side Matrix Population Associated with a 3 Zone Boundary Element Analysis

3.0 MATRIX CONDENSATION AND DESIGN SENSITIVITY ANALYSIS

As shown in references [1-3] design sensitivity analysis requires the computation of rates of change of node point displacements and tractions. Partial derivatives of these quantities with respect to design variable X_L can be obtained by implicit differentiation of the boundary element system equations as shown below.

$$\left[A\, x \;=\; b \right]_{,L} \tag{17}$$

$$A_{,L}\, x \;+\; A\, x_{,L} \;=\; b_{,L}$$

or

$$A\, x_{,L} \;=\; b_{,L} \;-\; A_{,L}\, x \tag{18}$$

Note that equation (18) is a system of equations with the same left hand side matrix as equation (4). Notice further that these equations can also symbolically stand for the reduced system of equations shown in equation (16). Therefore, the sensitivities of the nodal point unknowns at master degrees of freedom $(X_{,L})$ can be found without the need for and additional matrix factorizations, provided the factors of A are saved from the last analysis step. Of course, one needs to be able to form $b_{,L}$ and $A_{,L}$ in equation (18). For this to be true, recall from equation (16) that the matrix A is composed of reduced matrices $M_{j,LM}^{(i)}$. Consider the formation of the first entry of that equation.

$$\left[M_1 = A_{MM} - A_{MC} A_{CC}^{-1} A_{CM}\right]_{,L} \qquad (19)$$

$$M_{1,L} = A_{MM,L} - A_{MC,L} A_{CC}^{-1} A_{CM} - A_{MC} \left[A_{CC}^{-1} A_{MC}\right]_{,L} \qquad (20)$$

Inspection of the last term of equation (20) reveals the apparent need to obtain the sensitivity of the inverse of A_{CC}. This would clearly be a computationally burdensome task. However, one can obtain the required last term in equation (20) without matrix inversion by considering the product $A_{CC}^{-1} A_{MC}$.

$$A_{CC}^{-1} A_{MC} = D$$

$$A_{MC} = A_{CC} D$$

$$A_{MC,L} = A_{CC,L} D + A_{CC} D_{,L}$$

or (21)

$$A_{CC} D_{,L} = A_{MC,L} - A_{CC,L} D$$

This shows that the term $\left[A_{CC}^{-1} A_{MC}\right]_{,L}$ (which is $D_{,L}$) can be formed by a forward reduction and backward substitution of the right hand side vector shown in equation (21) using the triangular factors of A_{CC} which have already been computed in the matrix condensation process. The other entries of $A_{,L}$ can be obtained in an analogous fashion. Thus, by adopting a procedure which follows the approach outlined, one can develop a computationally efficient design sensitivity analysis that can work very effectively with reduced models. Notice that the process described above allows for the computation of traction and displacement sensitivities at master nodes. For sensitivities at condensed nodes, one would have to do the computations that result from differentiating an expansion equation like equation (7). Note further that for boundary element zones that did not contain points associated with objectives and constraints, the calculation of structural sensitivities at condensed nodes could be totally avoided.

4.0 SPARSITY EXPLOITATION

The matrices involved in design sensitivity analysis of objects with geometrically insensitive zones exhibit significant additional sparsity. This additional sparsity is beyond that which occurs in the analysis. At the matrix block level, sensitivities of geometrically insensitive zones need never be integrated, assembled or multiplied. Within each matrix block there can be additional sparsity. For boundary element zones that are only partially geometrically sensitive, the entries in the matrix blocks will still contain zero entries. A sparsity accounting can be easily performed to determine if load points or elements are geometrically sensitive. For insensitive cases, the numerical integration and assembly or contributions to the matrix block can be skipped.

5.0 RE-ANALYSIS TECHNIQUES

Re-analysis can refer to any attempt to look for computations in repetitive analyses which are the same from one analysis to the next. The values computed in the first analysis can then be saved and re-used in the subsequent analyses, thereby saving recomputations with identical results. Certain re-analysis techniques are quite straightforward. For example, a boundary element zone that has not changed from one analysis to the next, need not have its zone system matrices re-integrated the second time. Instead, these matrices can be saved and re-assembled into the overall system matrices along with the re-computed matrices associated with modified zones. Of course, this would not be possible without the multizone analysis capability. The concept of matrix condensation described is actually an elaboration on this idea.

Most re-analysis research has focused on the equation solving issue. This is because of the large portion of the total analysis cost usually associated with this task. To facilitate the discussion of the impact of re-analysis techniques for the computation of the triangular factors of a partially modified matrix, consider again the three zone boundary element model and its associated left hand side matrix shown in Figure 4. Suppose, for the sake of this discussion, that only zone three has changed from the first analysis to the second. The steps in the second sparse blocked triangular factorization process begin as described below. Note that the notation used in this discussion symbolizes all matrix blocks in the left hand side as A_{ij}, with i being the block row and j being the block column.

Factor: $A_{11} = L_{11} U_1$

Solve: $A_{11} D_{12} = A_{12}$

Form: $A_{i2} = A_{i2} - A_{i1} D_{12}$; $i = 2,3$

Solve: $A_{11} D_{13} = A_{13}$

Form: $A_{i3} = A_{i3} - A_{i1} D_{13}$; $i = 2,3$

Solve: $A_{11} D_{14} = A_{14}$

Form: $A_{i4} = A_{i4} - A_{i1} D_{14}$; $i = 2,3$

Solve: $A_{11} D_{17} = A_{17}$

Form: $A_{i7} = A_{i7} - A_{i1} D_{17}$; $i = 2,3$

Factor: $A_{22} = L_{22} U_{22}$

Solve: $A_{22} D_{23} = A_{23}$

Form: $A_{i3} = A_{i3} - A_{i2} D_{23}$; $i = 3,4,5,6$

Solve: $A_{22} D_{24} = A_{24}$

Form: $A_{i4} = A_{i4} - A_{i2} D_{24}$; $i = 3,4,5,6$

Solve: $A_{22} D_{27} = A_{27}$

Form: $A_{i7} = A_{i7} - A_{i2} D_{27}$; $i = 3,4,5,6$

Factor: $A_{33} L_{33} U_{33}$

Solve: $A_{33} D_{34} = A_{34}$

Form: $A_{i4} = A_{i4} - A_{i3} D_{34}$; $i = 4,5,6,7,8,9$

The important fact to notice is that it is not until the formation of A_{74} that any new entries in the matrix blocks occur.

That is to say, the matrix blocks processed before the formation of A_{74} will be exactly the same as those in the first

equation solve. Therefore, all of the steps preceeding the formation of A_{74} could have been avoided. Figure 4 points out the blocks of the left hand side matrix which can simply be saved and re-used, and those which must be recomputed. Of course, one observes that, if zone one is the modified zone instead of zone three, the matrix becomes "contaminated" immediately and the entire factorization must be done from the beginning. Simple zone renumbering could obviously be employed to force changing zones to be numbered last. This strategy, however, generally will be in conflict with zone numbering schemes to reduce fill-in and computation in the original analysis. Note further that even after A_{74} is formed, many additional computations can still be saved. If the D_{ij} matrices are saved for i = 3,4,..., 9 and j = 4,5,6, for example, they need not be recomputed.

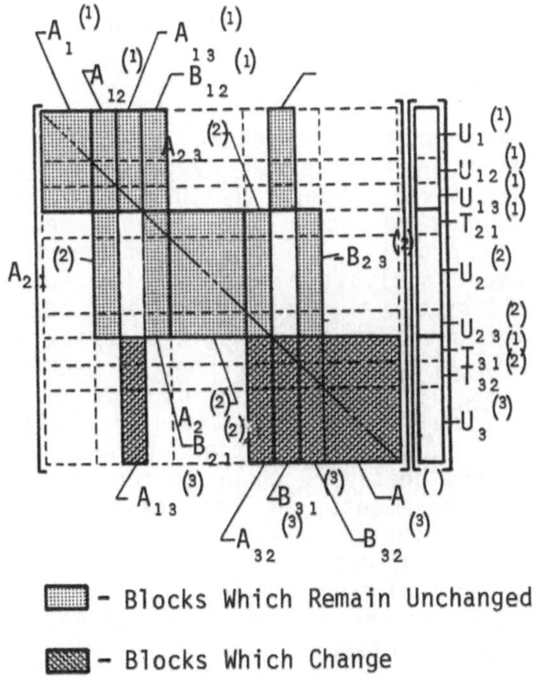

Figure 4: Changed and Unchanging Matrix Blocks in an Example Boundary Element Re-Analysis. Only last zone is changed.

6.0 REDUCED RE-ANALYSIS

Within the context of shape optimization using the matrix condensation concept, the issue of strategic numbering of the boundary element zones presents investigators with some significant opportunities. Normally, ordering the zones based upon re-analysis concerns will be in conflict with ordering the zones to reduce storage and computation in the original matrix factorization step. However, with matrix condensation techniques present, the overall system matrix to be factored is drastically reduced. Therefore, ordering for reduced computation in the factorization step should have only secondary influence, while ordering zones to populate the evolving system matrices to minimize effort in re-analysis computations producing the more significant overall economy.

Actual experience with the proposed combination of re-analysis and matrix condensation in the shape optimization context will be used to develop a logical basis and rules of thumb concerning the zone numbering issue.

7.0 CONCLUSIONS

Multi-zone boundary element analysis is a powerful tool to be incorporated in an overall system for shape optimization research. The novel concept that has been demonstrated to be viable in other references [1-3] is now being extended to a wider class of problems and greatly enhanced. The multi-zone boundary element analysis allows for the natural partitioning of the overall system matrices into blocks associated with each zone. The exploitation of dramatic sparsity in the sensitivities of matrix blocks corresponding to geometrically insensitive zones was discussed. The concept of boundary element substructuring (matrix condensation) was also shown to hold potential for substantial computational economy. Finally, re-analysis techniques were shown to represent another area for substantial improvements in computational performance when incorporated in overall shape optimization procedures. The fundamental research described in part in this paper is aimed at the development of more computationally efficient techniques that will allow for the use of shape optimization on orders of magnitude larger problems.

REFERENCES

1. Kane, J.H., 'Optimization of Continuum Structures Utilizing a Boundary Element Formulation,' Ph.D. Dissertation, University of Connecticut, Storrs, CT (1986).

2. Kane, J.H., 'Shape Optimization Utilizing a Boundary Element Formulation,' Proceedings BETECH86, Boundary Element

Technology Conference, MIT, Cambridge, MA (1986).

3. Kane, J.H. and Saigal, S., 'Design Sensitivity Analysis of Solids Using BEM,' Journal of Engineering Mechanics, ASCE, submitted for review.

4. Banerjee, P.K. and Butterfield, R., Boundary Elements in Engineering Science, McGraw-Hill, London (1981).

5. Banerjee, P.K. and Butterfield, R., Eds., Developments in Boundary Element Methods--Vol. I, Applied Science Publishers Ltd., England (1979).

6. Banerjee, P.K. and Shaw, R.P., Eds., Developments in Boundary Element Methods--Vol. II, Applied Science Publishers Ltd., England (1982).

7. Banerjee, P.K. and Muhkerjee, S., Eds., Developments in Boundary Element Methods--Vol. III, Applied Science Publishers Ltd., England (1974).

8. Brebbia, C.A. and Walker, S., Boundary Element Techniques in Engineering, Newnes-Butterworths, England (1980).

9. Brebbia, C.A., The Boundary Element Methods for Engineers, Pentech Press, England (1978).

10. Brebbia, C.A., Ed., Boundary Element Methods--Proceedings of the Third International Seminar, Springer-Verlag (1981).

11. Brebbia, C.A., Ed., Boundary Element Methods in Engineering--Proceedings of the Fourth International Seminar, Springer-Verlag (1982).

12. Cruse, T.A. and Rizzo, F.J., Eds. Boundary-Integral Equation Method: Computational Applications in Applied Mechanics, ASME (1975).

13. Crouch, S.L. and Starfield, A.M., Boundary Element Methods in Solid Mechanics, Allen and Unwin Ltd., London (1983).

AN EXPERIMENTAL AND THEORETICAL STUDY OF SOIL-STRUCTURE INTERACTION IN THE CASE OF A SHALLOW FOUNDATION MODEL

S. LABANIEH, M. BOULON
INSTITUT DE MECANIQUE DE GRENOBLE
B.P. n° 68
38402 - St Martin d'Hères Cédex
France

SUMMARY :

A shallow, circular, rigid and rough foundation model instrumented with eleven normal and tangential stress cells equiped with electrical strain gauges is first detailed. Two loading tests corresponding to two densities of the soil (Sand of Hostun) are then presented. Results of the interaction soil-structure and stresses and deformations in the mass of the soil are precised and analysed and the influence of the density is pointed out.

A numerical model using the finite element method with T3 elements and axisymetrical geometry is then presented. The soil behaviour considered is an isotropic hypoelasticity, function of the stress level. The incremental non linearity is solved by a classical initial stress method. Two extremal cases were calculated : smooth foundation and a rough one. Comparison with experimental results is made and results are analysed.

PART I : The shallow foundation model and experimental results

The shallow foundation model |1| is a circular (40 cm diameter), rigid and rough block of duralumin instrumented, along a diameter in a tunnel opened in its lower part, with eleven normal and tangential stress cells (fig. 1). Each of these cells has the form of I and is equiped with three pairs of electrical strain gauges (fig. 2), J3 for the normal stresses and J1 and J2 for the tangential ones. The response of J3 to normal stresses is directly obtained by an appropriate calibration. As for J1 and J2, their responses are due to the

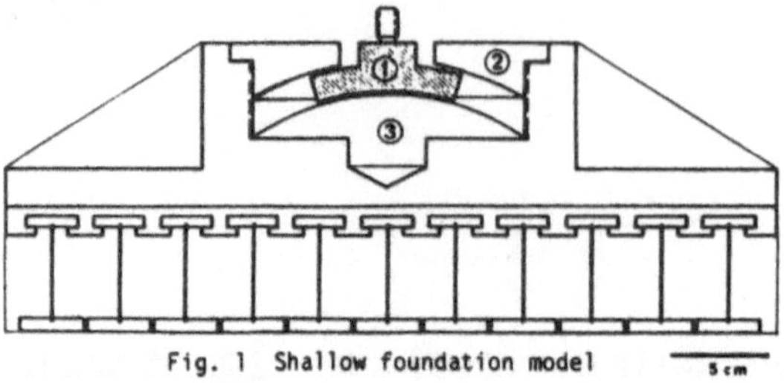

Fig. 1 Shallow foundation model

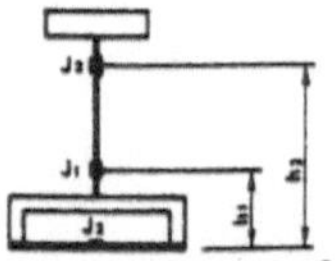

Fig. 2 Diagram of an I cell

application of a couple C - result of the normal stress gradient - and the moments Th1 and Th2 due to the tangential effort T. The difference between the indications of J2 and J1 gives T(h2-h1) and appropriate calibrations permit to determine for each I cell the distance (h2-h1) and then T. Total load and settlement are also measured by appropriate electrical cells. Measurements in the mass of the soil are made by means of total stress cells working on the back pressure principal for the normal stresses (Glötzl cells) and induction displacement cells (Bizon cells) for the deformations. These results should be considered rather qualitatively than quantitatively. Finally, the shallow foundation model is instrumented with a ball-joint having its center of rotation at the interface soil-structure. Thus it permits to avoid the perturbation of the measurement of the total effort due to moment rise which comes from the unavoidable non-homogeneous density of the soil.

Two loading tests were performed, with the sand of Hostun at two relative densities (Dr = 18 % and 81 %) in a 150 cm diameter and 200 cm deep chamber. The loading was realized by means of a mechanic jack at the rate of about 1 mm/mn. For the dense sand test, only the measurements of the soil-structure interaction were made. For the loose sand test the disposition of the Glötzl and Bizon cells are shown in fig. 3.

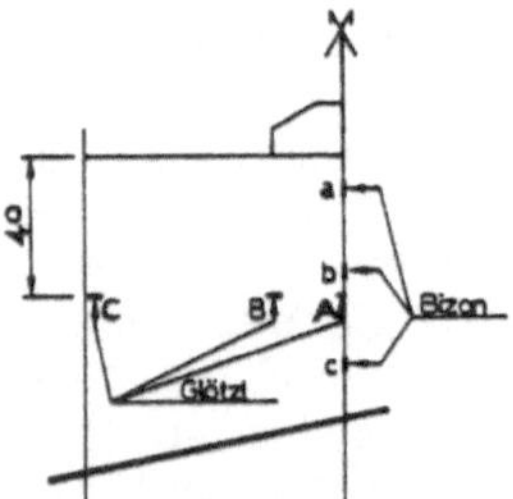

Fig. 3 Disposition of Glötzl and Bizon cells

In fig. 5 we present the total load -settlement curves for the two tests where we remark the large influence of the density. The dense sand test was stopped at 0.7 cm settlement as a security for the normall stress cells. The dense sand test curve being almost linear (small mobilized deformations), while loose sand test curve shows an asymtotic value giving experimental values of Nq and N |Le Gall 2| of about 94 and 47 respectively values corresponding to a friction angle comprised between 32° and 34°, correct values of for this density.

In fig. 6 we present the repartition of normal stresses at the interface soil-structure for the two tests for the precised respective settlements. The stresses represent in both cases the mean value given by symmetrical I cells. We remark that the repartition of normal stresses of the loose sand test is almost uniform while that of the dense sand test presents large variations density heterogeneities. The integration of these stresses at the end of each loading compared with the total load gives a difference less than 2 % in the case of loose sand test and less than 4 % in the case of dense sand test.

In fig. 7 we present the repartition of tangential stresses at the interface soil-structure for the two tests also and for the precised respective settlements. These stresses represent also the mean value given by symmetrical I cells. We remark also that the repartition of these stresses for the loose sand test is almost uniform while that of the dense sand test presents large variations. δ , the inclination of the stresses at the interface determined from the preceeding two figures varies from 11° to 41° for the loose sand test and from 3° to 32° for the dense sand test. The general tendancy for small values of δ is due to the large flexibility of the vertical plates of the I cells.

As for the evolution of vertical stresses in the soil (fig. 8) we note that for cell B and C there is almost no evolution, while cell A presents at the end of the settlement an asymtotic value which is about 96 % of the stress at the interface and at the center of the foundation model. If vertical stresses at 40cm in the axis beneath the foundation model presented an asymptotic value, deformations, 6cm higher (cell a), continued their evolution as fig. 9 shows.

PART II : Numerical modelling and experiment computation comparison.

The aim of this part of the communication is to test the accuracy of a simple numerical model in order to appraise the possible deviations from the experiment and the effect of some hypothesis, in the case where a prediction is searched. The classical finite element method is used, involving displacement unknowns, and deriving from the principle of the virtual work. Small displacements, strains and rotations are assumed ; consequently, the stress tensor is the so called Cauchy tensor, available for most of the civil engineering applications.

The spatial discretization, shown in figure 4, was carried out using "pairs of linear, axisymmetrical elements"

(T3). The rheological non linearities were based on the value of the stresses and strains at the gravity center of each pair; this technique gave us the same approximation level as an analysis performed with tetragonal elements (Q4), after a set of previous tests. The size of the element is obviously adapted to the strain/stress gradients.

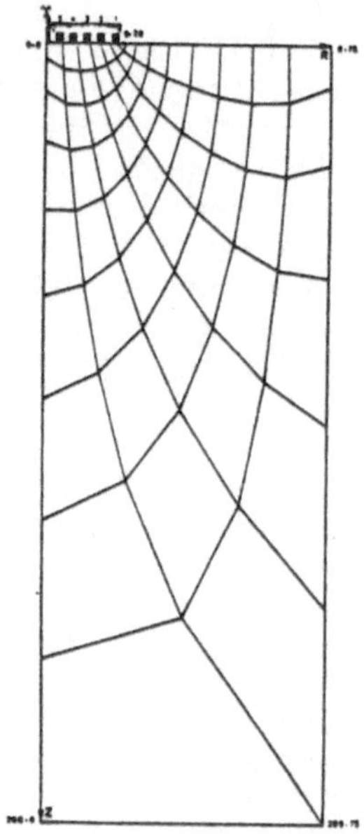

Fig. 4 Mesh of the experimental model

This choice, in full agreement with the common physical meaning, can sometimes introduce some instability in the computation ; in fact, one knows, that large elements disposed in the critical part of a domain (like near the corner of a rigid foundation) have a stabilizing effect on the possible divergence.

The constitutive equation modelling the behaviour of the sand, is an isotropic hypoelasticity |3| deriving from the DUNCAN's model, function of the mean stress level and of the deviatoric stress level for a given initial variation ; the mean stress level (σ i) is governing the "pseudo-Young modulus" $E^{\circ}_{\sigma_i}$ from an isotropic stress state, after the classical KONDNER's law :

$$E^{\circ}_{\sigma_i} = E^{\circ}_{réf} \left(\sigma_i / \sigma_{i\,réf} \right)^n$$

with $E^{\circ}_{réf}$: young modulus at the isotropic stress $\sigma_{i\,réf}$
n : exponent

The deviatoric stress level q = σ_1 - σ i acts through the "pseudo-modulus" $E^{q}_{\sigma_i}$ which decreases along an axisymmetrical biaxial stress path, from the isotropic state :

$$E^{q}_{\sigma_i} = E^{\circ}_{\sigma_i} \left(\frac{q_{lim} - q}{q} \right)^2$$

where q_{lim} is the limit deviatoric stress level, during plastic flow. The friction angle of the material is assumed to be constant, in order to simplify the model. Hence, the expression of q_{lim} :

$$q_{lim} = c + tg\phi \cdot \sigma_i$$

C and tg ϕ being respectively the effective cohesion and friction angle of the material. In the range of the small strains (engineering purpose) the "pseudo Poisson ratio" is also choosen as constant. As first step from this point, one have to generalize this element behaviour to paths starting from a non isotropic stress state ($\sigma_1 > \sigma_2 > \sigma_3$) ; The previous formulations remain available, assuming that the "intermediate stress in the limit state" (extrapolation of the actual load into the plastic flow) doesn't affect the phenomena. The second step of this generalization concerns the tridimensionnal loading : the principal axis of the actual stresses are taken as the local material axis. Moreover, a given tridimensionnal loading is considered as an interpolation between three element loading (see |3|). The simplicity of this model comes essentially from the assumption that parameters in compression and extension are the same, leading to a unique "tensorial zone". The identification laboratory tests were conventionnal triaxial tests in compression, in the range of isotropic pressure interesting our experiment (100 kPa). Therefore, it must be noted that very low pressures are reached in certain parts of the domain, during the loading ; not any theory could specify the true parameters of any model in this range of pressure !

Let be $\underline{\underline{D}}$ the local incremental behaviour matrix so as

$$\underline{\Delta\sigma} = \underline{\underline{D}} \cdot \underline{\Delta\varepsilon}$$

with $\underline{\Delta\sigma}$: incremental stresses, and $\underline{\Delta\varepsilon}$ incremental strains. The non linearity of $\underline{\underline{D}}$ is solved classically by iterations using the initial stress method. Applying this method, we had no problem excepted near the border of the foundation where horizontal tensile strains were observed. Small increments at the beginning of the numerical loading process were efficient to overcome this obstacle.

A mean number of 20 increments was necessary to reach a relative settlement W/B of 0.1.

The load-settlement curve is represented in figure 5, for loose and dense sand. Two extremal assumptions were considered for the calculations : smooth and rough foundation. One can observe first that the influence of this hypothesis is of importance ; the relative position of experiment and simulations shows clearly that a relative sliding occurs between soil and foundation, this sliding being not taken into account in the simulations ! The computation is in good agreement with the experiment for

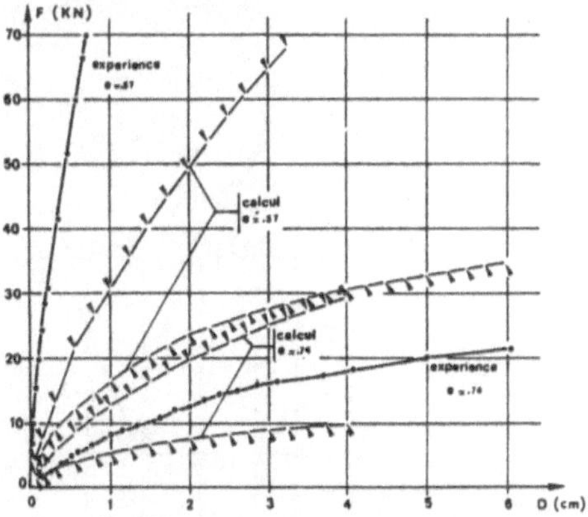

Fig. 5 Comparison computation-experience for total load-settlement curves

the loose sand. On the contrary a divergence is exhibited in the case of the dense sand. In our opinion, the reason does not come from the dilatancy in the mass of the soil, because the difference begins since the very small strains ; this deviation can arise both from the extension parameters, and from the dilatancy due to the relative sliding soil-foundation, these two aspects being not modelized.

The normal stress acting on the foundation is shown in figure 6, for two characteristic settlements (D = 0.57 cm for the dense sand ; D = 4.0 cm for the loose sand). The shear stress is plotted in figure 7 for the same displacements. The comments remain the same as for figure 5 ; the question of the soil foundation contact hypothesis is again asked.

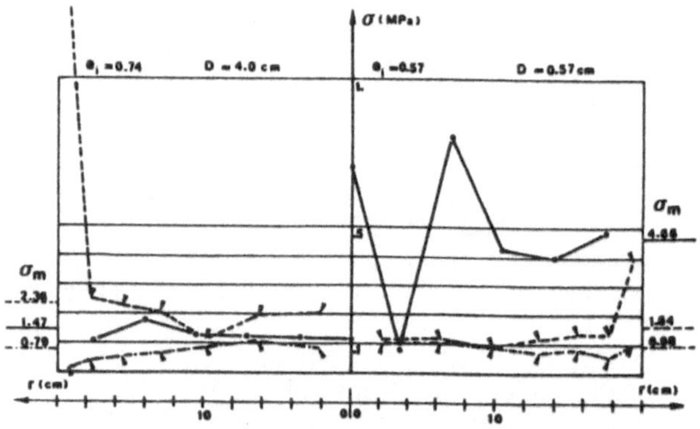

Fig. 6 Comparison computation-experience for the repartition of normal stresses at the interface soil-structure

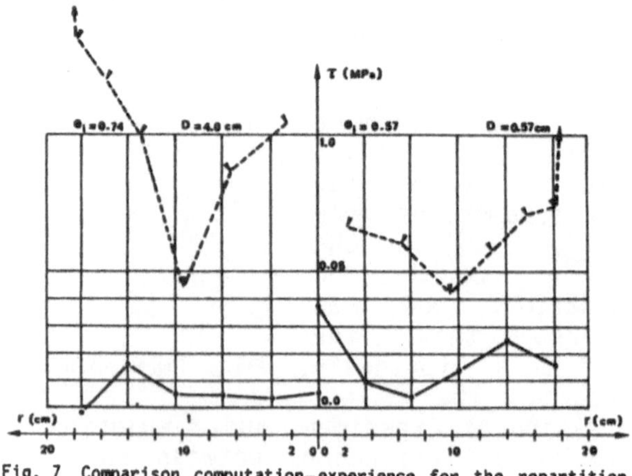

Fig. 7 Comparison computation-experience for the repartition of tangential stresses at the interface soil-structure

The normal stress in the soil at the relative depth of D/B = 1 on the axis of the foundation (A), under its border (B) and near the boundary of the chamber (C) are represented in figure 8. The agreement between experiment and simulation (rough foundation) is very good for (A). The influence of the contact conditions between the soil and the chambers boundary appears on the curve (C).

The vertical strains under the axis of the foundation at the depth 10 cm (a), 33 cm (b), and 60 cm (c) are shown in figure 9. The strains are more localized in the experiment than in the calculation (rough plate). The influence of the discretization (mesh) should be a governing factor of the result.

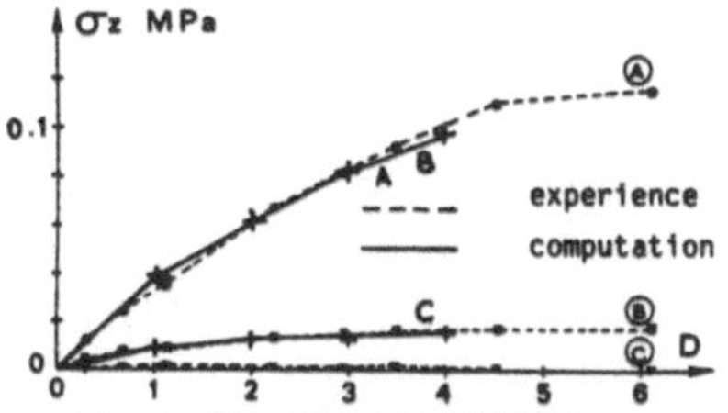

Fig. 8 Comparison computation-experience for the evolution of normal stresses in the mass of the soil (Glötzl cells)

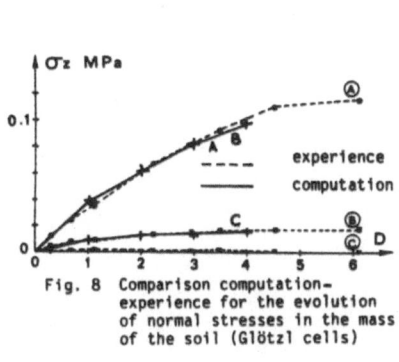

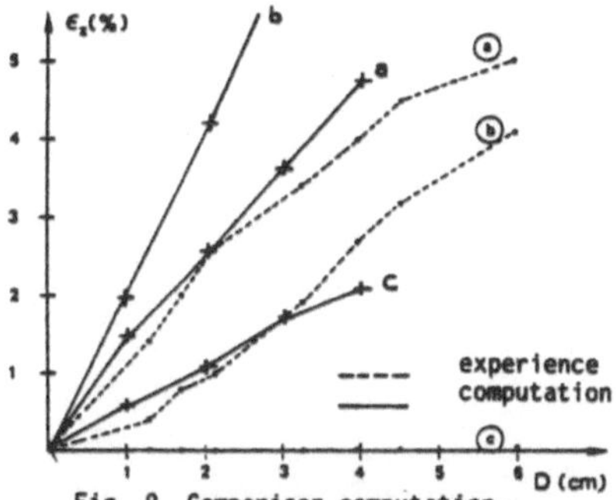

Fig. 9 Comparison computation-experience for the evolution of vertical deformations in the mass of the soil (Bizon cells)

CONCLUSION :

The experimental results show that the non-homogeneity of the soil affects largely the results (stress repartition curves), that stress cells should be of an important rigidity (tangential stresses) and that right under the shallow foundation model the soil does not form a rigid corner even at relative settlement W/B of 1.5 (fig.9)

From the numerical point of wiew, this example of comparison shows clearly that a numerical simulation needs not only a good finite element scheme and a non linear constitutive equation : firstly, the modelized behaviour should be validated on some different possible stress paths ; here, we know that the extension characteristics were ignored. Secondly, the questions of the soil-structure and of the soil-chambers boundary must be taken correctly into account.

REFERENCES :

1. LABANIEH S. , "Modélisations Non-Linéaires de la Rhéologie des Sables et Applications", Thèse de Doctorat d'Etat, Grenoble, 1984.
2. LE GALL Y., "Contribution à l'Etude de la Force Portante de Fondations Circulaires Peu Profondes", Thèse de Docteur-Ingénieur, E.N.P.C., Paris, 1981.
3. CHAU B., DARVE F., LANIER J., "Incrementally non linear and other constitutive relations : a comparative study for practical use", proc. of ECONMIG 86, vol 2, Stuttgart, september 1986.

D45/1

DYNAMIC ANALYSIS OF GENERAL THIN SHELLS

S.NAOMIS & P.C.M. LAU

Civil Engineering Department
University of Western Australia, Nedlands, W.A. 6009,
Australia.

Abstract
This paper presents a method for computing the natural frequencies and mode shapes of a general thin shell.The surface of a thin shell is described by a set of α-β curvilinear coordinate lines which may not be orthogonal.Only three translational components of displacement at intersections of two curvilinear lines are required for establishing the computational model.The method uses the large memory of a microcomputer to allow the process of formation of the stiffness and consistent mass matrices, and iterations of eigenvectors to be carried out entirely in the core memory.The method also uses curvilinear finite differences[1] to approximate the complete displacement and strain equations developed by Flugge [2] for thin shell structures. Numerical examples are presented to illustrate the applications and verify the acurracy of the method.

Introduction
In the dynamic analysis of a thin shell of arbitrary geometry,a high resolution discretisation is central to model accurately the local geometric changes and approximate sufficiently the displacement and deformation characteristics of the thin shell.This approach demands a large amount of computing storage for the stiffness and consistent mass matrices, and requires a long computing time for solving the related eigen-problem[3].One approach to reduce such a demand of computing effort is to reduce the number of degrees of freedom required at a node for the numerical formulation by including only the translational displacements and excluding the rotational displacements, so as to reduce the total computing storage requirement.The energy based finite difference method uses such an approach and one such method was applied to solve vibration and non-linear problems of shells of revolution [4].The search of improvement of computing strategy along this line leads to the development of curvilinear finite differences which relaxes the regular grid restriction and makes the finite difference code readily programmable.This method was applied to solve some orthotropic plate bending problems[5] and non-linear thin shell problems[6].

General theory:
The principal of virtual displacement statement for a thin shell (Fig.1) undergoing free vibration can be expressed in the form:

$$\int_{v} \delta(\eta_{\alpha\beta})^{t}(\sigma^{\alpha\beta})\, d\underline{v} + \int_{v} \delta(v_{i})^{t}\rho(\ddot{v}^{i})\, d\underline{v} = 0 \qquad (1)$$

where α,β are curvilinear coordinate lines,$\{\sigma^{\alpha\beta}\}$ contains the contravariant stress components, $\{\eta_{\alpha\beta}\}$ the covariant components of the general strain tensor , $\{v_i\}$ and $\{v^i\}$ the covariant and contravariant components of the general displacement components respectively.

These three dimensional stress and strain components can be reduced into a set of pseudo two dimensional expressions by applying the Kirchoff's hypothesis to simplify the displacements and deformation equations of a thin shell [2] .The modified equations for strain components are as follows :

$$\eta_{\alpha\beta} = \mu^{\gamma}_{\alpha}\mu^{\delta}_{\beta}\epsilon_{\gamma\delta} - 1/2\, z\,\kappa_{\gamma\delta}(\delta^{\gamma}_{\alpha}\mu^{\delta}_{\beta} + \delta^{\gamma}_{\beta}\mu^{\delta}_{\alpha}) \qquad (2a)$$

$$\epsilon_{\alpha\beta} = 1/2\,(u_{\alpha}\|_{\beta} + u_{\beta}\|_{\alpha} - 2\,w\,b_{\alpha\beta}) \qquad (2b)$$

$$\kappa_{\alpha\beta} = u_{\gamma}\,b^{\gamma}_{\beta}\|_{\alpha} + u_{\gamma}\|_{\alpha}\,b^{\gamma}_{\beta} - u_{\beta}\|_{\gamma}\,b^{\gamma}_{\alpha} + w\|_{\alpha\beta} + w\,b^{\gamma}_{\alpha}\,b_{\gamma\beta} \qquad (2c)$$

$$\mu^{\beta}_{\alpha} = \delta^{\beta}_{\alpha} - z\,b^{\beta}_{\alpha} \qquad (2d)$$

where $\varepsilon_{\gamma\delta}$ is the in-plane middle surface strain tensor, $\kappa_{\gamma\delta}$ is the change in middle surface curvature tensor , $\mu_\alpha^{\ \gamma}$ is termed the shift tensor, $b_\alpha^{\ \beta}$ denotes the mixed-variant components of the curvature tensor and $\delta_\alpha^{\ \beta}$ is the Kronecker delta. In addition, the notation ' $\|$ ' denotes the covariant derivative of a tensor with respect to the α–β coordinate system.

Numerical approximation

The method consists of discretising the middle surface of a thin shell into computational grids using an assembly of nine node curvilinear meshes .The values of x,y and z coordinates are assigned to every node.Within a curvilinear mesh (Fig. 3), the x,y and z coordinates of these nine nodes are used to approximate the surface of the thin shell in this local region.The approximated surface is quadratic and is suitable to model a local curved surface covering a part of the thin shell.In this way,the local changes in the geometry of the surface can be modelled without using a complicated surface equation to represent the entire surface of the thin shell . The x,y and z coordinates of the nodes are used to compute the surface metric tensors, curvature tensors, Christoffel symbols and their pertinent derivatives using the curvilinear finite differences [6,7].These geometric quantities are central in the following description of the deflection and deformation characteristics of a thin shell.

The displacement of a node on the middle surface (Fig. 2) can be defined by the three unknown covariant components (u_1, u_2, w) and a set of base vectors.Within a nine node curvilinear mesh, the displacements at any point can be approximated by the equation:

$$\{u\}_i = \{u_1\ u_2\ w\}^t = [\phi_i^{\ u}]\{U_i\} \tag{3}$$

where $[\phi_i^{\ u}]$ contains the interpolation coefficients for a given (α, β) and $\{U_i\}$ contains the three unknown covariant displacement components assigned to each of the nine nodes within the mesh. By differentiating the above expression with respect to the α–β coordinate system, similar expressions can be derived which define the local partial derivatives of the middle surface displacements at each node in terms of the vector $\{U_i\}$.

i.e.

$$\{u_{\alpha,\beta}\}_i = \{u_{1,1}\ u_{1,2}\ u_{2,1}\ u_{2,2}\}^t = [\phi_i^{\ u\alpha,\beta}]\{U_i\} \tag{4a}$$

$$\{w_{,\alpha}\}_i = \{w_{,1}\ w_{,2}\}^t = [\phi_i^{\ w,\alpha}]\{U_i\} \tag{4b}$$

$$\{w_{,\alpha\beta}\}_i = \{w_{,11}\ w_{,12}\ w_{,21}\ w_{,22}\}^t = [\phi_i^{\ w,\alpha\beta}]\{U_i\} \tag{4c}$$

At a point z distance above the middle surface (Fig. 2), the covariant components of the general surface displacement, v_i , can be expressed in terms of the displacement of the middle surface by

$$v_\alpha = u_\gamma \mu_\delta^\gamma \mu_\alpha^\delta - z\, w_{,\delta} \mu_\alpha^\delta \tag{5a}$$

$$w_3 = w \tag{5b}$$

Expanding the above equations and assembling them in matrix form yields :

$$\{v_i\} = [D_{cv}]\{U_i\} \tag{6a}$$

where

$$[D_{cv}] = [\,[A][\phi_i^{\ u}] - z[B][\phi_i^{\ w,\alpha}]\,] \tag{6b}$$

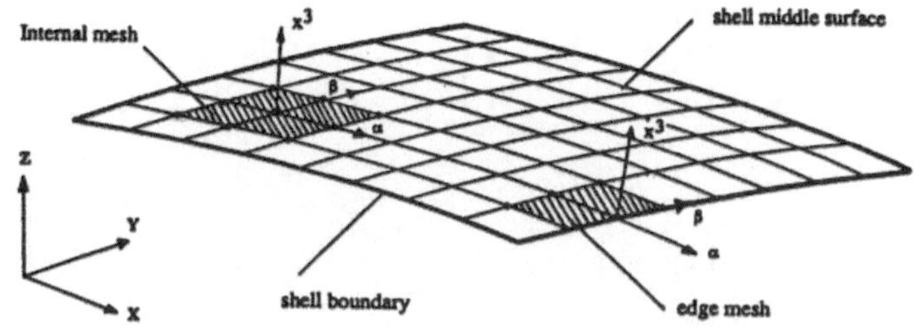

Figure 1. Discretization of a shell's middle surface.

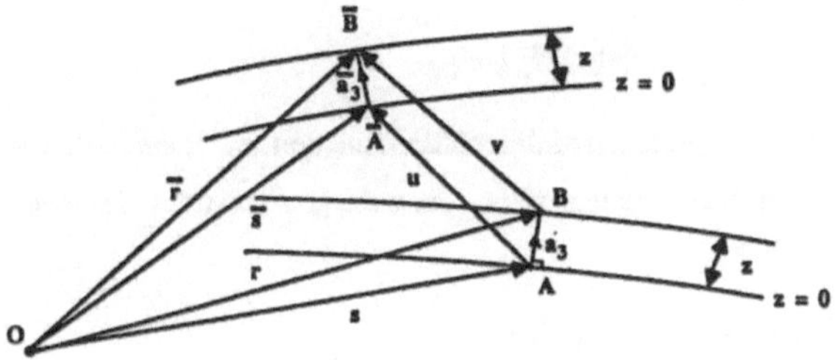

Figure 2. Section through a shell before and after deformation.

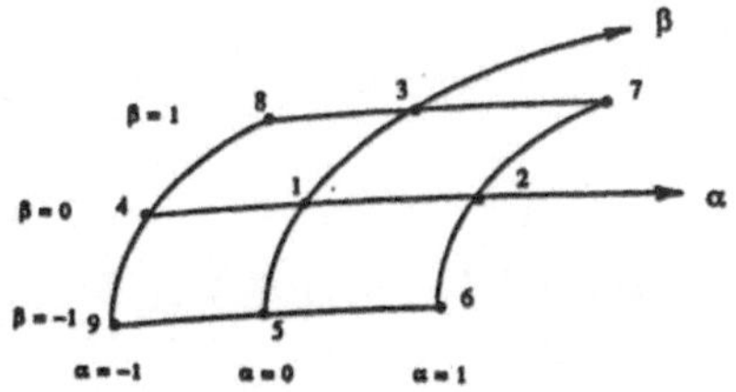

Figure 3. 9 node curvilinear mesh.

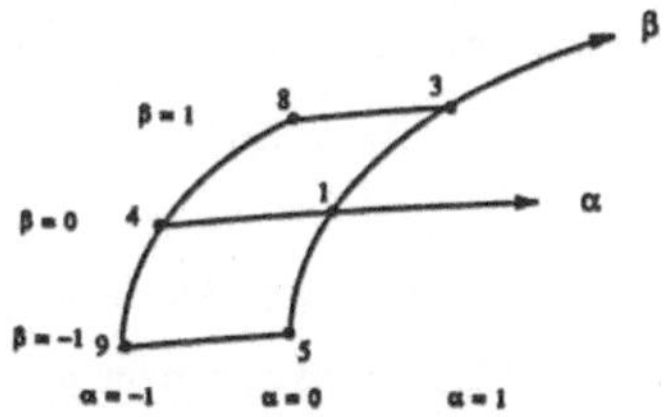

Figure 4. Typical 6 node edge mesh.

$$[A] = \begin{bmatrix} \mu^1_\gamma \mu^\gamma_1 & \mu^2_\gamma \mu^\gamma_1 & 0 \\ \mu^1_\gamma \mu^\gamma_2 & \mu^2_\gamma \mu^\gamma_2 & 0 \\ 0 & 0 & 1 \end{bmatrix} \qquad (6c)$$

and

$$[B] = \begin{bmatrix} \mu^1_1 & \mu^2_1 \\ \mu^1_2 & \mu^2_2 \\ 0 & 0 \end{bmatrix} \qquad (6d)$$

The contravariant components of the displacement vector $\{ v_i \}$ are

$$v^\alpha = v_\gamma \, g^{\gamma\alpha} \qquad (7a)$$

$$w^3 = w \qquad (7b)$$

i.e.

$$\{ v^i \} = [D^{cn}] \{ U_i \} \qquad (7c)$$

Using a quadratic displacement interpolation function $\{\phi_i{}^u\}$, the middle surface strain tensor $(\varepsilon_{\alpha\beta})$,the change in curvature tensor $(\kappa_{\alpha\beta})$ and the general strain tensor $(\eta_{\alpha\beta})$ can be defined in the following manner.

$$\{\varepsilon_{\alpha\beta}\} = [E] \{U_i\} \qquad (8a)$$

$$\{\kappa_{\alpha\beta}\} = [K] \{U_i\} \qquad (8b)$$

$$\{\eta_{\alpha\beta}\} = [H] \{U_i\} \qquad (8c)$$

For a homogeneous and isotropic material, the contravariant components of the stress can be determined from the equation:

$$\sigma^{\alpha\beta} = \frac{E}{(1-\nu^2)} [(1-\nu) \delta^\xi_\gamma \lambda^\alpha_\mu + \nu \delta^\alpha_\gamma \lambda^\xi_\mu] \lambda^\gamma_\beta \lambda^\beta_\sigma \lambda^\delta_\nu a^{\mu\nu} a^{\rho\sigma} \eta_{\xi\delta} \qquad (9a)$$

which can be rewritten in the form

$$\{ \sigma^{\alpha\beta} \} = [\Sigma] \{ U_i \} \qquad (9b)$$

The details of derivations of the elements of the matrices $[E]$, $[K]$, $[D_{cv}]$, $[D^{cn}]$, $[H]$ and $[\Sigma]$ are given in reference (7).

Substituting these equations into the principal of virtual displacement to obtain

$$\int_V \delta\{U_i\}^t [H]^t [\Sigma] \{U_i\} \, dV + \int_V \delta\{U_i\}^t [D_{cv}]^t \rho [D^{cn}] \{U_i\} \, dV = 0 \qquad (10a)$$

i.e.

$$\delta\{U_i\}^t [K] \{U_i\} + \delta\{U_i\}^t [M] \{U_i\} = \{0\} \qquad (10b)$$

where

$$[K] = \text{stiffness matrix} = \int_v [H]^t [\Sigma] \, dV \tag{10c}$$

$$[M] = \text{consistent mass matrix} = \int_v [D_{cv}]^t \rho [D^{cn}] \, dV \tag{10d}$$

For an arbitrary virtual displacement, $\delta\{U_i\}^t$,

$$[K]\{U_i\} + [M]\{\ddot{U}_i\} = \{0\} \tag{11}$$

The local stiffness and consistent mass matrices defined above can be computed by assuming a constant energy for the region surrounding a given node.These local stiffness and mass matrices can be assembled separately to form the total stiffness and mass matrices of the thin shell respectively.

Boundary conditions

The restraints imposed along a boundary are assumed to be composed of a combination of force and displacement constraints.The procedure is best illustrated by considering an example of a fixed edge boundary.

At a fixed edge, the following boundary conditions exist:

$$u_1 = u_2 = w = 0 \quad \text{and} \quad w_{,\alpha} = 0 \tag{12}$$

The displacement constraints (u_1, u_2, w) are implemented by modifying the shell's stiffness and mass matrices in the conventional manner. The zero slope condition ($w_{,\alpha} = 0$) is imposed by modifying the displacement interpolation functions associated with the mesh point under consideration.

At an edge, a 6 node mesh (Fig. 4) is used as a basis for constructing the interpolation functions. For each restraint imposed by a particular condition at the boundary, an unknown displacement is introduced at an external imaginary node. This node is positioned at the local coordinates $\alpha = \pm 1, \beta = 0$ or $\alpha = 0, \beta = \pm 1$,depending on the orientation of the boundary mesh.

Adopting the correct displacement interpolation functions it becomes possible to write the boundary conditions in the general form:

$$\{BC\} = [\phi_{bc}]\{U_i\} \tag{13a}$$

expanding this equation yields:

$$\{BC\} = [\phi_{r1}]\{U_{r1}\} + [\phi_i]\{U_{im}\} + [\phi_{r2}]\{U_{r2}\} \tag{13b}$$

For the boundary conditions under consideration,

$$\{BC\} = \{0\} \tag{13c}$$

which allows $\{U_{im}\}$ to be written in the form

$$\{U_{im}\} = -[\phi_i]^{-1}[\phi_{r1}]\{U_{r1}\} - [\phi_i]^{-1}[\phi_{r2}]\{U_{r2}\} \tag{13d}$$

The above equation relates the displacements associated with the ficticious nodes $\{U_{im}\}$ to the real unknown displacements and can be back substituted into the assumed displacement interpolation functions to remove the ficticious displacements.

After modifying the stiffness and consistent mass matrices, the equations of motion describing the free vibration of a thin shell structure are then solved using the subspace iteration technique to obtain its eigenvalues and corresponding eigenvectors.

Computer program implementation
The method outlined above was employed in the construction of a general shell analysis computer program. This program was developed using the 'Light-Speed C' compiler on the 'Apple Macintosh XL' microcomputer and later ported to an IBM RT (model 1510 with 3MBytes RAM)for generating the results of the illustrative examples .The computer program was written entirely in the C language to take advantages of the availabilty of quality C compilers and to use pointer arithmetic to speed up the index manipulations of the sparse stiffness and consistent mass matrices.

Numerical examples :
The first six circular frequencies and mode shapes of a clamped cylindrical shell panel and a spherical cap were computed using a number of mesh discretisations ranging from 7*7to 21*21. The most refined case required a total of 1.44 MBytes of memory to store 360K elements of the 1323 by 135 stiffness and consistent matrices.This amount is well within the available 3MBytes of RAM. The computation of subspace iterations was carried out in the in-core memory without using disk storage facilities.

Clamped cylindrical shell panel
The first example considered is a cylindrical panel clamped at each boundary. The dimensions and material properties are given in figure (5) and are identical to the ones given by Olsen & Lindberg [8].There is no exact solution to this problem and the tabulated results given in table (1) are in good agreement with those given by the finite element analysis of Olsen & Lindberg [8].

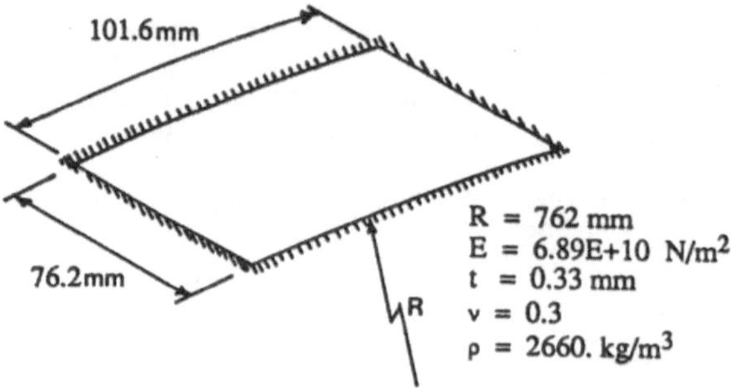

Figure 5. Clamped cylindrical shell panel.

Table 1 Natural frequency (Hz)

mode	7*7	9*9	13*13	17*17	21*21	Finite element [Ref.8]
1	793	815	840	852	857	869
2	849	881	920	936	943	957
3	1130	1167	1219	1244	1258	1287
4	1194	1268	1317	1335	1345	1362
5	1210	1287	1359	1390	1406	1437
6	1211	1322	1532	1627	1676	1752

Spherical cap on a square base
A shallow spherical shell depicted in figure (6) was analysed using a number of different meshes.

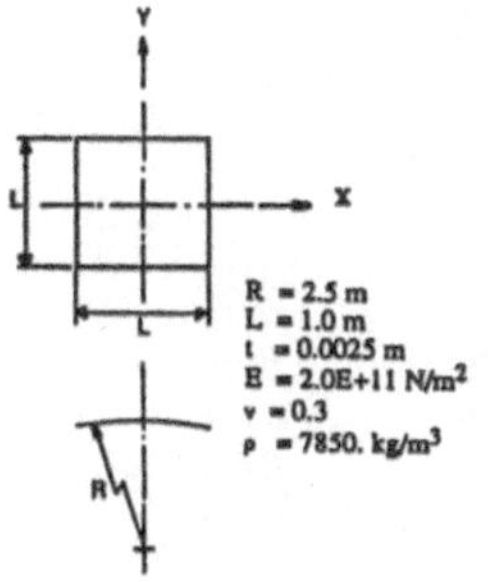

Figure 6. Spherical cap on a square base.

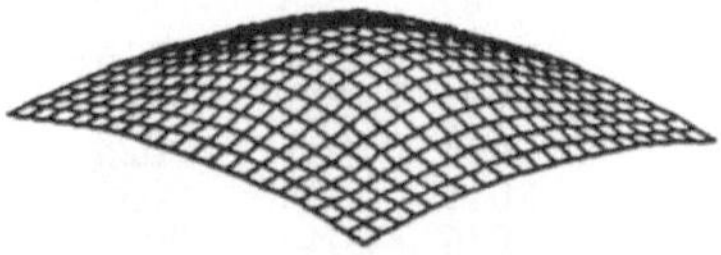

Figure 7. Spherical Cap - 1st mode.

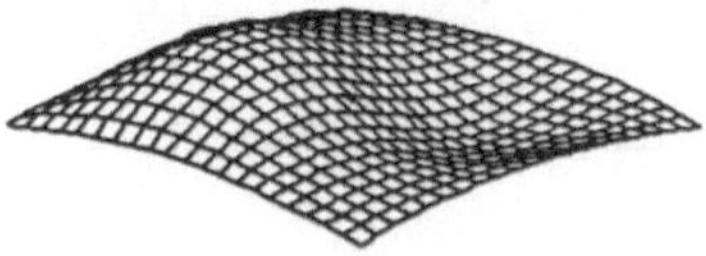

Figure 8. Spherical Cap - 2nd mode.

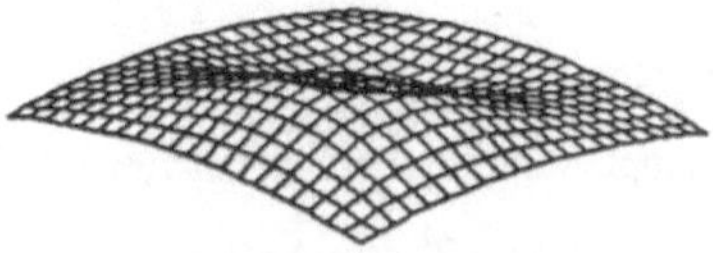

Figure 9. Spherical Cap - 3rd mode.

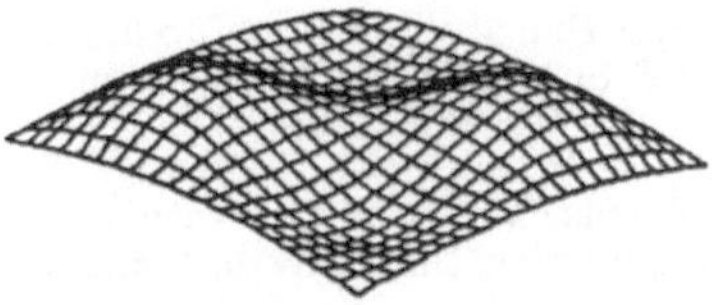

Figure 10. Spherical Cap - 4th mode.

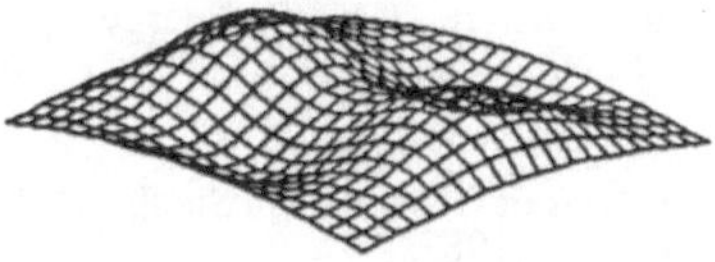

Figure 11. Spherical Cap - 5th mode.

The shell is freely supported on a boundary of square planform .The displacements w & u_α are constrained to zero along the edges x= -L/2 and x = L/2, and w & u_β are zero along y= -L/2 and y = L/2. The above configuration allows an exact solution for this problem to be obtained from shell theory [9].
The numerical results for the first six modes of vibration are presented in table (2) and compared well with the analytical solution. Figures (7) - (11) show the respective mode shapes for the first 5 nodes.

Table 2 Circular frequency (radians / second)

mode	7*7	9*9	13*13	17*17	21*21	Exact solution
1	2018	2017	2016	2016	2016	2020
2	2024	2025	2025	2026	2026	2028
3	2024	2025	2025	2026	2026	2028
4	2034	2036	2038	2039	2039	2041
5	2036	2044	2048	2050	2051	2054
6	2039	2046	2051	2053	2053	2054

Concluding remarks

A method for computing the natural frequencies and mode shapes of thin shells has been described. The tensor equations of Flugge's thin shell theory are approximated by curvilinear finite differences.The three displacements per node approach is computationally efficient in reducing the in-core memory required by the stiffness and mass matrices.This allows an analysis of a finer mesh model to be carried out in a microcomputer with a large memory. The proposed method is systematic, readily programmable and capable of representing the geometric and deformation characteristics of a general shell surface.

References

1. Lau ,P.C.M. 'Numerical solution of Poisson's equation using curvilinear finite differences', Appl. Math. Modelling,Vol 1,pp349-350,(1977)

2. Flugge,W. - Tensor Analysis and Continuum Mechanics, Springer-Verlag, Berlin, 1972.

3. Zienkiewicz, O.C. - The Finite Element Method, 3 rd ed., McGraw-Hill, 1977.

4. Bushnell, D. 'Analysis of buckling and vibration of ring-stiffened,segmented shells of revolution' Int.J. Solids & Structures,Vol.6,pp.157-181,(1970)

5. Christensen,P.J. & Lau ,P.C.M. 'A microcomputer based technique for the analysis of orthotropic thin plates' ,Proc. NUMETA 85, pp.965-972,(1985)

6. Kwok,S.K. 'Geometrically Nonlinear Analysis of General Thin Shells using a Curvilinear Finite Difference (CFD) Energy Approach', Computers & Structures Vol. 20, No. 4, pp. 683-697, 1985.

7. Naomis,S. 'The Application of the Curvilinear Finite Difference Method to the Dynamic Analysis of General Thin Shells', Department of Civil Engineering, University of Western Australia, 1987.

8. Olsen, M.D. & Lindberg,G.M. 'Dynamic Analysis of Shallow Shells with a doubly-curved triangular finite element',J. Sound & Vibration Vol 19(3),pp.299-318,(1971)

9. Kraus, H. - Thin Elastic Shells: An Introduction to the Theoretical Foundations and the Analysis of their Static and Dynamic Behavior, Wiley, New-York, 1967.

Acknowledgement

The authors would like to thank Mr.P Grove of IBM(Australia) in Perth who provided the IBM RT computing environment for obtaining the results of the illustrative examples.

D46/1

A SIMPLIFIED ANALYSIS OF TWO PLATE BENDING ELEMENTS — THE MITC4 AND MITC9 ELEMENTS

by

Klaus-Jürgen Bathe
Massachusetts Institute of Technology
Cambridge, MA 02139

&

Franco Brezzi
Università di Pavia
I.A.N. del C.N.R.
27100 Pavia, Italy

ABSTRACT

We consider the convergence behavior of two mixed-interpolated plate bending elements -- the MITC4 element which has already found wide use and a new element, the MITC9 element. A simplified analysis is given that renders valuable insight into the predictive capabilities of these elements.

1. INTRODUCTION

In this paper we analyze some finite element approximations of Mindlin-Reissner moderately thick plates. One element is the four-node element MITC4 [1,2] that was already analyzed from the mathematical point of view in [3]. The other element is a new element that we call MITC9, which is presently under testing and shows much promise.(*)

The type of analysis that we are carrying out here is in some sense a simplified one. In order to study the shear-locking phenomenon, we consider a sequence of plate bending problems $\{P_t\}$ with a thickness t going to zero, and the corresponding sequence $\{P_{th}\}$ of discretized problems. Now, instead of studying the convergence of P_{th} to P_t for positive t, we consider just the two limit problems P_{oh} and P_o, and we analyze convergence and error estimates only for this case. It is clear that, if P_{th} displays a "good behaviour" uniformly in t, then P_{oh} must also behave properly. Since the converse is not true, our analysis is not complete. However, we conjecture that the good behaviour for $t = 0$ is a very reasonable test that can be of great help in designing a new element. On the other hand, a comparison of the analysis of the MITC4 element in [3] (where the general case t >

(*)MITC4 denotes our element based on mixed-interpolated tensorial components using 4 nodes, and similarly for the abbreviations MITC8 and MITC9 [4].

0 was considered) and the present analysis shows clearly that the study of the limit case alone is considerably simpler. In particular the relationship with the analysis of some incompressible fluid elements can be established much more clearly.

For the sake of simplicity we consider only uniform decompositions of a square plate. However, it will be clear from the analysis that, at least for the MITC9 element, the arguments also hold for the general case.

2. THE SEQUENCE OF PROBLEMS AND THE LIMIT PROBLEM

We consider the spaces: $\underline{\Theta} = (H_0^1(\Omega))^2$ and $W = H_0^1(\Omega)$ and a load function f given in $L^2(\Omega)$. The sequence of problems under consideration is:

$$P_t \quad \operatorname*{Inf}_{\underline{\theta}\in\underline{\Theta},\, w\in W} \frac{t^3}{2} a(\underline{\theta},\underline{\theta}) + \frac{\lambda t}{2} \|\underline{\theta}-\underline{\nabla} w\|_0^2 - t^3(f,w)$$

where $\frac{t^3}{2} a(\underline{\theta},\underline{\theta})$ is the bending internal energy, λ includes the shear correction factor and $\| \ \|_0$ and $(\ , \ )$ represent respectively the norm and the inner product in $L^2(\Omega)$.

Assume now that we are given finite element subspaces $\underline{\Theta}_h \subset \underline{\Theta}$ and $W_h \subset W$. The corresponding discretized problem is described by

$$\tilde{P}_{th} \quad \operatorname*{Inf}_{\underline{\theta}_h\in\underline{\Theta}_h,\, w_h\in W_h} \frac{t^3}{2} a(\underline{\theta}_h,\underline{\theta}_h) + \frac{\lambda t}{2} \|\underline{\theta}_h-\underline{\nabla} w_h\|_0^2 - t^3(f,w_h).$$

In general, $\tilde{P}_{th}$ "locks" for small t. A common procedure is to reduce, somehow, the influence of the shear energy. We consider here the case in which the reduction is carried out in the following way: we assume that we are given a third finite element space, $\underline{\Gamma}_h$, and a linear operator R which takes values in $\underline{\Gamma}_h$. Then we use $\|R(\underline{\theta}_h-\underline{\nabla} w_h)\|_0^2$ instead of $\|\underline{\theta}_h-\underline{\nabla} w_h\|_0^2$ in the shear energy. For the sake of simplicity we shall assume that:

$$R\, \underline{\nabla} w_h = \underline{\nabla} w_h \text{ for all } w_h\in W_h \tag{1}$$

so that the discretized problem takes its final form

$$P_{th} \qquad \operatorname*{Inf}_{\underline{\theta}_h \in \underline{\Theta}_h, w_h \in W_h} \frac{t^3}{2} a(\underline{\theta}_h, \underline{\theta}_h) + \frac{\lambda t}{2} \| R\underline{\theta}_h - \underline{\nabla} w_h \|_0^2 - t^3 (f, w_h).$$

Setting

$$\underline{\gamma} = \lambda t^{-2}(\underline{\theta} - \underline{\nabla} w) \text{ and } \underline{\gamma}_h = \lambda t^{-2}(R\underline{\theta}_h - \underline{\nabla} w_h) \qquad (2)$$

the Euler equations of P_t and P_{th} are respectively

$$\begin{aligned} & a(\underline{\theta}, \underline{\eta}) + (\underline{\gamma}, \underline{\eta} - \underline{\nabla}\zeta) = (f, \zeta) \quad \forall\, \underline{\eta} \in \underline{\Theta},\ \forall\, \zeta \in W \\ & \underline{\gamma} = \lambda t^{-2} (\underline{\theta} - \underline{\nabla} w) \end{aligned} \qquad (3)$$

and

$$\begin{aligned} & a(\underline{\theta}_h, \underline{\eta}) + (\underline{\gamma}_h, R\underline{\eta} - \underline{\nabla}\zeta) = (f, \zeta) \quad \forall\, \underline{\eta} \in \underline{\Theta}_h,\ \forall\, \zeta \in W_h \\ & \underline{\gamma}_h = \lambda t^{-2} (R\underline{\theta}_h - \underline{\nabla} w_h) \end{aligned} \qquad (4)$$

From now on we will limit ourselves to the analysis of the limit problems

$$\begin{aligned} & a(\underline{\theta}, \underline{\eta}) + (\underline{\gamma}, \underline{\eta} - \underline{\nabla}\zeta) = (f, \zeta) \quad \forall\, \underline{\eta} \in \underline{\Theta},\ \forall\, \zeta \in W \\ & \underline{\theta} = \underline{\nabla} w \end{aligned} \qquad (5)$$

and

$$\begin{aligned} & a(\underline{\theta}_h, \underline{\eta}) + (\underline{\gamma}_h, R\underline{\eta} - \underline{\nabla}\zeta) = (f, \zeta) \quad \forall\, \underline{\eta} \in \underline{\Theta}_h,\ \forall\, \zeta \in W_h \\ & R\underline{\theta}_h = \underline{\nabla} w_h\ . \end{aligned} \qquad (6)$$

REMARK It is not difficult to show that (5) and (6) are the limit problems of (3) and (4) respectively, see for instance [3]. In particular the limit w will be the solution corresponding to the Kirchhoff model. Note also that the limit $\underline{\gamma}_h$ that appears in (6) will still belong to $R(\underline{\Theta}_h) - \underline{\nabla}(W_h)$. Although we are not studying here the convergence of $\underline{\gamma}_h$ to $\underline{\gamma}$, the results given in [5] with the discussion below give some insight into the behavior of $\underline{\gamma}_h$.

3. THE FINITE ELEMENT DISCRETIZATIONS

Following the discussion of the previous section, the finite element discretization is characterized by the choice of the finite element spaces $\underline{\Theta}_h, W_h, \underline{\Gamma}_h$ and by the choice of the linear operator R. Note that these choices are not independent from each other since we assumed (1) to be satisfied. We introduce now the two choices that we consider in this paper.

The MITC4 Element [1,2]

We set

$$\underline{\Theta}_h = \{\underline{\eta} \mid \underline{\eta} \in (H_0^1(\Omega))^2, \underline{\eta}|_K \in (Q_1)^2 \ \forall K\} \tag{7}$$

$$W_h = \{\zeta \mid \zeta \in H_0^1(\Omega), \zeta|_K \in Q_1 \ \forall K\} \tag{8}$$

where, here and in the following, Q_1 is the set of polynomials of degree ≤ 1 in each variable and K is the current element in the discretization (we recall that we assumed a uniform decomposition of a square Ω). The space $\underline{\Gamma}_h$ is given by

$$\underline{\Gamma}_h = \{\underline{\delta} \mid \underline{\delta}|_K \in TR(K) \ \forall K, \ \underline{\delta}\cdot\underline{\tau} \text{ continuous at the interelement boundaries}\} \tag{9}$$

where $\underline{\tau}$ is the tangential unit vector to each edge of each element and

$$TR(K) = \{\underline{\delta} \mid \delta_1 = a_1 + b_1 y, \ \delta_2 = a_2 + b_2 x\} \tag{10}$$

is a sort of "rotated Raviart-Thomas" space. We have finally to introduce the reduction operator R. We describe its action on the current element: for $\underline{\eta}$ smooth in K, $R\underline{\eta}|_K$ is the unique element in TR(K) that satisfies

$$\int_e (\underline{\eta} - R\underline{\eta})\cdot\underline{\tau} \, ds = 0 \text{ for all edges e of } K. \tag{11}$$

Note that if $\underline{\eta} \in \underline{Q}_1$ then (11) is satisfied if and only if $\underline{\eta}\cdot\underline{\tau} = R(\underline{\eta})\cdot\underline{\tau}$ at the midpoints of each edge. Hence clearly (1) holds.

The MITC9 Element

We introduce now a new element. We set

$$\underline{\Theta}_h = \{\underline{\eta} \mid \underline{\eta} \in (H_0^1(\Omega))^2, \underline{\eta}|_K \in (Q_2)^2 \ \forall\ K\} \tag{12}$$

$$W_h = \{\zeta \mid \zeta \in H_0^1(\Omega),\ \zeta|_K \in Q_2^r \ \forall\ K\} \tag{13}$$

where Q_2 is the space of polynomials of degree ≤ 2 in each variable (corresponding to a 9 node element) and Q_2^r is its usual serendipity reduction (corresponding to an 8 node element). In order to introduce the space $\underline{\Gamma}_h$ we define first the space of polynomials

$$G = \{\underline{\delta} \mid \delta_1 = a_1 + b_1 x + c_1 y + d_1\, x\, y + e_1 y^2, \tag{14}$$

$$\delta_2 = a_2 + b_2 x + c_2 y + d_2\, x\, y + e_2 x^2\}$$

which is some kind of rotated Brezzi-Douglas-Fortin-Marini space. Note that if $\zeta \in Q_2^r$ then $\underline{\nabla}\zeta \in G$. This is the main reason why W_h has been discretized with the interpolations of 8-node elements instead of 9-node elements. (But could we use a larger space G ? We shall deal with this question briefly later on.) We introduce now the space $\underline{\Gamma}_h$:

$$\underline{\Gamma}_h = \{\underline{\delta} \mid \underline{\delta}|_K \in G\ \forall\ K,\ \underline{\delta}\cdot\underline{\tau} \text{ continuous at the interelement boundaries}\}. \tag{15}$$

Further, we define the action of the reduction operator R on the current element K in the following way: for $\underline{\eta}$ smooth in K, $R\underline{\eta}|_K$ is the unique element in G that satisfies

$$\int_e (\underline{\eta}-R\underline{\eta})\cdot\underline{\tau}\ p_1(s)\, ds = 0 \quad \forall\ e \text{ an edge of } K,\ \forall\ p_1(s) \text{ polynomial of degree} \leq 1 \text{ on } e \tag{16}$$

$$\int_K (\underline{\eta}-R\underline{\eta})\,dx\, dy = 0. \tag{17}$$

Here again (1) is satisfied. Note also that if $\underline{\eta} \in \underline{Q}_2$ then (16) holds if and only if $\underline{\eta}\cdot\underline{\tau} = (R\underline{\eta})\cdot\underline{\tau}$ at the two Gauss points of each edge.

REMARKS

We could think of using $\underline{\eta} = R\underline{\eta}$ at the center of the element instead of (17). However, our proof is then not applicable, although numerical experiments may show good element behavior even in this case.

It is clear that for a general decomposition R should be defined by covariant interpolations (see [1,2,4]).

4. THE ERROR ANALYSIS

It is convenient to recall the definition of the differential operators

$$\varphi \to \underline{\mathrm{rot}}(\varphi) = (-\partial\varphi/\partial y,\ \partial\varphi/\partial x)$$

and

$$\underline{\varphi} = (\varphi_1,\varphi_2) \to \mathrm{rot}\underline{\varphi} = (\partial\varphi_1/\partial y - \partial\varphi_2/\partial x).$$

We now look for a "pressure space" Q_h made of discontinuous finite element functions(**) such that, for all $\underline{\eta} \in \underline{\Theta}$, we have

$$(\mathrm{rot}\ \underline{\eta}, q_h) = (\mathrm{rot}(R\underline{\eta}), q_h) \quad \forall\ q_h \in Q_h \tag{18}$$

and

$$\mathrm{rot}(\underline{\Gamma}_h) \subseteq Q_h. \tag{19}$$

Conditions (18), (19) are strictly related to the so-called "commuting diagram property" of Douglas and Roberts [6] that is used in the study of mixed methods for elliptic equations. It is easy to check that (18) and (19) hold if we take for the MITC4 element

$$Q_h = \{q \mid q|_K \in P_0\ \forall\ K\} \tag{20}$$

and for the MITC9 element

$$Q_h = \{q \mid q|_K \in P_1\ \forall\ K\}. \tag{21}$$

(**) This space corresponds to the pressure space in incompressible solutions.

In both cases P_k denotes the set of polynomials of total degree $\leq k$: hence Q_h has local dimension 1 in the MITC4 case and 3 in the MITC9 case.

In order to analyze the error between $\underline{\theta}$ and $\underline{\theta}_h$ in (5) - (6) (and as a consequence the error between w and w_h) we want to build a pair $\hat{\underline{\theta}}, \hat{w}$ in $\underline{\Theta}_h \times W_h$ such that $\|\underline{\theta}-\hat{\underline{\theta}}\|_1$ is optimally small and

$$R\hat{\underline{\theta}} = \underline{\nabla}\hat{w}. \tag{22}$$

Condition (22) implies

$$\text{rot } R\hat{\underline{\theta}} = 0 \tag{23}$$

which, in its turn, using (18), (19) is equivalent to

$$(\text{rot } \hat{\underline{\theta}}, q_h) = 0 \quad \forall\, q_h \in Q_h. \tag{24}$$

A possible way of constructing $\hat{\underline{\theta}}$ is the following. For $\underline{\theta}$ given in $(H_0^1(\Omega))^2$ and satisfying $\text{rot}\underline{\theta} = 0$, consider the problem:

$$\begin{aligned}
&\text{Find } \underline{\beta},\ p \in \underline{\Theta} \times L^2(\Omega) \text{ such that}\\
&a(\underline{\beta},\underline{\eta}) + (p,\ \text{rot}\underline{\eta}) = a(\underline{\theta},\underline{\eta}) \quad \forall\, \underline{\eta} \in \underline{\Theta}\\
&(q, \text{rot}\underline{\beta}) = 0 \quad \forall\, q \in L^2(\Omega)
\end{aligned} \tag{25}$$

and its approximation

$$\begin{aligned}
&\text{Find } \hat{\underline{\theta}},\ p_h \in \underline{\Theta}_h \times Q_h \text{ such that}\\
&a(\hat{\underline{\theta}},\underline{\eta}) + (p_h, \text{rot}\underline{\eta}) = a(\underline{\theta},\eta) \quad \forall\, \underline{\eta} \in \underline{\Theta}_h\\
&(q, \text{rot}\hat{\underline{\theta}}) = 0 \quad \forall\, q \in Q_h
\end{aligned} \tag{26}$$

Note that (25) is a kind of Stokes problem and its solution is given by $\underline{\beta} = \underline{\theta}$, $p = 0$. If the pair $\underline{\Theta}_h$, Q_h is a suitable finite element discretization for the Stokes problem one might expect to have optimal error bounds for $\hat{\underline{\theta}}-\underline{\theta}$. For instance, in the case of the MITC4 element the pair $\underline{\Theta}_h$, Q_h is the classical bilinear velocities-constant pressure (or Q_1-P_0) element, and we know that with minor assumptions on the decomposition that are surely satisfied in the present case:

$$\|\underline{\theta}-\hat{\underline{\theta}}\|_1 \leq c\, h\, \|\underline{\theta}\|_2 \quad \text{for } Q_1 - P_0 \text{ element.} \tag{27}$$

On the other hand in the case of the MITC9 element the pair $\underline{\Theta}_h$, Q_h is the biquadratic velocities and linear pressure (the Q_2-P_1) element and for a general decomposition:

$$\|\underline{\theta}-\hat{\underline{\theta}}\|_1 \leq c\, h^2\, \|\underline{\theta}\|_3 \quad \text{for } Q_2 - P_1 \text{ element.} \tag{28}$$

Note on the other hand that once $\hat{\underline{\theta}}$ satisfying (23) has been found, then one can uniquely determine the $\hat{w} \in W$ that satisfies (22). It is easy to check that in our two cases such a $\hat{w}$ is an element of W_h.

We are now ready for proving error estimates. We set

$$\underline{\delta} = \underline{\theta}_h - \hat{\underline{\theta}}\ ;\quad \xi = w_h - \hat{w} \tag{29}$$

and we note that

$$R\underline{\delta} = \underline{\nabla}\xi. \tag{30}$$

Now we have

$$\begin{aligned}
\alpha\|\underline{\delta}\|_1^2 &\leq a(\underline{\delta},\underline{\delta}) = a(\underline{\theta}_h-\underline{\theta},\underline{\delta}) + a(\underline{\theta}-\hat{\underline{\theta}},\underline{\delta}) = \\
&= -(\underline{\gamma}_h, R\underline{\delta}) + (\underline{\gamma},\underline{\delta}) + a(\underline{\theta}-\hat{\underline{\theta}},\underline{\delta}) = \\
&= (\underline{\gamma},\underline{\delta}-R\underline{\delta}) - (\underline{\gamma}_h-\underline{\gamma}, R\underline{\delta}) + a(\underline{\theta}-\hat{\underline{\theta}},\underline{\delta}) = \\
&= (\underline{\gamma},\underline{\delta}-R\underline{\delta}) - (\underline{\gamma}_h-\underline{\gamma}, \underline{\nabla}\xi) + a(\underline{\theta}-\hat{\underline{\theta}},\underline{\delta}) = \\
&= (\underline{\gamma},\underline{\delta}-R\underline{\delta}) + a(\underline{\theta}-\hat{\underline{\theta}},\underline{\delta}) \leq \\
&\leq \left[\sup_{\underline{\beta} \in \underline{\Theta}_h} \frac{(\underline{\gamma},\underline{\beta}-R\underline{\beta})}{\|\underline{\beta}\|_1} + c\, \|\underline{\theta}-\hat{\underline{\theta}}\|_1 \right] \|\underline{\delta}\|_1
\end{aligned} \tag{31}$$

which implies

$$\|\underline{\delta}\|_1 \leq c \left\{ \|\underline{\theta}-\hat{\underline{\theta}}\|_1 + \sup_{\underline{\beta} \in \underline{\Theta}_h} (\underline{\gamma},\underline{\beta}-R\underline{\beta})/\|\underline{\beta}\|_1 \right\} \tag{32}$$

In the case of the MITC4 element we have

$$|(\underline{\gamma},\underline{\beta}-R\underline{\beta})| \leq \|\underline{\gamma}\|_0 \ \|\underline{\beta}-R\underline{\beta}\|_0 \leq c\ h\ \|\underline{\gamma}\|_0\ \|\underline{\beta}\|_1 \tag{33}$$

and using (32), (33) and (27) we obtain

$$\|\underline{\delta}\|_1 \leq c\ h\ (\|\underline{\theta}\|_2 + \|\underline{\gamma}\|_0) \tag{34}$$

and finally from (34), (29), (27) and the triangle inequality:

$$\|\underline{\theta}-\underline{\theta}_h\|_1 \leq c\ h(\|\underline{\theta}\|_2+\|\underline{\gamma}\|_0) \qquad \text{for MITC4} \tag{35}$$

Let us consider now the case of the MITC9 element, and set $\hat{\underline{\gamma}}$ = mean value of $\underline{\gamma}$ in each K. Using (17) we have

$$(\underline{\gamma},\underline{\beta}-R\underline{\beta}) = (\underline{\gamma}-\hat{\underline{\gamma}},\underline{\beta}-R\underline{\beta}) \tag{36}$$

so that

$$\begin{aligned} |(\underline{\gamma},\underline{\beta}-R\underline{\beta})| &\leq \|\underline{\gamma}-\hat{\underline{\gamma}}\|_0\ \|\underline{\beta}-R\underline{\beta}\|_0 \leq \\ &\leq c\ h^2\|\underline{\gamma}\|_1\ \|\underline{\beta}\|_1 \end{aligned} \tag{37}$$

and from (32), (37) and (28)

$$\|\underline{\delta}\|_1 < c\ h^2(\|\underline{\theta}\|_3 + \|\underline{\gamma}\|_1) \tag{38}$$

so that from (38), (29), (28) and the triangle inequality

$$\|\underline{\theta}-\underline{\theta}_h\|_1 \leq c\ h^2(\|\underline{\theta}\|_3 + \|\underline{\gamma}\|_1) \qquad \text{for MITC9} \tag{39}$$

Finally we want to estimate $w-w_h$. Since $\underline{\nabla} w_h = R\underline{\theta}_h$ we have

$$\underline{\nabla}(w-w_h) = \underline{\theta} - R\underline{\theta}_h = (\underline{\theta}-R\underline{\theta}) + R(\underline{\theta}-\underline{\theta}_h) \tag{40}$$

It is easy to check that

$$\|\underline{\theta}-R\underline{\theta}\|_0 \leq c\ h\|\underline{\theta}\|_1 \qquad \text{for MITC4} \tag{41}$$

$$\|\underline{\theta}-R\underline{\theta}\|_0 \leq c\ h^2\|\underline{\theta}\|_2 \qquad \text{for MITC9} \tag{42}$$

while in both cases

$$\|R(\underline{\theta}-\underline{\theta}_h)\|_0 \le c\ \|\underline{\theta}-\underline{\theta}_h\|_1. \qquad (43)$$

Therefore from (40), (41), (43) and (35):

$$\|\underline{\nabla}w - \underline{\nabla}w_h\|_0 \le c\ h\ (\|\underline{\theta}\|_2 + \|\underline{\gamma}\|_0)\ \text{for MITC4} \qquad (44)$$

and from (40), (42), (43) and (39)

$$\|\underline{\nabla}w-\underline{\nabla}w_h\|_0 \le c\ h^2\ (\|\underline{\theta}\|_3 + \|\underline{\gamma}\|_1)\ \ \text{for MITC9.} \qquad (45)$$

REMARKS

Hinton and Huang [7] suggested other constructions of mixed-interpolated elements and gave interesting numerical results.

The use of 9 nodes to describe W_h, considering our theory, would require the $\underline{\Gamma}_h$ to be of the form

$$(a_1 + b_1x + c_1y + d_1\ x\ y + e_1y^2 + f_1\ x\ y^2,$$

$$a_2 + b_2x + c_2y + d_2\ x\ y + e_2x^2 + f_2\ x^2y)$$

in each K. Then, since we need (19), we would have Q_h made of bilinear (instead of linear) functions in each element. Hence the pair $\underline{\theta}_h$, Q_h will be of the type (Q_2-Q_1) which is not as good as the (Q_2-P_1) pair [8].

ACKNOWLEDGMENT

F. Brezzi was partially supported by M.P.I. 40%.

REFERENCES

1. E. Dvorkin and K. J. Bathe, "A Continuum Mechanics Based Four-Node Shell Element for General Nonlinear Analysis, *J. Engineering Computations*, *1*, 77-88, 1984.

2. K. J. Bathe and E. Dvorkin, "A Four-Node Plate Bending Element Based on Mindlin/Reissner Plate Theory and a Mixed Interpolation", *Int. J. Num. Meth. in Eng.*, *21*, 367-383, 1985.

3. K. J. Bathe and F. Brezzi, "On the Convergence of a Four-Node Plate Bending Element Based on Mindlin/Reissner Plate Theory and a Mixed Interpolation", *Proceedings Conf. on Mathematics of Finite Elements and Applications* V, Academic Press, (J. R. Whiteman, ed.), 491-503, 1985.

4. K. J. Bathe and E. Dvorkin, "A Formulation of General Shell Elements -- The Use of Mixed Interpolation of Tensorial Components", *Int. J. Num. Meth. in Eng.*, 22, 697-722, 1986.

5. F. Brezzi and K. J. Bathe, "Studies of Finite Element Procedures -- The Inf-Sup Condition, Equivalent Forms and Applications", in Reliability of Methods for Engineering Analysis, (K. J. Bathe and D. R. J. Owen, eds.), Pineridge Press, 1986.

6. J. Douglas, Jr. and J. E. Roberts, "Global Estimates for Mixed Methods for Second-Order Elliptic Equations", *Math. of Comp.*, 44, 39-52, 1985.

7. E. Hinton and H. C. Huang, "A Family of Quadrilateral Mindlin Plate Elements with Substitute Shear Strain Fields", *J. Computers & Structures*, 23, No. 3, 409-431, 1986.

8. T. Sussman and K. J. Bathe, "A Finite Element Formulation for Nonlinear Incompressible Elastic and Inelastic Analysis", *J. Computers & Structures*, 26, No. 1/2, 1987.

High-speed Loading Analysis of Reinforced Concrete Columns

Shigekatsu ICHIHASHI KOZO KEIKAKU Eng. Inc., Nippon Holstein Kaĩkan, 4-38-13, Honcho, Nakano-ku, Tokyo, 164, JAPAN
Akira WADA Associate Prof. Dept. of Arch. and Build., Tokyo Inst. of Tech., 2-12-1, Ookayama, Meguro-ku, Tokyo, 152, JAPAN

1. Summary

There are two concepts that buildings would be fractured by large earthquakes. One is that buildings would be yielded and fractured gradually by repeated forces of earthquakes. Other is that some parts of buildins would be fractured instantly with impact or high-speed forces occured in buildings by earthquakes. With latter concept, the series of experiments for the fractures of reinforced concrete columns have been done by installing the large actuator with static loadings and high-speed forces. At the same time,static analyses and dynamic analyses for high-speed forces have been performed about experiment modelswith the implicit nonlinear finite element method. For the analytical models, discreted models between steels and concrete elements have been used without considering bond effect and changing the geometrical configurations. Both results were compared.

2. Abstract

There are a lot of unclear facts relating to how the structural members will be fractured by external dynamic forces, such as earthquakes. There are two concepts for understanding the mechanism about fractures of members. One is that structural members will be gradually fractured with repeated deformations by the vibration of the seismic force and the other is that the building will be destroyed instantly by the large high-speed impact forces of earthquakes. We have studied the latter concept. We have taken up the column made with reinforced concrete (RC) as the object of study. It is evident that RC members are made of composite materials, so the behavior of RC members is quite complex. The stress-strain relations of RC members vary with the loading speed and are closely associated with progressive micro-cracking. The width of crack, and fracture propagation also depend on the loading speed. The material characteristics of steel bars that are used in RC members also change with the loading speed. Loading speeds affect the interaction between the concrete materials and the reinforcing bars, such as slip behaviors. A lot of researchers have been developing and studing the numerical techniques for the dynamic response analysis including the impact problems, and nonlinear response analysis. However, it is very important to compare the results from numerical analysis to those from experiment. As a result, we have confidence to apply the numerical analysis to the design. In this paper, we performed the actural experiment for the reinforced concrete column. We also performed the numerical analysis for this experimental model and compared the results.

3. Experiment

3.1 Exprimental model and material properties

We show the model geometry and positions of steel bars in Fig. 1 and material properties in Table 1 and 2. We use the RC column of the building as the experimental model. This RC column has the stags at each end parts as Fig.1. We performes the high-speed compression loading that direction was inclined 20° from the center line of the column in Fig.1. On this loading, we can see the behavior of the RC column on the mixed loading conditions that there are axial force, shear force and bending moment simultaneously.

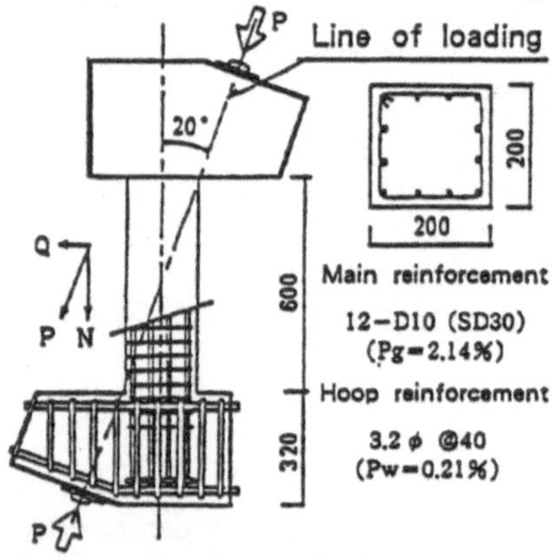

Fig. 1 Specimen & bar arrangement

Table 1 Raw material test of steel

Specimen	Yield stress (kg/mm^2)		max.stress (kg/mm^2)	Strain rate (%)
	Upper	Lower		
D-10	39.9	38.7	55.7	17.1
3.2 φ	35.1	34.3	43.8	29.8

Table 2 Raw material test of concrete

Specimen	Compressive Strength (kg/cm^2)		Tensile Strength (kg/cm^2)
	28days	43days	
RC20	112.0	119.0	11.4 (47days)

We used SD30 as steel bars, and normal concrete with combined flyash. At the experiment, material age of the concrete was 43 days.

3.2 Loading plan

We adopted the one way loading by using the dynamic testing actuator (Maximum static load : ±110 ton, Maximum dynamic load : ±75 ton) shown in Fig.2.

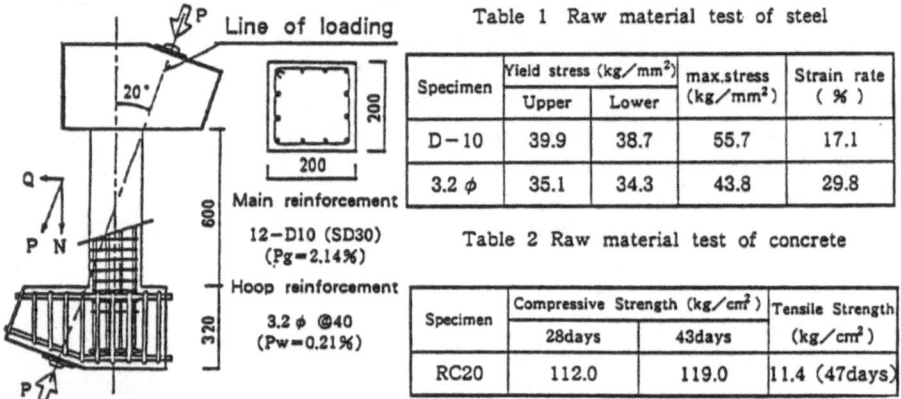

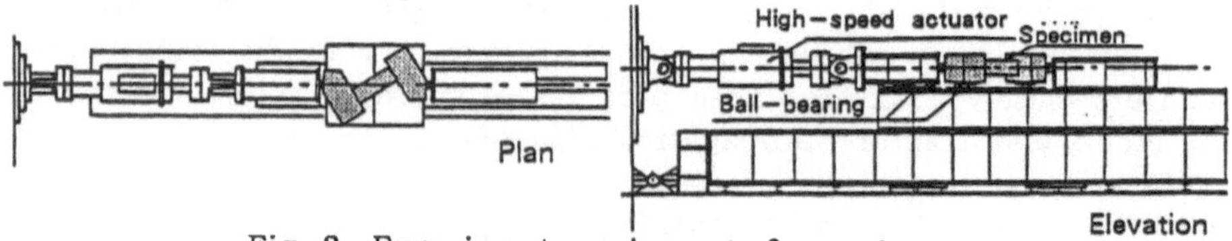

Fig. 2 Experiment equipment & specimen

As the boundary conditions, we set the pin which moved along the loading direction at the loading point, and also, we set the pin which couldn't move the horizontal direction at the reaction point. We used the ball bearings under the specimen. By using those bearings, specimem could move in any direction freely on the horizontal plane without the friction between the specimem and plate. We performed the experiments for two specimens which had made on the same conditons. One was static testing, and another was dynamic high speed testing. About the dynamic testing, we used 22.5 cm/sec as the loading speed. We selected this value by considering the response of relative story velocity in the lowest column of the general RC building that has 3 floors. The specimens' sizes were ⅓ model of the actual building. The variance of the dynamic loading is shown in Fig.3 from load−cell data.

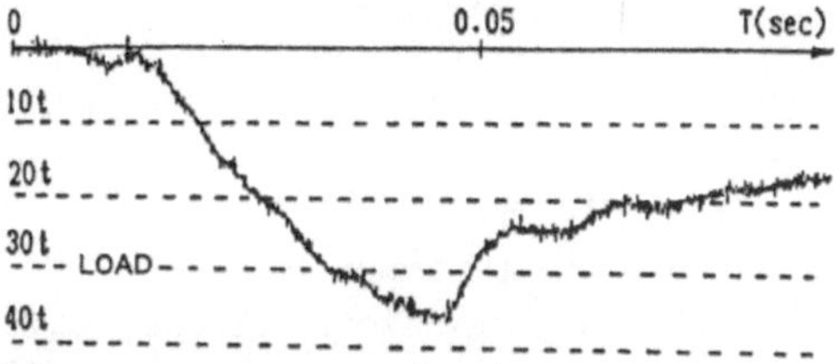

Fig.3 Dynamic loading

3.3 Measurement method

Measurement for displacements was performed with the measurement jig that moved with the specimen as one body shown in Fig.4. Relative displacement and axial displacement were measured by it.

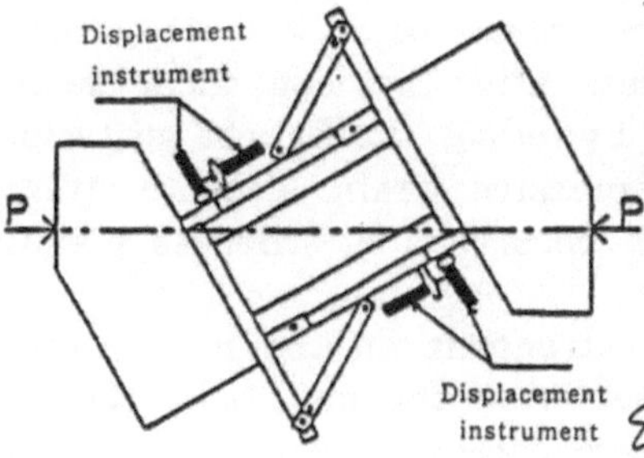

Fig.4 Measurement jigs

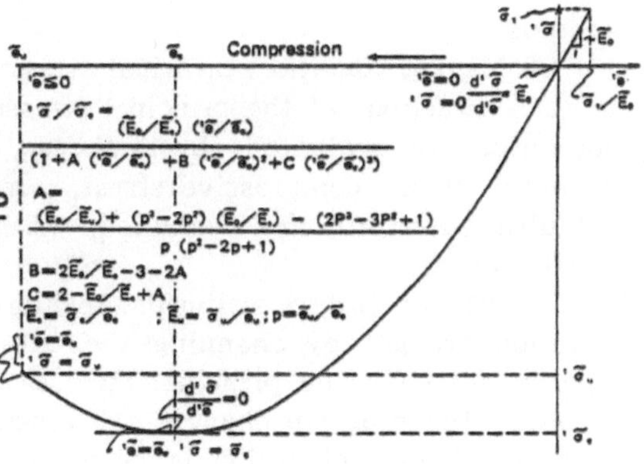

Fig.5 Concrete materials

4. Nonlinear Analysis method for the RC members.

We used the discrete model about the the steel bars and concrete without bond effect, and interaction for each other. We also assumped the both materials as hypo−elastic. The method by K. J. Bathe was used, because it agreed with the results of experiments by Kupfer, Khan, Saugy, and Liu. The relation between uniaxial forces and uniaxial strain is shown in Fig.5.

$\tilde{E}_0$: initial elastic modulus
$\tilde{e}_c$: strain equivalent to $\tilde{\sigma}_c$ (maximum compression stress)
$\tilde{e}_u$: ultimate strain equivalent to $\tilde{\sigma}_u$ (ultimate stress)
$\sim$ means the values from uniaxial test.

If ${}^tE$ is considered the equivalent elastic modulus at time t, we can get the next equation for the parts in ${}^t\tilde{e} \leqq 0$.

$$ {}^tE = \frac{\tilde{E}_0 [1 - B ({}^t\tilde{e}/\tilde{e}_c)^2 - 2C ({}^t\tilde{e}/\tilde{e}_c)^3]}{[1 + A ({}^t\tilde{e}/\tilde{e}_c) + B ({}^t\tilde{e}/\tilde{e}_c)^2 + C ({}^t\tilde{e}/\tilde{e}_c)^3]^2} \quad \cdots\cdots\cdots\cdots\cdots (1) $$

In the numerical analysis. Poisson's ratio should be constant. The relation between stress and strain should be changed on the loading and unloading process according to the multi−direction stress conditions. Loading function t_f is expressed as followings.

$$ t_f = t_s + 3\alpha^t \sigma_m \cdots\cdots\cdots\cdots\cdots\cdots\cdots\cdots\cdots\cdots (2) $$

t : time, ${}^t\sigma_m = {}^t\sigma_{ii}/3$, $t_s = (\frac{1}{2}\, {}^tS_{ij}\, {}^tS_{ij})^{\frac{1}{2}}$
${}^tS_{ij} = {}^t\sigma_{ij} - {}^t\sigma_m \delta_{ij}$, ${}^t\sigma_{ii}$: principal stress
${}^tS_{ij}$: deviatoric stress t_s : effective deviatoric stress
α : 0 or small negative value

In the case of loading, t_f is greater than equal f_{max}, and in the case of unloading t_f is small than equal f_{max} . f_{max} means the maximum value in both loading and unloading. On unloading condition, concrete is considered as isoropic material and also concrete keeps initial elastic modulus.

On loading condition, principal stresses are evaluated, and slso strains for each directions of the principal stresses. After that equivalent elastic moduluses are evaluated. In this case, by using the failure criterion on multi−axises, compressive stress, compressive strain, ultimate stress, and ultimate strain are evaluated. With recorrection, stiffness matrix is calculated.

About the failure criterion, tensile strengths on the multi−axial condition are got by changing the values from the uni−axial tension experiment. If tensile principal stresses are exceeded over those tensile strengths, the program changes the tensile stiffness as 1/200 of tensile strength. Stress relaxation is performed. About the shear strength, 1/2 of maximum shear strength. Similarly, for compressive failures, the values are evaluated by using the curves issued with Kupfer, and others. If the compressive stresses are exceeded over the values on the failure enveloped curves, stresses are relaxed. And this element has no stiffness. When tensile failure occured, crack in the element could be seen. However on this element compressive stress acts, crack would be closed. In this analysis, program considers about it. These treatments are performed on integral points of the element.

5. Numerical Analysis Model

In this experiment, two same specimens were made for both static and dynamic experiments. For the numerical analysis model, two dimensional stress problem model was evaluated as Fig.6. By considering the experimental boundary conditions, boundary condition was set like as Fig.6. The parts of the enveloped dash line are shown the main steel bars.

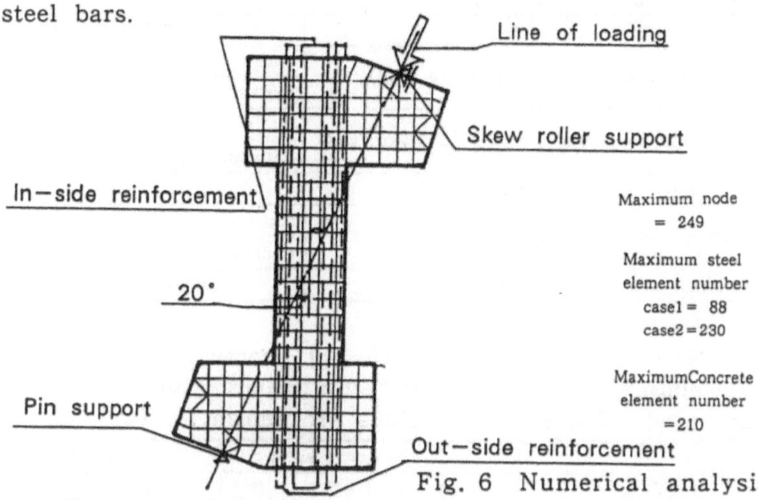

Fig. 6 Numerical analysis model

The material properties of steel bars and concrete were evaluated by the raw material testings. Each values are shown in table 3. About the steel bars, setional areas of outer main steel bars, inner main steel bars, and hoop bars were different. Other properties of steel bars were same.

material properties	values	
Section area of inner main steel bar (cm^2)	3.1416	
Section area of outer main steel bar (cm^2)	1.5708	
Section area of hoop bar (cm^2)	0.1680	
Density (ρ) (g/cm^3)	7.86	
Initial elastic modulus (kg/cm^2)	2.1×10^6	
Yield stress (kg/cm^2)	static	3.93×10^3
	dynamic	5.0×10^3
Tangent modulus after yield point (kg/cm^2)	2.1×10^5	

Table 3 Material properties of steel bars

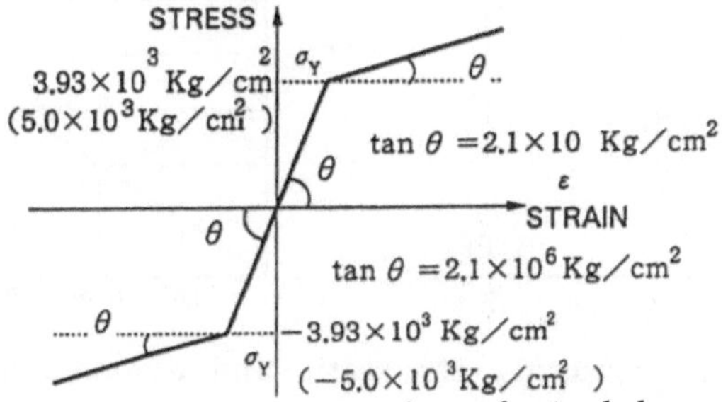

Fig.7 Material properties of steel bars

Material properties of steel bars are shown in Table 3 and Fig. 7. The material properties of concrete were evaluated from raw material testings. These values are shown in Table 4.

material properties	values	
Density (ρ) (g/c㎥)	2.4	
Initial elastic modulus ($\tilde{E}_c$) (kg/c㎡)	2.1×10^5	
Poisson's ratio (ν)	0.166667	
Uniaxial maximum tensile stress ($\tilde{\sigma}_t$) (kg/c㎡)	static	11.4
	dynamic	50.0,* 30.0**
Uniaxial maximum compressive stress ($\tilde{\sigma}_c$) (kg/c㎡)	static	119.0
	dynamic	150.0,* 180.0**
Uniaxial maximum compressive strain ($\tilde{e}_c$)	0.002	
Uniaxial ultimae compressive stress ($\tilde{\sigma}_u$)	static	108.0
	dynamic	135.0,* 162.0**
Uniaxial ultimate compressive strain	0.003	

* : case 1 ** : case 2

Table 4 Material properties of concrete

In this calculation, failure criterion factor values issued by Kupfer and others are used. These factor values are shown in Table 5.

ratio of principal stress	enveloped number					
	I	II	III	IV	V	VI
$\tilde{\sigma}1p/\tilde{\sigma}c$	0.0	0.25	0.5	0.75	1.0	1.2
$\tilde{\sigma}3p/\tilde{\sigma}c$ ($\tilde{\sigma}1p=\tilde{\sigma}2p$)	1.0	1.35	1.75	2.15	2.5	2.8
$\tilde{\sigma}3p/\tilde{\sigma}c$ ($\tilde{\sigma}2p=\beta\tilde{\sigma}3p$)	1.25	1.7	2.1	2.55	2.95	3.3
$\tilde{\sigma}3p/\tilde{\sigma}c$ ($\tilde{\sigma}2p=\tilde{\sigma}3p$)	1.2	1.6	2.0	2.4	2.8	3.1

Table 5 Triaxial failure criterion enveloped number

In static analysis, maximum load 27.2 ton was performed to specimen. Number of loading steps were 5 for numerical analysis. Loading diagram is shown in Fig.8.

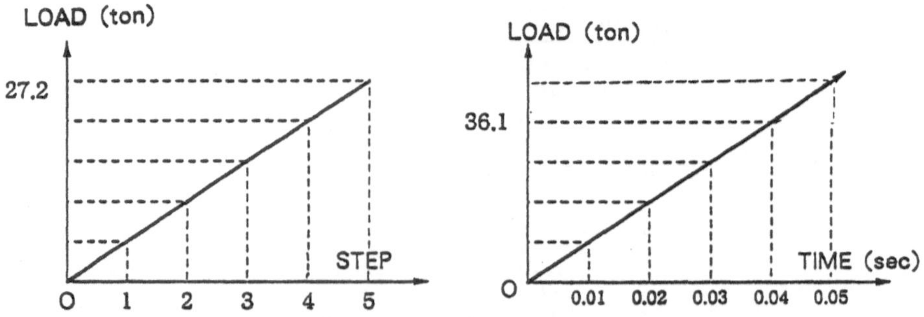

Fig.8 Load condition for static and dynamic analysis

The dynamic response results were got with the load−cell at the tip of actuator. We can see those data in Fig.3. By referencing with those data, we evaluated the high−speed dynamic conditions as concentrated load in Fig.8. The dynamic response analysis was performed by using those data. The maximum load was 36.1 ton.

The analytical time step (Δt) was evaluated with Courant criterion. First of all, propagating speeds (C_e) of steel bar and concrete were caluculated with elastic modulus and density. Nextly, time step (Δt) was calculated with minimum element length ($\Delta \ell$) and Courant condition ($\Delta t \leqq \Delta \ell / c_e$). Lumped mass concept was used. In this non−linear dynamic numerical analysis, $\Delta t = 8 \times 10^{-6}$ and $\Delta t = 8 \times 10^{-2}$ were used. At the case of $\Delta t = 8 \times 10^{-6}$, and the parameter case 2, we could get almost exact results.

6. Comparison between numerical analysis and experiment results.

In this part, comparison for the results between numerical analysis and experiment is discussed. By comparing the relative displacement between numerical analysis (solid line) and experiment (dash line), the good agreement for static loading could be seen in Fig.9.

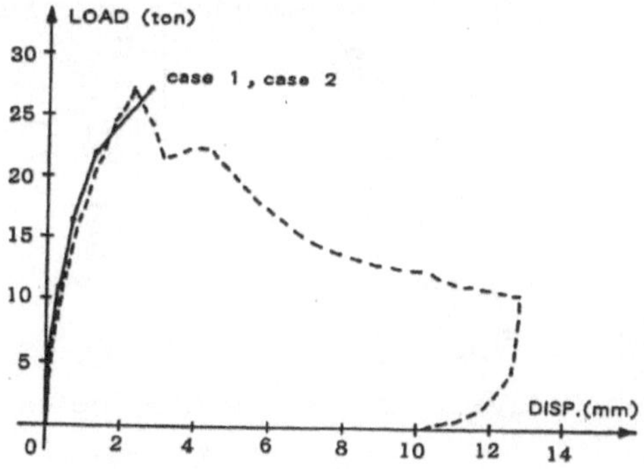

Fig.9 Load−relative displacement diagram

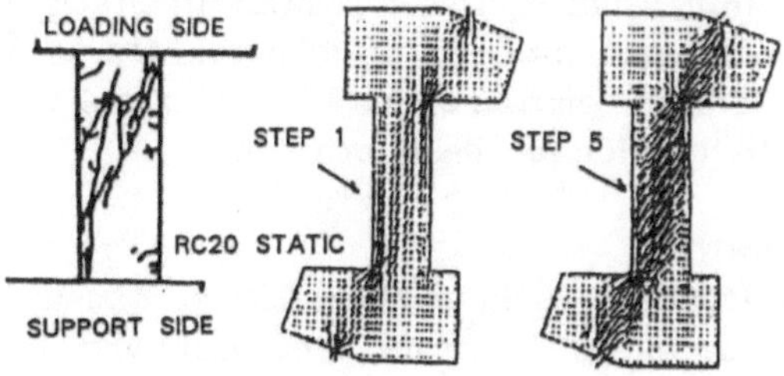

Fig.10 Last crack failure and stress distribution

In Fig.11, high−speed dynamic analysis results (solid line) and experiment results (dash line) are shown. By comparing the both results, we can see quicker failure in the numerical analysis than experiment. For this reason, the numerical analysis was performed without bond effect and geometrical remeshing. Last crack failure and stress distribution diagram are shown in Fig.12.

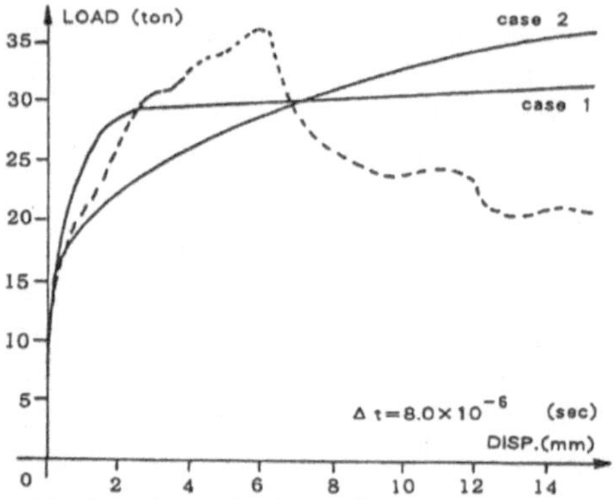

Fig.11 Load−relative displacement diagram

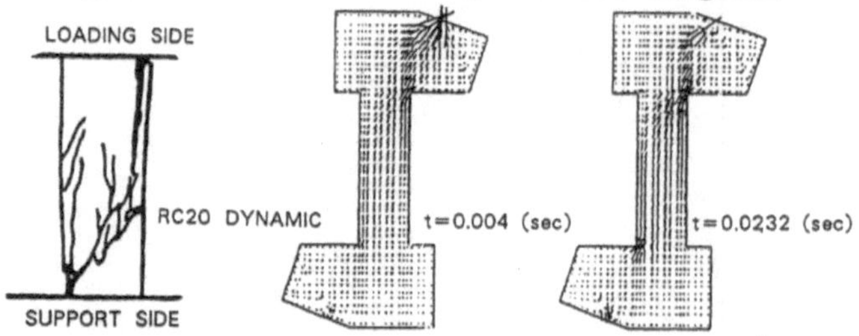

Fig.12 Last crack failure and stress distribution diagram

7. Conclusion

It is very important to compare the experimental results to numerical analysis results. By this method, there was good agreement with experiment and numerical analysis results in static problem. However for high−speed dynamic problem, it is necessary to insert the concept for bond effect, friction problem, contact technique for crack modelling. We will have to challenge these problems.

8. Acknowlegemnt

We thanks Hazama−Gumi to make the specimen. Also we are very happy to get a lot of advice from Prof. Kokusho, the Director of the Research Laboratory of Engineering Materials, Tokyo Institute of Technology.

9. Reference

K.J.Bathe "Finite Element Procedures in Engineering Analysis"

ANALYSIS OF FOOTING BEHAVIOUR ON HOMOGENEOUS AND LAYERED SOILS

AL-MUKHTAR M. , ROBINET J.C. , SHAHROUR I.
Université des Sciences et Techniques de Lille - EUDIL
59655 Villeneuve d'Ascq Cedex - FRANCE

Abstract

In the first part of this paper we present footings tests performed in calibrated chamber on homogenous and layered samples. These tests are then compared with known experimental and analytical data on footings behaviour.

In the second part we present the simulation of experimental tests on homogeneous and layered samplę by Lade's elastoplastic model.This model has been introduced in a finite element program .

1.-INTRODUCTION

Design of foundations on layered soils is of great importance in practice . This problem has been under investigation for the last years ; Hanna and Meyerhof (1979-1980) have carried experimental and analytical studies on foundations on two layered soils and proposed some formulas for thier design ; Griffith (1982) analysed this problem by means of elasto-plastic models introduced in a finite element program.

The civil engineering laboratory of Lille has carried out footings tests on cylindrical samples placed in a triaxial cell, these tests were performed under various initial and limit conditions . In this paper we present experimental tests on two layered soils and the predictions obtained by Lade's elastoplastic model .

2. EXPERIMENTAL APPARATUS

Tests were performed on cylindrical samples of 18 cm in diameter and height set in a triaxial cell . The lower base of the triaxial cell was modified to allow the driving into the sample of a rigid footing of 5 cm in diameter as shown in figure 1 . During the test the applied load, the footing penetration and the volume change are mesured.

Tests can be carried out under drained or undrained conditions on samples submitted initially to isotropic or deviatotic stress state .

3. EXPERIMENTAL RESULTS

3.1 Tests on homogeneous samples

Penetration tests were performed on Hostun sand ; in figure 2 we give the results obtained on loose sand for various initial isotropic confinement, this figure shows that the applied load increases with the confinement pressure.

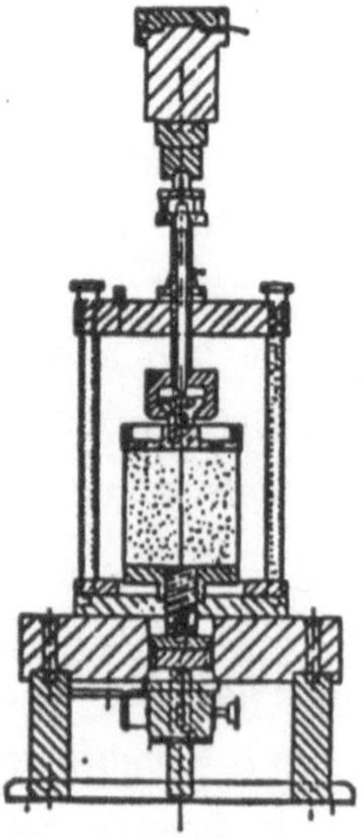

Figure 1 : Testing Device

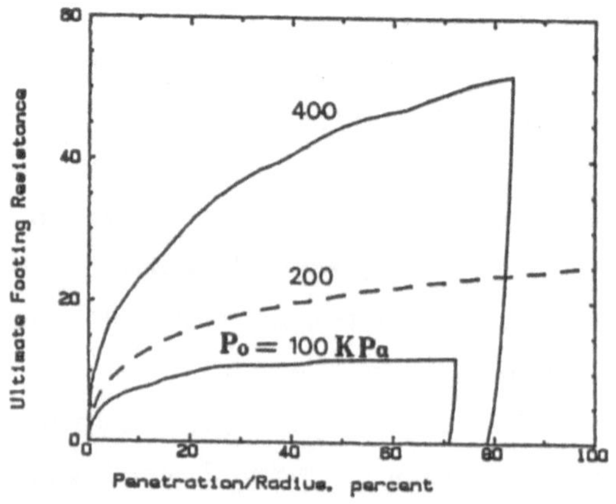

Figure 2 : Penetration tests on loose sand

In figure 3 we give the experimental ultimate loads in terms of the confinement pressure ; on the same figure we give the ultimate loads obtained by Terzaghi's expressions (1943) for generalised and local shear failure . This figure shows that in our tests a mechanism of local failure is expected to occur . The difference between the experimental results and Terzaghi's theoritical ones can be explained by the scale effet Vesic (1969) and the decrease of the confinement load on the lower side of the sample (surcharge q) during the test .

3.2 Tests on two layered samples

In this section we present tests performed on two layered samples, the lower layer which is in contact with the footing is constituted of dense sand, the second one is made of loose sand as shown in figure 4 . Tests were performed for various thickness of the dense layer (ratio H/B varies between 0 and 3.6).

In figure 5 and 6 we give the obtained results for various values of H/B . These figures show that the footing resistance increases with the ratio H/B .

Figure 7 shows that the ultimate footings resistance increases when the the ratio H/B rises from 0 to 2 and then it remains constant. These results agree qualitatively well with those preposed by Hanna and Meyerhof (1979) given in figure 8 .

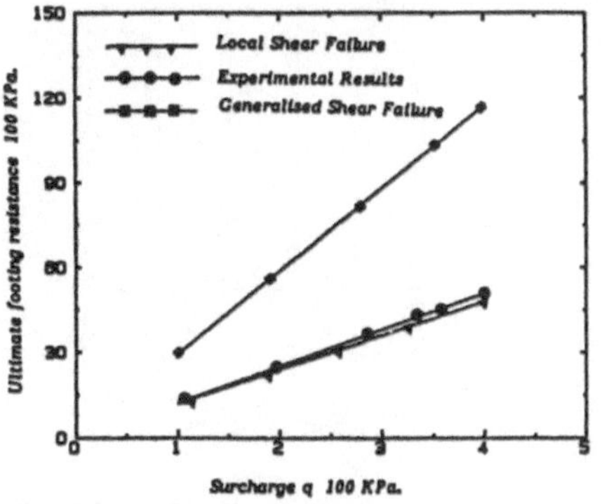

Figure 3 : Experimental and theoretical bearing capacity for loose sand

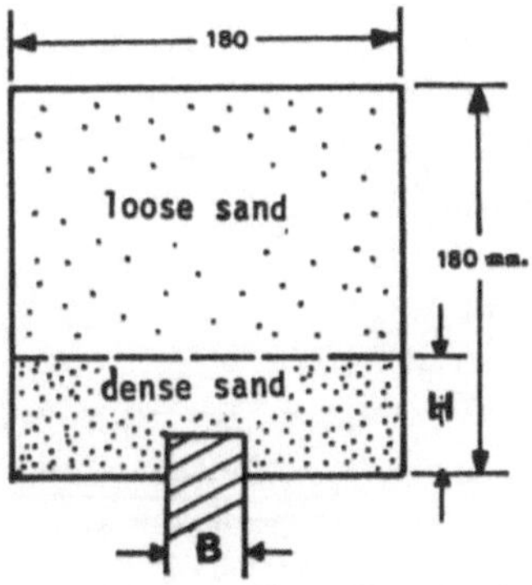

Figure 4 : Two-layer sample

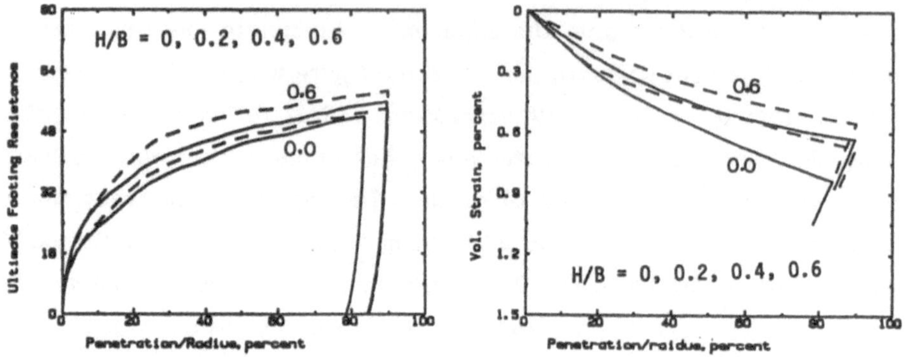

Figure 5 : Penetration tests for low values of H/B

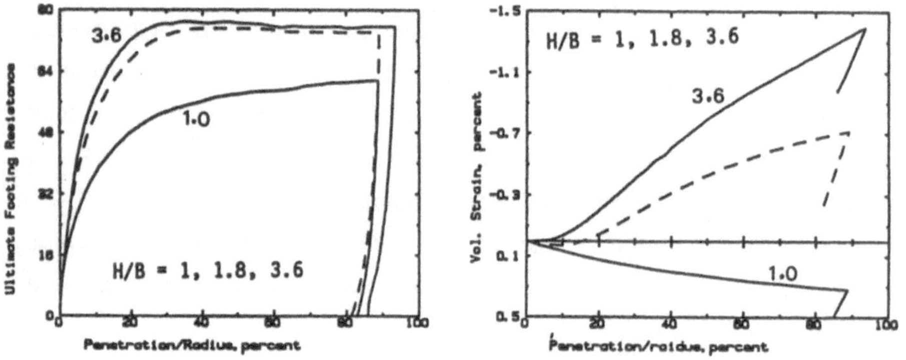

Figure 6 : Penetration tests for high values of H/B

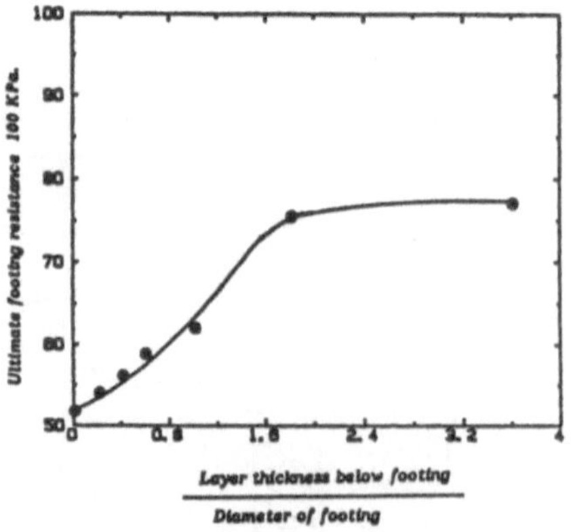

Figure 7 : Experimental bearing capacity for two layer sand

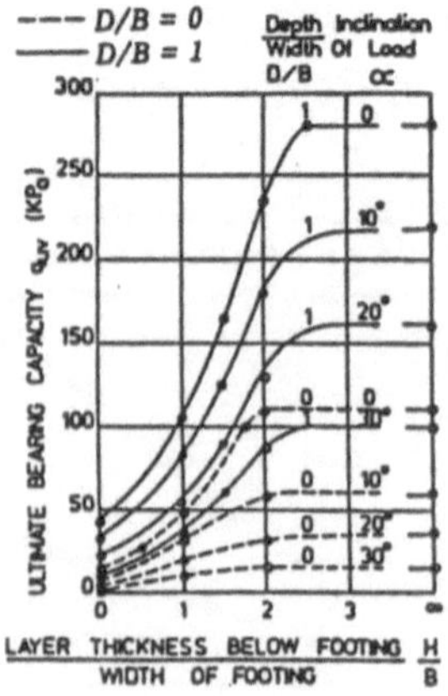

Figure 8 : Bearing capacity for two layer sand (Meyerhof)

4. NUMERICAL PREDICTION OF PENETRATIONS TESTS

Penetrations tests were carried out in order to validate soils costitutive models ; in this section we present predictions obtained by Lade's elasto-plastic model .

4.1 Model presentation

Lade (1975) has proposed a non associative elasto-plastic model for soils . The elastic part of this model is isotropic and non linear Young's modulus is given by :

$$E = E_0 \, (P/P_a)^n$$

P and P_a represent respectively the mean and atmospheric pressure, E_0 and n are two model parameters ; Poisson's ratio (N_0) is assumed to be constant .

The yield criterion is given by the following expression :

$$F = f + K\,(W_p)$$

where $f = I1^3 \,/\, I3$

$$K\,(W_p) = ft + W_p \,/\, (a + d\, W_p)$$

I1 and I3 are the first and third invariants of stress tensor,
W_p represent the plastic work, a and d are coffecients given by :

$$a = m\, P_a \,(P \,/\, P_a)^l$$

$$d = rf \,/\, (K1 - ft)$$

where m, l, rf, K1, and ft are model parameters .

The plastic potential is expressed by :

$$g = I1^3 \,(f - 27)^c \qquad c = 1/(1-A)$$

A is a model parameter .

Model parameters have been determined for Hostun sand from triaxial tests AL-Mukhtar (1987) . In the following table we give the values of these parameters for loose and dense sand .

Sand	N_0	E_0	n	m	l	rf	ft	A	K1
Loose	0.2	430	0.47	0.0012	1.04	0.95	30	0.35	48
Dense	0.2	540	0.5	0.0002	1.57	0.89	34	0.43	60

4.2 Footings tests predictions

Lade's model has been introduced in a finite element program based on Newton-Raphson modified method . 25 eight-nodes isoparametric elements mesh was used in numerical test, loads were applied by imposing a vertical displacement at nodes under the footing .

We consider that the sample is submitted in each point to a given initial pressure ; on the boundary of the sample we impose the conditions shown in figure 9 .

The modelisation of the two layered samples was obtained by giving to the elements in the mesh the appropriate model's parameter .

As convergence became difficult for high penetrations, predictions were carried up only for footings penetrations of nearly 20 % of its radius .

$u = 0$ on AE
$v = 0$ on AB
$v = \bar{v}$ on ED
$p = 0$ on BC U CD

where

u : the horizontal displacement
v : the vertical displacement
$\bar{v}$: the footing penetration
p : the exterior pressure

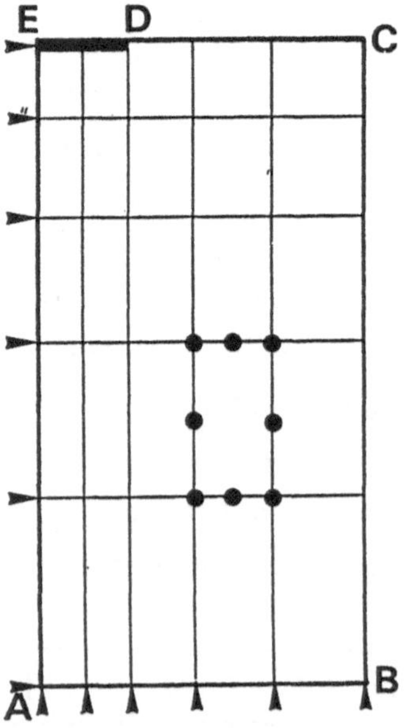

Figure 9 : Finite elements mesh and boundary conditions

In figure 10a and 10b we give the obtained results for low values of ratio H/B ; this figure shows that Lade's model gives satisfactory results for loose sand and describes the increase of the footing resistance with H/B .

In figure 11a and 11b we give the obtained results for high values of this ratio ; we observe that, in this case the model underestimates the footing resistance .

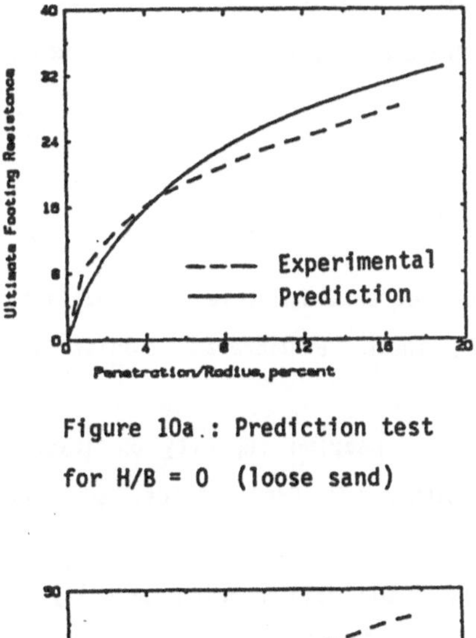

Figure 10a.: Prediction test for H/B = 0 (loose sand)

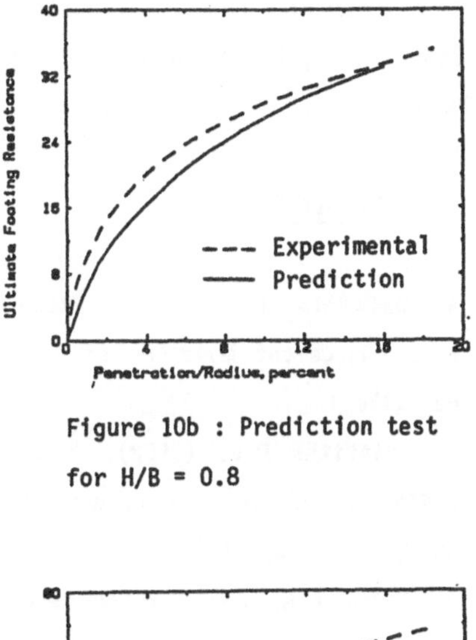

Figure 10b : Prediction test for H/B = 0.8

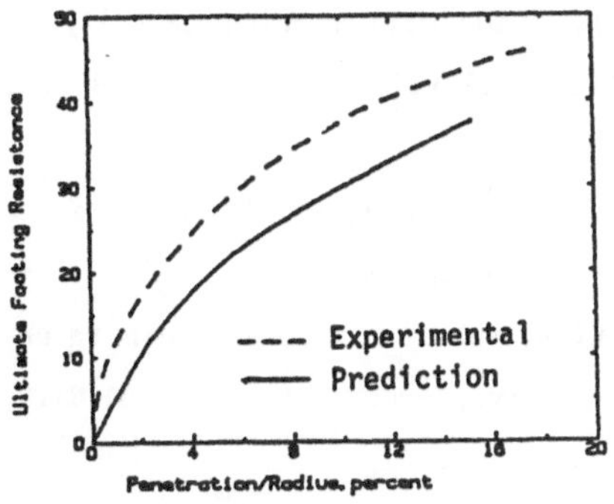

Figure 11a : Prediction test for H/B = 2

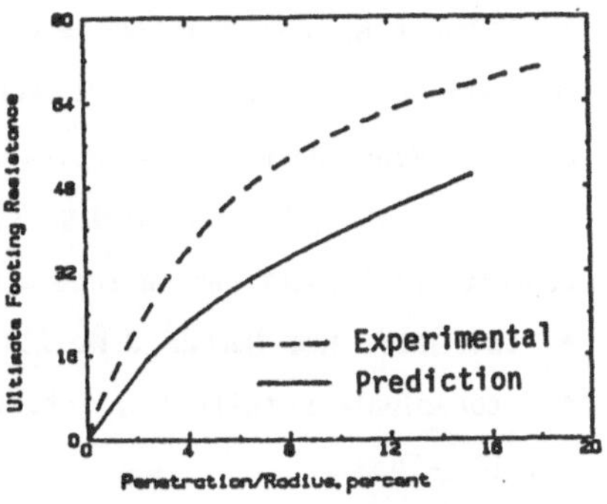

Figure 11b : Prediction test for H/B = 3.6 (dense sand)

CONCLUSION

In this paper we have presented footings tests performed in calibrated chamber on homogeneous and two layered samples.

The obtained ultimate capacity is lower than that proposed by Terzaghi and Hanna and Meyerhof ; this difference is owing to scale effects and to the decrease of the applied confinement load (surcharge) on the sample during the test.

Predictions of these tests were carried with Lade's elastoplastic model ; predictions show that the model gives satisfactory results for loose sand and describes the increase of the ultimate footing resistance when the thickness of the dense layer rises, but underestimates the applied load in the case of dense sand.

REFERENCES

1. **AL-Mukhtar M.** (1987), " Etude théorique et expérimentale des fondations sous chargement monotons et cycliques ", Thése de Doctorat - Université de Lille 1

2. **Griffiths D.V.** (1982), " Computation of bearing capacity on layered soils ", 4th Int. Conf. on numerical methods in Geomechanics, Edmonton 1982, Vol 1, pp. 163170 .

3. **Hanna A.M. and Meyerhof G.G.** (1979), " Ultimate bearing capacity of foundations on a three-layer soils with special reference to layered sand Canadian Geot. Jour. no.16, pp. 412-414 .

4. **Hanna A.M. and Meyerhof G.G.** (1980), " Design charts of ultimate bearing capacity of foundations on sand overlying soft clay " .

5. **Lade P.V. and Duncan J.M.** (1975) " Elasto-plastic stress-strain theory for cohesionless soils " , Int. Geo. Eng. Div., Proc. ASCE, Vol.101,GT10, pp. 1037-1053 .

6. **Vesic A.S.** (1969) " Effects of scale and compressibility on bearing capacity of surface foundations ", 7th Int. Conf. on Soil Mechanics and foundation Engineering, Mexico City, Vol. III, pp. 270-272 .

FRACTURE MECHANICS FOR DELAMINATION ON COMPOSITE STRUCTURES IN COMPRESSION

H. CHAOUK and G. P. STEVEN
The University of Sydney
N.S.W. 2006. Australia.

SUMMARY

In Graphite/Epoxy composites delamination of the plies under compressive loading is a serious failure mode. When the delamination is near the surface the lay-up of the surface plies has a significant effect on delamination growth. Linear elastic fracture mechanics together with 3D FE modelling is used to evaluate the components of the energy release rate for specimens with artificially introduced delaminations. The results are compared with experimental observations.

1. INTRODUCTION

Delamination, or interlaminar cracking, is one of the frequently encountered types of damage in advanced materials [1]. This phenomenon is unique to composites and bonded joints and is not found in metals. It usually takes the form of separation of plies and is generally initiated at geometric boundaries or defects such as voids, microcracks and foreign inclusions. Delamination causes structural degradation, stiffness reduction and leads ultimately to failure at stresses well below the design levels for an undamaged laminate. Therefore understanding of some of the basic mechanics of delamination is of great importance in designing for and maintaining the structural integrity of advanced composite materials and structures. By regarding a delamination region as a plane crack in a three dimensional body, concepts of fracture mechanics can be employed to study the growth of delaminations and their behaviour under load. Also since the common matrix materials such as epoxy resins are relatively brittle at normal temperatures, linear elastic fracture mechanics is often adopted as a first modelling strategy. The analysis of the stress state in the proximity of the delamination can thus be seen as an important contribution to the understanding of delamination behaviour.

Earlier studies [1] have used 2D finite element analysis and others [2-5] have had laminate in tension, which is not considered to be the most severe loading case. These investigations have also concentrated merely on the stress analysis around the delamination. The authors have recently [6] presented some 3D finite element analysis for a near surface delamination under compression, which demonstrated the importance of near surface lay up and the significance of the local inter-laminar stresses. It was shown both theoretically and experimentally that the near surface lay up can influence whether delamination occurs or not and the rate at which delamination will spread under repeated load. In this paper we study the edge delamination under compression loading for eight lay up sequences and show how the near surface lay up can influence delamination behaviour and therefore growth. It will be shown how near surface lay up can change the nature of the membrane to coupling matrix for the surface delaminated plies. In compression the membrane bending coupling can either tend to close or open up the delamination.

Fracture mechanics concepts to do with linear elastic energy release rates are employed for each of the eight lay ups. A virtual crack extension method together with a 3D anisotropic finite element model can determine the energy associated with any mode of crack loading, be it mode 1 or 2 or 3. The various energy release rates G_1, G_2 or G_3 are related to the type of near surface lay up which in turn relates to delamination behaviour.

Experimental studies reported in [8] show that edge delaminations behave differently according to near surface lay up sequence. Combining the theoretical energy release rates with experimentally observed growth laws gives us the starting point for some delamination growth laws in real structures.

2. BASIC FORMULATION

From the fundamental theory for anisotropic laminates, in the case of both inplane and out-of-plane forces $\{N\}$ and $\{M\}$, the stiffness matrix of a laminate is of the form

$$\begin{bmatrix} N_x \\ N_y \\ N_{xy} \end{bmatrix} = \begin{bmatrix} A_{11} & A_{12} & A_{16} \\ A_{12} & A_{22} & A_{26} \\ A_{16} & A_{26} & A_{66} \end{bmatrix} \begin{bmatrix} \varepsilon_x{}^0 \\ \varepsilon_y{}^0 \\ \gamma^0_{xy} \end{bmatrix} + \begin{bmatrix} B_{11} & B_{12} & B_{16} \\ B_{12} & B_{22} & B_{26} \\ B_{16} & B_{26} & B_{66} \end{bmatrix} \begin{bmatrix} K_x \\ K_y \\ K_{xy} \end{bmatrix}$$

$$\begin{bmatrix} M_x \\ M_y \\ M_{xy} \end{bmatrix} = \begin{bmatrix} B_{11} & B_{12} & B_{16} \\ B_{12} & B_{22} & B_{26} \\ B_{16} & B_{26} & B_{66} \end{bmatrix} \begin{bmatrix} \varepsilon^0{}_x \\ \varepsilon^0{}_y \\ \gamma^0{}_{xy} \end{bmatrix} + \begin{bmatrix} D_{11} & D_{12} & D_{16} \\ D_{12} & D_{22} & D_{26} \\ D_{16} & D_{26} & D_{66} \end{bmatrix} \begin{bmatrix} K_x \\ K_y \\ K_{xy} \end{bmatrix}$$

where $\{\varepsilon^0\}_x$ and $\{K\}$ are the lowest midplane strains and curvatures.

In this notation, (A) is called the extensional stiffness matrix, (B) is called the coupling stiffness matrix and (D) is the bending stiffness matrix. These matrices are related to ply properties and positions in the laminations as shown in [6]. It was also shown in [6] that stretching a lamination, that has non zero term (B), will produce bending and/or twisting in addition to extensional and shear deformation, and that a proper choice of ply orientation can give appropriate terms for the (B) matrix which can increase the damage tolerance of the laminate for its service loads. In general a decrease in damage tolerance comes about because the sign of the terms in the (B) matrix are such that membrane compressive loading can cause the surface plies above a delamination to curve outwards, thus tending to increase the delamination.

3. NUMERICAL ANALYSIS

In this paper, the laminates considered are always modelled as a set of orthotropic 20 nodes brick elements as shown in Figure 1 and the STRAND5 finite element program is used. STRAND5 has been developed at the Department of Aeronautical Engineering, The University of Sydney over the last ten years and has been extensively benchmarked.

Each ply is modelled separately around the delaminated region (i.e. 2 plies above, and 2 plies below crack centreline), and the rest of the plies are then modelled as one super brick element with appropriate properties using the principle outlined in [7]. The laminates considered are as shown in Table 1. Indeed the numerical model was carefully developed so as to easily vary the ply lay up and the location of delamination between the plies (i.e. through the thickness). The laminates were loaded by a uniform compressive stress. Constraints were

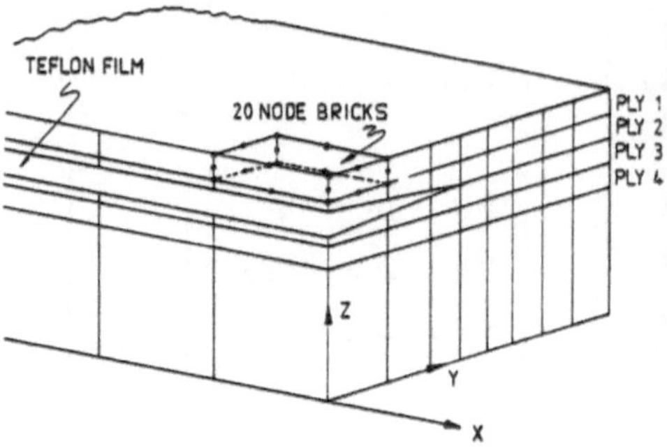

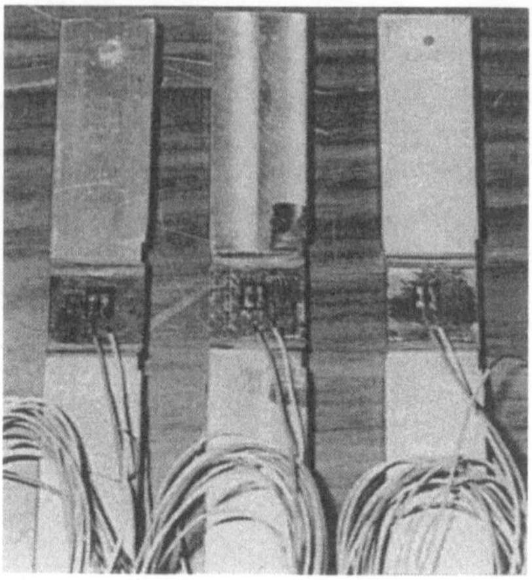

Figure 1 Finite Element Representation.   Figure 2.

PLY NO.	CONFIGURATION							
	1	2	3	4	5	6	7	8
1	0	0	0	45	45	-45	90	90
2	45	0	90	-45	-45	45	0	45
3	-45	45	45	0	90	0	45	-45
4	90	-45	-45	90	0	90	-45	0
5	0	0	0	0	0	0	0	0
6	45	45	45	45	45	45	45	45
7	-45	-45	-45	-45	-45	-45	-45	-45
8	0	0	0	0	0	0	0	0
9	45	45	45	45	45	45	45	45
10	-45	-45	-45	-45	-45	-45	-45	-45
11	0	0	0	0	0	0	0	0
	SYMMETRIC							

TABLE 1.

induced by means of beams elements in those areas of the delamination where closure was found to occur. Such constraint beam elements had zero extensional strength for crack opening and a very high strength for closing.

Since the contact between the two surfaces is modelled by this non-linear constraint it is necessary to use a Newton-Raphson solution procedure for the finite element modelling.

For the Newton-Raphson iterative cycle the displacement field $\{\delta\}$ at the $(n+1)^{th}$ iteration is given by

$$\{\delta\}^{n+1} = \{\delta\}^n + [k_T{}^n]^{g-1} \{\{F\}^g - [k_S{}^n]^g\{\delta\}^n\}.$$

In this $[k_T{}^n]^g$ and $[k_S{}^n]^g$ are the global tangent and secant stiffness matrices at the $(n)^{th}$ iteration and $\{F\}^g$ the applied loads. It is to be noted that since only the contact constraint beams have the non-linear behaviour, it is only necessary to assemble the stiffness matrix for the 3D laminate once. Using the subscripts 3D and b for the laminate and the beams respectively:

$$[k_T]^g = [k_{3D}]^g + [k_{T,b}]^g$$

$$[k_S]^g = [k_{3D}]^g + [k_{S,b}]^g.$$

Thus it can be seen that the iterative cycle involves only a small amount of additional numerical effort and indeed since the contact constraint has two essentially linear portions, convergence is always achieved in only two or three iterations.

The super-brick element, which allows for an asymmetric lay up, is carefully modelled by calculating its appropriate modulus and substituting it in the three-dimensional finite element so that correct representation is achieved.

Only half of each tested specimen was modelled and the boundary constraintsapplied were:

$u_x = 0$ for all nodes on specimen centreline (see Figure 1).

$v_x = w_z = 0$ for only the four corner nodes on the loading surface where the load is transferred through the tabs to the specimen's testing section (see Figure 2) see specimen description later.

4. FRACTURE MECHANICS

The principles of linear elastic fracture mechanics are well established and testing in finite element modelling, see [9] and [10], and need not be elaborated on here. The purpose of this investigation is to relate the near surface ply lay up to the composition of the energy release rate when a delamination grows. The energy release rate G contains terms from the three fracture modes. Mode 1 is an opening mode, Mode 2 is due to movement normal to the crack front and Mode 3 is due to movement parallel to the crack front. The associated energy release rates are denoted as G_1, G_2 and G_3 and combine to give a total release rate of $G = (G_1{}^2 + G_2{}^2 + G_3{}^2)^{\frac{1}{2}}$.

A virtual crack extension method is used to determine G_1, G_2 and G_3 for any of the FE models of the configurations in Table 1. To apply this method the delamination front is allowed to advance by a small amount. The displacement components (u, v and w) of the nodes which lay on the initial crack front are recorded. A subsequent FE analysis applies kinematic constraints such that these displacements are returned to being zero and the energy associated with each component of the displacement gives the corresponding mode of energy release G_1, G_2 and G_3.

5. EXPERIMENTAL ANALYSIS

The specimens used in this investigation were fabricated from unidirectional AS4/3501-6 graphite/epoxy tapes. Panels with symmetrical 0/45/-45/90/0/45/-45/0/45/-45/0 s and 45/-45/-/90/0/45/-45/0/45/-45/0 s lay up were prepared and cured according to the manufacturer's recommendations. Delaminations were introduced by inserting teflon films of different sizes between the second and third ply as shown in figure 1. Individual specimens were cut using a diamond head saw and bonded to aluminium tabs using FM300 hot cured adhesives. To ensure no failure at laminate/tab attachment the aluminium was carefully etched beforehand.

The specimens were then tested using a modified Iitri compression jig mounted on a 1195 series Instron machine. Strain gauges on both sides of each specimen as well as a clip gauge were used to accurately monitor the delamination behaviour in each laminate.

6. RESULTS AND DISCUSSION

Table 2 gives the results of our analyses. From this table some significant conclusions may be reached.

a) Configurations 1, 2 and 3 have Mode 1 as the dominant term. The ratio G_3/G_1 is small with configuration 2 having the largest G_1/G. These lay ups all exhibit opening behaviour. This is seen in the deformed FE model shown in Figure 3.
b) Configurations 4, 5 and 6 have Mode 2 as the dominant term with a lower G value overall. All these models exhibit almost total crack closing behaviour which would be expected to reduce significantly opening Mode 1 behaviour. Non-linear analysis was necessary for all these lay-ups and a sample of a deformed FE model is shown in Figure 4.
c) Configurations 7 and 8 exhibit partial crack closure near the free edge and have a more balanced distribution of G_1, G_2 and G_3. Number 8 has similar ratios to number 5 but its deformation and stresses are totally different.

When examining the stresses near the crack tips for these various lay ups [8] it has been observed that there are signif-

CONFIG-URATION	NEAR SURF-ACE LAY UP	$\frac{G_1}{G}$	$\frac{G_2}{G}$	$\frac{G_3}{G}$	$\frac{G_2}{G_1}$	$\frac{G_3}{G_1}$	G_0 J/M^2
1	0/45/-45	0.5142	0.438	0.048	0.851	0.0933	93.1
2	0/0/45	0.81659	0.0896	0.0939	0.058	0.1147	96.3
3	0/90/45	0.6269	0.3131	0.05985	0.499	0.0954	114.88
4	+45/-45/0	0.069	0.8913	0.0391	12.9	0.56	32.5
5	45/-45/90	0.1278	0.633	0.239	4.95	1.87	40.0
6	-45/45/0	0.058	0.868	0.078	15.0	1.27	29.7
7	90/0/45	0.3289	0.4362	0.2348	1.326	0.713	92.12
8	90/45/-45	0.1436	0.6438	0.2125	4.483	1.479	55.7

TABLE 2: Results of virtual crack extension method on 3D modelling of delamination growth.

For material properties

E_{11} = 145.5 GPA $\quad v_1$ = 60%
E_{22} = 13.79 GPA
G_{12} = 5.86 GPA $\quad \nu_{12}$ = 0.3

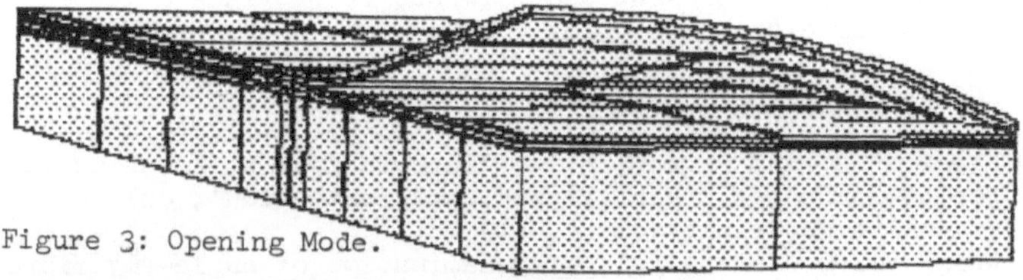

Figure 3: Opening Mode.

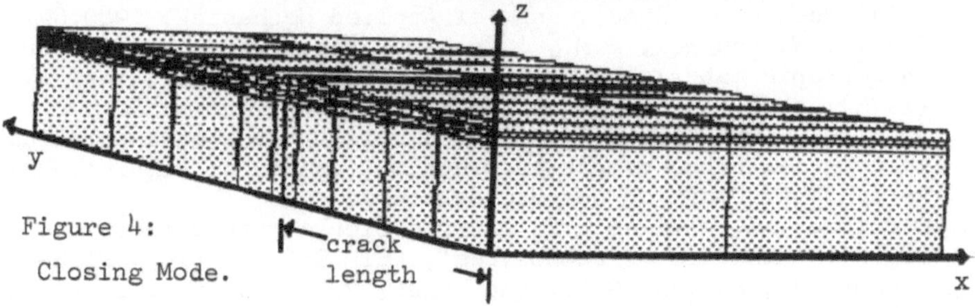

Figure 4:
Closing Mode.

icantly different features between the three groups a) b) and c) above. The experimental studies undertaken and reported in (6) bear out totally the analytically observed phenomena and provide us with some justification for our further work towards determining growth laws and failure criteria in debonded composites under compression loading.

To summarise the physical features that emerge from this study it can be said that laminates with 0° surface plies would give rise to Mode 1 and Mode 2 failure behaviour whereas laminates with 90° surface plies increase the influence of Modes 2 and 3. Laminates with 45°/-45° surface plies do not experience much Mode 1 and are dominated by Mode 2. In designing damage tolerant composite structures such information is important.

7. CONCLUDING REMARKS

In all the models studied herein the delamination has been between the 2nd and 3rd plies. Further work with other lay ups and delamination positions has still to be done. Eventually it is hoped that a sufficient body of knowledge can be established to be able to present criteria for growth in any type and form of delamination.

Our work demonstrates that the F.E. computer modelling is capable of reproducing some of the phenomena observed experimentally and thus provides a basis for developing fracture parameters based on modelling rather than experiment. We appreciate that compression testing of strips with Teflon inserts does not relate well to the actual structural environment where delamination damage may be present and grow but it provides a laboratory situation for close controllable study.

8. REFERENCES

1. Pipes, R.B. and Pagano, N.J., 'Interlaminar stresses in composite laminates under uniform axial extension', Jnl. of Composite Material, vol. 4, (1970).
2. Wang, S.S., 'An analysis of delamination of angle-ply fibre reinforced composites', Jnl. of Applied Mechanics, 1980,69-70
3. Tsai, S.W. and Wui, E.M., 'A general theory of strength for anisotropic materials', Jnl. of Composite Material, vol. 15, (1971), 58-80.
4. Hong, C.S. and Kim, K.S., 'An analysis of free edge delamination in laminated composite under uniform axial strain', Proc. Fourth Int. Conf. on Composite Materials.
5. O'Brien, T.K., 'Mixed mode energy release rate effects on edge delamination of composites', ASTM STP 836 (1984),125-142
6. Chaouk, H., Steven, G.P., and Jones, R., 'Closure effects in delaminated graphite epoxy laminate under compression', To be published in the Jnl. of Composite Structures, (1986).
7. Jones, R.J., Callinan, R., Teh, K.K., and Brown, K.C., 'Analysis of multilayer laminates using three dimensional super-elements', Int. Jnl. Num. Meth. in Engng., vol. 20.
8. Chaouk, H., and Steven, G.P., 'Finite Element studies of near surface delamination in laminated composites under compression', Proc. 5th Int. Conf. on FE in Australia, Aug. 86, Melbourne University.

9. Brock, D., *Elementary Engineering Fracture Mechanics*, Martinos Nijhoff, 1982.
10. Pian, T.H.H., 'Crack Elements', Proc. World Con. FE in Struct. Mech., vol. 1, Bournemouth, (1975), F1-F39.

A NEW MULTIVARIABLE FINITE ELEMENT ALGORITHM AND A BREAKABLE ELEMENT ALGORITHM FOR ELASTO-PLASTIC FRACTURE ANALYSIS

J.Y.Zhang
Dept. of Mechanics, Tianjin University, Tianjin, China
T.R.Hsu
Dept. of Mech. Engng. University of Manitoba, Winnipeg, Manitoba, Canada

SUMMARY

A new multivariable finite element algorithm is formulated on the basis of the generalized virtual work principles. The algorithm possesses the advantages of the traditional hybrid finite element[1] and quasi-conforming element[2] but it has more flexibility and makes the hybrid concept and quasi-conforming concept to be easily generalized to different problems in the continuum mechanics. This algorithm has been implemented in a quadrilateral planar hybrid element with finite displacement consideration and strain singular terms for the application to elasto-plastic fracture analysis.

The "breakable element" algorithm was proposed by one of the authors (T.R.H)[3] for producing smoother crack growth. Some improvements of the algorithm have been made in present paper. The multivariable finite element algorithm mentioned above with the breakable element algorithm was successfully used to predict the stable growth of a line crack in a large thin aluminum plate. The rupture effective strain was used as a fracture criterion.

1. INTRODUCTION

The hybrid model of finite element method has found ample applications in solid mechanics since it was first introduced by professor Pian in 1964. The essence of this model is that more than one kind of variable functions can be treated as primary variables. Thus among all of the requirements the variables should satisfy in solid mechanics, one can selectively make some of them to be exactly satisfied, but the rest of them to be approximately satisfied in a sense of an energy principle, so the balance between equilibrium requirements and compatibility requirements can be kept in order to make the model avoiding over-stiff and over-compliant, as well as it is easy to follow certain physical variation of some variables. These characteristics mentioned above can be used to improve the accuracy of both stresses and displacements[4],[5]. The conventional method to derive a hybrid finite element is based on a modified variational principles or generalized variational principles. The authors found that one can bring the advantages of hybrid concept and quasi-conforming concept into full play and simplify the derivations of hydrid model by using a multivariable finite element algorithm formulated on the basis of the generalized virtual work principles.

Special technique is required when using the finite element method for the simulation of crack growth in a ductile material. Proper numerical procedure has to be developed to describe this type of fracture. Two such procedures has been used by many researchers. These are: (1) the use of "dynamic crack tip elements" such as described in [6], and (2) the nodal relaxation process [7]. The former method requires the adjustment of nodal coordinates of moving crack tip and those at the neighborring nodes. Excessive iteration is often required to achieve accurate result. Once the crack tip has travelled by one-element length, re-numbering of all elements in the model has to be carried out. The nodal relaxation method, although appears to be relatively simpler in the computation, the crack tip requires to "jump" from one end of the element to the other, which can not be regarded as realistic. An alternative approach called "Breakable element" algorithm allows the crack tip to move within one element, yet no re-numbering of nodes or elements is required during the entire process. This algorithm has been incorporated into a code named TEPSA[8], but the computed results came under the influence of the size of breakable elements. The authors found that this disadvantage can be overcome by adjusting the mechanical propertis of the breakable elements.

2. MULTIVARIABLE FINITE ELEMENT ALGORITHM BASED ON THE GENERALIZED VIRTUAL WORK PRINCIPLES

Different mixed-hybrid finite element model can be formulated by this algorithm. A quadrilateral planar element similar to the hybrid-stress element was formulated as an example, as well as the geometric nonlinearity and the singular terms were considered for the application to elasto-plastic fracture analysis.

The algorithm consists of two steps. As the first step,

the generalized virtual stress principle is used to express the compatibility requirements in an element. By assuming an element of V^e with boundary S^e, for the finite displacement case the corresponding generalized virtual work equation can be expressed as[9]:

$$\int_{V^e} (e_{ij}\,\delta\sigma_{ij} - u_{i,j}\,\delta\sigma_{ij} - 1/2\,u_{k,i}u_{k,j}\,\delta\sigma_{ij})\,dV = \int_{S^e} (\tilde{u}_i - u_i)\,\delta(\sigma_{ij}n_j + \sigma_{jk}u_{i,k}n_j)\,dS \qquad (1)$$

where u_k, e_{ij}, σ_{ij} and $\delta\sigma_{ij}$ are the respective displacement, strain, stress and virtual stress components in an element. $\tilde{u}_k$ are the displacement components on the boundary of the element. n_i are the direction cosine with the corrdinates x_i. A distinct advantage of the generalized virtual work equation is that there is no subsidiary condition for the element variables.

For the general elements through comparison the following expressions are employed (Fig.1):

$$\{e\} = \begin{Bmatrix} e_x \\ e_y \\ e_{xy} \end{Bmatrix} = [P] \begin{Bmatrix} \alpha_1 \\ \alpha_2 \\ \vdots \\ \alpha_5 \end{Bmatrix} = [P]\{\alpha\} \qquad (2)$$

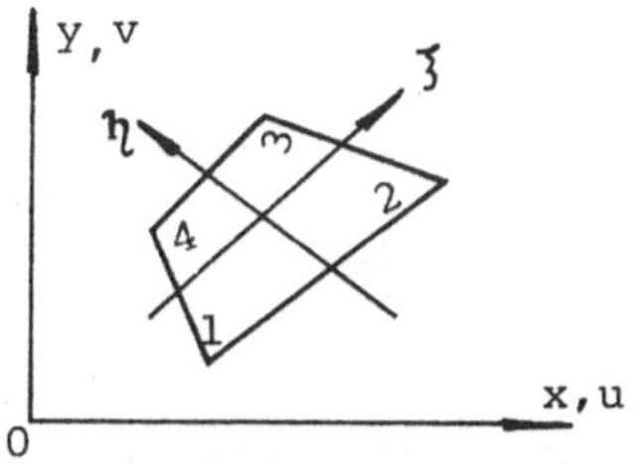

Fig.1 Quadrilateral Element

$$\{\delta\sigma\} = \begin{Bmatrix} \delta\sigma_x \\ \delta\sigma_y \\ \delta\sigma_{xy} \end{Bmatrix} = [P] \begin{Bmatrix} \beta_1 \\ \beta_2 \\ \vdots \\ \beta_5 \end{Bmatrix} = [P]\{\beta\} \qquad (3)$$

$$\{U\} = \{\tilde{U}\} = \begin{Bmatrix} u \\ v \end{Bmatrix} = \sum_{i=1}^{4} N_i \begin{Bmatrix} u_i \\ v_i \end{Bmatrix} \qquad (4)$$

where $\{e\}$, $\{\delta\sigma\}$ and $\{U\}$ are repective strain colum matrix, virtual stress column matrix and displacement column matrix in an element, $\{\tilde{U}\}$ is the displacement column matrix on the boundary of the element, $\{\alpha\}$ are the unknown strain parameters, $\{\beta\}$ are the virtual stress parameters, u_i, v_i are the nodal displacements and

$$[P] = \begin{bmatrix} 1 & y & 0 & 0 & 0 \\ 0 & 0 & 1 & x & 0 \\ 0 & 0 & 0 & 0 & 1 \end{bmatrix} \qquad (5)$$

$$N_i = 1/4(1+\xi_i\xi)(1+\eta_i\eta) \qquad (6)$$

For the singluar elements, we take 11 parameters for $\{\alpha\}$ and $\{\beta\}$, and

$$[P] = \begin{bmatrix} 1 & y & x/r^Q & y/r^Q & 0 & 0 & 0 & 0 & 0 & 0 & 0 \\ 0 & 0 & 0 & 0 & 1 & x & x/r^Q & y/r^Q & 0 & 0 & 0 \\ 0 & 0 & 0 & 0 & 0 & 0 & 0 & 0 & 1 & x/r^Q & y/r^Q \end{bmatrix} \quad (7)$$

moreover, on the sides(i-j) intersecting with the crack tip, U should take the form:

$$\{U\} = \begin{Bmatrix} u \\ v \end{Bmatrix} = \begin{bmatrix} \dfrac{r_i^b - r^b}{r_j^b - r_i^b} & \dfrac{r_j^b - r^b}{r_i^b - r_j^b} & 0 & 0 \\ 0 & 0 & \dfrac{r_j^b - r^b}{r_j^b - r_i^b} & \dfrac{r_i^b - r^b}{r_i^b - r_j^b} \end{bmatrix} \begin{Bmatrix} u_i \\ u_j \\ v_i \\ v_j \end{Bmatrix} \quad (8)$$

Where r is the distance of the point to the crack tip, Q and b are indexes describing the severity of the strain singularity.

In terms of the generalized virtual stress principle, the unknown strain components can be related to nodal displacements $\{q\}$ as:

$$\{e\} = [B]\{q\} \quad (9)$$

It should be noted that owing to the geometric nonlinearity matrix [B] still contain nodal displacements. Furthermor, with the aid of the constitutive relations the stress components in an element can be expressed by the nodal displacements of the element as:

$$\{\sigma\} = [C]\{e\} = [C][B]\{q\} \quad (10)$$

Where [C] can either be the elasticity or elastic-plasticity matrix of the material.

As the second step, the virtual displacement principle can be used in real structure to express the equilibrium requirements that the stress field (in Eq. (10) should satisfy. As a result, the simultaneous equations used for solving nodal displacements can be established. When having nodal displacements the stress components in an element can be calculated by Eq.(10).

3. "BREAKABLE ELEMENTS" ALGORITHM

We give the main points of the algorithm for a center-cracked plate as follows:

(1) To position the "brealable elements" along the expected path of crack axtension as illustrated in Fig.2. In order to realize the boundary condition at symmetric

line a-a we set the Young's modulus E_y of breakable elements to 10^{10} times larger than real one, E_x to zero.

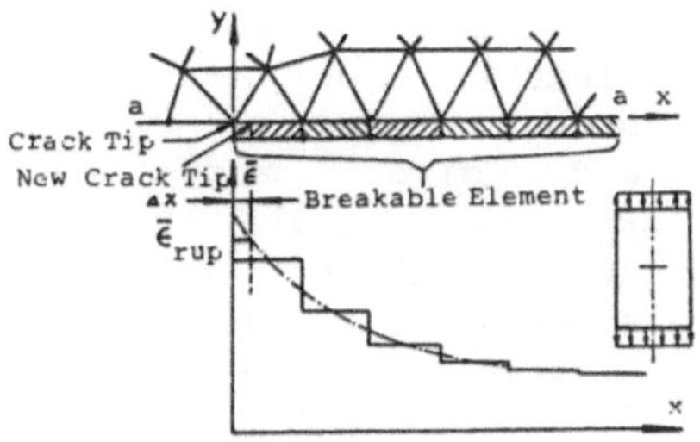

Fig.2 Breakable Element Algorithm

(2) Upon completion of elasto-plastic stress analysis at a given load step, the effective strains $\bar{\epsilon}$ in the elements in front of the crack tip are extrapolated as a smooth curve toward the crack tip using a least squares fitting technique. If the extrapolated effective strain at the crack tip, $\bar{\epsilon}_{ext}$ does not exceed the assigned fracture criterion, rupture effective strain $\bar{\epsilon}_{rup}$, then the stress analysis proceeds with the application of the next load increment.

(3) If $\bar{\epsilon}_{ext}$ exceeds $\bar{\epsilon}_{rup}$, the crack growth process begins and the crack tip takes new position. Fig.2 illustrates schematically how to determine the amount of crack extension, Δx. When an element begins to break, it Takes the real Young's modulus, but zero Young's mudulus for the broken part Δx.

(4) Once the crack front has passed through a crack tip element, a nodal force relaxation procedure is implemented in order to redistribute the stress field previonsly supported by the crack tip element before the element breaks.

4. NUMERICAL RESULTS AND CONCLUTIONS

The singular hybrid finite element and the "breakable element" algorithm stated above were used to predict the stable growth of a line crack in a thin aluminium plate. Fig.3 shows the curves of crack growth. the conforming quadrilateral finite element has already existed in the TEPSA code. Fig.4 shows the profiles of moving crack. Some parameters (e.g. COA, CTOA) in the crack tip can be calculated from the profiles of the crack.

It has been seen from numerical results that the predicted growth of the crack under monotonically increasing mechanical loads by the "breakable element" algorithm showed good correlations with those observed by experiments. However, the conforming finite element shows distinct over-stiff

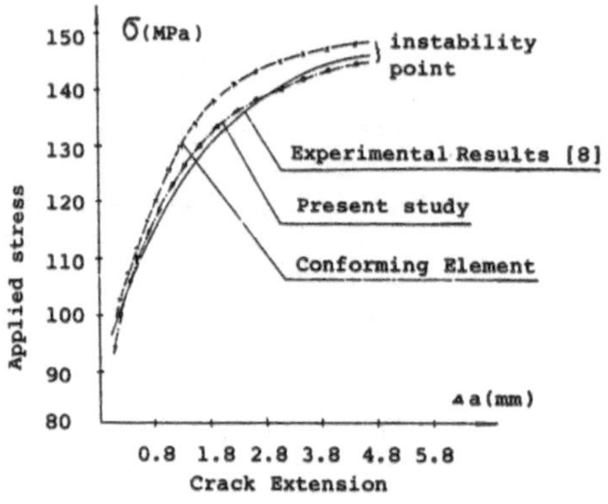

Fig.3 Curves of Crack Growth

character, it gives higher instability point (Fig.3), beyond which no further load was needed to maintain further growth, but the singular hybrid finite element derived above showes nice character and gives much better results. The unique advantages of using this element and the breakable element algorithm for the detailed description of the crack profiles, stresses and strains at any given stage have been also demonstrated.

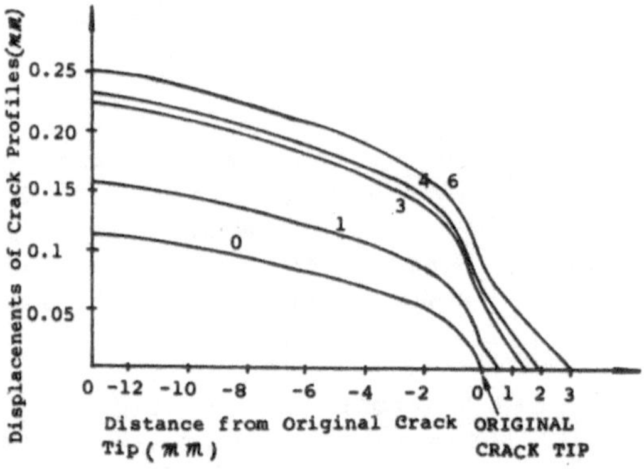

Fig.4 Profiles of Moving Crack

5. REFFERENCE

1. T.H.H.Pian and P.Tong, Basis of finite element methods for solid continua, Int. J. Num. Meth. Eng. 1, 3-28, 1969.
2. Tang Limin, Chen Wanji, Liu Yingxi, Quasi-conforming element analysis, J. Dalian Institute of Technology, 2, 1980.
3. T.R.Hsu and A.W.M. Bertels, Propagation and opening of a through crack in a pipe subject to combined cyclic thermomechanical loading, J.of Pressure Vessel Technology, Vol.98, 17-25 (1976).
4. T.H.H. Pian and Z.Tian, Hybrid solid element with a traction-free cylindrical surface, The Applied Mechanics Division (ASME), Vol.73, pp.69-75, 1985.
5. W.H.Chen and C.W. Wu, on elastodunamic fracture mechanics

analysis of Bi-material structures using finite element method, Engineering Fracture Mechanics, Vol.15, No. 1-2, pp.155-168, 1981.
6. T.Nishioka and S.N.Atluri, Numerical modeling of dynamic crack propagation in finite bodies by moving singular elements, J. of Appl. Mech., v.47, 570-576, (1980).
7. Bjorn Brickstad, A viscoplastic analysis of rapid crack propagation experiments in steel, J.Mech. Phys. Solids, Vol.31, 1983, pp.307-327.
8. T.R.Hsu, "The finite element method in thermomechanics", George Allen & Unwin lstd, England, 1986.
9. J.Y.Zhang and I.Q.Wang, The virtual stress principle in finite displacement theory and its application in hybrid finite element, proceedings of the International Conference on Nonlinear Mechanics, ShangHai, China, 1985.

PRELIMINARY SEISMIC ANALYSIS AND DESIGN OF LIQUID STORAGE TANKS

R. C. BARROS

Assoc. Professor, Departamento de Engenharia Civil, Faculdade de Engenharia, Universidade do Porto
Porto 4099, Portugal

1. SUMMARY

The behavior of thin shell cylindrical steel liquid storage tanks anchored to the foundation is studied when these structures are subjected to earthquake horizontal ground motions. Different analytical procedures are applied for the analyses of typical liquid storage tanks that suffered damage during the earthquake that occured in San Juan (Argentina) in November 1977. These procedures consider the generalized tank-fluid system as rigid or as a system with distributed flexibility. Results from the analyses using different methods are presented and compared, and some possible reasons for the inadequate observed behavior are detected. An improved procedure for the comprehensive seismic resistant design of cylindrical tanks is summarized. Finally, some recommendations and research needs to improve the design and construction of this type of liquid storage tanks are suggested.

2. ANALYSIS OF THE RESPONSE OF TANKS TO HORIZONTAL EARTHQUAKE MOTIONS

2.1 - Brief review of the state-of-the-art

The earliest studies of dynamic effects in cylindrical vessels were concerned with the earthquake response of bottom-supported tanks. This work was carried out in 1940's and 1950's by Jacobsen [1], Jacobsen and Ayre [2] among others. In these investigations the vessel was considered rigid and the dynamic response of the contained liquid was sought. Rigorous mathematical treatment of this problem was developed for the case of small liquid displacements. It involves finding an expression for the liquid velocity potential which satisfies Laplace's equation and appropriate boundary conditions at the liquid surface and at the shell-liquid interface. This velocity potential function may be separated into two parts, one treating the liquid as if it was constrained by a rigid immovable membrane at its surface, and the other, accounting for the response of the liquid free surface (sloshing oscillations). Dynamic effects associated with the rigidly constrained liquid are termed 'impulsive' responses, while dynamic effects associated with the free sloshing of the liquid are termed 'convective' responses. When a tank containing a liquid at rest is subjected to an earthquake, the subsequent ground motions cause the tank to accelerate in the horizontal plane as well as in the vertical direction. Figure 1 clearly visualizes the two responses associated with horizontal ground motions.

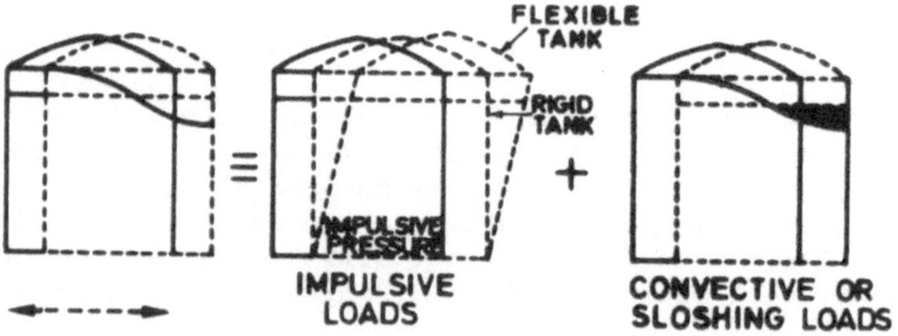

Figure 1. Response of a tank to earthquake horizontal ground motion

For structural analysis purposes, it is usually most convenient to deal with an equivalent mechanical analogue with discrete degrees of freedom consisting of spring mass systems and a mass rigidly attached to the container walls. In civil engineering applications, it is a common practice to further simplify the mechanical model considering only one sloshing mass, that is, representing the free surface motion by the first sloshing mode. This simplified model, developed by Housner [3] in 1963, has been widely applied in the prediction of seismic effects in liquid storage containers, with the introduction of some corrections [4].

In the early 1970's there was a resurgence of scientific interest in the seismic response of tanks, set forth by progress in the dynamic analysis of liquid-filled shells of the aerospace vehicles. The tank shell was regarded as a flexible structure anchored to the foundation and restrained against cross-section distortions. In 1973 Veletsos [5] used a less rigorous but simpler approach to the same problem. The tank was modelled to behave as a cantilevered beam with shear and bending flexibili ties.

Since 1974 research has been conducted at the University of California, Massachusetts Institute of Technology, Rice University, California Institute of Technology, Southwest Research Institute, and at other institutions around the world. The available literature has been continuing to report on analytical methods for the solution of a variety of dynamic problems involving cylindrical shells filled with liquid [6, 7, 8, 9, 10], as a result of such research. Nevertheless, usually they constitute extremely sophisticated analyses based on methodologies that restrict their immediate practical use, preventing them from being applied on a short term basis at engineering design offices.

2.2 - Housner - Epstein method for rigid tanks

The method available for the analysis of response of rigid tanks is attributed to Housner [3]. In a latter work Epstein [11] summarized the main formulas developed by Housner and suggested a procedure for their use in analyzing shallow and tall tanks. A simple mechanical analogue is used in which resultant effects of impulsive pressures can be simulated by considering some portion of the liquid mass, m_o, rigidly attached to the container. The height at which m_o must be attached to the container is labelled as h_o^o or h_o, depending on whether pressures on the tank bottom are considered or not. The effective convective mass m_1 is assumed suspended between the container walls by means of springs at a height labelled as h_1^o or h_1, depending on whether the overturning moment includes or excludes the effect of pressure acting on the tank bottom. When the tank analyzed has a ratio of height to base radius greater than 1.5 (ie, $\alpha = H/R > 1.5$) it falls under the category of tall slender tanks, in which case Epstein introduced an additional mass, m_o', in Housner's mechanical model, accounting for a mass of liquid below 1.5 R from free surface and considered rigidly attached to the container at the height h_o'.

For circular cylindrical tanks with base radius R and height H_s, filled up to a height H with a liquid of mass density ρ, a summary of equations simplified and implemented by Epstein [11] for shallow and tall tanks, is given in Table I.

	Shallow Tank	Tall Tank
h_o	$\frac{3}{8}$ H	$(1 - \frac{15}{16\alpha})$ H
h_o^o	$\frac{1}{8}$ $[\frac{4\,(\sqrt{3}/\alpha)}{th\,(\sqrt{3}/\alpha)}]$ H	$(1 - \frac{0.630}{\alpha})$ H
h_1	$[1 - \frac{ch\,(1.84\alpha) - 1}{(1.84\alpha)\,sh\,(1.84\alpha)}]$ H	$(1 - \frac{0.479}{\alpha})$ H
h_1^o	$[1 - \frac{ch\,(1.84\alpha) - 2.01}{(1.84\alpha)\,sh\,(1.84\alpha)}]$ H	$(1 - \frac{0.409}{\alpha})$ H
h'_o	—	$(0.5 - \frac{0.75}{\alpha})$ H
T	$2\pi\,[\frac{R}{1.84g\,th\,(1.84\alpha)}]^{1/2}$	$4.65\,(R/g)^{1/2}$
$P_o = m_o A_o$	$\frac{\alpha}{\sqrt{3}}\,th\,(\frac{\sqrt{3}}{\alpha})\,(\pi R^2 H\rho) A_o$	$3.34\,\rho R^3 A_o$
$P_1 = m_1 A$	$\frac{0.318}{\alpha}\,th\,(1.84\alpha)\,(\pi R^2 H\rho)\,A$	$0.991\,\rho R^3\,A$
$P'_o = m'_o A_o$	—	$(1 - \frac{1.5}{\alpha})\,(\pi R^2 H\rho)\,A_o$
V	$P_o + P_1 + W_T\,A_o/g$	$(P_o + P'_o) + P_1 + W_T A_o/g$
BM	$P_o h_o + P_1 h_1 + W_T\,A_o\,h_T/g$	$(P_o h_o + P'_o h'_o) + P_1 h_1 +$ $+ W_T A_o h_T/g$
OTM	$P_o h_o^o + P_1 h_1^o + W_T\,A_o\,h_T/g$	$P_o h_o^o + P'_o h'_o + P_1 h_1^o +$ $+ W_T\,A_o\,h_T/g$

Table I. Summary of equations for shallow and tall tanks

In Table I, T is the period of the oscillating liquid; A_o and A are respectively the maximum ground acceleration and maximum acceleration of the sloshing mass; P_o and P'_o are the maximum impulsive forces on the rigid tank; P_1 is the maximum convective force on the tank; V is a conservative estimated of the base shear, given by the sum of impulsive, convective and tank weight, W_T, terms; BM is a conservative estimate of the maximum bending moment caused by the liquid; h_T is the height from the base to the center of gravity of the tank; and OTM is a conservative estimate of the overturning moment caused by the liquid.

2.3 - Contribution of higher sloshing modes in the analysis of rigid tanks

Veletsos and Yang [12] and Sogabe, Shigeta, Shibata [13] among others, considered a series of convective terms representing the modal contributions of the portion of the liquid in sloshing motion. A summary of equations using this method is given below:

$$\omega_k = [\lambda_k \, (g/R) \, \text{th} \, (\lambda_k \, H/R)]^{1/2} \tag{1}$$

$$m = \rho\pi R^2 H = m_o + \sum_{k=1}^{\infty} m_k \tag{2}$$

$$\frac{m_k}{m} = \frac{2}{\lambda_k^2 - 1} \, \frac{R}{\lambda_k \, H} \, \text{th} \, (\lambda_k \, H/R) \tag{3}$$

$$h_k = H - (R/\lambda_k) \, \text{th} \, (\lambda_k \, H/2R) \tag{4}$$

$$h_o = \frac{1}{m_o} \left(\frac{H}{2} m - \sum_{k=1}^{\infty} m_k \, h_k\right) \tag{5}$$

$$V = m_o \, A_o + \sum_{k=1}^{\infty} m_k \, A_k \tag{6}$$

$$BM = m_o \, A_o \, h_o + \sum_{k=1}^{\infty} m_k \, A_k \, h_k \tag{7}$$

$$M_b = m \frac{R^2}{H} \{[\frac{1}{4} - \sum_{k=1}^{\infty} \frac{2}{\lambda_k^2 \, (\lambda_k^2 - 1)} \, \frac{1}{\text{ch} \, (\lambda_k \, H/R)}] \, A_o + $$

$$+ \sum_{k=1}^{\infty} \frac{2}{\lambda_k^2 \, (\lambda_k^2 - 1)} \, \frac{A_k}{\text{ch} \, (\lambda_k \, H/R)}\} \tag{8}$$

$$OTM = BM + M_b \tag{9}$$

in which λ_k are the zeros of the first derivative of the Bessel function of first kind and first order; ω_k is the kth sloshing frequency of the liquid in radians/s; m represents the total mass of the liquid equal to the sum of the impulsive mass m_o, rigidly attached to the tank at height h_o, and the convective masses m_k, attached to the tank by springs at heights h_k; and M_b is an additional base moment given by the sum of impulsive and convective terms, obtained integrating the hydrodynamic pressure distribution $p(r,\theta)$ at the tank bottom.

2.4 - Veletsos method for flexible tanks

Sufficient tank flexibility, specially in the case of slender tall tanks, can significantly undermine Housner's assumption of tank rigidity. A method presented by Veletsos et al. [5,12] accounts for tank flexibility along the height, with no distortion of its cross section. The tank-fluid system is idealized as a SDOF system with distributed generalized flexibility, vibrating in a prescribed vibration mode ψ (z). Only impulsive effects are affected by tank flexibility. A brief summary of the main alterations introduced by the method is given below:

$$\omega^2 = \frac{K^*}{m^*} = \frac{\int_0^H EI(z)\,\psi(z)\,dz}{\int_0^H \mu_T(z)\,\psi^2(z)\,dz} \quad \text{with} \left\langle \begin{array}{l} \psi_C(z) = 1 - \cos\frac{\pi z}{2H} \\ \psi_A(z) = \sin\frac{\pi z}{2H} \end{array} \right. \tag{10}$$

$$b_o(z) = \frac{4}{\pi} \sum_{i=1}^{\infty} \frac{\delta_i}{2i-1}\,\varepsilon_i \cos\left[(2i-1)\frac{\pi z}{2H}\right] \tag{11}$$

$$\delta_i = \frac{1}{H}\int_0^H \psi(z) \cos\left[(2i-1)\frac{\pi z}{2H}\right] dz \tag{12}$$

$$\varepsilon_i = \frac{I_1\left[(2i-1)\frac{\pi R}{2H}\right]}{I_1'\left[(2i-1)\frac{\pi R}{2H}\right]} \tag{13}$$

$$V^{(i)} = C_p\, m_{x1}\, A_o^* \tag{14}$$

$$m_{x1} = m\,\frac{8H}{\pi^2 R} \sum_{i=1}^{\infty} \frac{(-1)^{i+1}}{(2i-1)^2}\,\delta_i \varepsilon_i \tag{15}$$

$$BM^{(i)} = \left[\int_0^{2\pi}\int_0^H z\, p^{(i)}(z,\theta,t)\cos\theta\, dz\, R d\theta\right]_{max} =$$

$$= \pi\, C_p\, A_o^*\, \rho H R \int_0^H z\, b_o(z)\, dz \tag{16}$$

$$M_b^{(i)} = \left[\int_0^{2\pi}\int_0^R p^{(i)}(0,\theta,t)\, r^2 \cos\theta\, dr\, d\theta\right]_{max} =$$

$$= \pi\,\frac{R^3}{3}\, C_p\, A_o^*\, \rho\, H\, b_o(0) \tag{17}$$

$$M_b^{(c)} = \pi\,\rho\,\frac{R^4}{3} \sum_{k=1}^{\infty} \frac{2}{\lambda_k^2 - 1}\,\frac{A_k}{ch(\lambda_k H/R)} \tag{18}$$

in which EI (z) is the flexural stiffness of the tank; ω is the circular natural frequency of vibration of the generalized SDOF tank-fluid system, with a total mass per unit of height $\mu_T(z)$; b_o (z) is a pressure distribution function; I_1 e I_1' are respectively the modified Bessel function of first order and its first derivative; A_o is the spectral value of the pseudo-acceleration corresponding to the natural frequency ω; C_p is a participation factor given by a radio of effective masses [5,12]; $V^{(i)}$ is the absolute maximum value of the impulsive component of the hydrodynamic base shear; and the superscripts (i) and (c) label accordingly the absolute maximum values of the impulsive and convective components of BM and M_b.

3. NUMERICAL RESULTS

3.1 - Description of the tanks analyzed

Considerable damage to wine storage tanks occured in the city of San Juan during the 1977 earthquake. Some of this damage was caused by earth settlement and by liquefaction of the subsoil resulting from the earthquake [14]. However, this paper will only be concentrating on the damage caused by direct structural action induced by the earthquake ground motion. In order to verify if the available analytical methods and their results could predict the damage observed to wine tanks, the tank-fluid systems, shown in Figures 2 and 3, were investigated. They constitute typical circular cylindrical tanks of A36 steel, one is a shallow tank and the other a tall slender tank.

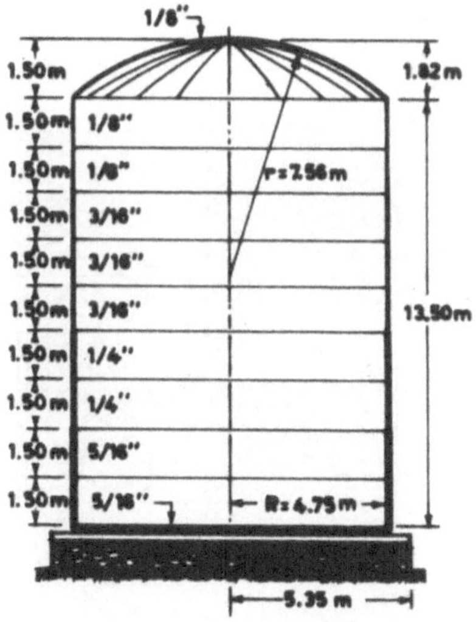

Figure 2. Tank of 1,100,000 liters

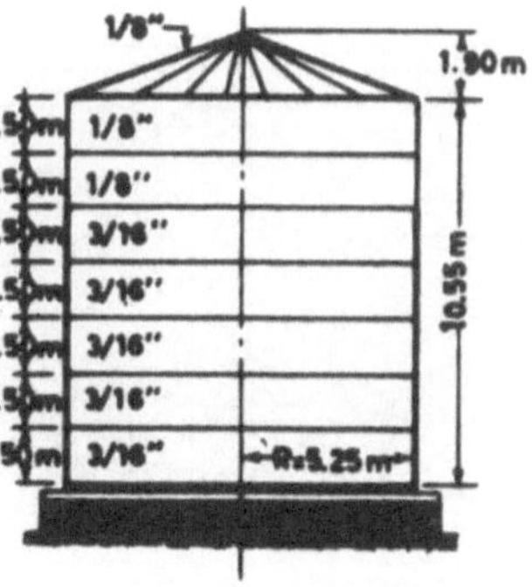

Figure 3. Tank of 1,000,000 liters

For the purpose of the analyses, the tanks were modeled to have a uniform wall thickness equal to the weighted average of the different wall thicknesses, and filled with liquid up to the height of the roof. Because of the form of the roof an equivalent cylindrical volume was calculated, establishing new heights of the tanks. The upper surface of the liquid is considered to be free. The fluid is incompressible and inviscid, and only linear effects are investigated. The tanks are presumed to be fixed at the base and excited by horizontal components of earthquake ground motion. Complete fixity of the tanks is achieved by anchor bolts, which would restrain the cylindrical shell from lifting off the foundation. A detail of the foundation and anchorage was presented in reference [15], emphasizing the way the anchor bolts were connected to the tank shell.

3.2 - Design loads

In general, damage to tank accessories can usually be avoided by proper detailing. Nevertheless damage to shell itself, like the elephant's foot bulge, can only be avoided through a comprehensive understanding of both structural response and buckling phenomenon. Therefore, there is a need to quantify earthquake induced forces in the tank. Under the present circumstances, only the primary loads due to the weight of the tank and contained liquid as well as the secondary loads due to the earthquake ground motion were considered. The dynamic analysis of the tall tank was based on the earthquake elastic response spectra shown in Figure 4 for both south and east directions, corresponding to the San Juan earthquake with ground peak accelerations of 0.195 g in both directions.

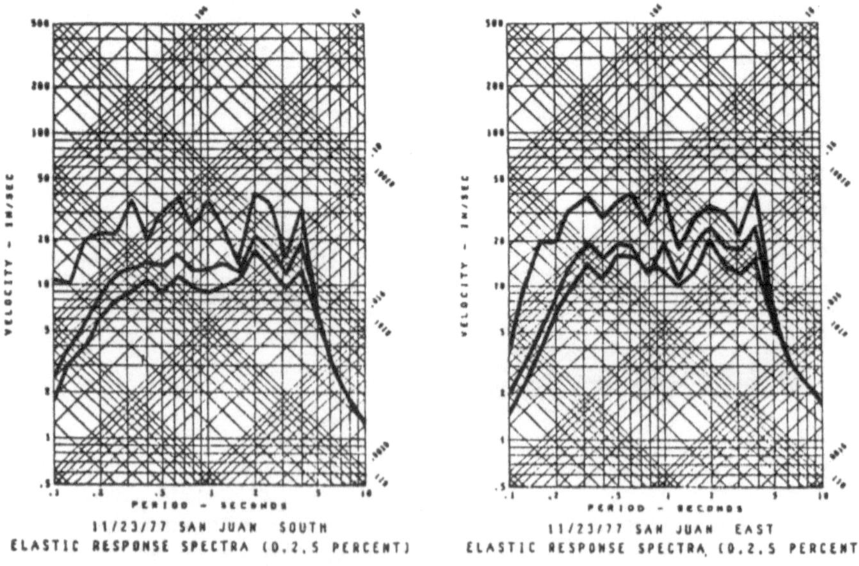

Figure 4. San Juan earthquake spectra

The dynamic analysis of the shallow tank was carried out using a smooth linear elastic response spectrum, corresponding to a peak ground acceleration of 0.33 g, obtained modifying the one proposed by Newmark and Rosenblueth [4] according to the guidelines of the Nuclear Regulatory Commission, NRC Guide 1.60.

3.3 - Discussion of results

For the tall tank, the values obtained by the different methodologies for the base shear, bending moment at bottom (supported by the shell) and overturning moment (supported by the anchor bolts), in both earthquake horizontal directions, are added vectorially giving the estimated results compared in Table II, extracted from reference [15]. Also registered on the table are impulsive and convective shear and moment resultants of the south and east components.

Table II. Comparison of results for tall tank

METHOD OF ANALYSIS		$V^{(i)}$	$V^{(c)}$	V (KN)	$BM^{(i)}$	$BM^{(c)}$	BM (KM.m)	OTM (KN.m)
Rigid tank (Housner, Epstein)		2190	82	2321	12657	918	13913	15354
Rigid tank (Veletsos et al. and Sogabe et al.)		2159	186	2391	12747	2061	15123	16198
Flexible tank (Veletsos)	ψ_C	2754	186	2932	22685	2061	24665	25143
	ψ_A	5194	186	5376	37409	2061	39424	40951

Table III compares the values of the earthquake generalized forces on the shallow tank obtained by different methods, with those that can be obtained using the Uniform Building Code (UBC' 76). Comparing the values obtained using UBC 76 with those obtained for flexible tanks (with prescribed generalized mode ψ_A), it can be seen that even with K = 2 the code values underestimate the seismic generalized forces in about 60%. However, even if the code values were multiplied by $\sqrt{2}$ (considering the simultaneous action of two equal horizontal components of ground motion) the resultant values would underestimate the base shear and bending moment at the base in more than 45%.

The results reveal the fact that assuming the tank to be rigid would significantly underestimate the seismic generalized forces. The procedure for selecting the prescribed shape of vibration [5] revealed that the given tanks deflected more like ψ_A (z), as justified by Barros [15] calculating the tank deflections including flexural and shearing deformations. Considering the results for flexible tank analyses, when comparing the com pressive stresses at the toe of the tanks with the buckling criteria proposed by Marshall [16] for fabricated cylinders, it indicates that the tank shell would buckle [15]. Observed damage in San Juan confirmed the occurrence of this kind of failure in many of the damaged tanks of this capacity [14].

Table III. Comparison of results for shallow tank

METHOD OF ANALYSIS		$V^{(i)}$	$V^{(c)}$	V (KN)	$BM^{(i)}$	$BM^{(c)}$	BM (KN.m)	OTM (KN.M)
UBC 76 (K = 2)		-	-	2547	-	-	13449	-
Rigid tank (Veletsos et al.)		3200	429	3629	14672	3416	18088	20743
Flexible tank (Veletsos et al.)	ψ_C	3499	429	3928	21178	3416	24594	26080
	ψ_A	6077	429	6506	32424	3416	35840	39524

4. COMPREHENSIVE DESIGN AND RECOMMENDATIONS

Although the tank fabrication industry has a valuable design code for welded steel tanks for oil storage (API Standard 650), it is worth while to mention a few concepts that should be present in a comprehensive design. The different aspects of the comprehensive conceptual design should contemplate: (i) Definition of the structural environment and establishment of the design excitations and design criteria; (ii) Selection of the tank configuration and structural system; (iii) Prediction of the tank mechanical behavior, with due emphasis on first preliminary design, modelling of the tank, structural analysis, stress analysis and check of preliminary design in both hoop tension and allowable buckling stress; (iv) Final detailing and construction aspects of unanchored and anchored tanks, tank roof and attached piping.

In anchored tanks, past performance shows a potential for tearing of the shell caused by the attached anchor bolts. To prevent this situation, the anchor bolt assemblies should be designed so that the bolts yield before any anchorage or shell attachment failure. Also, anchor bolts can be used to dissipate energy in a large earthquake.

To develop a better design procedure for cylindrical liquid storage tanks a few suggestions for needed research are mentioned: (i) Reliable method of dynamic analysis to include the effects of vertical excitations; (ii) Influence and importance of initial out-of-roundness in the tank responses; (iii) Parametric study on the effects of base flexibility and of tank movements; (iv) State of stress and uplift resistance of unanchored tanks; (v) Realistic shell buckling criteria for the actual seismic response of the liquid-filled tank; (vi) Effects of appendage to the shell; (vii) Inelastic behavior of the anchorage.

5. REFERENCES

[1] Jacobsen, L.S., 'Impulsive Hydrodynamics of Fluid inside a Cylindrical Tank and of Fluid surrounding a Cylindrical Pier', Bulletin of the Seismological Society of America, Vol. 39 (1949), 189-204.

[2] Jacobsen, L.S., and Ayre, R.S. 'Hydrodynamic Experiments with Rigid Cylindrical Tanks subjected to Transient Motions', Bulletin of the Seismological Society of America, Vol. 41 (1951), 313-346.

[3] Housner, G.W., 'The Dynamic Behavior of Water Tanks', Bulletin of the Seismological Society of America, Vol. 53 (1963), 381-387.

[4] Newmark, N.M. and Rosenblueth, E. - Fundamentals of Earthquake Engineering, Prentice-Hall, Inc., Englewood Cliffs, N.J., 1971.

[5] Veletsos, A.S., 'Seismic effects in flexible liquid storage tanks', Proc. 5th World Conf. Earthq. Engrg., Rome, Italy, Vol. 1, 630-639, 1974.

[6] Shaaban, S.H. and Nash, W.A., 'Finite element analysis of a seismically excited cylindrical storage tank, ground supported, an partially filled with liquid', Dept. of Civil Engrg., Univ. of Massachusetts, 1976.

[7] Parkus, H., 'Modes and Frequencies of Vibrating Liquid-filled Cylindrical Tanks', Int. J. Engrg. Sci., Vol. 20, No.2 (1982), 319-326.

[8] Committee on Gas and Liquid Fuel Lifelines, - Guidelines for the Seismic Design of Oil and Gas Pipeline Systems, ASCE Technical Council on Lifeline Earthquake Engineering, New York, 1984.

[9] Manos, G.C., 'Dynamic response of a broad storage tank model under a variety of simulated earthquake motions', Proc. 3rd U.S. Nat. Conf. Earthq. Engrg., Charleston, South Carolina, Vol. III, sec. 10/2131-2142, 1986.

[10] Peek, R., Jennings, P.C. and Babcock, C.D., 'The Preuplift method for anchoring fluid storage tanks', Proc. 3rd U.S. Nat. Conf. Earthq. Engrg., Charleston, South Carolina. Vol. III, sec. 10/2155-2166, 1986.

[11] Epstein, H.I., 'Seismic design of liquid-storage tanks', Journal of the Structural Division, ASCE, Vol. 102, No. ST9 (1976), 1659-1673.

[12] Veletsos, A.S. and Yang, J.Y., 'Earthquake response of liquid storage tanks', Advances in Civil Engineering through Engineering Mechanics, ASCE, 1, 1977.

[13] Sogabe, Shigeta and Shibata, 'A fundamental study on the aseismic design of liquid storages', Report of the Institute of Industrial Science, Univ. of Tokyo, Vol. 26, No. 7, 1977 (in Japanese).

[14] Carneiro de Barros, R., 'Projecto Anti-Sísmico de Tanques Metálicos de Armazenamento de Líquidos', Relatório interno, Depto. Engª Civil, F.E.U.P., Porto, 1986.

[15] Carneiro de Barros, R., 'Dynamic Analysis of Liquid Storage Tanks', Proc. 8th European Conf. Earthq. Engrg, Lisbon, Portugal, Vol. 3, sec. 6.8/81-88, 1986.

[16] Johnston, B.G., ed., - Guide to Stability Design Criteria for Metal Structures, ch. 10, John Wiley & Sons, New York, 1976.

OPTIMAL DESIGN OF SEMICONDUCTOR COMPONENTS

J.B. Waddell and J. Middleton
Department of Civil Engineering
University College of Swansea
U.K.

Abstract

A numerical model for the analysis and design optimisation of semiconductor devices under reverse-bias off-state conditions is investigated. The finite element method is used for the analysis phase and the design objective is the minimisation of the peak surface electric field in order to encourage bulk breakdown of the device. The optimal design process consists of sequential redesign using sensitivity analysis and an adaptive technique for design variable assignation is introduced.

Two termination techniques are considered, these bearing the shape of the surface termination profile (for non-planar devices), and the lateral variation of doping density across a material region adjacent to the surface (for planar devices). Results are presented to show the effectiveness of the technique in reducing the peak surface field.

1. Introduction

Numerical modelling of semiconductor devices is reviewed in the paper by Engl et al [1], and examples of the application of the finite element method to off-state semiconductors are given in references [2] and [3]. Such methods allow the accurate prediction of device response and due to versatility in application they can be used for a wide range of device designs and shape configurations.

Due to the adverse conditions that are inherent at the surface of semiconductors it is important that correct surface termination designs are used. It has been recommended in the paper by Baliga [4] that the peak surface field ϵ_{max} should be reduced to at least 50 percent of the bulk peak field value thereby encouraging bulk breakdown. This gives improved device behaviour regardless of the conditions that exist at the surface of the device. It has been shown recently [5] that sensitivity analysis can be accurately applied to non-linear device response and mathematical models developed in preparation for optimisation procedures. Using the proposition described above optimal device design to minimise the peak surface electric field is applied to the surface shape termination of non-planar devices and to the doping density profile of planar devices. Here the device design variables are the set of pre-selected nodal coordinates of the elements which define the surface termination shape for non-planar problems, or doping density values at specified nodes for the planar problem.

A technique is also introduced which allows device design variables to be added adaptively such that accurate modelling and smooth profiles can be produced. This method is evoked if the reduction in the surface field, ϵ_{max}, is considered small between two consecutive designs.

2. Governing Device Equations

The potential distribution across a two-dimensional semiconductor under reverse-bias off state condition can be expressed as

$$\frac{\partial^2 \psi}{\partial x^2} + \frac{\partial^2 \psi}{\partial y^2} = - \frac{P(\psi,x,y,)}{k\epsilon_o} \tag{1}$$

The complete space charge equation is given as follows and allows the accurate prediction of the potential ψ and electric field ϵ at the depletion edges.

$$p(\Psi,x,y) = q\left[(p+N_d) - (n+N_a)\right] \tag{2}$$

If the charge in the dielectric regions is assumed zero, equation (1) reduces to the well known Laplace equation

$$k\epsilon_o \left[\frac{\partial^2 \Psi}{\partial x^2} + \frac{\partial^2 \Psi}{\partial y^2}\right] = 0 \tag{3}$$

Then using a Galerkin weighted residual approach [6] the above equations can be assembled into the matrix form

$$[C]\ \{\Psi\} = \{Q(\Psi)\} \tag{4}$$

where [C] is the constant capacitance matrix, $\{\Psi\}$ is the potential vector, and $\{Q(\Psi)\}$ is the highly non-linear charge vector.

3. The Optimisation Model

The mathematical model developed for the optimisation phase is based on accurate design sensitivities

$$\left[\frac{\partial \epsilon_q}{\partial x_1}, \frac{\partial \epsilon_q}{\partial x_2}, \ldots \frac{\partial \epsilon_q}{\partial x_n} \right]^T$$

which are determined from resolutions of equation (4). This sensitivity analysis, although applied to a highly non-linear problem has been shown to be capable of predicting accurate device response and is fully described in [5].

The design objective is the minimisation of the maximum surface electric field, ϵ_{max}, and this may be written in discretised form as

$$\text{Min } Z(\underline{x}) \cong \text{Min } \epsilon_q(\underline{x})_{max} \quad q=1, \ldots p$$
$$\text{subject to } g_r(x) \leqq 0 \qquad r=, \ldots m \tag{5}$$

where $\underline{x} = (x_1, \ldots x_n)^T$ is the design vector

$\underline{\epsilon}_q(\underline{x}) = (\epsilon_1, \ldots \epsilon_p)$ are the discrete set of surface electric field values

$g_r(x) = (g_1(x) \ldots g_m(x))$ are prescribed design constraints

Equations (5) can be approximated by using first order Taylor series in which the only unknowns are the design derivatives

$$\frac{\partial \epsilon_{max}}{\partial x} \text{ and } \frac{\partial g_r}{\partial x}.$$

These can be calculated explicitly by using the sensitivity analysis described in reference [5].

3.1 Move Limits and Constraint Control

A problem often ecountered when dealing with the optimisation of field problems is that the peak field may increase at some undefined position due to the linearisation of the design model. This behaviour however, can be controlled by applying a set of constraints which prevent the surface field, at any location along the termination profile from exceeding $\epsilon_q(x^k)_{max}$. This is applied at all nodes along the profile which have a surface field value of $\geqslant 0.3\epsilon_q(x^k)_{max}$. These applied constraints enforce an upper bound on the surface fields sampled at the Gauss points by restricting the field at any point to be less than $\epsilon_q(x^k)_{max} - \Upsilon$, where Υ is typically preset to 0.6V/μm.

In order to control the accuracy of the desing model a move limit method is used. If the move limit remains constant throughout the design process this can result in slow convergence or inaccurate modelling of the system. In order to predict a suitable size for the move limits it is possible that they can be linked to the magnitude of the design sensitivities. This can be imposed by applying a parabolic relationship which effectively scales up values of the design vector which are less than unity and scales down values of the design vector which are greater than unity.

4. Design Variable Definition

(i) Non-Planar Device

The mesh perturbation for this device is shown in Figure (1). Selected nodes on the termination surface are chosen as design variables with all other nodes being described by linear interpolation. This allows arbitrary surface termination to be achieved and ease of adaption in the number of variables to be included in the model. Typically 5 variables are used initially with up to 25 variables being active near the optional solution.

(ii) Planar Device

In this problem the design variables are the doping density values at preselected design nodes along the surface termination region, as shown in Figure (2). Again a linear interpolation procedure is used between nodes to determine the doping profile and arbitrary lateral variation of doping density within the material region is achieved by the inclusion of addition variables (nodes).

5. Examples

(i) Surface Termination Profile Optimisation

Consider the diffused n-p junction shown meshed in Figure (3) which has a vertical termination surface. The maximum field was initially 216 kV/cm and 4 design variables were used to describe the profile. The progress of field reduction, design steps and the number of design variables used are shown in Figure (4). The final surface shape is shown in Figure (5) and the respective surface field for each design step is shown in Figure (6). At the converged solution 25 design variables were active and the reduction in ϵ_{max} was from 216 kV/cm to 50 kV/cm.

(ii) Doping Profile Optimisation

The mesh for the doping density problem is shown in Figure (7). Initially 4 design nodes were allocated with the same doping concentration which produced a maximum field value of 473 kV/cm. After 15 design steps this value was reduced to 80.2 kV/cm. The resulting change in doping density and the corresponding surface field profiles are shown in Figure (8). As can be seen from the figure the optimal field shape is flat throughout the material region which is very encouraging indeed from the design and performance viewpoint.

6. Conclusions

A method for the optimal design of off-state semiconductors with arbitrary surface and doping density profiles has been implemented. A 4 noded bilinear element was used for the solution procedure since it is highly resistant to error under severe geometrical distortions. Two different examples were successfully optimised and accurate device response obtained at the end of each design step. Each device was discretised into approximately 2500 elements with solution times being approximately 8,000 c.p.u. seconds on a CDC Cyber 176.

References

[1] W.L. Engl, H.K. Dirks and B. Meinerzhagen, "Device modelling", Proc. IEEE, vol. 71, no. 1, pp. 10-33, 1983.

[2] J.J. Barnes and R.J. Lomax, "Finite element methods in semiconductor device simulation", IEEE Trans., Ed-24, pp. 1082-1089, 1977.

[3] SWANOFF 2: A 2-dimensional off-state finite element package, Dept of Electrical and Electronic Engineering, University College of Swansea, Swansea, U.K.

[4] B.J. Baliga, "High-voltage termination techniques - a comparative review", IEE Proc., vol. 129, pt. 1, no. 5, pp. 173-179, 1982.

[5] J. Middleton and J.B. Waddell, "Sensitivity analysis and design optimisation of bevelled p-n junctions", IEE Trans. on Electron Devices, Feb. 1987

[6] E. Hinton and D.R.J. Owen, "An introduction to finite element computations", Pineridge Press, 1979.

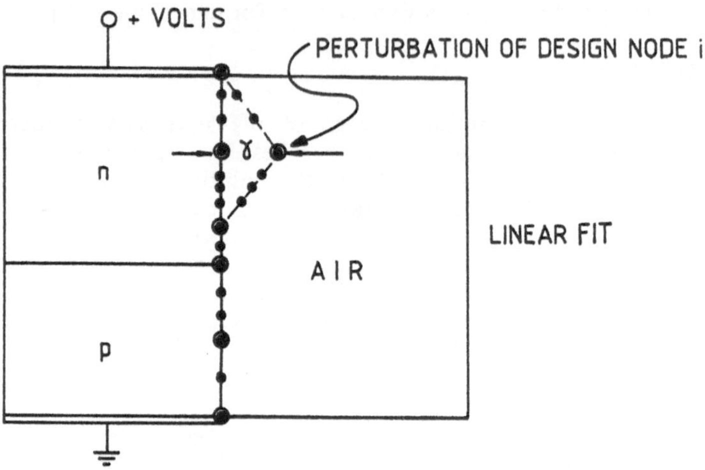
+ VOLTS
PERTURBATION OF DESIGN NODE i
γ
n
p
AIR
LINEAR FIT

fig.1.

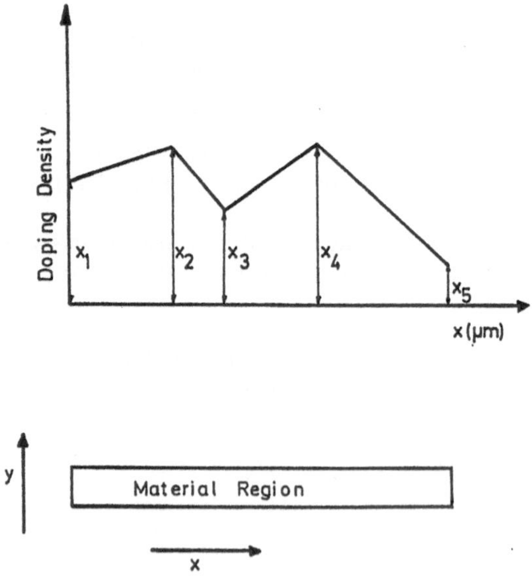
Doping Density
x1
x2
x3
x4
x5
x (μm)
y
Material Region
x

fig.2.

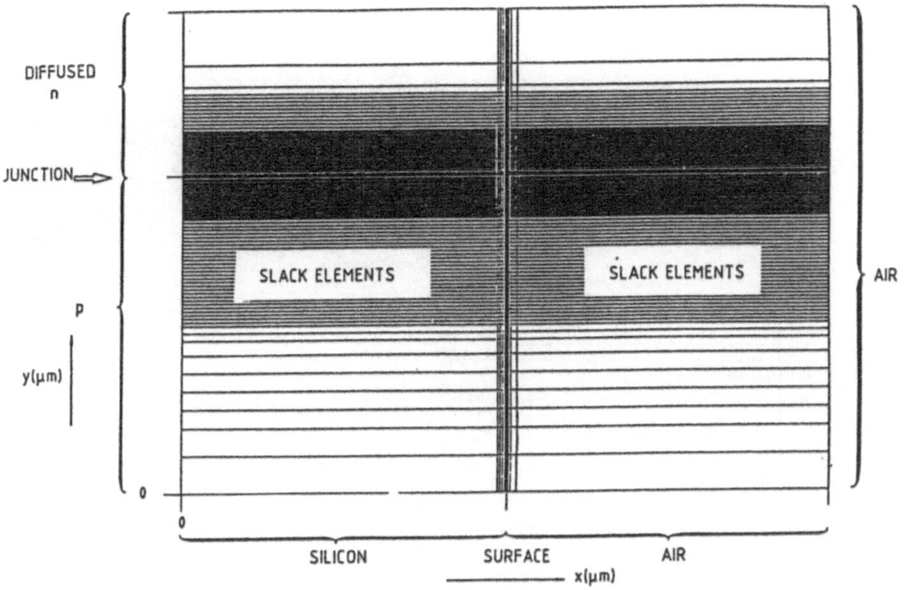

fig.3. Finite element mesh for diffused n-p junction. Device discretised into 2121 linear quadrilateral elements.

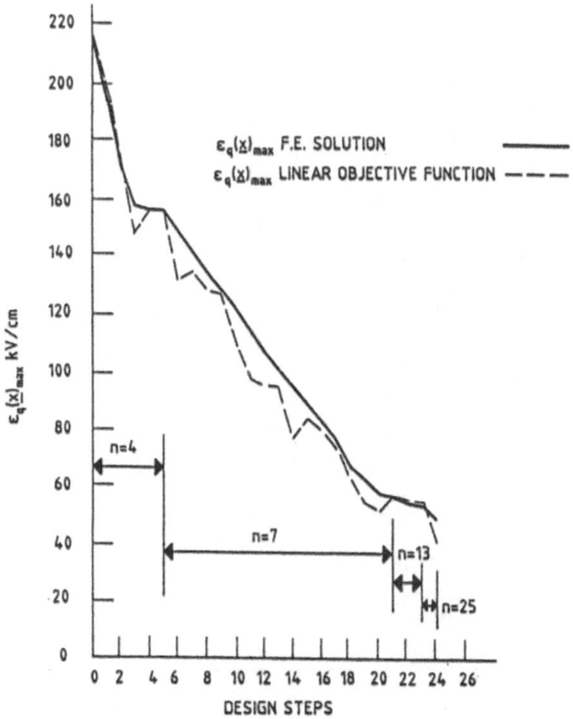

fig.4.

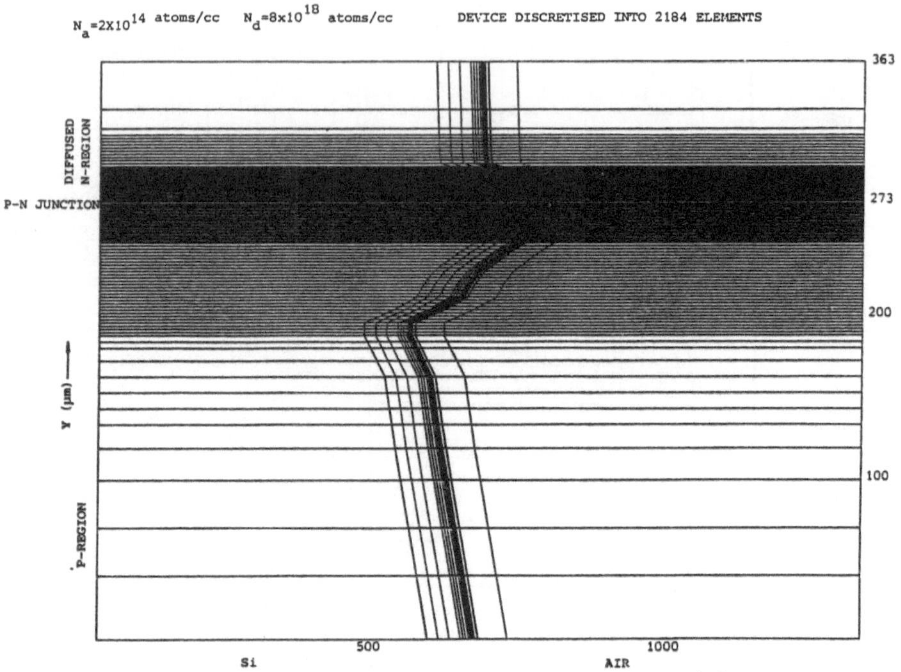

fig.5. FINAL DESIGN SHAPE OF P-N JUNCTION

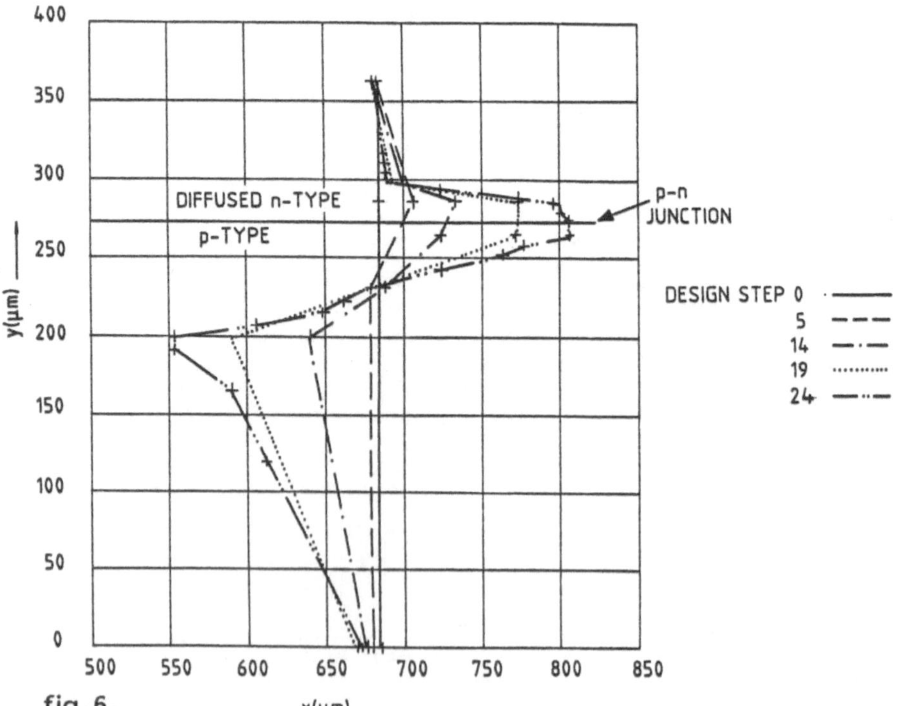

fig.6.

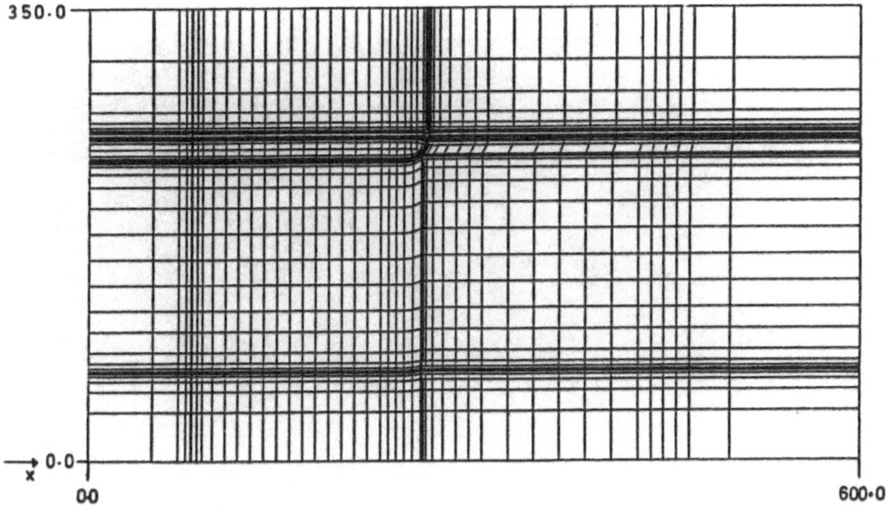
350.0
0.0
x
0.0
600.0

fig. 7.

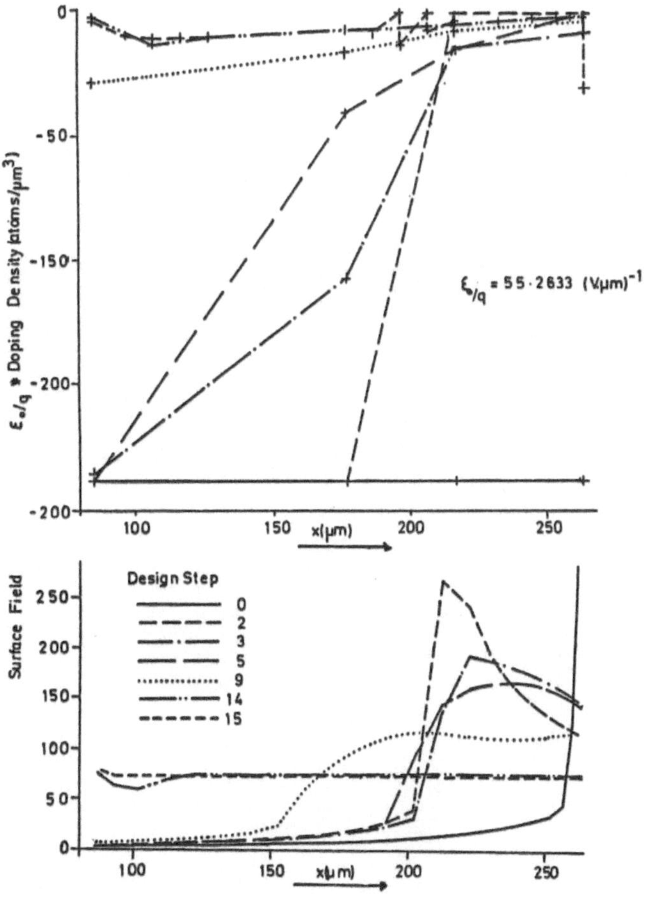
ε_o/q = 55.2633 (V.μm)$^{-1}$
ε_o/q * Doping Density (atoms/μm^3)
x(μm)
Surface Field
Design Step
0
2
3
5
9
14
15
x(μm)

fig. 8.

REFLECTIONS ON FINITE ELEMENT PLATE ANALYSIS

E. Hinton

University College of Swansea

SUMMARY

Mindlin/Reissner (M/R) formulations have been increasingly adopted in the finite element analysis of plates because they provide a more accurate, versatile and convenient basis for the development of finite elements. This paper addresses two questions: (a) 'Are there any inherent limitations in the use of M/R formulations?' and (b) 'Are there any attractive alternatives?'

1. INTRODUCTION

The last decade or so has seen a growing movement away from the use of classical, Kirchhoff formulations [1] in the finite element analysis of plates. Mindlin/Reissner (M/R) formulations [2,3] have been increasingly adopted because they are more accurate, versatile and convenient than their Kirchhoff counterparts, especially in the finite element context. The author has been working actively in this area for some time and believes that it is now useful to assess this trend of adopting M/R elements.

2. DIFFICULTIES IN THE USE OF MINDLIN/REISSNER FINITE ELEMENTS

2.1 Basic problems

'Are there any inherent difficulties in the use of M/R formulations in the finite element analysis of plates?' In a joint paper with H.C. Huang [4] at

the last NUMETA conference, the author considered problems associated with the development of a reliable M/R plate element capable of passing the widely accepted benchmark tests?' However, there also exists a major problem concerned with modelling the distribution of shear forces and twisting moments.

There appear to be two main difficulties in sampling stress resultants for thin plates obtained by M/R elements:

(1) The first of difficulties emanates from the fact that very steep gradients of shear forces and twisting moments are predicted by M/R theory on the plate boundaries for certain types of support conditions. Unless a very fine mesh is used near the plate boundary, the finite element solution may fail to capture the peak values of the stress resultants and there will be a local disruption of the stress resultants.

(2) There is a tendency for the shear forces obtained by conventional displacement based M/R plate elements to oscillate dramatically even when there are no steep gradients of shear forces present.

2.2 Stress resultant sampling

To illustrate the phenomena described in item (2) above consider the solution of a uniformly loaded, thin (h/a=0.001), simply supported, square plate using a 2×2 mesh of 9-noded, selectively integrated elements [5] in a symmetric quadrant.

If the shear forces are evaluated at the nodal points in the conventional way with subsequent nodal averaging, then the results shown in Table 1 are obtained. The values in parentheses given for comparison, are obtained using local (bilinear) smoothing [6]. The smoothed values are in very close agreement with the exact solution. The highly oscillatory behaviour of the unsmoothed shear forces obtained from the finite element solution is clearly seen. Even when moderately thick plates are analysed by Mindlin elements the shear forces are unreliable. When the problem described above is solved again for a plate of thickness-to-span ratio h/a=0.1, although the oscillations in the unsmoothed values are not excessive as in the thin plate case they are still unreliable.

Table 1 Shear force distribution for a uniformly loaded, simply supported, thin (h/a=0.001), square plate using a 2×2 mesh of selectively elements in a symmetric quadrant – smoothed and unsmoothed values

			Q_y		
$x =$	$y = 0$	$y = a/8$	$y = a/4$	$y = 3a/8$	$y = a/2$
$a/2$	2253.20	-1126.20	2253.00	-211.10	422.33
	(0.3376)	(0.02354)	(0.1356)	(0.0677)	(-0.0026)
3a/8	2100.40	-1049.8	2100.20	-188.24	376.606
	(0.3119)	(0.2147)	(0.1199)	(0.0599)	(-0.0026)
a/4	1644.10	-821.76	1643.90	-133.81	267.720
	(0.2866)	(0.1909)	(0.0991)	(0.0504)	(-0.0022)
a/8	901.62	-450.60	901.470	-68.49	137.042
	(0.1683)	(0.1058)	(0.0486)	(0.0261)	(-0.0017)
0	0.0000	0.0000	0.0000	0.0000	0.0000
	(0.0373)	(0.0164)	(0.0013)	(0.0024)	(-0.0023)

2.3 Shear forces and twisting moments on plate edges

Using the segmentation method, Kant and Hinton [7] have shown that a 'boundary layer' of shear forces and twisting moments exists for plate problems with certain types of support conditions. The size of the boundary layer is directly related to the plate thickness and this causes acute problems for finite element analysis of thin plate situations. In attempting to reproduce steep gradients with M/R finite elements, the shear forces and twisting moments oscillate violently in such regions and also disturb the shear force and twisting moment distribution in adjacent regions of the plate. This is to be expected – at least, by experienced analysts. However, a structural engineer not well versed in the vagaries of the finite element method is likely to lose confidence in finite element simulation tools when faced with highly uneven distributions of shear forces and twisting moments.

To illustrate this type of behaviour, Kant and Hinton [7] used the finite segment method to analyse a series of square plates with two opposite edges simply supported and with various boundary conditions on the two remaining edges. They focussed attention on the edge shear forces and twisting moments for plates of thickness-to-span ratio h/a=0.02 and Poisson's ratio $\nu = 0.3$ which are subjected to a uniformly distributed load of intensity q. One of these problems is re-analysed here using a graded mesh of Huang/Hinton, Mindlin plate elements (referred to hereafter as QUAD9* element). See Figure 1. This study is used to illustrate the accuracy of QUAD9* elements in representing the shear forces and twisting moments on the edges of the plates. It also highlights the potential difficulties in representing these stress resultants.

2.3.1 Plate with two opposite edges simply supported and two edges free (SFSF)

Figures 2 and 3 show the variations of the stress resultants $M_{xy}(x,0)$ and $Q_y(x,0)$ respectively. Figure 2 shows the variation of M_{xy} at the supported edge along the line $y = 0$. Mindlin plate theory predicts a zero value for M_{xy} at the plate corner whereas Kirchhoff plate theory does not. Again, the steep gradient of M_{xy} near the plate corner is well represented by the QUAD9* solution.

Also at the supported edge along the line $y = 0$, the variation of Q_y is shown in Figure 3. Mindlin plate theory predicts a very high value at the corner of the plate whereas Kirchhoff plate theory does not. The QUAD9* solution is in very good agreement with the Mindlin solution.

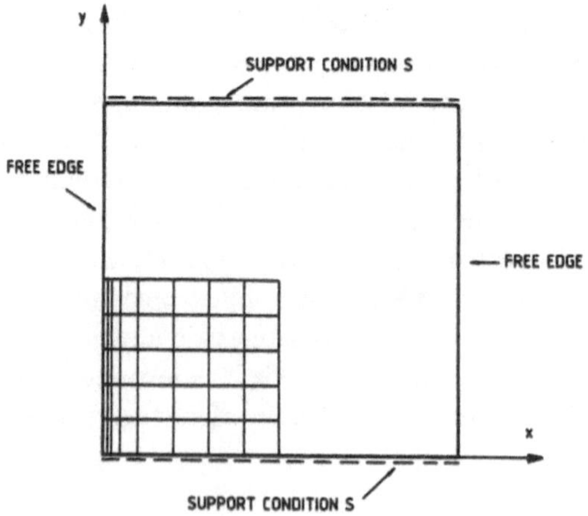

Figure 1 Finite element mesh for a SFSF plate quadrant

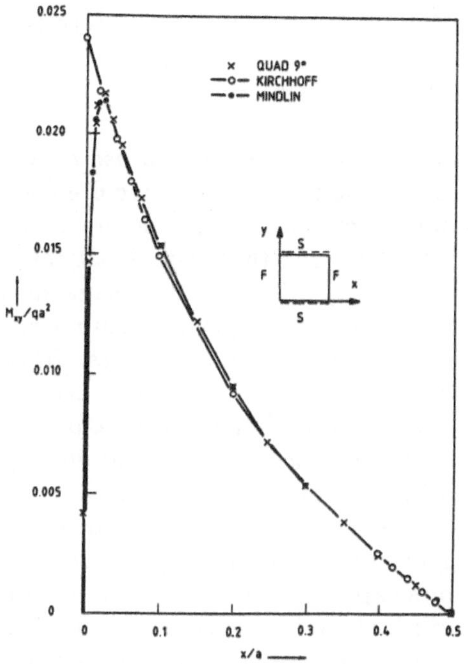

Figure 2 Variation of M_{xy} along the line $y = 0$ for the uniformly loaded square plate with SFSF boundary conditions

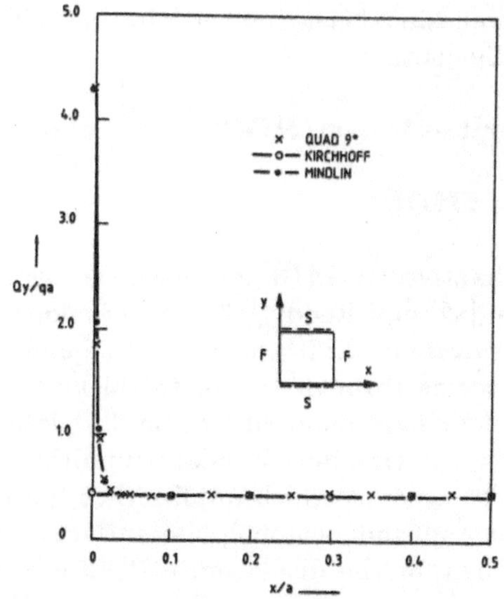

Figure 3 Variation of Q_y along the line $y = 0$ for the uniformly loaded square plate with SFSF boundary conditions

3. ALTERNATIVE PLATE FORMULATIONS

3.1 Three dimensional formulations

'Are there any attractive alternatives to the use of M/R formulations?' The author has considered the use of fully three dimensional formulations for the non-linear analysis of reinforced concrete plate and shell structures [9]. The motivation here is that with M/R formulations for such a problem, quadratic layered, Mindlin plate elements with as many as 72 sampling/quadrature points are typically required in the evaluation of the tangential stiffness matrix and residual forces. However, in the three dimensional formulation, 20-noded brick elements with a 15-point quadrature [10] can be successfully used. In a series of experiments on uniformly loaded square plates in the elastic range, carried out by the author [9], it was concluded that locking problems rather than ill-conditioning limit the use of such elements to element ratios of less than 25 to 1. Twenty seven noded brick elements with $3 \times 3 \times 3$ quadrature/stress sampling can also be used. In the author's experience no locking occurs for aspect ratios beyond even 100 to 1. With these integration rules for the 20 and 27 noded brick elements, no mechanisms are present and further tests on standard, non-linear reinforced concrete plate problems demonstrate the high accuracy and economy of these three dimensional bricks. In other words, three dimensional elements are worthy of attention for plate analysis. It should be noted that three dimensional elements with incompatible modes should not be forgotten in this context.

3.1 Higher order plate theory (HOPT) formulation

3.2.1 Kinematics of HOPT

Other attractive alternatives to M/R formulations exist. For example, Murthy [11], Levinson [12] and Reddy [13] have developed a higher order plate theory which is novel in that it accounts for a quadratic distribution of transverse shear strains through the plate thickness and thus shear modification factors (which are required by Mindlin plate theory) are not needed. In this theory, the traction boundary conditions on the faces of the plate are taken into account which implies that the transverse shear stresses vanish on the top and bottom plate surfaces. To achieve this condition, a special form for the displacement field is required in which the in-plane displacements u_1 and u_2 in the plane of the plate (i.e. in the x and y directions) are expanded as cubic functions through the plate thickness. This results in displacement fields of the form

$$u_1 = z(\psi_x - 4z^2(w_{,x} + \psi_x)/3h^2) \tag{1a}$$

$$u_2 = z(\psi_y - 4z^2(w_{,y} + \psi_y)/3h^2) \tag{1b}$$

$$u_3 = w \tag{1c}$$

in which w denotes the lateral displacement of the plate midsurface and ψ_x and ψ_y are the rotations of the normal to the middle plane in the xz- and yz-planes respectively.

3.3 Modified higher order theory (HOT) formulation

3.3.1 Introduction

Ren and Hinton [14] have developed a finite element based on HOPT which unfortunately proved to be unsatisfactory. However, they subsequently made some minor modifications to the theory regarding the choice of the main variables in the original displacement field and thereby produced a better element. These modifications are now briefly summarised – the modified formulation is hereafter called HOT is then used as a basis for the development of the modified Razzaque-Irons triangular plate element [15].

3.3.2 Kinematics of HOT

The assumptions made in HOT are identical to those of HOPT – the only differences are in the choice of the main variables. The new displacement field is of the form

$$u_1 = z(\phi_x - w_{,x} - 4z^2\phi_x/3h^2) \tag{2a}$$

$$u_1 = z(\phi_y - w_{,x} - 4z^2\phi_y/3h^2) \tag{2b}$$

$$u_3 = w \tag{2c}$$

Note that ϕ_x and ϕ_y are everywhere constrained to zero in the plate, then HOT reduces to Kirchhoff plate theory.

3.4 The modified Razzaque-Irons triangle

3.4.1 Introduction

A finite element formulation based on HOT requires a representation for the lateral displacement w, based on either (a) shape functions with C(1) continuity or (b) non-conforming shape functions which lead to patch test satisfaction in the Kirchhoff sense. However, for the parameters ϕ_x andϕ_y only C(0) continuity is required. Therefore, for a three-noded triangular element, Kirchhoff-type shape functions may be used to describe w whereas only linear shape functions are needed for the description of ϕ_x and ϕ_y. Further details are presented in Reference 16.

3.4.2 Examples of the use of the new element

Consideration is now given to the analysis of uniformly loaded, square plates with two sets of boundary conditions – SSSS: simply supported all round and SFSF: two opposite edges simply supported and the remaining edges free. Table 2 contains non-dimensionalised central deflections and bending moments and maximum shear forces. For both thick (h/a=0.1) and thin (h/a=0.01) cases, the results of the use of the modified Razzaque-Irons plate element (HOT) are compared with results obtained using finite element analyses based on Mindlin plate theory (MPT). In both cases fine meshes of (16 $\times$ 16) elements were used in a symmetric quadrant. The results are in excellent agreement. For the plate with SFSF boundary conditions, a boundary layer of shear forces and twisting moments adjacent to the free edge again develops as with MPT formulations.

ACNOWLEDGEMENTS

The author gratefully acknowledges the collaboration and helpful discussions with the following former students and colleagues: M. Cervera, O. Hassan, H.C. Huang, L. Iossifidis, T. Kant and J.C. Ren.

REFERENCES

1. S.P. Timoshenko and S. Woinowsky-Krieger, *Theory of Plates and Shells*, 2nd edn., McGraw-Hill, New York, 1961.

2. R.D. Mindlin, Influence of rotary inertia and shear in flexural motions of isotropic elastic plates, *J. Appl. Mech.* **18**, 1031-1036, 1951.

Table 2 Comparison of solutions for uniformly loaded, square plate with SSSS and SFSF boundary conditions

solution method	boundary condition	h/a	wD/qa^4 $(\times 10^{-6})$	M_x/qa^2 $(\times 10^{-5})$	M_y/qa^2 $(\times 10^{-5})$	Q_{max}/qa $(\times 10^{-4})$
HOT	SSSS	0.01	4064	4794	4794	3370
MPT	SSSS	0.01	4064	4794	4794	3380 mid-edge
HOT	SSSS	0.1	4273	4795	4795	3420
MPT	SSSS	0.1	4273	4795	4795	3380 mid-edge
HOT	SFSF	0.01	13090	12230	26870	44710
MPT	SFSF	0.01	13090	12250	27190	34670 corner
HOT	SFSF	0.1	13460	12210	25530	13820
MPT	SFSF	0.1	13460	12280	25830	12257 corner

3. E.Reissner, The effect of transverse shear deformation on the bending of elastic plates. *J. Appl. Mech.* **12**, 69-76, 1945.

4. H.C. Huang and E. Hinton, An improved Lagrangian 9-node Mindlin plate element. In *Proc. Int. Conf. on Numerical methods in engineering: theory and applications - NUMETA 85*, Swansea (Eds. J. Middleton and G.N. Pande), Balkema, Rotterdam, 707-713, 1985.

5. T.J.R. Hughes, M. Cohen and M. Haroun, Reduced and selective integration technique in the finite element analysis of plates, *Nucl. Eng. Design* **46**, 203-222, 1978.

6. E. Hinton and J. Campbell, Local and global smoothing of discontinuous finite element functions using a least squares method. *Int. J. Num. Meth. Engng.* **8**, 461-480, 1974.

7. T. Kant and E. Hinton, Mindlin plate analysis by the segmentation method, *J. Eng. Mech. Div., ASCE* **109**, 537-556, 1983.

8. H.C. Huang and E. Hinton, A 9-node Lagrangian Mindlin plate element with enhanced shear interpolation, *Eng. Comput.* **1**, 369-379,1984.

9. M. Cervera, E. Hinton and O. Hassan, Nonlinear analysis of reinforced concrete plate and shell structures using 20-noded isoparametric brick elements, *Computers and Structures*, (to be published).

10. B.M. Irons Quadrature rules for brick-based finite elements, *Int. J. Num. Meth. Engng.* **3**, 293-294, 1971.

11. M.V.V. Murthy, An improved transverse shear deformation theory for laminated anisotropic plates, *NASA Technical paper 1903*, 1981.

12. M. Levinson, An aacurate simple theory of the statics and dynamics of elastic plates, *Mech. Res. Commun.* **7**, 343-350,1980.

13. J.N. Reddy A simple higher-order theory for laminated composite plates, *J. Appl. Mech.* **51**, 745-752, 1984.

14. J.G. Ren and E. Hinton, The finite element analysis of homogeneous and laminated composite plates using a simple higher-order theory, *Commun. Appl. Numer. Meth.* **2**, 217-228, 1986.

15. B.M. Irons and A. Razzaque, Shape function formulation for elements other than displacement models, In *Proc. Int. Conf. on Variational Methods in Engineering*, 4/59-71, Univ. of Southampton, 1972.

16. E. Hinton, L. Iossifidis and J.G. Ren, Higher-order plate analysis using the modified Razzaque-Irons triangle *Computers and Structures* (to be published).

FINITE ELEMENT MODEL FOR LAYERED PLATES/SHELLS

M G RAJENDRAN Ph D RESEARCH SCHOLAR

S RAJASEKARAN PROFESSOR

K JAWAHAR REDDY
S VALLIAMMAI GRADUATE STUDENTS

PSG COLLEGE OF TECHNOLOGY
COIMBATORE, TAMIL NADU, INDIA

SUMMARY

The sixteen term complete 3rd order Hermitian polynomial initiated by Bogner, Fox and Schmit [3] is used to develop the finite element model for static and classical buckling analyses of layered rectangular plates made of composite materials. The concept of Isoparametric elements with independent rotational and transverse degrees of freedom originally introduced by Ahmad, Irons and Zienkiewicz [1], is extended to develop the finite element models for static and classical buckling analyses of layered axisymmetric shells as well as layered plate/shell. Computer programs for the above models suitable for microcomputer have been developed and some practical problems have been solved and compared with the available results.

1 INTRODUCTION

The analysis of multilayered plates/shells made of composite materials, using classical lamination theory (viz, Kirchhoff hypothesis for plates and Kirchhoff-Love hypothesis for shells) neglect the transverse shear effects. Finite element model is developed in section 2 for layered rectangular plate using sixteen term 3rd order Hermitian polynomial. Finite Element Models are also developed in section 3 for layered axisymmetric as well as general plate/shell using degeneration concept which takes care of transverse shear effects.

2 LAYERED RECTANGULAR PLATE

The composite plates, in general, do not have a neutral midsurface due to the presence of layers of different materials and fibre orientations. As is well known in the composite literature, the displacements of any point can be defined in terms of the displacements and slopes of the corresponding point on the reference surface.

$$u = u(x,y,z) = u_o(x,y) - z\, w,_x \quad (1 - a)$$

$$v = v(x,y,z) = v_o(x,y) - z\, w,_y \quad (1 - b)$$

$$w = w(x,y,z) = w_o(x,y) \quad (1 - c)$$

where x,y,z is the Right Handed Rectangular coordinates with z downwards.

u_o,v_o,w_o - displacements of reference surface in the coordinate direction.

u,v,w - displacements of the point under consideration.

, - denotes differentiation.

The lateral displacement of any point can be interpolated using the 3rd order sixteen term hermitian polynomials 'f' used by Bogner etal [3], for the 4 noded rectangular plate element with 4 degrees of freedom at each node (viz, the lateral displacement, its derivatives and the twist) and the in plane displacements of that point are linearly interpolated in terms of shape functions 'N'.

Now the strain displacement relation can be written as:

$$\begin{Bmatrix} x \\ y \\ xy \end{Bmatrix} = \begin{bmatrix} N,_x & O \\ O & N,_y \\ N,_y & N,_x \end{bmatrix} \begin{Bmatrix} \underline{u} \\ \underline{v} \end{Bmatrix} - z \begin{Bmatrix} f,_{xx} \\ f,_{yy} \\ 2f,_{xy} \end{Bmatrix} \{\underline{w}\} \quad (2)$$

Applying the principle of minimum potential energy for stable equilibrium, and by summing up the stiffness contributions using layer-wise integration the equilibrium equation can be obtained as

$$\begin{bmatrix} k_{PP} & k_{PB} \\ k_{BP} & k_{BB} \end{bmatrix} \begin{Bmatrix} \begin{Bmatrix} \underline{u} \\ \underline{v} \end{Bmatrix} \\ \{\underline{w}\} \end{Bmatrix} = \begin{Bmatrix} \{R\} \\ \{Q\} \end{Bmatrix} \qquad (3)$$

k_{PP} = Membrane stiffness matrix

k_{BB} = Bending stiffness matrix

k_{PB} = $k_{BP}{}^{T}$ = Coupling stiffness matrix

= 0 for plate of isotropic material or symmetric geometry and material properties.

Likewise, the geometric stiffness matrix can be formulated using the strain energy or work done due to the in plane forces.

The convergence and suitability of the layered model is verified by performing static and buckling analyses of a clamped isotropic plate and six-layer isotropic plate. The results are compared with Timoshenko's solution [5,6] in Table 1.

Table 1:

Results of static analysis (maximum central deflection by Timoshenko is 0.03693 cm) and classical buckling analysis (Nx = Ny by Timoshenko is 730.9254 N/mm). Layered plate 32 cm x 32 cm x 0.25 cm; udl = 0.04 N/mm^2; Young's Modulus = 1×10^6 N/mm^2 and Poisson's ratio = 0.3 - (1/4 symmetry used)

No of Elements	Static analysis: Isotropic	Static analysis: Layered Isotropic	Buckling analysis: Isotropic	Buckling analysis: Layered Isotropic
1	.03883	.03883	739.5074	739.6338
2	.03790	.03790	737.5473	737.6728
4	.03768	.03767	734.6674	734.7906
8	.03745	.03744	733.1963	733.3201

The layered model is then checked for maximum central deflection in cm, of five layer symmetric orthotropic plate with Timoshenko's [5] solution for that particular material which satisfies the relation $H = \sqrt{Dx\ Dy} = D1 + 2Dxy$. The results are shown in Table 2.

Table 2:

Results of static analysis of simply supported orthotropic plate 11.892 cm x 10.0 cm x 0.02 cm; udl = 0.1 N/mm^2; $E_L = 4 \times 10^6$ N/mm^2; $E_T = 2 \times 10^6$ N/mm^2; $G_{LT} = 1.2018$ N/mm^2 and $\nu_{LT} = 0.25$ (whole plate considered).

No of Elements	Orthotropic	Layered Orthotropic	Timoshenko [5]
4	3.26 049	3.25 853	2.95 710
8	3.10 516	3.09 968	

Static and classical buckling analyses of layered anisotropic rectangular plate are performed with material properties as given in Table 3 and the results are shown in Table 4.

Table 3:

Material Properties of composites.

Name of Composite	E_L N/mm^2	E_T N/mm^2	ν_{LT}	G_{LT} N/mm^2	Material identification
Glass epoxy	0.386×10^6	0.0827×10^6	0.26	0.0414×10^6	A
Kelvar epoxy	0.76×10^6	0.055×10^6	0.34	0.0230×10^6	B
Graphite epoxy	1.81×10^6	0.103×10^6	0.28	0.0717×10^6	C
Boron epoxy	2.04×10^6	0.185×10^6	0.23	0.0559×10^6	D

Table 4:

Results of static and classical buckling analyses of multilayered anisotropic clamped rectangular plate 32 cm x 32 cm x 0.26 cm thick; udl = 0.04 N/mm^2 - whole plate discretized into 4 elements

Configuration - symmetric with respect to midsurface- 0° orientation		Maximum Central static deflection in cm	Nx = Ny in N/mm as per classical buckling analysis
Top	A - 0.04 cm	0.205 113	198.8085
	B - 0.04 cm		
	C - 0.04 cm		
Middle	D - 0.02 cm		
	C - 0.04 cm		
	B - 0.04 cm		
Bottom	A - 0.04 cm		

3 LAYERED MODEL FOR SHELL

3.1 Formulation

The degenerated shell elements introduced by Ahmad etal [1,2] have been successfully applied for axisymmetric shells as well as general plate/shells.

Using the principle of virtual work and using Gauss Quadrature numerical integration technique, the stiffness matrix can be computed as

$$[K] = \sum_{k=1}^{\text{No of layers}} \int_{Vol} B_k^T \; D_k \; B_k \quad u \quad dv$$

where B_k is the matrix relating local strains to global displacement, D_k is the material property matrix obtained by applying appropriate transformation to the elasticity matrix in the material coordinates.

In a similar manner, the geometric stiffness matrix for the layered axisymmetric shell as well as general plate/shell can be formulated.

The coordinate system for the layered axisymmetric shell are shown in Figure 1 and 2. Following Ramm [4], the global displacements of any point are interpolated in terms of midsurface nodal displacements and the stiffness matrix formulated. Gauss Quadrature numerical integration is used to evaluate the stiffness contribution by each layer. In a similar manner, the stiffness matrix for general layered plate/shell can be formulated, following the procedures given by Ahmad etal [1,2].

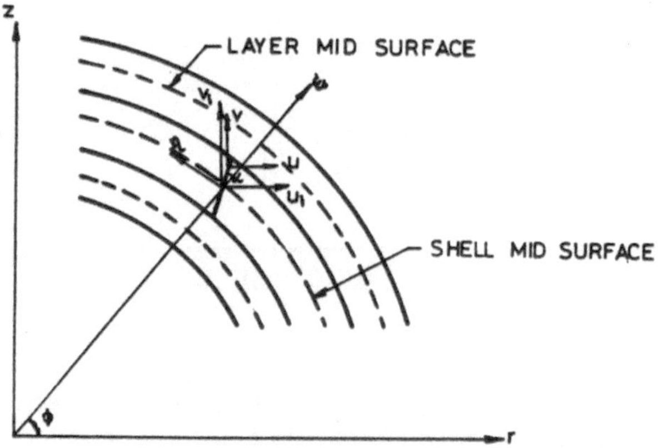

Fig: 1 GLOBAL DISPLACEMENTS IN AN AXI-SYMMETRIC LAYERED SHELL

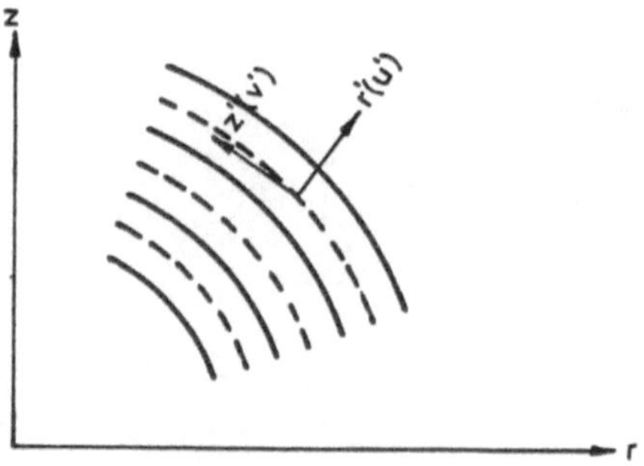

Fig: 2 LOCAL ORTHOGONAL CO-ORDINATES

3.2 Layered axisymmetric shell

The suitability of the model has been checked by performing static analysis of a cylinder with internal radius 50 cm, thickness 0.1 cm and uniform internal pressure 0.1 N/mm^2. The cylinder was made of 3 equal layers of isotropic material with $E = 0.3 \times 10^6$ N/mm^2 and $\nu = 0.3$. The cylinder is modelled into 8 elements and the hoop stress at the mid-surface is obtained as 50 N/mm^2 as against the closed form solution of 50.3 N/mm^2. The model is also checked for classical axial buckling load for a hinged cylinder of radius 5.125 cm, height 10.0 cm and thickness 0.081 cm with $E = 0.28 \times 10^7$ N/mm^2 and $\nu = 0.3$. The axial buckling load obtained was 69608.15 N/mm^2 which fairly agrees with the analytical solution 69000 N/mm^2 of Timoshenko [6].

3.3 Layered Plate/Shell

Plate being a particular form of shell without initial curvature, the development of a general layered model with different material properties and fibre orientation is under development.

4 CONCLUSION

The layered model for rectangular plate using 16 term hermitian polynomial for transverse displacement and linear model for in-plane displacements is found to give acceptable results for static and classical buckling analyses.

The layered model for axisymmetric shell using the degeneration concept is also found to be suitable. The formulation for the layered model for a general plate/ shell using degeneration concept is under development.

5 ACKNOWLEDGEMENT

The research work reported in this paper is a part of the Ph D Research work of the Senior Author. The Authors wish to thank Dr A Shanmugasundaram, Principal, PSG College of Technology for providing necessary facilities to carry out the Research. The senior author thanks the Government of India for financial assistance through the UGC Research Scheme.

6 REFERENCES

1 Ahmad, S., Irons, B.M and Zienkiewicz, O.C., 'Curved thick shell membrane Elements with particular reference to Axisymmetric problems'. Proc. (Second) Conf. on Matrix Methods in Structural Mechanics, Wright Patterson A F B, Ohio, December 1969.

2 Ahmad, S., Irons, B M. and Zienkiewicz, O.C., 'Analysis of thick and thin shell structures by curved finite elements', International Journal for Numerical Methods in Engineering, Vol.2, 1970, pp 419 - 451.

3 Bogner, F.K., Fox, R.L and Schmit, L.A., 'The Generation of Interelement compatible stiffness and mass matrices by the use of Interpolation formulae'. Proc. (First) Conf. on Matrix Methods in Structural Mechanics, Dayton, Wright - Patterson A F B, Ohio, Oct.1968.

4 Ramm, E. and Stegmuller, H., 'The displacement finite element method in non-linear buckling analysis, of shells', Buckling of Shells, A state of the Art colloquium, Vol.2, May 6-7, 1982, Stuttgart.

5 Timoshenko, S. and Woinowsky Kreiger, S., 'Theory of Plates and Shells', Mc Graw Hill Book Company Inc, Second Edition, 1959.

6 Timoshenko, S.P. and Gere, J.M., 'Theory of Elastic Stability', Mc Graw Hill Book Company Limited, Second Edition, 1961.

AN APPROACH TO CORRECT ELASTO-VISCOPLASTIC STRESS PREDICTIONS

T. Rodic[+], Owen D.R.J, Damjanic F.

Dept.of Civil Engineering, University of Wales, Swansea, U.K

SUMMARY:

When high strain rate processes are simulated by standard elasto-viscoplastic algorithms it is necessary to apply short time steps to avoid the overestimation of stresses and therefore heat generation due to plastic work. In many cases time step limitations lead to unaccaptably expensive computation. In this paper a simple approach is proposed which allows the use of larger time steps.

1. INTRODUCTION

In the modelling of hot working processes it is essential that accurate values of strain, strain rate and temperature are predicted not only for steady state conditions but also throughout the transient response, i.e. at each time station, in order to obtain a realistic simulation. Many hot working processes operate under high strain rates which cause high stress rates when the material behaviour is elastic. Using standard elasto - viscoplastic solution procedures this produces overestimation of the stresses over the time step in which yielding occurs. An overestimation of the stresses and heat generation due to plastic work gives an error in the temperature field prediction and to avoid large discrepancies small time steps have to be used. However, in many cases this time step limitation leads to unaccaptably expensive computation and in this paper a simple approach is described which allows the use of arbitrarily large time steps. The approach is based on an assumed relation between the effective stress and the equivalent strain under particular loading conditions. The examples presented illustrate the applicability of the approach.

+ Currently research scholar and member of "RSS Akcija 2000" FNT,E.Kardelj University of Ljubljana

2. BASIC CONCEPTS OF ELASTO-VISCOPLASTICITY

Details of the standard elasto-viscoplastic algorithm can be found in Ref.[1] and only the essential expressions are reproduced here. The total strain rate is separated into elastic and viscoplastic components

$$\left\{ \dot{\epsilon} \right\} = \left\{ \dot{\epsilon}_e \right\} + \left\{ \dot{\epsilon}_{vp} \right\} \tag{1}$$

The elastic strain rate $\{\dot{\epsilon}_e\}$ obeys Hooke's law and $\{\dot{\epsilon}_{vp}\}$ is expressed by the viscoplastic flow rule for associated plasticity. This gives the governing equation

$$\left\{ \dot{\epsilon} \right\} = \left[E \right]^{-1} \left\{ \dot{\sigma} \right\} + \gamma <\Phi(F)> \frac{\partial F}{\partial\{\sigma\}} \tag{2}$$

where γ is a fluidity parameter and $<\Phi>$ is taken as non-zero for positive values of the yield function, F, only.

$$F(\{\sigma\},\chi) = f(\{\sigma\}) - \sigma o(\chi) = 0 \tag{3}$$

The uniaxial yield stress is denoted by σo and χ is a hardening parameter. The rate equations (1) and (2) can be written in an incremental form to give the stress increment occuring in a time step $\Delta t_n = t_{n+1} - t_n$ as

$$\left\{ \Delta\sigma^n \right\} = \left[D \right] \left[\left[B \right] \left\{ \Delta U^n \right\} - \left\{ \dot{\epsilon}_{vp}^n \right\} \Delta t_n \right] \tag{4}$$

where

$$\left[D \right] = \left[\left[E \right]^{-1} - \Theta \, \Delta t_n \left[\frac{\partial\{\dot{\epsilon}_{vp}^n\}}{\partial\{\sigma\}} \right] \right]^{-1} \tag{5}$$

and in which [B] is the usual strain-displacement matrix. During the n-th time step most of the plastic work is converted into heat. The average heat generation rate due to the plastic work which enters the thermal analysis [2] is computed as

$$Q = (1-f) \left\{ \sigma^{n+\alpha} \right\}^T \frac{1}{\Delta t_n} \left[\left\{ \epsilon_{vp}^{n+1} \right\} - \left\{ \epsilon_{vp}^n \right\} \right] \tag{6}$$

with

$$\left\{ \sigma^{n+\alpha} \right\} = \left\{ \sigma^n \right\} + \alpha \left\{ \sigma^{n+1} \right\} \tag{7}$$

where $0 \leq \alpha \leq 1$

3. ILLUSTRATIVE EXAMPLE

The problem under present consideration is illustrated by the solution of the simple unidirectional tension test described in Fig.1. Results obtained by the standard elasto-viscoplastic procedure (Ref.[1]) are compared with the analytical solution. The main steps of the analytical solution are given below.

Thermal b.c....insulated
$\rho x Cp = 6.24\ N/mm^2K$

Young's modulus...$E=200\ N/mm^2$
Poisson's ratio.........$\nu=0,3$
yield stress......$\sigma o=10\ N/mm^2$
hardening parameter......$H'=0$
fluidity parameter....$\gamma=0.2/s$
Von Mises yielding function F
linear Φ function.........$n=1$
X-velocity.........$V=0,2$ mm/s

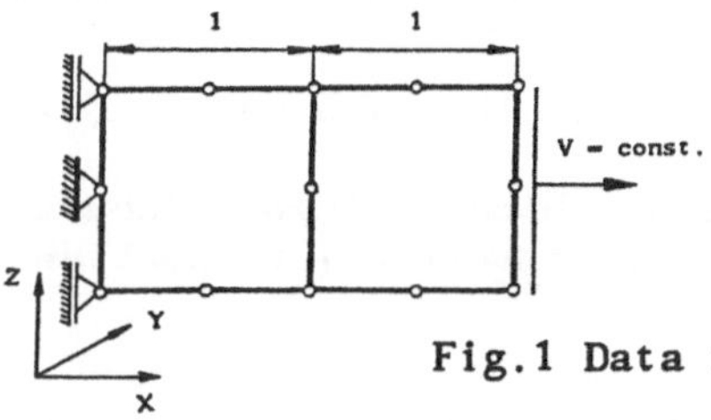

Fig.1 Data for uniaxial tension

The coordinate system X-Y-Z coincides with the direction of the principal stresses $\sigma_1,\sigma_2,\sigma_3$ and the governing equation (2) can be written as

$$\begin{Bmatrix} \dot\epsilon_1 \\ \dot\epsilon_2 \\ \dot\epsilon_3 \end{Bmatrix} = \begin{bmatrix} & & \\ & E & \\ & & \end{bmatrix} \begin{Bmatrix} \dot\sigma_2 \\ 0 \\ 0 \end{Bmatrix} + \gamma < \Phi > \frac{\partial F}{\partial\{\sigma\}} \tag{8}$$

If a linear function Φ is used then

$$\Phi = \frac{\bar\sigma - \sigma o}{\sigma o} \tag{9}$$

Note that the effective stress $\bar\sigma$ is equal to σ_1 in this case. When the Von Mises yield criterion $f=\sqrt{(3J'_2)}$ is employed then

$$\frac{\partial F}{\partial\{\sigma\}} = \frac{\partial F}{\partial J'_2}\,\frac{\partial J'_2}{\partial\{\sigma\}} = \frac{\sqrt{3}}{2\sqrt{J'_2}} \begin{Bmatrix} S_1 \\ S_2 \\ S_3 \end{Bmatrix} \tag{10}$$

where the second deviatoric stress invariant is

$$J'_2 = \frac{1}{6}\left[(\sigma_1-\sigma_2)^2 + (\sigma_2-\sigma_3)^2 + (\sigma_3-\sigma_1)^2 \right] = \frac{\sigma_1^2}{3} \tag{11}$$

and the components of the deviatoric stresses are

$$S_1 = \frac{2\,\sigma_1}{3} \quad ; \quad S_2 = -\frac{\sigma_1}{3} \quad ; \quad S_3 = -\frac{\sigma_1}{3} \tag{12}$$

Substituting (12) and (11) into (10) results in

$$\frac{\partial F}{\partial\{\sigma\}} = \begin{Bmatrix} 1 \\ -1/2 \\ -1/2 \end{Bmatrix} \tag{13}$$

Since the total strain rate in the X-direction is constant

$$\dot\epsilon_1 = V/L = C \tag{14}$$

and consequently the following differential equation can be obtained from the first row of (8)

$$C = E^{-1}\,\dot{\bar\sigma} + \gamma < \Phi(\bar\sigma)> \tag{15}$$

The solution of differential equation (15) is

$$\bar{\sigma} = E\,C\,t \qquad \text{when } \bar{\sigma} < \sigma o \tag{16}$$

and

$$\bar{\sigma} = \sigma o + \frac{C}{\gamma}\,\sigma o \left[\, 1 - \exp(-E\gamma\tau/\sigma o) \,\right] \qquad \text{when } \bar{\sigma} \geq \sigma o \tag{17}$$

where τ is <t-to> and to is the time at which yielding occurs. A fraction (1-f) of the plastic work is converted into heat and leads to a change of temperature

$$\Delta T(t) = \frac{(1-f)}{\rho\ Cp} \int_0^t \bar{\sigma}(t)\ \dot{\epsilon}_{vp1}(t)\ dt \tag{18}$$

in which f is the fraction of plastic work stored in the material, ρ is the material density and Cp is the specific heat capacity. The value of $d\epsilon_{vp1}/dt$ can be derived from the following equation:

$$\dot{\epsilon}_{vp1} = \dot{\epsilon}_1 - \dot{\epsilon}_{e1} = C - E^{-1}\,\dot{\bar{\sigma}} = C\left[\, 1 - \exp(-E\gamma\tau/\sigma o) \,\right] \tag{19}$$

where the value $d\bar{\sigma}/dt$ is obtained from the time derivative of (17). Substitution of equations (19) and (17) into (18) gives the change of the temperature with respect to time as

$$\Delta T = \frac{\sigma o\ C}{\rho\ Cp}\left\{(1+C/\gamma)\tau + \frac{\sigma o}{E\gamma}\,e^{\frac{-E\gamma\tau}{\sigma o}}\left[\,1 + \frac{2C}{\gamma}\left[1 - \frac{1}{4}\,e^{\frac{-E\gamma\tau}{\sigma o}}\right]\right] - \frac{\sigma o}{E\gamma}\left[1 - \frac{3C}{2\gamma}\right]\right\} \tag{20}$$

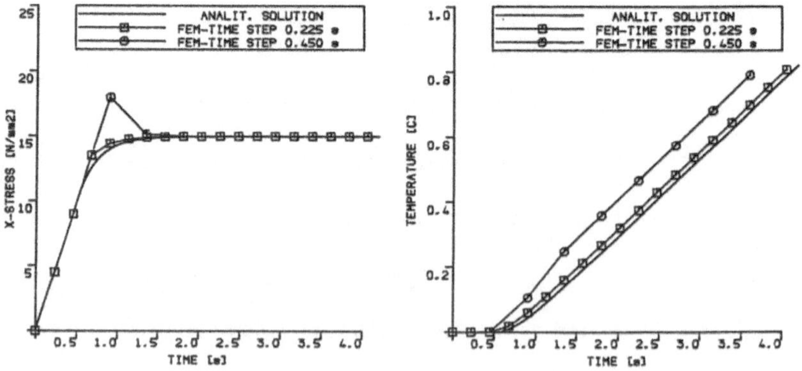

a) stress history b) temperature history

Fig.2 Results obtained by standard el.-viscoplastic algorithm

The stress and temperature histories obtained numerically are compared with the analytical values in Fig.2. It is seen from Fig.2(a) that the stress prediction is overestimated at the end of the time step in which yielding occurs.This directly causes the overestimation of the predicted temperature due to plastic work as shown in Fig.2(b).

4. THE PROBLEM RELATING TO METAL FORMING PROCESSES

As seen from the previous example, reasonably accurate results can be obtained by using a time step such that the elastic stress increments $\Delta\sigma_e$ are smaller than the difference

$$\Delta\bar{\sigma}e < \bar{\sigma}s - \sigma o \qquad (21)$$

where $\bar{\sigma}s$ represents the steady state flow stress. In many hot working forming processes the strain rate $d\epsilon/dt$ reach values as high as 100/s, the Young's modulus E and the difference $\sigma s - \sigma o$ are of the order 120 000 N/mm^2 and 200 N/mm^2 respectively. According to limitation (21) the time increment must satisfy the following inequality:

$$\Delta t < \frac{\bar{\sigma}s - \sigma o}{E\ \dot{\bar{\epsilon}}} = \frac{200\ N/mm^2}{120\ 000\ N/mm^2\ 100/s} = 0{,}17\ 10^{-4}\ s \qquad (22)$$

Supposing that only one stroke (duration 0.03 s) of the radial forging process [3] is analysed, over 1000 increments would be required to obtain an accurate solution. A simple approach which allows larger time steps is described in the following section.

5. STRESS CORRECTION ALGORITHM

Consider that the material behaviour at the Gauss point is elastic at time tn (Fig.3). The stress increment in the next time step is calculated according to (4) as

$$\left\{ \Delta\sigma_e^n \right\} = \left[E \right] \left[B \right] \left\{ \Delta U^n \right\} \qquad (23)$$

where the subscript ,e, denotes that elastic behaviour is assumed. The stess increment can now be added to the previous state of stress to give

$$\left\{ \sigma_e^{n+1} \right\} = \left\{ \sigma^n \right\} + \left\{ \Delta\sigma_e^n \right\} \qquad (24)$$

Employing the Von Mises expression, the effective stress can be calculated as

$$\bar{\sigma}_e^{n+1} = \sqrt{J_2'} \qquad (25)$$

Now we can check whether or not the Gauss point has yielded during this time step. If it has, than an effective procedure has to be undertaken to eliminate the remaining portion ,R, of the overestimated stresses (Fig.3).

$$R = \frac{\bar{\sigma}_e^{n+1} - \bar{\sigma}^{n+1}}{\bar{\sigma}_e^{n+1} - \bar{\sigma}^{n}} \qquad (26)$$

The main question now is: HOW TO DEFINE σ^{n+1}?

Recalling equation (15), the term C represents the total strain in the X-direction.The principal strain rates under uniaxial tension are then

$$\dot{\epsilon}_1 = C \quad ; \quad \dot{\epsilon}_2 = -q\, C \quad ; \quad \dot{\epsilon}_3 = -q\, C \tag{27}$$

where q is a Poisson's ratio dependant on the current state of stress. The ratio of transverse contraction $\dot{\epsilon}_q$ to longitudinal extension rate $\dot{\epsilon}_1$ under simple unidirectional tension can be defined according to equation (2) as

$$q = \frac{\dot{\epsilon}q}{\dot{\epsilon}_1} = \frac{\nu\ \dot{\bar{\sigma}}/E + \gamma\ \langle\Phi(\bar{\sigma})\rangle\ 1/2}{\dot{\bar{\sigma}}/E + \gamma\ \langle\Phi(\bar{\sigma})\rangle} \tag{28}$$

where ,ν, is the elastic Poisson's ratio.Employing (27), C can be expessed in terms of the equivalent strain rate as

$$C = \frac{\dot{\bar{\epsilon}}}{\surd(2/3)\,[1+q^2]^{1/2}} \tag{29}$$

Substitution of C from (29) into (15) results in

$$\dot{\bar{\sigma}} + E\left[\ \gamma\ \langle\Phi(\bar{\sigma})\rangle - B(\dot{\bar{\sigma}},\bar{\sigma},\dot{\bar{\epsilon}})\right] = 0 \tag{30}$$

where

$$B(\dot{\bar{\sigma}},\bar{\sigma},\dot{\bar{\epsilon}}) = \frac{\dot{\bar{\epsilon}}}{\left[\ \frac{2}{3}\ \left[\ 1 + q^2(\dot{\bar{\sigma}},\bar{\sigma},\dot{\bar{\epsilon}})\ \right]\right]^{1/2}} \tag{31}$$

Differential equation (30) gives a relationship between effective stress and equivalent strain under uniaxial tension when the Von Mises yielding function is considered. This relationship is assumed to be valid for general states of stress and strain. The average equivalent strain rate within the n-th time step is

$$\dot{\bar{\epsilon}}^n = \frac{1}{\Delta t_n}\left[\ \frac{2}{3}\ \{\Delta\epsilon^n\}^T\{\Delta\epsilon^n\}\ \right]^{1/2} \tag{32}$$

in which

$$\left\{\ \Delta\epsilon^n\right\} = \left[\quad B\quad \right]\ \left\{\ \Delta U^n\right\} \tag{33}$$

Solution for $\bar{\sigma}(t)$ from (30) under a constant equivalent strain rate with initial condition

$$\bar{\sigma} = \sigma o \qquad \text{at } \tau = 0 \tag{34}$$

provides an estimated change of effective stress with respect to time and at time station t_{n+1} gives

$$\bar{\sigma}^{n+1} = \bar{\sigma}(\tau_o) \tag{35}$$

where τ_o is defined according to Fig.3 as

$$\tau_o = R\,\Delta tn \tag{36}$$

Stress vector $\{\sigma^{n+1}\}$ is therefore required to satisfy

$$P(\{\sigma\}) = f(\{\sigma\}) - \bar{\sigma}^{n+1} = 0 \tag{38}$$

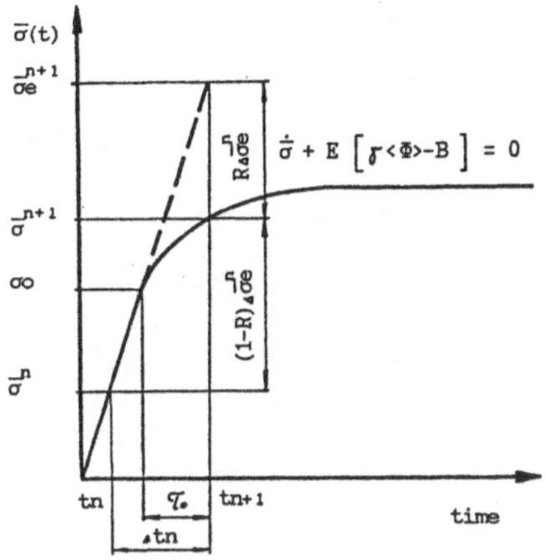

Fig.3: $\bar{\sigma}(t)$ under $\dot{\bar{\epsilon}}$=const.

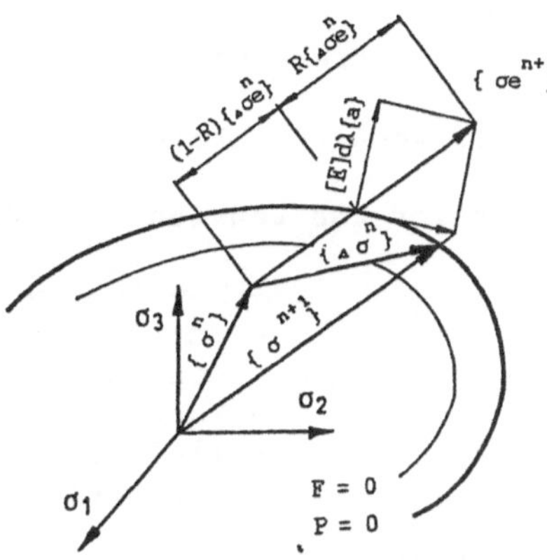

Fig.4:Stress vector correction

Since $\{\sigma e^{n+1}\}$ has a higher potential (see Fig.3,4), the remaining portion R of the elastic stress increment has to be eliminated in some way. One possible way is graphically presented in Fig.4. The procedure is similar to that normally used in plasticity to bring the stress vector back to the yield surface. A detailed description of such a procedure is provided in Ref.[1].The only difference in this case is that the yield surface is replaced by the current stress potential. The analytical solution of the non-linear differential equation (30) for various functions Φ generally results in mathematical difficulties. However, (30) can always be solved numerically and the Runge-Kutta approach [4] has been implemented in the present work. The part τo of the time step Δtn in which yielding is registered, is divided into several time increments and the B function is evaluated at the begining of each increment.

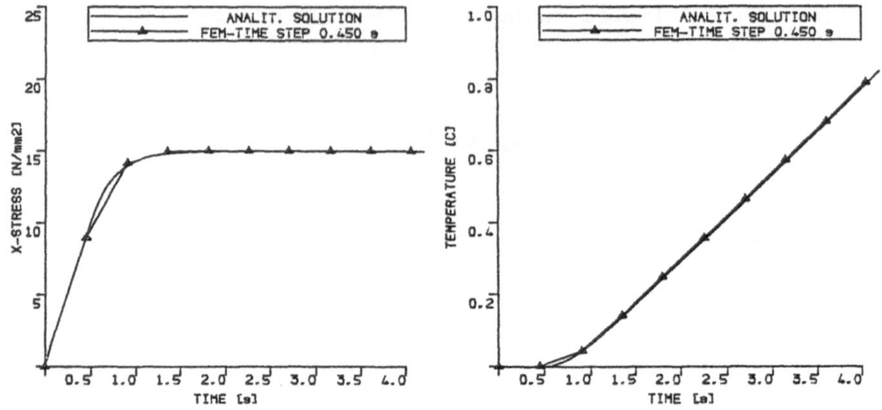

a) stress history b) temperature history

Fig.5 Results obtained by use of the stress correct. algorithm

Data for the tension of a X10CrNi18.8 steel specimen

- X-velocity.....................21.256 mm/s
- Initial temperature................ 1100 C
- density x specific heat cap.....6,24 N/mm^2K
- Young's modulus............E =120 000 N/mm^2
- fluidity function..........γ = A exp(-Q/RT)
- Φ-function................$\Phi = [\sinh(\beta\sigma)]^n$

A=1.2 10^{12}/s; β=0.0012 mm^2/N; Q=351 kJ/mol; n=3.5; R=8.31J/molK

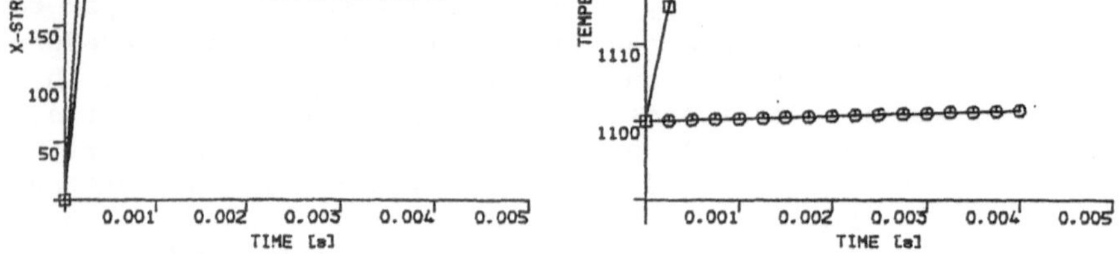

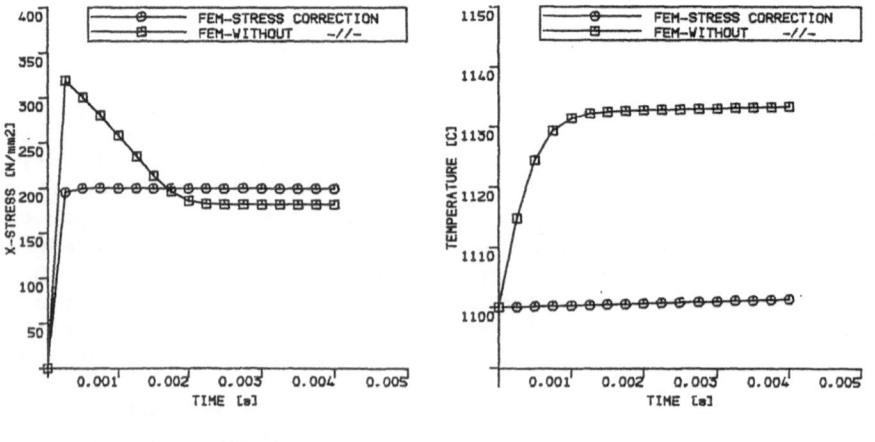

a) stress history b) temperature history

Fig.6 Tensile loading of X10CrNi18.8 steel analysed by both algorithms

6. RESULTS

The example from Section 3 shown in Fig.1 is reanalysed using the stress correction algorithm. The results (Fig.5) are in good agreement with the analytical values even for the large time step employed. The same example is then analysed using the material properties [5] of a steel with German designation X10CrNi18.8. The Young's modulus and strain rate are both high and therefore the suitability of the algorithm is even more apparent.

7. REFERENCE

[1] OWEN D.R.J and HINTON E.: Finite Elements in Plasticity. Pineridge Press, Swansea, U.K., (1980)

[2] DAMJANIC F. and OWEN D.R.J.: Practical Considerations for Thermal Transient Finite Element Analysis Using Isopara - metric Elements.Nucl.Engng.Design, Vol.69,pp.109-126,(1982)

[3] RODIC T., STOK B., GOLOGRANC F. and OWEN D.R.J: Finite Element Modelling of a Radial Forging Process. 2nd International Conference on Technology of Plasticity, Stuttgart, (1987)

[4] BRAUN M.: Differential Equations and Their Applications. 2nd Edition. Springer-Verlag, New York Inc., (1978)

[5] KRENGEL R.: Anpassung von Warmwalz-Stichplanen an Veranderungen der Walzguttemperatur Institut fur Werkstoffunformung, T.U. Clausthal, (1985)

FINITE ELEMENT ANALYSIS OF A VISCOELASTIC SOLID SLIDING OVER RIGID TRIANGULAR ASPERITIES

N. Purushothaman and I.D. Moore
University of Newcastle
Newcastle, N.S.W., 2308, AUSTRALIA

SUMMARY

A three dimensional viscoelastic finite element procedure has recently been developed as a part of an investigation of the skid resistance of wet pavement. The procedure is described briefly, and example calculations show how it is used to examine a viscoelastic solid sliding over a rigid triangular asperity. A comparison is made between theoretical predictions of hysteretic friction and measurements obtained in the laboratory. These indicate that the procedure successfully predicts the influence of sliding speed, contact pressure, asperity, geometry and three dimensional effects, on hysteretic friction.

1. INTRODUCTION

The skid resistance of wet pavement has a significent effect on road safety. Studies of skid resistance are usually experimental in nature. However theoretical studies of skid resistance are also valuable, and various solutions (e.g. [3, 4]) to idealised sliding contact problems have been reported. A three dimensional viscoelastic finite element theory and its application to the study of skid resistance are described in this paper.

Firstly the idealised sliding contact problem is defined, and the numerical procedure which has been developed is outlined. Some typical solutions are then examined, and comparisons are made between theoretical predictions of sliding contact behaviour, and experimental measurements.

2. STATEMENT OF PROBLEM

A vehicle skidding across a rough wet pavement represents a

complex mechanical system, and the task of modelling it is daunting indeed. In the present study all consideration of vehicle suspension, pavement deformation and surface water drain age is put aside. Furthermore it is assumed that dynamic (or inertial) effects can be neglected, as well as the influence of changes in geometry on rubber stiffness, [3].

In order to make reasonable comparisons with experimentally measured behaviour, the rubber block of length 76mm, width 24mm and depth 5mm which is attached to the pendulum skid resistance testing devices being used, must be modelled, Figure 1. The block has an aluminium plate bonded to its top surface, so requires some form of three dimensional analysis.

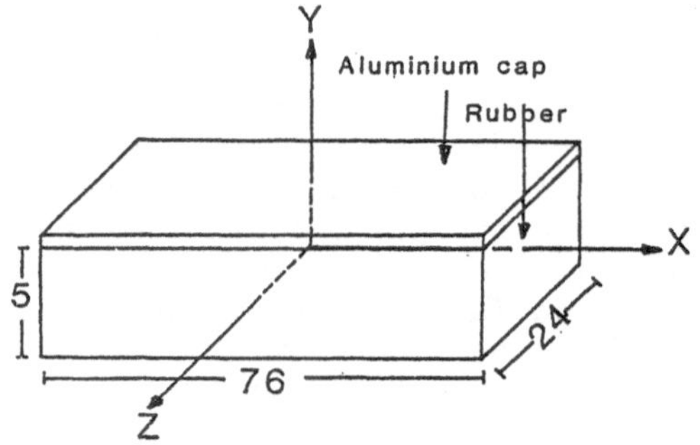

Figure 1. Rubber Block Used in Pendulum Tester

The surface of a road pavement has a complex geometry, featuring asperities of many different scales which all contribute to skid resistance. The asperities of different scales will be considered separately in this study, and for each scale the profile will be idealised as a series of smooth equally spaced prismatic asperities of triangular cross-section. The objective is to calculate the coefficient of friction associated with the rubber hysteresis (the rubber is stiffer in loading than unloading and a net horizontal force develops across the asperity).

3. FINITE ELEMENT ANALYSIS

Three aspects of the numerical analysis are worthy of mention: (1) the three dimensional model; (2) the implementation of viscoelastic material behaviour; (3) the treatment of rubber-asperity interaction.

3.1. Simplified Three Dimensional Analysis

Generally plane strain or plane stress conditions are assumed for this type of analysis [3,4]. However the three-dimensional nature of the rubber block attached to the pendulums has already been emphasised. In order to avoid the substantial costs associated with conventional three-dimensional analysis, an alternative strategy has been developed.

For the rubber block shown in Figure 1, it is assumed that displacements in the direction of sliding u_x and vertically u_y do not vary across the block, i.e.

$$u_x = u_x(x,y); \quad u_y = u_y(x,y) \tag{1}$$

and that lateral displacements u_z are symmetric about the centre line z = 0,

$$u_z = u_z(x,y,z) = - u_z(x,y,-z) \tag{2}$$

The displacement functions employed are

$$(u_x, u_y) = \sum_{i=1}^{n} N_i(x,y)\ (u_x,u_y)_i \tag{3}$$

$$u_z = \sum_{i=1}^{n} N_i(x,y)\ \frac{2z}{W}\ u_{zi}$$

where N_i is the shape function for node i of a two-dimensional element with n nodes, W is the width of the rubber block, u_{xi} and u_{yi} are the x and y displacements of nodal "line" i and u_{zi} is the z displacement at the external face (z = W/z) of nodal "line" i.

Warping across the block is neglected, and the importance of this has been examined elsewhere [1] by comparing theoretical predictions based on equations (1), (2), (3) with direct laboratory measurement of static rubber block response. The simplified theory was found to yield satisfactory estimates of rubber block behaviour, so that warping effects are not significant.

3.2. Viscoelastic Material Response

Schapery [3] demonstrated that for a periodic disturbance travelling at constant speed, it is straightforward to use the correspondence principle to calculate viscoelastic material response once the discrete Fourier series is used to decompose all degrees of freedom into harmonic components. If the y,z planes containing finite element nodes are numbered consecutively in the x direction, and the "long" rubber block is responding to a periodic disturbance of period N, then only one typical wavelength of the response needs to be considered (neglecting end effects for the block of finite length). Applying the discrete Fourier Series transformation to nodal displacements $\underset{\sim}{\delta}_k$ and forces $\underset{\sim}{f}_k$ (for y,z plane number k)

$$(\underset{\sim}{\delta},\ \underset{\sim}{f})_k = \frac{1}{\sqrt{N}} \sum_{j=1}^{1} (\underset{\sim}{\Delta},\ \underset{\sim}{F})_j \exp\ (i\ \alpha\ jk) \tag{4}$$

where $\alpha = 2\pi/N$ and the coefficients $\underset{\sim}{\Delta}_j$ and $\underset{\sim}{F}_j$ are given by

$$(\underset{\sim}{\Delta},\ \underset{\sim}{F})_j = \frac{1}{\sqrt{N}} \sum_{j=1}^{N} (\underset{\sim}{\delta},\ \underset{\sim}{f})_k \exp\ (-i\alpha\ jk) \tag{5}$$

For an elastic material, stiffness equations can be derived relating coefficients for mode j, [2] viz:

$$\underset{\sim}{F}_j = K_j\ \underset{\sim}{\Delta}_j \tag{6}$$

Now the simple rheological model shown in Figure 2 is characterized by

$$\sigma(E_1^{-1} + E_2^{-1}) + \dot{\sigma}(\eta E_1^{-1}) = \varepsilon + \dot{\varepsilon}\ (\eta E_2) \tag{7}$$

where σ is stress, ε is strain, E_1 and E_2 are "spring" stiffnesses, η is "dashpot" viscosity and $(\dot{\ })$ denotes differentiation with respect to time. Consider the harmonic contributions to stress and strain for sliding speed C, and wavelength L, $(\sigma_j,\ \varepsilon_j)\ \exp\ [i\alpha j(k + CtN/L)]$.

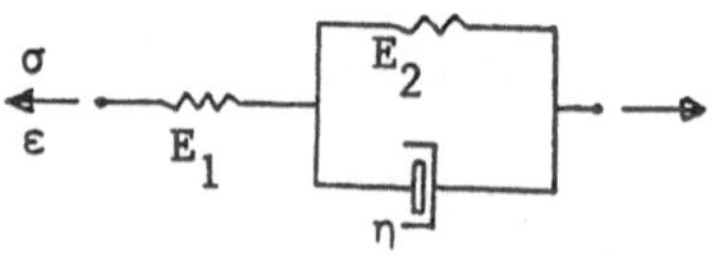

Figure 2. Rheological Model used for Viscoelastic Solid

Substitution into equation (7) yields the pseudo-static relationship

$$\sigma_j = E^* \varepsilon_j \tag{8}$$

where $E^*(E_1, E_2, \eta, j)$ is given by

$$E^* = \frac{[E_1E_2(E_1+E_2) + E_1(2\pi\eta j\ C/L)^2] + i[E_1{}^2 2\pi\eta j\ C/L]}{(E_1+E_2)^2 + (2\pi\eta j\ C/L)^2} \tag{9}$$

It is not difficult to obtain E^* for other rheological models, nor to show that a viscoelastic finite element solution can be obtained for each node j by calculating K_j for unit Young's modulus, followed by solution of

$$\underset{\sim}{F}_j = E^* K_j\ \underset{\sim}{\Delta}_j \tag{10}$$

3.3 Contact Analysis

Analysis of the contact between the rubber block and the triangular asperities involves the use of prescribed displacements at contact points and forces of zero value elsewhere. Unfortunately, for each location in the y,z plane a solution based on the discrete Fourier Series transformation must have the same type of boundary condition at all positions in the x direction.

An iterative procedure using only prescribed forces along the contact surface is therefore employed: (1) accumulated displacement is compared with total prescribed displacement to obtain the incremental deformation required; (2) elastic analysis is used to predict the increments in contact force associated with those incremental deformations; (3) viscoelastic analysis yields the actual deformations associated with those forces, and accumulated displacement is updated.

4. RESULTS OF SLIDING CONTACT ANALYSIS

4.1 Material Properties and Mesh Design

All solutions reported in this section are for Poisson's ration 0.47, static Young's modulus $E_\infty = E_1E_2/(E_1 + E_2)$ = 3.29 MPa and "spring" ratio E_1/E_2 = 5 (these are the rubber block properties as measured). Eight noded rectangular elements are extended into the third dimension using the procedure outlined in section 3.1.

4.2 Deformations and Contact Pressures

Figure 3 shows the shape of the contact surface of the rubber for a triangular asperity of slope 0.36 and spacing L=4mm apex to apex. For this low sliding speed ($\eta C/L = 0.025$) the deformation profile is almost symmetric about the apex, and hysteretic friction is small $\mu_h = 0.014$. For higher speeds the profile is significantly less symmetric and hysteretic friction increases (e.g. for $\eta C/L = 2.5$, $\mu_h = 0.19$)

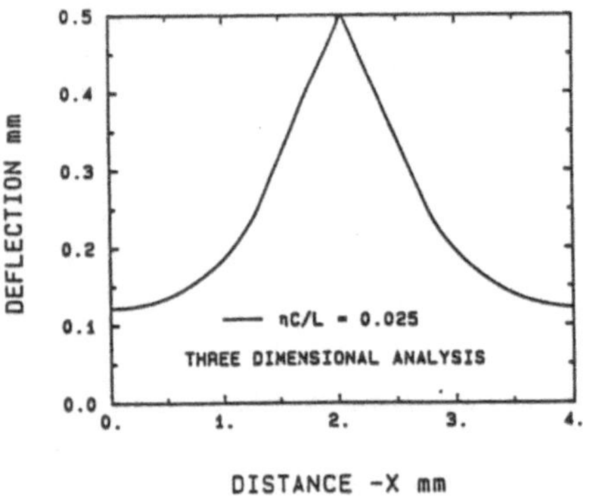

Figure 3. Contact Shape - Triangle Slope 0.36, $\eta C/L = 0.025$, $\mu = 0.014$

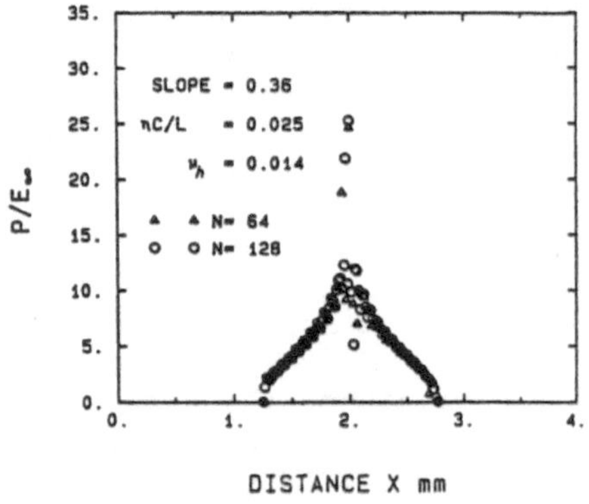

Figure 4. Contact Pressures - Triangle slope 0.36, $\eta c/L = 0.025$.

The distribution of contact pressures at nodes is shown in Figure 4. Again the distribution for this low speed is nearly symmetric, but is distinctly nonsymmetric for high velocity. Solutions are shown for two discretisations: 32 elements (N=64) and 64 elements (N=128) along the contact surface.

4.3 Comparison with Experimental Measurements

Three brass plates were accurately milled to form a set of equally spaced traingular prisms with side slopes 0.25, 0.36 and 0.5 and apex to apex distance 4 mm. The British Portable Pendulum Tester (BPT) and a Variable Speed Pendulum Friction Tester (VSPFT) developed at Newcastle University were used to measure hysteretic friction for each of the three profiles. The performance of the theory was then examined by making comparisons with the experimental measurements.

Figure 5 shows measurements of hysteretic friction μ_h obtained using the British Portable Tester. Theoretical predictions are shown for BPT sliding speed of 10km/h. Clearly the numerical procedure is providing reasonable estimates of hysteretic friction. The prediction is excellent for triangle slope 0.25, but friction is overestimated for the steeper triangles.

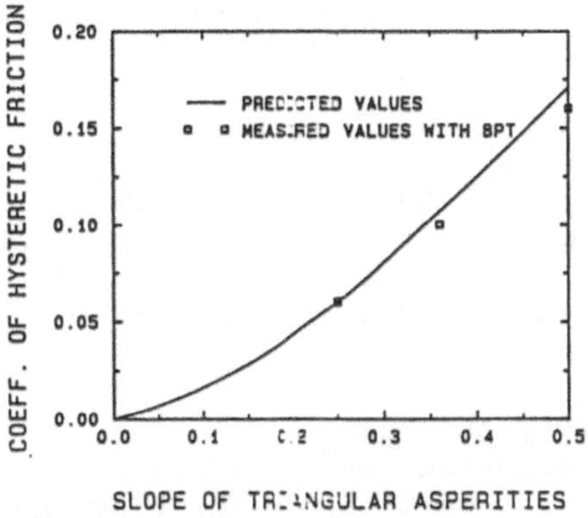

Figure 5. Hysteretic friction Measurements from BPT - C = 10 km/h.

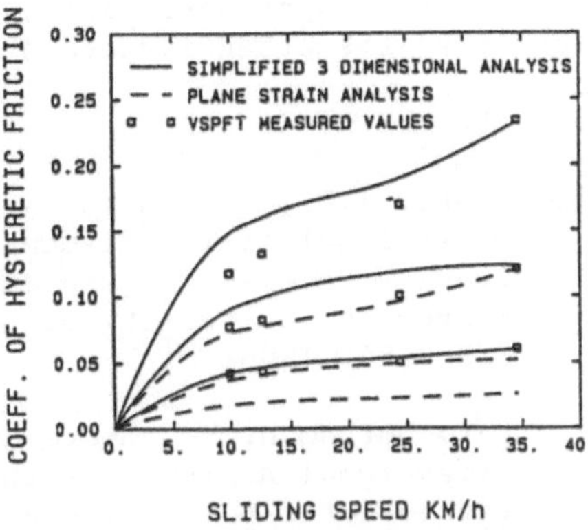

Figure 6. Hysteretic Friction Measurements from VSPFT - Triangle slopes 0.25, 0.36, 0.5.

Figure 6 shows friction measurements obtained using the Variable Speed Pendulum Friction Tester. VSPFT measurements for each of the triangle slopes were made for sliding speeds of 10km/h, 12.6km/h, 24.4km/h and 34.6km/h. Theoretical predictions obtained using three-dimensional and plane strain sliding contact analysis are also shown.

At a sliding speed of 10km/h (the same as for BPT) the trends are similar to those shown in Figure 5. For higher sliding velocities there is an improvement in the predictions made for steep triangles. The three-dimensional analysis seems capable of predicting the influence of sliding speed, triangle slope and contact pressure (contact pressures are different for BPT and VSPFT so that results for 10km/h are not the same in Figures 5 and 6). Two-dimensional analysis is significantly less satisfactory.

5. CONCLUSIONS

A finite element solution to a three-dimensional viscoelastic sliding contact problem has been described, and its application to studies of pavement skid resistance has been considered. The simplified three dimensional theory yields reasonable predictions of test equipment response without the need for substantial increases in solution cost above that associated with two dimensional analysis. The use of the correspondence principle after harmonic decomposition using discrete Fourier Series transformation, yields an efficient solution of the viscoelastic material problem.

The analysis was used to determine the hysteretic friction associated with rubber sliding on rigid triangular asperities.Comparisons with experimental measurements confirmed the effectiveness of the numerical model.

6. REFERENCES

1. Moore, I.D., Purushothaman, N. and Heaton, B.S., 'Three Dimensional Finite Element Study of the Skid Resistance of Grooved Pavements', Department of Civil Engineering and Surveying, University of Newcastle, Research Report 012-8-86, 1986.

2. Purushothaman, N., Moore, I.D., and Heaton, B.S. 'Three Dimensional Analysis of the Behaviour of Elastic and Viscoelastic Solids', Department of Civil Engineering and Serveying, University of Newcastle, Research Report 013-8-86, 1986.

3. Schapery, R.A., 'Analysis of Rubber Friction by Fast Fourier Transform, Tire Science and Technology', Vol.6, 2, pp.89-113, 1978.

4. Yandell, W.O., 'The Measurment of Surface Textures of Stones with Particular Regard to the Effect on the Frictional Properties of Road Surfaces', Ph.D. Thesis, University of New South Wales, Australia, 1970.

NON-LINEAR ANALYSIS OF ARCH DAMS

C. Pina and R. Câmara

Laboratório Nacional de Engenharia Civil
Lisbon, Portugal

SUMMARY

The evaluation of safety conditions of arch dams calls for the verification of the global rupture conditions of significant volumes of the structure, for scenarios of material strength decrease and/or actions increase. For the study of these scenarios the material non-linear behaviour must be considered.

Shell models of arch dams developed with an elastoplastic behaviour of the concrete are presented in this paper. In the analysis of these models the finite element method is used with a stress-transfer technique to consider the material non-linear behaviour.

Examples of the use of these models in the static and dynamic analysis of some arch dams are briefly presented and confronted with the results of three-dimensional scaled models.

1. INTRODUCTION

In the study of arch dams, at the design phase, at the construction phase or at the phase of interpretation of already built structures, it becomes necessary to use models in order to assess the safety and deterioration conditions. These models are essentially based on assumptions related with the actions, the structural behaviour and the boundary conditions.

Linear elastic material behaviour is one assumption usually adopted in arch dams in order to analyse the combination of actions that are representative of normal operating conditions (dead weight, water and temperature actions, and also the action of small or moderate intensity earhquakes). The safety criteria that usually correspond to these scenarios are the limitation of the stresses to low values (1.0 MPa for maximum tensile stress and 8.0 MPa for maximum compressive stress) which clearly justifies the linearity hypothesis assumed. However, the observation of arch dams shows that beyond the elastic (reversible) effect, a non-elastic (irreversible) effect remains, related, among other causes, with the structure deterioration (local ruptures).

On the other hand, the evaluation of safety conditions of arch dams calls for the verification of the global rupture conditions of significant volumes of the structure, for scenarios of material strength decrease and/or actions increase (including strong intensity earthquakes). For the study of these scenarios where the stress level reaches the material strength values the use of non-linear models is obviously required.

Three-dimensional scaled models were initially used in the analysis of global rupture conditions of arch dams [1] ; more recently however they are used together with mathematical methods [2,3] .

In this paper, a shell model, with an elastoplastic behaviour, is presented, and the main features of the static and dynamic analysis of this model by the finite element method are referred to. Examples of the use of this model in the analysis of arch dams and the comparison of the results obtained with experimental and mathematical methods are also presented.

2. MODEL

In the model adopted the analysed structure is assumed to have a thin shell structural behaviour. The structure is idealized as an assembly of a set of continuous and homogeneous blocks, possibly, linked by joints.

Time-dependent actions must respect the equilibrium conditions of the applied forces ($\underline{F}(t)$) and forces associated with inertia, damping and deformability. The equations of equilibrium may be expressed in function of displacements ($\underline{U}(t)$) and their time derivatives ($\dot{\underline{U}}(t)$, $\ddot{\underline{U}}(t)$) by means of:

$$\underline{M}\,\ddot{\underline{U}}(t) + \underline{C}\,\dot{\underline{U}}(t) + \underline{K}(U(t))\,\underline{U}(t) = \underline{F}(t) \qquad (1)$$

In the above equation $\underline{M}$, $\underline{C}$ and $\underline{K}$ are the mass, the damping and stiffness matrices.

The effects of the vibration of the water in the reservoir can be roughly introduced in this equation by adding to the structure mass matrix another diagonal matrix defined by Westergaard solution [4] , so avoiding the solution of the coupled problem [3] .

For static actions represented by applied forces ($\underline{F}$) The equations of equilibrium reduce to

$$\underline{K}(U)\ \underline{U} = \underline{F} \tag{2}$$

The behaviour of the material (of blocks and joints) was supposed to be elastoplastic, with peak and residual resistances. Linear elastic behaviour has an upper limit for the tensile rupture stress

$$\sigma \leqslant f_t \tag{3}$$

and is limited by a Mohr-Coulomb criterion for shear rupture with compressive normal stress

$$\tau \leqslant - \sigma \text{ tg } \emptyset + C \tag{4}$$

being C the cohesion and $\emptyset$ the angle of internal friction(fig.1).

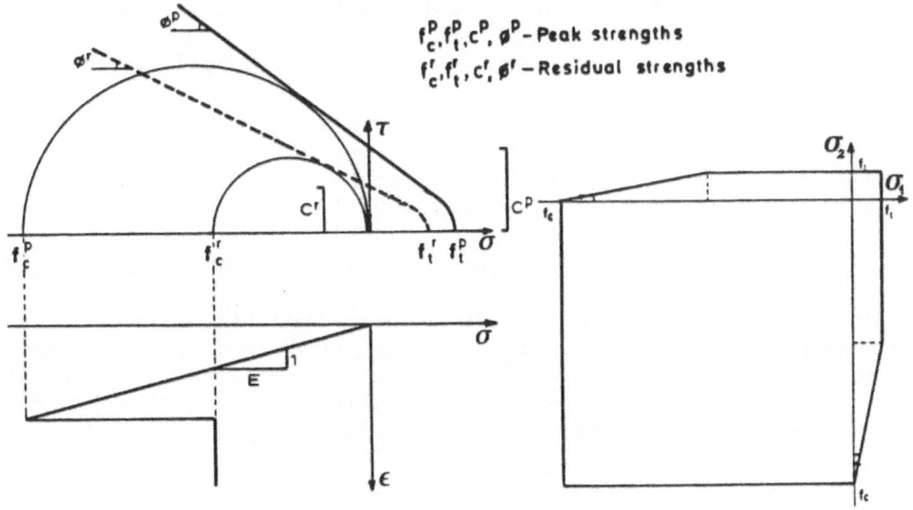

Fig.1 Failure criterion.

When the above conditions for peak strengths do not hold, the behaviour of the material becomes perfectly plastic, and the same conditions for residual values will be fulfilled.

It should be noted that, as a thin shell model was used, a principal stress is assumed to be zero in a normal direction to the middle surface, so the maximum normal stress is always a tensile stress or, at least, zero.

The boundary conditions may be of the two following types:

i) rigid supports in any direction; and ii) elastic supports, when the force-displacement relationships on the boundary surface are defined by means of the well-known Vogt's coefficients [5] .

3. METHODS

The finite element method was used to analyse the above model; as field variables, displacements within the element and on the boundary were considered.

The structure is divided into a set of triangular flat elements, with nodal points at vertices and with displacement interpolation functions not time-dependent, linear for components parallel to the middle plane and cubic for the component perpendicular to the middle plane [6,7] .

A stress-transfer technique is used to take into account the non-linear behaviour of the materials. This technique allows to write the first term of equation (1) in function of the stiffness of the structure that corresponds to the elastic behaviour:

$$\underline{M}\,\underline{\ddot{U}}(t) + \underline{C}\,\underline{\dot{U}}(t) + \underline{K}_o\,\underline{U}(t) = \underline{F}(t) + \left[\underline{K}_o - \underline{K}(U(t))\right]\underline{U}(t) \qquad (5)$$

The solution of equation (5) in a time t_o may be obtained by an iterative scheme, where the difference of the fictitious stresses (corresponding to elastic behaviour) and the real stresses are represented by equivalent forces:

$$\underline{M}\,\underline{\ddot{U}}^i(t_o) + \underline{C}\,\underline{\dot{U}}^i(t_o) + \underline{K}_o\,\underline{U}(t_o) = \underline{F}(t_o) + \underline{\varphi}(U^{i-1}(t_o)) \qquad (6)$$

As for time-dependent loads the equation (1) always has a solution for a specific instant t_o, the unbalanced forces always converge to finite values. For static problems when the failure of the structure is attained the solution of equation (2) does not exist, so the iterative scheme will diverge.

The method of Newwark was used for time integration:

$$\underline{A}\,\underline{U}^i(t_j) = \underline{F}(t_j) + \underline{F}_o(t_{j-1}) + \underline{\varphi}(U^{i-1}(t_j)) \qquad (7)$$

being:

$$\underline{A} = \frac{1}{b\,\Delta t^2}\underline{M} + \frac{a}{b\,\Delta t}\underline{C} + \underline{K}_o \qquad (8)$$

$$\underline{F}_o(t_{j-1}) = \left(\frac{1}{b\,\Delta t^2}\underline{M} + \frac{a}{b\,\Delta t}\underline{C}\right)\underline{U}(t_{j-1}) + \left[\frac{1}{b\,\Delta t}\underline{M} + \left(\frac{a}{b} - 1\right)\underline{C}\right]\underline{\dot{U}}(t_{j-1}) + \left[\left(\frac{1}{2b} - 1\right)\underline{M} + \left(\frac{a}{2b} - 1\right)\Delta t\,\underline{C}\right]\underline{\ddot{U}}(t_{j-1}) \qquad (9)$$

$$\dot{\underline{U}}(t_j) = -\frac{a}{b\Delta t}\underline{U}(t_{j-1}) + \left(1-\frac{a}{b}\right)\dot{\underline{U}}(t_{j-1}) + \left(1-\frac{a}{2b}\right)\Delta t\,\ddot{\underline{U}}(t_{j-1}) + \frac{a}{b\Delta t}\underline{U}(t_j) \tag{10}$$

$$\ddot{\underline{U}}(t_j) = -\frac{1}{b\Delta t^2}\underline{U}(t_{j-1}) - \frac{1}{b\Delta t}\dot{\underline{U}}(t_{j-1}) - \left(\frac{1}{2b}-1\right)\ddot{\underline{U}}(t_{j-1}) + \frac{1}{b\Delta t^2}\underline{U}(t_j) \tag{11}$$

Where a and b are constants and $t_j = t_{j-1} + \Delta t$.

4. EXAMPLES

4.1.Experimental studies up to failure of arch dams have been traditionally made by increasing proportionally the dead weight and the water load [1] to simulate a uniform decrease of material strengths. In order to compare the results obtained by the finite element models developed with those previously obtained, a similar failure scenario was considered in the mathematical study up to failure of a Portuguese arch dam (fig.2) [8] .

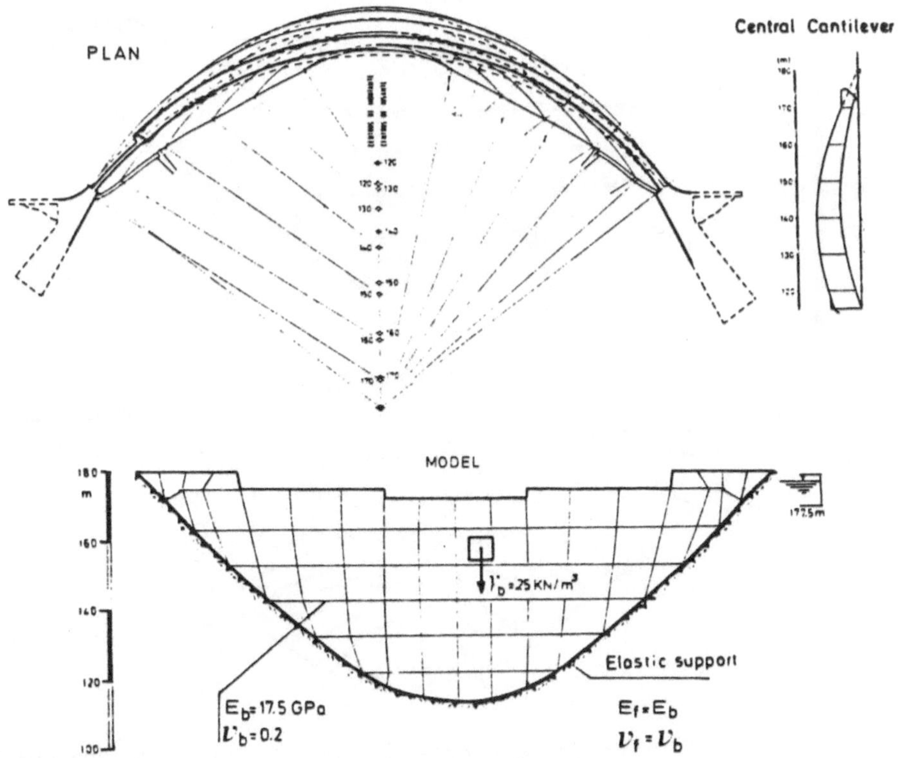

Fig.2 Bouçã arch dam. Finite element model.

The dead weight of concrete was represented by vertical body forces (γ_c = 25 KN/m^3) assuming a single phase of construction for all the structure. Water load was represented by the hydrostatic pressure (γ_w = 10 KN/m^3) acting on the upstream face of the dam with the water at 177.5 m.

Concrete was assumed to be homogeneous, isotropic and with an elastoplastic behaviour represented in the linear elastic phase, by the Young's modulus E_c = 17.5 GPa and the Poisson's ratio ν = 0.2; this phase was limited by the tensile uniaxial strength f_t = 2.0 MPa and by a Mohr-Coulomb criterion with a cohesion C = 5.55 MPa and an internal friction angle $\emptyset$ = 55º. The residual values were assumed to be f_t= 0,C = 4.75 MPa and $\emptyset$ = 55º. The deformability of the foundation was represented by Vogt's coefficients assuming that the Young's modulus and the Poisson's ratio are the same as those of the concrete.

The study was made first applying the forces representative of the dead weight and the hydrostatic pressure and then increasing proportionally these forces up to the failure of the dam. The

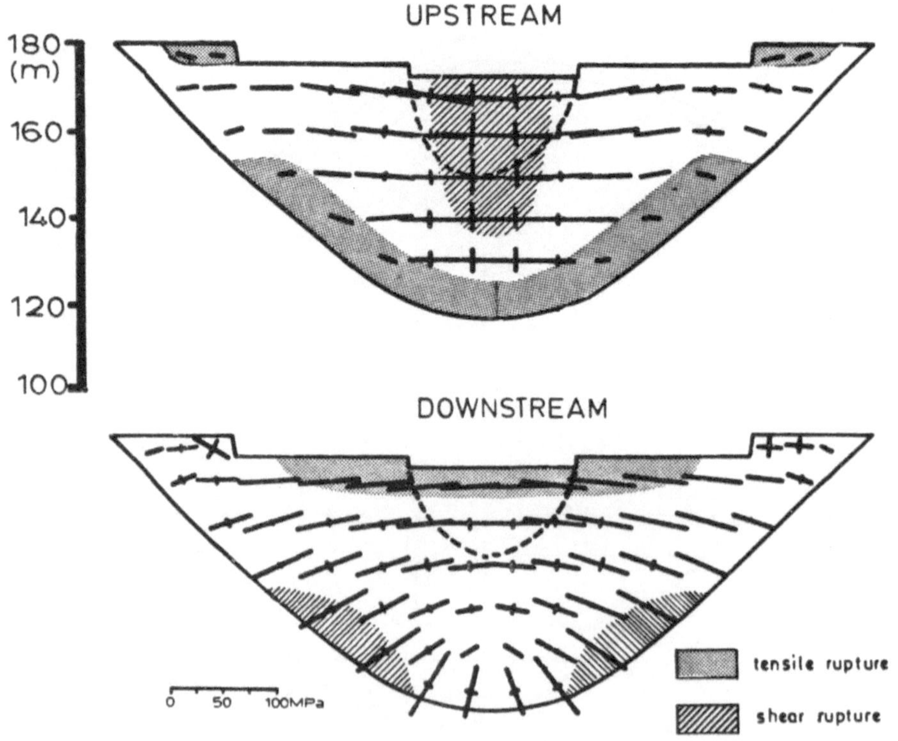

Fig.3 Principal stresses and craked zones on upstream and down--stream faces close to the failure (λ= 9.3).

failure occurs for loads about 9.3 times higher than those initially applied. Similar results were obtained in tests of three-dimensional scale models of the same dam with a similar material behaviour [9,1] .

Fig.3 shows the principal stresses on the upstream and downstream faces for a load level near the rupture, the zones being marked where rupture have been attained and also the cross-section in the central zone of the dam where the failure will occur when the compressive strength is reached all over that continuous surface. In fig.4 same strains measured in the test of the experimental model are compared with those obtained in the analytical model, and a good agreement can be noted.

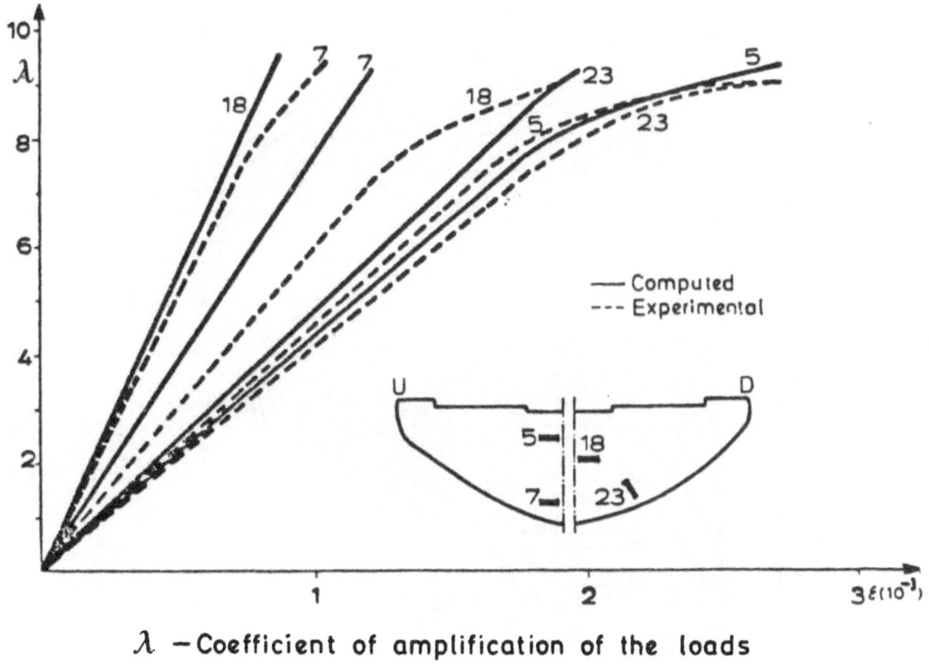

Fig.4 Evolution up to failure of strains measured in experimental test and computed in finite element model.

4.2.The models and methods described above were also used in the dynamic analysis of another Portuguese arch dam (fig.5) [3] .A study was made of the response to an earthquake with a peak acceleration of 0.7 g and a duration of 10 s, assuming full reservoir and an initial state of stress due to dead weight and water pressure.

The dam was discretized by thin shell elements supported by a homogeneous foundation, whose deformability was represented by Vogt's coefficients. Both concrete and rock mass were assumed to have dynamic values for the Young's modulus and for strengths 50% higher than the static values ($E_c = E_r$ = 30 GPa; f_t = 4.5 MPa; C = 7.125 MPa; Ø = 55°). The effect of the vibration of the water in the reservoir was represented by a mass added to the structure mass and the damping was assumed to be proportional to the mass matrix $\underline{C} = 2.5\underline{M}$.

A step-by-step time analysis was made for the strong earthquake above referred to, applying on the support surface of the dam a uniform acceleration with a horizontal component (upstream -downstream) and a vertical component, being $a_h = 2/3\ a_v$.

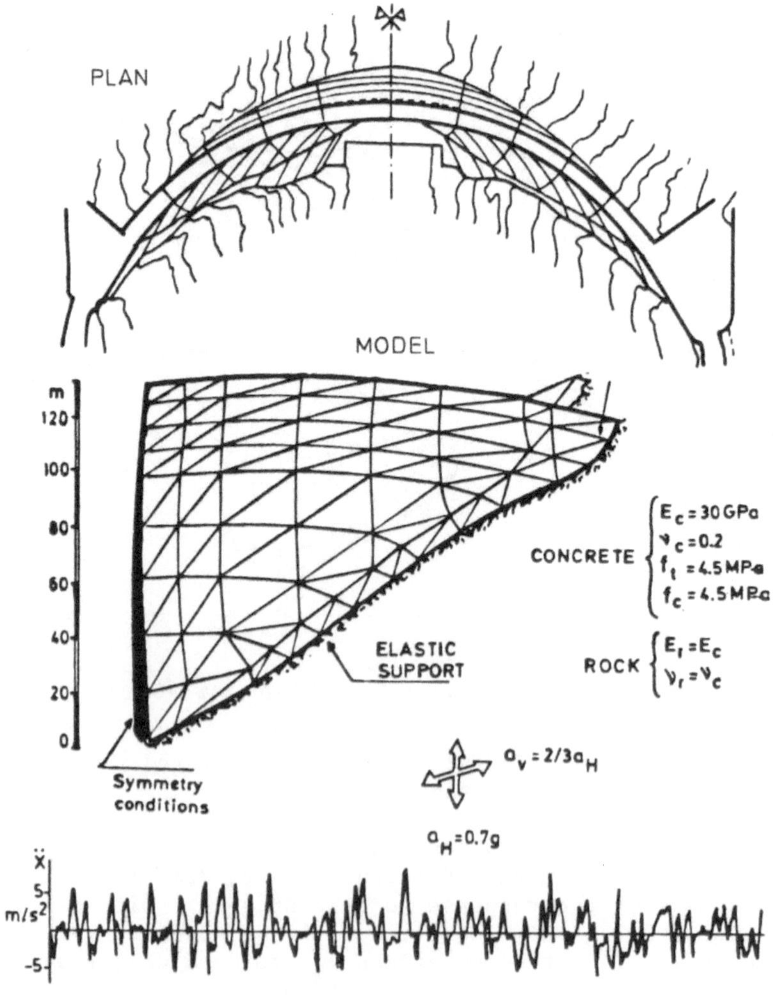

Fig.5 Cabril arch dam. Dynamic model.

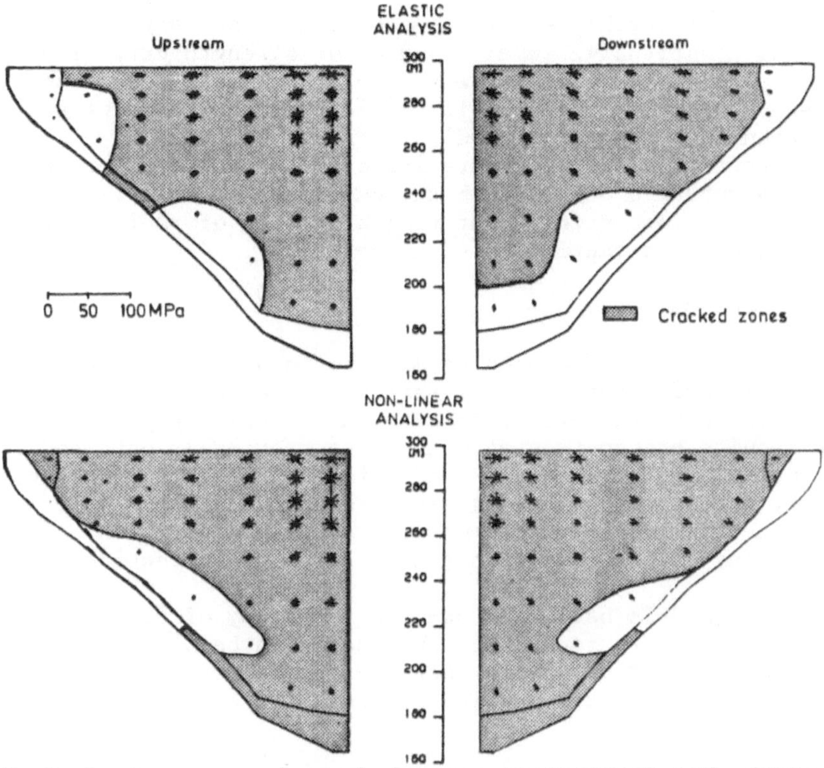

Fig.6 Peak stresses and cracked zones on upstream and down-stream faces due to elastic and non-linear analysis.

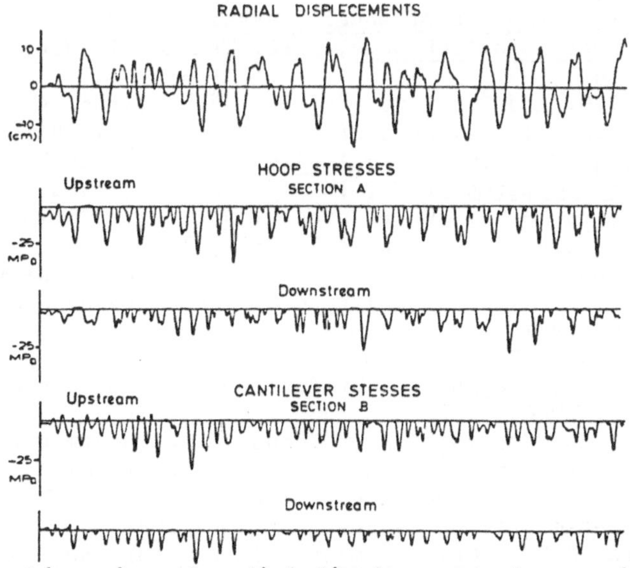

Fig.7 Time histories of radial displacement, hoop and cantilever stresses.

For this earthquake, it is possible to satisfy the equilibrium and constitutive equations and the strength criterion assumed, with a peak hoop stress of -37 MPa and a peak cantilever stress of -29 MPa, being still valid the assumption of small displacements (0.16 m of peak displacement). In fig.6 the peak stresses at the dam faces are indicated and also the zones where tensile ruptures occur for both non-linear and elastic analysis. Fig.7 shows some significant time histories of radial displacements, hoop and cantilever stresses.

5. CONCLUSION

Shell models developed at LNEC for non-linear analysis of arch dams are referred to in this paper. An example of static analysis up to failure of an arch dam is briefly presented and the results are confronted with those previously obtained in experimental tests. A dynamic analysis of an arch dam subjected to a very strong earthquake is also presented. Both examples show that the failure of arch dams will usually occur when compressive strength is attained in a continuous cross-section.

These models have been used in the study of other dams and for other failure scenarios. They can also be used for studies of other thin structures - representing joints possibly - and loaded by body forces, surface forces, temperature variations, or seismic actions.

REFERENCES

1. Rocha, M. - The use of Model Tests in the Determination of the Safety Factor in Arch Dams, LNEC, Technical paper nº433, Lisbon, 1974.
2. Pedro, J.O. and Pina, C., "Finite element analysis of dams up to failure", Trans.Int.Conf. FEICOM-85,vol.1, Bombay, 1985.
3. Câmara, R. and Pina, C. - Analise dinâmica não-linear de estruturas laminares, LNEC, Internal Report, 1986.
4. Westergaard, H.M., "Water pressures on dams during earthquakes". Trans.ASCE, V98, paper nº1835, 1933.
5. U.S. Bureau of Reclamation - Treatise on Dams. Denver, 1956.
6. Zienkiewicz, O.C. - The Finite Element Method, McGraw-Hill, New York, 1971.
7. Pedro, J.O. - Finite element stress analysis of plates, shells and massive structures, CEB International Course on Structural Concrete, LNEC, 1973.
8. Pina, C. and Baptista, A., "Estudo do comportamento até à rotura de barragens abóbada. Comparação de resultados obtidos por métodos analíticos e experimentais", Trans.Nat.Conf. Análise Experimental de Tensões, Lisbon, 1986.
9. Ferreira, M.E. - Comportamento até à rotura de modelos de barragens abóbada, LNEC, Lisbon, 1964.

2-D MOVING GRID FEM FOR DIFFUSION PROBLEMS WITH CHEMICAL REACTIONS

P.GERREKENS,M.HOGGE
Aerospace Laboratory of the University of Liège,Belgium

SUMMARY

Deforming finite element techniques are presented in the context of coupled heat conduction,gas diffusion, pyrolysis and surface ablation of charring materials. The physics of the phenomena and their mathematical model are briefly described. Effects of the deforming grid on conventional space Galerkin formulation is recalled and an original 2-D mesh deformation is proposed. An artificial diffusivity technique is retained to avoid spatial oscillations of density. The proposed numerical methods are finally applied to a plastic heat shield in a typical steady re-entry situation.

1.INTRODUCTION

This paper deals with the numerical prediction of the thermal behaviour of charring materials. Such materials may be used for ablative heat shields of re-entry bodies. When they are heated above a certain temperature, thermochemical decomposition occurs, which forms a porous char layer and gases that flow to the heated surface. Ablation of the char at the surface may occur subsequently if the external heat flux is sufficient. Surface ablation partially removes the charred material and implies a recession of the boundary [1,5,6].

Although fixed grid FEM were used to treat this class of problems [1], a deforming mesh is more convenient: it allows to track the moving boundary, to work with a constant number of nodes and, last but not least, to define the nodes positions according to the instantaneous needs of space discretization [6].

A deforming FEM was already proposed for ablation of non-charring materials [2]. In this paper, we show how it can be extended to ablation problems with pyrolysis.

2.MATHEMATICAL MODEL OF COUPLED HEAT CONDUCTION, PYROLYSIS, GAS DIFFUSION AND SURFACE ABLATION

The following PDE's govern the evolution of temperature T, gas pressure p and solid density ρ in a medium undergoing pyrolysis [1,6]:

heat equation:

$$H_p \, A \exp(-E/RT) \, (\rho - \rho^c)^n + \rho c \, \dot{T} = \partial_i (k_{ij} \, \partial_j T) + K_p \, \partial_i p \, \partial_i H^g \quad (1)$$

gas diffusion equation:

$$\partial_i (K_p \, \partial_i p) = -A \exp(-E/RT) \, (\rho - \rho^c)^n \quad (2)$$

chemical kinetic equation:

$$\dot{\rho} = -A \exp(-E/RT) \, (\rho - \rho^c)^n \quad (3)$$

Equations (1) and (3) are subject to the initial conditions

$$T = T_o(x_i) \text{ and } \rho = \rho^v \quad \text{at } t=0 \, .$$

$H_p(T,p,\rho)$ is the heat of pyrolysis, A,E,n are constant parameters, R is the universal gas constant, $k_{ij}(T,\rho)$ and $\rho c(T,\rho)$ are the solid conductivity coefficients and heat capacity, $K_p(T,p,\rho)$ is a mass flow conductivity for gas diffusion, ρ^v and ρ^c are the densities of virgin and charred materials.

boundary conditions and surface ablation

We shall consider here that the boundary is subject to aerodynamic and radiative heating as well as to chemical ablation.The surface recession rate $\dot{s}$ depends therefore on various parameters such as temperature, pressure...

$$\dot{s} = f(T,p) \quad (4)$$

and determines the normal velocity of any point $X_i(t)$ of the boundary

$$\dot{X}_i = -\dot{s}\, n_i \tag{5}$$

The heat balance at the boundary is thus written

$$k_{ij}\, \partial_j T\, n_i = q_e - \rho H_c\, f(T,p) \tag{6}$$

where q_e is the total external heat flux and H_c is the heat of chemical ablation.

The boundary conditions for pressure may be of two kinds: either pressure is prescribed by the external aerodynamic flow

$$p = p_e \qquad \text{on } S_p \tag{7}$$

or the gas mass flux vanishes, a condition which may be expressed in terms of the pressure gradient as follows

$$\partial_i p\, n_i = 0 \qquad \text{on } S_m \tag{8}$$

3.SPACE DISCRETIZATION

Space discretization is realized according to Galerkin's method. If a moving grid is used, the shape functions ψ_k will depend on time. The unknown fields $(T, p, \rho, \dot{s})$ are interpolated between their nodal values (T_k, p_k, ρ_k, v_k)

$$\begin{aligned}
T(x_i,t) &= \psi_k(x_i,t)\, T_k(t) && \text{(a)}\\
p(x_i,t) &= \psi_k(x_i,t)\, p_k(t) && \text{(b)}\\
\rho(x_i,t) &= \psi_k(x_i,t)\, \rho_k(t) && \text{(c)}\\
\dot{s}(x_i,t) &= \psi_k(x_i,t)\, v_k(t) && \text{(d)}
\end{aligned} \tag{9}$$

where the last equation only concerns the ablative boundary.

The time derivatives $\dot{T}$ and $\dot{\rho}$ may be written in the form

$$\begin{aligned}
\dot{T} &= \frac{DT}{Dt} - u_k\, \partial_k T && \text{(a)}\\
\dot{\rho} &= \frac{D\rho}{Dt} - u_k\, \partial_k \rho && \text{(b)}
\end{aligned} \tag{10}$$

where D/Dt denotes the time derivative at a moving point of the mesh and $u_k\,\partial_k$ is a "convective" operator accounting for mesh deformation with velocity u_k [2,3,6].

The shape functions ψ_k do not vary with time at a given moving point of the mesh, hence

$$\frac{D\psi_k}{Dt} = 0 \tag{11}$$

By introducing the discretizations (9) in the expressions (10) of $\dot{T}$ and $\dot{\rho}$, one obtains

$$\begin{aligned} \dot{T} &= \psi_k\ \dot{T}_k - u_k\ \partial_k\psi_l\ \ T_l & \text{(a)} \\ \dot{\rho} &= \psi_k\ \dot{\rho}_k - u_k\ \partial_k\psi_l\ \ \rho_l & \text{(b)} \end{aligned} \tag{12}$$

Application of Galerkin's method to the various field equations with the approximations (9) and (12) then leads to the following weak forms :

1)for the heat equation (1) with boundary condition (6)

$$\underset{\sim}{C}_{TT}\ \dot{\underset{\sim}{T}} + (\underset{\sim}{K}_{TT}-\underset{\sim}{K}^{*}_{TT})\ \underset{\sim}{T} + \underset{\sim}{K}_{Tp}\ \underset{\sim}{p} = \underset{\sim}{g}_T \tag{13}$$

where $C_{TT,kl}=\int_V \rho c\ \psi_k\psi_l\ dV$

$K_{TT,kl}=\int_V k_{ij}\ \partial_i\psi_k\ \ \partial_j\psi_l\ dV$

$K^{*}_{TT,kl}=\int_V \rho c\ u_m\,\partial_m\psi_l\ \psi_k\ dV$

$K_{Tp,kl}=\int_V K_p\ \partial_i\psi_l\ \partial_i H^g\ \psi_k\ dV$

$g_{T,k}=-\int_V H_p\ A\ \exp(-E/RT)\ (\rho-\rho^c)^n\ \psi_k\ dV$

$\qquad + \int_S q_e\ \ \psi_k\ dS$

2)for the diffusion equation (2) with boundary condition (7)

$$\underset{\sim}{K}_{pp}\ \underset{\sim}{p} = \underset{\sim}{g}_p \tag{14}$$

where $K_{pp,kl}=\int_V K_p\ \ \partial_i\psi_k\ \ \partial_i\psi_l\ dV$

$g_{p,k}=\int_V A\ \exp(-E/RT)\ (\rho-\rho^c)^n\ \psi_k\ dV$

3)for the chemical kinetic equation (3) multiplied by H_p

$$\underset{\sim}{C}_{\rho\rho}\,\dot{\underset{\sim}{\rho}} - \underset{\sim}{K}^{*}_{\rho\rho}\,\underset{\sim}{\rho} = \underset{\sim}{g}_{\rho} \tag{15}$$

where $C_{\rho\rho,kl}=\int_V H_p\ \psi_k\ \psi_l\ dV$

$K^{*}_{\rho\rho,kl}=\int_V H_p\ u_m\ \partial_m\psi_l\ \psi_k\ dV$

$g_{\rho,k}=-\int_V H_p\ A\ \exp(-E/RT)\ (\rho-\rho^c)^n\ \psi_k\ dV$

4)for the surface ablation equation (4) multiplied by ρ

$$\underset{\sim}{M}\ \underset{\sim}{v} = \underset{\sim}{g} \tag{16}$$

where $M_{kl}=\int_S \rho\,\psi_k\ \psi_l\ dS$; $g_k=\int_S \rho\ f(T,p)\ \psi_k\ dS$

4.TWO-DIMENSIONAL GRID DEFORMATION

Grid deformation is developed with a double objective:

1)track the moving boundary due to ablation

2)ensure a sufficient accuracy of space discretization at the interior of the domain.

We suppose that the boundary consists of (fig.1) :

1)an ablative surface S_A

2)a non-ablative surface S_{NA}

3)straight boundaries S_S for which the direction of ablation is prescribed (e.g.,a symmetry axis).

All nodes of the finite element mesh are put on a set of straight "master lines" linking the ablative boundary S_A to the non-ablative one, S_{NA}.

At a given time the ablative surface S_A has moved to S_A' (fig.2). The new positions of the nodes are defined as follows.Let NN be the total number of nodes on a line, and P_i (i=1,NN) their initial positions with $P_1\in S_A$ and $P_{NN}\in S_{NA}$. The node P_1 moves to P_1' according to equation (5), while the node P_{NN} remains fixed. This determines the new position of the master line $P_1'P_{NN}$. We then decide to put an important number N_{PD} of nodes into the thermal boundary layer, the estimation of which is proportional to the heat penetration depth $\delta(t)$, say r $\delta(t)$. The remaining nodes are distributed in the complementary part of the line.This is a manner to control the mesh density within the boundary layer at each time. As soon as a uniform node distribution is restored on a line, its deformation no longer depends on the boundary layer estimation, but only consists in a simple rotation and subsequent contraction.

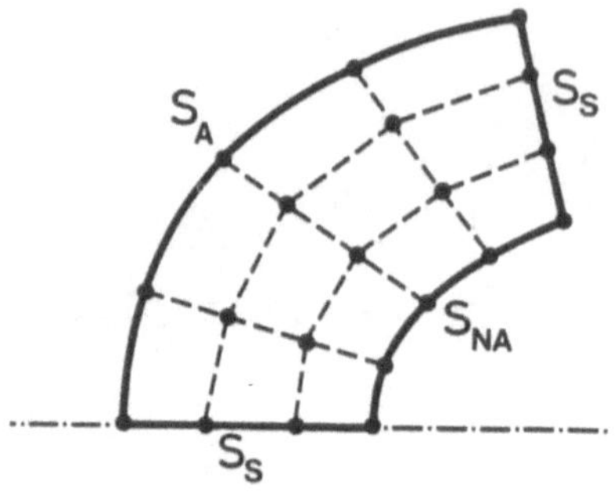

Fig. 1

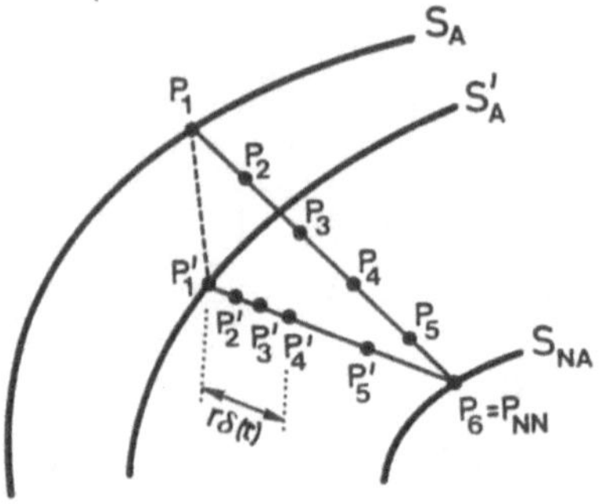

Fig. 2

Along each master line, the penetration depth is estimated by the differential equation

$$\frac{d}{dt}\left(\frac{\delta^2}{\eta_w}\right) = 1 - \frac{\dot{s}\delta}{\eta_w} \qquad (17)$$

where η_w is the thermal diffusivity at the moving boundary. This equation has been obtained by applying the heat balance integral method [4] to a 1-D ablation problem without pyrolysis. It allows for a sufficiently accurate prediction of the boundary layer if an adequate proportionality factor r is chosen. Numerical experiments have shown that this factor should vary from 3 to 2 for pyrolysis problems [6].

5. TIME INTEGRATION AND ITERATIVE PROCEDURE

Time integration of the spatially discretized field equations (13) and (15) is realized by using the one-step backward difference scheme. The same scheme is used to integrate equation (5) for the boundary nodes and equation (17) for each master line, so as to estimate the mesh deformation velocity u_k from the displacements of the interior nodes.

The resulting non-linear problem is solved at each time step by a mixed secant-tangent method. At each iteration, one first determines the values T_k, p_k, ρ_k, from which the ablation rates v_k and the mesh deformation are then derived.

6. ARTIFICIAL DIFFUSION FOR THE CHEMICAL KINETIC EQUATION

The density variation generally occurs in a very small part of the domain, the so-called pyrolysis zone, which is an order of magnitude smaller than the thermal boundary layer. This zone may thus be much thinner than the elements size (unless an extremely rich space discretization is used) and the numerical solution will exhibit spatial oscillations. To avoid this inaccuracy, we add an artificial diffusive term to the chemical kinetic equation (3), which becomes, after multiplication by H_p :

$$H_p \dot{\rho} = -A\, H_p \exp(-E/RT)\, (\rho - \rho^c)^n + \partial_i (H_p \nu\, \partial_i \rho) \qquad (18)$$

where ν denotes the artificial diffusivity. It is chosen in order to avoid spatial oscillations for the used space and time discretization. From simplified 1-D considerations [6], one may obtain the following expression for ν :

$$\nu = \frac{L^2}{6\,\Delta t} + \frac{uL}{2} \qquad (19)$$

where L is the element length, Δt the time step and u the deformation velocity. This expression may also be used for the multidimensional case if we define the length L and the velocity u in the direction of the density gradient

$$L = \frac{\rho_{max} - \rho_{min}}{\sqrt{\overline{\partial_i \rho}\;\overline{\partial_i \rho}}} \quad ; \quad u = \frac{|u_i\,\overline{\partial_i \rho}|}{\sqrt{\overline{\partial_i \rho}\;\overline{\partial_i \rho}}}$$

where ρ_{max} , ρ_{min} are the maximum and minimum values of ρ at the element's nodes, and $\overline{\partial_i \rho}$ is the mean density gradient on the element.

7. NUMERICAL ILLUSTRATIONS

We present some results of two-dimensional pyrolysis of a typical heat shield material, a low conductivity reinforced plastic. The external heat flux mainly consists in aerodynamic heating. The reduction of this flux, due to mass injection into the environment, is also taken into account. The complete data (material and heating) cannot be mentioned here.

Pyrolysis and ablation of a plastic heat shield in a steady re-entry configuration

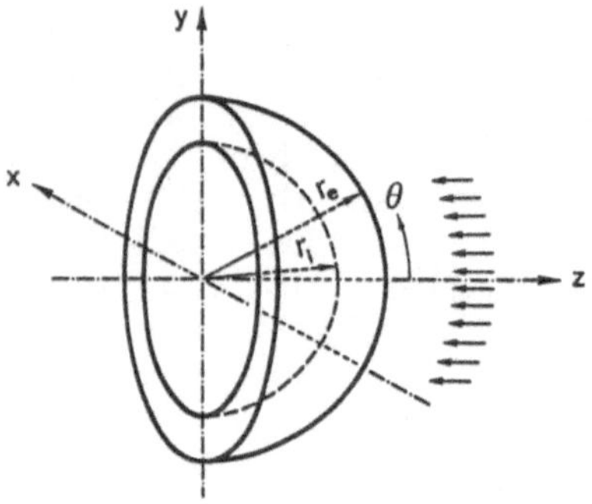

Fig. 3

The heat shield consists in a hole sphere of external and internal radii 8 cm and 7 cm (fig.3). The external spherical surface is subject to a convective heating resulting from a steady aerodynamic flow at zero incidence, so as to radiative cooling. The other faces are insulated and impermeable to the pyrolysis gases.

Time steps elected for the response integration are 20 times 0.05 s, 6 times 0.5 s and 10 times 1 s.

The mesh consists in 100 quadrilateral 4-node elements; its evolution is shown on figure 4. At t=5 s, it mainly deforms according to the expansion of the boundary layer, that consists of the finely discretized region. At t=11 s, we discern two deformation modes: near the symmetry axis, ablation causes a sensitive surface recession; far from the stagnation point, no surface ablation occurs, but the grid still deforms with the penetration depth. Finally, at t=15 s, a quasi uniform mesh is restored and surface recession becomes relatively important near the axis.

Finally, figure 5 presents the isotherms, isobars and isovalues of density at t=15 s. Heat conduction reveals to be essentially radial; this is no longer true for gas diffusion, that also occurs in the azimutal direction. Finally, the isovalues of density clearly show the location of the pyrolysis zone.

REFERENCES

1 CHIN, J.H., "Charring Ablation by Finite Elements", Numerical Methods in Thermal Problems II, pp. 218-229, Pineridge Press, Swansea, 1981.

2 HOGGE, M. and GERREKENS, P., "Two-Dimensional Finite Element Methods for Surface Ablation", AIAA Journal, Vol. 23, n°3, pp. 465-472, 1985.

3 LYNCH, D.R., "A Unified Approach to Simulation on Deforming Elements with Application to Phase Change Problems", Journal of Computational Physics, Vol. 47, pp. 387-411, 1982.

4 GOODMAN, T.R., "The Heat Balance Integral and its Application to Problem Involving a Change of Phase", Trans. ASME, Vol. 80, pp. 335-342, 1958.

5 HURWICZ, H. and ROGAN, J.E., "Ablation and High-Temperature Thermal Protection Systems", Handbook of Heat Transfer, Ed. Rohsenow W.M. & Hartnett J.P., Mc Graw Hill, chp. 16 & 19, 1973.

6 GERREKENS, P., " Modélisation par éléments finis des phénomènes d'ablation thermique avec pyrolyse", Ph.D. thesis, 1987, University of Liège.

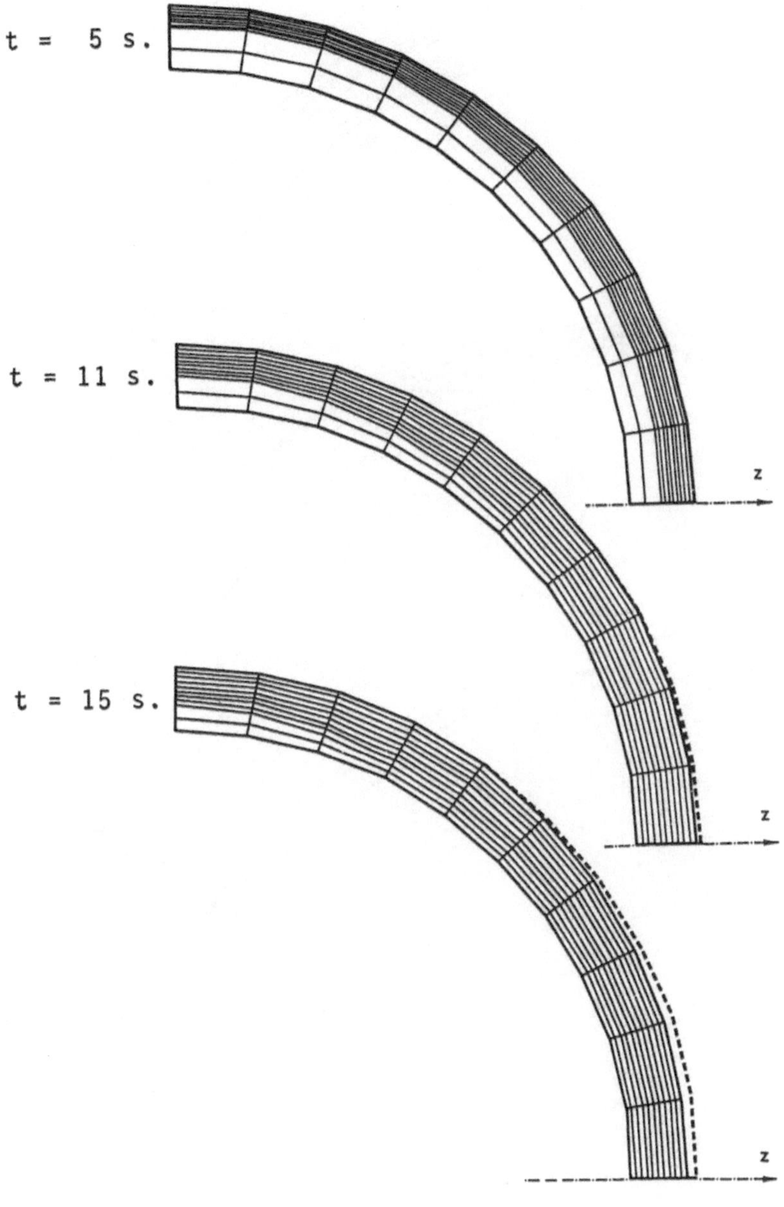

Fig. 4 : Mesh Evolution

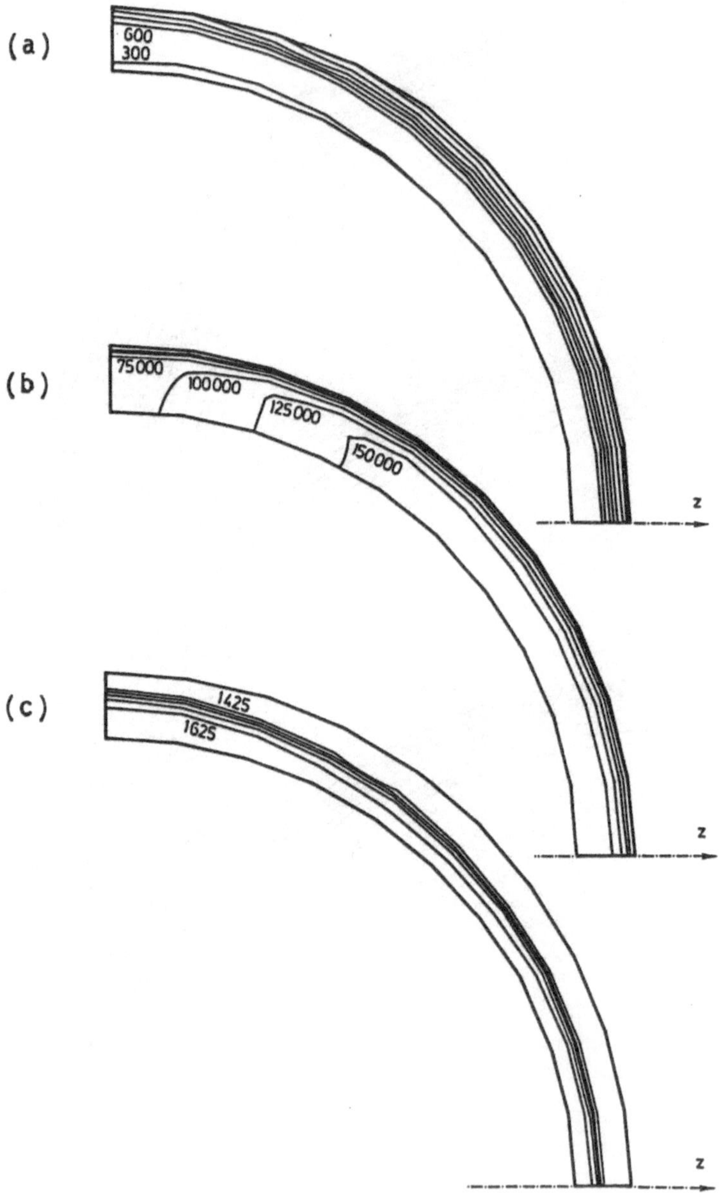

Fig. 5 : Isotherms (a), isobars (b) and isovalues of density (c) at t = 15 s.

SECTION S : DEVELOPMENTS IN ENGINEERING SOFTWARE

Are High Degree Elements Preferable? Some Aspects of the h and h-p Version of the Finite Element Method

Ivo Babuška
Institute for Physical Sciences
and Technology
University of Maryland

ABSTRACT

The paper discusses some aspects of the comparison betweem low and high degree elements especially in the relation to the p and h-p versions.

1. INTRODUCTION

The finite element method has become the main tool in computational mechanics. The success is manifested by the development of over five hundred user-oriented finite element program systems [19]. The literature on the subject is overwhelming and to date there are over two hundred monographs and conference proceedings [25]. Nevertheless, almost all results relate to the *h-version* of the finite element method, which fixes the degree of the elements p (usually p = 1,2), and achieves the desired accuracy with a sufficiently fine mesh. Recently, the *p and h-p version* was developed when the desired accuracy is obtained, either by increasing the degree p of elements or by simultaneous increase of the degrees and the mesh refinement. The commercial code PROBE (Noetic Tech. Inst., St. Louis, MO, USA) treating 2-dimensional problems and research program STRIPE (Aeronaut. Res. Inst. of Sweden), treating 3-dimensional problems based on p and h-p version are

Partially supported by the Office of Naval Research under Contract #N00014-85-K-0169 and by the National Science Foundation grant #DMS-85-16191.

available. For surveys we refer to [1],[23]. The references give the basic results for the h and h-p version available today.

2) THE MODEL PROBLEM

We will consider - for simplicity - the problem for one second-order elliptic partial differential equation. Nevertheless, we will illustrate the results also by numerical examples of elasticity.

Let $\Omega \subset \mathbb{R}^2$ and Γ be the boundary of Ω, $\Gamma = \bigcup_{i-1}^{M} \bar{\Gamma}_i$ where Γ_i are simple arcs. Assume that $\Gamma = \bar{\Gamma}^{(0)} \cup \bar{\Gamma}^{(1)}$, $H_0^1(\Omega) = \{u \in H^1(\Omega) \mid u = 0 \text{ on } \Gamma^{(0)}\}$. We shall consider the problem of minimizing quadratic functional

$$(2.1) \qquad \Phi(u) = B(u,u) - 2\int_{\Omega} fu\,dx - 2\int_{\Gamma^{(1)}} g^{(1)} u\,ds$$

over all functions $u \in H^1(\Omega)$ such that $u = g^{(0)}$ on $\Gamma^{(0)}$, where

$$(2.2) \qquad B(u,u) = \int_{\Omega}\left[\sum_{i,j=1}^{2} a_{i,j} \frac{\partial u}{\partial x_j}\frac{\partial u}{\partial x_i} + \sum_{i=1}^{2} b_i \left[\frac{\partial u}{\partial x_i} - u\right]^2 + cu^2\right]dx.$$

We will assume that $a_{i,j} = a_{j,i}$, b_i, $c>0$, and

$$(2.3) \qquad \sum_{i,j=1}^{2} a_{i,j}\, \xi_i \xi_j \geq \gamma(\xi_1^2 + \xi_2^2), \quad \gamma > 0.$$

The class of problems under consideration is given by a characterization of the set of input data, i.e. Ω, $a_{i,j}$, b_i, c, f, $g^{(0)}$, $g^{(1)}$. We shall assume that the input data depend continuously on a parameter t, $0 < t < 1$, so that for $t = 0$ the data are singular.

Let us show some examples:

a) Domain Ω: Consider the class of domains

$$\Omega(t) = \{x_1, x_2 \mid |x_1| < 1, |x_2| < t\}.$$

Obviously, for $t = 0$ the domain degenerates into straight line. Plates and shells could be interpreted in this framework.

b) Coefficients $a_{i,j}$, b_j, c: Let $a_{11} = a_{22} = (1+\frac{1}{t})$, $a_{12} = t(1-\frac{1}{t})$, $b_j = \frac{1}{t}$, $c = 0$. This problem is of the same type as the problem of nearly incompressible material, Reissner-Mindlin plate, etc.

Let us remark that one parameter family of the domains could be related to the one parameter family of coefficients by a transformation of coordinates.

c) Functions $f, g^{(0)}$, $g^{(1)}$: Consider the case when $\Gamma^{(1)} = \{x_1, x_2 | -1 < x_1 < 1,\ x_2 = 0\}$ and $g^{(1)}(t) = t/(x^2+t^2)$. Obviously, the "load" converges to the Dirac function (concentrated load) for which the solution (although it exists in $L_2(\Omega)$) does not belong to $H^{(1)}(\Omega)$.

3. REGULAR CLASS OF PROBLEMS WITH PIECEWISE ANALYTIC DATA

Let us assume that $\Gamma^{(i)}$ are analytic arcs,

$$|a_{i,j}|,\ |b_i|,\ |c| < C,\ \gamma > \alpha > 0,\ f$$

are analytic on $\bar{\Omega}$, $g^{(j)}$ are analytic on $\bar{\Gamma}^{(j)}$, $g^{(0)}$ is continuous on $\Gamma^{(0)}$. Then the solution of the problem is analytic on $\bar{\Omega} - \bigcup A_i$ (A_i, A_{i-1} are the end points of Γ_i) and in the neighborhood of every vertex A_i,

$$(3.1)\quad |D^{\alpha}u| \le Cd^k k!\ (r_i^{-1-k+\beta_i}),\ |\alpha| = k,\ k = 1,2,\ldots,$$

where (r_i, θ_i) are polar coordinates with the origin in A_i and $0 < \beta_i < 1$ depend on the type of the boundary conditions on both sides of A_i, the internal angle of the domain in A_i and on the coefficients $a_{i,j}$. For details see [3],[4],[5]. Then there exists a sequence of meshes and elements with increasing degrees such that for the error $e = u - u_{FE}$ (u_{FE} is the finite element solution) we have

$$\|e\|_E = [B(e,e)]^{1/2} \le Ce^{-\bar{\gamma}\sqrt[3]{N}},\ \bar{\gamma} > 0$$

where N is the number of degrees of freedom. For the proof and details see [4],[5],[21],[22].

Example: Consider the elasticity problem on a L-shaped domain with exact solution being the singular corner solution. Fig. 3.1 shows the domain and a (geometric) mesh with two layers. Fig. 3.2 shows the error of the p-version with various number of layers, performance of the h-version with the uniform mesh and p-version on the uniform mesh. The envelope gives the performance of the h-p version.

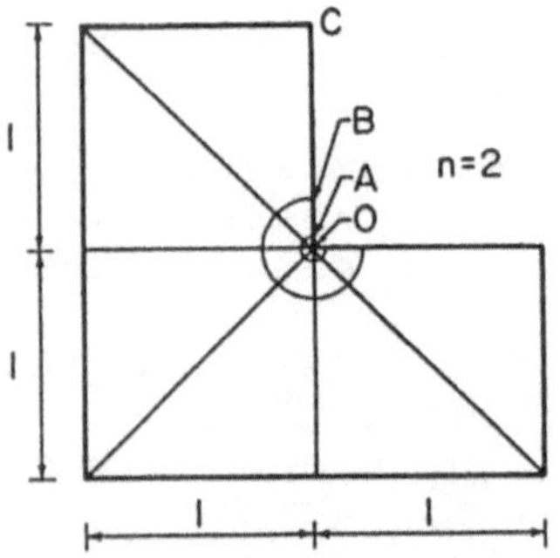

Fig. 3.1. L-shaped domain and the geometric mesh

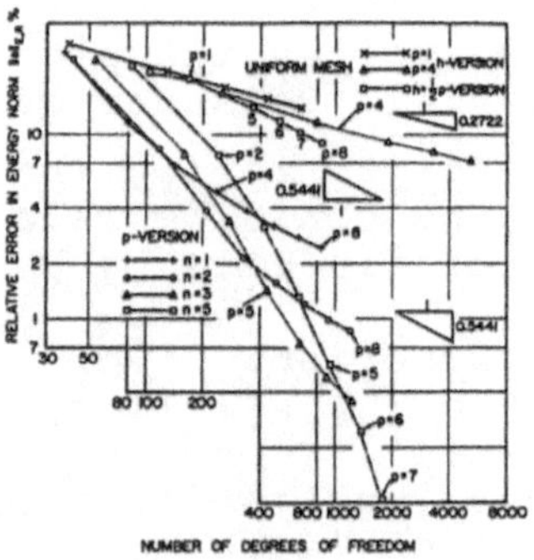

Fig. 3.2 Performance of various versions of the finite element method.

4. CLASS OF PROBLEMS WITH GENERAL g,f

Let us assume that $u \in H^k(\Omega)$ (and no information is given where the solution could be singular). Consider the uniform mesh. Then (see [9],[10],[11],[12].)

$$(4.1) \qquad \|e\|_E \le C(h^{\mu-1}/p^{k-1})\|u\|_{H^k(\Omega)} \quad \mu = \min(p+1,k)$$

Because N $h^{-2}p^2$, we have $\|e\|_E \le CN^{-(\mu-1)/2}/p^{k-\mu}$ and hence with respect to N, higher degree elements are preferable. We refer to [8] for more detailed analyses of computational complexity.

5. POLLUTION PROBLEM

Let us consider the limiting case t=0 when the function $g^{(1)}$ will be Dirac function (concentrated load). Then the solution exists but has infinite energy on Ω. Nevertheless, it has finite energy on $\Omega^* \subset \Omega$ when the area around the concentrated load is excluded. Hence we can be interested in the error measured in the energy norm on Ω^*. For the h-version it is known (applying the theory of interior estimates) that $\|e\|_{E(\Omega^*)} \longrightarrow$ on Ω^* as $h \to 0$. For the p-version this is not the case, and $\|e\|_{E(\Omega^*)} \longrightarrow 0$ as $p \to \infty$, hence pollution effect occurs. Nevertheless, refinement of the mesh i.e., the h-p version removes the pollution effect. For more see [6].

6. THE PROBLEM OF NEARLY DEGENERATED DIFFERENTIAL EQUATIONS

Let us consider the case that

$$\Omega = \{x_1, x_2 \mid |x_i| < 1,\ i = 1,2\},$$

$$(6.1)\qquad B_1(u,u) = \int_\Omega \left[\left(\frac{\partial u}{\partial x_1}\right)^2 + \frac{1}{t}\left(\frac{\partial u}{\partial x_2}\right)^2\right] dx$$

$$(6.2)\qquad B_2(u,u) = \int_\Omega \left(\frac{\partial u}{\partial x_i} + \frac{\partial u}{\partial x_2}\right)^2 + \frac{1}{t}\left(\frac{\partial u}{\partial x_i} - \frac{\partial u}{\partial x_2}\right)^2 dx$$

$g^{(0)} = 0$, $g^{(1)}$ be such that the exact solution is

$$u_1 = \sin x_1\ e^{-\sqrt{t}\cdot x_2},\quad u_2 = \sin(x_1 + x_2)\ e^{\sqrt{t}(x_1 - x_2)}.$$

Consider the uniform square mesh with $h = \frac{1}{2}, \frac{1}{4}$.
Tables 6.1 and 6.2 show the relative error $\|e\|_{ER}$ in % in dependence on h,p and t and the number of degrees of freedom N.

t	h	p = 1		p = 2		p =3		p =4	
		N	$\|e\|_{ER}$	N	$\|e\|_{ER}$	N	$\|e\|_{ER}$	N	$\|e\|_{ER}$
10^{-1}	1/2	8	7.79	20	.81	32	.03	48	.0
	1/4	24	3.94	64	.28	104	.01	160	.0
10^{-3}	1/2	8	7.42	20	.80	32	.02	48	.0
	1/4	24	3.76	64	.20	104	.0	160	.0
10^{-6}	1/2	8	7.42	20	.78	32	.01	48	.0
	1/4	24	3.70	64	.20	104	.0	160	.0

Table 6.1. Error for the Problem (6.1)

t	h	p = 1		p = 2		p =3		p =4	
		N	$\|e\|_{ER}$	N	$\|e\|_{ER}$	N	$\|e\|_{ER}$	N	$\|e\|_{ER}$
10^{-1}	1/2	8	16.07	20	.85	32	0.09	48	.02
	1/4	24	8.71	64	.22	104	0.03	160	.01
10^{-3}	1/2	8	21.36	20	3.35	32	0.31	48	.03
	1/4	24	18.96	64	1.67	104	0.07	160	.01
10^{-6}	1/2	8	21.50	20	3.72	32	0.42	48	.05
	1/4	24	19.30	64	3.66	104	0.36	160	.04

Table 6.2. Error for the Problem (6.2)

For the analyses of the problem of nearly incompressible material we refer to [13],and [32].

7. PROBLEM OF NEARLY DEGENERATED DOMAINS

Let us consider the elasticity problem on $\Omega(t) = \{x_1, x_2 \mid |x_1| < 1, |x_2| < t/2\}$ uniformly loaded on $\Gamma_1 = \{x_1, x_2 \mid |x_1| < 1, x_2 = t/2\}$ and clamped at $x_1 = \pm 1$. Then the singularity of the solution is due to the boundary layer at $x \quad \pm 1$, in the corners of the domain and in general a locking effect occurs. For $p > 4$ the locking effect essentially disappears and both the boundary layer and the corner singularity can be dealt with effectively by a proper mesh refinement.

CONCLUSION

The theory and numerical experience show that higher order elements seems to be highly preferable, especially when the error has to be in the range of 1-2%. Higher order elements are more robust than the lower ones and lead to smaller computer time to achieve this accuracy. Theory, numerical and industrial experience are given in the cited literature.

REFERENCES

[1] Babuška, I., The p and h-p versions of the finite element methods. The state of the Art. To appear in Proc. of the Workshop on Finite Element Method, Nasa Langley, ed. R. Voigt, Springer 1987 (Tech. Note BN-1056 IPST, Univ. of Maryland, 1986).

[2] Babuška, I., Dorr, M. R., Error Estimates for the Combined h and p-version of the Finite Element Method, Numer. Math 37 (1981), 257-277.

[3] Babuška, I., Guo, B., Regularity of the solution of elliptic problems with precise analytic data, Part I: Boundary value problem for linear elliptic equations of second order, Tech. Note BN-1047, IPSTUM 1986.

[4] Babuška, I. Guo, B., The h-p Version of the Finite Element Method for Domains with Curved Boundaries, To appear in SIAM J. Numer. Anal., 1987 (Tech. Note BN-1057, IPSTUM 1987).

[5] Babuška, I., Guo, B., The h-p version of the finite element method for problems with non-homogenous essential boundary conditions. To appear.

[6] Babuška, I., Oh, Hae-Soo, Pollution problems for the p-Version and the h-p Version of Finite Element Method, To appear in Communications in Applied Numerical Methods, 1987 (Tech. Note BN-1054, IPSTUM 1986).

[7] Babuška, I., Rank, E., An Expert-System-Like Feedback Approach in the h-p version of the Finite Element Method. To appear in Finite Elements in Analyses and Design 1987 (Tech. Note BN-1048, IPSTUM 1986).

[8] Babuška, I., Scapolla, T., Computational aspects of the h,p and h-p versions of the finite element method. To appear in Proc. of the 6th IMAC Symposium 1987, (Tech. Note BN-1061, IPSTUM 1987).

[9] Babuška, I., Suri, M., The optimal convergence date of the p-version of the finite element method. To appear in SIAM J. Numer. Anal., 1987 (Tech. Note BN-1045, IPSTUM 1985).

[10] Babuška, I., Suri, M., The h-p version of the Finite Element Method with quasi uniform meshes. To appear in Math. Modeling Numer. Anal. (RAIRO) 1987 (Tech. Note BN-1046, IPSTUM 1986).

[11] Babuška, I., Suri, M., The treatment of non-homogenous Direchlet boundary condition by the p-version of the finite element method, Tech. Note BN-1063, IPSTUM 1987.

[12] Babuška, I., Suri, M., The p-version of the finite method for constraint boundary conditions, Tech. Note BN-1064, IPSTUM 1987.

[13] Babuška, I., Szabo, B. A., Rates of Convergence of the Finite Element Method, Int'l. J. Numer. Math. Engr. 18 (1982), 323-341.

[14] Babuška, I., Szabo, B. A., Katz, I. N., The p-Version of the Finite Element Method, SIAM J. Numer. Anal. 18 (1981), 512-545.

[15] Barnhart, M. A., Eisenmann, J. R., Analysis of a Stiffened Plate Detail Using p-Version and h-Version Finite Element Techniques, Paper presented at the First World Congress on Computational Mechanics, Sept. 22-26, 1986, Univ. of Texas at Austin.

[16] Dorr, M. R., The approximation theory for the p-version of the finite element method, SIAM J. Numer. Anal. 21 (1984), 1180-1207.

[17] Dorr, M. R., The Approximation of Solutions of Elliptic Boundary Value Problems Via the p-Version of the Finite Element Method, SIAM J. Numer. Anal. 23 (1986), 58-77.

[18] Dunavant, D. A., Szabo, B. A., A-posteriori Error Indicators for the p-Version of the Finite Element Method, Int'l. J. Numer. Engr. 10 (1983), 1851-1870.

[19] Frederiksson, B. and Mackerle, I., Structural Mechanics: Finite Element Computer Programs, new up-to-date forth edition, Advanced Engineering Corp., Linkoping, Sweden, 1984.

[20] Gui, W., Babuška, I., The h,p and h-p versions of the finite element method in one dimension. Part I: The error analyses of the p-version, Part II: The error analyses of the h and the h-p versions, Part III: The adaptive h-p version, Numer. Math. 49 (1986), 577-612, 613-657, 659-683.

[21] Guo, B., Babuška, I., The h-p version of the finite element method. Part 1: The basic approximation results, Part 2: General results and approximations, Computational Mechanics 1 (1986), 21-44, 203-220.

[22] Guo, B., The h-p version of the finite element method for elliptic order of 2k-order. To appear.

[23] Guo, B., Babuška, I., The theory and practice of the h-p version of the finite element method. To appear in Proc. of the 6th IMAC Symposium 1987, (Tech. Note BN-1062, IPSTUM 1987).

[24] Katz, I. N., Wang, E. W., The p-Version of the Finite Element Method for Problems Requiring

C^1-continuity, SIAM J. Numer. Anal. 22 (1985), 1082-1106.

[25] Noor, A. K., Books and Monographs on Finite Element Technology, Finite Elements in Analysis and Design, 1 (1985), 101-111.

[26] Rank, E., Babuška, I., An Expert System for the Optimal Mesh Design in the h-p Version of the Finite Element Method. To appear in Int'l, J. Numer. Meths. Engrg. 1987 (Tech. Note BN-1053, IPSTUM 1986).

[27] Szabo, B. A., Some recent development in finite element analyses, Comp. Math. Appl. 5 (1979), 99-115.

[28] Szabo, B. A., PROBE: Theoretical Manual Noetic Tech. St. Louis, 1985.

[29] Szabo, B. A., Mesh design for the p-Version of the Finite Element method, Comp. Math. Appl. Math. Engrg 55 (1986), 181-197.

[30] Szabo, B. A., Estimation and Control of Errors Based on P-convergence, in Accuracy Estimates and Refinements in Finite Element Computations, I. Babuška, J. Gago and E. R. de A Oliveira, O. C. Zienkiewicz eds, J. Wiley, 1986, 61-70.

[31] Szabo, B. A., Implementation of a finite element software systems with h and p-extension capabilities, Finite Elements in Analysis and Design 2 (1986), 177-194.

[32] Vogelius, M., An Analyses of the p-Versions of the Finite Element Method for nearly Incompressible Materials - Uniformly Valid, Optimal Estimate, Numer. Math. 41 (1983), 39-53.

[33] Wang, D. W., Katz, I. N., Szabo, B. A., h and p-version Finite Element Analyses of a Rhombic Plate Int'l.,J. Numer. Meths. Engrg. 20 (1984), 1309-1405.

HANDICRAFT IN FINITE ELEMENTS

J. Blaauwendraad, A.W.M. Kok

Delft University of Technology
P.O. Box 5048, 2600 GA DELFT, The Netherlands

SUMMARY

The subject of this paper is a general purpose finite element program, yet designed for specialistic problems. In two respects the new program TILLY differs form existing ones. Firstly we use discrete elements based on the concept of generalized deformations. Secondly we apply the Kok-γ integration method to solve the equations. This method is believed to be an improvement with respect to the widely applied Newmark-β method. The ultimate goal of TILLY is not primarily friendliness in use, but rather freedom in applicability. The user is supposed to contribute substantially in the input; he is repaid for his efforts by big freedom in the modelling of problems. The Kok-γ method has specific advantages in nonlinear dynamics. High accuracy can be achieved without the need of extremely small time steps, because it is possible to iterate within each time step. A complete analogy with the Newton-Raphson technique for statics is shown. An application on a personal computer demonstrates the usefulness of the program.

1. SPECIFICATIONS OF THE PROGRAM

Do we really need another finite element program? No and yes. We do not need a new FEM package for the design and analysis of structures in the consulting engineers practise. A sufficient number of user-friendly programs serves this market. However, the answer is less self-evident for research-workers. It is admittedly clear that many research problems of an advanced character are solved by existing FEM programs which offer strong capabilities for nonlinear analysis. At the same time, paradoxical enough, the power of FEM programs also imposes rigid restrictions. The restrictions have different sources:

* The best programs offer the user a library of elements, a library of constitutive models and a library of solvers. The

aim is to permit all possible combinations and so to place at the disposal of users a wide range of options. In reality a restricted number of combinations is applied in industry, so the owner or distributor of the program focusses on these options. This means necessarily that other options either are not present, or have not been tested intensively.

* Advanced nonlinear programs apply the technique of assumed displacement fields and make use of a numerical integration procedure (Gauss points) to calculate the stiffness matrix. It can be a constraint to insert new elements if such a numerical integration scheme is an essential building-brick in the package.
* Big programs can be maintained only if a restricted number of authorized people has permission to change or extend the software. This can be a hindrance for universities if the goal is also to train students in development of finite element programs.
* Big programs require at least 32 bits microprocessors, which are not always available for all research-workers and research-students.
* Research in structural engineering and structural mechanics often is a process of interaction between experimental investigation and numerical modelling. On the one hand finite element programs do support the tests, but on the other hand tests provide new information about materials or structural components to feed into the program. Only in the case that the tests have been prepared in a way which takes account of the requirements of FEM programs, we can use the test results in the program. Very often this is not the case. Sometimes only in a late phase of an experiment the need for analysis becomes evident. Another time the results do not fit in the more or less rigid scheme of a finite element program. This is particularly the case if complete structural components are tested such as beams, columns and beam-column connections (possible result: moment-curvature diagram).
* General purpose finite element packages tend to focus on big problems and high numbers of degrees of freedom. In structural research we do not always need big models, but rather very simple ones, if possible even one-dimensional. However, the constitutive behaviour may be unexpected, for instance for a new type of material. Experience learns that it may be difficult to utilize a strong bruiser for light (yet special) tasks.

We conclude that room is left for small and special "handicraft shops" next to the big and general "supermarkets". The program TILLY aims to be such a handicraft shop.

From the beginning we adopt the viewpoint that the user is a research-oriented engineer with sufficient knowledge of structural mechanics. The all-important objective is freedom in applicability. User-friendliness is no (strong) requirement.

If necessary a research-worker is willing to prepare a laborious input if his efforts are repaid by big freedom in the modelling of the problem under consideration.

Big freedom in modelling is achieved if one gives up the privilege of the conventional FEM programs that they completely calculate the stiffness matrix, damping matrix and mass matrix for the user. In the conventional programs the user has to specify the geometry of the elements and to provide material data, and the program does all the rest. After all, this is a major source of rigidity. To support this statement we go through the procedure once more. In case of the element stiffness matrix the procedure is as follows. The vector $u(x,y,z)$ for the displacements field is connected to the vector d of the degrees of freedom of an element by:

$$u(x,y,z) = B(x,y,z)\ d \tag{1}$$

in which $B(x,y,z)$ is the matrix of shape functions. From this we derive the vector $\varepsilon(x,y,z)$ for the deformations (strains, curvatures, etc.):

$$\varepsilon(x,y,z) = D(x,y,z)\ d \tag{2}$$

The constitutive relation is

$$\sigma(x,y,z) = S_\varepsilon(x,y,z)\ \varepsilon(x,y,z) \tag{3}$$

in which $S_\varepsilon(x,y,z)$ is the rigidity matrix. The wanted element stiffness matrix is:

$$S = \iiint_V D^T(x,y,z)\ S_\varepsilon(x,y,z)\ D(x,y,z)\ dV \tag{8}$$

The work is kept away from the user, except that he has to specify the coordinates of the element nodes and some material constants. The integrations are done in a numerical way.

In TILLY we replace this procedure by one in which two parts are distincted in the vector d

$$d = d_r + e_g \tag{5}$$

Herein the vector d_r contains the full number of rigid body displacements of the element and e_g is the vector of generalized deformations. Now we use the relations:

$$e_g = D_g d \qquad\qquad \sigma_g = S_g e_g \tag{6}$$

and the element stiffness matrix is calculated from:

$$S = D_g^T S_g D_g \tag{7}$$

Note that the matrices D_g and S_g are not functions of x, y and z. In stead they contain real numbers.
The user of TILLY is asked to provide in the input the matrices D_g and S_g. Many, many engineering problems can be modelled by a number of single spring elements, damper elements and lumped masses. In that case S_g is just one single stiffness number (and also C_g and M_g) and D_g contains only one row. Any configu-

ration of such single elements is permitted, and the only restriction is a limited imagination of the user. For beam elements e_g consists of two generalized deformations, and so on.

The name TILLY is composed of the first character of a list of specifications. This list is:

* **T**ransient and static analysis: Both dynamic and rheologic transient processes have to be simulated by the model.
* **I**ncremental loading and initial strains: The load and initial strains are always applied step-wise in time; even for static calculations one has to introduce one or more time steps.
* **L**inear and nonlinear behaviour: In due time both material nonlinearity and geometrical nonlinearity will be covered. Material nonlinearity may be plasticity and fracturing, hardening and softening.
* **L**umped masses, springs and dampers: In due time to be extended to elements of two and more generalized deformations.
* **Y**oung and aging materials: Material properties can be constant in time, but also dependency of time must be included. Material stiffness and damping data may increase in time (for instance: young concrete) or may deteriorate (for instance: damaging by cyclic loading).

It has been chosen consciously to focus at the start on elements of one generalized deformation and material nonlinearity only. It is remarkably how many structural engineering problems can be modelled already by this research tool. A major advantage is also that students are forced to take an active part in making a numerical model. And the program can run on their personal computer (Olivetti M24)!

Because of the kind of elements that can be used in TILLY, the name of "discrete elements" is chosen. The most needed discrete element is an elasto-plastic spring element with proper loading and unloading. Also an elastic-fracture element is most useful.

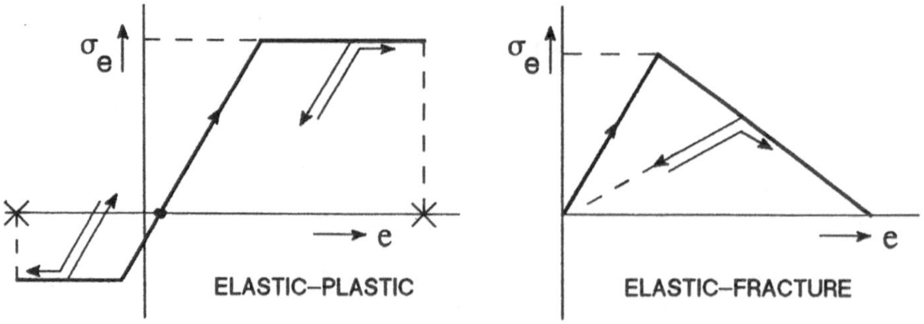

Fig. 1. Two spring types. Left: Elastic-plastic spring with different yield levels and different ultimate strains (present in TILLY). Right: Elastic-fracture spring with softening (being implemented).

2. THE KOK-γ METHOD

In general each finite element program has to solve a system of equations of the type:

$$S\,d + D\,\dot{d} + M\,\ddot{d} = f \qquad (8)$$

Herein f is the time-dependent load vector. In the widely used Newmark-β integration method time steps Δt are applied and the displacements d_{i+1} at time t_{i+1} are calculated from the earlier found displacements d_{i-1} and d_i.
One of the authors, Kok, published in 1981 a new presentation of the method, in which he developed an analogy between dynamic and static analysis [1]. He introduced momentums (pulses) for dynamic analysis and showed that they play a similar role versus displacements as forces do versus displacements in statics. Above that, he held out a prospective of an advantage above the conventional time stepping procedure in case of nonlinear dynamics. It has been decided to implement this method in TILLY.

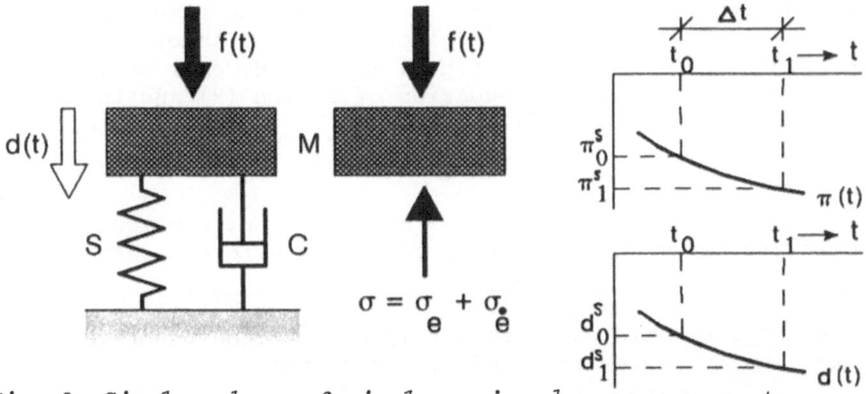

Fig. 2. Simple scheme of single spring-damper-mass system.

In the restricted space of this paper the method is clarified using a single spring-damper-mass system (see fig. 2). A more detailed explanation is found in [1].

The (generalized) deformation of the spring is e and the deformation rate of the damper is $\dot{e}$. The (generalized) force in the spring is σ_e and in the damper $\sigma_{\dot{e}}$, making together σ. We introduce the symbol π for the momentum of the mass, which is Mv. Herein v is the velocity.

According to Newton's first law the equation of motion is:

$$\dot{\pi} = f - \sigma_e - \sigma_{\dot{e}} \qquad (9)$$

The constitutive relations are

$$\pi = M\,v, \qquad \sigma_e = S\,e, \qquad \sigma_{\dot{e}} = C\,\dot{e} \qquad (10)$$

Finally we have three kinematic conditions:

$$v = \dot{d}, \qquad e = d, \qquad \dot{e} = \dot{d} \qquad (11)$$

The method takes account of time steps Δt. In a section between two time steps (intervals) we denote the momentum by π^s and the displacement by d^s. Let us consider the interval between time t_0 and time t_1. For this time step we have two dynamic boundary conditions with respect to time:

$$\begin{aligned} \pi_0^s - \pi &= 0 \qquad \text{at } t = t_0 \\ \pi_1^s - \pi &= 0 \qquad \text{at } t = t_1 \end{aligned} \tag{12}$$

and two kinematic boundary conditions with respect to time:

$$\begin{aligned} d_0^s - d &= 0 \qquad \text{at } t = t_0 \\ d_1^s - d &= 0 \qquad \text{at } t = t_1 \end{aligned} \tag{13}$$

At $t = t_0$ the initial momentum π_0^s and the initial displacement d_0^s are known. At $t = t_1$ the end momentum π_1^s and the end displacement d_1^s are unknown. These two quantities will be calculated in the time step under consideration. After that, they will act as the initial quantities in the next time step, and so on. The constitutive relations (10), the kinematic conditions (11) and the kinematic boundary displacements (13) should be satisfied exactly in the pure stiffness method. However it is permitted to approximate the equation of motion (9) and the dynamic boundary conditions (12) by imposing a Galerkin condition:

$$G = \int_{\Delta t} \delta d\,(f - \sigma_e - \sigma_{\dot{e}} - \dot{\pi})dt + \delta d\,(\pi_0^s - \pi)_{t_0} - \delta d\,(\pi_1^s - \pi)_{t_1} = 0 \tag{14}$$

Using partial integration of the momentum contribution, with respect to time, this condition can be transformed into:

$$G = \int_{\Delta t} \{-\delta e\,(\sigma_e + \sigma_{\dot{e}}) + \delta v\,\pi + \delta d\,f\}dt + \delta d_0 \pi_0^s - \delta d_1 \pi_1^s = 0 \tag{15}$$

Using the kinematic conditions (11) and the constitutive relations (10) the Galerking condition becomes:

$$G = \int_{\Delta t} (-\delta d\,K\,d - \delta d\,C\,\dot{d} + \delta\dot{d}\,M\,\dot{d} + \delta d\,f)dt + \delta d_0 \pi_0^s - \delta d_1 \pi_1^s = 0 \tag{16}$$

Next we choose a linear distribution in time for the displacement d. For the kinetic energy and the damper energy this must be a continuous field at $t = t_0$ and $t = t_1$. For the spring energy we are free to adopt a non-continuous distribution. Both distributions are covered by one expression:

$$d(t) = (\tfrac{1}{2} - \gamma\tfrac{t}{\Delta t})d_0 + (\tfrac{1}{2} + \gamma\tfrac{t}{\Delta t})d_1 \tag{17}$$

The origin of the t-axis is in the mid of the interval Δt. For $\gamma = 1$ the distribution is continuous in time, for other values of γ it is not.

Executing the integral over time the Galerking condition becomes:

$$G = -\{\delta d_0 \quad \delta d_1\} \begin{bmatrix} H_{00} & H_{01} \\ H_{10} & H_{11} \end{bmatrix} \begin{Bmatrix} d_0 \\ d_1 \end{Bmatrix} + \{\delta d_0 \quad \delta d_1\} \begin{Bmatrix} \mu_0 \\ \mu_1 \end{Bmatrix} = 0 \tag{18}$$

in which:

$$
\begin{aligned}
H_{00} &= -\frac{1}{\Delta t} M - \frac{1}{2}C + \left(\frac{1}{4} + \frac{\gamma^2}{12}\right) \Delta t\, S \\
H_{01} &= \frac{1}{\Delta t} M + \frac{1}{2}C + \left(\frac{1}{4} - \frac{\gamma^2}{12}\right) \Delta t\, S \\
H_{10} &= \frac{1}{\Delta t} M - \frac{1}{2}C + \left(\frac{1}{4} - \frac{\gamma^2}{12}\right) \Delta t\, S \\
H_{11} &= -\frac{1}{\Delta t} M + \frac{1}{2}C + \left(\frac{1}{4} + \frac{\gamma^2}{12}\right) \Delta t\, S
\end{aligned} \tag{19}
$$

and

$$
\begin{aligned}
\mu_0 &= \pi_0^s + 1/3\Delta t\, f_0 + 1/6\,\Delta t\, f_1 \\
\mu_1 &= -\pi_1^s + 1/6\,\Delta t\, f_0 + 1/3\,\Delta t\, f_1
\end{aligned} \tag{20}
$$

Herein it is assumed that f varies linearly from f_0 at time t_0 to f_1 at time t_1.
Requiring $G=0$ for any value of δd_0 and δd_1 unequal zero results in:

$$
\begin{bmatrix} H_{00} & H_{01} \\ H_{10} & H_{11} \end{bmatrix} \begin{Bmatrix} d_0 \\ d_1 \end{Bmatrix} = \begin{Bmatrix} \mu_0 \\ \mu_1 \end{Bmatrix} \tag{21}
$$

In the first line of these equations only d_1 is unknown, so we can calculate:

$$
d_1 = H_{01}^{-1} (\mu_0 - H_{00} d_0) \tag{22}
$$

Now in the second line only μ_1 is unknown, so we find:

$$
\mu_1 = H_{10} d_0 + H_{11} d_1 \tag{23}
$$

From μ_1 we calculate π_1^s. Now we shift d_1 to d_0 and π_1^s to π_0^s and can start the calculation for the next time step.

If the calculation has been accurate, equation (15) must be satisfied. At the end of each time step we can check the accuracy. In general the equation (15) will hold for momentums $\tilde{\pi}_0^s$ and $\tilde{\pi}_1^s$ which are slightly different from π_0^s and π_1^s. Now we can improve d_1 by Δd_1 which is calculated from $\Delta\mu = \pi_0^s - \tilde{\pi}_0^s$:

$$
\Delta d_1 = H_{01}^{-1}\, \Delta\mu \tag{24}
$$

So the obvious advantage is, that we have got at our disposal a criterion to check the results for the loss of accuracy. This criterion is to be compared with the accuracy check for reactions and forces in statics.

Precisely the same scheme holds for nonlinear dynamics. The stresses σ_e and $\dot{\sigma}_e$ in (15) are analysed, taking account of the non-linearities. Now the resulting $\Delta\mu_0$ is not only due to loss in accuracy but also to the nonlinear behaviour. We have to iterate until $\Delta\mu_0$ is sufficiently small. Fig. 3 illustrates the procedure and clearly shows the complete analogy with the Newton-Raphson technique for nonlinear statics. As it has been drawn in this figure, a new H_{01} has to be derived and inverted (or decomposed) for each iteration. We could also use the same

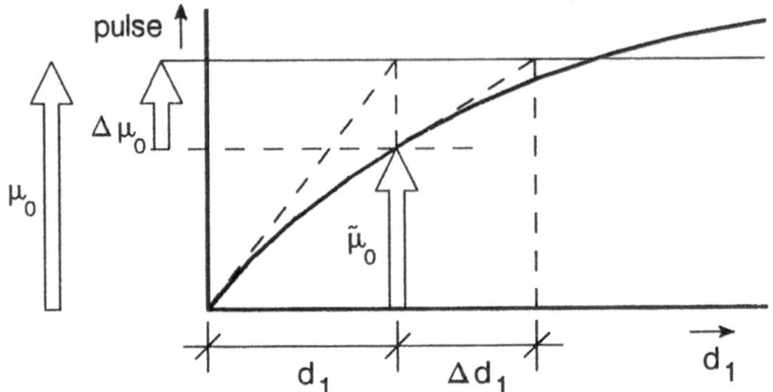

Fig. 3. Improvement of solution by iterations within a time step

H_{01} for all iterations, at the cost of more iterations.
It is the aim of the method, that no smaller time steps are required for nonlinear dynamics than for linear dynamics. At the worst more iterations per time step will be necessary.

In this paper the method has been explained for a simple system of one degree of freedom, because all attention has been drawn to the integration procedure in the time domain. The reader easily understands that a similar approach holds for systems of more degrees of freedom. In that ease π, d and f are vectors and S, M and C are the well-known matrices of the conventional stiffness method. For linear dynamics the method has already been used a number of years in the Delft-version of ICES-STRUDL [3].

3. APPLICATION

For the purpose of demonstration an analysis has been made for a simply supported beam of reinforced concrete which was lifted up over some height at the left support and was then dropped freely. Experimental results are available from Switserland [4]. Time t = 0 is taken when the beam just touches the support. So at t = 0 an initial velocity field is known and of course dead weight has to be considered. We use a model of 21 rigid bars which are connected by 20 rotational springs (fig. 4). The spring characteristics are nonlinear, to achieve the best possible simulation of the moment-curvature diagram (M–κ). The springs must have unloading branches, for many of them initially yield but transfer moderate moments later on.

Two models have been applied and in both cases two parallel springs are used which together produce the required M–κ diagram. In the first model both parallel springs are elasto-plastic springs. This model is correct for increasing loading but will behave too stiff for unloading. In the second model one spring is elasto-plastic and one is an elastic-fracture spring (fig. 5). Account is taken of proper material strength data in regard of the occurring strain rate. Fig. 6 shows the

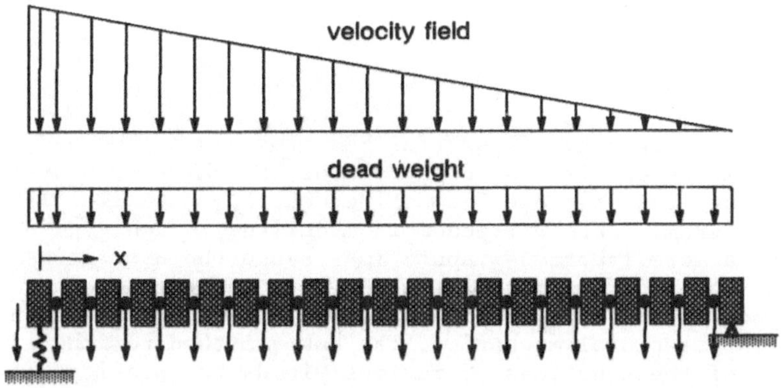

Fig. 4. Discretization of beam, loading, initial velocity field.

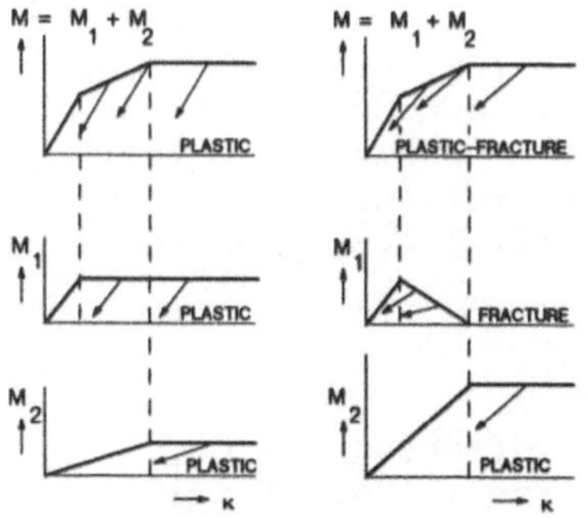

*Fig. 5. **Decomposition** of constitutive model in two components. Left: two elastic-plastic springs representing the plastic model. Right: one elastic-fracture spring and one elastic-plastic spring **representing the** plastic-fracture model.*

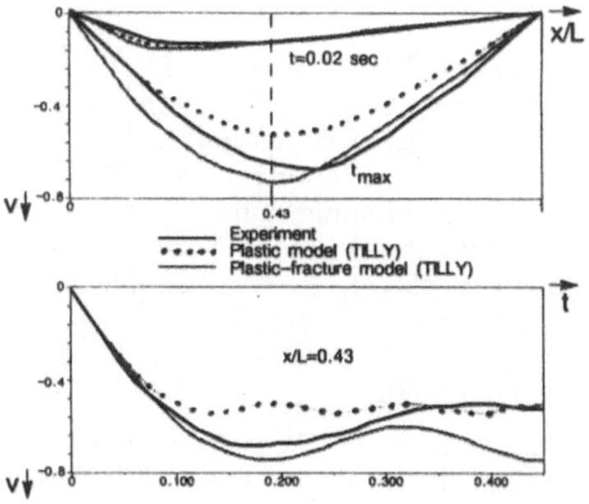

Fig. 6. Comparison of experimental data and numerical results.

results. The double-plastic model is too stiff, but the plastic -fracture model appears to be more usefull. The reader is referred to [5] for more details.

4. CONCLUSION

Much of the rigidity of powerful FEM programs can be avoided if we utilize generalized deformations or (which is more or less equivalent) the Discrete Element Method. So, actually, TILLY is a DEM program. The new freedom in modelling is achieved at the cost of a more laborious input. In a sense the user gives in some of the benefactions of the FEM, but he enriches in another way. And the new type of input challenges him to call upon his skill in structural mechanics. The Kok-γ method for the integration of the equations of motions yields two profits. It offers a possibility to judge and (if necessary) to improve the numerical accuracy and it presents an accurate scheme for non-linear dynamics. One does not need extremely small time increments, because it is possible to iterate within a time step.

5. ACKNOWLEDGEMENT

The authors are greatly indebted to Mr. P. Willems who translated the ideas on the discrete element method and the integration technique into operable software in the scope of his masters thesis. They also owe Mr. L.J. Sluys and Mr. A.G.J. Heinsbroek many thanks, who kindly ran TILLY to analyse the falling beam problem. Finally they express their gratitude towards the secretary Ms. R.W. Griffioen for her dutiful typing of the manuscript.

6. REFERENCES

1 Kok, A.W.M.: Pulses in Finite Elements. Proceedings First Internat.Conf. on Computing in Civil Engr. 1981, N.Y., 1981, p. 286-301.

2 Willems, P.: TILLY, the theoretical background of a Discrete Element Program (in Dutch). Delft University of Technology, Dept. of Civ. Engr., 1985.
Willems, P.: User guide of TILLY
Willems, P.: Programmers guide of TILLY.

3 Kok, A.W.M.: Theoretical Manual for STRUDL, Eurocopi dept. A, JRC Ispra, 1981.

4 Amman , W.: Stahlbeton- und Spannbeton tragwerke unter stossartiger Belastung, ETH Zürich, dissertation, 1983.

5 Blaauwendraad, J.; Sluys, L.J.; Heinsbroek, A.G.J.: DEM and Beam Dynamics, an Application of TILLY. To be published in proceedings of IABSE Colloquium on Computational Mechanics of Concrete Structures, Delft 1987.

ADAPTIVE TECHNIQUES IN FINITE ELEMENT ANALYSIS

J. Z. Zhu and O. C. Zienkiewicz
Department of Civil Engineering
University College of Swansea, U.K.

A. W. Craig
Department of Mathematical Science
University of Durham, U.K.

Summary

An effective h-version finite element adaptive strategy combined with mesh regeneration is presented, which is based on the error estimator developed in [1]. The rate of convergence of the adaptive procedure has been tested for two examples. A very strong convergence has been observed. Unlike some existing h-version adaptive procedures, a nearly optimal mesh and predicted accuracy can be obtained in one or two steps of mesh regeneration.

1. Introduction

In the last few years considerable effort has been devoted to the development of the theory and the implementation of the adaptive procedures and a-posteriori error estimation for the finite element method. Various approaches to achieve a-posteriori error estimates with consequent adaptive process have been developed [e.g. 2-8] to use effectively finite element codes for practical engineering analysis. However, the cost of computation associated with error estimation and the difficulty of computer implementation of the adaptive algorithm are still the main problems.

A recently developed h-version error estimator and adaptive strategy with mesh regeneration techniques overcome these difficulties. These allow both the local and global energy norm errors to be evaluated in a simple way for a variety of currently used elements [1]. A fully or partially automated process can be made available within an existing code structure. The reliability of the error estimator has been demonstrated both numerically [1] and mathematically [9].

As many authors have pointed out [2,6,8], when adaptive processes are used in the mesh refinement, the maximal convergence rate for the type of element used can be achieved, providing the error is uniformly distributed and this rate only depends on the element used. At that stage the solution provides a nearly optimal error. Much mathematical and numerical research work has been reported for the convergence rate of h-version quadrilateral and triangular element with successive adaptive refinement procedures [e.g. 8]. The convergence of the mesh regeneration adaptive procedure with linear and quadratic triangular elements has only been reported recently [10].

In this paper we shall present an adaptive refinement strategy which is based on the error estimates developed in [1] for h-refinement triangular elements. By coupling the adaptive refinement strategy with an automatic mesh generator, the computational efficiency of adaptive process has been much improved by reducing the number of adaptive iterations. At the same time the maximum rate of convergence is also achieved, and therefore a nearly optimal control of error.

2. Error measures

In order to keep the discussion in focus, we shall be concerned only with problems of linear elasticity which may be posed by the differential equilibrium equation

$$\underline{Lu} - \underline{q} = \underline{S}^T \underline{DSu} - \underline{q} = 0 \qquad (1)$$

in a domain Ω, with boundary conditions

$$\underline{u} = \bar{\underline{u}} \qquad \text{on } \partial\Omega_1 \qquad (2a)$$

$$\underline{S}_n^T \underline{DSu} = \bar{\underline{t}} \qquad \text{on } \partial\Omega_2 \qquad (2b)$$

where $\underline{S}$ is the strain operator defined by

$$\underline{\varepsilon} = \underline{Su} \qquad (3)$$

$\underline{D}$ is the elasticity matrix giving stresses as

$$\underline{\sigma} = \underline{D}\,\underline{\varepsilon} \qquad (4)$$

The finite element solution $\hat{\underline{u}}$ can be obtained by a standard Galerkin process (or equivalently by minimising the potential energy expression) with respect to a set of functions that can be written as

$$\hat{\underline{u}} = \sum_{i=1}^{n} \underline{u}_i \, \underline{N}_i \qquad (5)$$

where $\underline{N}_i$ are the trial functions, constructed from the element shape functions in such a way that the appropriate continuity and boundary conditions are satisfied.

The error is the difference between the exact and finite element approximate solution of (1), thus

$$\underline{e} = \underline{u} - \underline{\hat{u}} \tag{6}$$

The error in energy norm can be written now as

$$||\underline{e}|| = [\int_\Omega (\underline{S}\underline{e})^T \underline{D}(\underline{S}\underline{e})d\Omega]^{\frac{1}{2}} \tag{7a}$$

$$= [\int_\Omega (\underline{\sigma} - \underline{\hat{\sigma}})^T \underline{D}^{-1} (\underline{\sigma} - \underline{\hat{\sigma}})d\Omega]^{\frac{1}{2}} \tag{7b}$$

where

$$\underline{\hat{\sigma}} = \underline{DS\hat{u}} \tag{8}$$

For the usual C_o continuous approximation, the stress field is discontinuous and inaccurate. A more accurate and continuous stress field can be achieved by projection writing

$$\underline{\sigma}^* = \underline{N}\ \underline{\bar{\sigma}}^* \tag{9}$$

and requiring that

$$\int_\Omega \underline{N}^T(\underline{\sigma}^* - \underline{\hat{\sigma}})d\Omega = 0 \tag{10}$$

Then an error estimate can be achieved simply by equation (7b)

$$||\underline{\bar{e}}|| = [\int_\Omega (\underline{\sigma}^* - \underline{\hat{\sigma}})^T \underline{D}^{-1} (\underline{\sigma}^* - \underline{\hat{\sigma}})\ d\Omega]^{\frac{1}{2}} \tag{11}$$

The effectivity of such an error estimator is high as shown in [1] both globally and at the element level.

3. Adaptive mesh refinement

Unlike in the successive mesh enrichment, authors have presented a technique attempting to achieve the permissible error value in each subdomain by predicting directly a new mesh size [1,10]. If we assume that the error is equally distributed on each element with expected value $||\underline{e}||_i$ and the error in a particular element of size h of an existing subdivision is estimated to be ρ_i, we can predict the local element size required by assuming a convergence rate of order p, i.e.

$$h = h_i/\xi^{1/p} \tag{12}$$

where

$$\xi = \rho_i / ||\underline{e}||_j \tag{13}$$

Obviously, if the rate of convergence is exactly as assumed and the error is exactly measured, the new element sizes predicted should give a mesh the solution yielding the desired accuracy in a most economical manner. Generally we shall assume that p is equal to the polynomial order of the trial function used, although this simple convergence rule is, of course, not valid near singularities.

For the elements near to singular points, the predicted element size can be calculated by

$$h = h_i / \xi^{1/\lambda} \tag{14}$$

where

$$0 < \lambda < 1 \tag{15}$$

is the a-priori convergence rate of finite element solution for the problems with singularities. In many cases, λ is known. For most plane elastic problems with re-entrant corners we have [11]

$$0.5 \leq \lambda < 1.0 \tag{16}$$

The range of λ is fairly small, and in practice one can always choose $\lambda = 0.5$ for convenience, which is corresponding to the line crack in an elastic body. The possible over-refinement caused by smaller λ is local and only around the singularities.

We note that the statements made on convergence rates are global, but numerical experience has shown that they give reasonable approximation to the local convergence of the finite element solution.

Indeed, the success of implementation of the procedures depends on availability of automatic mesh generators specified. One such generator has been recently developed on triangle basis and proceeds by direct construction of triangular mesh size locally specified starting from a suitably modelled boundary and a background grid. For a full description of the mesh generator we refer to [12] and for the detail of combination of the adaptive strategy and mesh generator we refer to [10].

4. h-convergence

It is well known that finite element solutions minimize the error in energy norm. For the h-refinment, when a sequence of mesh is obtained through uniform refinement, the error is bounded by

$$||\underline{e}|| \leq CN^{-\frac{1}{2}\min(p,\lambda)} \tag{17}$$

where $||e||$ is the error in energy norm as we defined before, C is positive constant, N is the number of degrees of freedom, p is the polynomial degree of shape functions being used, and λ represents the smoothness of the exact solution. The exponent of N is called the asymptotic rate of convergence and obviously it depends on the exact solution of the problem and the finite element meshes.

If the sequence of meshes is obtained in such a way that the error is approximately equally distributed on each element, the meshes are called nearly optimal and the estimate of error is

$$||e|| \leq CN^{-p/2} \tag{18}$$

therefore the rate of convergence is independent of the problem being solved, and an optimal convergence rate is then achieved.

5. Examples

The quadratic 6 node triangular element has been used to check the convergence rate of the adaptive strategy proposed. $\lambda = 0.5$ is used in the examples at the elements surrounding the singularities. For the testing of 3-node element, we refer to [10]. In the latter we seek directly to obtain a 5% accuracy. The relative error η is shown at every refinement for the examples.

As first example, a short cantilever beam is considered, the geometric definition of the problem and the load condition are shown in Figure 1. Plane strain conditions and Poisson's ratio of 0.3 are assumed.

The problem has also been solved by Szabo to test the convergence rate of the h and p versions of finite element method [13].

The rate of convergence is shown in Figure 2 for both uniform mesh refinement and adaptive refinement. It is seen that the theoretically predicted convergent rate is realized in the case of both 6-node and 9-node elements when the mesh is uniform. The convergence of the adaptive refinement is very strong and the desired 5 percent accuracy is obtained in two steps. Even the average rate of convergence has reached the maximum value according to (18).

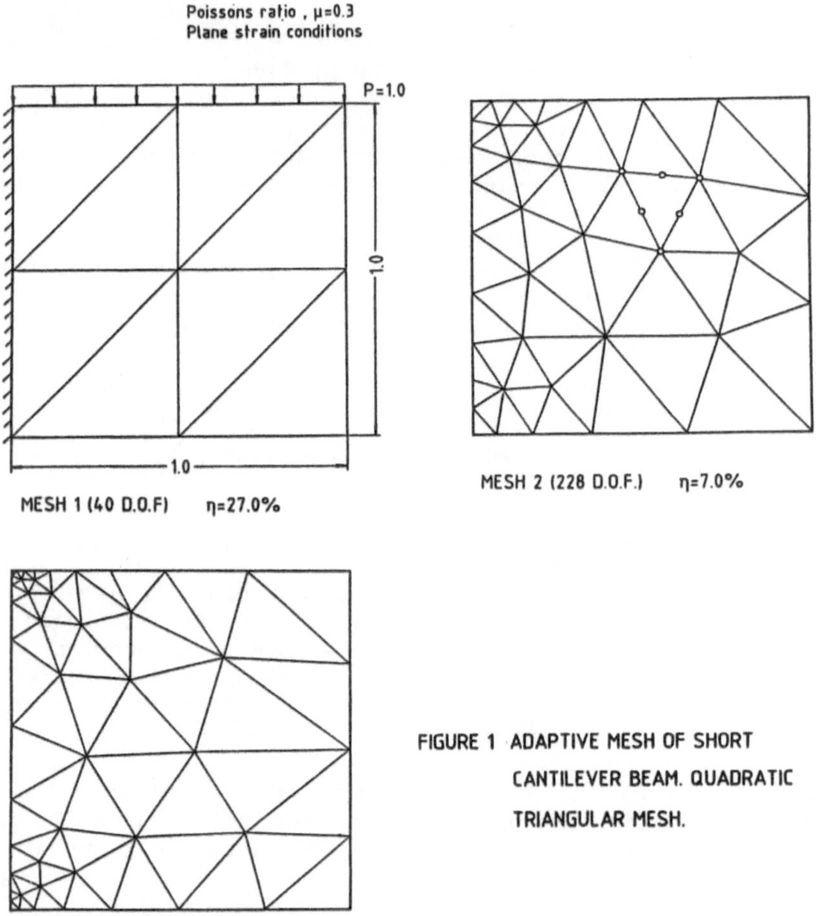

FIGURE 1 ADAPTIVE MESH OF SHORT CANTILEVER BEAM. QUADRATIC TRIANGULAR MESH.

We note that at this stage there is almost no element needed to be refined by following the strategy of (12) and (14). In fact, most elements need a slight derefinement.

The second example studied is an L-shaped domain conforming to plane stress conditions. Figure 3 shows the geometrical dimensions and the load distribution. The problem has also been tested by 6-node and 9-node elements using uniform refinement.

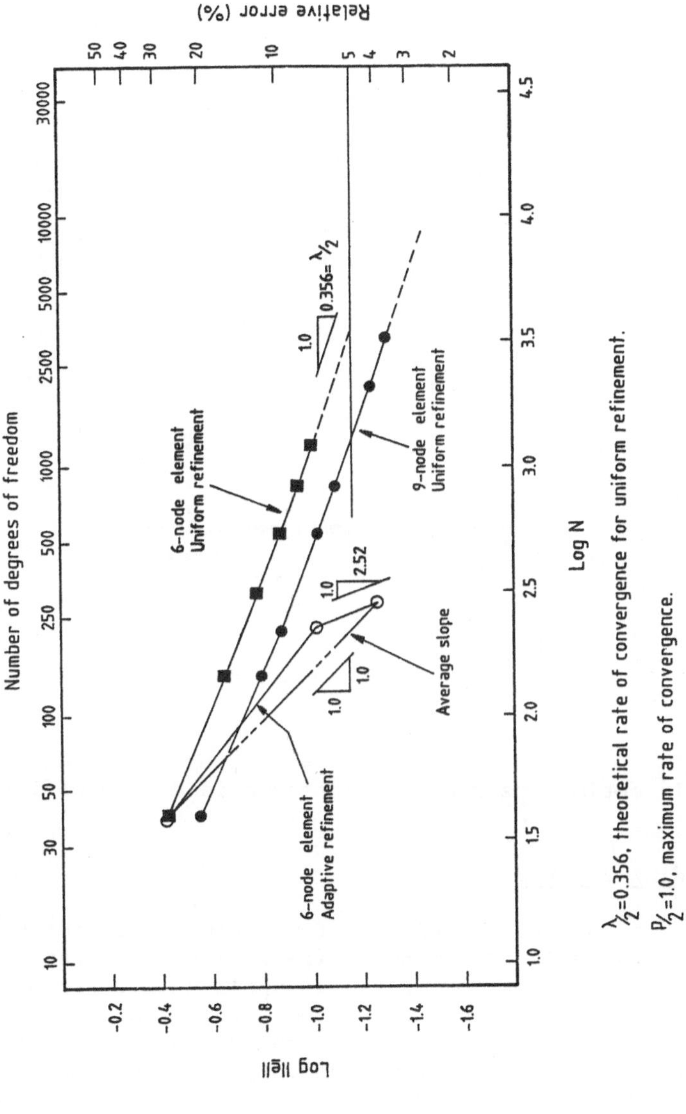

Figure 2. Experimental rates of convergence for cantilever beam. Performance of h refinement of quadratic elements. Adaptive refinement aims at 5% accuracy.

Once again, strong convergence of adaptive procedures is observed (Figure 4). For this problem, the predicted 5 percent accuracy has been achieved in one step. The maximum theoretical convergence rate is also obtained, and hence a nearly optimal mesh. The second step done here is carried out to ensure local refinement.

The adaptive procedure with mesh regeneration is not only easy to implement but also cost-effective. The strong convergence behaviour of the procedure has shown that the predicted accuracy

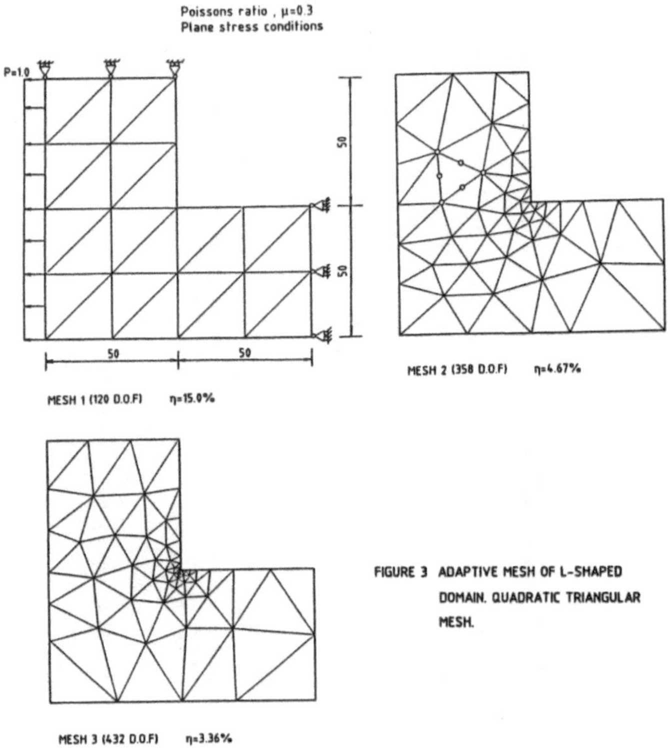

FIGURE 3 ADAPTIVE MESH OF L-SHAPED DOMAIN. QUADRATIC TRIANGULAR MESH.

and a nearly optimal mesh can be achieved in a very efficient manner.

6. Concluding remarks

We have shown the effectivity of the simple refinement strategy suggested in this paper. The refinement strategy combined with the error estimator (7) is applicable to problems other than stress analysis alone. The method can be used for almost any linear finite element discretization and, in fact, the development of non-linear methodology is in progress.

It is of interest to note that if we change to p-refinement after a nearly optimal h-version mesh has been achieved, much stronger convergence can be reached, because the mesh around the singularities will be well designed [13]. The implementation of this type of h and then p version refinement is only a simple combination of conventional h and p version.

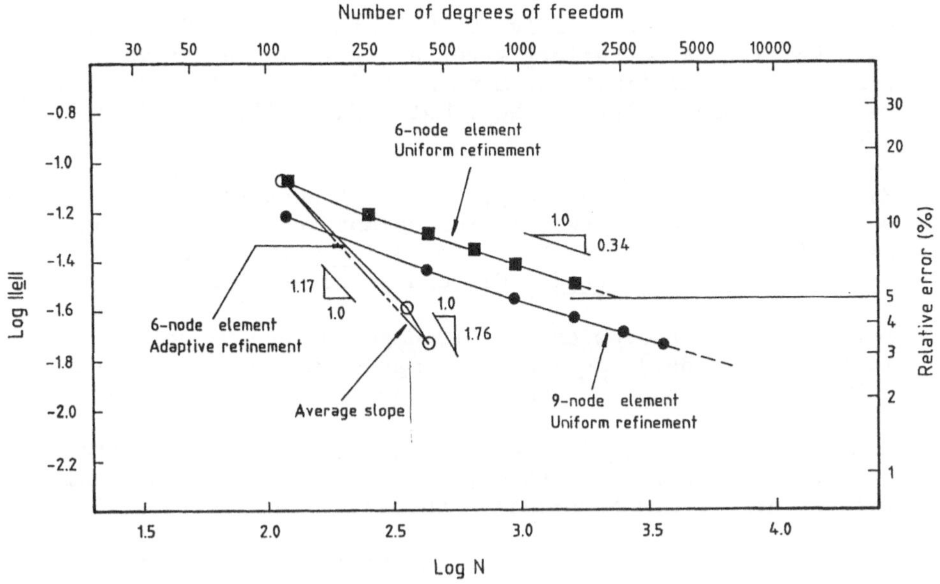

$p/2 = 1.0$, maximum rate of convergence.

Figure 4. Experimental rates of convergence for L-shaped region. Performance of h refinement of quadratic elements. Adaptive refinement aims at 5% accuracy.

7. References

1. Zienkiewicz, O. C. and Zhu, J. Z., 'A simple error estimator and adaptive procedure for practical engineering analysis', Int. J. Num. Meth. Eng., V.24, pp.337-357, 1987.

2. Babuska, I. and Rheinboldt, W. C., 'Error estimates for adaptive finite element computations', SIAM J. Num. Analysis, 15 (4), pp.736-754, 1978.

3. Babuska, I. and Rheinboldt, W. C., 'Adaptive approaches and reliability estimates in finite element analysis', Comp. Meth. Appl. Mech. Eng., 17/18, pp.519-540, 1979.

4. Zienkiewicz, O. C. and Gago, J. P. De S. R. and Kelly, D. W., 'The hierarchical concept in finite element analysis', Computers and Structures, 16, pp.53-65, 1983.

5. Kelly, D. W., Gago, J. P. De S. R., Zienkiewicz, O. C. and Babuska, I., 'A-posteriori error analysis and adaptive

processes in the finite element method: Pt.I - error analysis', Int. J. Num. Meth. Eng., 19, pp.1593-1619, 1983.

6. Gago, J. P. De S. R., Kelly, D. W , Zienkiewicz, O. C. and Babuska, I., 'A-posteriori error analysis and adaptive processes in the finite element method: Part II - Adaptive mesh refinement', Int. J. Num. Meth. Eng., 19, pp.1621-1656, 1983.

7. Zienkiewicz, O. C. and Craig, A. W., 'A-posteriori error estimation and adaptive mesh refinement in finite element method'. in Griffiths, D. F. (ed.) The Mathematical Basis of Finite Element Methods, Clarendeon Press, Oxford, 1984.

8. Babuska, I., Zienkiewicz, O. C., Gago, J. and Oliveira, E. R. de A. (eds.), Accuracy Estimates and Adaptive Refinement in Finite Element Computerations, John Wiley & Sons, 1986.

9. Rank, E. and Zienkiewicz, O. C., 'A simple error estimator in the finite element method', (to be published in Communications in Applied Numerical Methods).

10. Zhu, J. Z., 'Error Estimation, Adaptivity and Multigrid Techniques in the Finite Element Method', Ph.D. Thesis, University of Wales, Swansea, February 1987.

11. Szabo, B. A. and Babuska, I., 'Computation of the amplitude of stress singular terms for cracks and re-entrant corners', Report WU/CCM-86/1, Center for Computational Mechanics, Washington University, 1986.

12. Peraire, J., Vahdati, M., Morgan, K. and Zienkiewicz, O. C., 'Adaptive remeshing for compressible flow computations', to be published J. Comp. Physics.

13. Szabo, B. A., 'Estimation and control of error based on p-convergence', in Accuracy Estimates and Adaptivity for Fintie Elements, ed. by Babuska, I et al., pp.61-78, 1986.

ASPECTS ON METHODOLOGY FOR FE-PROGRAM DEVELOPMENT

Harald Tägnfors

Department of Structural Mechanics, Chalmers University of Technology, Göteborg, Sweden.

SUMMARY

There is a great need for a systematic approach to the development of Finite Element (FE) programs. This paper describes the reasons for such an approach and gives examples on general methods which have been developed for certain steps in the program development process. A selection of methods, along with rules how to apply them, constitutes a methodology. The special features in the FE-method which have to be considered when selecting methods in a methodology for FE-program development are discussed as well as the possibilities to support the methods with tools. The author also presents an architectural design of a FE-program with modules in seven hierarchical levels, where control of calculation can be switched between the Program Developer and User in a flexible way.

1 INTRODUCTION

Special purpose and general purpose FE-programs are widely used to solve engineering problems. Programs of these types, many of them designed in the 1960's and developed in a traditional way, have often shown to be expensive to maintain, Schrem[1] , and they are not well suited for further development except for small modifications within their range of application. By use of techniques developed since the beginning of the 1970's it is possible to develop programs which are more reliable and adaptable to new User requirements, concerning both the range of application and the execution environment.

Main steps in software development in general are: specification, architectural design, detailed design, implementation and maintenance. These steps are also applicable in the development of FE-programs. When large FE systems are developed, like SITU, Tägnfors[2] , FINITE, Dodds and Lopez[3] and NICE, Felippa and Stanley[4] the steps are applicable both in the design of the system and in the development of the application programs but the meaning of the steps are different.

2 DESIGN METHODOLOGY

During the last ten years some research has concerned methodologies for design and development of software for finite element calculations. The goal has been to provide Program Developers with a good methodology, with concepts and tools, by which flexible and safe programs easily can be developed. One example of a methodology is given in the handbook by Bell et al.[5], where the Program Developer is guided, step by step, in each phase of the program development. A methodology for design of more general engineering software is given by Rzevski[6]. The methodology comprises definitions, design principles and design guidelines. Rzevski and Wells[7] describe an interactive computer-aided system for software design. All presented methodologies propose modular programs where most of the modules can be used in several programs.

2.1 Specification

A functional description of the system shall be developed and constraints on the structure shall be specified. A general method for specification is Structured Systems Analysis (SSA), Gane and Sarson[8], which is intended to be used before a design is made with so-called Structured Design (SD), which is a technique for architectural design. The need for advanced methods for specification of a FE-program to be developed is however limited. The effort can be concentrated on the architectural and detailed design steps. Important specification decisions are:

- Range of application. FE-methods to be used, preferably represented as a sequence of matrix operations, have to be specified as well as the size of problems to be solved. These decisions will influence the methods to be chosen for data management and solution of systems of equations.

- Tools for application program development. The use of a System Specification Language (SSL), a Program Specification Language (PSL) and Data Management System (DMS) will facilitate application program development, see below, and the characteristics of them have to be specified.

- Program execution. The ways in which the User shall interact with the system must be specified, as well as how results are to be presented. A rapidly changing computational environment shall be taken into consideration.

2.2 Architectural design

In the architectural design step, a structure of a system which satisfies the requirements in the specification is sought. The Finite Element Analysis can in principle be described as a sequence of operations on different types of data. The data structure can therefore be derived from the structure of operations. The program structure is defined by the operations which shall be performed on the data structure. The architectural design can be made in a systematic way by Structured Design, where data flow diagrams are transformed into program structures, Yourdon and Constantine[9]. In composite/structured designed introduced by Myers[10] there are possibilities to evaluate the quality of the design by the introduction of the concepts of cohesion and coupling. A widely used method for design of commercial programs is the Jackson Design Method, Jackson[11]. A number of diagrams with input data structures and output data structures are drawn. Operations to be performed are listed and inserted in a program structure, by which the program can be implemented. A technique with emphasis on representation of the design is HIPO (Hierarchical Input Process Output), Stay[12]. In a diagram for each of the hierarchically organized modules the calculation is described and input and output items are specified. At the design of a FE-system it is important to provide for the use of program parts in many different applications.

When the data and program structures have been designed, different alternatives concerning control should be considered. Two extreme designs are pictured in Figure 2.1. Either the Programmer steers the calculations completely by implementation of a control structure within the program, Figure 2.1 a), or the User (representing the physical problem at hand), Figure 2.1 b), steers the calculations by commands. Parts of the program to be used for a certain physical problem with one chosen method are denoted by digits indicating the sequential order in which calculations are performed.

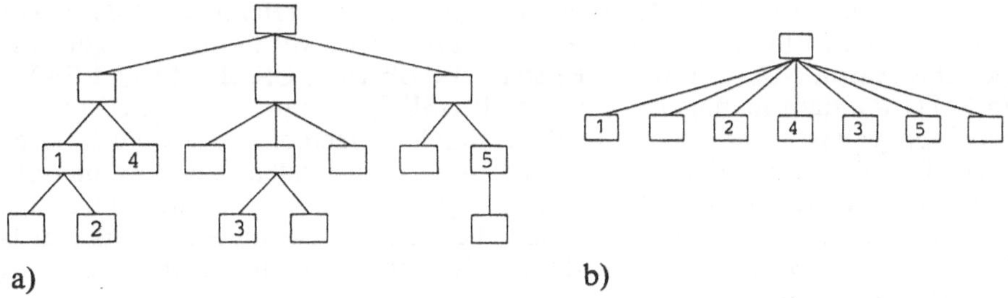

Figure 2.1. a) Program with Programmer control
b) Program with User control

A solution where the Programmer makes one part of the control, and the User the other, is the best in most situations. It is very important for a flexible system that there are possibilities to easily change the steering function from one to the other. For example, in many FE-programs stresses are calculated and printed on a lineprinter. In a CAD environment the User want to select stresses to be calculated and store them in a data base where they can be retrieved for display in the CAD-system. Thus there is a need for the User to enter the program at a certain point and perform a single function. This is often difficult in a program with a conventional design where the control is within the FORTRAN-program.

In the following a modular design will be discussed and for reasons of clearity modules at the lowest levels are described first. A module at level 1 (M1), which is taken as the lowest level, can be defined as in Figure 2.2, where notations according to the Jackson Structured Design is used. Module M1 may contain sequences (concatenations) of calculations including calls for modules at the same level (not the same module), selections of alternative calculations due to control data, and iterative calculations. The module is implemented as a subroutine in FORTRAN, and may have control data and calculation data as input and output. Other data which are produced and used in the module are never referenced in the environment and the same input will always produce the same output. Schrem[13] defines this type of module to be a functional module. If the module performs a single function it has the best possible cohesion, functional strength, Myers[10] and is well suited to be stored in a subroutine library for use in several modules. Smith[14] proposes that libraries with FE-modules at this level should be used for FE-program development together with general mathematical libraries.

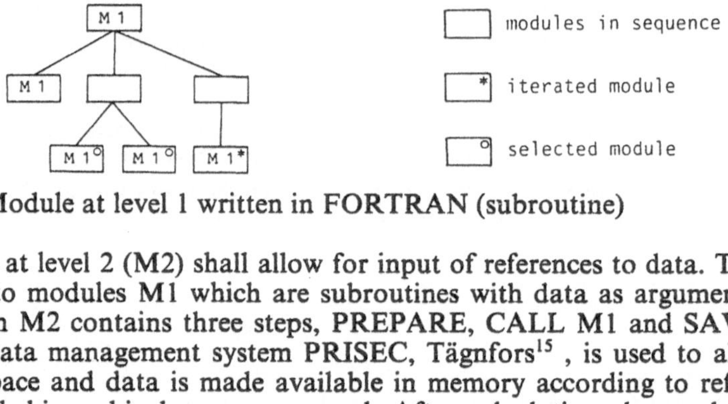

Figure 2.2. Module at level 1 written in FORTRAN (subroutine)

Modules at level 2 (M2) shall allow for input of references to data. This is in contrast to modules M1 which are subroutines with data as arguments. For this reason M2 contains three steps, PREPARE, CALL M1 and SAVE. In SITU the data management system PRISEC, Tägnfors[15] , is used to allocate storage space and data is made available in memory according to references for which hierarchical names are used. After calculations by modules M1, data are stored in the data base, Figure 2.3. Such modules are termed functional units according to Schrem[13] because the calculated results can be different for successive entries with the same input data due to the interaction with the environment.

Modules at level 3 (M3) can be built up of an arbitrary combination of modules M2, Figure 2.3. Modules M1, M2 and M3 are implemented as subroutines and can, by the Programmer, be included in a FORTRAN-program by call-statements. The introduction of modules M2, where references to data, (not the data themselves) are passed to the module, gives the possibility to enter the modules from a language above FORTRAN, a Program Specification Language (PSL), Tägnfors[2] . If the PSL, like FORTRAN, see Figure 2.2, has the three structural components: sequence, selection and iteration, then any programming problem can be solved, Böhm and Jacopini[16] . The inclusion of these three components in the PSL gives the Programmer full freedom to choose language (FORTRAN or PSL) for a certain module. SITU-PSL, which has these three components, is used to specify modules to be entered for a command. In Figure 2.3 this module is denoted M4. Data to be used are specified as names of matrices, arrays, numbers or character strings in a hierarchical data structure, or scalar variables.

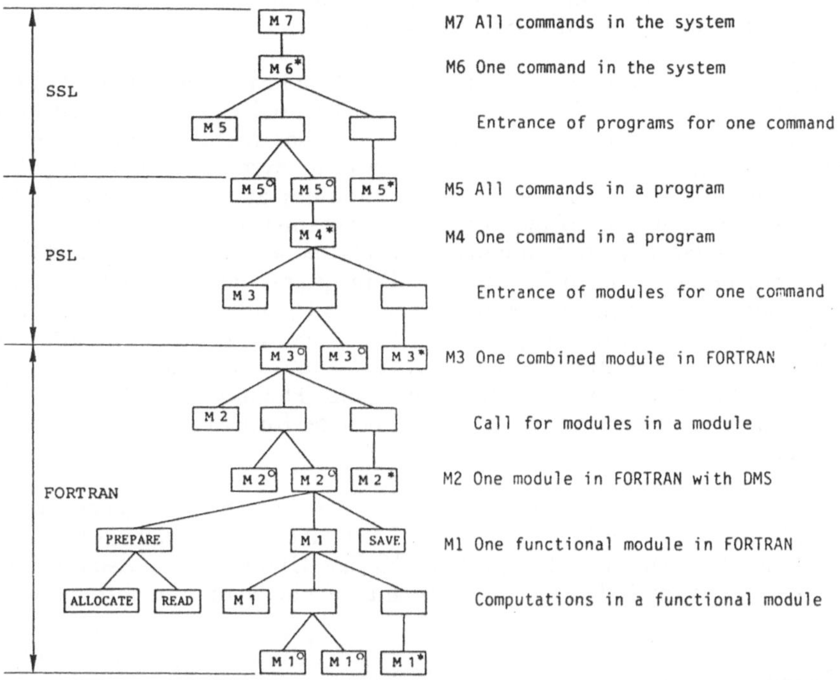

Figure 2.3. Architectural design with hierarchical levels of modules

In POLO, with its FE-program FINITE, Dodds and Lopez[3] , an hierarcic data structure can be defined in a language called F. In the language G items in the data structure can be specified and modules which use these data can be entered.

Within SITU-PSL a sequence of subsequent commands can be specified to be executed for a command (macro command). For example, for the command ALL in the command sequence DISPLAY INPUT ALL, the sequence of all commands for input display can be specified.

A module is defined by Parnas[17] to be a program part having well defined interfaces that hide a design decision. A selection of other definitions can be found in Roos et al.[18] . By the concept of abstraction, Dijkstra in Dahl et al.[19] , a sequence of calls for modules M4 in Fig. 2.3 can be regarded as a module. It may be called a logical module because it is composed due to commands given by the User (explicitly or implicitly by commands generated by PSL for other commands).

For the same reasons the module M5 can be defined to be a logical module performing calculations within a program. The sequence of commands (with data) which are used at the execution of a program may be given by the User or they may emerge from calculations in a module and are then provided to the program by the SSL, see below.

When the system is distributed on several main programs the selection of programs to be executed and data(files) to be used can be specified in a System Specification Language (SSL). The SSL has to be a language with possibilities to enter commands to the Operating System (OS). If the SSL allows for all structural component types, modules at level 6 (M6) can be designed as a combination of modules M5 (programs). For SITU running on a mainframe the terminal system language is chosen as SSL. The introduction of SSL is not necessary in SITU because all programs can be included in the PSL. It is however convenient for the Developers to allow for use of different versions of the system, which is possible if the system is replicated in different programs. Users utilize SSL to specify programs and program versions to be used as well as to define data files to be used, etc. The module at level 7 (M7) can be defined to be the logical module which performs calculations within the system.

NICE, Felippa[20] , is designed to be used in a network system, and a Command Language Interface Program (CLIP) covers the options in both the PSL and the SSL. The use of a modern OS like UNIX gives possibilities to include features in the SSL which are difficult to achieve on a mainframe with an oldfashioned OS.

2.3 Detailed design

The modules from the architectural design shall be made detailed. Algorithms and data structures shall be specified for each module as well as interfaces between modules. The methods used at architectural design can be used also at detailed design but the purpose of this design is to produce a representation which can be implemented in a program. Rules according to Structured Programming (SP) shall be followed. The design can preferably be represented in pseudocode because then it can easily be written and modified by an ordinary text editor. Tools to construct the program code automatically from the design are useful. The data structure is designed and represented in detail.

2.4 Implementation

The design shall be realized in an executable form. It includes programming, testing of parts, and integration of parts for further tests until the whole system is completed and tested. The methodology shall specify layout of subroutines, naming conventions etc. and provide tools for scanning and administration of source code, etc. Evaluation of performance and modifications shall be made so that the system conforms to the specification.

2.5 Maintenance

Maintenance after the system is finished includes identification and correction of misbehaviour, addition of new functions and tuning the system when required resources seem to be unfavourable. Such maintenance represents up till now, an extremely heavy part of the total program cost during the life time. Through the methodology discussed above it will hopefully be reduced to a great extent.

3 CONCLUSIONS

A methodology for software engineering has proven to be indispensible in several presented FE-projects. Different architectural solutions have been discussed and the possibilites to distribute the control of the calculations between the Programmer and the User is demonstrated in a modular design with several levels of modules. The use of a PSL and a SSL facilitates program development and use and give both the Program Developer and the User possibilities to control the calculation flow in an easy way.

4 REFERENCES

[1] Schrem, E., 'Trends and aspects of the development of large finite element software systems', Comp. & Struct., 10 (1979), 419-426.

[2] Tägnfors, H., 'Architecture of an adaptive and interactive FEM-program for CAD-purposes', NUMETA 85, Proc. Int. Conf. on Num. Meth. in Eng., Swansea, 1029-1038, 1985.

[3] Dodds Jr., R.H. and Lopez, L.A., 'Software virtual machines for development of finite element systems', Eng. Comp. 3 (1986), 18-26.

[4] Felippa, C.A. and Stanley, G.M., 'NICE: a utility architecture for computational mechanics', Europe - US Symposium, Finit Element Methods for Nonlinear Problems, Trondheim, Norway, 1985.

[5] Bell, K., Carr, A. and Sylvertsen, T.G., *Handbook of computer programming - Design, development and maintenance of engineering (Fortran) software, an internal reference manual,* SINTEF report no. STF71 A83005, 1983.

[6] Rzevski, G., 'On the design of engineering software', Proc. First Int. Conf. on Eng. Software, Southampton, 428-440, 1979.

[7] Rzevski, G. and Wells, M., 'Computer-aided design of engineering software', Proc. Second Int. Conf. on Eng. Software, London, 86-90, 1981.

[8] Gane, C. and Sarson, T., *Structured systems analysis,* Prentice-Hall, Englewood Cliffs, New Jersey, 1979.

[9] Yourdon, E. and Constantine, L.L. *Structured design,* Prentice-Hall, Englewood Cliffs, New Jersey, 1979.

[10] Myers, G.J. *Composite/Structured design,* Van Nostrand Reinhold, New York, 1978.

[11] Jackson, M.A., *Principles of program design,* Academic Press, London, 1975.

[12] Stay, J.F., 'HIPO and integrated program design', IBM Systems Journal, 15 (1976), 143-154.

[13] Schrem, E., 'Functional software design and its graphical representation', Comp. & Struct. 8 (1978), 491-502.

[14] Smith, I.M., 'Adaptability of truly modular software', Eng. Comp., 1 (1984), 25-35.

[15] Tägnfors, H., 'Data management techniques in engineering computer programs', SEAS, proceedings anniversary meeting, Oxford, (1983), 634-643.

[16] Böhm, T.L. and Jacopini, G., 'Flow diagrams, turing machines and languages with only two formation rules', Comm. ACM. 9 (1966), 366-371.

[17] Parnas, D.L., 'On the criteria to be used in decomposing systems into modules', Comm. ACM, 15 (1972), 1053-1058.

[18] Ross, D.T., Goodenough, J.B. and Irvine, C.A., 'Software engineering: Process, principles and goals', IEEE, Computer, May 1975.

[19] Dahl, O.-J., Dijkstra, E.W. and Hoare, C.A.R., 'Structured programming', Academic Press, London, 1972.

[20] Felippa, C.A., 'Architecture of a distributed analysis network for computational mechanics', Comp. & Struct. 13 (1981), 405-413.

THE SIGNIFICANCE AND PRACTICE OF RANK ESTIMATION IN STRUCTURAL DYNAMICS IDENTIFICATION ALGORITHMS

John Brandon
Department of Mechanical and Manufacturing Systems Engineering
University of Wales Institute of Science and Technology
PO Box 25, Cardiff, CF1 3XE, United Kingdom

1. INTRODUCTION

It is common practice in structural dynamics to assume that a structure can be adequately described in terms of the linear n dimensional matrix equation:

$$\underset{\sim}{M}\,\underline{\ddot{x}} + \underset{\sim}{K}\,\underline{x} = \underline{f} \qquad \ldots (1)$$

The inclusion of damping is usual, but the analysis presented here can readily be extended. There are instances of different assumptions, most notably that of Baruh and Meirovitch [1] who use a continuum representation. These are, however, quite rare and will not be considered here. Neither is it intended to undertake a comprehensive survey of identification algorithms, but rather to assess some typical well known algorithms to demonstrate the need for rank estimation. The reader is referred to the recent paper by Snoeys et al [2] for a brief survey of current identification algorithms. More general bibliographies have been presented by Mitchell and Mitchell [3] and Allemang [4]. There is a strong case to be made, however, for a survey providing a detailed categorisation of algorithms, similar to the work by Eykhoff [5] who has classified identification methods in control engineering. Similarly, it is not intended to provide a thorough survey of the literature on rank estimation, but rather to highlight a small number of pertinent references.

For the purposes of this paper a pseudo-algorithm is constructed. This has properties of the practically implemented algorithms discussed but is deficient in the practical programming aspects of current software.

The statement of the problem in equation (1) is not of

great use to the designer. More interesting is the solution to the inverse problem:

$$\underline{x} = \underset{\sim}{R}\ \underline{f} \qquad \ldots (2)$$

where the receptance matrix $\underset{\sim}{R}$ is a function of frequency. (It is not unusual to discount the effects of the initial conditions and consider the solution to be sufficiently described by the particular integral of (1).)

The receptance matrix $\underset{\sim}{R}$ is often constructed analytically using the spectral decomposition of the quadratic eigenproblem represented by the homogeneous form of equation (1):

$$(p_i^{\,2}\ M + K)\ \underline{q}_i = 0 \qquad \ldots (3)$$

leading to the spectral decomposition:

$$\underset{\sim}{R} = \sum_{i=1}^{n} \frac{\underline{q}_i\ \underline{q}_i^{\,t}}{(p_{ex}^{\,2} - p_i^{\,2})\, m_i}$$

where $m_i = \underline{q}_i^{\,t}\ \underset{\sim}{M}\ \underline{q}_i$ and p_{ex} is the excitation frequency. The receptance matrix comprises the point to point force-displacement transfer functions. These are typically evaluated by the special purpose experimental instrumentation hardware and will be assumed already to be available prior to the application of the identification algorithms discussed here.

Whilst in theory the identification can be achieved by evaluating the full receptance matrix at a small number of frequencies, this entails serious practical difficulties (particularly the necessity to provide a force excitation at each point of the structure included in the model). Traditionally only a single point is excited, with responses measured all model points, giving a single column of the receptance matrix for each excitation frequency. (This is called in the control literature a Single Input - Multiple Output (SIMO) model.) Recent publications describe multiple excitation configurations (MIMO), see for example Allemang [6].

2. THE PHYSICAL SIGNIFICANCE OF RANK

The observations of the column of the receptance matrix may be stated in the form:

$$\underset{\sim}{Y} = \underset{\sim}{Q}\ \underset{\sim}{F} \qquad \ldots (5)$$

where $\underset{\sim}{Y}$ n x m is the matrix whose columns are the observations of the column of the receptance matrix, $\underset{\sim}{Q}$ n x n is the matrix of the eigenvectors of equation (3) (the modes of vibration) and $\tilde{F}$ n x m (with m > n in general) is a matrix containing only the frequency dependent element of the denominator, ie $F_{ij} = 1/(p_{exj}{}^2 - p_i{}^2)$.

It should be noted that the columns of $\tilde{Q}$ are scaled in terms of the experimental configuration. Of particular importance is the possibility that the excitation point may coincide with a node of a modal vector $\underline{q}_i$. This will result in the annihilation of a column of $\tilde{Q}$ in its image in the observation matrix $\tilde{Y}$, which in consequence will be rank deficient.

In practice, the observation matrix $\tilde{Y}$ will be subject to the effects of noise. The spectral decomposition shows that the contribution of any given mode is strongly dependent on the frequency of excitation and thus the noise may overwhelm the contribution of modal vectors when the excitation frequency p_{ex} and natural frequency p_i are widely separated.

A third potential source of corruption of the observation matrix $\tilde{Y}$ is the possibility of the resonant characteristics of the instrumentation system itself being mistaken for structural properties.

All of the above sources of error lead to uncertainties in the measurements which can in turn affect the perceived rank of the observation matrix.

3. A PSEUDO-ALGORITHM

This is loosely based on the method of Van Loon [7] which is widely quoted in the literature. He describes a procedure:

(i) initial estimates are derived for the natural frequencies, ie the p_i by graphical methods
(ii) the modal vectors $\underline{q}_i$ are calculated by a non-iterative process
(iii) the use of an iterative gradient type non-linear method to refine both the estimates of natural frequencies p_i and modal vectors $\underline{q}_i$.

Thus the analyst makes decisions about the rank of the observation matrix in stages (i) and (ii). Stage (i) allows the construction of the $\tilde{F}$ matrix and the linear least squares process (ii) can be expressed in terms of the equation:

$$\tilde{Q}^* = \tilde{Y}\,\tilde{F}^g \qquad \ldots (6)$$

where $\tilde{Q}^*$ is the identified modal matrix and $\tilde{F}^g$ is the Moore-Penrose generalised inverse of $\tilde{F}$. The quality of the inversion can then be assessed, eg by evaluating the residual $||\underset{\sim}{y} - \underset{\sim}{Q}^* \underset{\sim}{F}||$, the most suitable norm (considering the use of the M-P inverse) being $||.||_2$. Common practice in structural dynamics is the use of the restricted (ie full rank) M-P inverse $\underset{\sim}{F}^g = \underset{\sim}{F}^t (\underset{\sim}{F} \underset{\sim}{F}^t)^{-1}$.

As has been observed by Noble [8], this has the undesirable effect of squaring the condition number of inversion. It is widely agreed that the singular value decomposition, whilst expensive to compute, gives a reliable solution procedure for the M-P inverse, and perhaps more importantly gives insight into the quality of the data, (see Golub and Kahan [9]).

4. THEORETICAL WORK ON RANK ESTIMATION

Wedin [10] considered rank estimation particularly in terms of "non-linear least squares problems ... solved iteratively by successive approximations with linear problems" and is thus appropriate here. Wedin considers both the effect of inappropriate estimates of rank and the adverse effects of the scaling of $\underset{\sim}{Q}$. If the rank is r then the criterion for the bounds for the perturbation of the residual depends on the ratio s_r/s_{r+1} where s_i are the ordered singular values of $\underset{\sim}{F}$. Elsewhere Wedin [11] shows that if $||\underset{\sim}{B}-\underset{\sim}{A}||$ is small and rank $(\underset{\sim}{A}) \neq$ rank $(\underset{\sim}{B})$ then $||\underset{\sim}{B}^g-\underset{\sim}{A}^g||$ is always large. Equivalently, Noble [8] states "if a matrix $\underset{\sim}{B}=\underset{\sim}{A}+\underset{\sim}{E}$ is near $\underset{\sim}{A}$, but has rank greater than $\underset{\sim}{A}$, its generalised inverse can be larger and completely different from $\underset{\sim}{A}^g$, and the smaller $\underset{\sim}{E}$, the worse the trouble can be".

Stewart [12] discriminates between three cases of close matrices. Firstly, the case when the rank differs gives $||\underset{\sim}{B}||_2 \gtrsim 1/||\underset{\sim}{E}||_2$. The equal rank case is well behaved when the row spaces of B and A have the same span, called the acute case, the appropriate condition being $||\underset{\sim}{A} \underset{\sim}{A}^g - \underset{\sim}{B} \underset{\sim}{B}^g||_2 < 1$. In the full rank case where A and B are not acute then $||\underset{\sim}{B}^g-\underset{\sim}{A}^g||_2 \gtrsim 1/||\underset{\sim}{E}||_2$.

A measure for estimation of rank in least squares analysis has been suggested by Lawson and Hanson [13]. "We define pseudo rank to be the rank of the matrix $\underset{\sim}{A}^*$ that replaces $\underset{\sim}{A}$ as the result of a specific computational algorithm. Note that pseudo rank is not a unique property of the matrix $\underset{\sim}{A}$ but also depends on other factors such as ... algorithm ... tolerances ... and ... round off errors." Noble [8] defines a different and more specific measure, epsilon rank, in terms of the ordinate of the smallest singular value greater than a preset tolerance.

5. PRACTICAL APPROACHES TO RANK ESTIMATION

Perhaps the simplest method in use is due originally to Kennedy and Pancu [14]. This method, originally graphically based, depends on the assumption that the response of a mode dominates in the region of its natural frequency. This can be seen by consideration of the spectral decomposition. The method is only suitable for well spaced modes, although it has been claimed that methods proposed by Brandon and Cowley [15] and Elliot and Mitchell [16] improve the resolution of close modes. This method is a rank one method and often is used as the preliminary step to the multi-mode algorithms such as that of Van Loon (ie step (i) of the method described above). Goyder [17] uses what is essentially a rank one method, but has evolved the imaginative technique of using each mode as it is identified to deflate the observation matrix, identifying a multi-mode system as a superposition of individually modelled single degree of freedom systems. The individual modes are identified by partitioning the observation matrix $\underset{\sim}{Y}$, using only the columns corresponding to excitation frequencies close to the resonant frequency of the suspected mode. Ewins and Gleeson [18] assume that the effects of adjacent modes are sufficiently significant to affect the quality of the single rank fit and have developed from Goyder's work a rank three method with a similar partitioning strategy, which they describe as "windowing". In this case the successive partitions of $\underset{\sim}{Y}$ must overlap.

At the other extreme there have been a number of methods developed recently which assume the data matrix to be of full rank. Perhaps the most widely known of these methods is due to Ibrahim [19]. The justification for the method suggests that measurement noise and numerical truncations are sufficient to complete the (grossly deficient) rank of the measurement matrix. This method depends, therefore, for its operation on the inversion of a nominally singular matrix, which in Lawson and Hanson's terms has full pseudo rank! Whilst in general such a procedure will give meaningless results, subsequent processing enables the extraction of valid modal data. The method has been demonstrated to be effective on practical engineering structures (Ibrahim and Pappa [20]) and can be explained in terms of the disjointness of the subspaces containing the structural data and the noise (see Brandon [21]).

Between the two extremes come the methods, such as that of Van Loon [7] where all of the data is processed simultaneously using an estimate of rank provided by the analyst, based usually on graphical interpretation of the driving point transfer function. This estimate is implicit, being based on the number of identified resonances supplied.

The method of Van Loon has been developed by Mergeay [22] to allow the rank of the observation matrix to be determined incrementally by considering the size of the residual. The observed behaviour of the residual as shown in Fig 1 agrees closely with Noble's description (p289 [8]) of an ill conditioned system: "... if p_1 ... p_k are of order of magnitude unity and p_{k+1} drops suddenly to a small multiple of 10^{-t} on a t-decimal machine, we can assume the rank is k; if the number decrease gradually to a small multiple of 10^{-t} the problem is ill-conditioned".

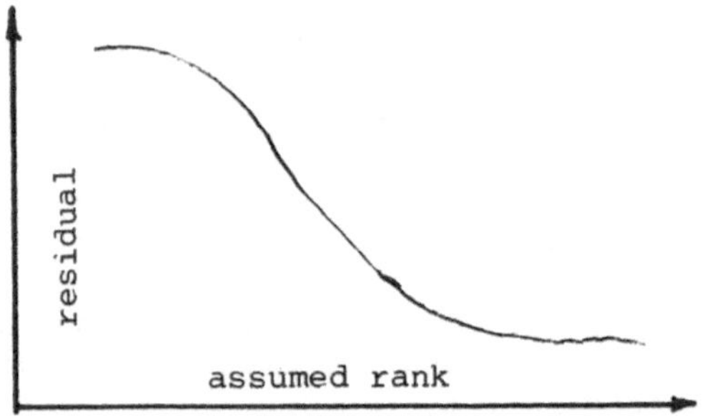

Fig 1. Variation of Residual with Assumed Rank (After Mergeay [22])

It is common practice in experimental tests to prefer to include "ghost" modes, (ie characteristics constructed solely from the noise) rather than exclude faintly measured genuine structural properties, relying on intuition and experience to reject the spurious information (see Ewins, p189 [23]). Thus in general the policy would be adopted of overestimating the rank of the observation matrix. The adverse consequences of this strategy are discussed by Turunen [24], whose conclusions suggest that the assumptions of Ewins on the effectiveness of using experience must be questioned, since the algorithms studied may give modes which are strongly biased, combined with low variance, making results repeatable but not accurate. The identification of an excess number of modes is not restricted to simultaneous multi-mode algorithms, however.

Even if the rank was to be correctly identified, the analyst could still run into serious difficulties due to poor condition. This corresponds to Stewart's criterion for the non-acute case, potentially suppressing an actual structural mode, and replacing it with one constructed from the noise.

6. A PROPOSED RANK ESTIMATOR

The estimator proposed by Noble [8] depends on the explicit evaluation of the singular value decomposition, which may be impractical to apply. In addition, the setting of the epsilon tolerance in advance prevents the adaptation of the rank estimator to characteristics of the data.

Since it may be expected in general that the identification may be ill-conditioned, due both to the separation of the highest and lowest frequencies and also due to the scaling of the modes, then the analysis of Wedin [10] should be considered. This suggests that the ratio of successive singular values of the error matrix should be the criterion in assigning rank. Thus apparently the appropriate criterion should be based on the point of inflexion of Fig 1.

7. CONCLUSIONS

A general overview of the practice of rank estimation in structural dynamics has been presented. This has been assessed in terms of relevant work in numerical analysis.

There is a strong case for a taxonometric study of existing methods similar to the recent studies in the field of control engineering.

A rank estimation criterion, which it is believed will be acceptable to both the engineering and numerical analysis communities, has been proposed.

8. REFERENCES

1. Baruh, H and Meirovitch, L, 'Parameter Identification in Distributed Systems', Jnl Sound & Vibration, 101, 4 (1985) 551-564.
2. Snoeys, R, Sas, P, Heylen, W and Van der Auweraer, H, 'Trends in Experimental Modal Analysis', Mechanical Systems & Signal Processing, 1, 1 (1987) 5-27.
3. Mitchell, L D and Mitchell, L D, 'Modal analysis bibliography - an update 1980-3', 2nd Intl Modal Analysis Conf, Orlando, Fa, 1098-1114, 1984.
4. Allemang, R J, 'Experimental modal analysis bibliography', 2nd Intl Modal Analysis Conf, Orlando, Fa, 1085-1097, 1984.
5. Eykhoff, P, 'On the coherence among the multitude of system identification methods', IFAC Symp on Identification and System Parameter Identification, Washington, 1982.
6. Allemang, R J - Investigation of Some Multiple Input/Output Frequency Response Modal Analysis Techniques, PhD, University of Cincinnati, 1980.
7. Van Loon, P - Modal Parameters of Mechanical Structures, PhD, University of Leuven, 1974.

8. Noble, B - Methods for Computing the Moore-Penrose Generalised Inverse and Related Matters in _Generalised Inverses and Applications_, Ed Nashed, M Z, Academic, 245-301, 1976.
9. Golub, G and Kahan, W, 'Calculating the Singular Values and Pseudo-Inverse of a Matrix', SIAM Jnl Numerical Analysis, 2(B) (1965) 205-224.
10. Wedin, P-A, 'On the Almost Rank Deficient Case of the Least Squares Problem', BIT 13 (1973) 344-354.
11. Wedin, P-A, 'Perturbation Theory for Pseudo-Inverses', BIT, 18 (1973) 217-232.
12. Stewart, G W, 'On the Perturbation of Pseudo-Inverses, Projections and Linear Least Squares Problems', SIAM Review, 19 (1977) 634-662.
13. Lawson, C L and Hanson, R J - _Solving Least Squares Problems_, Prentice-Hall, 1974.
14. Kennedy, C C and Pancu, C D P, 'Use of Vectors in Vibration Measurement and Analysis', Jnl Aeronautical Sciences, 14 (1947) 603-625.
15. Brandon, J A and Cowley, A, 'A Weighted Least Squares Method for Fitting Circles to Frequency Response Data', Jnl Sound & Vibration, 89 (1983) 419-424.
16. Elliot, K B and Mitchell, L D, 'Improved frequency response function circle fits, in modal testing and modal refinement', ASME, AMD-59, 63-75, 1983.
17. Goyder, H G D, 'Methods and Application of Structural Modelling from Measured Structural Frequency Response Data', Jnl Sound & Vibration, 68 (1980) 209-230.
18. Ewins, D J and Gleeson, P T, 'A Method for Modal Identification of Lightly Damped Structures', Jnl Sound & Vibration, 84 (1982) 57-79.
19. Ibrahim, S R amd Mikulcik, 'The Experimental Determination of Vibration Parameters from Time Responses', Shock & Vibration Bulletin, 46 (1976) 187-196.
20. Ibrahim, S R and Pappa, R S, 'Large Modal Survey Testing Using the Ibrahim Time Domain Identification Technique', Jnl Spacecraft, 19 (1982) 459-465.
21. Brandon, J A, 'Subspace behaviour in Ibrahim's time domain algorithm', 5th Intl Modal Analysis Conf, Imperial College, London, April 1987.
22. Mergeay, M, 'Theoretical background of curve fitting methods used by modal analysis', 6th Modal Analysis Sem, University of Leuven, September 1981.
23. Ewins, D J - _Modal Testing: Theory and Practice_, Research Studies Press, 1984.
24. Turunen, R, 'Statistical performance of modal parameter estimation methods', 5th Intl Modal Analysis Conf, London, April 1987.

THE USE OF A TENSION PARAMETER IN SURFACE MODELING

Da-Pan Chen, Professor
Tser-Liang Lin, Graduate Student

Department of Mechanical Engineering, National Chiao Tung University, Hsinchu, Taiwan, Republic of China.

ABSTRACT

In this paper, the Nu-spline surface is derived and used to model several complicated machine parts. A Nu-spline is essentially a conventional cubic spline with tension control. Tensions applied at the joints of the curve segments enable the curve to turn around sharp corners without undesirable oscillations. The Nu-spline surface thus can turn around sharp edges without undesirable effects. It is derived on the basis to satisfy the G^2-continuity. Several restrictions are imposed to the surface to make it practicable.

1. INTRODUCTION

The idea of applying tension in a spline to pull out unwanted oscillations was developed by Schweikert [1] in 1966. This spline under tension was formulated with the analogy to an elastic beam bent by lateral loading under the simultaneous action of axial force. The axial force, when applied as a tension, tends to flaten the beam and thus can be utilized to eliminate the unwanted oscillations in spline curves. This idea was further extended and generalized by Nielson [2] and Barsky [3].

The practical use of spline curves with tension control is in the geometric modeling of objects with a complex surface. Numerous machine parts, such as the internal combustion engine piston, and a cam mounted on a circular shaft, possess complex surface. The modeling of such parts requires a spline curve to turn around square corners without the undesirable oscillation. A spline curve with tension control serves this purpose nicely.

In this paper, the use of a tension parameter in complex

surface modeling is explored as to its extent and limits. The satisfaction of the geometric continuity across the boundary oi adjacent surface patches is also discussed.

2. THE GEOMETRIC CONTINUITY AND BARSKY'S Nu-SPLINE

Barsky [3], in 1984, extended the Nielson Nu-spline from a polynomial basis to the form using Hermite basis functions. This relates the Nu-spline to the conventional cubic interpolatory spline, i.e., Barsky's Nu-spline is the cubic interpolatory spline with tension control. The difference between Barsky's Nu-spline and the conventional cubic spline is in the type of continuity for the second parametric derivative. For the derivation of Barsky's Nu-spline, we begin with the conventional cubic spline in the form of

$$Q_i(u) = (1-3u^2+2u^3)P_{i-1} + (3u^2-2u^3)P_i + (u-2u^2+u^3)P^1_{i-1} + (-u^2+u^3)P^1_i \tag{1}$$

where $Q_i(u)$ is the vector function of the i-th segment of the curve, P_i is the position vector of the i-th data point, and P^1_i is the first parametric derivative vector of the curve at the point P_i. The four functions of u on the right-hand side of the above equation are recognized as the Hermite basis functions. In matrix form, the cubic spline can be expressed as

$$Q_i(u) = [1\ u\ u^2\ u^3]\begin{bmatrix} 1 & 0 & 0 & 0 \\ 0 & 0 & 1 & 0 \\ -3 & 3 & -2 & -1 \\ 2 & -2 & 1 & 1 \end{bmatrix}\begin{bmatrix} P_{i-1} \\ P_i \\ P^1_{i-1} \\ P^1_i \end{bmatrix} \tag{2}$$

$$= U\ C\ P_i$$

For composite cubic spline, the continuity equation for the second parametric derivative vector at the joining point of two curve segments is

$$Q^{(2)}_{i+1}(u_i) = Q^{(2)}_i(u_i) \tag{3}$$

or, after the normalization of the parameter,

$$Q^{(2)}_{i+1}(0) = Q^{(2)}_i(1) \tag{4}$$

or composite Nu-spline, the continuity equation takes the form

$$Q_{i+1}^{(2)}(0) - Q_i^{(2)}(1) = t_i\, Q_{i+1}^{(1)}(0) = t_i\, Q_i^{(1)}(1) \qquad (5)$$

where $Q_i^{(1)}(u)$ and $Q_i^{(2)}(u)$ are the first and second parametric derivatives of the vector function $Q_i(u)$, and t_i is the tension value applied at point P_i. It can be seen from (5) that, when $t_i = 0$, the continuity condition for the composite Nu-spline reduces to that of the composite cubic spline. Equation (5) has the meaning that the vector difference of the second parametric derivative to the right and to the left of the joint is proportional to the first parametric derivative vector at the joint, and the proportional constant is equal to the tension value applied there. This type of continuity is referred to as a geometric second derivative continuity, or a G^2-continuity, which implies that the curvature of a Nu-spline is continuous.

Equations (1) and (5) produce a set of simultaneous equations for the determination of the first parametric derivative vectors at internal joints

$$P_{i-1}^1 + \left(4 + \frac{t_i}{2}\right) P_i^1 + P_{i+1}^1 = 3(P_{i+1} - P_{i-1}), \qquad i = 2, 3, 4, \ldots, n-1 \qquad (6)$$

Different end conditions, such as free ends, clamped ends, and cyclic ends, can be used to specify the two end-point first parametric derivative vectors P_0^1 and P_n^1. Thus, together with the known position vectors P_i, each segment of the composite curve can be plotted according to equation (1). Since each P_i^1 at internal joints, is used by the two joining curve segments at that point, the composite Nu-spline do have first parametric derivative continuity.

Examples of Barsky's Nu-spline are given in Fig. 1. In the first plot are curves fitting through four data points with cyclic end conditions. It can be seen that as the tension values at data points increase, the original visually circular curve with zero tensions is gradually flatened and becomes the visually squarish curve with all tension values equal to a hundred. In the second plot are curves fitting through five data points with free-free end conditions. It can be seen that, when the applied tension values all equal to a hundred, the curve turns around the sharp angles at the internal points with almost straight edges. In the third plot, an L-shaped curve is attempted. With zero tensions, the curve goes through the points in a smooth curvy pattern. As the tensions increased to a thousand, we obtain straight connections where it is needed.

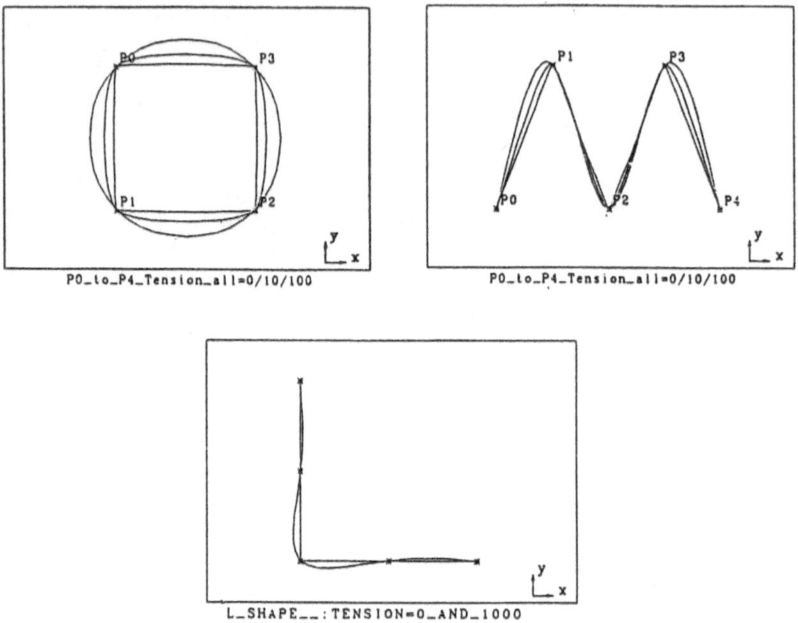

Fig. 1 Examples of Barsky's Nu-Spline

3. THE Nu-SPLINE SURFACES

For the development of the Nu-spline surfaces, we again begin with the formulation of the bicubic parametric surface. The tensor product equation of the bicubic parametric surface is

$$Q_{ij}(u,v) = U\,C\,B_{ij}C^{T}V \tag{7}$$

where $V^T = [1\ v\ v^2\ v^3]$

and

$$B_{ij} = \begin{bmatrix} P_{i-1,j-1} & P_{i-1,j} & P^{01}_{i-1,j-1} & P^{01}_{i-1,j} \\ P_{i,j-1} & P_{i,j} & P^{01}_{i,j-1} & P^{01}_{i,j} \\ P^{10}_{i-1,j-1} & P^{10}_{i-1,j} & P^{11}_{i-1,j-1} & P^{11}_{i-1,j} \\ P^{10}_{i,j-1} & P^{10}_{i,j} & P^{11}_{i,j-1} & P^{11}_{i,j} \end{bmatrix}$$

Referring to the sketch shown in Fig. 2, P_{ij} is the position vector of the top-right corner point of surface patch Q_{ij}, P^{10}_{ij} and P^{01}_{ij} are the first partial parametric derivative vector in u and v-direction, and P^{11}_{ij} is the cross-derivative vector of the point P_{ij}

For the bicubic parametric surface, the continuity

equations for the second partial derivative across the patch boundaries in u and v-direction are

$$Q^{(20)}_{i+1,j}\ (0,v) = Q^{(20)}_{ij}\ (1,v)$$
$$Q^{(02)}_{i,j+1}\ (u,0) = Q^{(02)}_{ij}\ (u,1) \qquad (8)$$

Now, for the bivariate Nu-spline surface, the continuity equations for the second partial derivatives take the form

$$Q^{(20)}_{i+1,j}(0,v) - Q^{(20)}_{ij}(1,v) = \alpha_{ij}(v)Q^{(10)}_{ij}(1,v) \qquad (9a)$$

$$Q^{(02)}_{i,j+1}(u,0) - Q^{(02)}_{ij}(u,1) = \beta_{ij}(u)Q^{(01)}_{ij}(u,1) \qquad (9b)$$

with $\alpha_{ij}(v)$ denoting the u-direction tension across the boundary between patches Q_{ij} and $Q_{i+1,j}$, and $\beta_{ij}(u)$ denoting the v-direction tension across the boundary between patches Q_{ij} and $Q_{i,j+1}$. Note that, in general, α_{ij} are functions of v depending on the u-direction tensions applied at points $P_{i,j-1}$ and P_{ij}, and β_{ij} are functions of u depending on the v-direction tensions applied at points $P_{i-1,j}$ and P_{ij}.

Equation (7) and (9a) produce, after dropping the common factor C^TV,

$$[0026]CB_{ij} + \alpha_{ij}(v)[0123]CB_{ij} = [0020]CB_{i+1,j} \qquad (10)$$

And, equation (7) and (9b) produce, after dropping UC,

$$B_{ij}C^T[0026]^T + \beta_{ij}(u)B_{ij}C^T[0123]^T = B_{i,j+1}C^T[0020]^T \qquad (11)$$

Equation (10) and (11) each contains four scalar equations. They can be summarized as the following four equations, after eliminating the repeatness in the indexing,

$$P^{10}_{i-1,j}+(4+\alpha_{ij}/2)P^{10}_{ij}+P^{10}_{i+1,j} = 3(P_{i+1,j}-P_{i-1,j}) \qquad (12a)$$

$$P^{11}_{i-1,j}+(4+\alpha_{ij}/2)P^{11}_{ij}+P^{11}_{i+1,j} = 3(P^{01}_{i+1,j}-P^{01}_{i-1,j}) \qquad (12b)$$

$$P^{01}_{i,j-1}+(4+\beta_{ij}/2)P^{01}_{ij}+P^{01}_{i,j+1} = 3(P_{i,j+1}-P_{i,j-1}) \qquad (12c)$$

$$P^{11}_{i,j-1}+(4+\beta_{ij}/2)P^{11}_{ij}+P^{11}_{i,j+1} = 3(P^{10}_{i,j+1}-P^{10}_{i,j-1}) \qquad (12d)$$

These equations, together with appropriate boundary edge

conditions, can be used to determine the u-direction and the v-direction first partial parametric derivative vectors and the cross-derivative vectors for all internal points of the composite surface. The only requirement to be met is on the choice of the tension distribution functions $\alpha_{ij}(v)$ and $\beta_{ij}(u)$. It can be seen from equations (12b) and (12d) that, in order to have unique solutions of P_{ij}^{11}, the tension distribution functions must satisfy certain restrictions.

4. MODELS OF MECHANICAL COMPONENTS

For the practical purpose of mechanical component modeling, we make the following restrictions on the use of Nu-spline surface:

(i) All the tension distribution functions, including $\alpha_{ij}(v)$ and $\beta_{ij}(u)$, are constants.

(ii) All the cross-derivative vectors on the modeling surface are zero.

The first restriction condition implies that the u-direction tension along a v-curve on the surface is a constant and the v-direction tension along a u-curve is a constant. The second condition seems to impose a heavy restriction on the modeling and renders the surfaces to have only G^1-continuity. But, considering that most mechanical part surfaces do not have twist and if we carefully place all the surface fitting points along lines of curvature of the surface, the cross-derivative

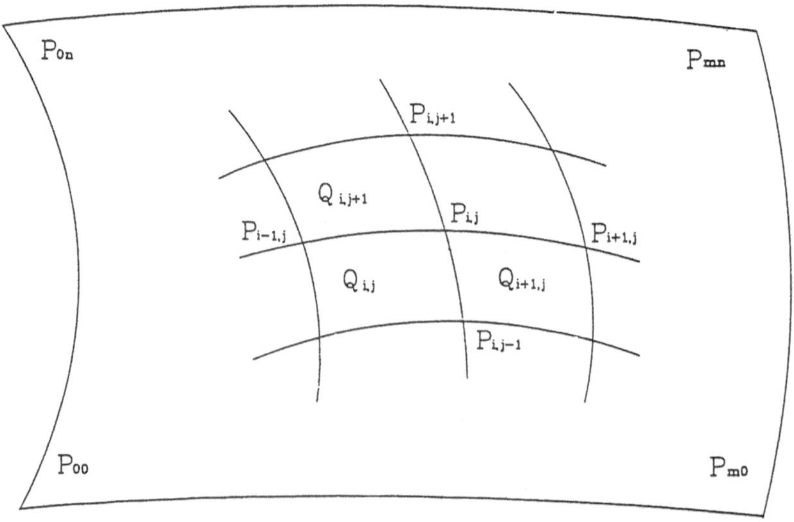

Fig. 2 The Points and Patches on a Nu-spline Surface

vectors will naturally be zero. Hence, the G^2-continuity will be preserved for most of the cases.

Three surface models of mechanical parts are given in the following figures. The object modeled in Fig. 3 is a cam mounted on a shaft. The object in Fig. 4 is a piston of internal combustion engine and that in Fig. 5 is the inner ring of a ball bearing. All models are constructed with tensions applied at points along the u-curves which run parallel to the object axis. Tensions are applied at points where the curves make a right-angle turning. The tension values used are all

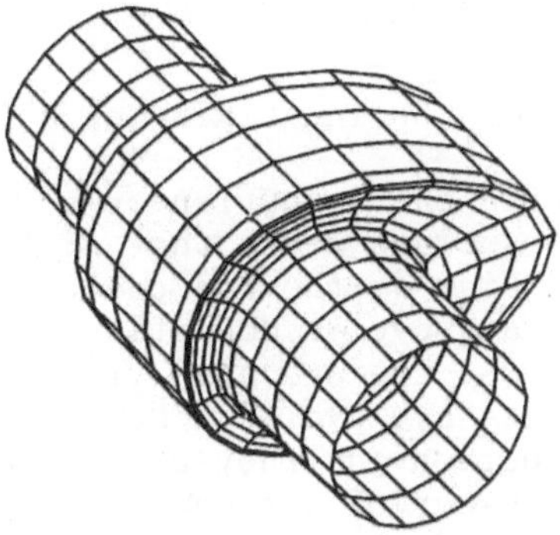

Fig. 3 Model of a Cam Mounted on a Shaft.

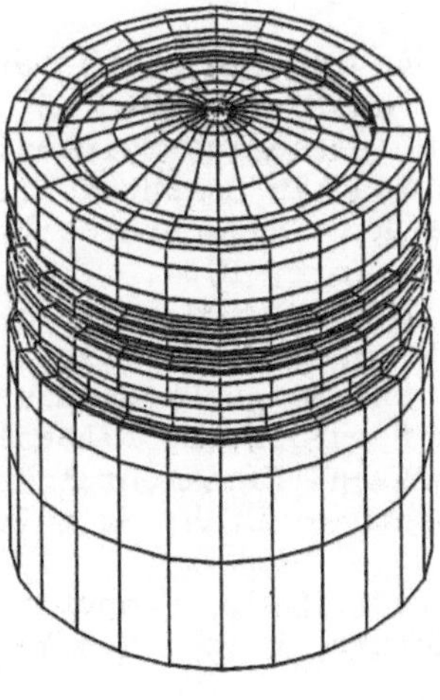

Fig. 4 Model of the Piston of Internal Combustion Engine.

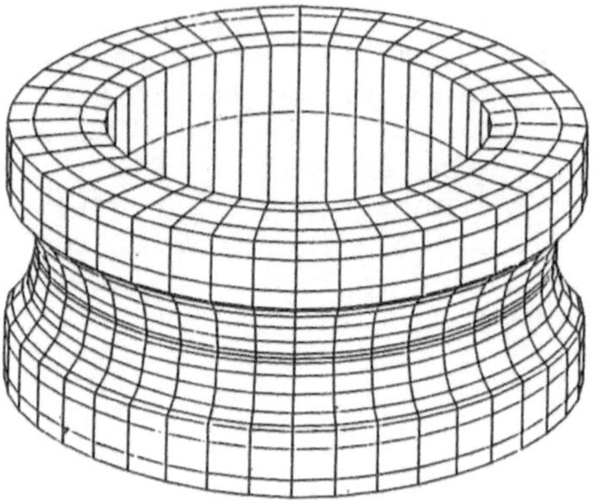

Fig. 5 Model of the Inner Ring of a Ball Bearing.

equal to a hundred with the exception of a fifty at those points where the cam surface meets that of the shaft. For the models in Fig. 3 and Fig. 4, the end conditions for the u-curves are free-free and cyclic conditions for the v-curves. For the model in Fig. 5, the end conditions for both the u-curves and the v-curves are cyclic. Notice that tensions are applied in the u-direction and they are of the same constant value along a single v-curve. It can be seen from the models that all the u-curves and v-curves are lines of curvature on the modeled surfaces. Hence, the two restriction conditions set up earlier in the section are all met.

REFERENCES

1. Schweikert, D.G., An interpolation curve using a spline in tension, J. Math. Phys., 45, 1966, 312-317.
2. Nielson, G.M., Some piecewise polynomial alternatives to splines under tension, in Computer Aided Geometric Design, (R.E. Barnhill and R.F. Riesenfeld, Eds.) pp. 209-235, Academic Press, New York, 1974.
3. Barsky, B.A., Exponential and polynomial methods for applying tension to an interpolating spline curve, Computer Vision, Graphics, and Image Processing, 27, 1984, pp. 1-18.
4. Barsky, B.A., The Beta-Spline: A Local Representation Based on Shape Parameters and Fundamental Geometric Measures, Ph. D. Thesis, University of Utah, Salt Lake City, Utah, December, 1981.
5. Faux, I.D., and Pratt, M.J., Computational Geometry for Design and Manufacture, Wiley, New York, 1979.

DIRECT DESIGN VERSUS RANGE SELECTION ALGORITHMS USED IN MECHANICAL COMPONENT SOFTWARE

Dr J Vogwell

University of Bath, UK

SUMMARY

The conventional approach to selecting catalogue components (such as gears, bearings, springs, etc) tends to be an iterative or direct process when tackled manually. This usually involves making a subjective selection of an item at the design stage, then performing some analysis to confirm suitability in a given application.

Computerising both the analysis and the selection procedure, it has been shown, can lead to significant benefits including a much improved choice of component if a range of solutions is generated. However, the formulation of the design problem may require that a radically different approach to the manual method be adopted and this necessitates using efficient computer codes.

This paper describes some numerical techniques which have been used developing bearing and gear selection software for use by practising mechanical engineering designers. Of particular significance has been the speed of response of such routines and so efficiency of the numerical methods used has been of vital importance.

1. INTRODUCTION

Engineering design is unlike most other scientific disciplines in that there is rarely, if ever, a unique solution to a problem. Indeed, there are usually many alternatives which satisfy a specification and the version which is chosen is often arrived at quite arbitrarily. This is because the design function characteristically involves identifying a solution (usually from a vast range of alternatives) then performing some analysis -either directly or by iteration [Ref 2]. The first suitable solution found, therefore, is unlikely to be the best available.

Also, the design process is such that some degree of originality (perhaps inventiveness) may be required or alternatively involve the selecting suppliers' components from catalogues. This latter aspect of design is becoming increasingly important in the mechanical element design field because considerable advantage can be gained in terms of standardisation and subsequent cost reduction from using readily available components. Consequently, the statement **'if you can buy it - don't design and make it'** [Ref 3] is very true and adopting such a strategy necessitates using an efficient selection and analysis verification procedure.

2. THE NEED FOR AUTOMATIC SELECTION

Manual component selection techniques are frequently cumbersome, time consuming, prone to error and difficult to follow, especially when using an unfamiliar catalogue. It is highly probable that the design iteration loop will be traversed more than once before a suitable solution is obtained, so any form of optimisation is highly unlikely, especially across a range of manufacturers.

Work carried out at Bath University's Mechanical Design Department has shown that analysis selection software with data files containing details of components is best configured to generate a range of all acceptable solutions. This leads to many advantages for the engineering designer in terms of the final selection made and likely benefits include:-

a) an improved solution is identified - especially if an optimisation algorithm is used to sort in order of priority.
b) the selection process is rapidly speeded up.
c) different types of solution are considered (such as bearing types) - which may not otherwise have been thought suitable and so worth trying using a manual method.
d) encourages the use of catalogue components - often the difference in performance between a mathematical optimum design (for example a gear) and a carefully selected catalogue version can be very minor.
e) improved reliability - because of the relative complexity and tediousness of manual selection procedures, a computer aided solution is less prone to human error.
f) improved presentation - having made a selection by computer, details of the component can be available in hard copy form

3. RANGE SELECTION - A PLAIN BEARING EXAMPLE

To illustrate how displaying a range of solutions can aid the designer when selecting a component consider the case of a plain bearing. In a typical application a designer would primarily be concerned with selecting the material type and geometric sizes (such as bore diameter and length) to satisfy an operating load and rotational speed. The principal constraints used for judging suitability would be acceptable pressure stress, peripheral velocity and pressure-velocity product (since heat generation and dissipation is critical).

Conventionally an iterative approach would be taken in which a likely component is chosen then some analysis completed checking functional suitability. The procedure would be repeated if necessary until a satisfactory bearing is obtained. Taking a range approach by computer provides a clearer picture of the situation and so can influence the solution chosen. Figure 1 below displays a typical design range for a bearing application and illustrates the constraint boundaries and also includes a vendors components superimposed.

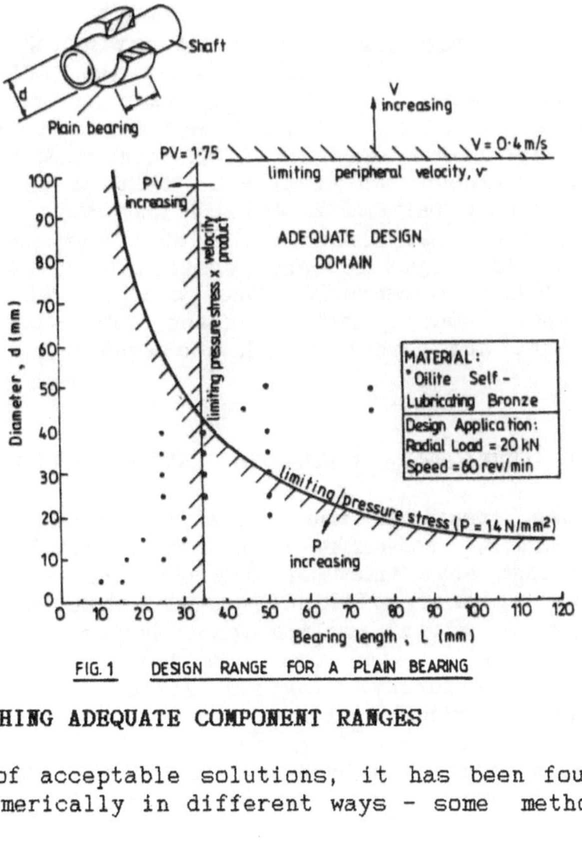

FIG. 1 DESIGN RANGE FOR A PLAIN BEARING

4. ESTABLISHING ADEQUATE COMPONENT RANGES

The range of acceptable solutions, it has been found, can be achieved numerically in different ways - some methods proving more suited to particular component types than others. With some items component details are required for the verifying analysis whereas with others, performance requirements appropriate to an application can be determined initially then a search can be carried out to identify suitable versions.

4.1 Direct Graphical Display. With relatively simple components, such as plain bearings, the 'Adequate Design Domain' [Ref 4] can be obtained if constraint boundaries can be described mathematically and then displayed graphically. If there are just two variables the catalogue versions can be shown directly as shown in Figure 1. With plain bearings the constraint boundaries are dictated by the bearing material as well as operating conditions.

4.2 Analysing Each Version. Frequently the contraint boundaries are displayed in empirical chart form and so do not enable use of convenient curve fitting methods. Also, there may be many variables and a two-dimensional graphics display is impractical. In such circumstances it is necessary to consider, in turn, the suitability of every alternative and this requires a considerable amount of computation time.

4.3 Direct Search Techniques. When performance requirements can be established prior to considering the suitability of individual components, a search can then be carried out to identify adequate ones. Different search techniques can be used to suit the characteristics of the component concerned and the information known. An example is that of rolling element bearings in which a particular dynamic load rating is necessary to fulfill desired load, life and reliability requirements. The magnitude of the dynamic and static load ratings being determined before commencing the search. Also, limiting operating speed and geometric sizes influence acceptable solution ranges and direct search techniques can likewise be applied.

5. SEARCHING TECHNIQUES - A ROLLING BEARING EXAMPLE

In catalogues, bearings are normally listed in ascending geometrical order, increasing with bore diameter. Fortunately the load ratings also increase with size and a direct search technique is the most efficient way of establishing suitable bearings for a given application. Having identified the transitional bearing (the smallest bearing at which load ratings are satisfactory), so all larger bearings will be suitable from load considerations.

Various search techniques have been tried for identifying the transitional bearing and perhaps the most obvious is the widely used binary method [Ref 5 and others]. However, as speed of response is a critical factor in choosing a search technique, a more efficient method (in terms of bearings retrieved) uses the principle of convergent linear interpolation. A comparison of a search of 400 bearings using the binary method would require that 10 bearings would need to be sought whereas using linear interpolation this is reduced to just 5 as illustated in Figure 2. This is a significant time saving when considering the number of such searches which may be required.

FIG. 2 COMPARISON OF SEARCH TECHNIQUE EFFICIENCY

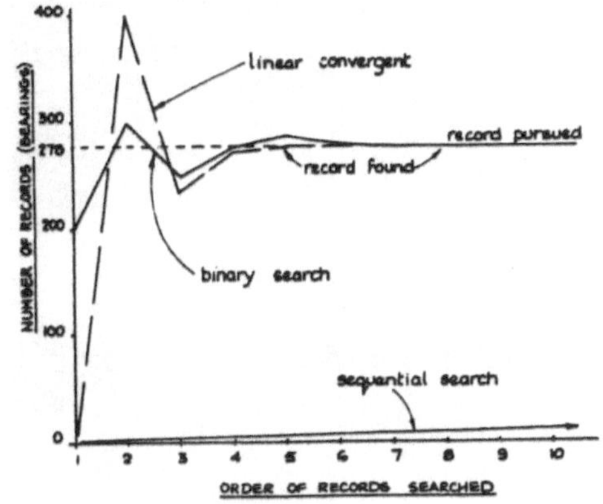

5.1 Convergent-Linear Interpolation Search

This method requires that both first and last bearings are searched - although even this is unnecessary as the salient information can be kept in current memory (as it is frequently used) thus making the technique even more efficient! A linear relationship is assumed to exist between load rating and bearing as shown in Figure 3. The next bearing to be sought is that one whose record is proportionally displaced from the end boundaries as the bearing rating is displaced from the boundary values. Using knowledge of its actual load rating and using one of the previous boundary values the procedure can be repeated to determine a new estimated bearing record to search. Repeating the procedure will quickly converge upon the pursued bearing as shown in Figure 3.

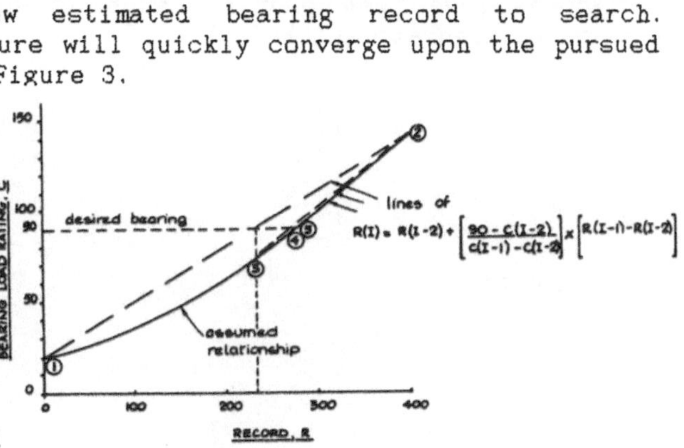

FIG. 3 SEQUENCE OF RECORDS SEARCHED USING LINEAR CONVERGENT TECHNIQUE

6. HANDLING ASSEMBLIES

When a satisfactory assembly of catalogue components is required, the number of possible combinations is greatly increased compared to considering items separately. Consequently, the technique used for identifying solutions becomes even more critical if the speed of response is to be acceptable. Selecting mating gear tooth numbers to produce a desired reduction ratio for a multiple reduction gearbox arrangement, as shown in Figure 4, is a typical example.

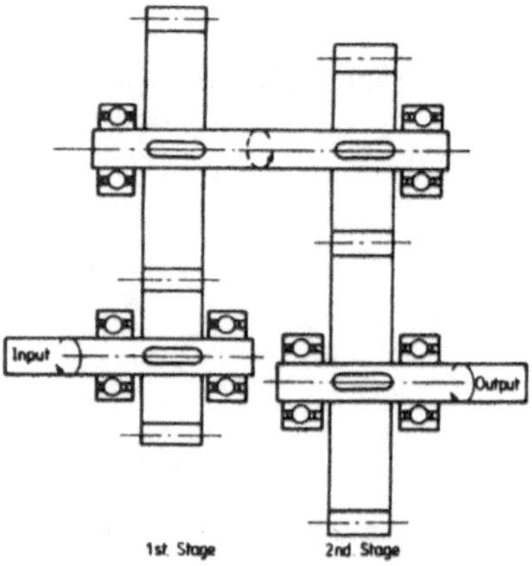

FIG. 4 A TWO-STAGE GEAR REDUCTION UNIT

6.1 Establishing Tooth Numbers in Gear Sets

Various direct design methods exist for establishing the optimum distribution of reductions. Some advocate an equal sharing of the reduction between stages such that :-

$$R_1 = R_2 = \sqrt{R} \qquad \text{---------------(1)}$$

where R is the total reduction and R_1 & R_2 are the first and second reductions respectively. Alternatively, empirical breakdowns have been proposed such as solving the following simultaneous equations [Ref 1]:-

$$R_n * R_{n-1} * R_{n-2} * \ldots * R_1 = R \qquad \text{----------(2)}$$

$$R_n^n + R_n^{n-1} + R_n^{n-2} + \ldots + R_n = R \qquad \text{----------(3)}$$

For a double reduction gearbox these would simplify to :-

$$R_1 * R_2 = R \qquad \text{---------------(4)}$$

$$R_2 + R_2^2 = R \qquad \text{---------------(5)}$$

For most target reductions either of these approaches would lead to individual, non-integer ratios which could not be obtained using standard gears - and most probably not exactly anyway since gears must have an integer number of teeth. Any errors may be cumulative thus possibly magnifying the overall inaccuracy of the reduction.
Seeking a target reduction of 30, for example, using the simultaneous equation method leads to first and second stage reductions of 6 and 5 respectively. Most target reductions, however, will not result in integer sub-ratios and an equal sharing of the reduction clearly does not in this example.

TASK - To select standard gears to give a double stage reduction of 30 : 1

Typical Table of Manufacturers Standard Gears

No. of teeth	12	13	14	15	16	18	20	22	24	25	28	30	32
No. of teeth	35	36	40	48	50	54	60	64	70	72	80	84	-

	Combinational	Sub-system Isolation	Sub-optimisation
Target Ratio per Stage	$R_1 \times R_2 = 30$	$R_1 = R_2 = \sqrt{30}$	$R_1 = \sqrt{30}$ $R_2 = 30/R_1$ actual
Optimised Selection	72:12 60:12	72 : 13 72 : 13	72 : 13 70 : 13
Actual Reduction Ratio	30 : 1	30·67 : 1	29·82 : 1
Permutations Computed	25^4 (= 390625)	25^2 (= 625)	$25^2 \times 2$ (= 1250)

FIG. 5 COMPARISON OF VARIOUS OPTIMISATION TECHNIQUES

Having determined target sub-ratios the next objective is to select gear tooth numbers such that ratios are best achieved. Various numerical means of doing this are possible with differing advantages as illustrated in Figure 5.

6.1.1 Combinational - if the distribution of reductions between stages is not critical and the overall optimum arrangement is sought, then it is necessary to consider every combination. Although the optimum solution(s) will be identified using this approach the computing task can be enormous as indicated by the simple example given in Figure 5. This is based upon a modest range of components (just 25 standard gears) and will be much greater for more components and larger multiple gearsets.

6.1.2 Sub-System Isolation - if individual target sub-ratios are established (by whatever means) and treated independantly in a combinational manner the computing task is much reduced. However, the resulting accuracy may suffer as shown in Figure 5 for the simple example given.

6.1.3 Recursive Sub-Optimisation - a compromise between accuracy of identified solution and computing time is accomplished when second and subsequent gear set ratios are recalculated as a consequence of actually achieved previous ratios. Should a tolerance on target ratio be acceptable (as is usual) this method can be formulated to display a range of solutions thus identifing the optimum yet at an efficient usage of computer time.

7. CONCLUSIONS

The overall aim of the work completed (and ideas described in this paper) has been to give an insight into the problems associated with the computer-aided selection of catalogue components and discuss how they are overcome numerically. Taking a range approach has been a vital strategy in the formulation of the software codes used in this work.

Computerising the selection of catalogue components is undoubtedly a much ignored area. It seems quite absurd that having used sophisticated CAD equipment on expensive computer systems, vital components are selected on a largely ad-hoc basis by manually searching through catalogues.

Efficient numerical methods have a crucial role to play in the computerising process. It is essential that software can manipulate the vast amounts of data and handle the necessary analyses at response rates which are acceptable to users - that is the practising engineering designer.

8. REFERENCES

Paper

1 Savage, M Coy J.J. and Townsend D.P. - Optimal Tooth Numbers for Compact Spur Gear Sets, Journal of Mechanical Design, Transactions of the ASME, Volume 104, 1982

Book

2 Besant, C and Lui,C.W.K. - Computer Aided Design and Manufacture, Ellis-Horwood Int. 1985

3 Deutschman,Michels and Wilson,- Machine Design (Theory and Practise), Collier Macmillan

4 Johnson, R.C. - The Optimum Design of Mechanical Elements, Wiley 1981

5 Martin J. - Principles of Database Organisation, Prentice Hall 1982

ANALOG/HYBRID AND DIGITAL SIMULATUIONS IN CIVIL ENGINEERING

HAMDY YOUSSEF

Concordia University
Department of Civil Engineering
1455 De Maisounneuve Blvd. West
Montreal Quebec H3G 1M8 CANADA

1. ABSTRACT

The objective of the present paper is to illustrate the utilization of Analog/Hybrid/Digital Simulation for solving engineering physical problems with special emphasis on Civil Engineering applications. In this respect the simulation solution of the viscou damping non-linear differential equation is presented utilizing the facilities at Concordia University, Montreal, Canada. The differential equation describe the water/earth dam interaction. Both the Simulation language [MIMIC-developed by the USA Air Force] and the Control Data Analog Computer EAI 680 were utilized for the solution. Utilization of the MIMIC Simulation language and the Analog computer present advantages for solving the differential equations which describe most of the engineering phenomena.

2. INTRODUCTION

Many fundamental laws of nature are conveniently expressed mathematically by differential equations. It is thus important for scientists and engineers to understand and solve these equation Proficiency in formulating differential equations, given a physical solution, requires a through knowledge of the appropriate field, mathematical skill and common sense. Once the differential equation is obtained the next step is to solve. Analytical methods are mostly available for linear differential equations with constant coefficients. For non-linear differential equations which have varying coefficients it is necessary to use computers.

There are mainly two types of computers, namely, digital and analog computers, the hybrid system provides a linkage between the two types. Digital computer is based in converging the problem variables to binary numerical system, and the solutions are obtaine in discrete numerical approximate forms utilizing finite element or finite difference methods. Analog computers are based on utilizing physical electronic componenents which permits parallel and continuous solutions. Both computers have some advantages and some limitations of their utilizations [Youssef 1976a]. For more complex problems, for example for design and implementation of electrical power system load prediction model [Youssef 1975b] both computer are utilized having the hybrid system with Analog/ Digital and Digital/Analog convertors to transmite the intermediate results to obtain the optimum design. Due to the space limitations for the present paper, only an example of solution of one Civil Engineering problem is summarized. The solution is given for the motion of the interface between the water face in a reservoir and the water motion inside a porous media [water/earth dam interaction]

3. PHYSICAL CIVIL ENGINEERING SYSTEM

Porous structures, such as rockfill dams and rubblemound breakwaters as well as embankments are frequently subjected to wave motion. When water waves interface with a rockfill earth dam, the resultant unsteady, internal flow is often one of the non-Darcy flow, Dracos [1969]. Thus the governing equation is non-linear, and the solution becomes more difficult than Darcy flow. The water earth dam interaction is simulated by a sine-wave motion Oscillating along a mean water level with amplitude [A_o] as shown in Fig. [1]. The interface between the water motion in the reservoir and the phreatic line within the porous media is described by the movement of the outcrop point. In case that the inside and outside water levels are not coincide a challenging problem arises at the interface boundary. Figure [1] presents the movements of the free water level in the reservoir and the outcrop point at the interface with the porous media, by the dotted and solid lines with elevations of [Y_f and Y_c] respectively. The two wave motions, coincide in the first phase up to a time $t=t_1$, then diverge with approximate straight line relationship up to a time $t=t_2$ where the maximum deviation [ΔY_p] is achieved, then the two curves descriping the water motion converge again and the amplitude [$\Delta Y \to 0$] where the two curves coincide again in the rising phase with maximum amplitude of [A_o]. The sine wave motion of the impact wave is described by the following equation [Dracous 1969]:

$$Y_f = h_o + A_o \sin \omega t \qquad [1]$$

where h_o is the mean water level, A_o is the maximum amplitude and ω is the frequency of Oscillation [$\omega = 2\,\pi/T$] and [T] is the period of motion. [t] is the time. By differentiating equation [1]:

$$\dot{Y}_f = V_f = A_o \cos \omega t \qquad [1.1]$$

is the velocity of motion of the free water surface.

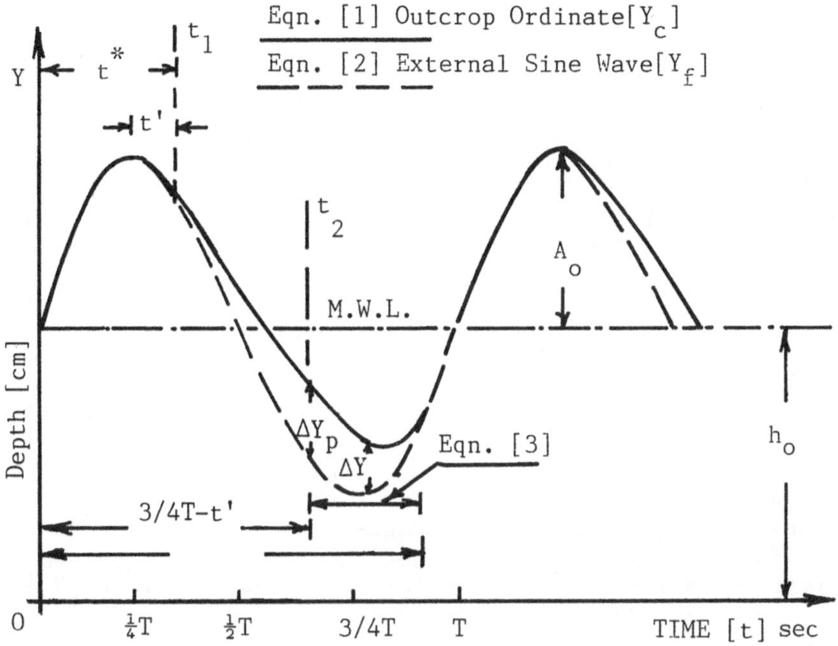

Fig. [1]. Movements of the Free Water Wave And The OutCrop Point[3]

The motion of the outcrop point are described by [3]:

For the slow drop case:

$$Y_c = h_o + A_o \sin [\omega t\ [T/4 + t']] - V_* [t-[T/4 + t']] \qquad [2]$$

where the maximum internal fall velocity $[V_*]$ is given by:

$$V_* = - [a/[2bm]] + \tfrac{1}{2}[\ [[a/[bm]]^2 + 4[[\sin^2\theta]/[m^2 b]]]^{\frac{1}{2}} \qquad [2\text{-}1]$$

where [a and b] are the Darcy and non Darcy's terms in Forchheimer's [5] equation, [m] is the prosity, $[\theta]$ angle of inclination of the porous structure with the horizontal line, and $[t']$ is the time during which the external and internal movements are identical for the start of the falling phase. The fast drop case is described by the following non-linear viscous damping differential equation:

$$\ddot{Y}_c = [g \sin^2\theta/\Delta Y_p][h_o + A_c \sin [\omega t - Y_c] - gm\ [a+mb|\dot{Y}_c|\ ]\ \dot{Y}_c \qquad [3]$$

where $[A_c]$ is the amplitude of the outcrop point movement from the mean-water level, $[Y_c]$ is the elevation of the outcrop point movement from the mean-water level line, $[Y_c]$ is the elevation of the outcrop point motion $[Y_c = A_c + h_o]$, $[\ \dot{Y}_c\ ]$ is the derivative of Y_c and [g] is the gravitational acceleration.

Equation [3] satisfies the following initial conditions at the time [t]:

$t_i = 3T/4 - t'$ [4]

and

$Y_{ci} = h_o + A_o \sin\omega t^* - V_* [T/2 - 2t']$ [5]

where $[t^*]$ is the time during which the external and internal movements are identical [Fig. 2], where:

$t' = t^* - T/4$ [6]

$t^* = T/[2\pi] \cos^{-1} [-W\sin^2\theta]$ [7]

$W = K_* T/[2\pi\ m\ A_o]$=Dimensional Coefficient [8]

where:

$K_* = 1/[a + bm|V_*|]$ [9]

K_* is defining the conductivity coefficient for Darcy flow. The following two equations are defining the two cases of motions:

[1] For fast drop case:

$W\sin^2\theta < 1$ [10]

[2] For slow drop case:

$W\sin^2 \geqq 1$ [11]

Referring to figure [1]; the motion of the outcrop point is described by the following four stages:

[I] for $W\sin^2 \geqq 1$, t^* either does not exist or it is equal to [T/2] this represents a slow drop case for which the elevation $[Y_c]$ of the outcrop point can be calculated from Eqn. [1].

[II] For $W\sin^2 < 1$, the movement of the outcrop point undergoes various phases as follows:

[i] $o \leqq t \leqq t^*$; $[1/W]\cos \omega t > -\sin^2\theta$ [12]

From the time 0 to [T/4] there is a rising phase, followed by a slow drop phase during the interval of time indicated by [t']. For this stage, the movement of the outcrop point is governed by Equation [1].

[ii] $t^* < t \leqq [3T/4 - t']$; $[1/W] \cos\omega\ t > -\sin^2\theta$ [13]

This indicates a fast drop stage; therefore, the outcrop point fal at its maximum rate $[V_*]$ and Equation [2] is applied.

[iii] $[3T/4 - t'] < t \leqq t_c$

At $t = 3T/4 - t'$, the outcrop point curve will have reached its peak deviation $[\Delta Y_p]$ from the free water level curve; thereafter the curve starts to converge until the differences are diminishes, this indicates a deceleration motion, for the outcrop point. Assuming that the deceleration varies linearly with $[\Delta Y]$ and referring to Fig. [1] the driving force, per unit mass, can be expressed by Eqn. [3] which solved for $[Y_c]$ in terms of the values at the two previous time stages. The time $[t_c]$ is the time at which the inside and ouside water surfaces rejoin, and marks the upper limits for the damping face.

[IV] $t_c \leqq t \leqq T$

This stage represents another rising phase during which the movement of the outcrop point is again described by Eqn. [1].

4. UTILIZATION OF THE DIGITAL/ANALOG SIMULATION LANGUAGE [MIMIC]

MIMIC is a digital computer language and utilized to solve systems of ordinary differential equations. MIMIC uses a Control Data 6000 Series computer with a minimum memory size of 32K. MIMIC is an extension of the design concept of MIDAS [Modified Integration Digital Analog Simulator], developed by the US Air Force. MIMIC is more powerfull, versatile, and efficient than its predecessor [MIDAS] it retains also, programming simplicity. MIMIC provides the engineers with a simple and effective means of communicating with a digital computer, and it is flexible enough to appeal to the professional analog or digital programmers [8].

Working from the mathematical description of the problem as previously described, MIMIC statments was written to describe the system. MIMIC is frequently referred to as a digital/analog simulator or a continuous system simulator. These lables are utilized because they are descriptive of the most important feature of a program written in MIMIC language [Parallelism]. MIMIC is a parallel language since the statments defining a program can be written in any order. This parallism is in contradiction to the more general programming language, as FORTRAN, where the essence of the language is the order in which the statments are executed. Since, MIMIC does not necessitate the ordering of the statments; it is not possible to make an error by writting them in the wrong order. Another advantages, which is less obvious but equally important, is that MIMIC program can be made to resemble closely the original problem statment. The parallism of MIMIC is achieved by a sorting algorithm in the MIMIC program [8].

A variable step integration routine is used to perform all integrations. The step size is automatically varied by the requirements of the problem to ensure that the integration errors does not exceed a given bound at any one time. This feature generally, but not always, relieves the user from any further concern about selecting integration parameters. The MIMIC program for the above mentioned mathematical models descriping the water/earth dam interaction is written considering [a,b and m] as constant and [θ, h_o, A_o and T as parameters]. The following intermediate results were obtained from the [MIMIC] solution: $V_*=14.40$, $W\sin^2\theta=0.3944$, $t'=0.072$, $\dot{Y}_{ci}=-13.26$, $K_*=8.937$, $t^*=0.447$, $\Delta Y_p=11.613$, $Y_{ci}=39.44$, $k_1=[g\sin^2/\Delta Y_p]=42.23$, $\omega=[2/T]=4.19$, $K_3=gm=330.60$, $K_3K_4=2.341$, $aK_3=3.31$, $K_4=mb=0.0071$, $A_oK_1=449.75$, $h_oK_1=1604.74$. These obtained intermediate results enabled writting the equations descriping the wave motion of the outcrop point [Equations 1,2 and 3] in the form:

$$Y_f = 38 + 10.65 \sin [4.19t] \quad [1']$$
$$Y_c^f = 54.61 - 14.4t \quad [2']$$

$\ddot{Y}_c$= 1604.74 + 449.75 sin 4.19t - 42.23 Y_c -3.31 $\dot{Y}$ -2.341|$\dot{Y}$|Y [3']
Furthermore, from Eqn. [2] we express the constant term as k_5, since $k_5 = h_o + A_o$ sin[[T/4+t']] + V_* [T/4 +t'] And K_5=54.61

5. SOLUTION UTILIZING ANALOG COMPUTER

Analog computer as the name implies uses physical models equivelent to the given equation and the model is subjected to the inputs in the given physical system. For instance an electrical analog computer can use a voltage to represent the elevation of the outcrop point [Y_c] another voltage to represents the velocity [$\dot{Y}_c$], and time to represent the independent variable [t] in Eqauation [3]. The solution is obtained as a continuously varying voltage which is a function of time and may be displayed on an Oscilloscope or strip chart recorder as well as X-Y plotter. The analog computer at Concordia University [EAI680] were utilized by the author for solving the non-linear viscous damping differintial equation descriping the stage of movement of the outcrop point. The basic idea used in programming or setting up an analog computer to solve a differential equation is based on utilization of electronic circuits. Each unit of circuits is set-up to perform a mathematical function. Connections on voltages fed to each of them and yield appropriate output voltages. For instance, if a Voltage [V(t)] is fed to an integrator unit the output voltage would be ∫V(t)dt. There are other units which add voltage, divide one input by another and so on. These units are properly interconnected to solve differential equations.

To ensure a one to one correspondance between the variables of the original physical system and the variables of the analog computer. For instance, in simulating the motion of the outcrop point at the interface of the free water board and the rockfill dams. One to one correspondance may be maintained between the force per unit mass, $\ddot{Y}$ and the input voltage, the constant [gam/[2bm]] and the setting voltage to the potentiometer, and similary for all the parameters in Eqn. [3]. It is thus easy to change the values of θ, h_o, A_o, T, a,m, ...etc. in the analog set-up and observe almost instantly their effect on the motion of the outcrop point. This flexibility allows considerable innovations in design at small costs. The use of the analog computer in solving the physical engineering problems requires scaling of the equations variables amplitudes to adjust for the computer voltages, the computer time is real time [when β=1 is utilized] to speed or slow up obtaining the results, time scalling is required. Then following the amplitude and time scalling the program [patch pannel] is checked by arbitrary variables on the potentiometers and following the circuit to check the theoretical outputs of the amplifiers, these values is then checked against the outputs of the measured ones on the analog computer, this process called [Static check]. Figure [2] presents the overall

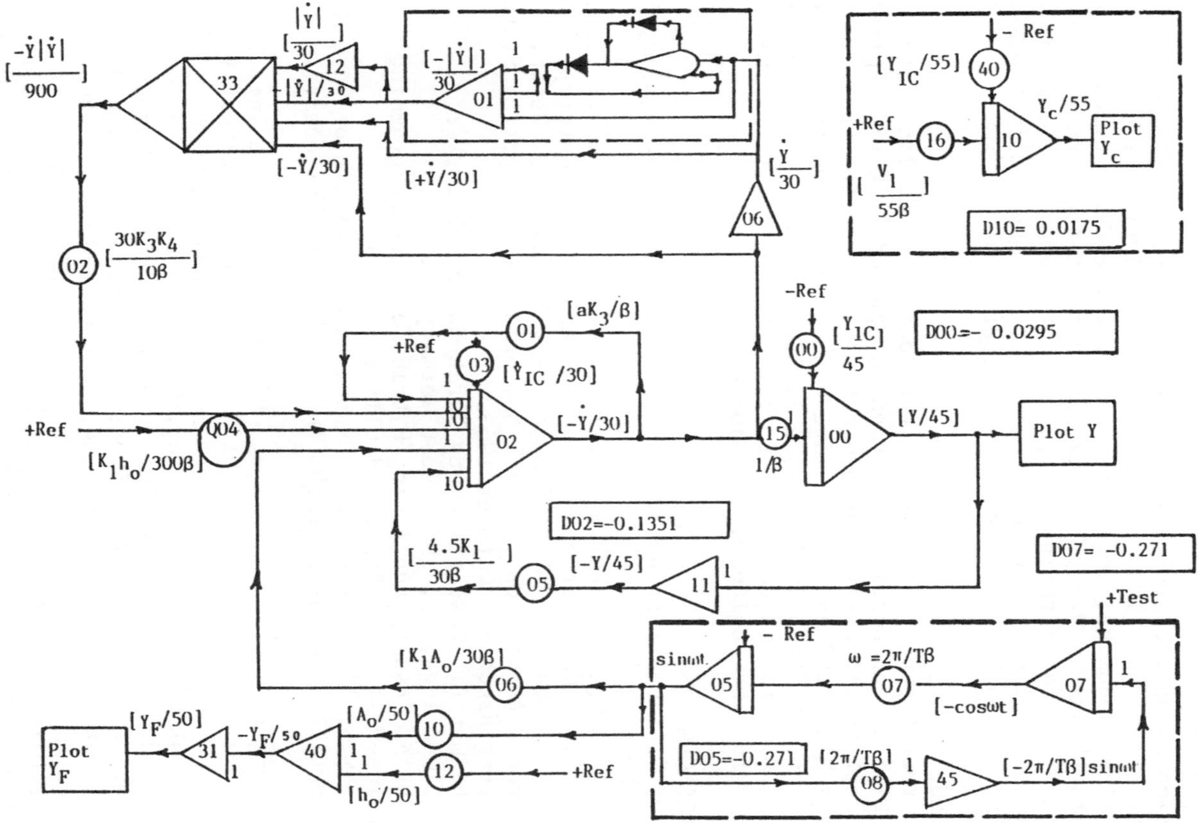

Fig. [2] CIRCUIT DIAGRAM WITH STATIC CHECK VALUES FOR WAVE MOTION SIMULATION [9]

analog computer circuits with static check values [9].for solving the three differential equations describing the wave motion of the outcrop point [3]. Detailed description of the solution is given by the author [9] and the concept of the analog/Digital/Hybrid simulation as applied to Civil Engineering Physical phenomena is given by the author [10]. For optimization solution of more complex engineering phenomena, the author utilized the three types of computers [Youssef 1976]. The simulation techniques are of primary importance for facilitating and providing economized solutions for most of the engineering phenomenon in nature, especially those described by non-linear differential equations.

6. CONCLUSIONS

The present paper illustrates the concept of utilizing the Analog/Digital and Hybrid simulation as applied in Civil Engineering The concept is illustrated by presenting solution for specific examp utilizing MIMIC and Analog Simulation techniques. The theme of the paper is to motivate Civil Engineering Scients and Researchers for utilization of the three types of computers to achieve ultimate solutions for the engineering phenomenon.

7. REFERENCES

1. Beckey G.A. and Karplus W.J. 'Hybrid Computation'John Wiley & Sons, New York, N.Y.
2. Chow V.T. 'Open Channel Hydraulics',McGraw-Hill Book Co.,USA.
3. Dracous T. ' Calculations of The Movement of The Outcrop Point' Proc. XIII Congress of IAHR,Paper D-2, Kyoto, 1969, Japan.
4. Electronic Associates Inc. EAI 680 Scientific Computing System, Reference Handbook, October 1968.
5. Forchheimer P. 'Wasserbewegung Durch Boden', Zeit Ver. Deutchland Ing., 45, 1901, West Germany.
6. Hausner A. 'Analog and Analog/Hybrid Computer Programming', Prentice-Hall Inc., Englewood Cliffs, N.J. 1971.
7. Nasser M.S. 'Theoretical and Experimental Analysis of Wave Motion In Rockfill Structures' Ph.D. Thesis, Windsor University, 1974, Canada.
8. MIMIC Digital Simul. Language, CDC600, Control Data Inc. 1973.
9. Youssef H. 'Analog And Digital Simulation of Non-Linear Viscous Damping Equation of Wave Motion'Research Report,December 1975, Department of Civil Engineering, Concordia University, Canada.
10. Youssef H. 'Analog/Hybrid/Full Hybrid and Digital Computers Simulation' February 27,1976, Seminar Notes, Concordia University Department of Civil Engineering, Montreal, Quebec, Canada.
11. Youssef H. 'Design and Implementation of Electrical Power System Load Prediction Model' Research Report [pp. 70],April 22,1976, Concordia University, Department of Civil Engineering, Montreal, Quebec, Canada.
12. Weyrick R.C. 'Fundamentals of Analog Computers' Printice-Hall Inc., Engelwood Cliffs,N.J. 1969.

CURVE DESIGN USING HIERARCHICAL FINITE ELEMENT FORMS

S. Virtanen
Tampere University of Technology
P.O. Box 527, SF-33101 Tampere, FINLAND

SUMMARY

Hierarchical elements are used in curve design problem. Three different p-refinement versions are compared using error in curve length and error in function value as error measure. A circular arc is used as a test example. The most promising way to add more degrees of freedom seems to be to increase the number of nodal parameters in the nodes where no boundary conditions are given.

1. INTRODUCTION

The creation of different kind of curves plays an important role in computer aided design. In existing CAD-systems only pointwise coordinate information is used. In many applications, however, there are also other types of geometrical requirements. For example the enclosed area and/or the first and/or second moment of the enclosed area with respect to the given line is fixed in earlier stages of design process.

When using the present interpolation algorithims of the existing CAD-systems the above mentioned requirements are achieved only by a length trial and error procedure. But when the curve creation task is formulated as a constrained minimisation problem, the requirements are fullfilled automatically. This minimisation problem is solved by FEM. Examples of this kind of approach are given in references [1-4].

The hierarchical finite element forms are used nowadays in many linear problems because of their advantages in equation conditioning, estimating the errors existing at any given stage of the subdivision and calculating error indicators if successive refinement is sougth. But in non-linear problems only a few results have been obtained.

In this work we use p-adaptive hierarchical line elements in the curve design problem, which is formulated as a constrained minimisation problem. The functional under minimisation is the length of the curve

$$L = \int_a^b \sqrt{1 + (y_{,x})^2}\, dx \tag{1}$$

and the constraints are the area between the curve and x-axis and/or the first and second moment of the area with respect to y-axis. Because the functional is not quadratic, this leads to non-linear equations which are solved iteratively.

2. FINITE ELEMENT DISCRETISATION OF THE PROBLEM

For generality and for avoiding problems with infinite slopes, the curve is described parametrically. Thus the FEM representation for the curve in x,y-plane is

$$x = \underset{\sim}{N}\, \underset{\sim}{\hat{x}} \equiv \sum_j N_j\, \hat{x}_j \tag{2}$$

$$y = \underset{\sim}{N}\, \underset{\sim}{\hat{y}} \equiv \sum_j N_j\, \hat{y}_j \tag{3}$$

where $\underset{\sim}{N}$ is the shape function matrix and $\underset{\sim}{\hat{x}}$ and $\underset{\sim}{\hat{y}}$ are lists of nodal parameters. Shape functions N_j are in each element functions of parameter ξ, $\xi \in [-1,1]$, and the derivatives in the nodal parameter lists are w.r.t. parameter ξ. The area enclosed by the curve $y = y(x)$ and the x-axis in the interval $[a,b]$ is

$$A = \int_a^b y\, dx \tag{4}$$

when equations (2) and (3) are substituted into (4) it becomes a linear equation

$$\underset{\sim}{G}\, \underset{\sim}{\hat{y}} = A \quad , \tag{5}$$

when equation (3) is substituted into equation (1) and the integration is thought to have been carried out, the length L is controlled by $\underset{\sim}{\hat{y}}$, which now contains the only unknowns, formally

$$L = L(\underset{\sim}{\hat{y}}) \quad . \tag{6}$$

We now have to find $\underset{\sim}{\hat{y}}$ which minimises (6) and satisfies (5) plus possible pointwise constraints. In using the Lagrangian procedure we first write a modified function

$$L^* = L(\underset{\sim}{\hat{y}}) + \lambda(\underset{\sim}{G}\,\underset{\sim}{\hat{y}} - A) \quad . \tag{7}$$

Demanding L^* to be stationary w.r.t. $\underset{\sim}{\hat{y}}$ and λ the equations obtained are

$$\frac{\partial L}{\partial \underset{\sim}{\hat{y}}} + \lambda\, \underset{\sim}{G}^T = \underset{\sim}{0} \tag{8}$$

$$\underset{\sim}{G}\hat{\underset{\sim}{y}} - A = 0 \quad . \tag{9}$$

From equation (1) we see that the typical element of the vector $\partial L/\partial\hat{\underset{\sim}{y}}$ is (before going into parametric from)

$$\frac{\partial L}{\partial \hat{y}_i} = \int_a^b \frac{1}{\sqrt{1+y^2_{,x}}} \; y_{,x} \frac{\partial y_{,x}}{\partial \hat{y}_i} \; dx \quad . \tag{10}$$

From equations (3) and (10) we see that if the square-root term would be nonexistent, right hand side of (10) would be linear in $\hat{y}_j$. Because other terms in equations (8) and (9) are linear in $\hat{y}_j$, we can try to find the solution iteratively by approximating during the p:th iteration cycle the denominator by the values of the previous cycle. Hence we obtain for the p:th iteration cycle

$$\frac{\partial L^{(p)}}{\partial \hat{y}_i} = \sum_j K_{ij}^{(p)} \; \hat{y}_j^{(p)} \tag{11}$$

where the typical element contribution is

$$K_{ij}^{e(p)} = \int_{-1}^{1} \frac{1}{\sqrt{1+(y_{,x}^{(p-1)})^2}} \quad N_{i,x} \, N_{j,x} \, x_{,\xi} \; dt \tag{12}$$

The derivative $y_{,x}^{(p-1)}$ is calculated from the results of the previous cycle.

Hence, during iteration cycle p we have to solve linear system of equations (vide equations (8) and (9))

$$\begin{bmatrix} \underset{\sim}{K}^{(p)} & \underset{\sim}{G}^T \\ \underset{\sim}{G} & 0 \end{bmatrix} \begin{bmatrix} \underset{\sim}{y}^{(p)} \\ \lambda^{(p)} \end{bmatrix} = \begin{bmatrix} \underset{\sim}{0} \\ A \end{bmatrix} \quad . \tag{13}$$

However, before solution the given values for elements of $\hat{\underset{\sim}{y}}$ must be inserted. Convergence of the iteration is followed by calculating after each cycle the current value for the length L. Iteration is stopped when the improvement (relative shortening) is under preset limit.

3. HIERARCHICAL ELEMENTS

Because for curve design the continuity of the function itself and its first derivative is required the element type used is one-dimensional Hermitian isoparametric element with two nodes. The nodal parameter vector for a typical element e is

$$\hat{\underset{\sim}{x}}^e = [\hat{x}_1 \; (\hat{x}_{,\xi})_1 \quad \hat{x}_2 \; (\hat{x}_{,\xi})_2]^T \tag{14}$$

and an analogous expresson for $\hat{\underset{\sim}{y}}^e$. The interpolating functions
are the first order Hermitian polynomials H^1_{ij} ($i=0,1$; $j=1,2$).

The hierarchical, additional degrees of freedom are introduced in two alternative ways. One way is to use Legendre-type
surplus functions S_i which have zero values and slopes at nodes,
i.e. same as Delpak and Peshkam used in reference [5]. The
surplus function of the i:th order is defined as

$$S_i = \frac{d^i}{d\xi^i} (\xi^2-1)^{i+2} \tag{15}$$

These new degrees of freedom can be condensed easily at the element level and the size of the matrices in e.g. (13) do not
increase. The other way to make hierarchical elements is to use
more nodal parameters. These new parameters are no longer values
of the function or its derivatives at nodal points but are the
departures of the new solution due to refinement. For example
if we take the first hierarchical function

$$N_5 = \frac{1}{16} (1-\xi)^3 (1+\xi)^2 \tag{16}$$

then the corresponding nodal parameter is

$$a_5 = (\hat{x}_{,\xi\xi})_1 + \frac{3}{2} (\hat{x}_1-\hat{x}_2) + 2 (\hat{x}_{,\xi})_1 + (\hat{x}_{,\xi})_2 \tag{17}$$

Usually we do not need this kind of equations between parameters
a_i and derivatives $\hat{x}_{,\xi}$, $\hat{x}_{,\xi\xi}$ etc. because we usually do not give
boundary conditions to the higher derivatives.

One drawback of this kind of elements, where we increase the
number of nodal parameters, is that these new degrees of freedom are shared by two elements. So we can't condense them at the
element level but they must be kept along when the e.g. (13) are
solved. Also the possible different number of unknowns in each
node causes a little extra work in renumbering the elements of
the matrices in e.g. (13).

4. NUMERICAL EXAMPLE

The test problem is a modification of the classical Dido's
problem, 'among curves of given length, find the one that en-
closes largest area together with the straight line connecting
the endpoints'. The solution of this is simply a circular arc.
According to the reciprocity law of isoperimetric problems the
solution is the same as in our problem, 'find the shortest curve
enclosing a given area'. The example is calculated by computer
programs developed according to the above formulation. We start
with a mesh of two elements and three nodes (Fig. 1) and use
four different kind of refinement strategies. In the end nodes
only the function value is fixed to zero as a boundary condition.

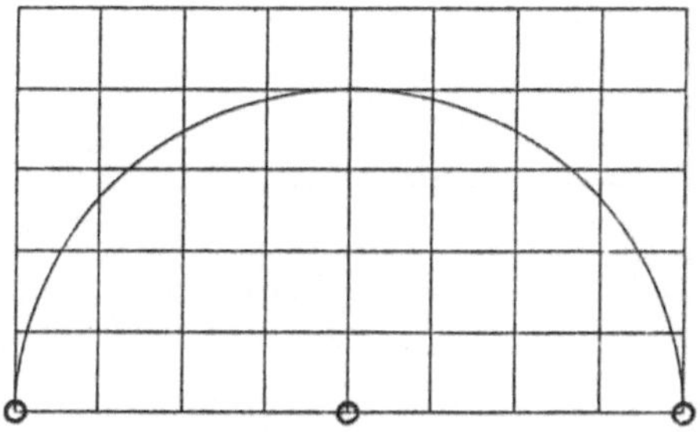

Figure 1. Circular arc with infinite slopes at ends.

(a) In h version we have used uniform mesh refinement with h values $\frac{1}{3}$, $\frac{1}{4}$, $\frac{1}{6}$, $\frac{1}{8}$ and $\frac{1}{16}$
(b) In the first p version we increase the number of nodal parameters at first to three then to four and lastly to five.
(c) The second p version is like the former one but now we increase the number of parameters only in the middle node, not in those nodes where the boundary conditions are given.
(d) In the third p version we use surplus functions defined by e.g. (15). Index i gets the values from zero to four.

The area constraint is the area of a half circel. The iteration is stopped when the relative change in the curve length is below 10^{-4}.

The rates of convergence are shown in Figure 2. The relative error is

$$e_{r1} \doteq \frac{|L-L_n|}{L} \tag{18}$$

where L is the exact length of the circular arc and L_n the computed length of the curve. N is the number of degrees of freedom.

The performance of the h version is linear in log-log scale. In the p version (b) the first increase (2→3) in the number of nodal parameters per node does not improve the result but after that the convergence rate is about twice that of the h method. The reason of this behaviour is not yet understood but it has something to do with the boundary conditions and the infinite slopes at ends because in p version (c) this effect is avoided. The version (d) first converges but when p>5 the results became a little poorer. To compare the results we have also used another error measure. Let

$$e_{r2} = \max_i \frac{|y_i-\hat{y}_i|}{y_i} \tag{19}$$

where y_i is the analytical function value at the node i and $\hat{y}_i$ is the calculated value. The rates of convergence using this error are shown in Figure 3.

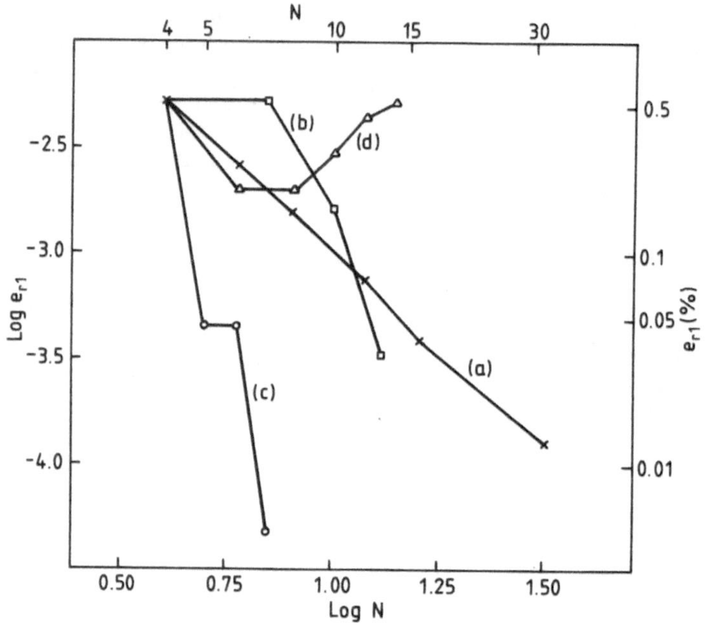

Figure 2. Convergence rates of the h and p extensions respect to error e_{r1}. Letters (a) ... (d) refer to the different refinement strategies mentioned above.

With h refinement the error in function value is about the same regardless of the number of elements. The (b) version behaves almost like in figure 2. With the version (c) the start is similar in both figures but for some reason the last point (p=9) makes an exception. The version (d) gives the best results but when p>5 the convergence behaviour is the same as in figure 2. According to results of both figures the p-version (c) seems to be the best alternative. We must, however, keep in mind that in this example the number of degrees of freedom is quite small and perhaps the asymptotic behaviour of these refinements is not yet shown. More examples are needed.

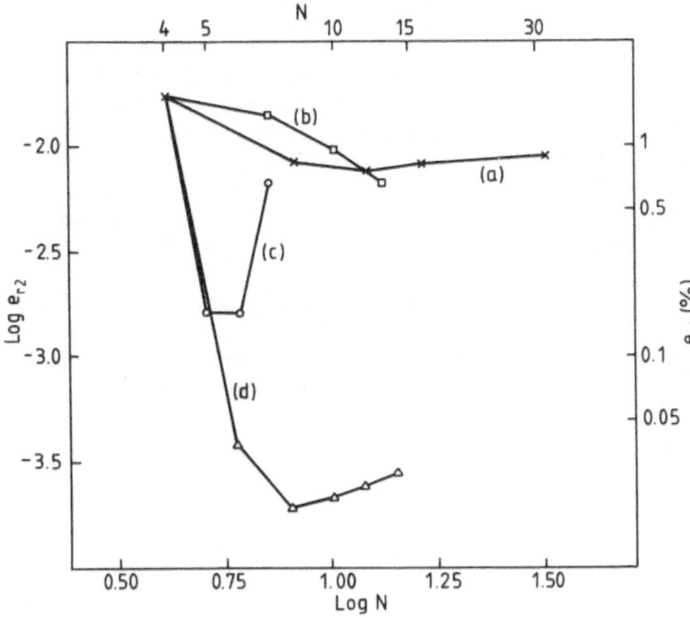

Figure 3. Convergence rates of the h and p extensions respect to error e_{r2}.

5. CONCLUSIONS

In this paper we have used p-refinement in a curve design problem. The use of more nodal parameters seems to be very efficient way to reduce the error. Also the use of internal surplus functions gives very accurate function values. This encourages further research to find reliable correction indicators and to obtain an accurate solution quickly and cheaply.

6. ACKNOWLEDGEMENTS

The financial supprot received from the Academy of Finland and the Foundation of Tampere Research is gratefully acknowledged.

7. REFERENCES

1 A. Pramila, 'Ship Hull Surface Using Finite Elements', International Shipbuilding Progress, Vol. 25 (1978) No 4, pp. 97 - 107.

2 A. Pramila 'A Novel Finite Element Method for Preliminary Design of Ship Hull Form and for Calculating Hydrostatic

Particulars', International Shipbuilding Progress, Vol 26 (1979) No 6 and 7, pp. 116 - 158.

3 A. Pramila, 'A New Curve and Surface Design Method Employing Hemitian Isoparametric Finite Elements', Proceedings of the 2nd International Conference on Engineering Software, London, March 1981, pp. 735 - 746.

4 A. Pramila and S. Virtanen, 'Surfaces of minimum area by FEM', Int. j. numer. methods eng., 23, 1669 - 1677 (1986).

5 R. Delpak and V. Peshkam, 'A study of the influence of hierarchical nodes on the performance of selected parametric elements', Int. j. numer. methods eng., 22, 153 - 171 (1986).

INTEGRATION OF FEM, OPTIMIZATION, AND CAD ON MICROCOMPUTERS

Gu Yuanxian, Cheng Gengdong

Research Institute of Engineering Mechanics,
Dalian Institute of Technology, Dalian, China.

SUMMARY

A integrated CAD system, MicroComputer-Aided engineering structural Design System (MCADS), has been introduced. The idea is to combine finite element method (FEM), structural optimization, and techniques of computer-aided design (CAD) for producing more powerful computer-based design systems. The emphasis is put on interfaces of FEM-CAD, FEM-optimization, and CAD-optimization. Particularly, a quasi-analytic scheme is available for general sensitivity analysis with various constraints, variables, and elements. It easies the way to develop structural optimization software which is based on existing FEM system and has a equivalent ability of complicated problem handling to later one.

1. INTRODUCTION

The FEM has now become the most powerful computational tool for engineering analysis and been widely used. The structural optimization, appeared at almost the same time with FEM, has not such achievement yet. In spite of numerous researches on theory and algorithm, applicable optimization programs for complicated problems of real life and nice applications in industry are very fewer. From the engineer point of view, no matter analysis or optimization, it is neccessary to transform computation data into forms suitable for design, e.g. graphs, and take advantages of designer. But this had not got enough attention in past.

CAD has got great achievements recent years due to the rapid developments of computer hardware and software, especially due to computer graphics and microcomputers. In general, CAD is a technology to assist engineers during whole process of design by means of computer facilities. It should include FEM and optimization as tools of analysis and synthesis. However, principle functions of CAD systems up to now are geometric modelling, graphical viewing, engineering drawing, and man-computer interaction. Although some of them have the function of FEM analysis, but it is still short of real optimization ability. It is a challenge for computational mechanics to combine traditional FEM and optimization with new CAD techniques. In fact, there is indeed a such trend since this decade [1-3].

The most important things for this combination are interfaces of FEM-CAD, FEM-optimization, and CAD-optimization. The FEM-CAD interface generates FE model from geometric modelling, and represents computational data by means of graphical viewing. There are already many works, and difficulties lie in 3D cases. The FEM-optimization interface is mainly design sensitivity analysis for general cases, i.e. various types of design variables, constraints and objective functions, finite elements, etc. This is the key for practicalization of optimization technology. The section 3 will discuss it in detail. The CAD-optimization interface includes state check using graphics facility, interactive design modification and parameter adjustment during iteration, as well as shape description and mesh refinement on changable boundary for structural shape optimization. Additionally, effective and reliable database management is the foundation of integrated CAD systems. Section 5 represents an approach successfully used in MCADS. Because of increasing performance/cost ratio and extensive usage of microcomputers, it is not only possible but also significant to develop integrated CAD systems on them.

2. THE GENERAL PICTURE OF MCADS

MCADS is developed from a microcomputer program system of FEM analysis, DDJ-W[4], with extensions of graphics and interaction facilities, pre- and post-processing for FEM, sensitivity analysis, and design optimization. DDJ-W is a general FEM program for structural static, dynamic, and stability analysis. Its element library involves plenty of element types such as 3D bars, 3D beams, bricks, various isoparametric membranes, plates and shells, and has a convenient interface for adding new elements. It has rich functions to facilitate the modelling of complicated structures and boundary conditions. Its data organization and strategy of equations solving are designed skillfully so as to effectively analyze large scale structures, e.g. more than 1000 nodes or several thousands degrees of displacement freedom, on microcomputers such as IBM-PC. DDJ-W has many users in Chinese industry. All the functions of DDJ-W have been remained to give MCADS a powerful structural analysis ability.

MCADS is composed of the following seven subsystems:

JINEGS, IGSS, DDJ-W, MESHG, GRAPH, OPTMOD, and OPTSOV. The system structure is shown by Fig.1.

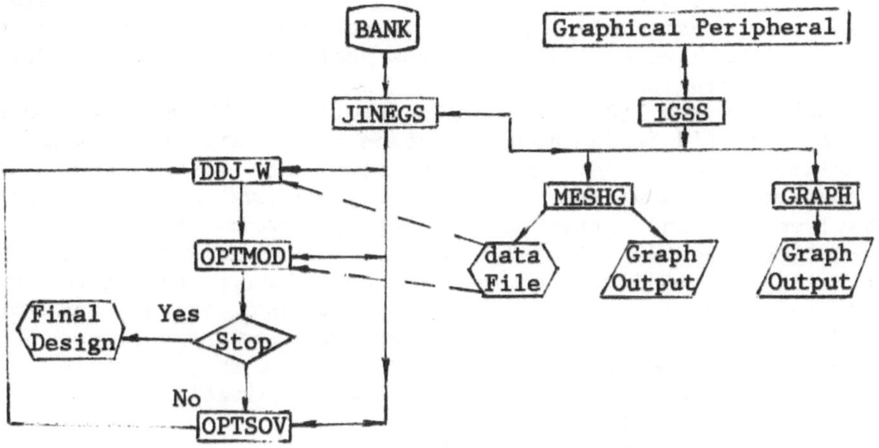

Fig.1. Structure of MCADS

The hardware environment needed by MCADS are microcomputer compatible with IBM-PC possessing 512KB memory and following options: hard disk for midium and large problem more than 100 nodes in FE model, graphics terminal for graph display, plotter or dot-matrix printer for graph drawing. MCADS produces only one direct file, named BANK, on disk as its physical database to support each subsystem executing independently during whole running process. The operation procedure is very convenient without such confusion caused by many data files.

JINEGS is a file system for memory and data management, and IGSS is an interactive graphics support system supplying basic procedures for the programming of graphics and interaction functions. They are fundamental support subsystems of MCADS.

MESHG is an interactive graphical system used for modelling of FE and shape optimization. First, it has geometric modelling ability with simple wire-frame model for describing the shape of structure and definning the requirement of FE mesh. According to this description and definition, MESHG then is able to generate FE mesh and some data useful for shape optimization automatically. In structural shape optimization, the initial mesh model and the produced data, linking initial and generated mesh, are used to define design model and refine FE mesh during iteration. MESHG can describe structural boundaries with quadratic curves and surfaces, and B-Spline curves. It is available for 2D and 3D continuum structures and complicated structures of multi-type elements. So it facilitates data pre-processing enormously. Besides, MESHG can display and draw structural graph in any stage of MCADS running for checking FE mesh and results of shape optimization.

GRAPH is a data post-processing system for assistance of structural analysis and optimization by means of various graphs display and drawing. For instance, structural outline graph, mesh graph, deformation graph, vibrational and instable mode graph, stress contour graph, stress vector graph, etc., with or without hidden line deleting.

OPTMOD is a FEM-optimization interface. It has two main functions: (1) design model creation and modification, and (2) sensitivity analysis. The design modelling includes input of optimization data after initial design analysis, selection of displacement and stress constraints at each iteration, and manual modification on design model at designer's will. User is permited to modify, delete, and add design variables and constraint conditions, replace optimization algorithm, and stop iteration. In this way, their activity and experience are combined with MCADS. Sensitivity analysis is versatile for general elements and variables, and suitable for static response at present.

Upon the foundation of OPTMOD, OPTSOV is expected to be a library of optimiziton algorithms. It has quadratic and linear programming solvers for the present. These two algorithms have been tuned with some practical techniques.

MCADS is application-oriented and user-friendly program, and is capable of solving as large scale problems as about 1000 nodes or several thousands degrees of displacement freedom in FE model and hundreds of variables and constraints in optimization model, without difficulty in memory and storage (The only difficulty is processing speed of microcomputers).

3. DESIGN SENSITIVITY ANALYSIS WITH QAM

In comparison with commercial FE softwares, the principle weakness of structural optimization systems is lack of the ability of complicated problem modelling, which results from few type of finite elements, design variables, and constraints. The conventional method of optimization programming is to develop program from the very beginning, and to derive analytic sensitivity formulations for limited cases. This scheme is computationally efficient, but is tremendously difficult and man power consuming as the type of element, variable, and constraint increases. One attractive approach is to treat existing FEM program as a black box and develop optimization system based on it [2,5]. Optimization systems developed with this 'black box' approach possess same modelling ability as parent FEM program. The key of FEM-optimization interface in this approach is sensitivity analysis, some similar schemes have been proposed [6-8].

A quasi-analytic method (QAM)[7,8] which takes both advantages of analytic and finite difference scheme, has been used in MCADS for general sensitivity analysis. Let us denote structural response as $R=R(X,U)$, where $U=(u_1,u_2,\ldots,u_n)$, $X=(x_1, x_2,\ldots,x_m)$ are nodal displacement and design variable vectors. The equation of FEM is

$$KU=P \qquad (1)$$

The K is structural stiffness matrix, P is load vector. For a given proper perturbance vector $\Delta X=(0,0,\ldots,\Delta x_i,\ldots,0)$ of design variable, the sensitivity of response R can be computed as follows.

1.Compute the difference-approximation of pseudo-load vector Q_i

$$Q_i=[P(X+\Delta X)-P(X)-K(X+\Delta X)U+K(X)U]/\Delta x_i \qquad (2)$$

2. Solve displacement derivative $U_i=\partial U/\partial x_i$ from equation:

$$KU_i=Q_i \qquad (3)$$

3. Evaluate first-order approximation of displacement at design $X+\Delta X$

$$U(X+\Delta X)=U(X)+U_i\Delta x_i \qquad (4)$$

4. Get response sensitivity $R_i=\partial R/\partial x_i$ by difference

$$R_i=[R(X+\Delta X,U(X+\Delta X))-R(X,U)]/\Delta x_i \qquad (5)$$

The local differences are employed in formulas (2)(4)(5) to replace analytic derivatives of P,K, and R with respect to x_i, while analytic formulation (3) is still kept. All the computations can be done by calling procedures (or subroutines) of parent FEM program without knowing its details.

Obviously, the QAM is programming-oriented and much more easier to implement than pure analytic method. It is also versatile for any finite element and structural response, provided they are included in parent FEM program, and any design variable so long as the relationship between variable and FE model could be determined, e.g. arbitrary control parameter of boundary shape can be taken as design variable for shape optimization. The computational efficiency of QAM is close to analytic method, because about the same time is needed for computing of P, K in formulation (2) and their analytic derivatives. The accuracy of QAM has been studied in [7], better accuracy was gained with proper step length of difference. MCADS has employed QAM for size and shape optimizations to arrive at the destination--optimize whatever can be analyzed.

4. STRUCTURAL OPTIMIZATION FACILITY

The optimization modelling of MCADS presently consists of: (1) Size variables(cross-section area of bar, cross-section sizes of beam, thickness of membrane, plate, and shell) and shape variables. (2) constraints of displacement, stress, and resultant inner-force, with or without assigned load case, element, and node, as well as structural weight. (3) Objective function of weight, stress, or displacement. A user interface for adding new types of variable and constraint is provided.

The program execution is automatic or interactive. During interactive iteration, user can examine the state of design by means of data output and graph display or drawing, modify, delete, and add variables or constraints, change algorithm, and stop iteration whenever necessary. The design state and sensitivity of each iteration has been stored in database for later investigation.

The method of shape optimization in MCADS is similar to that of V.Braibant and C.Fleury [1,3]. Initial mesh of MESHG is used to defined 'design element', and control nodes of interpolation curves or surfaces are shape variables. For 2D or 3D continuum structures, shape variable controls coordinate of boundary and inner nodes of design elements, and mesh of design elements is refined by node moving to avoid element distortion. Design element boundaries are described with quadratic curves, quadratic surfaces, and B-Spline curves.

5. MEMORY AND DATABASE MANAGEMENT

Efficient and reliable management of memory and database is extraordinary important for the development of large scale integrated systems (particularly, based on microcomputers with memory and word-length limitations), the 'Software Virtual Machine' concept is benefitial for it. The JINEGS[4], as a such software virtual machine, has been used in MCADS at higher level for the management of memory and engineering database.

JINEGS is a file system embedded into application program to help application programmer dynamically manages data area in memory, data external storage in direct file BANK, and input/ouput interchange. JINEGS organizes data in terms of file and record two level data structure, a file contains any number of records and record contains data. A so-called management-type record is the outstanding feature of JINEGS. It keeps trace not the data itself(stored in BANK all the time) but the management informations of data or subordinate files. These informations are: the index for equal-length data, the index and the length for unequal-length data, and indexes(3 integers) for subordinate file. By means of management-type record, multi-level file organization and effective management of unstructured data are implemented, this is significant for engineering softwares. All files are stored in BANK without limit on number, and loaded into memory for using, on-line files are less than 20. The operation is logically according to numbers of files and records. So JINEGS helps application programmer to use memory space as economically as possible, and to programming clearly in logic.

JINEGS has a mechanism to calculate and keep the address of its files, records, and data controled by management-type records in BANK automatically, so that application programmer need not pay attention to details of physical storage structure and access path of data. It is designed available for 16-bit microcomputer (with 20MB hard disk) and 2-byte integer. JINEGS has also a virtual memory mechanism for data interchange between memory and BANK, which uses a part of high address memory space as data buffer and the algorithm of LRU(least recently used) page replacement for fixed-size page management. It improves execution efficiency of application program notably. JINEGS supplies a set procedures to be embedded into application program, and a series of error detecting facilities. These error detecting facilities are very helpful in quality assurance of software systems.

6. EXAMPLES

Example 1. Ten bar truss, a classical test problem[2], is designed with displacement and stress constraints and variables of cross-sectional areas. After 12 iterations, MCADS have got an optimal design with weight of 5063(lbs), which is very close to the standard result.

Example 2. A truss shown in Fig.2. is consisted of 8 nodes and 16 bars. The material parameters are as follows: Young's modulus is $2*10^7 t/m^2$, allowable stress is $50000t/m^2$, specific weight is $5t/m^3$. The cross-sectional areas are $F1=...=F6=0.003m^2$, $F7=F8=F9=0.002m^2$, $F10=...=F16=0.001m^2$, and weight is 0.3932t. Size and shape optimization are carried out to minimize truss weight in three cases:(1) Design variables are $F7=F8=F9\geqslant 0.001m^2$ and $F11=...=F16\geqslant 0.0005m^2$. (2) Four design variables are node coordinates Y1, Y3, Y5, and Y7. Their lower and up bounds are 2.0m and 3.2m. (3) Six variables are Y1, Y3, Y5, Y7, F7=F8=F9, and F11=...=F16. The limit values are same as above. The displacement of any node is limited 0.02m in all cases. Results of optimizations are listed in table 1.

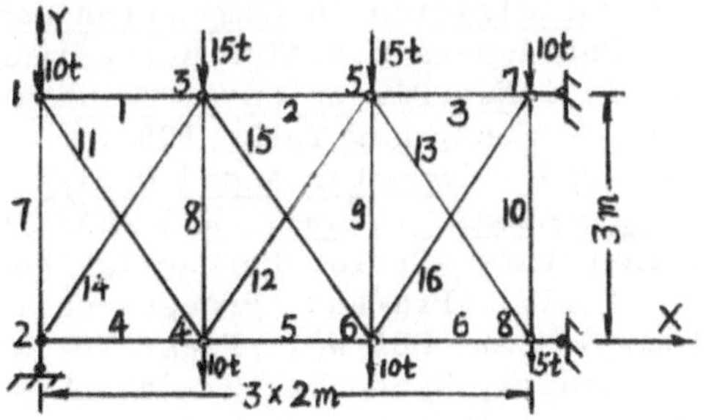

Fig.2

Table 1.

Result	case 1.	case 2.	case 3.
F7-F9	0.00100		0.0010
F11-F16	0.00106		0.0010
Y1		2.000	2.000
Y3		2.055	2.401
Y5		2.601	2.650
Y7		2.850	2.949
Weight	0.3546	0.3552	0.3276
Iterations	6	12	14

Example 3. A square membrane with a square hole at centre is stressed by uniform tensile loads along its four edges, Fig.3. is a quarter of it. The aim is to minimize the maximum Von Mises stress, which is measured at nodes after stress smoothing, by means of changing boundary shape of hole. The 'design element' is divided into quadrilateral elements and fixed part is divided into triangular elements. The changable boundary is defined as quadratic curve controled by nodes 1, 2, 3. The design variables are coordinates of control nodes: Y1, X2, Y2, X3. The initial design has stress concentration at corners of hole. After 7 iterations, the result shown in Fig.4. has been obtained, which has reduced maximum stress from 149.146 to 106.107 and equalized stresses along hole boundary.

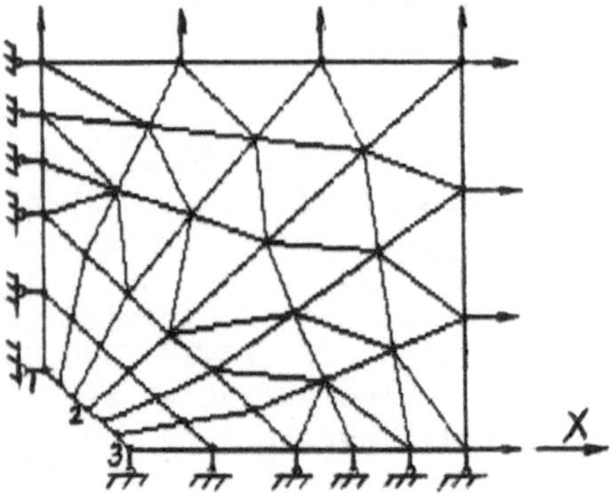

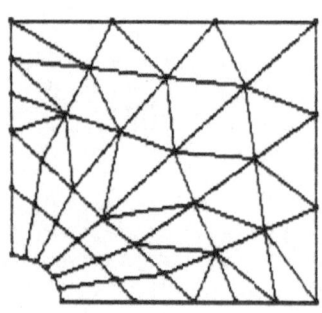

REFERENCE

1. Gero,J.S.(eds) - Optimization in Computer-Aided Design, Elsevier Science Publishers B.V, The Netherlands, 1985.
2. Atrek,E. et al(edi) - New Direction in Optimum Structural Design, John Wiley & Sons, New York, 1984.
3. Mota Soares,C.A.(eds) - Computer Aided Optimal Design, structural and mechanical systems, NATO/NASA/NSF/USAF Advanced Study Institute, Troia, Portugal, 1986.
4. Zhong Wanxie, et al. 'Pre/Post Processing in Structural Micro Computing System DDJ-W', First World Congress on Computational Mechanics, September 22-26, 1986, Austin, USA.
5. Sobieski,J., 'From a 'black box' to a programming system', Foundations of Structural Optimization: An Unified Approach, Ed by A.J.Morris, John Wiley & Sons, 1982.
6. Iott,J., Haftka,R.T. and Adelman,H.M.,'Selecting Step Sizes in Sensitivity Analysis by Finite Differences', NASA Technical Memorandum, 86382, 1985.
7. Cheng Gengdong and Liu Yingwei, 'New Computation Scheme of Sensitivity Analysis', Engineering Optimization, to appear.
8. Wang Liu, 'Shape Optimization of Revolutional Bodies and Sensitivity Analysis Method', Master thesis, Dalian Institute of Technology, China, December, 1984.

AN ADAPTIVE HP-VERSION IN THE FINITE ELEMENT METHOD

ERNST RANK

Fachgebiet El. Rechnen im konstruktiven Ingenieurbau, Technische Universität München, Arcisstr. 21, D-8000 München 2, Germany
now :
SIEMENS AG, Corporate Research and Technology, Otto-Hahn-Ring 6, D-8000 München 83

SUMMARY

Adaptivity is now widely accepted in finite element methods. Most adaptive codes refine the finited element mesh locally controlled by some *a posteriori* estimation. In this paper an adaptive hp-version is presented. The algorithm increases the polynomial degree p and refines the finite element mesh ,i.e. decreases the local mesh-width h. Numerical examples show that even in presence of singularities in the exact solution exponential rate of convergence is obtained.

1. INTRODUCTION

There are three ways to achieve convergence in the finite element method: the h-version improves the accuracy of an approximation by refining the mesh and using shape functions of usually low degree. The p-version uses a fixed mesh but increases the polynomial degree of the shape functions to improve its accuracy and to obtain convergence. This method has been analysed during the past 5 to ten years,[1,2,3] and there has been some very promising software development[4,5] which proves the superiority of the p-version over the h-version. A combination of the h- and p-version, i.e. a simultaneous local mesh refinement and increase of the polynomial degree is called hp-version. It has been shown theoretically [2,6] and practically [7,8] that exponential rate of convergence in the energy norm can by achieved by this method even in cases with singularities in the exact solution. In a prototype finite element expert system[8] optimal combinations of mesh and polynomial degree are predicted from a starting computation with low polynomial degree on a coarse mesh. Using this prediction the user constructs with the help of this expert system a mesh-degree-combination which yields the desired accuracy at minimal computational cost.

In this paper an alternative to the expert system mentioned above will be presented. The algorithm described below is fully adaptive, i.e. starts on a coarse mesh with low polynomial degree and refines in several cycles completely automatically using *a posteriori* estimations of the distribution of the error in energy norm on every mesh. In contrast to [7,8] the polynomial degree of the shape functions needs not be constant over the entire mesh, i.e. every element can have polynomial order which is adjusted to yield the desired accuracy with minimal cost.

2. P- AND HP-VERSIONS IN FEM

As model problem, consider

$$-\Delta u = f \quad \text{in } \Omega \in \mathbb{R}^2$$
$$u = u_0 \quad \text{on } \Gamma_1 \quad ; \quad \frac{\partial u}{\partial n} = g_0 \quad \text{on } \Gamma_2 \tag{1}$$

The smoothness of the solution u of (1) depends on the shape of the boundary and on f, u_0 and g_0. Assume first that u is analytic up to the boundary of Ω. Then the error $\|e\|_E := \|U-u\|_E$ of an approximation U to u in the p-version, i.e. an increase of the polynomial degree p of the elements on a fixed mesh converges exponentially in the energy norm.

$$\|e\|_E \le C\, e^{-\alpha N(p)^{1/2}} \qquad p \longrightarrow \infty \tag{2}$$

C,α, are positive constants, N(p) is the number of degrees of freedom depending on the polynomial degree p. If there are reentrant corners in Ω or if there is a sudden change of the boundary condition then the exact solution can be written in the form

$$u = u_0 + \sum_{i=1}^{\infty} C_i r^{\lambda_i} g_i(\theta) \tag{3}$$

C_i are stress intensity factors,u_0 and g_i smooth functions, (r,θ) polar coordinates centered at the singularity and λ_i ordered increasingly. On a mesh as shown in figure 1a the p-version converges as in figure 2 curve (a). An exponential preassymptotic range is observed (curved down) and assymtotically the convergence is levelling off to an algebraic rate (straight line) which is governed by the power of the singularity, i.e. by λ_1. If a geometrically refined mesh as in figure 1b or 1c is used, similar S-curves can be observed (curves (b) and (c) in figure 2), yet shifted compared to that of mesh 1a. An optimal hp-version of the FEM 'switches' now from one mesh to a geometrically refined one just at the intersection points of the convergence curves always staying on the lower left envelope of the curves. This envelope itself is 'bent down', i.e. shows exponential convergence rate in the energy norm. This behaviour has been proven theoretically in [2,6] and shown numerically in [7] and [8]. Moreover it has been proven [2] that the optimal geometric progression factor for the hp-version is independent of the strength of the singularity and should be chosen as .15, yielding a very strong grading toward the singular point. Yet the optimal combination of number of refinement layers and polynomial degree depends on the stress intensity factors C_i and the exponents λ_i.

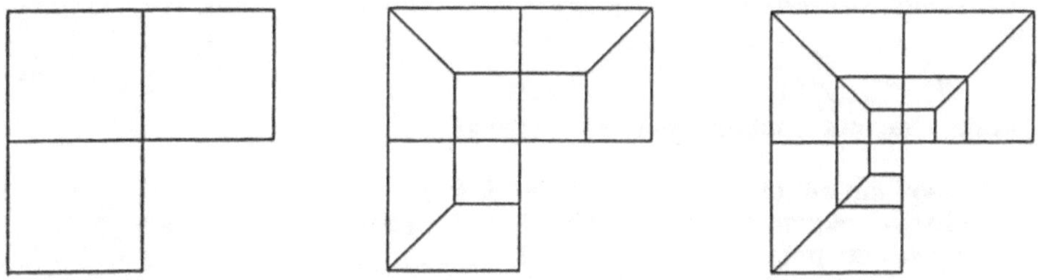

Figure 1a-c: Mesh with 0,1 and 2 refinement layers

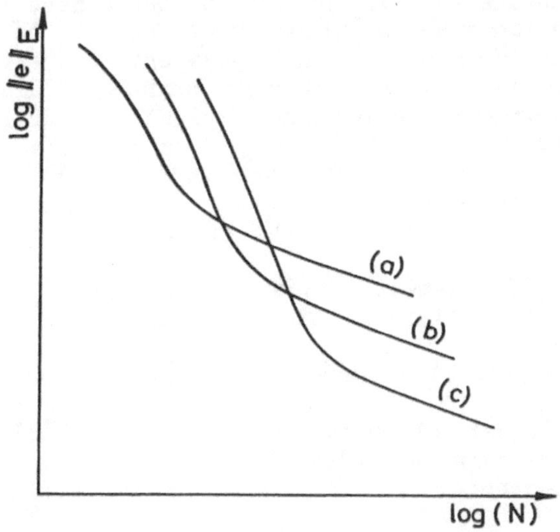

Figure 2 : Convergence of a p-version FEM for different meshes

3. AN ADAPTIVE HP-VERSION

The core of the adaptive hp-version is the p-version finite element code for linear potential, elasticity and Reissner-Mindlin-plate problems which is presented in [5]. The polynomial degree p can be varied freely over the mesh in all variables. Error indicators and estimators are similar to those presented in [9]. On the edges Γ_i of an element i the jumps $J_x(U)$ and $J_y(U)$ of the derivatives of the approximate solution U in x- and y-direction are computed and integrated to the error indicator

$$\lambda_i^2 = \frac{h}{24\, p} \int_{\Gamma_i} J_x(U)^2 + J_y(U)^2 \, d\Gamma \qquad (4)$$

h is the diameter of the element, p the polynomial degree. The error estimator for the error in energy norm is then defined as

$$\eta^2 := \sum \lambda_i^2 \qquad (5)$$

where the sum ranges over all elements.

Now an adaptive hp-version can be defined. As the goal is to achieve exponentional rate of convergence, the strategy is to increase the polynomial degree p in smooth parts of the solution and to refine geometrically at singularities. The basic algorithm has the following form:

Step 1: Choose a basic mesh, which is just fine enough to describe geometry, boundary conditions and loads of the problem.

Step 2 : Seperate the elements of the basic mesh into two parts, those, where the exact solution is expected to be smooth (called non-critical elements) and those adjacent to a singular point of the exact solution, e.g. reentrant corners, points of change of boundary conditions etc.(called critical elements).

Step 3: Assign polynomial degree p=1 to each element.

Step 4: Perform a FEM-computation and compute error indicators for each element.

If the accuracy estimated by the error estimator is sufficient, STOP.

Step 5: For each element decide if the error indicator is above a prespecified level, i.e. if the accuracy has to be improved. If yes, then

for noncritical elements increase the polynomial degree by 1,

for critical elements refine geometrically towards the singularity in this element.

Step 6: Goto step 4.

4. NUMERICAL EXAMPLES

In two numerical examples the behaviour of various extension strategies will be compared. The uniform h-version (marked as 'H 2' in the plots) refines, starting from the basic meshes uniformly and uses polynomial degree p=1 on all elements. The uniform p-version ('P 2' in the plots) uses the basic mesh and increases the p-degree uniformly over the mesh. The adaptive h-version [10] ('H 1') uses elements of degree 1 and refines locally controlled by the error indicators (4). In the adaptive p-version ('P 1') the basic mesh is unchanged but the polynomial degree is increased adaptively over the mesh, controlled again by the error indicators (4). The adaptive hp-version was run in two variants. One increases the polynomial degree uniformly over the mesh and refines locally at the singularities ('HP 2'). The other ('HP 1') varies the polynomial degree over the mesh and refines locally as defined in the algorithm of chapter 3. For the hp-versions a list of possible singularities, i.e. points of change of boundary conditions and reentrant corners was provided as input data to the program.

Example 1. As domain of computation the rectangle $\Omega = (-50,0)\times(-7,0)$ was chosen with the boundary conditions

$$u(-50,y) = u_0 \quad \text{for } -7 \le y \le 0$$

$$u(0,y) = 0 \quad \text{for } -7 \le y \le -3.5 \quad ; \quad \frac{\partial u}{\partial n} = 0 \text{ elsewhere.}$$

u_0 was chosen so that the exact solution could be computed analytically. Due to the change of boundary conditions at the point (0,-3.5) the exact solution shows a singularity of order

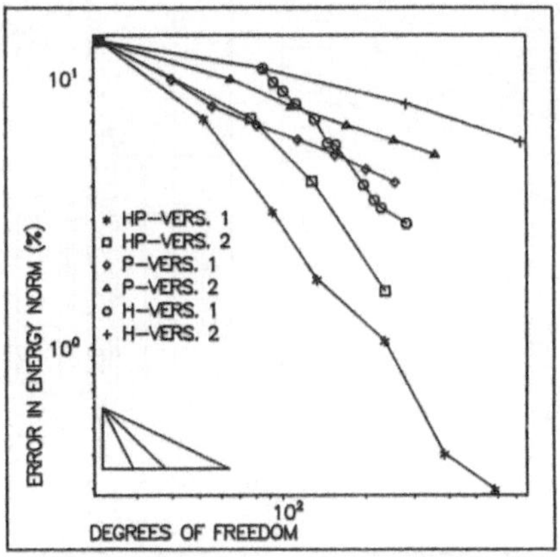

Figure 3 : Error in energy norm for example 1

$r^{-1/2}$ in the flux. On the other hand the exact solution is extremely smooth (essentially linear) in the rest of the domain.

Figure (3) shows the convergence in energy norm for the extension processes described above. The two adaptive hp-versions show superior accuracy, with only 600 degrees of freedom an error of less than .3 % is achieved. There is also a significant difference between the hp-version with uniform p-degree ('HP 2') and with variable p-degree ('HP 1'). This is due to the large smooth part of the solution where HP 1 'wastes' degrees of freedom whereas HP 2 uses only linear or quadratic elements in this part of the domain. Both HP 1 and HP 2 show the desired exponential rate of convergence in energy norm.

Example 2. The domain of computation with equipotential lines for example 2 is shown in figure (4). As the exact solution for this example is not known, the exact energy was estimated by extrapolation from very fine meshes with high polynomial degree. Figure (5) shows an adaptively refined h-version mesh for linear elements and figure (6) gives the mesh constructed by the adaptive hp-code. The different 'strengthes' of the various singularities are reflected in the different number of refinement layers at these points. Due to the strong geometric refinement towards the singularities not all refinement layers can be seen in the plots. For example at the change of boundary conditions at the lower boundary of the domain there are 5 refinement layers towards the singular point.

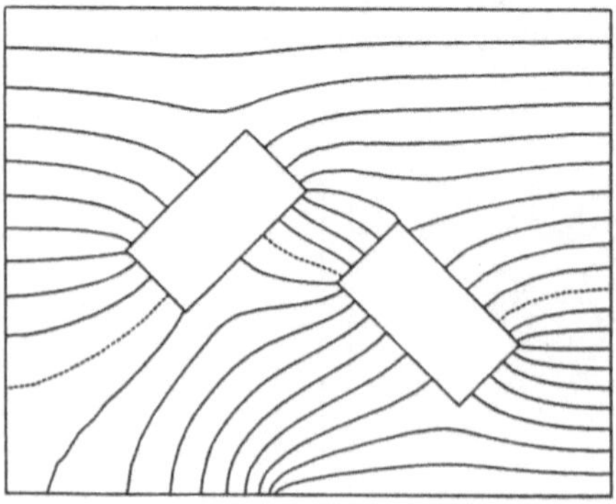

Figure 4 : Domain of computation with equipotential lines for example 2

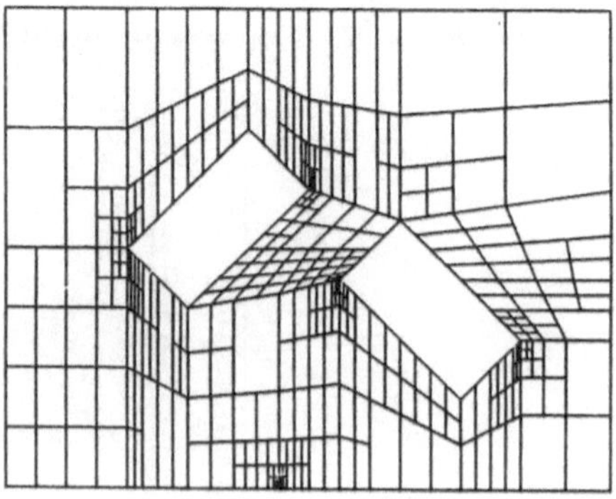

Figure 5 : Adaptively refined mesh (h-version)

In figure (7) the convergence for the extension processes is plotted. Essentially the same behaviour can be observed as in example 1, yet now there is nearly no difference between adaptive and non-adaptive p-versions and between HP 1 and HP 2. This is due to the fact that nearly the whole domain is under the influence of one of the 9 singularities leading to a nearly uniform optimal p-distribution. Again the convergence curves for the hp-versions are 'bent down' showing the exponentially decreasing error.

The effectivity index $\theta = \eta / \|e\|$, which gives a measure of the quality of the error estimator (5) is for all examples reasonably close to 1. For example HP 2 in problem 2 yielded an index of 1.17 for 38 dofs and .96 for 1104 degrees of freedom.

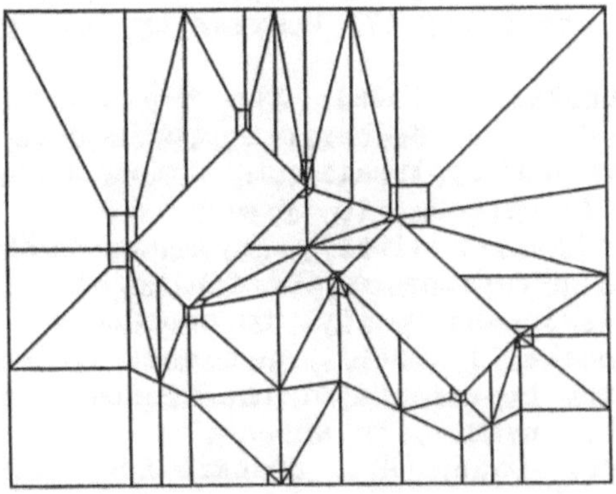

Figure 6 : Adaptively refined mesh (hp-version)

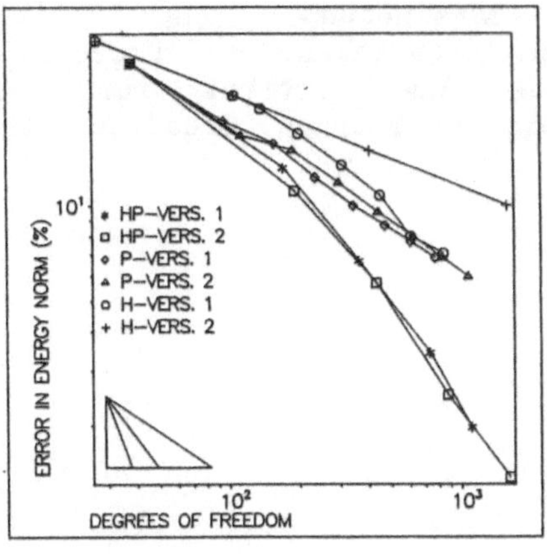

Figure 7 : Convergence in energy norm for example 2

References

/1/ Babuska,I., Szabo,B.A., Katz,I.N. (1981), The p-version of the finite element method, SIAM J. Numer. Anal. 18,No 5,515-545

/2/ Gui,W., Babuska,I. (1986), The h, p and h-p versions of the finite element method in 1 dimension. Part 1: The error analysis of the p-version. Part 2: The error analysis of the h- and h-p versions. Part 3: The adaptive h-p version, Numerische Mathematik, to appear.

/3/ Szabo,B. (1986), Estimation and control of error based on p-convergence. in : Accuaracy estimates and adaptive refinements in finite element computations, Babuska, I., Zienkiewicz, O.C., Gago, J., de Oliveira, E.R. (eds), John Wiley & Sons

/4/ Szabo, B.A., PROBE: Theoretical Manual, NOETIC Technology Corporation, 7980 Clayton Road, Suite 205, St. Louis, MO 63117

/5/ Bellmann, J. (1987), Convergence of hierarchical finite elements, Proc. of NUMETA 87, University College, Swansea, July 6-10th, 1987

/6/ Guo,B., Babuska,I. (1986), The h-p version of the finite element method. Part 1: The basic approximation results. Part 2: General results and applications, Computational Mechanics 1, Part 1:, pp 21-41, Part 2 : to appear.

/7/ Babuska,I., Rank,E. (1986), An expert-system-like feedback approach in the hp-version of the finite element method, Finite Elements in Analysis and Design, to appear.

/8/ Rank,E., Babuska,I. (1986), An expert system for the optimal mesh design in the hp-version of the finite element method, Int. J. for Num. Meth. in Eng., to appear.

/9/ Kelly, D.W., Gago, R., Zienkiewicz, O.C., Babuska, I. (1983), A posteriori error analysis and adaptive refinement processes in the finite element method: Part I: Error analysis, Int. J. f. Num. Meth. in Eng., Vol 19, pp 1593-1619

/10/ Rank, E. (1985), A-posteriori-Fehlerabschätzungen und adaptive Netzverfeinerung für Finite-Element- und Randintegralelement-Verfahren. Doctorial Dissertation, Mitteilungen aus dem Institut für Bauingenieurwesen I, Technische Universität München. Grundmann, Knittel, (Hrsg.)

A FINITE ELEMENT TRANSITIONAL MESH GENERATION TECHNIQUE

L. Carter Wellford, Jr.
M. R. Gorman
University of Southern California
Los Angeles, California, 90089, U.S.A.

SUMMARY

Transitional mesh generation procedures involve the definition of meshes which provide a changing element density from one part of the mesh to another. In this work a transitional mesh generation procedure based on blending and sweeping function is introduced. This procedure is used to generate typical transitional meshes. These transitional meshes show that the developed procedure produces a quad-dominated mesh.

1. INTRODUCTION

It has been clearly understood, since the beginning phases of finite element research, that mesh generation was a critical procedure in proving the ultimate usefulness of the method. Various mesh generation techniques were proposed by early investigators. They included a Laplacian smoothing procedure [1] and a mapping procedure [2] based on isoparametric or other mappings. At this point in time the mapping methods have been widely implemented. However, the development of the Quadtree procedures [3] may eventually represent a challenge to the dominance of the mapping type generators.

In this paper a technique is introduced which is designed to make the mapping type generators more versatile. This technique allows the generation of transitional meshes on the basic patches of the geometric model. The transitional meshes allow an arbitrary change in element density from side to side in the basic patch. However, the procedure has the property that it results in a quad-dominated mesh.

2. A TRANSITIONAL PATCH

Consider a square patch parameterized by coordinate ξ and η. Let the sides of the parent patch be labeled f_{ik} where i is the direction number and k is the side number in that direction. Let the number of points on the side f_{ik} of the patch be denoted N_{ik}. The initial mesh is created from a grid of horizontal and vertical lines. The maximum number of vertical lines K_1 is equal to the maximum number of points on a horizontal face of the parent patch. The maximum number of horizontal lines K_2 is equal to the maximum number of points on a vertical face of the parent patch. Thus

$$K_i = \underset{k=1,2}{\mathrm{MAX}} (N_{ik}) \; .$$

Let KS_1 be the side number with the maximum number of horizontal points. Let KS_2 be the side number with the maximum number of vertical points

$$K_{si} = \begin{cases} 1 \text{ IF } N_{i1} = K_i \\ 2 \text{ IF } N_{i2} > N_{i1} \end{cases} \; .$$

Let $S^{(i,k)}$ be a local coordinate on the side ik of the parent patch. The $S^{(i,k)}$ local coordinates can be identified with the (ξ,η) parent coordinate on specific sides. In fact

$$\begin{aligned} S^{(1,1)} &= S^{(1,2)} = \xi \\ S^{(2,1)} &= S^{(2,2)} = \eta \end{aligned} \; . \tag{1}$$

Let the nodal points on the side ik be denoted $S_j^{(i,k)}$, $j = 1,\ldots N_{ik}$. Let a series of nodeless functions $\tilde{\phi}^{(i,k)}$ be defined as follows

$$\tilde{\phi}^{(i,k)} = (S^{(i,k)} - S_1^{(i,k)})(S^{(i,k)} - S_2^{(i,k)})\ldots(S^{(i,k)} - S_{N_{ik}}^{(i,k)}) \tag{2}$$

Let a set of <u>sweeping functions</u> $\phi^{(i,k)}$ be defined as follows

$$\phi^{(i,k)} = C_{ik}\tilde{\phi}^{(i,k)}$$

where if $R^{(i,k)}$ is the value of $S^{(i,k)}$ at the peak of the curve $\tilde{\phi}^{(i,k)}$ closest to the point $S^{(i,k)} = 0$, then

$$C_{ik} = \frac{1}{\tilde{\phi}^{(i,k)}(R^{(i,k)})} \; . \tag{3}$$

Then

$$\phi^{(i,k)}(S^{(i,k)}) = C_{ik} \prod_{j=1}^{N_{ik}} (S^{(i,k)} - S_j^{(i,k)}) \quad . \qquad (4)$$

The locations of the nodes in the parent element block can be defined as the zeros of certain function. In particular, the node of the tentative parent mesh can be defined as the zeros of the following tentative node definition function $X_1(\xi,\eta)$:

$$X_1 = \phi^{(1,KS_1)} S^{(1,KS_2)^2} + \phi^{(2,KS_2)} S^{(2,KS_2)^2} \quad . \qquad (5)$$

Because of the definition of the boundary coordinates $S^{(i,k)}$ in (1), this tentative node definition function can be shown to take the following form:

$$X_1(\xi,\eta) = \phi^{(1,KS_1)^2}(\xi) + \phi^{(2,KS_2)}(\eta)^2 \quad . \qquad (6)$$

In addition to the tentative node definition function, a final node definition function $X_2(\xi,\eta)$ can be defined. This function is more complicated than X_1 and requires the introduction of certain blending functions. Let $\beta_i(S)$ be a set of blending functions. In particular, let us define functions capable of blending the mesh from one side to another. Then the β_i functions are

$$\beta_1(S) = \frac{1}{2}(1-S)$$

$$\beta_2(S) = \frac{1}{2}(1+S) \quad .$$

The final node definition function can then be shown to take the following form:

$$X_2(\xi,\eta) = \left(\sum_{k=1}^{2} \beta_k(\eta)\phi^{(1,k)}(\xi)\right)^2 + \left(\sum_{k=1}^{2} \beta_k(\xi)\phi^{(2,k)}(\eta)\right)^2 \quad . \qquad (7)$$

The final node definition function has zeros at the positions of the nodes on the faces of the parent patch.

In order to move or sweep the nodes from their original or tentative positions to their final positions, a continuation procedure based on a parameter α may be used. Associated with this continuation parameter is a <u>continuation node definition function</u> X_3 defined as follows:

$$X_3(\xi,\eta) = [(1-\alpha)\phi^{(1,KS_1)}(\xi) + \alpha \sum_{k=1}^{2} \beta_k(\eta)\, \phi^{(1,k)}(\xi)]^2$$
$$+ [(1-\alpha)\phi^{(2,KS_2)}(\eta) + \alpha \sum_{k=1}^{2} \beta_k(\xi)\, \phi^{(2,k)}(\eta)]^2 \quad . \tag{8}$$

When α equals zero, the continuation node definition function is the same as the tentative node definition function X_1. When α equals one, the continuation node definition function is the same as the final node definition function X_2. As α is varied from 0 to 1, the nodes of the mesh, represented by zeros of $X_3(\xi,\eta)$, move from their initial positions to their final positions.

4. NUMERICAL RESULTS

To demonstrate the mesh generation procedures proposed in this paper, various transitional meshes were defined. These meshes are pictured in Figures 1 to 3.

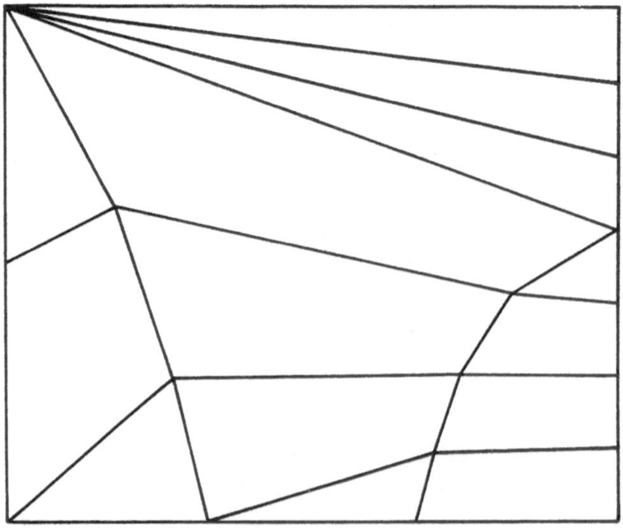

Figure 1. 3/7/1/2 Transition

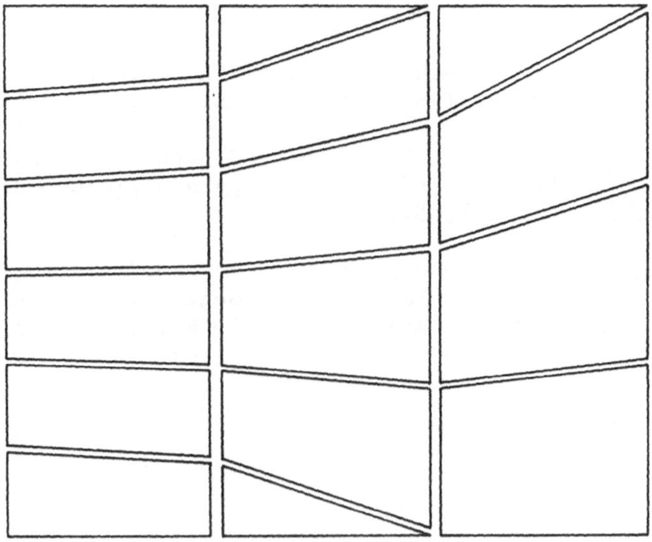

Figure 2. 3/3/3/6 Transition

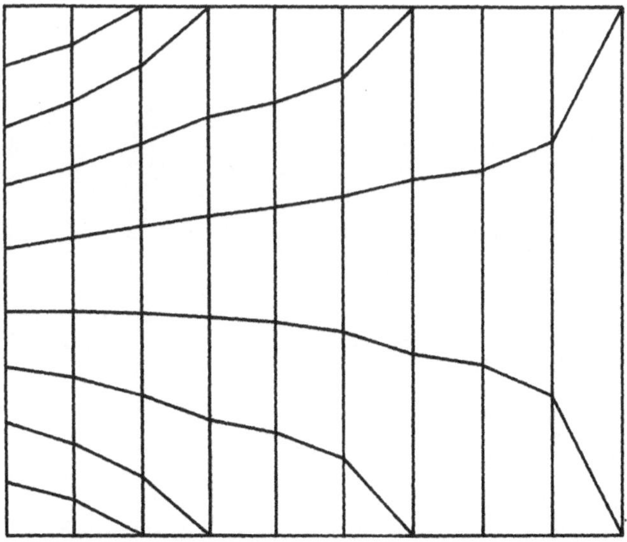

Figure 3. 9/1/9/9 Transition

REFERENCES

1. Buell, W. R., "Mesh Generation - A Survey," Journal of Engineering for Industry, A.S.M.E. (1973), 332-346.

2. Zienkiewicz, O. C., "An Automated Generation Scheme for Plane and Curved Surfaces by Isoparametric Coordinates," I.J.N.M.E. (1971), 519-538.

3. Yerry, M. A. and Shepherd, M. S., "A Modified-Quadtree Approach to Finite Element Mesh Generation," IEEE Computer Graphics and Applications, Vol. 3, No. 1, (1983), 39-46.

EXPERT SYSTEM FOR MATERIAL SELECTION

James K. Blundell and R. Bryan Greenway -- University of Missouri-Columbia/Kansas City
Design Productivity Center 4747 Troost Ave Kansas City, MO 64110

1. The Design Process

There are complex demands made upon the practicing design engineer. The advance of technology has compressed product life cycles, and if there are inefficiencies in the development process, then the product may be obsolete even before production begins. Hence, design decisions must be made promptly and intelligently.

In order to improve the designer's performance, investigation of the design process has become a top priority of researchers and, although no universal methodology has been discovered, some relevant facts have been noted. For example, the design process is not algorithmic; rather, it is the manipulation of a collection of discrete conceptual elements until a perceived need is satisfied. The models of the design process[1-7] indicate a series of intermediate steps in which the design of a system is reduced to the specification of individual components, each of which presents the designer with individual functional and manufacturing requirements. The conceptual elements referred to above may be broadly classified as geometry, material selection and manufacturing process selection and these provide the basic ingredients. Designing is the integration of these elements into a form which satisfies all the needs. This synthesis requires deductive thinking and trade offs in the decision. The element of design which includes conceptualizing and synthesis is best performed from a foundation of experience.

The other sphere of design is concerned with analysis. Analysis involves the rationalization of the design into a set of manageable parts. Neophyte engineers are usually more competent and comfortable with analysis than synthesis because it comprises most of their engineering education.

From the above,it might appear that the design process is sequential; however, within each step, the procedure is less easily defined as all three conceptual elements must play interactive roles in the decision making process.

Crucial to the development of a successful design is information, consisting of data and knowledge. It is needed at each step, becoming more specific as the process moves from concept design to detail design. Data is exemplified by material properties. Knowledge consists of analytical tools and design experience.

Appplying the correct type and depth of information at appropriate points in the design process is crucial and sub-optimal designs result if any one of the conceptual elements is over- or under-emphasized.

Greater complexity is incurred when part interactions are considered in their integration

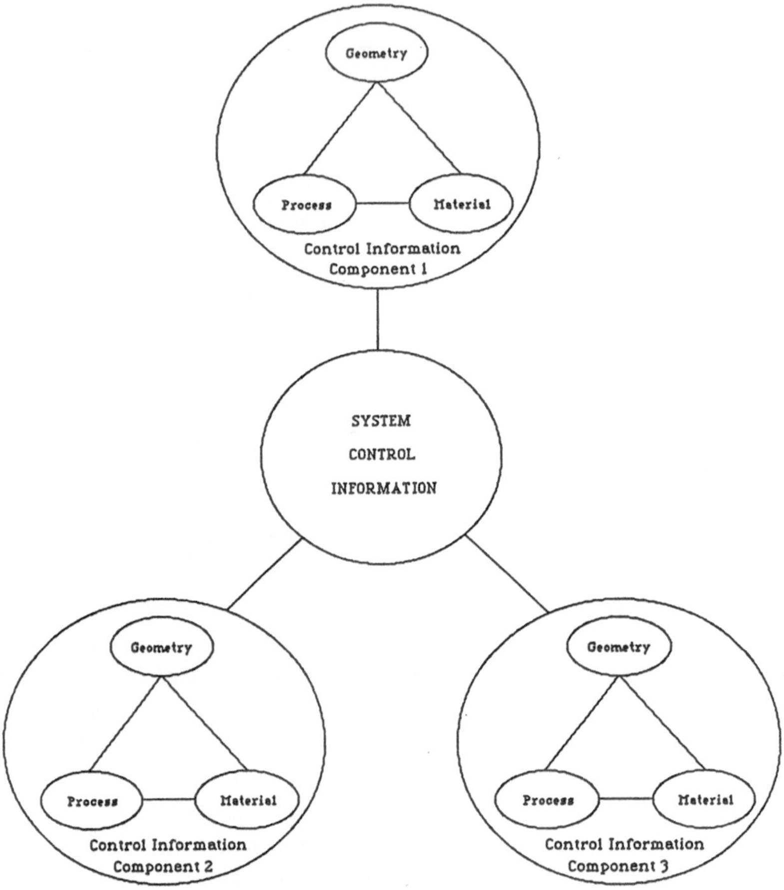

FIGURE 1 Design Control Diagram

into an assembled design. Under "real-world" design conditions, control information is also needed as such information will direct the design within the parametric constraints. Such information would include functional requirements, for example, static loads, torque transmission, operating temperatures or corrosion resistance. In addition, there may be size restrictions, price limitations or aesthetic demands. The control information list will then be used to drive the individual and collective component designs. The control information list will not be stable since if requirements are found to be too restrictive or costly then values in the information list may be modified either for individual components or for system assemblies. (Figure 1)

Computer Aided Design requires numerical and symbolic manipulations which must be complemented by appropriate levels of knowledge and expertise. This implies that expert systems must be judiciously integrated with numerical analyses to provide a flexible design environment.

2. Expert Systems for Design

An expert system is an integrated, humanlike strategy for problem solving with a large database of factual knowledge and experience relevant to the problem. Expert systems operate best when the thinking is reasoning not calculating and reasoning power not mathematics is what sets experts apart from novices.

In general, expert systems lend themselves to "combinatorial" problems where a number of possibilities must be evaluated. These types of problems can become unwieldy and the expert system can use heuristics to prune the number of search paths. A second strength is the ability to sort through and interpret large amounts of data or information.

Attempts to formalize the process of design in an expert system might suggest that rules-based systems may not adequately capture the process of synthesis. However,as described in the following sections, it may still be a viable and relevant approach providing the design and manufacturing philosophies are thoroughly understood and the database requiremnets are adequately addressed.

Providing a design environment that offers completeness without unwieldiness indicates a probable relevancy of expert systems. However, when coupled with the need to compress the cycle of innovation, expert systems may be vital in systematically ensuring optimal design solutions in accelerated time frames.

The problems faced by designers include:-

- No simple, consistant design routine;
- Integrating functional design within manufacturing constraints;
- Performing synthesis without experience;
- Incomplete knowledge of all available materials and manufacturing methods;
- Detailing the design too early;
- Converting functional requirements to material properties too early thereby forcing the design down sub-optimal paths.

Application of expert knowledge is needed at key points in the design process to prevent such problems from arising.[9] Ideally, a group of design specialists would work together to convert a list of specifications for a component into an optimal detail design. A flow chart of a design process which incorporates a hierarchical expert system has been developed.[10] This establishes a hierarchy of control over the system by the use of a "meta-rule" philosophy. Meta-rules give the system information about how to use what it already knows. There could be an arbitrary number of levels of control, each of which controls the use of knowledge at the next lower level. Meta-rules become useful when there are large stores of domain-specific knowledge such as in engineering design. By supplying strategic information to the system, meta-rules make possible a finer degree of control over the use of knowledge in the system. In other words, we can gain leverage at higher levels by using heuristics that guide the use of heuristics and rather than adding more rules to improve performance, more information can be added at the next higher level concerning the effective use of existing rules.

Globally optimal design is only possible when manufacturing and design functional requirements are merged into a single list of constraints before any conceptual designing takes place. Inherently, there will be conflicting design and manufacturing requirements such as strength versus machinability. At this point, expert conflict resolution takes place in order to determine the needed compromises and define the final product constraints.

If an expert system for design is to be created then, as in all computerization exercises, it is imperative that the current process be thoroughly understood before the system is specified. For the design process , it is important to identify what the system should be capable of doing and to then specify the best means of realising those system goals. Fundamentally, an expert system may not be desirable if it implies the existence of an " Engineering Oracle". Perhaps, an "Advice and Consultation" system is conceptually more viable as this implies greater flexibility for the designer and grater emphasis on personal creativity.

Once the system understands the primitive size and shape, it can perform a feasibility study. The required material and processing properties are derived from a translation of the functional requirements through the meta-rules and conflict resolution expert. The conceptual idea is checked against the required material characteristics and manufacturing restrictions to show plausibility. If the idea does not prove feasibile, the process reverts to the conceptual design stage and an iterative approach is taken.

The expert system would also be capable of relaxing constraints and considering options which might have functional viability but which technically violated original constraints. Parts quality and performance might be increased with insignificant impacts on costs until some manufacturing constraint were violated which might project the part into a more precise and incrementally more expensive process. The "rate of satisficing" may thereby be a potential benefit of such a knowledge system.

A merit ranking of feasible materials and compatible manufacturing methods is then derived. From this list of possibilities, a detail design can be produced using the optimal materials/process selection. Optimality may be developed from a "criterion of excellence " [11] or a Principal Design Equation [4] depending upon the part type. The best selection is reduced, by the engineer, to a description for manufacture which includes all dimensions, tolerances, and specifications.

3. Expert Systems for Materials Selection

In determining the format of an expert system, the material selection process may be designed differently than if the system were analysing engineering data such as material properties. Such systems, described in Dieter [1],are useful but require direct input from the designer to specify what material properties are to be evaluated and what their relative merits might be. It may be beyond the ability of a neophyte designer to adequately weight material properties for a design and may exceed the expertise of an experienced designer to consistantly rank material properties so that the design and manufacturing requirements are optimally met and that the specification falls within the design and manufacturing philosophies of the organization.

In the material selection process, several levels of expert system may be required. The information required at conceptualization is generally broad-based and should emphasise material specifications that will best accomplish the functional and manufacturing requirements within the design and manufacturing philosophies of the system. At the preliminary design stage, the widest data base of materials and processes should be considered whereas in detail design more intricate specifications for the most likely material-process combinations can be included.

Existing computer aided materials and/or process programs have had measures of success in translating the process or parts of the process to an interactive environment.[12,13], however certain limitations can arise.[10]

Expertise in material selection demands several knowledge bases in a hierarchy of distinct but integrated expert systems. Meta-rules will determine the relative importance of the design and manufacturing requirements and should present the designer with a weighted ranking.

The expert system can secondly assist in the translation of functional requirements to material properties and can resolve conflicts where design and manufacturing objectives may be incompatible.

In the material database, a wide range of performance characteristics should be available. Apart from traditional engineering properties such as:-

- Physical and Chemical
- Mechanical
- Geometric

others of importance are:-

- Material Availability
- Standard sizes
- Property variability
- Cost
- Consumer Appeal

A third knowledge base is for manufacturing processes and this should contain process data which relates to materials, geometry, batch sizes and quality.

Material selection usually does not enter into the problem until the designer has started detailing the design. Candidate materials should play an early role. The optimal material is directly related to the manufacturing process as well as the part geometry; therefore, at every decision point, a selection of possible materials should be retained.

Most materials selection methods match a material to the functional requirement alone. But, as stated previously, this will not lead to a truly optimal solution. A material must not only satisfy all the functional requirements but must also be manufacturable into the desired geometry by the company, economical (since the material cost composes a major portion of total cost), and obtainable within a reasonable time frame since timing is always critical.

The first step in choosing a material is to define the specific functional requirements for the part. Will the part be loaded? If so, will the load be static or dynamic? Does the environment contain any corrosive elements? What is the operating temperature? A variety of questions must be posed in order to develop a complete functional profile. The list of material properties can be quite extensive, as shown in Table 1.

The objective is to include only those properties which are relevant. If more than one property is to be considered, then the relevant properties must also be prioritized and weighted according to importance. Each property must also be normalized in order to put each on an equal level before weighting factors are applied. Once normalization and weighting have taken place, then an optimization equation can be developed.

TABLE 1 Example material performance characteristics

Physical props.:	Mechanical props.:	Thermal props.:
density	hardness	conductivity
viscosity	modulus of elasticity	specific heat
porosity	Poisson's ratio	absorptivity
permeability	yield strength	emissivity
reflectivity	tensile strength	flammability
transparency	fatigue properties	
dimensional stability	impact strength	**Chemical props.:**
	fracture toughness	corrosion resistance
Electrical props.:	creep rate	oxidation
conductivity	damping properties	thermal stability
dielectric strength	wear properties	stress corrosion
hysteresis		hydraulic permeability
	Fabrication props.:	
	Castability	
	heat treatability	
	hardenability	
	formability	
	machinability	
	weldability	

The best way to show the importance of normalization is with an example: An automobile drive shaft is to be designed on the basis of light weight, maximum stiffness, maximum strength, and minimum cost per pound. A simple relationship might be:

$$F_1 = E + ts + 1/d + 1/m\$ \quad (1)$$

where F = factor of optimization
E = Young's Modulus (stiffness)
ts = tensile strength
d = density
m$ = cost per pound

This equation incorporates all the factors to be maximized with the reciprocal of all the factors to be minimized; hence, the result gives a maximum F value to those materials which have the highest values in the maximize portion and the lowest values in the minimize portion. In this case, the material with the best combination of high stiffness and strength and low cost and density will have the highest F value.

For example, five materials could be in the list of candidates. These materials have all been found to be readily available, environmentally and functionally suitable and manufacturable by the company. Table 2 below lists their relavent properties and some selection factors (which are explained in later).

Calculated F_1 values using equation (1) for the materials are listed in Table 2 From these calculations, tungsten carbide (WC) would seemingly be the best choice. But also notice that none of the other properties had much effect on the total value of F. Hence, the selection was based entirely on the value of Young's modulus (E). In order to consider the properties on a more equal basis, each property must be normalized to some maximum value. Equation (2), shown below, performs a normalization on the values in equation (1).

$$F_2 = E/E_{max} + ts/ts_{max} + d_{min}/d + m\$_{min}/m\$ \qquad (2)$$

F_2 values calculated by using equation (2) are shown in Table 2.

Young's Modulus and tensile strength are now less dominant. Normalization has succeeded in putting all the properties on an equal basis. But , what if it is desired to consider the properties on an unequal basis? For example, cost and light weight may be much more important to the

TABLE 2 Example materials and their properties

Material	E (X10^6)	Density	T.S. (X10^3)	Cost	F_1	F_2	F_3
Al 6061-T6	10	0.098	45	3.51	10,045,010	1.602	0.462
CS 1020	30	0.284	50	0.95	30,050,004	1.944	0.587
Ti	45	0.163	63	22.00	45,063,006	1.486	0.311
WC	82	0.510	215	13.50	82,215,002	2.262	0.386
SS 440A	24	0.280	107	3.62	29,107,004	1.464	0.345

designer than strength and stiffness. To accomplish this, weighting factors can be tagged to relevant properties in order to reflect the desired emphasis. The final optimizing relation is shown in equation (3).

$$F_3 = w_E\,(E/E_{max}) + w_{ts}(ts/ts_{max}) + w_d(d_{min}/d) + w_{m\$}(m\$_{min}/m\$) \quad (3)$$

To continue the example, the following weighting factors are attached: E=0.1, ts = 0.2, d=0.3, m\$ = 0.4. Now F values can be recalculated according to the prioritization given above. The optimization factor now represents the value judgement of the designer. Carbon steel, which an experienced designer might intuitively have chosen, has the best F value. Aluminum follows fairly close behind and may also inspire further investigation.

The technique used above is very simple and has the same fundamental entities:

1. <u>Completely define the functions of the part</u>.

2. Completely define the part's operating environment.
3. Determine the quantities to be produced.
4. Define the required time frame and materials available.
5. Convert the overall requirements of the part into the associated material properties.
6. Search a database. Examine the available materials for possible candidates and subject the likely candidates to the test. If no possible candidates are found, then go back and try to relax some of the requirements. This process is especially suitable for computerization.

Another approach to computer aided materials selection in demonstrated by Hanley and Hobson. [14] They developed two different methods for use in the evaluating process of their program. One was based on a geometic model of a polygon shown in Fig. 3.

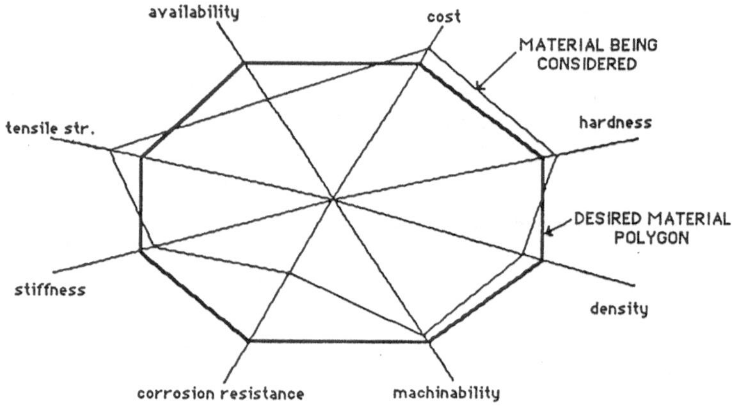

FIGURE 3 Polygon representation of geometrical material selection procedure

The closer the candidate polygon's size and shape is to the desired polygon, the better the chances of that material being selected. The equations used for the geometrical comparison are:

$$M.W.C. = \sum_{i=1}^{n} (\beta_i X_i / Y_i) / \sum_{i=1}^{n} (\beta_i)$$

$$B.F. = \sqrt{\sum_{i=1}^{n} \left(\frac{X_i}{Y_i} - M.W.C.\right)^2}$$

$$d = \sqrt{(1 - M.W.C.)^2 + (B.F.)^2}$$

Where M.W.C. - Mean Weighted Characteristic
β - weight factor for property
X - material property value for candidate material

Y - desired property value
n - number of material properties
B. F. - Balance Factor
d - selection factor

The Mean Weighted Characteristic (M.W.C.) is the normalized measure of how a particular material compares to the target material. The Balance Factor (B.F.) is the root-mean-square measure of the deviation of the candidate material's polygon to that of the desired. The selection factor (d) incorporates both of the previous factors and should, when minimized, provide the material with the best comparitive polygon.

The second method used is more straight forward. It is based on a simple algegraic equation which measures the per unit deviations of the candidate material property value from that of the desired value. The equation used to make the measurement is:

$$Z = \sum_{i=1}^{n} \beta_i \left| \frac{X_i}{Y_i} - 1 \right|$$

where
Z - selection factor
X - candidate material property value
Y - desired property value
ß - weight factor
n - number of properties

A minimum selection factor value (Z) will give the material with the least amount of total deviation from the desired values. This method also includes some constraint equations which can be used depending on whether the candidate properties must fall inside a maximum/minimum window or simply be compared to a target value. The constraining equations are:

$X_i/Y_i > 1$ for an upper limit on property i

$X_i/Y_i < 1$ for a lower limit on property i

$X_i/Y_i = 1$ for a targer value for property i

A system called MAPS [12] has been developed by a group lead by W.R.D. Wilson. Several years of work have been put into the three versions of MAPS (Material And Process Selection). The MAPS program was designed for general use and includes important aspects of manufacturing considerations in materials selection. The MAPS system is based around the premise that certain characteristics of a part; such as the number to be made, size, and geometry; restrict the available number of choices for manufacturing processes and materials.

The desired part is described to the computer via a 12-digit classification code. The code could be considered a special type of group technology classification. The first 5 digits of the code give information which allow for pruning of possible manufacturing processes (batch size, physical size, shape, tolerances, and surface roughness). The next three digits are used to describe environmental conditions during the foreseen operation of the part. This allows for pruning of candidate materials. The final four digits are used only if there are many possible material/process combinations.

There have been other attempts at computerized material selection, but the previous two examples are representative of the systems that already exist. The relative success of both systems should support the use of analytical models of a desired material and the use of "figure of merit" equations for material selection.

4. An Expert System for Materials Selection

This section describes a program for material selection that has been developed at the Design Productivity Center. Early in the development stages, much effort went into deciding the general format for such a program and it's supporting materials database. The following guidelines were decided upon:

1 - The system should be easy to use.

2 - The system should be as generic as possible, i.e. not slanted toward any specific type of component design.

3 - The system should provided adequate guidance for the neophyte designer.

4 - The system should be made as flexible as possible as possible, i.e. allow the operator to perform any type of material search desired.

5 - Develop a foundation for a more comprehensive system which will include materialproperty adjustments for changing environments and manufacturing and geometrical considerations.

6 - The system should provide quick and usable results.

7 - The system should be able to communicate why the particular materials were chosen.

8 - The database should contain several hundred materials with representatives from the major groups: ferrous and non-ferrous metals, plastics, ceramics, and composites.

9 - The properties given for each material in the database should include not only mechanical and physical, but also electrical, thermal, and commercial considerations.

10 - The database should be easy to update and expandible and be in a format which is accessible by different systems.

In an attempt to conform with the previously stated guidelines the anatomy shown in Figure 4 was developed for the system.

The system is comprised of several modular components, each of which has a definite

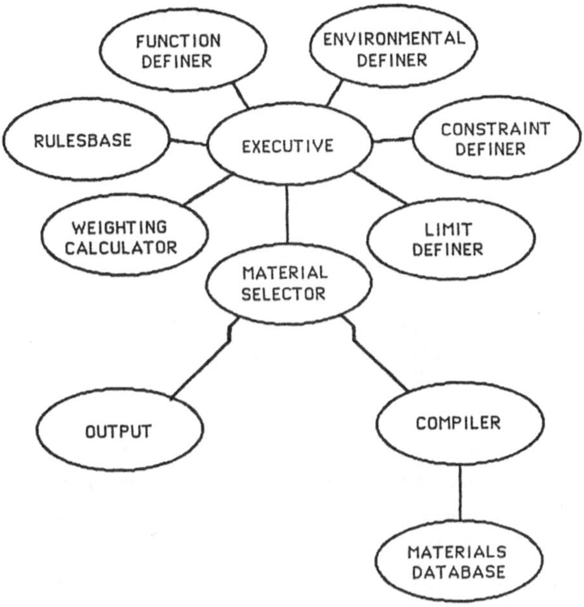

FIGURE 4 Relationship structure of the material selection program

function which can be accessed at any time. The entire process is monitored and controlled by an executive element. From here, the user can randomly visit any of the other elements. The four definition elements are used to describe the function, environment, and other desired characteristics of the part being designed and to set property value limits which must be met. The rulesbase is used to translate the information provided by the definer elements into weighted material properties. The weight calculator is then used to determine the coefficients of an optimization equation used in the material selector element.

The four definer elements are used to establish the parameters. The first visited is usually the function definer. Here the user selects one or more verb/noun combinations that best describe the basic function of the part. The functions described here are those which the part must perform in order not to fail. A part can ususlly be described with one or two basic functions. A typical function definer menu is shown below.

MATCH A VERB WITH A NOUN TO GIVE A FUNCTION

NOUNS	**VERBS**
A. CONTAIN	A. LOAD
B. SUPPORT	B. PRESSURE
C. MOVE	C. VIBRATION
D. ABSORB	D. TORQUE
E. DAMPEN	E. ELECTRICITY
F. INSULATE	F. HEAT
G. CONDUCT	G. ENERGY
H. TRANSMIT	

The user, for example, might choose "CONTAIN PRESSURE" as the function for an air tank. If the material chosen does not allow the tank to hold pressure then the tank should fail. Material properties derived from the basic function(s) will, hence be considered most important, because without these, the part will not function.

The user may choose one or more elements from the environment menu shown below.

ENVIRONMENTAL EFFECTS

SELECT ALL THAT APPLY:

A. DYNAMIC LOAD
B. IMPACT LOAD
C. DYNAMIC SURFACE CONTACT
D. HIGH TEMPERATURE ENVIRONMENT
E. CORROSIVE ENVIRONMENT

The selected environment(s) will correspond to an emphasis on material properties that will aid the part's survival. For example, if "CORROSIVE ENVIRONMENT" were chosen, then the system's response would be to apply a heavy weighting factor to corrosion resistance in the desired material profile.

The program also allows the imput of certain design guidelines through the constraint definer. This element of the system allows the operator, for example, to steer the selection towards a light-weight material or an inexpensive material. The choices provided are shown below.

CONSTRAINT SELECTION

SELECT ALL THAT APPLY:
A. MINIMIZE WEIGHT
B. MINIMIZE COST
C. MAXIMIZE AVAILABILITY
D. MAZIMIZE MACHINABILITY
E. MAXIMIZE FORMABILITY
F. PART TO BE MASSED PRODUCED

As with the function and environment definers, the chosen constraints are used to activate certain material properties for use in the screening process.

The final definer element, the limit definer, differs from the previous three. The definer allows the user to set a minimum value, maximum value, or both to any or all of the material properties. The limit values are not used in the optimization equation, but are instead used to screen materials. That is, the materials are first compared to the limits, if they violate them, the material is no longer considered. This can eliminate the possibility of having materials score well using the optimization equation, but falling short in one critical area. If any specific limit values are known by the designer, they should be inserted into the parameter listing. This will help the system more efficiently score and prioritize candidate materials.

The Executive, now knowing the part parameters, can use the rulesbase to translate the parameters into weighted materials properties corresponding to the needs of the part. The rules base consists of IF-THEN type statements, rules, which perform the translation. The idea is to incorporate expertise into the rules thereby giving the user the advantage of past experience (not necessarily his/her own).

The output from the rulesbase can be reviewed and changed very easily and quickly. This allows the operator to also include his/her own experience or intuition. Rules can be editted, added, or deleted to improve the system's performance.

Once the rulesbase has decided what materials properties are important, given each a weighting factor. The weighting calculation element then determines the coefficients for the optimization equation shown below:

$$F = \sum_{i=1}^{n} W.F._i \, (N.P.V.)_i$$

where F = optimization factor to be maximized

W.F. = weighting factor coefficient for material property i ; determined by weighting calculation element

N.P.V. = normalized property value

n = number of properties for each material; in this case n=25

The optimization factor (F) is calculated for each material using the sum of the products of the weighting factor multiplied by the normalized property value. For example if tensile strength had a weighting coefficient of 0.7, machinability had a coefficient of 0.3, and the rest of the properties had coefficients of zero, then the F value for a typical aluminum with the N.P.V. for tensile strength equal to 60 and the N.P.V. for machinability equal to 85 would be calculated as:

$$F = 0.7(60) + 0.3(85) = 67.5$$

The individual weighting factors for use in the optimization equation as coefficients, as stated before, is calculated by the weight calculation element using the information passed to it by the rulesbase. Each coefficient is scaled to a factor between 0 and 1 depending on its weighting factor and the weighting factors of all of the other properties. The equation used is :

$$W.C._i = W.F._i \, / \, TOT. W.F.$$

where $W.C._i$ = weighting coefficient used in optimization equation for material property i

$W.F._i$ = weighting factor given by rulesbase or operator input for material property i

TOT.W.F. = sum total of W.F.'s for all properties

Once the selection procedure is finished, the program lists the prioritized top n materials, where n equals the number of materials that the operator decided to prioritize.

Appendix 1 contains the results of a number of selections for commonly encountered engineering components. The materials database currently contains over 500 materials falling under the classification scheme in Figure 5.

5. Conclusions and Recommendations

The system developed can benefit the designer since:

1) It provides a rapid search of a large database for the materials which best matches the

desired characteristics.

2) It gives the novice a good starting point and leads him/her through the process of material selection in a fashion which is easily understood.
3) It helps prevent the possibility of the best material being omitted.
4) It provides enough flexibility that the knowledgable user can execute any search.
5) The program provides unlimited access materials.

The system is beneficial, but the results should be clinically analysed. The computer gives advice only, not absolute answers. The system is designed to emulate an expert designer who is no more than a trained and experienced human, and as we all know, humans make mistakes. Therefore, we must remember that the computer is not infallible! The system was designed as a tool to be used to free the designer from doing the menial task of searching a materials database. There is no desire (nor is it desirable) to build a system which will dominate the design process. Human conceptualization has yet to be adequately modeled and replicated and, until this is done, humans must spearhead the design process.

Operating the system was educational in itself as two very interesting trends were discovered. Firstly, it seemed that the part function was less influential than both the environment and applied constraints. The latter parameters provided more guidance when choosing and rank-ordering the best materials. This became very obvious during testing when the weight factors were interactively adjusted.

The second discovery concerned limiting values. Usually, the optimization equation is insufficient to perform the selection. Strategically assigned limiting values can play a pivotal role in pruning materials that are deficient in one or more areas. For example, an epoxy/carbon fiber composite may be poor in temperature resistance, but do so well as far as strength and cost that it becomes a high scoring material and therefore would be recommended even though it can not withstand the operating temperature. This trend can be eliminated by using a lower limit value for maximum usable temperature. The system would then ignore all materials that violate this limit. Hence, it is recommended that if any limiting values are known, they should be applied without hesitation.

In addition, many new requirements for an enhanced system were discovered. These are listed here in the form of recommendations.

- The materials database is incomplete, not in the number of materials, but within each material and it would be helpful if a method for calculating missing values were developed. This procedure could then be used during compilation of the database to create a compiled version with no missing values. Possible methods for calculating machinability and formability from other material properties are known to exist.[20]

- A data structure that would allow quick and efficient search strategies should be used to help speed of the search process. As the materials database grows and the amount of information to be calculated and checked for each material increases, this will be become more and more important. A hierarchical type structure will probably be beneficial.

- It may be desirable to modify material properties dynamically during the search according the type of operating environment. For example, tensile strength, creep strength, and fatigue end limit are all a function of temperature. If the part were to operate in a high temperature environment, then an adjustment of some of the properties may be useful.

- A truly optimal material selection cannot realistically be made without including manufacturing and geometrical considerations. As the system exists now, the manufacturing parameters are left to the operator.

• The optimization equation used in the program seemed to perform adequately, but since it so easy to change this equation in the program, alternative methods (equations) should be examined. One possibility might be in using linear penalty functions for deviations from the desired values.

• It may also be desirable to include the option of inputting secondary part functions into the part profile. These secondary functions would carry lower weighting factors than primary functions, but could be useful at times. For example, an automobile's carburetor provides little besides an enclosed area for fuel/air mixture. It is, however, desirable to have a carburetor with a reasonable amount of strength for attachment of additional parts. Some type of strength function could be considered, but secondary.

• Finally, it is recommended that the system be transferred (reprogrammed) in an interrupt-type programming environment such as LOOPS,since this method (and design in general) is non-algorithmic and requires access to all data manipulation routines at all times. LOOPS is also based around the hierarchical type decision structure which could be heavily implemented.

6. Acknowledgements

The authors wish to express their thanks to the staff and the Industrial Advisory Board of the Design Productivity Center for their support and the Columbia and Kansas City campuses of the University of Missouri for provision of facilities.

7. References

1 Dieter, G.E. , *Engineering Design - A Materials and Processing Approach,* McGraw-Hill Book Company, New York, 1983.

2 French, M. J., *Engineering Design: The Conceptual Stage*, Heinemann Educational Books Ltd., London, 1971.

3 Hill, P. H., *The Science of Engineering Design,* Holt, Rinehart and Winston, Inc., New York, 1970.

4 Johnson, R. C., *Optimum Design of Mechanical Elements,* John Wiley & Sons, Inc., New York, 1980.

5 Love, S. F., *Planning and Creating Successful Engineered Designs,* Advanced Professional Development Incorporated, Los Angeles, 1980.

6 Vidosic, J. P., *Elements of Design Engineering,* John Wiley & Sons, Inc., New York, 1969.

7 Woodson, T. T., *Introduction to Engineering Design,* McGraw-Hill Book Company, New York, 1966.

8 Buchanan, B. G. and Shortliffe, E. H., *Rules Based Expert Systems - The MYCIN Experiments of the Stanford Heuristic Programming Project,* Addison-Wesley Publishing Co., Reading, MA, 1985.

9 Brown, D. C. and Chandrasakian, B., "An Approach to Expert Systems for Mechanical Design," *Trends and Applications,* National Bureau of Standards, Gaithersberg, MD, May 25-26, 1983.

10 Greenway, R.B. and Blundell, J.K. "An Expert System for Design and Manufacturing Integration" 12th Design Automation Conference,Columbus ,Ohio, October 5-8, 1986

11 Groover, M. P., *Automation, Production Systems and Computer-Aided Manufacturing,* Prentice-Hall, 1980.

12 Dargie, P. P., Parmeshwar, K. and Wilson, W. R. D., "MAPS-1: Computer Aided Design System for Preliminary Material and Manufacturing Process Selection," Transactions of ASME, Design Engineering Conference, Sept. 28 - Oct. 1, 1980, Paper No. 80- DET-51.

13 Dixon, J. R. and Simmons, M. K., "Expert Systems for Mechanical Design: A Program of Research," ASME Design Engineering Conference, Sept. 10-13, 1985, Paper No. 85-DET-78.

14 Hanley, D.P. and Hobson, E., "Computerized Material Selection,"Journal of Engineering Materials and Technology, October 1973.

15 Datsko, J., "Putting the 'Science' in Materials Science," Machine Design, October 24, 1985, pp 85-87.

16 Olsson, L., Bengtson, U., and Fischmeister, H., "Computer-Aided Materials Selection," Computers in Materials Technology, Proceedings of the International Conference held at the Institute of Technology, Linkoping University, Sweden, June 4-5, 1980.

17 Miaw, D.C. and Wilson, W.R.D., "Use of Figures of Merit in Computer-Aided Material Selection and Manufacturing Process Planning," Transactions of ASME, Vol. 104, October 1982.

A DYNAMICALLY PARTITIONED OUT-OF-CORE SKYLINE SOLVER FOR MICRO-COMPUTERS

A J du Toit* & W S Doyle**

*Research Assistant ** Assoc. Professor
Department of Civil Engineering, University of Cape Town

SUMMARY

In this paper we present improvements to the well known block partitioned form for the out-of-core solution of sets of linear equations. Our solver operates on a coefficient matrix which is stored out-of-core as a one dimensional array of values under the skyline.

The improvements pertain to the partitioning of the coefficient matrix into blocks of reduced and unreduced terms which vary as the decomposition of the matrix proceeds. This dynamic form of partitioning selects an optimum size of the blocks and leads to substantial savings in terms of execution time and data transfer from backup to storage.

The algorithm uses the Cholesky method for decomposition of the coefficient matrix but can be modified to accommodate Crout reduction for instance.

The solver was developed specifically for the micro-computer environment but the concepts are general and may be implemented on a full range of machines.

INTRODUCTION

We are primarily interested in the implementation of the finite element technique on micro-computers with limited in-core storage capabilities. A solver with an out of core capability is therefore essential.

The coefficient matrix, resulting from the linear elastic finite element analysis of structures, is sparse, symmetric and positive definite. These characteristics are effectively exploited in the development of an efficient algorithm for the solution of large sets of equations. The most well known and widely used of these algorithm is the fixed block partitioned form of solution [1].

Another extremely important aspect in the micro-computer environment in particular, is modularity in programming. We will not deal in depth with this topic, but do feel that a brief word is called for as it is fundamental to our approach.

MODULARITY

A modular program is usually one which consists of an assembly of logical units or sub-routines dealing with specific aspects of the algorithm in a logical and sequential manner. We have extended the concept of modularity so that subsections of a more complex package are entirely independent program modules.

This approach has already been adopted very successfully in the R.M. [2] suite of programs for instance. The advantages of this modular approach to programming are numerous and will not be discussed here. Suffice it to say that modularity is usually desirable in the micro-computer environment, especially if the program is long and complicated.

ALGORITHM

It is generally accepted that, for large sets of linear equations, some form of direct solution algorithm, based on Gauss elimination, is the most efficient. The most suitable algorithms are those which decompose a sparsely populated, square, symmetric coefficient matrix into a lower/upper triangular form [3]. This category of algorithm includes Crout reduction and the Cholesky method. The development of our solver is based on the latter.

Considering the nature of the solving algorithm and our aim toward modularity, the solution was divided into the following two distinctly separate program modules;

a) Module 1: The decompsotion of the coefficient matrix to the form $A = U^T U$. The decomposed form of the coefficient matrix (U) is stored as a one dimensional vector of elements below the skyline on auxiliary storage.

b) Module 2: The forward reduction and back substitution of the load vectors to obtain the nodal displacements. The objective is that different load vectors can be added and existing load vectors can be combined without rerunning moodule 1. These two modules are described in detail in the following sections.

STORAGE OF COEFFICIENT MATRIX AND LOAD VECTORS

The terms below the skyline are stored for each consecutive column as shown in figure 1.

The skyline storage scheme is more complex to set up and manipulate than the often used banded form, but is by far the most efficient when it comes to data transfer and in-core memory requirements.

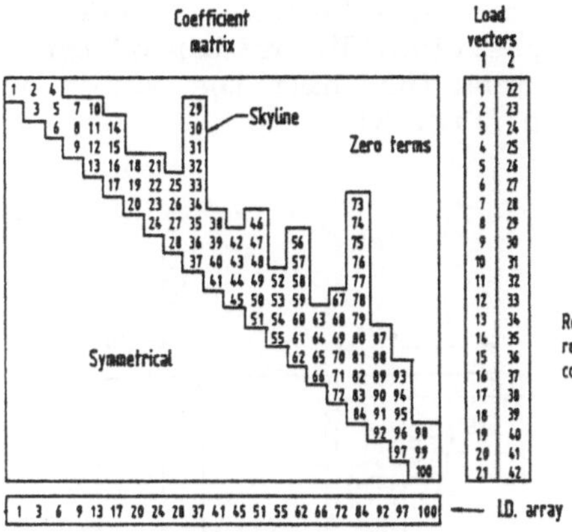

FIGURE 1

Skyline scheme of storage for the coefficient matrix

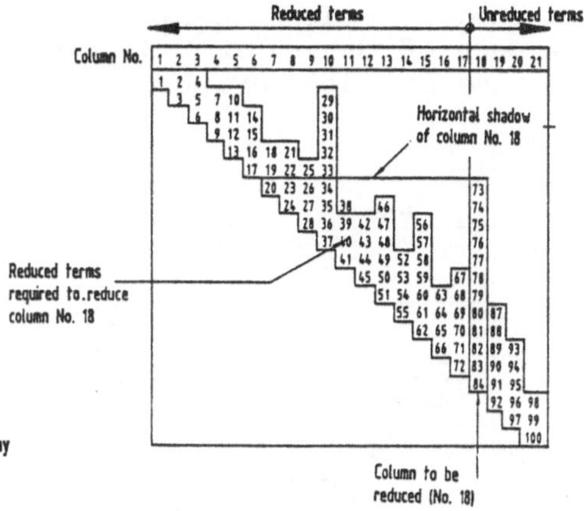

FIGURE 2

Terms required to reduce a particular column

In order to identify specific elements in the coefficient matrix from the one dimensional format of the skyline scheme, a diagonal address (ID) array is required. The ID array lists the addresses of the diagonal terms in the coefficient matrix (fig. 1). Load vectors are stored consecutively for each load case.

DYNAMIC V/S FIXED BLOCK PARTITIONING

This section deals with the program which decomposes the coefficient matrix to the form $A = U^T U$.

Our solver is based on the well known fixed partitioned form of solution [1]. The partitions are however not rigidly defined hence the term "dynamic" partitioning. A brief review of the fixed block partitioned form of the algorithm will be given in order to highlight the salient differences and point out the advantages of dynamic partitioning.

Both fixed block and dynamic block partitioning depend on the fundamental principle that the decomposition of elements in a particular column of the coefficient matrix only require those (already reduced) terms which fall within the horizontal shadow of the column under consideration (fig. 2). [3]

It is generally faster to work with whole columns at a time. This requires far less housekeeping and is more efficient in the transfer of data from backup storage.

With fixed block partitioning, the coefficient matrix is divided into blocks prior to decomposition. The number of terms in each of these blocks is always less than half the number of storage locations available in core (fig. 3).

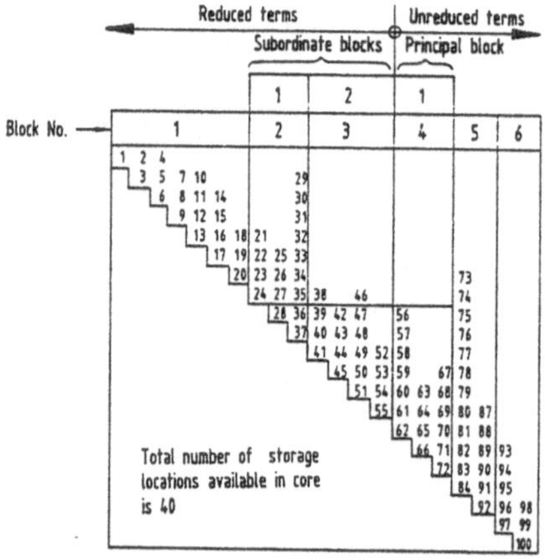

FIGURE 3

Block partitioning

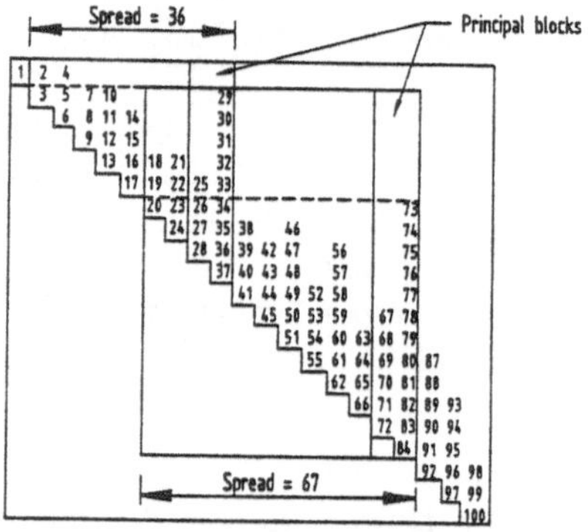

FIGURE 4

Spread of a principal block

Having partitioned the coefficient matrix, the solution proceeds as follows:

i) The first two blocks are transferred into core, reduced and these terms transfered to back-up storage.

ii) The next block, now termed the principle block, is then loaded into core.

iii) For the principle block, subordinate blocks are defined as those lying wholly or partially within the horizontal shadow of the principle block (fig. 3). These subordinate blocks (already reduced) are re-loaded into core (in ascending order) and the decomposition of the principle block proceeds.

iv) On completion of the decomposition, the principle block is transferred to back-up storage.

v) The next unreduced block now becomes the principle block and is loaded into core. The reduction procedure ((ii) to (iv)) is repeated.

Dynamic partitioning also approaches the problem in block form. The coefficient matrix is divided into blocks (principle and subordinate). The smallest block size for either the reduced or unreduced partition is the number of terms in the semi-bandwidth of the coefficient matrix.

We define the "spread" of any principle block in the coefficient matrix as the sum of the terms in the block and the terms in columns which fall wholly or partially within the horizontal shadow of this block (fig. 4). A parameter N defines the minimum size of any one unreduced block see later. With N set, partitioning of blocks then proceeds as outlined below.

Presume that the decomposition of the coefficient matrix has advanced some way along the matrix. Unreduced columns are added sequentially into the principle block until the total number of elements in the block (P) exceeds the minimum block size (N). ie: $\mathbf{P \geq N}$. The "spread" (S) of this block is then calculated (fig. 4).

If the total number of available in-core storage locations is M, three possibilities exist ie. $S > M$, $S = M$ and $S < M$ depending upon the value of S the decomposition proceeds as follows:

If $S > M$, fig. 5(a) the principle block is transferred to core. Subordinate blocks are partitioned to contain only whole columns, starting with the lowest required decomposed column. The subordinate blocks are sized such that at any one time the principle block and one subordinate block can reside in-core. Decomposition of the principle block then follows along the same lines as in fixed block partitioning.

FIGURE 5

Dynamic partitioning

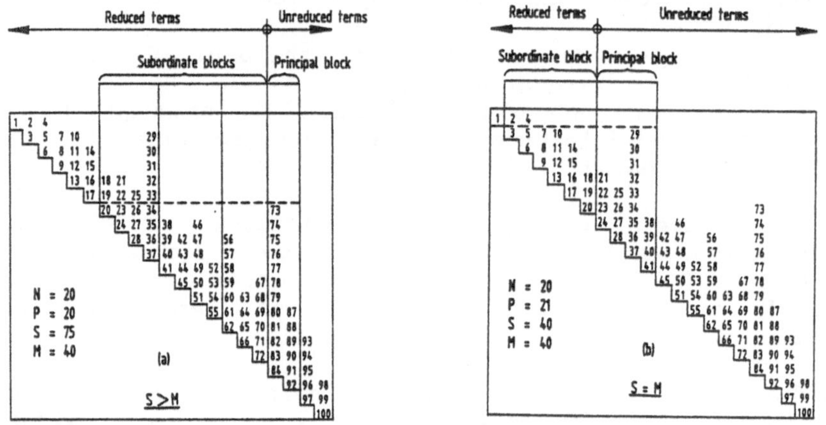

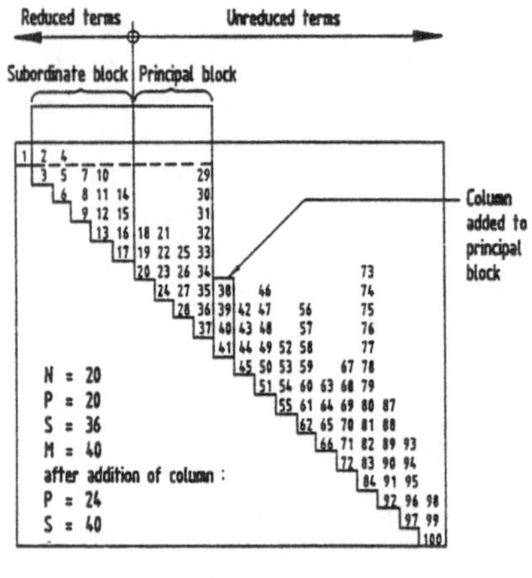

(c)

S<M

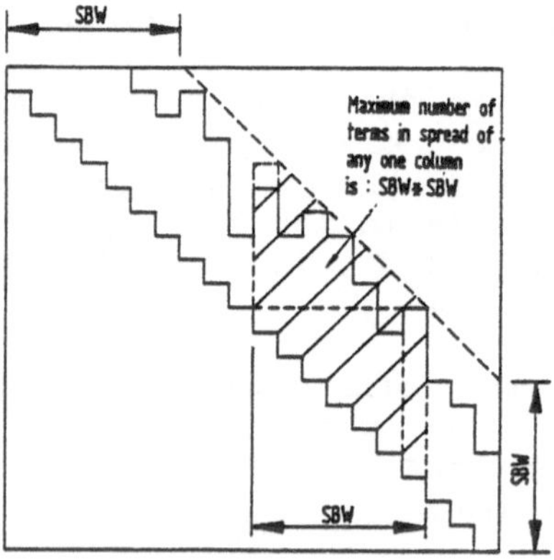

FIGURE 6

Maximum number of terms in spread of any column

If S = M, fig. 5(b) all the elements in the "spread" of the principle block are transferred to core, the principle block is decomposed and transferred to back-up storage.

If S < M, fig. 5(c) the possibility exists that the principle block can be increased in size. The next unreduced column is added to the principle block and the spread (S) is adjusted. Unreduced columns are added to the principle block until the addition of one more column would result in : S > M. The decomposition then proceeds along the same lines as in (b).

All the housekeeping and partitioning of principle and subordinate blocks is calculated using the ID array. As only integer arithmetic is involved, this is extremely fast and efficient.

MINIMUM BLOCK SIZE (N)

The selection of the parameter N, which determines the minimum size of the principle block, does affect the efficiency of the algorithm. For a matrix with a semi-bandwidth (SBW) and a machine with (M) available in-core storage locations. (N) must of course lie within the following limits:

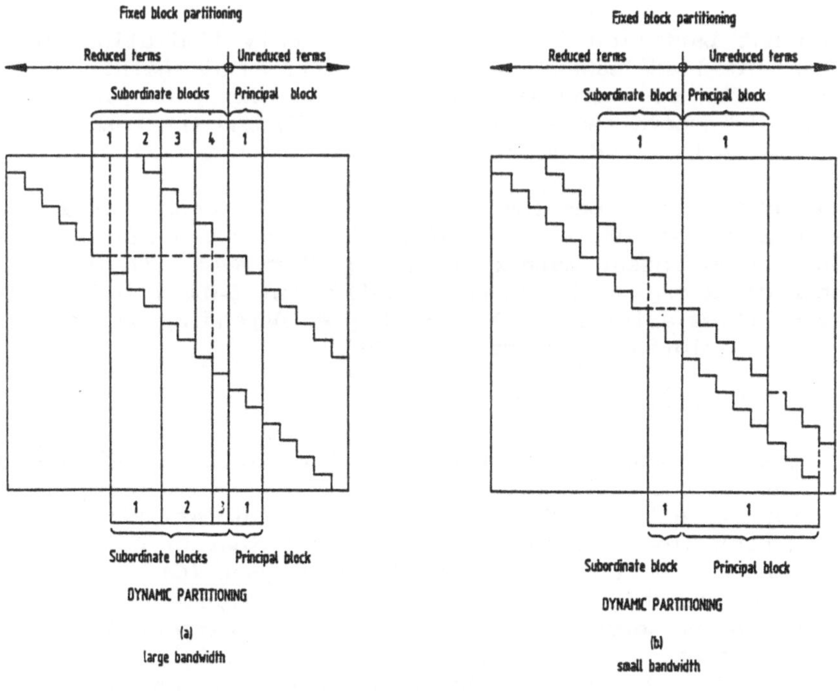

FIGURE 7

Dynamic vs block partitioning

$$SBW \leq N \leq M - SBW$$

In our algorithm we set N as follows (fig. 6):

a) If SBW*SBW > M/2 then N = M/2
b) If SBW*SBW < M/2 then N = M-SBW*SBW

It is clear that the narrower the bandwidth, the more efficient dynamic partitioning becomes. Fixed block partitioning on the other hand becomes progressively less efficient as the bandwidth reduces! (fig. 7). To this end we have also included a bandwidth reduction routine in our program [4].

FORWARD REDUCTION / BACK SUBSTITUTION

This section deals with the module which solves the system of linear equations $\mathbf{U}^{T}\mathbf{U}\,\mathbf{x} = \mathbf{b}$. The solution proceeds in two stages ie.:

a) Forward reduction : $U^{T} y = b$ - determine y
b) Back substitution : $U \; x = y$ - determine x

We do not partition load vectors. The only limitation on the number of load cases the solver will deal with, is the availability of auxiliary storage. Let the total number of available in-core storage locations be (M) and the total number of terms in the load vector (L).

Assuming that only one load vector will be processed at a time, the total number of in-core storage locations available for the reduced coefficient matrix is then: R = M-L . If the total number of terms in the reduced coefficient matrix is (C), two possibilities exist ie. C > R and C $\leq$ R. Depending on the value of R, the solution proceeds as follows:

a) C > R (fig. 8(a))

i) Transfer the first load vector (L) to be considered to core, now R = M - L..
ii) Partition the coefficient matrix into blocks. Block size $\leq$ R each block comprises the maximum number of columns that can reside in the remaining in core space (C).
iii) Load the blocks of the reduced coefficient matrix in ascending order into core and then perform a forward reduction.
iv) Load the blocks of the reduced coefficient matrix in decending order into core and then perform a back substitution.

v) Transfer the resulting displacement vector to auxiliary storage or print-out.

vi) Consider the next load case.

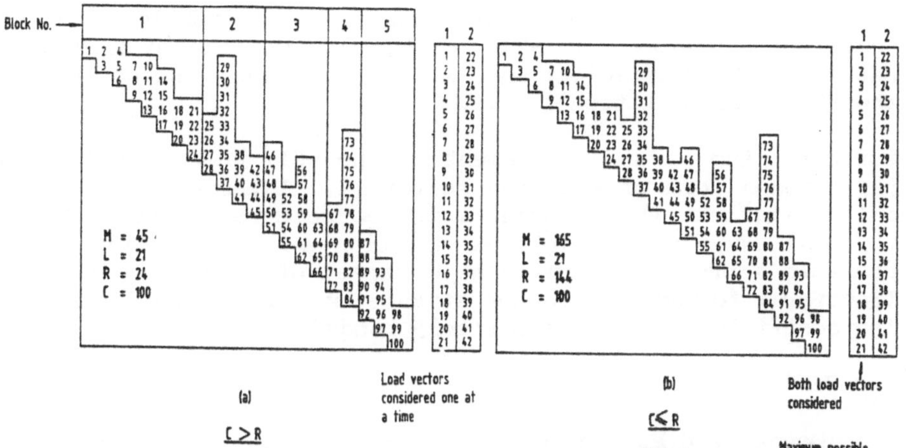

FIGURE 8

Forward reduction - Back substitution

b) C ≤ R (fig. 8(b))

In this instance the entire reduced coefficient matrix plus at least one load vector can reside in memory at any one time.

i) Transfer the entire reduced coefficient matrix into core. These terms will reside in core for the duration of the solution.

ii) Determine the number of load vectors that can reside in the remaining in-core space ((M-C)/L)).

iii) Load as many load vectors into core as possible or until the last load case is encountered.

iv) Perform a forward reduction and a back substitution on these load vectors.

v) Transfer the resulting displacement vectors to auxiliary storage or print-out.

Consider the next set of load vectors.

CONCLUSIONS

We have developed an out of core solver for large sets of simultaneous linear equations in matrix form. This development was activated by the need for an efficient out of core solver for the implementation of finite element techniques on micro-computers. Previous research in this field has on the whole concentrated on minimising in-core requirements. However, it is clear that the optimal use of in-core memory is more in keeping with modern trends in hardware development, and this was also our premise.

The algorithm is based on the lower/upper triangular form of Choleski reduction and takes account of the banded symmetric nature of the coefficient matrix most frequently encountered in the finite element technique.

Our algorithm is an improvement on the well known fixed block partitioned form of equation solver [1]. Dynamic partitioning only partitions the relevant matrices on consideration of the physical requirements of the solution as it proceeds. Most dramatic improvements in terms of speed of execution and data transferal to and from backup storage is achieved especially if the coefficient matrix is narrowly banded.

REFERENCES

1. Mondkar, D.P. and Powell, 'Large capacity equation solver for structural analysis', Comp. & Struc., 4, (1974), 699-728.

2. Pircher, H., R.M. Suite of Programs. Technische Datenuerbeitung, Austria.

3. Tewarson, R.P., Sparse matrices. Academic Press, 1973.

4. Collins, R.J., 'Bandwidth reduction by automatic renumbering', Int. J. Num. Meth. Engng., 6, (1973) 345-356.

A GENERALISED A-POSTERIORI ANALYSIS OF THE DISCRETIZATION ERROR IN NUMERICAL SOLUTIONS TO LINEAR PROBLEMS

D.W.KELLY
University Of New South Wales,
Kensington, NSW, 2033, Australia

SUMMARY

Aspects of a posteriori error analysis are discussed including the need to identify errors caused by roundoff and physical modelling approximations in addition to discretization. A general procedure is demonstrated on a one-dimensional example identifying exactly the error in stress and displacement. The extensions required to apply the procedures in two and three dimensions are then described.

1 INTRODUCTION

The problem of the validation of the results of a numerical solution is a complex one due to the variety of sources of error which can invalidate the solution. These include roundoff error in the mathematical computations, discretization errors sensitive to the design of the mesh and the number and type of elements used, physical approximations and errors in the data. Traditionally, new finite elements and finite element programs are subject to a set of benchmark tests which validate the theoretical development of the element and the computer code itself. However, it still remains to provide the user with a measure of the error in the solution which can indicate the effect, on nodal displacements and stress, of roundoff and discretization errors and those physical approximations that can be clearly identified.

Recent work [1] on the development of a posteriori error estimates has provided an error measure which can be applied to a solution from any source and indeed one application in [2] considers the "solution" given by a random number generator. Clearly this error measure is independent of the numerical procedure used to define the solution and will sense both the discretization error and roundoff error. The basis of the error analysis is described in Section 2 of this paper and considers applications to a simple problem in one dimension in Section 3. Extentions to problems in multiple dimensions are considered in Section 4.

The standard checks for roundoff error and poor discretization in finite element codes include the sensing of poor element geometry during assembly of the element matrices, diagonal decay during the reduction of the stiffness matrix and calculation of the residuals of the algebraic equations [3,4]. It should be noted however, that these procedures only serve to indicate the presence of an error, and do little to indicate, in absolute terms, the error in the final solution. An estimation of a measure of the error is the primary goal of the present research.

2 THE ERROR IN A NUMERICAL SOLUTION

Most boundary value problems in science and engineering are expressed mathematically by differential equations of the form

$$\mathcal{L}u + f = 0 \text{ over the domain } \Omega \tag{1}$$

subject to boundary conditions

$$u = \overline{u} \text{ on } \Gamma_u \tag{2}$$

and

$$\frac{\partial u}{\partial n} = \overline{\frac{\partial u}{\partial n}} \text{ on } \Gamma_n \tag{3}$$

An approximate solution u_h will not satisfy (1) exactly. Instead

$$\mathcal{L}u_h + f = R \neq 0 \tag{4}$$

where R is the residual in the differential equation. Equation (4) applies pointwise over the domain so the continuous variation of u_h must be defined. In the present work u_h is deemed to consist of linear interpolants between specified nodal values in one dimension and by bilinear interpolants between four and eight nodes in two and three dimensions. Then u_h is smooth (ie. continuous with continuous derivatives) over the subregions or elements, but has discontinuous derivatives across the boundaries between the elements. The residual R therefore has two components :- a regular residual r on each element, and a singular residual $\mathcal{J}$ on the boundaries between the elements.

The error in the solution is defined as:

$$e = u - u_h \tag{5}$$

Subtracting (4) from (1)

$$\mathcal{L}(u - u_h) = -R$$

or

$$\mathcal{L}e + R = 0 \tag{6}$$

so that the error is governed by the same operator as the original differential equation. Since $u_h = \overline{u}$ can be enforced exactly for $\overline{u}$ linear, the boundary conditions on e become

$$e = 0 \text{ on } \Gamma_u \tag{7}$$

and substituting from (5) into (3)

$$\frac{\partial e}{\partial n} = \overline{\frac{\partial u}{\partial n}} - \frac{\partial u_h}{\partial n} \text{ on } \Gamma_n \tag{8}$$

Clearly to determine the error means to determine the solution to (6) subject to the boundary conditions (7) and (8).

The natural distance measure in linear elasticity and potential flow is the energy norm which in one dimension is given by:

$$B(u,u) = \int_\Omega \frac{du}{dx}\frac{du}{dx}dx \tag{9}$$

substituting for u from (5)

$$B(u,u) = B(u_h,u_h) + 2B(u_h,e) + B(e,e). \tag{10}$$

Here

$$B(u_h,e) = B(u_h,u-u_h) = B(u,u_h) - B(u_h,u_h)$$

so that (10) can be written

$$B(u,u) = B(u_h,u_h) + 2\left[B(u,u_h) - B(u_h,u_h)\right] + B(e,e) \tag{11}$$

If the terms on the right-hand side of (11) are known exactly, then the energy norm of the exact solution can be calculated. Integrating by parts:

$$\begin{aligned} B(u,u_h) &= \int u_h \frac{du}{dn} d\Gamma - \int u_h \frac{d^2u}{dx^2} d\Omega \\ &= \left[u_h \frac{du}{dn}\right] + (f,u_h) \end{aligned} \tag{12}$$

It is therefore possible to evaluate $B(u,u_h)$ exactly for the standard problem in structural mechanics in which $u_h = 0$ on Γ_u so that the boundary integral contains contributions only from Γ_n on which du/dn is defined.

Finally $B(e,e)$ can be determined from e given by the solution of (6) subject to the boundary conditions (7) and (8). In general, if these equations cannot be solved to give an exact solution for e, then the analysis should be designed to ensure that an upper estimate $B_u(e,e)$ is obtained for $B(e,e)$. Then

$$B(e,e) \leq B_u(e,e) \tag{13}$$

and from (11)

$$B(u,u) \leq B(u_h,u_h) + 2\left[B(u,u_h) - B(u_h,u_h)\right] + B_u(e,e) \tag{14}$$

giving an upper bound on the energy of the exact solution. If $u_h \neq 0$ on Γ_u then du/dn in (12) must be estimated on Γ_n from the numerical solution u_h and the analysis, which was exact to this point, becomes approximate.

At this point we note that the origin of the solution u_h is arbitrary and can be provided by a numerical procedure such as finite elements (for which $B(u_h,e) = 0$ in (10)), finite difference, the boundary element method (so long as u_h at discrete points on the domain is evaluated), or indeed from an intelligent guess or random number generator. The latter clearly indicates that the effect of rounding errors will be sensed and correctly evaluated. Alternatively the procedure for deriving u_h could include physical modelling approximations which will be correctly sensed if the analysis carried out to estimate the error using (6), (7) and (8) does not contain these approximations.

3 AN EXAMPLE IN ONE DIMENSION

Let u be the axial displacement in a one-dimensional string subject to a body force b such as gravity. Then if E is Young's modulus and ε and σ are the axial strain and stress,

$$\varepsilon = \frac{du}{dx} \tag{15}$$

$$\sigma = E\varepsilon \tag{16}$$

and

$$\frac{d\sigma}{dx} = -b \tag{17}$$

Substituting (15) and (16) into (17) gives the governing differential equation

$$E\frac{d^2u}{dx^2} + b = 0 \tag{18}$$

where E has been assumed constant. If the string is of length 1,is fixed at $x = 0$ and free at $x = 1$, then the boundary conditions become

$$u = 0 \text{ at } x = 0 \tag{19}$$

and

$$\frac{du}{dx} = 0 \text{ at } x = 1. \tag{20}$$

Integrating (18) and substituting the appropriate boundary conditions gives the exact solution

$$u = \frac{b}{E}\left(x - \frac{x^2}{2}\right) \tag{21}$$

The solution is plotted in Figure 1 together with a proposal for the solution u_h. The aim of the following analysis is to determine the error $e = u - u_h$.

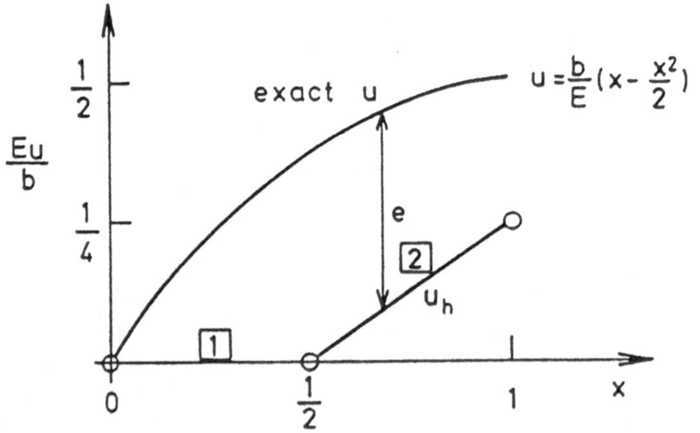

Figure 1. Solutions for the one dimensional example.

$u_h = 0$ at $0 \leq x \leq \frac{1}{2}$

$u_h = \frac{b}{E}(\frac{x}{2} - \frac{1}{4})$ at $\frac{1}{2} \leq x \leq 1$

Since u_h is linear on each element, the regular part of the residual

$$r = E\frac{d^2u_h}{dx^2} + b = b \tag{22}$$

and on the interface at $x = \frac{1}{2}$ the singular part is infinite such that

$$\mathcal{J} = \lim_{\Delta\to 0}\int_{\frac{1}{2}-\Delta}^{\frac{1}{2}+\Delta}(E\frac{d^2u_h}{dx^2} + b)dx$$

$$= E\frac{du_h}{dx}\bigg|_{x=\frac{1}{2}+\Delta} - E\,\frac{du_h}{dx}\bigg|_{x=\frac{1}{2}-\Delta} \tag{23}$$

$$= \sigma_h|_{\frac{1}{2}+\Delta} - \sigma_h|_{\frac{1}{2}-\Delta}$$

which is the external force per unit area which has to be applied to restore equilibrium at the interface. Finally

$$e = 0 \text{ at } x = 0 \tag{24}$$

and

$$\frac{de}{dn} = \frac{du}{dn} - \frac{du_h}{dn} = -\frac{du_h}{dn} \text{ at } x = 1 \tag{25}$$

To solve note that the error e is subject to the same differential equations as the original solution so that we can write

$$\overline{\varepsilon} = \frac{de}{dx} \tag{26}$$

$$\overline{\sigma} = E\overline{\varepsilon} \tag{27}$$

and

$$\frac{d\overline{\sigma}}{dx} = -r \tag{28}$$

Substituting $r = b$ and choosing to satisfy (28) exactly defines

$$\overline{\sigma} = -bx + c_1 \tag{29}$$

but at $x = 1$ (25) defines

$$\overline{\sigma}|_{x=1} = -E\frac{du_h}{dx} = -\frac{b}{2}$$

giving $c_1 = \frac{b}{2}$, so that on Element 2

$$\overline{\sigma}_2 = b(\frac{1}{2} - x) \tag{30}$$

On the interface (23) gives $\mathcal{J} = \frac{b}{2}$. Since on the interface u is continuous (23) can be expanded to

$$\mathcal{J} = E\left(\frac{du}{dx} - \frac{du_h}{dx}\bigg|_{\frac{1}{2}-\Delta}\right) - E\left(\frac{du}{dx} - \frac{du_h}{dx}\bigg|_{\frac{1}{2}+\Delta}\right)$$

$$= E\left.\frac{de}{dx}\right|_{\frac{1}{2}-\Delta} - E\left.\frac{de}{dx}\right|_{\frac{1}{2}+\Delta}$$

$$\text{thus}\quad \overline{\sigma}_1|_{\frac{1}{2}} = \mathcal{J} + \overline{\sigma}_2|_{\frac{1}{2}} = \frac{b}{2} + 0 = \frac{b}{2}$$

which can be used to define c_1 in expression (29) so that

$$\overline{\sigma}_1 = b(1-x) \tag{31}$$

Hence on Element 1,

$$\frac{de}{dx} = \frac{b}{E}(1-x)$$

so that

$$e = \frac{b}{E}\left(x - \frac{x^2}{2}\right) + c_2$$

But from (24) $c_2 = 0$ and $e = \frac{b}{E}\left(x - \frac{x^2}{2}\right)$ which , from (21) and the definition of u_h in Figure 1, is the exact solution. Similarly on Element 2

$$e = \frac{b}{E}\left(\frac{x}{2} - \frac{x^2}{2}\right) + c_3$$

From continuity and the expression for e on Element 1

$$e\left(\frac{1}{2}\right) = \frac{b}{E}\frac{3}{8}$$

so that on Element 2

$$e = \frac{b}{E}\left(\frac{x}{2} - \frac{x^2}{2} + \frac{1}{4}\right)$$

which again, is the exact solution for e.

To evaluate the bounds on the energy norm each term in (14) must be evaluated

$$\text{from (9)}\quad B(u_h, u_h) = \int_0^1 \left(\frac{du_h}{dx}\right)^2 dx = \frac{b^2}{8E^2}$$

$$\text{from (12)}\quad B(u, u_h) = \int_0^1 \frac{b}{E} u_h dx = \frac{b^2}{16E^2}$$

The upper bound $B_u(e,e)$ is defined from the complementary analysis which has been completed here. It follows that from (9)

$$B(e,e) = \int_0^1 \left(\frac{de}{dx}\right)^2 dx = \frac{1}{3}\frac{b^2}{E^2}$$

Substituting into (11)

$$B(u,u) = \frac{b^2}{e^2}\left(\frac{1}{8} + 2\left(\frac{1}{16} - \frac{1}{8}\right) + \frac{1}{3}\right)$$

$$= \frac{b^2}{E^2}\cdot\frac{1}{3} \tag{32}$$

A simple evaluation of $B(u,u)$ follows by substituting from (21) into (9) to give $B(u,u) = \frac{1}{3}\frac{b^2}{E^2}$. The upper bound is therefore exact which can be expected since the previous analysis has defined exactly the error.

4 EXTENSIONS FOR PROBLEMS IN MULTIPLE DIMENSIONS

The multi-dimensional equivalent of the analysis in Section 3 satisfying (28) a priori and preserving the upper bound in (13) is the complementary approach based on the superposition of equilibrating stress systems. The one dimensional example (28) enforces self-equilibration of the residual forces on each element since

$$\int_a^b \left(\frac{d\overline{\sigma}}{dx} + r\right) dx = \overline{\sigma}|_a^b + \int_a^b r dx = 0$$

In multiple dimensions

$$\int_\Gamma \overline{\sigma}_n ds + \int_\Omega r d\Omega = 0 \tag{33}$$

with the supplementary condition that across the interface

$$\overline{\sigma}^+ - \overline{\sigma}^- = \mathcal{J} \tag{34}$$

where $\mathcal{J}$ is the singular interface residual defined by (23). For the one dimensional example these self equilibrating systems took the form in Figure 2.

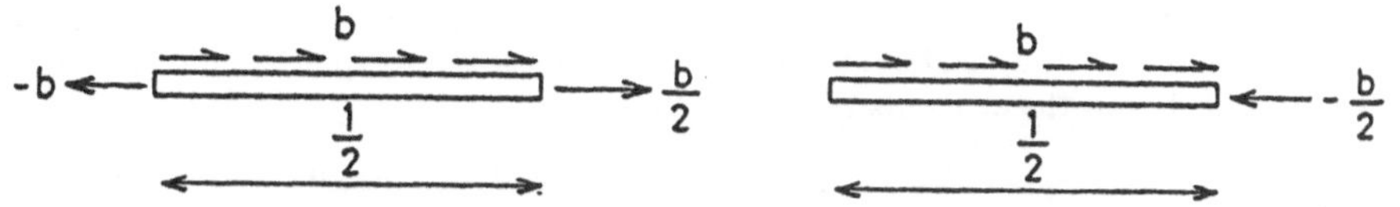

Figure 2: Self-equilibration of the residuals in the one dimensional example.

In two dimensions and plane stress (28) becomes

$$\frac{\partial \overline{\sigma}_x}{\partial x} + \frac{\partial \overline{\tau}_{xy}}{\partial y} + b_x = 0$$

$$\text{and} \quad \frac{\partial \overline{\sigma}_y}{\partial y} + \frac{\partial \overline{\tau}_{xy}}{\partial x} + b_y = 0$$

The basic equilibrating unit is depicted in Figure 3 with linear variation of normal and shear tractions if the variation of u_h on the element is assumed to be bilinear. Equilibrium now implies force equilibrium in each of the coordinate directions and moment equilibrium about an arbitrary point.

The major effort required in this analysis is to define the splitting of the singular interface residual implied by (34) such that each of the systems depicted in Figure 3

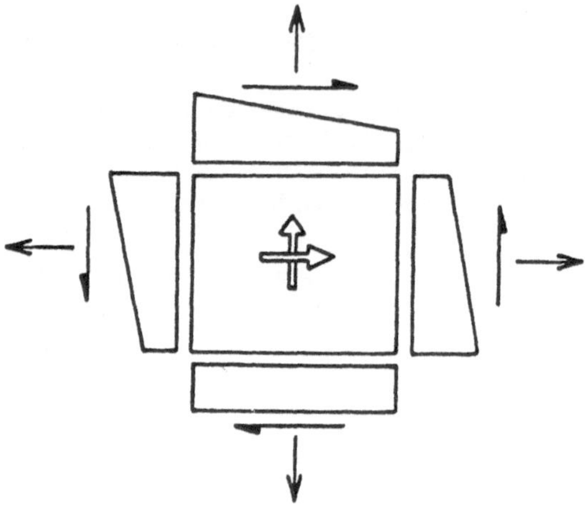

Figure 3: The residual system in two dimensions.

self-equilibrate. An initial guess (which is correct if u_h is a finite element solution and the error is uniformly distributed on a uniform mesh) is to split the interface residuals equally between the elements so that in (34)

$$\overline{\sigma}^{+} = c\mathcal{J} \text{ and } \overline{\sigma}^{-} = -(1-c)\mathcal{J} \text{ with } c = 0.5$$

In general however the equilibrating systems are not unique because, as each element is added to the general mesh depicted in Figure 4, it is possible to define the four splitting factors for the normal and shear tractions on the free edges to ensure the three conditions of equilibrium. The closest upper bound in (13) will define the minimum of a function of the form

$$\mathcal{I} = \alpha \sum_{\text{elements}} \left(\int_{\Gamma_i} \overline{\sigma}_n ds + \int_{\Omega_i} r d\Omega \right)^2 + \beta B_u(e,e) \tag{35}$$

In a proposal for a practical implementation of these procedures a threshhold value for an acceptable lack of equilibrium is defined and in the remaining regions of the mesh a search is executed to define the minimum of (35). Further we note that the evaluation of the energy of the error proceeds element by element so that

$$B(e,e) = \sum_{\text{elements}} B_i(e,e) \tag{36}$$

and it has been shown that in the optimal mesh $B_i(e,e)$ is uniform throughout the mesh [5]. The evaluation and plotting of $B_i(e,e)$ therefore provides a useful evaluation of the

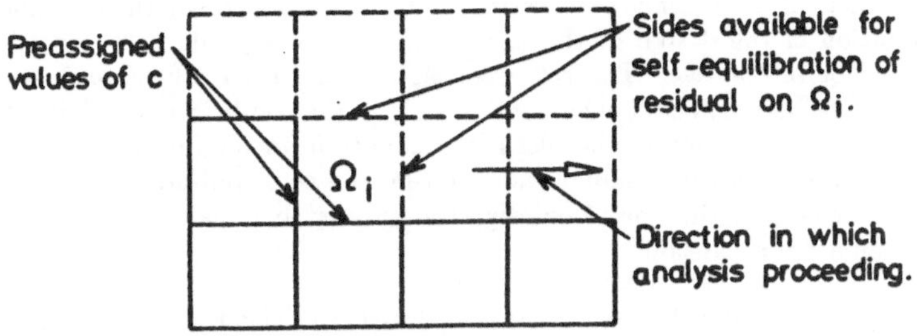

Figure 4: Equilibrium in a two dimensional mesh.

design of the mesh and the search proposed here should only be necessary for poorly designed meshes.

Finally the error measure $B(u,u) - B(u_h,u_h)$ in (14) needs to be converted to a qualitative estimate of the error in stress and displacement. As demonstrated in the one-dimensional example these can be determined directly. However, given the relative inexactness of the proposed practical implementation, the results presented in [1,2] scale the solution u_h from multiple meshes to the energy bound defined by

$$B(\overline{u},\overline{u}) = B(u_h,u_h) + 2\,[B(u,u_h) - B(u_h,u_h)] + B_u(e,e) \tag{37}$$

with

$$\overline{u} = u_h + c(u_h - u_{2h}) \tag{38}$$

For solutions produced by the finite element method which are free from rounding error

$$B(e,u_h) = 0$$

and (37) simplifies to

$$B(\overline{u},\overline{u}) = B(u_h,u_h) + B_u(e,e)$$

The solution u_{2h} is only needed to define the direction of the correction and can be defined by constraining the mesh in regions where the distribution of $B_i(e,e)$ in (36) indicates the mesh should be refined.

5 CONCLUSIONS

The analysis presented in Section 2 provides a general basis for analysis of the error from any source. In [1,2] the approach described in Section 4 has been applied to solutions produced by the finite element, finite difference and boundary element methods and in [2] is applied to the "solution" produced by a random number generator. The significance of the latter is that it indicates that the error caused by roundoff and truncation during the solution can also be sensed and correctly evaluated. In addition, if the differential equations defining (1)-(8) are an exact representation of the physics and theory, then the errors in u_h caused by physical and modelling approximations can also be assessed.

The practical significance of the error analysis is enhanced by the fact that it provides an evaluation of the design of the mesh essential for any adaptive procedure whether automated or interactive. The real expense of the error analysis proposed in Section 4 should only be significant when the design of the mesh is poor. The estimation of $B_i(e,e)$ to guide redesign of the mesh is adequate from the initial assumption of equal splitting of the interface residual and ignoring the self-equilibration requirement. The judicious choice may then be to redesign the mesh rather than execute a complete error analysis on a poor solution.

In any case the development of these procedures, and the relative inexpense of microcomputer implementation of the finite element method in particular should lead in the future to a greater effort to validate finite element solutions. Broadly based error estimates designed to trap the widest variety of sources of error will have a role to play in this future development of numerical methods.

ACKNOWLEDGEMENT

The author is grateful for the financial support provided for this work by the Australian Research Grants Scheme.

REFERENCES

1. Kelly, D.W., Mills, R.J., Reizes, J.A. and Miller, A.D., A Posteriori Estimates Of The Solution Error Caused By Discretization in the Finite Element, Finite Difference and Boundary Element Methods. Accepted for pubilcation Int.J.Num.Meth.Engng.

2. Mills, R.J. A Posteriori Error Estimates in Computational Fluid Mechanics, Ph.D. Thesis, University of New South Wales, Australia (1987)

3. Cook, R.D., Concepts and Applications of Finite Element Analysis., Second Edition J.Wiley., 1981.

4. Kelly, D.W. and Isles, J.D., Practical Error Checks That Can Be Implemented In Finite Element Analysis. Paper to be presented at the Fifth International Conference On Finite Elements in Australia, August 1987.

5. Babuška, I. and Rheinboldt, WC. Adaptive Approaches and Reliability Estimates in Finite Element Analysis. Comp. Meth. Applied Mech. Eng. 17/18 pp519-540, 1979.

ANALYSIS OF CANTILEVER SHEET PILING IN STRATIFIED COHESIVE SOILS

Dr. Jay S. DeNatale and Mr. German A. Ibarra-Encinas
The University of Arizona
Tucson, Arizona 85721, U.S.A.

Summary: Currently available equations for cantilever sheet pile design are restricted to homogeneous or presupposed homogeneous soils below the dredge line. Since the composition of natural soil deposits is often quite complex, an analytical formulation is developed to permit the determination of required piling penetration for the case of stratified foundation soils. A computer program SPILE is developed to solve the generalized equations for the required penetration depth and consequent maximum bending moment.

1. INTRODUCTION

Flexible lateral earth support systems are used in a variety of construction activities and geotechnical projects. According to Head and Wynne [2] cantilever sheet piling is now the most common type of flexible earth retaining system. However, all existing closed-form solutions for sheet piling analysis and design are restricted to homogeneous (or presupposed homogeneous) soils below the dredge line. To date, no solutions have been developed for analyzing cantilever systems in stratified foundation soils.

Since the structure of natural soil deposits is often quite complex, a method is developed which can account for strata of differing shear strength. The closed-form analytical formulation is based on conventional earth pressure theories, the assumption of a linear pressure distribution within each soil stratum, and the laws of static equilibrium.

2. DERIVATION OF GOVERNING EQUATIONS

2.1 Homogeneous Deposit Below The Dredge Line

A general cohesive soil derives its shear strength from both cohesive and a frictional components. However, when analyzing the short term stability of walls in cohesive soil, the frictional component is normally neglected, and a $\phi = 0$ shear strength characterization is generally used. Although such a characterization is assumed in the following derivations, the method could easily be extended to more general conditions.

The conventional method for analyzing cantilever sheet piling in homogeneous foundation soils is well defined and described in references [1] and [4]. The analysis is based on the assumption that the piling rotates about some "pivot point," causing active states of stress to develop in back of the wall above the pivot point and in front of the wall below the pivot point. The stability of the wall depends on the magnitude of the passive stresses that develop in front of the wall above the pivot point.

2.2 Two Strata Below The Dredge Line

The analysis of cantilever sheet piling in stratified foundation soils is conceptually identical to that for homogeneous foundation soils. The analysis is simplified if the lateral stresses due to the backfill above the dredge line is replaced with a single pressure resultant R_o acting at a distance y_o above the dredge line, as shown in Figure 1. The vertical effective stress at the dredge line level is defined as q_o. Referring to Figure 1, the assumed pressure distribution is as follows:

Zone	Lateral Stress on the Left Side	Lateral Stress on the Right Side	Net Stress
I - top	$2c_1$	$q_o - 2c_1$	$4c_1 - q_o$
I - bottom	$q_1 + 2c_1$	$q_o + q_1 - 2c_1$	$4c_1 - q_o$
II - top	$q_1 + 2c_2$	$q_o + q_1 - 2c_2$	$4c_2 - q_o$
II - bottom	$q_1 + q_2 - 2c_2$	$q_o + q_1 + q_2 + 2c_2$	$4c_2 + q_o$

where q_1 and q_2 are the vertical effective stress increments due to the weight of foundation strata 1 and 2, respectively.

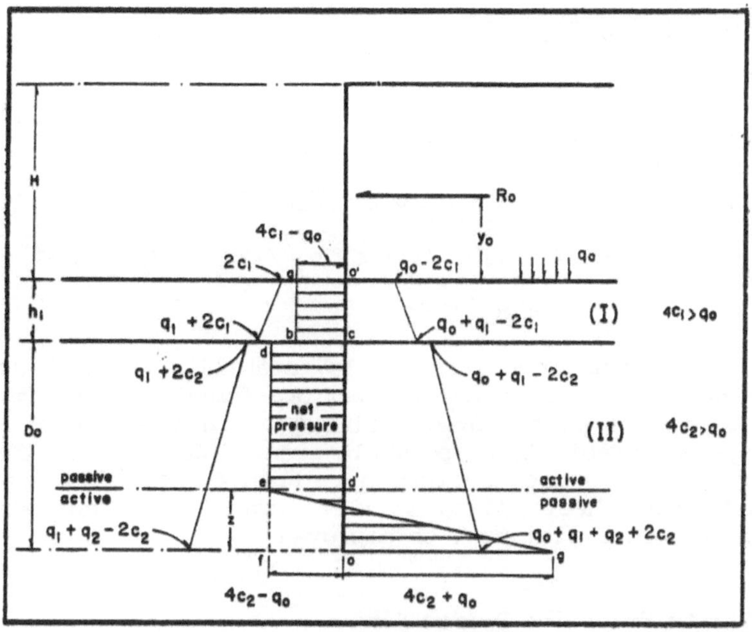

Figure 1. Net Earth Pressure Distribution For Cantilever Sheet Piling In A Two-Layer Foundation System (where $c_2 > c_1$).

The required depth of piling penetration below the dredge line is found by satisfying horizontal force equilibrium and moment equilibrium about the base of the wall. Summing forces in the horizontal direction yields

$$0 = R_o - s_1h_1 - s_2D_o + 4c_2z \quad \ldots\ldots\ldots\ldots \quad (1)$$

where $s_1 = 4c_1 - q_o$ and $s_2 = 4c_2 - q_o$, and where h_1, D_o and z are as shown in Figure 1. Equation (1) may be rearranged to permit evaluation of the unknown z

$$z = (s_1h_1 + s_2D_o - R_o)/4c_2 \quad \ldots\ldots\ldots\ldots \quad (2)$$

Summing moments about the base of the sheet piling provides a second independent equation

$$0 = R_o(yo+h_1+D_o) - s_1h_1[(h_1/2)+D_o] - s_2D_o^2/2 + 4c_2z^2/3 \quad \ldots\ldots \quad (3)$$

Finally, by substituting Equation (2) into Equation (3) it becomes possible to eliminate the first unknown z and generate a simple quadratic equation in the second unknown D_o

$$0 = AD_o^2 + BD_o + C \quad \ldots\ldots\ldots\ldots\ldots\ldots \quad (4)$$

where:

$$A = (s_2^2/12c_2) - s_2/2$$

$$B = R_o - s_1h_1 + s_1s_2h_1/6c_2 - R_os_2/6c_2$$

$$C = R_o(y_o+h_1) - s_1h_1^2/2 + s_1^2h_1^2/12c_2 - R_os_1h_1/6c_2 + R_o^2/12c_2$$

For homogeneous foundation soils, stability is possible only if the cohesion c is greater than one-fourth of the vertical effective stress at the dredge line q_o. In the case of a two-layer system, stability is possible even if $4c_1$ is less than q_o, so long as $4c_2$ is greater than q_o. When $4c_2$ is also less than q_o, cantilever sheet piling will be unstable and an alternate lateral support system must be employed.

2.3 N Strata Below The Dredge Line

The analysis for sheet piling which extends into N cohesive strata below the dredge line is a straightforward extension of the previously discussed case. A typical lateral stress distribution is depicted in Figure 2 below. Once again, consideration of force equilibrium in the horizontal direction leads to

$$0 = R_o - s_1h_1 - s_2h_2 - \ldots - s_nD_o + 4c_nz \quad \ldots\ldots\ldots \quad (5)$$

or

$$z = (s_1h_1 + s_2h_2 + \ldots + s_nD_o - R_o)/4c_n \quad \ldots\ldots\ldots \quad (6)$$

where:

$$s_1 = 4c_1 - q_o$$
$$s_2 = 4c_2 - q_o$$
$$\vdots$$
$$s_n = 4c_n - q_o$$

By summing moments about the base of the piling, substituting the right hand side of Equation (6) for z, and rearranging, the following quadratic equation may be constructed which permits direct solution for the critical unknown D_o:

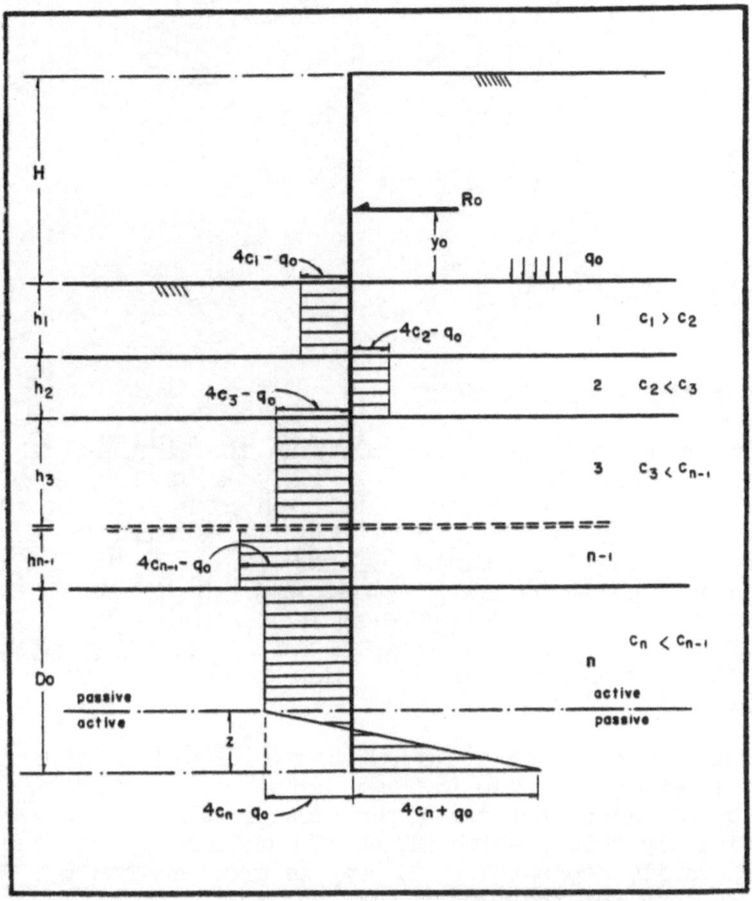

Figure 2. Net Earth Pressure Distribution For Cantilever Sheet Piling In A Multiple-Layer Foundation System.

$$0 = AD_o^2 + BD_o + C \quad \ldots\ldots\ldots\ldots\ldots\ldots\ldots\ldots \quad (7)$$

where:

$$A = s_n^2/2c_n - s_n/2$$

$$B = s_nT_1/6c_n - T_1$$

$$C = R_oL_o - s_1h_1L_1 - s_2h_2L_2 - \ldots$$
$$- s_{n-2}h_{n-2}L_{n-2} - s_{n-1}h_{n-1}L_{n-1}/2 - T_1^2/12c_n$$

$$T_1 = s_1h_1 + s_2h_2 + \dots + s_{n-1}h_{n-1} - R_o$$

$$L_o = y_o + h_1 + h_2 + \dots + h_{n-1}$$

$$L_1 = h_1/2 + h_2 + h_3 + \dots + h_{n-1}$$

$$L_2 = h_2/2 + h_3 + h_4 + \dots + h_{n-1}$$

$$\vdots$$

$$L_{n-2} = h_{n-2}/2 + h_{n-1}$$

3. THE COMPUTER PROGRAM SPILE

Computer program SPILE [3] contains the mathematical formulations outlined above for $\phi = 0$ analyses of cantilever sheet piling in stratified cohesive foundation soils. The program can identify both the required depth of piling penetration below the dredge line and the maximum bending moment. The program is written in FORTRAN 77 and compiled with the Microsoft FORTRAN compiler version 3.2. The program is executable on any IBM PC-compatible machine which uses a DOS 2.11 or later operating system. At least one double-sided disk drive and 128K of memory are required.

A flow chart which illustrates the program's structure is shown in Figure 3. The program computes a trial penetration depth by assuming that the first stratum below the dredge line is infinitely thick, which allows all underlying soil layers to be temporarily neglected. If $4c_1$ is greater than q_o, Equation (7) is solved for the penetration depth D_o. If D_o is less than the true thickness of the first stratum, the maximum bending moment is calculated and the program then stops. If D_o is greater than the true thickness of the first stratum, the properties and pressure within the next lower layer are taken into account. As before, this second stratum is assumed to be infinitely deep, and Equation (7) is used to compute D_o for the two-soil system. The iteration process is repeated, with an additional stratum being accounted for in each iteration, until the computed value of D_o is found to be less than the true thickness of the lowest layer in the system.

When the program incorporates layer j into the analysis and finds that $4c_j$ is not greater than q_o, the program immediately proceeds to the next iteration. If the program determines that the piling must extend into the final stratum N, and it is found that $4c_N$ is less than q_o, a stable cantilever design is not possible, and the analysis ends with a message to this effect.

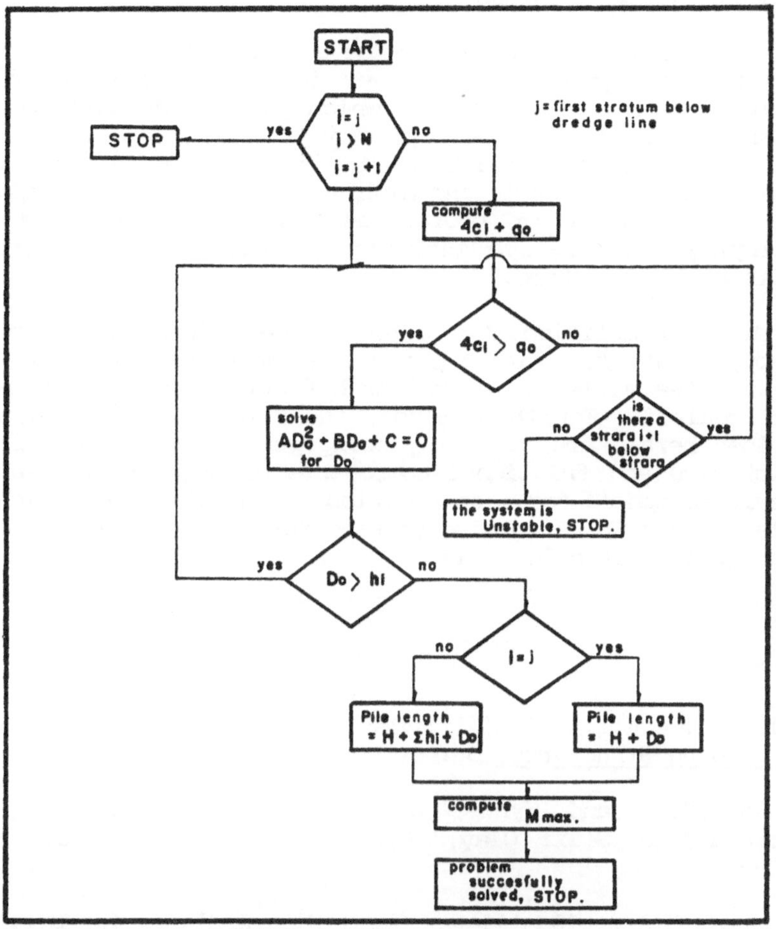

Figure 3. Flow Chart For Computer Program SPILE.

4. THE VERIFICATION PROGRAM

Program SPILE was used to analyze a wide variety of foundation problems. The results of these computer analyses agreed well with manually obtained solutions. Due to space limitations, specific example analyses will not be reviewed here, but a complete presentation of the entire testing program is presented in reference [3].

5. SUMMARY AND CONCLUSIONS

An analytical formulation is developed which permits a computer-aided analysis of cantilever sheet piling in stratified foundation soils. The equation which governs the required piling penetration depth is always quadratic, irrespective of the number of strata existing below the dredge line. The computer program SPILE is accurate and highly efficient. It provides the engineering student or professional with a means of rapidly identifying the relative influence of the various geometrical and material parameters.

Although simplified $\phi = 0$ shear strength characterizations are appropriate for some types of geotechnical analyses involving cohesive soils, more precise effective stress analyses are often required. For this reason, the mathematical formulations described herein are currently being generalized to permit both the cohesive and frictional components of shear strength to be properly accounted for. An additional generalization which will enable program SPILE to also be used for the analysis and design of anchored sheet pile walls.

6. REFERENCES

1. Clayton, C.R.I. and Milititsky, J. - Earth Pressure and Earth Retaining Structures, Surrey University Press, London, 1986.

2. Head, J.M. and Wynne, C.P., "Designing Retaining Walls Embedded in Stiff Clay," Ground Engineering, 18-3, (1985), 30-33.

3. Ibarra, G.A., "Cantilever Sheet Pile Analysis for Stratified Cohesive Soil Deposits," M.S. Thesis, Department of Civil Engineering and Engineering Mechanics, The University of Arizona, 1986.

4. United States Steel, Steel Sheet Piling Design Manual, 1969.

A GEOMETRICAL CONTINUOUS REMESHING PROCEDURE FOR APPLICATION TO FINITE ELEMENT CALCULATION OF NON-STEADY STATE FORMING PROCESSES

J.P. CESCUTTI and J.L. CHENOT

ECOLE DES MINES DE PARIS. CENTRE DE MISE EN FORME
SOPHIA ANTIPOLIS 06565 VALBONNE CEDEX . FRANCE
UA CNRS N° 852 et GRECO "GRANDES DEFORMATIONS ET ENDOMMAGEMENT"

ABSTRACT

Based on an incompressible visco-plastic formulation, a finite element program (FORGE2) is used to simulate isothermal forging. Two adaptative remeshing procedures are introduced, in order to achieve very large deformations. a static one and a continuous dynamic one. The two methods presented are applied to the plane strain upsetting and to the plane strain compression test. The results are compared and the advantages of continuous remeshing are discussed.

1. INTRODUCTION

The generation of solution-adaptative grids for the finite element method receives today a great deal of interest (/2/ to /9/). For simulating hot forging by an updated Lagrangian approach (see O.C. Zienkiewicz /10/, S. Kobayashi /11/ and J.L. Chenot /12/ works) the remeshing problem becomes crucial in order to allow large deformations. With an arbitrary Eulerian-Lagrangian approach, the problem is settled in terms of determining the remeshing velocity field. The purpose of this paper is to illustrate the interest of remeshing to achieve very large deformations.

Two broad categories can be distinguished among the solution-adaptative grid remeshing methods ; either equidistribution of some error measure or explicit use of variational principles. As for us, we follow the track of J.U. Brackbill /8/, J.T. Oden /7/ and R. Carcaillet /7/ in using variational principles, so we introduce a functional quantifying the grid quality, and the element volume adaptation to some solution quantity (as the deformation energy).

2. FINITE ELEMENT CALCULATION OF ISOTHERMAL FORGING PROCESSES

The material is assumed to obey a viscoplastic Norton-hoff law, and a similar viscoplastic friction law. We use a penalty approach to enforce the incompressibility requirement, so the associated functional is :

$$\Phi(\vec{v}) = \int_\Omega \frac{K}{m+1} (\sqrt{3}\,\dot{\bar{\varepsilon}})^{m+1} + \frac{1}{2}\rho \int_\Omega K(\operatorname{div}\vec{v})^2 + \int_{Sf} \frac{\alpha K}{p+1} \|\Delta\vec{v}_t\|^{p+1}$$

with
- $\dot{\bar{\varepsilon}}$: equivalent strain rate
- m : strain rate sensitivity
- K : material consistency
- α and p : parameters depending on the type of contact
- ρ : large positive constant ($\simeq 10^5$ to 10^7)
- Ω : workpiece domain
- $\vec{v}$: velocity field
- $\Delta\vec{v}_t$: relative velocity between die surface and workpiece
- Sf : friction surface

This formulation with the finite element discretization in space and time is now classical and may be found in /1/.

3. REMESHING ANALYSIS

3.1 Variational formulation

The present method is a natural extension to remeshing of recently published static grid generation and optimization concepts. We start from the variational method of Brackbill and Saltzman. They controlled the grid quality and grid adaptation by quantifying the mapping between the physical space (x,y) and the computational space (ξ,y) with the following integrals (in the physical space) :

smoothness : $I_s = \int_\Omega (\overrightarrow{\nabla_x\xi}^2 + \overrightarrow{\nabla_x\eta}^2)\, dx$

orthogonality . $I_o = \int_\Omega (\overrightarrow{\nabla_x\xi}.\overrightarrow{\nabla_x\eta})^2\, dx$

volume control . $I_w = \int_\Omega w^2(x,y)\, J\, dx$

where • ∇_x denotes the gradient operator in the physical space
• J the jacobian of the transformation $\vec{T}$.

$\vec{T}$: $\Omega \rightarrow \Omega$
$\vec{\xi}=(\xi,\eta) \rightarrow \vec{x}=(x,y)$

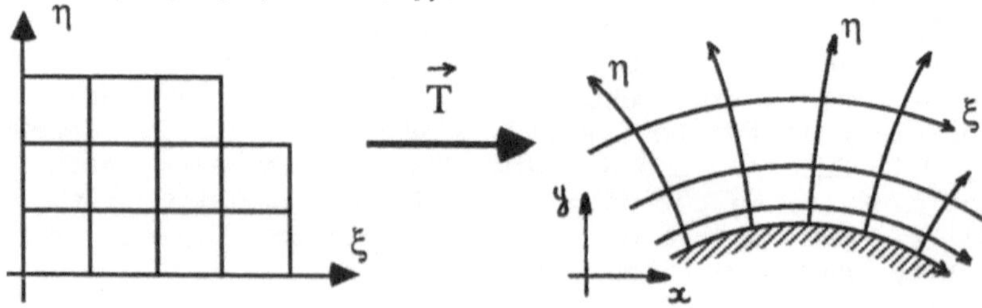

Figure 1

In our remeshing functional we only consider smoothness and volume control and we write the associated integrals I_s and I_w in the computational space

$$I_s = \int_\Omega (\overrightarrow{\nabla_\xi x}^2 + \overrightarrow{\nabla_\xi y}^2) \frac{d\vec{\xi}}{J} = \int_\Omega \|\nabla_\xi \vec{T}\|^2 \frac{d\vec{\xi}}{J}$$

$$I_w = \int_\Omega w^2(\xi,\eta)\, J^2\, d\vec{\xi}$$

To allow various kinds of mesh topology and various kinds of elements we replace on each element Ke, the restriction $\vec{T}/Ke$ of the global mapping $\vec{T}$, by the isoparametric mapping $\vec{\tau}_e$ between the reference element Ke and the real element Ω_e. We also replace I_w by a simplified element volume control. So we obtain the following functional $\pi(\vec{X})$ governing our remeshing procedures, depending on the vector $\vec{X}$ of the nodal position :

$$\pi(\vec{X}) = \sum_e \int_{Ke} \|\nabla_\xi \vec{\tau}_e\|^2 \frac{d\vec{\xi}}{Je} + \rho\, W_e \left(\int_{Ke} Je\, d\vec{\xi}\right)^2 \qquad (1)$$

$$\pi(\vec{X}) = \sum_e \int_{\Omega e} \frac{\|\nabla_\xi \vec{\tau}_e\|^2}{Je^2}\, d\vec{x} + \rho\, W_e \left(\int_{\Omega e} d\vec{x}\right)^2$$

where • $\vec{\tau}_e$: $Ke \rightarrow \Omega e$
$\vec{\xi} \rightarrow \vec{x} = \sum_k \vec{X}^k N^k(\vec{\xi})$

with $\vec{X}^k$ the coordinates of element nodes of Ωe

$N^k(\vec{\xi})$ the shape function in the reference space

• W_e : the weight of element Ωe ($W_e = \int_{Ke} w^2\, d\vec{\xi} / \int_{Ke} d\vec{\xi}$)

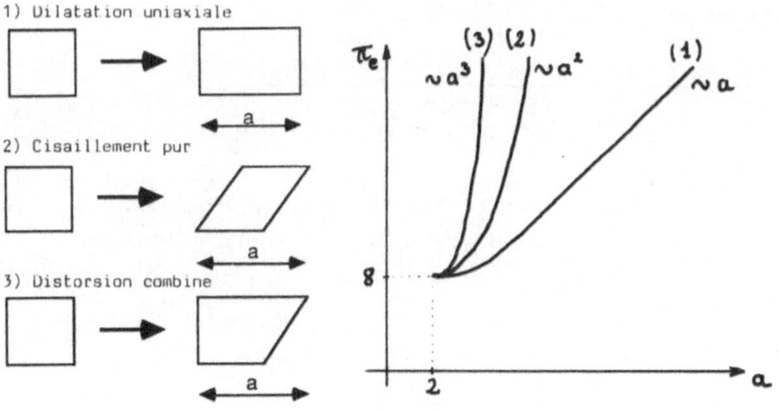

Figure 2

We can also interpret the first term of π as a measure of the geometrical grid quality by quantifying for each $\vec{\tau}_e$ its distance from similarity transformation ("it tries" to preserve the reference angles and proportions). Figure 2 shows the evolution of π, for a few distorsions of the Q_1 element. We notice different levels of selectivity.

3.2 Analytical π derivatives

Our remeshing process based on the minimization of the functional $\pi(\vec{X})$ will solve the necessary conditions of optimality :

$$\frac{\partial \pi}{\partial \vec{X}} = 0 \qquad (2)$$

with a Newton-Raphson method. So we need to calculate the first and the second derivatives of π with respect to the nodal positions. To do so, we use the ideas developed in derivation with respect to the domain theory /13/ : to calculate $\partial\pi/\partial X_i^\lambda$ we define the transformation $\vec{T}_t$.

$$\vec{T}_t : \Omega \rightarrow \Omega$$

$$\vec{x} \rightarrow \vec{x}_t = \vec{x} + t\, \psi^\lambda(\vec{x})\, \vec{e}_i$$

associated with the velocity field $\vec{W}(\vec{x})$.

$$\vec{W}(\vec{x}) = \psi^\lambda(\vec{x})\, \vec{e}_i$$

where : . $\psi^\lambda(\vec{x})$ is the shape function in the physical space
. $(\vec{e}_1,\vec{e}_2)$ are the base vectors of the physical space

Of course we have :

$$\frac{\partial \pi}{\partial X_i^\lambda} = \frac{d}{dt}\, \pi\, (\vec{T}_t(\vec{X}))\, \Big|_{t=0}$$

We have developed two procedures of remeshing : a geometrical one and a continuous dynamic one.

3.3 "Static" or "geometrical" remeshing

The problem is formulated as : knowing initial mesh $\vec{X}_0$, find $\vec{X}$ satisfying the optimality conditions (2) with the following contraints :

a) The boundary of the mesh $\vec{X}$ is prescribed to stay in the proximity of the Ω boundary (defined by the old mesh boundary or by a better approximation).

b) The points where the mechanical boundary conditions change must coïncide with a boundary node of $\vec{X}$.

After remeshing the nodal parameters are smoothed on the old grid and interpolated on the new one.

We developed the algorithm :

1) Evaluate an approximate solution with a Newton-Raphson scheme of the problem with the only constraint a) that we have chosen to write :

$$\Delta\vec{X}_b.\vec{n} = 0 \quad \text{on } \partial\Omega \quad (\vec{X}_b \text{ . boundary nodes}).$$

2) Correct the mesh to satisfy the constraint b)

3) Go on the Newton-Raphson iteration with the two constraints

4) Interpolate the nodal parameters of $\vec{X}_0$ on $\vec{X}$

It is quite easy to include this algorithm in the finite element simulation of forming process : we just have to

1) Compute the kinematic velocity : $\vec{V}$
2) Evaluate the new configuration : $\vec{X}_0(t+\Delta t) = \vec{X}(t)+\vec{V}.\Delta t$
3) Remesh on this configuration : $\vec{X}_0(t+\Delta t)\to..\to \vec{X}_n(t+\Delta t)$

3.4 "Dynamic" or "continuous" remeshing

In dynamic remeshing, we want to eliminate in the previous algorithm the treatment of the intermediate meshes $\vec{X}_0$, .., $\vec{X}_{n-1}$, so we wish to move the nodes with a remeshing velocity field which saves the mesh regularity. This field can be obtained by time differentiation of condition (2).

$$\frac{\partial^2\pi}{\partial\vec{X}^2} \, . \, \vec{V}_m = 0 \qquad (3)$$

where $\vec{V}_m$ is defined as the velocity of the mesh nodes, associated with the remeshing field :

$$\vec{v}_m(\vec{x}) = \sum_\lambda \vec{V}_m^{\,\lambda} \, \psi^\lambda(\vec{x})$$

To preserve a good geometrical evolution of the domain boundary there must be no material flux through this boundary, so we have the contraint :

$$\int_{\partial\Omega} \lfloor(\vec{v}-\vec{v}_m).\vec{n}\rfloor^2 = 0 \qquad (4)$$

To solve this new problem, we introduce an explicit time integration procedure as a first approximation, so the nodes are moved according to :

$$\vec{X}(t+\Delta t) = \vec{X}(t) + \vec{V}_m(t).\Delta t$$

and the nodal parameters Q are updated by the convective formulae (as in ALE method).

$$Q_{t+\Delta t} = Q_t + \overrightarrow{\text{Grad}}\, Q_t.(\vec{v}-\vec{v}_m)\,\Delta t$$

The dynamic remeshing algorithm used can be expressed as :

1) compute the kinematic velocity $\vec{V}$
2) write the boundary condition (4) on $\vec{V}_m$ as :
   - $\vec{V}_m = \vec{V}$ in a sharp angle of the mesh boundary
   - $\vec{V}_m.\vec{n} = \vec{V}.\vec{n}$ at other boundary nodes
3) solve $\frac{\partial^2\pi}{\partial\vec{X}^2} \, . \, \vec{V}_m = - \frac{1}{\Delta t}\frac{\partial\pi}{\partial\vec{X}}$ with these boundary conditions.
4) evaluate the new configuration deduced from the previous one with $\vec{V}_m$.

3.5 Rotation of special boundary elements

The remeshing difficulties that we had foreseen for quadrangular elements with two sides on the boundary, were confirmed by numerical results. In fact, these element do not have enough remeshing degrees of freedom, and they cannot avoid their transformations into triangles when they are flatted by a tool. But some cases of static remeshing applied to forging simulation show us how to treat this phenomenon the element is turned around its interior node (fig.3). They also gave indications for knowing when and how to perform this transformation. So in our remeshing procedures we force this transformation when a simple geometrical criterion is reached. This capability increases the efficiency of our remeshing procedure especially the dynamic one.

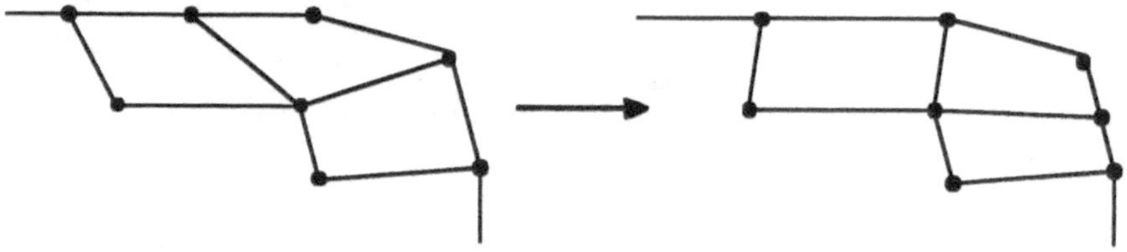

Figure 3 : rotation of an element around its interior node

4. NUMERICAL RESULTS

Most of our results are presented without the element volume control. Nevertheless, we present on figure 9 preliminary results with the adaptative term.

Figure 4 illustrates the possibilities of static remeshing with an exemple of grid generation improvement, for the case of forward extrusion. It includes a displacement of the nodes on the boundary.

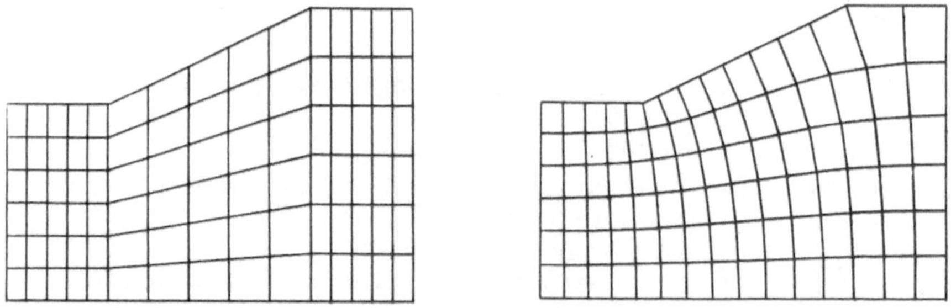

Figure 4 : grid improvement with static remeshing

The two methods presented are applied to the plane strain upsetting and the plane strain compression test, with a strain rate sensitivity m equal to 0,5 and the severe conditions for mesh evolution of sticking contact.

Figures 5 et 6 present the last step reached without remeshing for the two configurations.

Figure 7 a,b,c,d show some steps of the plane strain compression test with static remeshing. A reduction ratio of 50% is easily reached. Around 70% some geometrical approximations occur because of the element size. And we encounter remeshing instabilities near 90% of reduction due to the disproportionnal sizes of the element edges.

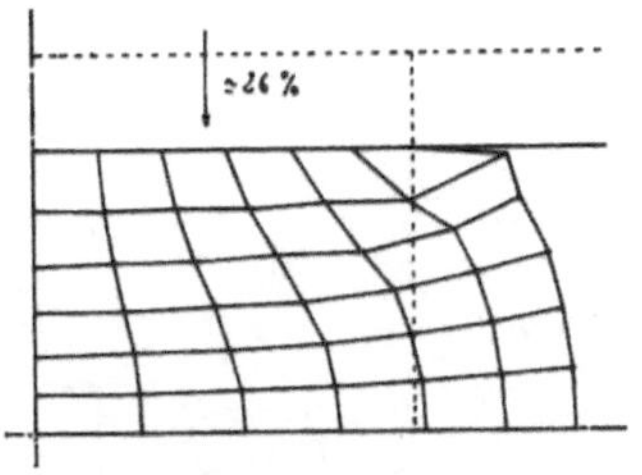

Fig.5: plane strain upsetting without remeshing

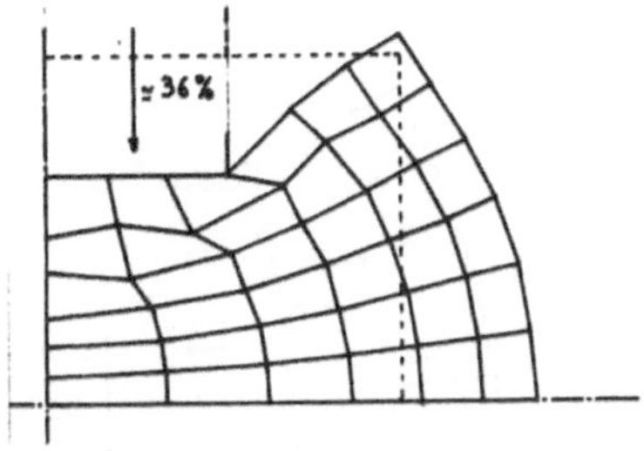

Fig.6: plane strain compression test without remeshing

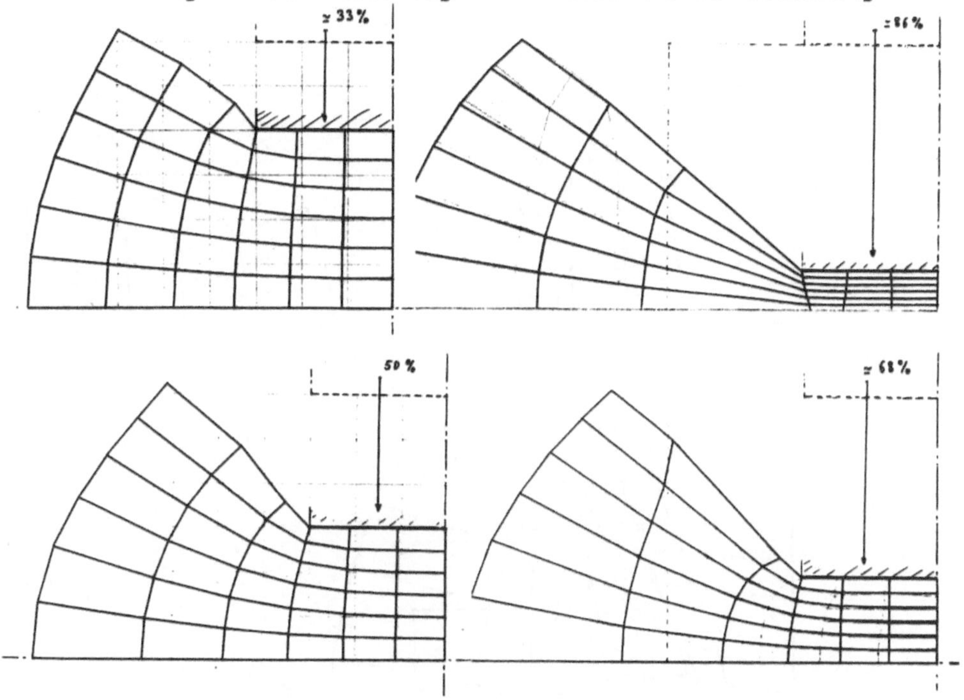

Fig. 7 a,b,c,d : Plane strain test with static remeshing

Figure 8 a,b,c,d show some steps of the same test case but with dynamic remeshing 70% of reduction ratio has been reached, but it is hard to go on, because dynamic remeshing is unable to do geometrical approximations as static one. If we introduce the adaptative term (equidistribution of the plastic deformation energy over the elements), we obtain a better shape of the workpiece (figure 9), but we can only reach a reduction ratio of 50%.

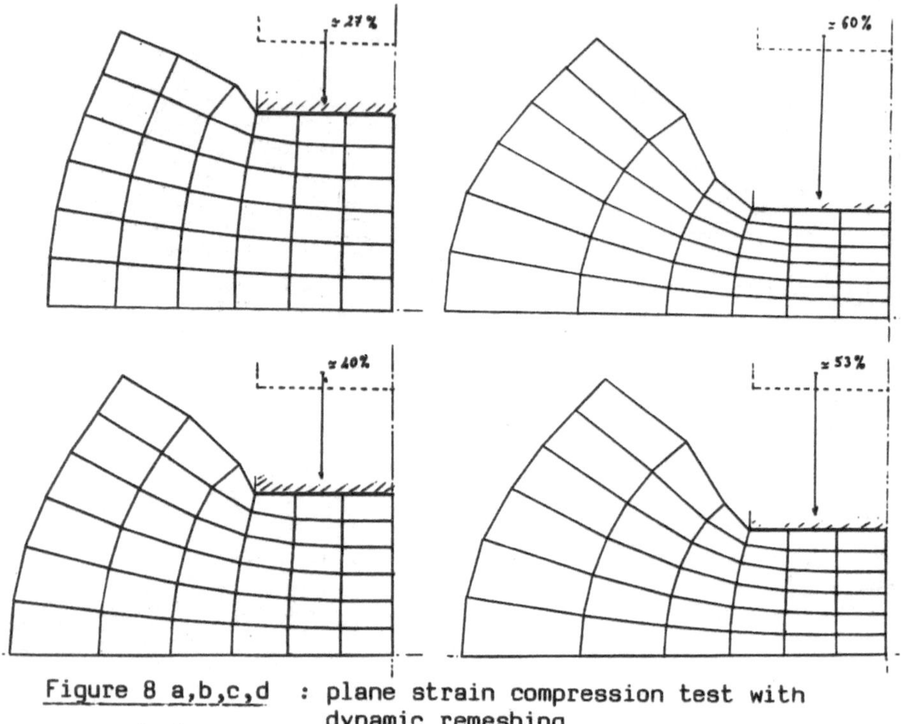

Figure 8 a,b,c,d : plane strain compression test with dynamic remeshing

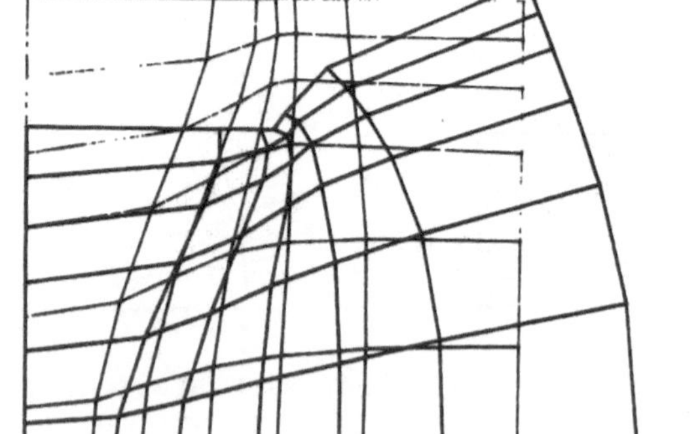

Figure 9 : plane strain compression test with adaptative dynamic remeshing

The testing of static remeshing with the plane strain upsetting gives almost the same results as previously. The rotation of the corner element allows to perform a reduction of 80% before encountering instabilities. With dynamic remeshing (figure 10 a,b,c), if we want to go as far, we have to do small steps, otherwise, with a constant step of about 3%

of reduction we reach a reduction of only 45%.

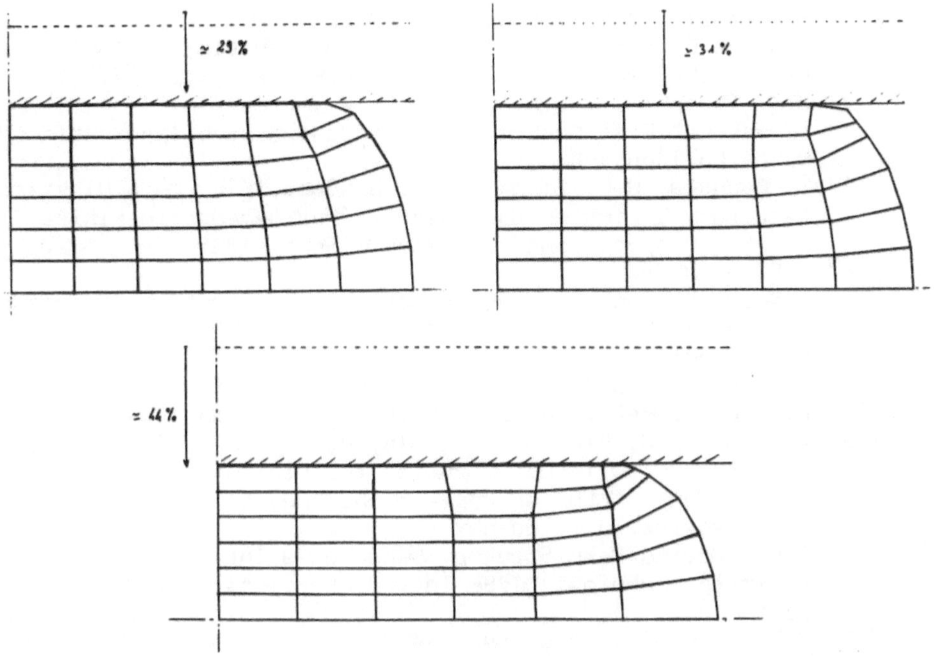

Figure 10a b c plane strain upsetting with dynamic remeshing a,b . just before and after the second rotation of corner element. c 15 th step just after the fourth rotation

CONCLUSION

Static remeshing with boundary nodes notion appears to be a powerful procedure of remeshing, but a little expensive in CPU time. With a good initial mesh topology, and the introduction of adaptative steps, dynamic remeshing seems to be an efficient choice. These good preliminary results allow us to consider the treatment of complex industrial workpieces.

REFERENCES

/ 1/ Y. Germain J.L Chenot P.E Mosser
"Finite Element Analysis of shaped lead-tin disk forgings" 1986 Num. Meth. in Ind. Form Proc pp. 271-275

/ 2/ J.F. Thompson, Z.U.A. Warsi, C. Wayne Mastin
"Numerical Grid Generation, Foundations and applications" North-Holland 1985.

/ 3/ I. Babuska, O.C. Zienkiewicz, J. Gago, E.R. de A.Oliviera
"Accuracy Estimates and Adaptive Refinements in Finite Element Computations" J. Wiley & Sons, 1985.

/ 4/ M.S. Shephard
"Finite Element Grid Optimization- A review" The American Soc. of Mech. Eng., pp. 1-13, 1979.

/ 5/ N. Kikuchi
"Adaptive Grid-Design Methods for Finite Element Analysis" Comp. Meth. in Appl. Mech. Eng. 55, pp 129-160, 1986

/ 6/ A.R. Diaz, N. Kikuchi, J.E. Taylor
"A method of Grid Optimization for Finite Element Methods" Comp. Meth. in Appl. Mech. Eng. 41, pp 29-45, 1983

/ 7/ L. Demkowicz, J.T. Oden
"On a Moving Mesh Strategy based on a Interpolation Error Estimate Technique" 1986 Int. J. Eng. Sci. Vol 24 N°1, pp 55-68.

/ 8/ J.U. Brackbill, J.S. Saltzman
"Adaptive Zoning for Singular Problems in two Dimensions" 1982 J. Comput. Physics 46, pp 342-368.

/ 9/ R. Carcaillet, G.S. Dulikravich, R. Kennon
"Generation of Solution-Adaptative Computational Grids Using Optimization" Comp. Meth. in Appl. Mech. Eng., 57 pp 279-295, 1986.

/10/ O.C. Zienkiewicz
"Flow Formulation for Numerical Solution of Forming Processes" 1984 Numerical Analysis of Forming Processes J. Wiley & Sons pp 1-44.

/11/ S. Kobayashi
"Thermoviscoplastic Analysis of Metal Forming Problems by the Finite Element Method" Numerical Analysis of Forming Processes J. Wiley & Sons, pp 45-69, 1984.

/12/ G. Surdon and J.L. Chenot
"FInite Element Calculation of Three-Dimensional Hot Forging" 1986 Num. Meth. in Ind. Form. Proc. pp 287-292.

/13/ J.P. Zolesio
"The material derivative for shape optimization" in E. Hang an J. Céa eds, optimization of distributed parameter structures (Sijthoff and Noorhoff, Alphen aan den Rijn, The Netherlands, 1981) 1089, 1151.

IMPROVED SYSTOLIC DESIGNS FOR THE ITERATIVE SOLUTION OF LINEAR SYSTEMS

D.J. Evans and K. Margaritis

Department of Computer Studies
University of Technology
Loughborough, Leicestershire, U.K.

ABSTRACT

Recent advances in VLSI technology have led to a rapid growth in the research into parallel computer architectures. A major attraction of the emerging VLSI technology is that it will give the scientists and engineers the potential to design, in a cost effective manner, closely related devices and algorithms tailored to particular problem classes (e.g. iterative solution of linear systems of equations is discussed in this paper).

An important development in the design of parallel computer architectures is the design of sophisticated microprocessors, (e.g. Transputer, Warp cell) that can be interconnected to form parallel processors, such as Wavefront Arrays, Systolic Arrays, etc.

The purpose of this paper is to consider the application of systolic array designs to problems arising in numerical computation, as encountered in many engineering domains. More specifically this paper describes systolic networks for the iterative solution of linear systems of equations using the Jacobi, Gauss-Seidel, JOR and SOR methods. Existing iterative systolic designs are modified to become more area-efficient, two new pipeline architectures are proposed based on an improved matrix-vector multiplication systolic array.

1. INTRODUCTION

Given a linear system of equations,

$$A\underline{x} = \underline{b} \ , \tag{1}$$

or

$$\sum_{j=1}^{n} a_{ij}x_j = b_i \ , \ i=1,2,\ldots,n$$

which without loss of generality has been so ordered that $a_{ii} \neq 0$, for all i. The ith equation can then be solved for x_i [J]:

$$x_i = \frac{1}{a_{il}} \left(-\sum_{i\neq j} a_{ij}x_j + b_i\right), \ i=1,2,\ldots,n; \tag{2}$$

the system (1) is then rewritten as,

$$\underline{x} = M\underline{x} + \underline{g} \ , \tag{3}$$

where,

$$m_{ij} = \begin{cases} 0, & i=j \\ -a_{ij}/a_{ii}, & i\neq j \end{cases} \ , \ g_i = b_i/a_{ii}, \ i=1,2,\ldots,n. \tag{4}$$

Equations (3)-(4) are the basis of many iterative methods for the solution of linear systems, e.g. [J],[F]:

(i) Jacobi method (J)

$$\underline{x}^{(k+1)} = M\underline{x}^{(k)} + \underline{g} \ , \tag{5}$$

or

$$\underline{x}_i^{(k+1)} = \frac{1}{a_{ii}} \left(-\sum_{i\neq j} a_{ij}x_j^{(k)} + b_i\right), \ i=1,2,\ldots,n$$

(ii) Gauss-Seidel method (GS)

$$\underline{x}^{(k+1)} = L\underline{x}^{(k+1)} + U\underline{x}^{(k)} + \underline{g} \tag{6}$$

or

$$x_i^{(k+1)} = \frac{1}{a_{ii}}\left(-\sum_{j=1}^{i-1} a_{ij}x_j^{(k+1)} - \sum_{j=i+1}^{n} a_{1j}x_j^{(k)} + b_i\right)$$

where L(U) is strictly lower (upper) triangular matrices derived from M, i.e. L+U=M.

(iii) Jacobi overrelaxation method (JOR),

$$\underline{x}^{(k+1)} = \omega(M\underline{x}^{(k)} + \underline{g}) + (1-\omega)\underline{x}^{(k)} \ , \tag{7}$$

or

$$x_i^{(k+1)} = \frac{\omega}{a_{ii}}\left(-\sum_{i\neq j} a_{ij}x_j^{(k)} + b_i\right) + (1-\omega)x_i^{(k)}$$

(iv) Successive overrelaxation method (SOR),

$$\underline{x}^{(k+1)} = \omega(L\underline{x}^{(k+1)} + U\underline{x}^{(k)} + \underline{g}) + (1-\omega)\underline{x}^{(k)} \tag{8}$$

or

$$x_i^{(k+1)} = \frac{\omega}{a_{ii}} \left(-\sum_{j=1}^{i-1} a_{ij}x_j^{(k+1)} - \sum_{j=i+1}^{n} a_{ij}x_j^{(k)} + b_i\right) + (1-\omega)x_i^{(k)}$$

Methods (5)-(8) have been considered for systolic implementation in [B], [BBD],[D]; Figures 1-4 illustrate linear arrays performing one iteration for methods (5)-(8) respectively. In the same Figures the data sequence format is shown as well as the synchronization delays that are necessary between two successive iterative steps.

In general, all the designs can be analysed in:two matrix-vector multiplication (mvm) arrays, performing the calculations $L\underline{x}+\underline{y}$ and $U\underline{x}+\underline{y}$, for L,U as defined in (4); and a special cell performing more complex computation as explained in Fig.5, for each one of Figures 1-4.

Table 1 summarises the area and time requirements for the iterative systolic designs implementing the methods (5)-(8), for k iterative steps, and for a banded matrix A, with bandwidth w=p+q-1. Notice that the time complexity for the special cell in the case of SOR is 1 (inner product step) IPS+1 ADD and therefore the time cycle for SOR must be prolonged.

Method	J,Eq.(5)	JOR,Eq.(7)	GS,Eq.(6)	SOR,Eq.(8)
Figures	1,5(a)	2,5(b)	3,5(c)	4,5(d)
Area	k((W-1)IPS+DIV)	k(wIPS+DIV)	k((w-1)IPS+DIV+ADD)	k(wIPS+DIV+ADD)
Time	2(n-1)+(k-1)(2p+1) +(w+1)	as for J	2(n-1)+(k-1)2p+p	as for GS

TABLE 1

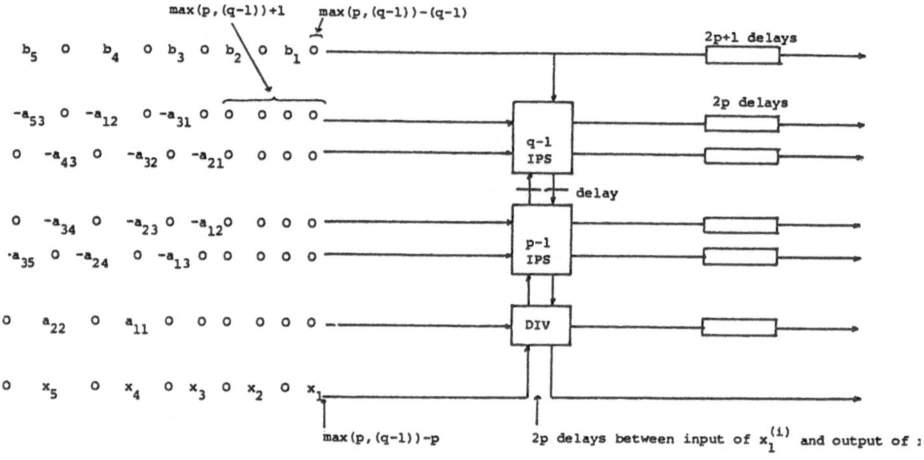

FIGURE 1: Array for J method; w=5, p=q=3, n=5.

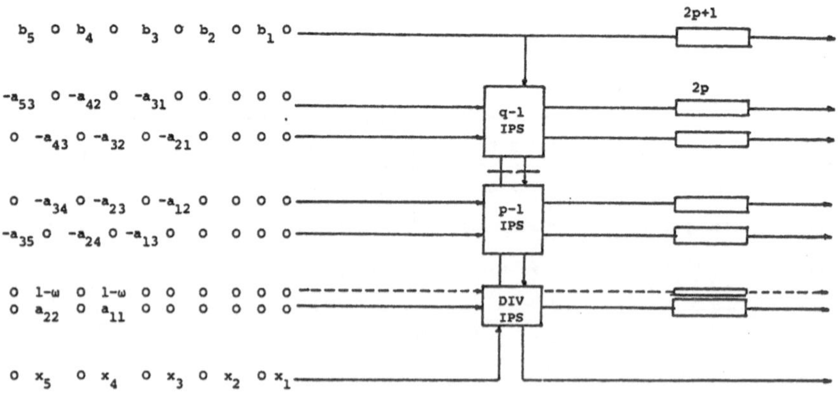

FIGURE 2: Array for JOR method

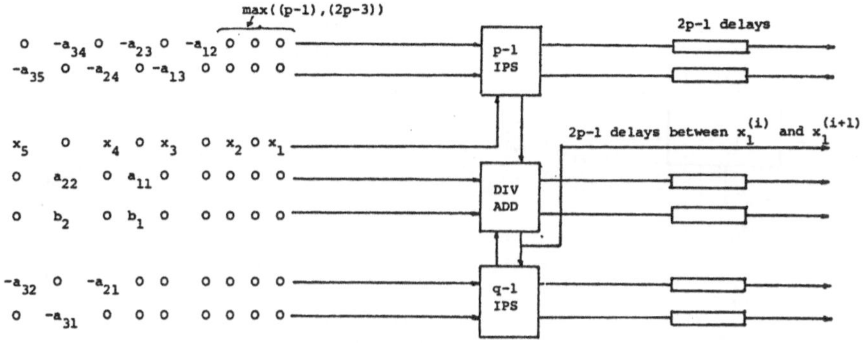

FIGURE 3: Array for GS method

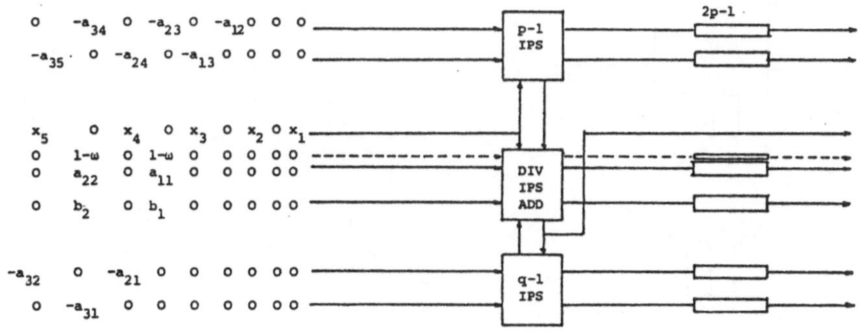

FIGURE 4: Array for SOR method

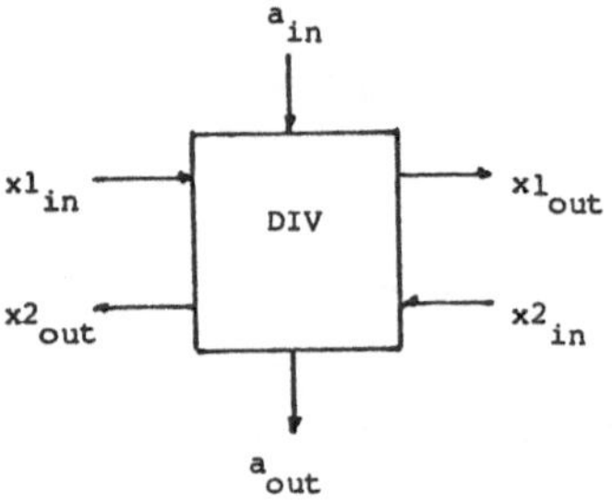

$x1_{out} := x1_{in}$

$x2_{out} := x2_{in}/a_{in}$

$a_{out} := a_{in}$

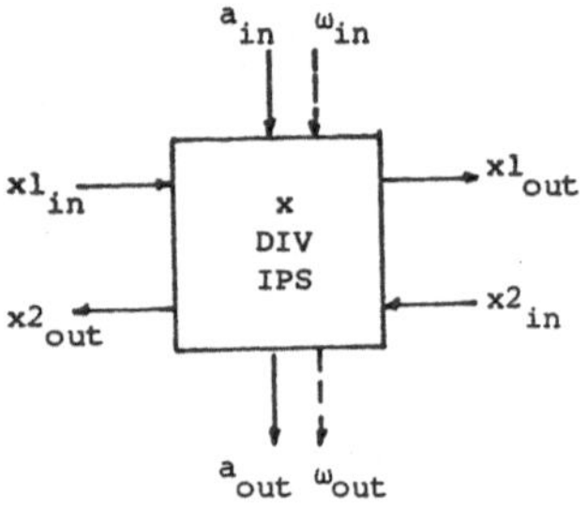

$x1_{out} := x1_{in}$

$x2_{out} := x2_{in}/a_{in} + x\omega_{in}$

($x := x1_{in}$ delayed 2p-1 cycles)

$a_{out} := a_{in}$

$\omega_{out} := \omega_{in}$

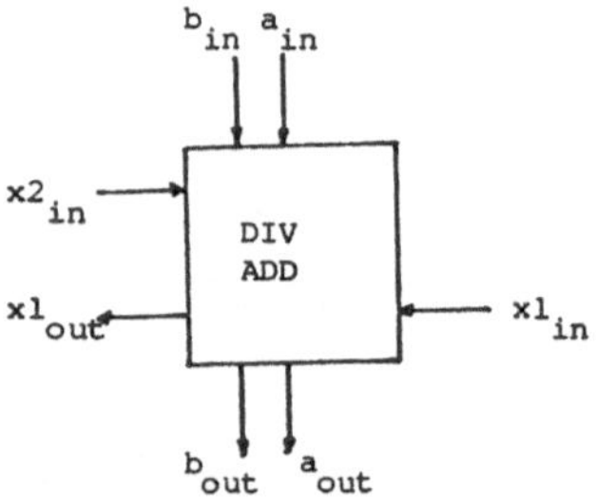

$x1_{out} := (x1_{in} + x2_{in} + b_{in})/a_{in}$

$a_{out} := a_{in}$

$b_{out} := b_{in}$

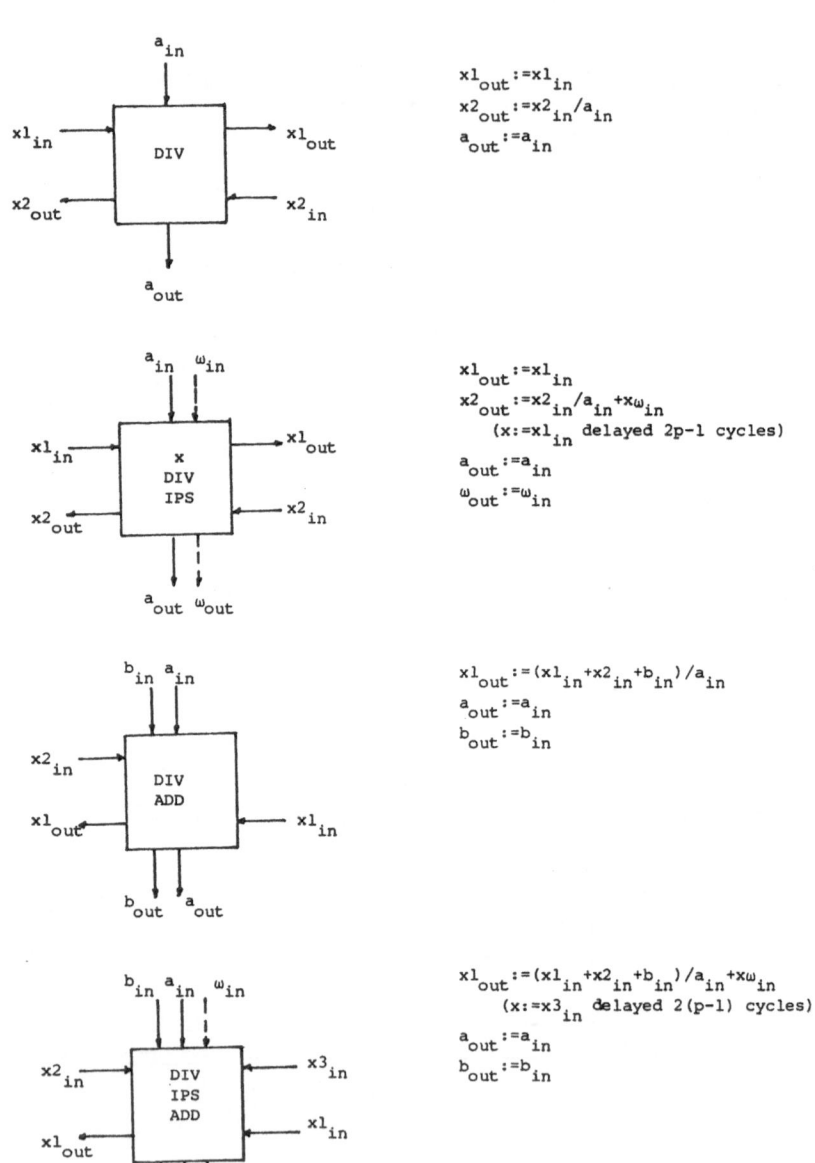

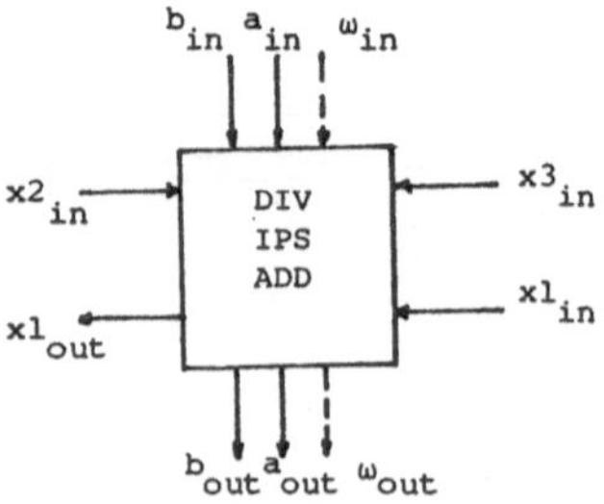

$x1_{out} := (x1_{in} + x2_{in} + b_{in})/a_{in} + x\omega_{in}$

($x := x3_{in}$ delayed 2(p-1) cycles)

$a_{out} := a_{in}$

$b_{out} := b_{in}$

<u>FIGURE 5</u>: Special cells for the arrays in Figs.1-4.

2. IMPROVED ITERATIVE SYSTOLIC DESIGNS

The following observations can be made on the designs of Figs.1-4:

(i) The quantities,

$$-\frac{a_{ij}}{a_{ii}}, \ \omega\frac{a_{ij}}{a_{ii}}, \ i\neq j \text{ and } \frac{b_i}{a_{ii}}, \ \omega\frac{b_i}{a_{ii}}$$

in eq.(5)-(8) are calculated repetitively in each step of the iterative process, while they need to be calculated only once in the beginning of the computation. Thus, the complex special cells will be avoided at the expense of some initial preprocessing.

(ii) For the JOR and SOR methods the addition of vectors $(1-\omega)\underline{x}^{(k)}$ can be interpreted as an additional inner product step in the mvm operations, i.e., (7) can be re-written as,

$$\underline{x}^{(k+1)} = (\omega L+(1-\omega)I+\omega U)\underline{x}^{(k)} + \omega\underline{g}\ , \tag{9}$$

and similarly for (8) we have,

$$\underline{x}^{(k+1)} = \omega L\underline{x}^{(k+1)} + ((1-\omega)I+ U)\underline{x}^{(k)} + \omega\underline{g}\ . \tag{10}$$

Thus, a more uniform implementation of the J, JOR and GS, SOR methods can be achieved since for $\omega=1$ $\omega L+(1-\omega)I+\omega U=\omega M$ and $(1-\omega)I+\omega U=U$.

Fig.6 and 7 illustrate the improved systolic arrays for one iterative step of methods J, JOR (9) and GS, SOR (10) respectively. The linear array of Fig.6 is a simple mvm array implementing a matrix-vector inner product step (mvips) as in (3). In Fig.7 there are again two mvm arrays for the $L\underline{x}+\underline{y}$ and $U\underline{x}+\underline{y}$ computations but now the middle cell is a simple 3-input adder:

The data sequence formats as well as the synchronization delays between two successive iterative steps are also shown in Fig.6 and 7. The preprocessing elements required for the formulation of the data sequences are shown in Fig.8 and 9, for the pipelines of Fig.6 and 7 respectively. The main diagonal of matrix A enters a divider cell calculating ω/a_{ii}; this quantity is propagated to the remaining cells while $(1-\omega)$ is the main diagonal element of the output matrix M. The other cells calculate $-(\omega/a_{ii})a_{ij}$, except for the cell corresponding to vector $\underline{b}$ that calculates $(\omega/a_{ii})b_i$.

Some reformatting delays are necessary for the output of the preprocessing elements to conform with the data sequences formats of the iterative systolic designs; these delays are shown in Fig.8 and 9.

The area requirements and the delays introduced by the preprocessing elements are summarized in Table 2, for a banded matrix A with bandwidth w=p+q-1.

Method	J,JOR	GS,SOR
Figure	8	9
Area	(w-1)IPS+DIV+MUL	as in J,JOR
Delay	4q-2	2q-1

TABLE 2

Table 3 summarizes the area and time requirements for the iterative systolic designs implementing methods (9) and (10); where k iterative steps are assumed.

Method	J,JOR Eq.(9)	GS,SOR Eq.(10)
Figures	6,8	7,9
Area	kwIPS+(w-1)IPS+DIV+MUL	k(wIPS+ADD)+(w-1)IPS+DIV+MUL
Time	2(n-1)+(k-1)(2p-1)+w+(4q-2)	2(n-1)+(k-1)2p+p+(2q-1)

TABLE 3

Comparing Tables 1 and 3 the following observations can be made:

(i) For J,JOR methods; the special cell in each iterative step, i.e. DIV or DIV+IPS is replaced by an IPS cell; furthermore a new array of ≅wIPS+DIV is introduced. Thus, the main pipeline is simpler and more regular as no special cells are required, at the expense of wIPS cells and 1 Divider. Especially for the J method, the IPS cell for the main diagonal for each mvm array (see Fig.6) can be replaced by a pair of delays (as in Fig.1) since 1-w=0. Finally, the computation times do not differ significantly.

(ii) For the GS,SOR methods: again the main pipeline is simpler and for the GS method the main diagonal cell can be removed (see Fig.7) giving better time and area results. Consequently, the computation time is increased by (2q-1); however notice that the SOR pipeline of Fig.4 has a longer time unit because of the complex operation in the special cell.

In general the improved designs offer simpler and more regular pipelines with approximately the same time complexity; furthermore some unification is achieved in the treatment of the different methods, i.e. the same systolic design can be used for both J and JOR, or for GS and SOR methods.

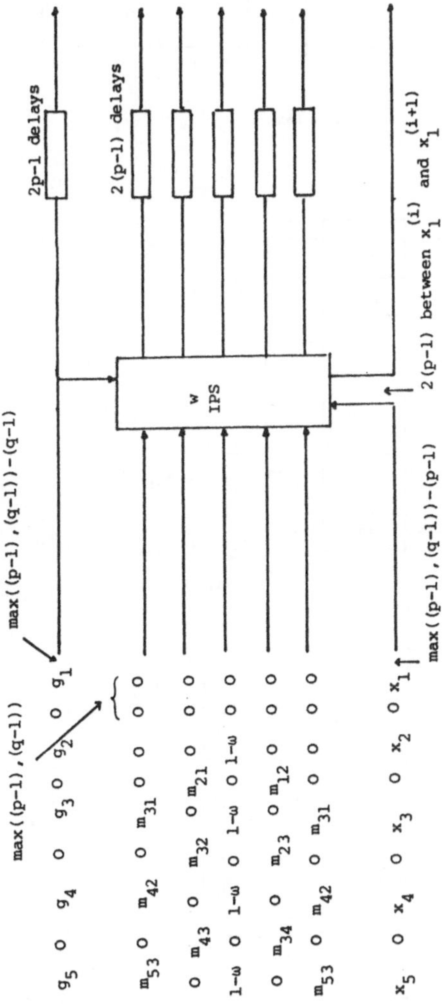

FIGURE 6: Modified array for J,JOR methods, $m_{ij} = -\omega \frac{a_{ij}}{a_{ii}}$, $g_i = \omega \frac{b_i}{a_{ii}}$

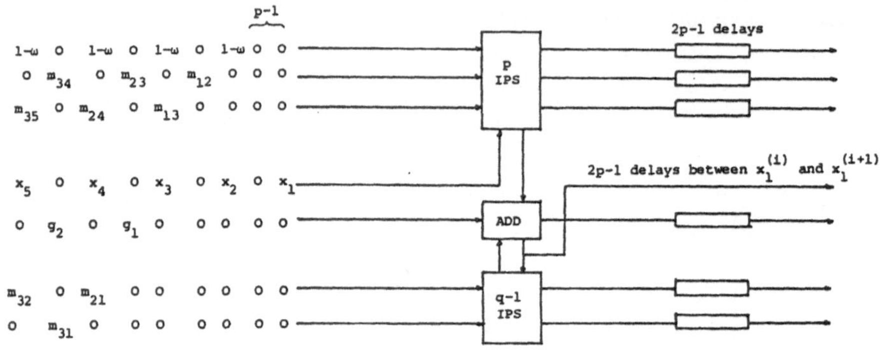

FIGURE 7: Modified array for GS,SOR methods

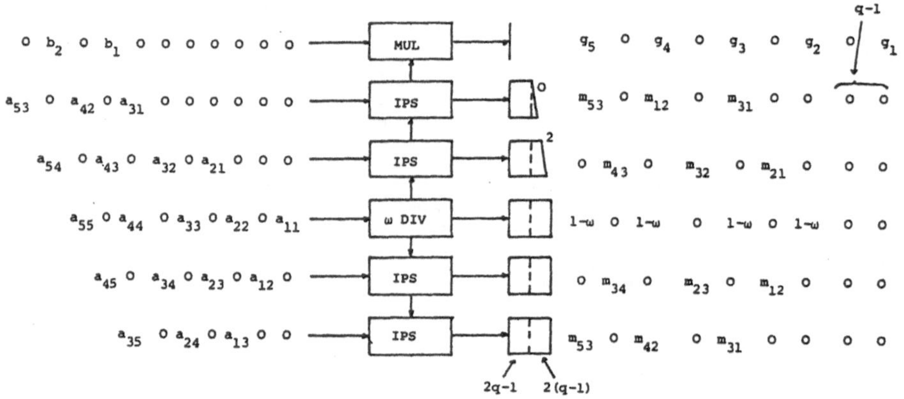

FIGURE 8: Preprocessor for J,JOR methods (Fig.6)

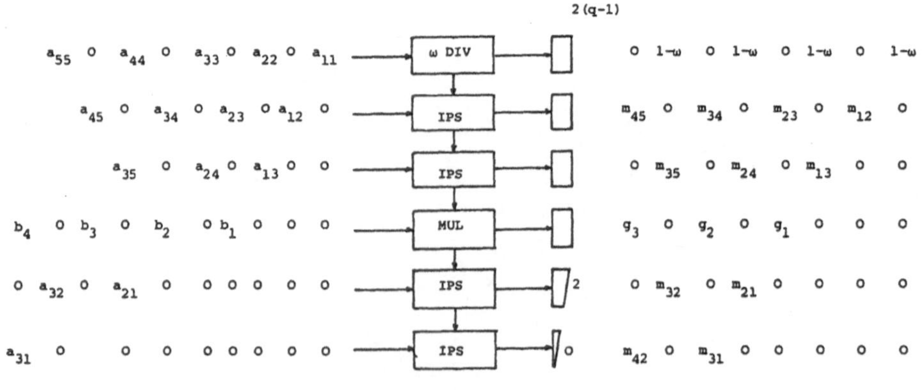

FIGURE 9: Preprocessor for GS,SOR methods (Fig.7)

3. UNIDIRECTIONAL MVM ARRAY FOR J,JOR METHODS

The iterative methods for the solution of linear systems discussed above can be analysed as a series of mvips. The basic building block of the iterative systolic networks of Figs.1-4 and 6,7 is the mvm array proposed in [L]. This array can be improved as discussed in [ME320] to give a systolic design with compact data sequence format, i.e. with no dummy elements between successive data items.

The computation time achieved by the mvm array in [EM320] is n+w+p-1 instead of 2n+w in [L]; furthermore there is an unidirectional data flow along the array and the processor utilization as well as the array throughput are also improved.

Fig.10 illustrates the configuration of a single iteration for the J,JOR method, i.e. the mvm array of Fig.10 corresponds to that of Fig.6.

The data sequence format as well as the synchronization delays are also shown in the figure; notice that the delay introduced by the modified mvm array is larger than that of that originally used in Fig.6: however the compactness of the data sequence gives a better overall computation time for the mvm array used in Fig.10.

The preprocessing elements of Fig.8 are modified as shown in Fig.11. In the same figure the delays required for the re-formatting of the data sequence before it enters the pipeline are also illustrated.

The area requirements for the preprocessor is the same as in Table 2; the delay which is introduced is now 2p IPS cycles.

The time requirements for the iterative systolic design using the improved mvm array are (n-1)+(k-1)(w+p-1)+w+2p IPS cycles; the area requirements are the same as in Table 3.

It is obvious that the computation time is reduced from $\cong$2n to $\cong$n, assuming $k,w \ll n$.

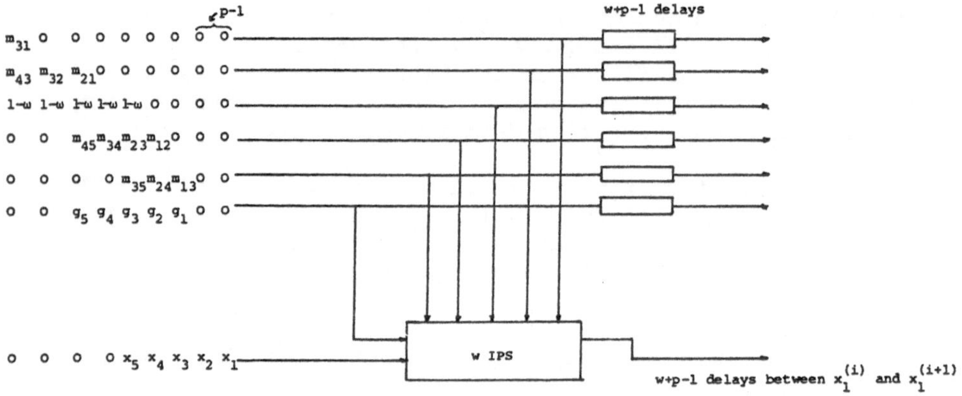

FIGURE 10: Array for J,JOR methods, using modified mvm array

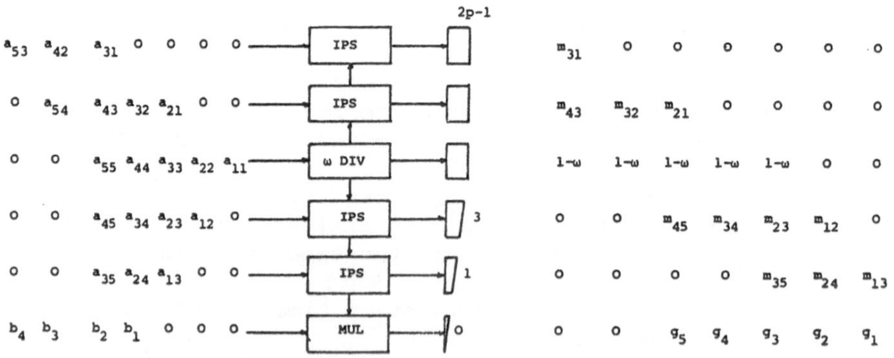

FIGURE 11: Preprocessor for J,JOR methods (Fig.10)

4. SYSTOLIC J,JOR METHODS USING MATRIX POWERS

Hotelling [H] has expressed the J method in a matrix format where the system (1) is written as an homogeneous system,

$$b_i x_0 - \sum_{j=1}^{n} a_{ij} x_j = 0 \quad , \; i=1,2,\ldots,n \tag{11}$$

by introducing $x_0=1$. The JOR-Hotelling method (JORH) can be similarly expressed as follows:

$$x_0^{(k+1)} = x_0^{(k)} = 1$$

$$x_i^{(k+1)} = \frac{\omega}{a_{ii}} (b_i - \sum_{j=1}^{n} a_{ij} x_j^{(k)}) + x_i^{(k)} \quad , \; i=1,2,\ldots,n \tag{12}$$

or

$$\underline{x}^{(k+1)} = T\underline{x}^{(k)}$$

where,

$$T = \begin{bmatrix} 1 & 0 & 0 & --- & 0 \\ \omega b_1/a_{11} & 1-\omega & -\omega a_{12}/a_{11} & --- & -\omega a_{1n}/a_{11} \\ \omega b_2/a_{22} & -\omega a_{21}/a_{22} & 1-\omega & --- & -\omega a_{2n}/a_{22} \\ \vdots & \vdots & \vdots & & \vdots \\ \omega b_n/a_{nn} & -\omega a_{n1}/a_{nn} & -\omega a_{n2}/a_{nn} & --- & 1-\omega \end{bmatrix} \tag{13}$$

The advantage of writing the JOR procedure in matrix form is that the sequence of matrices $T^2, T^4, T^6, T^8, \ldots, T^{2^k}$ can readily be obtained. Thus, four successive squarings will be equivalent to sixteen complete cycles of the JOR method. A further advantage is that the problem of solving a system of linear equations has been reduced to the problem of performing a sequence of matrix-matrix multiplications (mmm).

If the JORH process is convergent the first column of the matrix T will converge to the solution vector $\underline{x}$ [J].

The full JORH method is as follows:

(i) For the given system as in (1) and an overrelaxation factor ω, formulate T as in (13).

(ii) Apply k successive squarings on T, where k is

determined by the convergence rate.

(iii) Extract the first column of T^{2^k}, which is the solution vector $\underline{x}$.

From the description of the JORH method, it is evident that the main computational effort is the calculation of successive squarings of matrix T; systolic pipelines for successive matrix powers are described in [EMB284], using the unidirectional mmm systolic array of Fig.12 (see [LW]).

An important feature of the matrix T is that it can be partitioned as follows, (ω=1 for simplicity):

$$T = \left[\begin{array}{c|c} 1 & \underline{o}^T \\ \hline \underline{g} & M \end{array}\right] \tag{14}$$

It is also obvious that matrix M retains the sparsity characteristics of the original matrix A. Using the partitioned mmm formula [F]:

$$\left[\begin{array}{c|c} A_{11} & A_{12} \\ \hline A_{21} & A_{22} \end{array}\right] \cdot \left[\begin{array}{c|c} B_{11} & B_{12} \\ \hline B_{21} & B_{22} \end{array}\right] = \left[\begin{array}{c|c} A_{11}B_{11}+A_{12}B_{21} & A_{11}B_{12}+A_{12}B_{22} \\ \hline A_{21}B_{11}+A_{22}B_{21} & A_{21}B_{12}+A_{22}B_{22} \end{array}\right] \tag{15}$$

the matrix squaring for T can be written as:

$$\left[\begin{array}{c|c} 1 & \underline{o}^T \\ \hline \underline{g} & M \end{array}\right]^2 = \left[\begin{array}{c|c} 1 & \underline{o}^T \\ \hline \underline{g}+M\underline{g} & M^2 \end{array}\right] \tag{16}$$

Therefore, the squaring of the augmented matrix T is analysed in the mmm M and the matrix-vector inner product (mvips), $\underline{g}+M\underline{g}$. One block for a successive matrix squaring pipeline is given in Fig.13.

After a number of successive squarings using a pipeline with blocks as in Fig. 13, the output vector can be written as:

$$x = \underline{g}+M\underline{g}+M^2(\underline{g}+M\underline{g})+M^4(\underline{g}+M\underline{g}+M^2(\underline{g}+M\underline{g}))+ \ldots$$

$$= \underline{g}+M(\underline{g}+M(\underline{g}+\ldots+M\underline{g})\ldots) \tag{17}$$

The nested multiplication scheme allows for a computational network, similar to that shown in Fig.10: notice that 2^k-1 pipeline stages are required, instead of k as in Fig.13, where k is the number of successive squarings. However the network blocks have only mvips arrays instead of mmm and mvips arrays.

Table 4 summarises a comparison of the area and time requirements for the pipelines of Fig.13 and Fig.10 for k squaring stages.

For a banded matrix A the blocks of the pipeline of Fig. 13 increase in size since each matrix squaring operation doubles the bandwidth of matrix M. On the other hand, as there is no explicit powering-up of matrix M in the pipeline of Fig.10, all the blocks are identical, having area complexity w. Furthermore, the increasing bandwidth eliminates the computation time advantage of the pipeline of Fig. 13.

The matrix powering pipeline is interconnected with a preprocessor which produces matrix T implementing recurrences (13), i.e. in the form of matrix M and vector $\underline{g}$. This preprocessor is similar to the one described in Fig.11 and causes an additional delay of max((p+2),2q)-1 IPS cycles.

	Fig.13	Fig.10
Area	$\sum_{j=1}^{k-1}(2^{j-1}w-(2^{j-1}-1))^2+\sum_{j=1}^{k}(2^{j-1}w-(2^{j-1}-1))$	$(2^k-1)w$
Time	$\sum_{j=1}^{k}(2^{j-1}(w+p)-(2^{j-1}-1))-k$	$(2^k-1)(w+p-1)$

TABLE 4

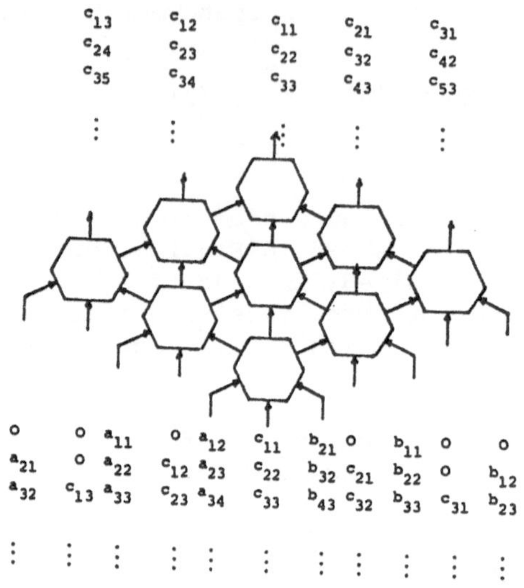

FIGURE 12: Unidirectional matrix-matrix multiplication systolic array

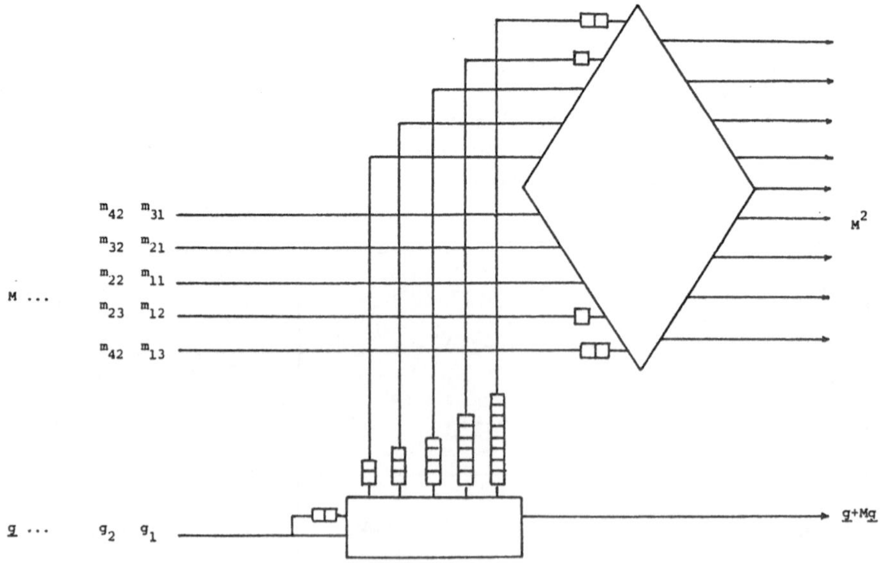

FIGURE 13: Pipeline block for matrix squaring and mvips

5. REFERENCES

[B] Brudaru, O., "Systolic Algorithms to Solve Linear Systems by Iteration Methods", An.St.Univ.Iasi, 31 (1985), 301-306.

[BBD] Berzins, M., Buckley, T.F. and Dew, P.M., "Systolic Matrix Iterative Algorithms", pp. 283-288, in Proc. of Int. Conf. 'Parallel Computing 83', Feilmeier, M., Joubert, J. and Schendel, U. (eds.), North Holland, 1984.

[D] Dew, P.M., "VLSI Architectures for Problems in Numerical Computation", pp.1-24 in 'Supercomputers and Parallel Computation', Paddon, D.J., (ed.), Oxford Univ. Press, 1984.

[EMB284] Evans, D.J., Margaritis, K. and Bekakos, M.P., "Systolic and Holographic Pyramidical Soft-Systolic Designs for Successive Matrix Powers", CS-284, LUT, June 1986 (to appear in Journal of Parallel Computing).

[F] Froberg, C.E., "Introduction to Numerical Analysis", Addison-Wesley, 1965.

[H] Hotelling, H. "Some New Methods in Matrix Calculation", Ann.Math.Stat., 4 (1943), 1-33.

[J] Jennings, W., "First Course in Numerical Methods", McMillan, 1965.

[L] Leiserson, C.E., "Area-Efficient VLSI Computation", Ph.D. Thesis, CMU, Oct. 1981.

[LW] Li, G.J. and Wah, B.W., "The Design of Optimal Systolic Arrays", IEEE Trans. Vol. C-34, No.1, Jan., 1985.

[ME320] Margaritis, K. and Evans, D.J., "Improved Systolic Matrix-Vector Multiplication", CS-320, LUT, Sept. 1986.

SUPERCONVERGENCE AND FINITE ELEMENT POST PROCESSING

G. F. Carey and R. J. MacKinnon

The University of Texas at Austin
Austin, Texas 78712

SUMMARY

We consider several practical and theoretical issues concerning point superconvergence of finite-element solutions and derivatives. This includes the development of special 'extraction' or post-processing formulas for improving solution accuracy (an example is flux post processing). Some new theoretical ideas and an analysis based on Taylor series expansions are introduced and lead to new superconvergence results and a class of post-processing formulas. This analysis method affords a more direct constructive approach to the subject and brings out some common misconceptions regarding superconvergence points.

1. INTRODUCTION

In finite-element calculations of boundary-value problems, it has been observed that there are special points where the solution or derivatives are exceptionally accurate (in certain instances, the point results may be exact). Moreover, mesh refinement experiments further reveal that the actual asymptotic rate of convergence at these superconvergence points is also superior to the well known global rates. In practice, we seek to exploit these superconvergence properties in interpreting and post processing the solution. The essential idea is to develop and analyse post-processing techniques for generating more accurate local solution and derivative (flux, stress) approximations.

Local superconvergence has been most extensively noted and analysed in one-dimension for two-point boundary-value problems. De Boor and Swartz [1973] found that, for a particular collocation method, the error at the inter-element nodes of certain spline approximations was of higher order in mesh size h than the global estimates. Douglas and Dupont [1974] showed that the Galerkin finite-element solution is also superconvergent at the inter-element nodes. This and other related studies led to the development of a more complete superconvergence analysis in one dimension (Dupont [1976]). Similar results have been demonstrated by Diaz [1977], Wheeler [1977] and Carey and Wheeler [1979] for C°-collocation-Galerkin finite-element methods.

In one dimension, the derivative for linear elements is simply the chord slope defined by the element end node values. If these nodal values are superconvergent, then, from the midpoint rule, we anticipate this chord slope would be a good approximation to the exact derivative at the element midpoint and this is consistent with experience. These observations and intuitive arguments have, in turn, led engineering analysts to infer superconvergence properties at corresponding Gauss points in higher dimensions (Barlow [1976]). For example, derivatives in two dimensions are commonly evaluated as centroid values for linear triangles and bilinear quadrilaterals. These computed values are then averaged or extrapolated to obtain nodal or boundary derivative approximations (e.g., see Hinton and Campbell [1974], Zienkiewicz [1977], Bathe [1982]).

In a more recent mathematical development, it was shown that the Green-Gauss integral identities lead to a class of post-processing formulas for the derivatives that yield superconvergent approximations in one dimension (Wheeler [1974], Dupont [1976], Carey [1982]). These ideas have been partially extended and applied to two-dimensional boundary-value problems (Carey, Seager and Chow [1985]).

In the present study, we develop a Taylor series approach to analyze superconvergent derivative points in one dimension and, more particularly, two dimensions. The approach applies also to three-dimensional elements. This development offers a very lucid interpretation of superconvergence and is easy to apply to elements in common use. From this analysis, we show, for instance, that several assumptions concerning derivative superconvergence points in two-dimensional elements are not valid. Further, we develop directly highly accurate post-processing formulas and estimates. We begin by briefly reviewing some fundamental global estimates and local superconvergence results.

2. GLOBAL AND LOCAL ESTIMATES

Consider a Galerkin finite-element approximation u_h to the solution u of an elliptic boundary-value problem of order $2m$. Let k be the element degree and h the mesh parameter. Then, under the standard assumptions of quasi-uniform refinement of the mesh and smoothness of the solution ($u \in H^r$), we have the basic global estimate of the error $e = u - u_h$,

$$\|e\|_m \leq Ch^{\mu} \|u\|_r \tag{1}$$

with $\mu = \min(k+1 - m, r - m)$ and $\|\cdot\|_m$ denoting the norm in H^m. Using the Aubin-Nitsche method, one can obtain estimates in H^s, $s < m$. For example, for second-order problems $m = 1$ and if the solution is sufficiently smooth ($r \geq k+1$), we obtain the global rate of convergence

$$|e|_s = O(h^{k+1-s}) \qquad s = 0, 1 \tag{2}$$

where $|\cdot|$ denotes the seminorm, e.g., in one dimension $|e|_0^2 = \int_a^b e^2 dx$, $|e|_1^2 = \int_a^b (e')^2 dx$ for solution and derivative, respectively.

To illustrate local superconvergence, let us consider the special example $-u'' = 1$, $u(0) = u(1) = 0$. The exact solution is a quadratic, and we find that the piecewise linear finite-element solution is *exact* at the nodes. That is, in this case the finite-element solution u_h is, in fact, the piecewise-linear nodal interpolant of u. This unusual result can be explained by noting that, in one dimension, the Green's function for this operator and a point source at node i is piecewise-linear and, hence, lies in the approximation space. We can use this property to show easily that the error e at the nodes is zero. This approach can be extended to more general 1-D situations by introducing a projection of the Green's function into the finite-element space to obtain the nodal estimate

$$|e(x_i)| \leq Ch^{2k} \|u\|_{k+1} \tag{3}$$

for sufficiently smooth solution ($r > k+1$) and data. (See Carey and Oden [1983] for further discussion.)

This property was computationally verified for the test case $-u'' + u = x$, $u(0) = u(1) = 0$: For uniform mesh refinement with a fixed interior node at $x = 1/3$ and a fixed inter-element node (knot) at $x = 1/2$ using C° cubic elements, we obtain numerically the rates $k+1 = 4$ and $2k = 6$, respectively.

Remark: A key concept in the error analysis is the use of negative norm estimates of the form $\|e\|_s \leq Ch^{2(k+1-m)} \|u\|_{k+1}$ for $s \leq 2m - k-1$ (see also Strang and Fix [1973]).

The derivative (flux, velocity, stress) is frequently of great practical interest, particularly at domain boundaries and interfaces. As observed in the introduction, the interior Gauss points in one dimension are superconvergence points for the derivative. More generally, one can use integration by parts (Green-Gauss identities) to develop formulas for superconvergent flux calculations. This leads directly to consideration of special post-processing schemes for derivatives. For example, in the previous test case, we have the identity

$$\int_0^1 (-u''+u - x)v \, dx - \int_0^1 (u'v'+uv - xv)dx + u'v\Big|_0^1 = 0 \qquad (4)$$

With $v = x$, this yields the post-processing formula

$$u^{*\prime}(1) = \int_0^1 (u_h'v'+u_h v - xv)dx$$

using finite-element solution u_h .

From Dupont [1976], the magnitude of the local derivative error $e' = u' - u^{*\prime}$ at $x = 1$ satisfies the estimate

$$|e'| \leq Ch^{r+k-1} \|u\|_r$$

with $2 \leq r \leq k+1$ so that the optimal result for the derivative is again $O(h^{2k})$.

This procedure can be conveniently used in practice with v chosen as the appropriate finite-element basis function and applied to subregions to extract more accurate interface and interior derivatives. For the previous example, a log-log plot of $e'(0.5)$ against h in a mesh refinement study with quadratic elements yields a rate of convergence (slope) of 4.

While the above approaches are fundamental, we now describe an alternative but related technique, based on Taylor series ideas, that is more explicit and constructive. Moreover, since this approach relies only on elementary analysis concepts, it is straightforward to understand and implement. Finally, the method directly leads to some new results concerning derivative superconvergence points in two dimensions. For brevity, here we shall consider only the derivative extraction problem for finite-element interpolation and summarize the main results. Similar results hold for the finite-element approximation to a boundary-value problem (Mackinnon and Carey [1987]).

3. TAYLOR SERIES ANALYSIS

For clarity of exposition, we begin with the one-dimensional case and the problem of derivative calculation from the finite-element nodal *interpolant* u_h on element Ω_e of degree $k = N-1$ with local nodes $x_1, x_2, \ldots, x_N$. Simply differentiating the element expansion

$$u_h'(\bar{x}) = \sum_{j=1}^{N} u_j \; \psi_j'(\bar{x}) \tag{5}$$

where u_j are the nodal values for element Ω_e, ψ_j are the element basis functions and $\bar{x}$ is an arbitrary point in the element. For example, if linear elements are used, then N=2 and we have $u_h'(\bar{x}) = (u_2 - u_1)/h$. Recalling the standard Taylor series argument for determining difference approximations to derivatives, at midpoint $\bar{x} = (x_1 + x_2)/2$,

$$u'(\bar{x}) = \frac{u_2 - u_1}{h} - \frac{2h^2}{3!} u''(\bar{x}) + O(h^4)$$

so that the finite-element result will be $O(h^2)$ superconvergent at the central Gauss point.

Similarly, if the element is quadratic $N = 3$ in (3) then at any interior point $\bar{x}$

$$u_h'(\bar{x}) = \sum_{j=1}^{3} u_j\psi_j'(\bar{x}) = \sum_{j=1}^{3} u_j\alpha_j \tag{6}$$

where we have introduced the notation $\alpha_j = \psi_j'(\bar{x})$. Setting $x_i - \bar{x} = \delta_i$, then

$$\alpha_1 = -\frac{(\delta_2 + \delta_3)}{2h^2} \quad , \quad \alpha_2 = \frac{(\delta_1 + \delta_3)}{h^2} \quad , \quad \alpha_3 = -\frac{(\delta_1 + \delta_2)}{2h^2} \tag{7}$$

Expanding u_1 , u_2 , u_3 in Taylor series about $\bar{x}$ and collecting terms, we have

$$\sum_{j=1}^{3} \alpha_j u_j = \left(\sum_{j=1}^{3} \alpha_j\right) u(\bar{x}) + \left(\sum_{j=1}^{3} \alpha_j\delta_j\right) u'(\bar{x}) + \left(\sum_{j=1}^{3} \alpha_j \frac{\delta_j^2}{2!}\right) u''(\bar{x}) + \left(\sum_{j=1}^{3} \alpha_j \frac{\delta_j^3}{3!}\right) u'''(\bar{x}) + \ldots \tag{8}$$

so that the derivative at $\bar{x}$ is defined by

$$\left(\sum_{j=1}^{3} \alpha_j\delta_j\right) u'(\bar{x}) = \sum_{j=1}^{3} \alpha_j u_j - \left(\sum_{j=1}^{3} \alpha_j\right) u(\bar{x}) - \left(\sum_{j=1}^{3} \alpha_j \frac{\delta_j^2}{2!}\right) u''(\bar{x}) - \left(\sum_{j=1}^{3} \alpha_j \frac{\delta_j^3}{3!}\right) u'''(\bar{x}) + \ldots \tag{9}$$

Since the basis is a complete quadratic, then it interpolates the functions 1 , x , x^2 exactly. This implies that $\sum_{j=1}^{3} \alpha_j = 0$, $\sum_{j=1}^{3} \alpha_j \delta_j = 1$, $\sum_{j=1}^{3} \alpha_j \delta_j^2 = 0$ and hence

$$u'(\bar{x}) = \sum_{j=1}^{3} \alpha_j u_j - \left(\sum_{j=1}^{3} \alpha_j \frac{\delta_j^3}{3!} \right) u'''(\bar{x}) - O(h^3) \tag{10}$$

Examining the leading term in the truncation error, we find that

$$\sum_{j=1}^{3} \alpha_j \delta_j^3 = 0$$

if $\bar{x}$ is chosen to be either of the two Gauss points (zeros of the shifted legendre polynomials) in the element interior; i.e., if we take $x_1 = -1$, $x_2 = 0$, $x_3 = 1$ for the reference interval $[-1,1]$, then $\bar{x} = \pm 1/\sqrt{3}$ are the special points at which (5) is exceptionally accurate. This procedure can be directly extended to one-dimensional elements of any degree $k = N-1$.

We next consider the two-dimensional case. Let α_j^x denote $\partial \psi_j / \partial x$ so that on differentiating the finite-element expansion, we have

$$\frac{\partial u_h}{\partial x} = \sum_{j=1}^{N} u_j \alpha_j^x \tag{11}$$

As in the one-dimensional analysis, we introduce Taylor series expansions for $u_j = u(x_j, y_j)$ in (11) about $x = \bar{x}$, $y = \bar{y}$ with $\delta_j^x = x_j - \bar{x}$, $\delta_j^y = y_j - \bar{y}$ to obtain,

$$\begin{aligned} \sum_{j=1}^{N} \alpha_j^x u_j = {} & \left(\sum_{j=1}^{N} \alpha_j^x \right) \bar{u} + \left(\sum_{j=1}^{N} \alpha_j^x \delta_j^x \right) \bar{u}_x + \left(\sum_{j=1}^{N} \alpha_j^x \delta_j^y \right) \bar{u}_y \\ & + \left[\sum_{j=1}^{N} \alpha_j^x (\delta_j^x)^2 / 2! \right] \bar{u}_{xx} \\ & + \left[\sum_{j=1}^{N} \alpha_j^x (\delta_j^y)^2 / 2! \right] \bar{u}_{yy} + \left[\sum_{j=1}^{N} \alpha_j^x \delta_j^x \delta_j^y \right] \bar{u}_{xy} + \ldots \end{aligned} \tag{12}$$

where $\bar{u} = u(\bar{x}, \bar{y})$, $\bar{u}_x = u_x(\bar{x}, \bar{y})$, and so on.

Using the completeness properties of the element basis of degree k, simplifying and regrouping terms yields the Taylor expansion for $\bar{u}_x$. In particular, for a linear basis $k=1$ and we obtain

$$\bar{u}_x = \sum_{j=1}^{N} \alpha_j^x u_j - \left[\sum_{j=1}^{N} \alpha_j^x (\delta_j^x)^2/2!\right]\bar{u}_{xx} - \left[\sum_{j=1}^{N} \alpha_j^x (\delta_j^y)^2/2!\right]\bar{u}_{yy}$$

$$- \left(\sum_{j=1}^{N} \alpha_j^x \delta_j^x \delta_j^y\right)\bar{u}_{xy} - O(h^2) \qquad (13)$$

The expression (11) will be an $O(h)$ approximation to u_x except at those points $(\bar{x},\bar{y})$ where the first-order terms are individually zero or collectively cancel. Examining this requirement, we find that this implies

$$\bar{x} = \frac{1}{2}\sum_{j=1}^{N} \alpha_j^x x_j^2 , \quad \bar{y} = \sum_{j=1}^{N} \alpha_j^x x_j y_j , \quad \sum_{j=1}^{N} \alpha_j^x y_j^2 = 0$$

The last relation in general will not hold! For example, for the triangle, we find that this condition holds only if one side is horizontal in which case the superconvergence point $(\bar{x},\bar{y})$ is the midpoint of that side. This result can be compared with the previous 1-D result.

Similar reasoning applies to the formula for u_y and the only superconvergence point is the midpoint of a vertical side. By employing a rotated coordinate system aligned along any given side, it follows that the tangential derivative is superconvergent at the midpoint (as one can also infer immediately from the 1-D case). The significant point is that u_x and u_y are not superconvergent at the interior Gauss point (centroid) of the triangle.

For the bilinear quadrilateral a similar analysis applies and a solution exists only if an associated consistency condition again holds

In the specific case of a rectangular element aligned with the axes, it follows that u_x is superconvergent on the <u>line</u> bisecting the horizontal sides and u_y on the <u>line</u> bisecting the vertical sides. Hence, the centroid is the superconvergent point in this instance. For quadrilateral elements, there will in general be no such superconvergence point in the interior and the situation reverts to one analogous to that of the triangle. A perturbation analysis indicates, however, that for small distortion from a rectangular shape the coefficients of the $O(h^2)$ terms in the Taylor series representation will be small.

If the solution u satisfies an additional relationship such as a differential equation, then this can be exploited to weaken the previous conditions. For example, if u satisfies $\Delta u = 0$, then for an equilateral triangle of side h we find that the exceptional points for u_x and u_y lie diametrically opposite on a circle of radius $h/2\sqrt{3}$ centered at the element centroid.

Remark. If the finite-element nodal solution is introduced on the right in our Taylor series representation, we obtain an approximate formula for derivative post processing. In the one-dimensional case, by appealing to the nodal superconvergence theory stated earlier and approximation of the integral form of the flux post-processing formula, we find that the error in this derivative approximation is $O(h^{k+1})$ for $k=1, 2$. It follows also that simply differentiating the finite-element solution yields an error at the Gauss points that is 'only' $O(h^{k+1})$ and $O(h^k)$ at other points.

We are also able to determine the "correction" needed to restore the $O(h^{2k})$ accuracy. Finally, the results in two dimensions are more limited and are presently under investigation. At this point, we have extended the analysis to linear 2-D elements.

ACKNOWLEDGEMENT

This research has been supported in part by the Office of Naval Research and the CEOGRR industrial associates.

REFERENCES

1. Bathe, K. J. - Finite Element Procedures in Engineering Analysis, Prentice-Hall, Englewood Cliffs, NJ, 1982.

2. Barlow, J., "Optimal Stress Locations in Finite Element Models," Int. J. Numer. Meth. Eng. 10, 243-251, 1976.

3. Carey, G. F., "Derivative Calculation from Finite Element Solutions," J. Comp. Meth. Appl. Mech. Eng. 35, 1-14, 1982.

4. Carey, G. F., S. S. Chow and M. Seager, "Approximate Boundary Flux Calculation," J. Comp. Meth. Appl. Mech. Eng. 50, 107-120, 1985.

5. Carey, G. F. and J. T. Oden - Finite Elements: A Second Course, Prentice-Hall, Englewood Cliffs, NJ, 1983.

6. Carey, G. F. and M. F. Wheeler, "C°-Collocation-Galerkin Methods," in Codes for BoundaryValue Problems in Ordinary Differential Equations, Lecture Notes in Computer Science, Springer-Verlag, 250-256, 1979

7. DeBoor, J. and B. Swartz, "Collocation at Gauss Points," SIAM J. Num. Anal. 10, 582-606, 1973.

8. Diaz, J., "A Collocation-Galerkin Method for the Two-Point Boundary-Value Problem Using Continuous Piecewise-Polynomial Spaces, SIAM J. Num. Anal. 14, 844-858, 1977.

9. Douglas, Jr., J. and T. Dupont, "Galerkin Approximations for the Two-Point Boundary-Value Problem Using Continuous, Piece-Wise Polynomial Spaces," Numer. Math., 22, 99-109, 1974.

10. Dupont, T., "A Unified Theory of Superconvergence for Galerkin Methods for Two-Point Boundary Problems," SIAM J. Numer. Anal., Vol. 13, No. 3, pp. 362-368, 1976.

11. Hinton, E. and J. S. Campbell, "Local and Global Smoothing of Discontinuous Element Functions Using a Least Squares Method," Int. J. Numer. Meth. Eng. 8, 461-480, 1974.

12. MacKinnon, R. J. and G. F. Carey, "Superconvergent Derivatives: A Taylor Series Analysis" (in preparation), 1987.

13. Strang, G. and G. Fix - An Analysis of the Finite Element Method, Prentice-Hall, 1973.

14. Wheeler, M. F., "A Galerkin Procedure for Estimating the Flux for Two-Point Problems," SIAM J. Numer. Anal. 11, 764-768, 1974.

15. Wheeler, M. F., "A C°-Collocation Finite-Element Method for Two-Point Boundary-Value Problems and One Space-Dimension Parabolic Problem, SIAM J. Num. Anal. 14, 71-90, 1977

16. Zienkiewicz, O. C., The Finite Element Method, 3rd. Ed., McGraw-Hill, London, 1977.

Automatic Generation of Shape Function Routines

Peter Bettess
Professor of Offshore Engineering
School of Marine Technology
University of Newcastle-upon-Tyne
Newcastle-upon-Tyne NE1 7RU

Jacqueline A Bettess
Assistant Director
Computer Centre
University of Durham
Durham DH1 3LE

Summary

In this paper we show how the use of an automatic symbolic algebraic manipulation system can greatly reduce the work involved in the generation of shape functions and their derivatives for finite element analyses. The system used is called REDUCE and is available for most IBM and VAX machines. REDUCE can carry out algebraic operations on rational functions accurately, no matter how complicated the expression becomes and also produce the output in the form of a FORTRAN program. The power of the system is demonstrated by giving the code required to generate most of the one and two dimensional families of C_0 continuous shape functions. The resulting FORTRAN code for the shape function and shape function derivatives are given for the 6 node triangular element. The advantages and disadvantages of using such a system are also considered.

1. Introduction

Computer programs for symbolic algebraic manipulation have been around for some time. They are however getting progressively more sophisticated. To date they do not seem to have been used extensively in the finite element world, although at least one usage has been reported [1]. The power of one such program entitled REDUCE which is available on the NUMAC system was recently exploited for the expansion of some algebra associated with Stokes fifth order waves, in connection with wave diffraction analysis. The amount of algebra was formidable, but it was handled very easily by the REDUCE program. The most attractive feature of the computer manipulation is not so much the time saved, as it takes some time to de-bug the input, but the confidence in the results.

Programs such as REDUCE[2] can perform addition, subtraction and multiplication of series, can operate on rational functions and can perform differentiation and integration of an ever increasing range of functions. REDUCE produces output either in a passable imitation of ordinary algebraic notation, so far as the line printer makes this possible, or in FORTRAN expressions. It also has a facilities for outputting strings. Thus the output from REDUCE can easily be a FORTRAN subroutine. In this paper we show how by using the FORTRAN capabilities and the power of algebraic manipulation of REDUCE, it is possible to generate from a fairly concise data, virtually all the one and two dimensional families of C_0 continuous shape functions. Moreover the confidence in their correctness is extremely high.

Appendix 1 gives a listing of the REDUCE input data used to generate the following shape functions.

(i) The one dimensional Lagrange shape functions, up to any predetermined order.
(ii) The two dimensional Lagrange shape functions, up to any predetermined order.
(iii) The quadratic and cubic serendipity shape functions.
(iv) The triangular element shape functions, up to any order.

The predetermined order is currently set to cubic and is only limited by the size of the arrays in the code. As can be seen, all the above shape function routines are generated from approximately 100 lines of input data. REDUCE has an algol type program structure.

2. Program

Arrays

X(i,j) holds the j th co-ordinate for the i th Lagrange polynomial, e.g.

X(1,1) = -1.0	X(1,2) = 1.0		
X(2,1) = -1.0	X(2,2) = 0.0	X(2,3) = 1.0	
X(3,1) = -1.0	X(3,2) = - 1/3	X(3,3) = 1/3	X(3,4) = 1

M(i,j,k) holds the k th term of the j th Lagrange polynomial, with unknown i th co-ordinate. So for example if the first co-ordinate was XI
M(1,1,1) = (1-XI)/2   M(1,1,2) = (1 + XI)/2

N(i) holds the shape function expression for the i th node

COORD (i) holds the i th co-ordinate name e.g.

COORD(1) = XI	COORD(2) = ET	or
COORD(1) = L1	COORD(2) = L2 etc.	depending upon the names selected.

SEM(4) stores some orientation information for the serendipity elements.

Executable statements

Lines

7-9 First all the X values are set up, for as many Lagrange polynomials are required.

10-13 The Lagrange polynomials are now set up, using C as the independent variable. This can be changed later quite freely.

14-16 The co-ordinate C is set in turn to each of the 3 chosen local co-ordinates, in this case XI, ET and ZE, which are stored in COORDS (i). The Lagrange polynomials, in terms of the three local co-ordinates are stored in M(i,j,k).

17-54 The procedure WRTSF which outputs the shape functions in FORTRAN is now input. The header statement writes out the subroutine name. This consists of SF, followed by the number of dimensions, a tag, for the element type (L lagrange, S serendipity, T triangle), and the number of nodes. Thus SF2S8 is the shape function for the 2 dimensional 8 node serendipity quadrilateral. The parameter list consists of the local co-ordinates, written out from COORD(i), and the arrays containing the shape function and shape function local derivatives, and their sizes, thus SF, ISF, SFDL, ISFDL, JSFDL. This is followed by write statements which produce the necessary array and type declarations, and some explanatory comment statements. The actual shape functions are simply written out in a do loop (line 50) and so are the shape function derivatives (line 53). The REDUCE code, if asked to write DF(A,B) will write the derivative of A with respect to B. So if variable B holds A**3 + 2*A**2+6, the output will be 3*A**2+4*A*. There is no need for any hand generation of the shape function derivatives or for any checking (except of the shape functions themselves).

55-59 The one dimensional lagrange shape functions are now generated. The variable C is reset to XI, the tag is set to L, and for each polynomial the number of nodes are calculated. The shape function is copied into N from L, and procedure WRTSF is invoked.

60-63 The two dimensional lagrange shape functions are now generated. In this case the method is as for one dimension, except that the two shape functions in XI and ET are multiplied together to form N. The node ordering is very straightforward and is shown in Figure 1.

64-88 Serendipity shape functions are now generated. The explanation given in Zienkiewicz [3], page 158 is followed precisely. The quadratic element is dealt with first. The shape functions for the midside nodes are first found, by multiplying together the appropriate quadratic and linear shape functions (lines 67-69). Then bi-linear shape functions are set up for each corner node and then modified by having half the two adjacent midside shape functions subtracted from them. (lines 70-75). The code thus exactly mirrors the theory.

In the cubic serendipity element the two nodes on each edge are first set up, by multiplying together the appropriate cubic and linear shape functions. (lines 78-81) Again corner nodes are initially set to be bi-linear and are modified by subtracting appropriate fractions of neighbouring edge nodes (lines 82-88). Both these shape functions are output by procedure WRTSF. Higher order serendipidity elements could be generated by the same method, but they do not seem very popular and have been omitted here.

89-104 The triangular elements are now generated. Zienkiewicz [3] explains how the triangular shape functions can be constructed from suitable lagrange polynomials in the local area co-ordinates L1, L2, and L3. In the present program these are simply set up in procedure TLAN, and multiplied together to form the element shape function. As before they are output by procedure WRTSF. The local coordinates are re-set, in array COORD, to L1, L2 and L3. In this case L3 has been redefined in terms of L1 and L2 so that there are only two independent variables (line 99). This can be easily changed to retain three independent variables if required. REDUCE can deal with all the necessary substitutions. Appendix 2 shows the code for the 6 node triangle elements as generated by REDUCE.

3. Advantages

1. The REDUCE code is extremely concise, in comparison with the FORTRAN code which it generates.
2. The REDUCE code is much closer to the theory and is easy to follow. There is no complicated algebra, since the program does that for you.
3. Changes of notation and variable names are trivially simple.
4. The co-ordinate basis can easily be changed. For example changing from the range (-1,+1), to (0,+1), would only require changing line 13.
5. Alterations in the theory can be made simply. For example changing to

keeping the three area co-ordinates as independent variables requires only the removal of line 145.

6. The extension to three dimensions is trivial, and is being carried out at the moment.
7. C_1 continuous shape functions also be dealt with.

4. Disadvantages

1. At present REDUCE only allows FORTRAN constants to be with or without a decimal point by changing a switch. In the shape function application we require the array subscripts to be without a decimal point and the constants in the expression to be with. This necessitates some post run editing.
2. The polynomial expressions produced by REDUCE are not optimised for fast execution.There is no optimal nesting, and a lot of exponentiation is carried out, so either an optimising compiler should be used, or some massaging of the code carried out, on a semi automatic basis.

5. Conclusions

A radical new approach to the generation of shape function routines has been presented. The use of symbolic manipulation codes, like REDUCE, for such tasks, simplifies the process and greatly reduces the opportunities for error. The authors believe that codes like REDUCE will become much more widely adopted for the automatic generation of reliable software for numerical algorithms. We think that this will be very much a growth area. For example REDUCE could be used to generate time stepping algorithms, infinite element mapping functions and other special finite element software.

We expect to see further developments. REDUCE can already produce output which is either algebraic or FORTRAN compatible. We expect the FORTRAN representation to be improved to include optimisation of expression evaluation. It is also possible that output in a type setter compatible format, eg TEX code could be produced, leading to the enticing prospect of no proof reading between algorithm formulation and published paper.

References

1. Nedergaard, H. and Pedersen, P.T., 'Analysis Procedure for Space Frames with Material and Geometrical Nonlinearities', Finite Element Methods for Non Linear Problems, eds. Bergan, P.G. and Bathe, K.J., Springer Verlag 1986.
2. Hearn, Anthony C., ed. 'REDUCE User's Manual', The Rand Corporation, Santa Monica, Ca 90406.
3. Zienkiewicz, O.C. 'The Finite Element Method', McGraw Hill, 3rd edition, 1977.

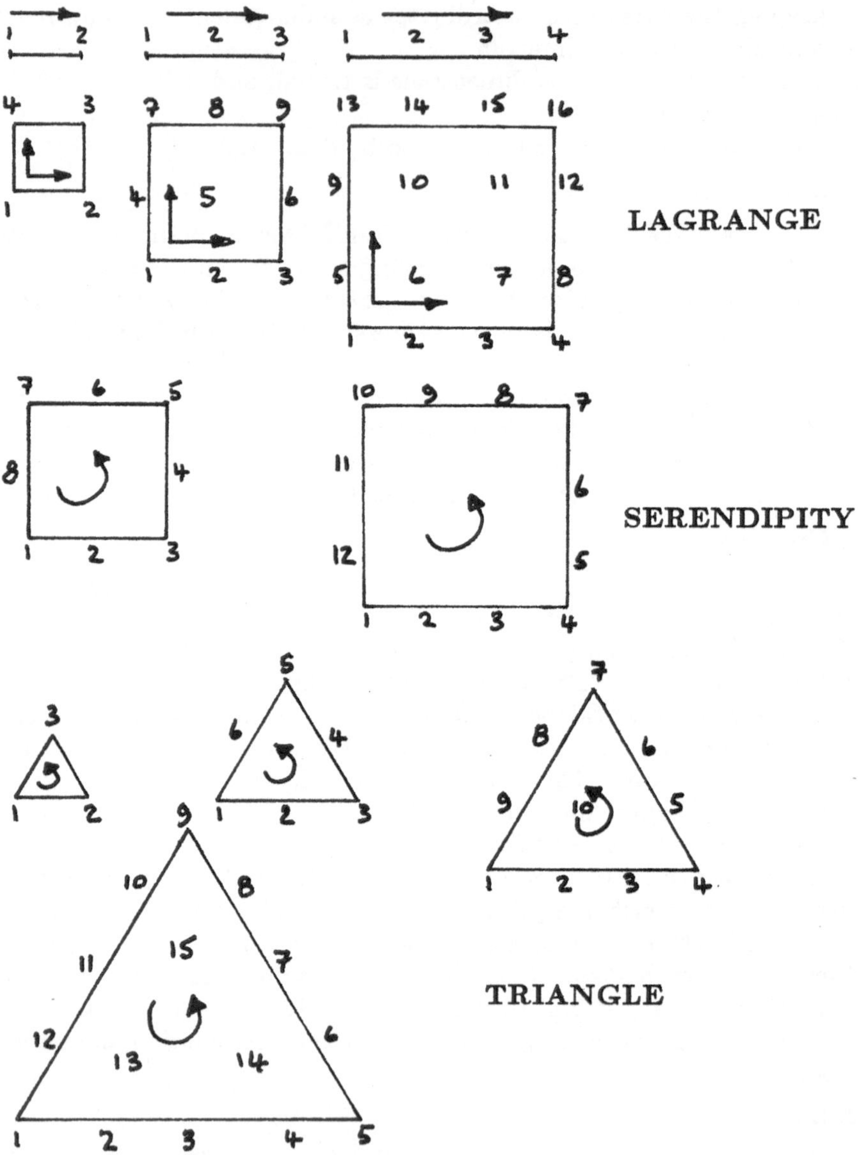

Figure 1 Node ordering

Appendix 1

Listing of REDUCE input data

```
OFF ECHO; OUT SFR; OFF OUTPUT; FORTWIDTH!* :=70;
COMMENT  Create shape functions for quadrilateral and triangular
two dimensional finite elements using REDUCE for algebraic
manipulations. Peter Bettess and Jacqueline A. Bettess, March 1987;
ARRAY  X(9,10), L(9,10), M(9,10,100), N(100), CX(10), LOC(10,100), COORD(10), SEM(20);
SEM(1):=1;SEM(2):=2;SEM(3):=2;SEM(4):=1;COORD(1):=XI; COORD(2):=ET;COORD(3):=ZE;
COMMENT  set up the co-ordinates.   X(I,J) holds the J th co-ordinate of the
I+1 th set of points, e.g. X(1,1) = -1,X(1,2) = +1. At present 4 sets are generated;UPL:=4;
FOR I:= 1:UPL DO BEGIN FOR J:=0:I DO BEGIN X(I,J+1):=-1+2*J/I END END;
COMMENT  now set up all the one dimensional Lagrange polynomials
in terms of the generic co-ordinate, C;
FOR I:=1:UPL DO BEGIN  FOR J:=1:I+1 DO BEGIN  L(I,J):=1;FOR K:= 1:I+1 DO BEGIN
IF K NEQ J THEN L(I,J) := L(I,J) * (C-X(I,K))/(X(I,J)-X(I,K)) END END END;
COMMENT  copy the lagrange shape functions into array M;
FOR K:=1:3 DO BEGIN C:=COORD(K);
FOR I:=1:UPL DO BEGIN  FOR J:=1:I+1 DO BEGIN M(K,I,J):=L(I,J) END END END;
COMMENT  define the procedure to write out the shape function routines;
PROCEDURE WRTSF(TAG,N,NODES,NDER,COORD); BEGIN ON FORT; OFF PERIOD;
IF NDER=1 THEN <<
WRITE"      SUBROUTINE SF",NDER,TAG,NODES,"(",COORD(1), SF, ISF, SFDL, ISFDL, JSFDL)">>;
IF NDER=2 THEN <<
WRITE"      SUBROUTINE SF",NDER,TAG,NODES,"(",COORD(1),", ",COORD(2),
   " ,SF, ISF, SFDL, ISFDL, JSFDL)">>;
ON PERIOD; WRITE"C"; WRITE"C***SHAPE FUNCTION SUBROUTINE";
WRITE"C***(c) Peter and Jacqueline A. Bettess, 1987"; WRITE"C";
WRITE"C-----------------------------------------------------------------";
WRITE"C PURPOSE";WRITE"C      Forms element shape function and derivatives";
WRITE"C";WRITE"C HISTORY";
WRITE"C       Written February 1987";
WRITE"C";WRITE"C ARGUMENTS IN";
WRITE"C       ",COORD(1),"     First local co-ordinate";
IF NDER>1 THEN WRITE"C      ",COORD(2),"     Second local co-ordinate";
IF NDER>2 THEN WRITE"C      ",COORD(3),"     Third  local co-ordinate";
WRITE"C       ISF    Dimension of shape function array, SF";
WRITE"C       ISFDL  1st dimension of shape function";
```

```
WRITE"C              derivative array, SFDL";
WRITE"C       JSFDL  2nd dimension of shape function";
WRITE"C              derivative array, SFDL";
WRITE"C";WRITE"C ARGUMENTS OUT";
WRITE"C       SF     Shape function array";
WRITE"C       SFDL   Array of shape function derivatives";
WRITE"C              with respect to local co-ordinates";WRITE"C";
WRITE"C**********************************************************************";
WRITE"      DOUBLE PRECISION  SF, SFDL";
IF NDER=1 THEN  WRITE"      DOUBLE PRECISION ",COORD(1);
IF NDER=2 THEN  WRITE"      DOUBLE PRECISION ",COORD(1),", ",COORD(2);
WRITE"      INTEGER  ISF, ISFDL, JSFDL";
WRITE"      DIMENSION  SF(ISF), SFDL(ISFDL,JSFDL)";
WRITE"C";WRITE"C***form the element shape functions"; WRITE"C";
FOR I:=1:NODES DO <<WRITE"      SF(",I,") = ",N(I)>>;
WRITE"C"; WRITE"C***form the shape function derivatives"; WRITE"C";
FOR KN:=1:NDER DO BEGIN
FOR I:=1:NODES DO <<WRITE"      SFDL(",KN,",",I,") = ",DF(N(I),COORD(KN))>>
END; WRITE"      RETURN"; WRITE"      END" END;
COMMENT  set up the 1 dimensional Lagrange shape functions
directions, using the generic form;
C:=XI; TAG:=L; NDER:=1;
FOR I:=1:UPL DO BEGIN  NODES:=I+1; FOR J:=1:NODES DO BEGIN N(J):=L(I,J) END;
WRTSF(TAG,N,NODES,NDER,COORD) END;
COMMENT  set up the 2 dimensional Lagrange shape functions;
NDER:=2; FOR I:=1:UPL DO BEGIN  NODES:=(I+1)*(I+1); K:=1;
FOR J:=1:I+1 DO BEGIN FOR P:=1:I+1 DO BEGIN
N(K):=M(1,I,P)*M(2,I,J); K:=K+1 END END; WRTSF(TAG,N,NODES,NDER,COORD) END;
COMMENT  set up the 2 dimensional Serendipity shape functions;
COMMENT  first, the 8 noded Serendipity element;
K:=1; L:=2; NODES:=8;
COMMENT  first set up the 4 mid side nodes;
FOR J:=1:4 DO BEGIN  NODENO:=J+J; LL:=SEM(J);
N(NODENO):=M(K,2,2)*M(L,1,LL); K:=3-K; L:=3-L END;
COMMENT now set up corner nodes;
```

Appendix 1 cont.

Listing of REDUCE input data

```
LL:=1; KK:=1; INK:=1;
FOR J:=1:4 DO BEGIN NODENO:=J+J-1; IF J>2 THEN LL:=2;
N(NODENO):=M(1,1,KK)*M(2,1,LL); NODM:=NODENO-1;
IF NODM<1 THEN NODM:=NODM+8; N(NODENO):=N(NODENO)-N(NODENO+1)/2-N(NODM)/2;
KK:=KK+INK; INK:=INK-1 END; TAG:=S; NDER:=2; WRTSF(TAG,N,NODES,NDER,COORD);
COMMENT  second, the 12 noded Serendipity element;
K:=1; L:=2; NODES:=12;
COMMENT  first set up the 8 mid side nodes; P1:=3;P2:=2;
FOR P:=1:2 DO BEGIN FOR Q:=1:2 DO BEGIN J:=(P-1)*2+Q;
NODENO:=J+J+J; LL:=SEM(J);  N(NODENO):=M(K,3,P1)*M(L,1,LL);
N(NODENO-1):=M(K,3,P2)*M(L,1,LL); K:=3-K; L:=3-L END; P1:=2; P2:=3 END;
COMMENT now set up corner nodes;
LL:=1; KK:=1; INK:=1; FOR J:=1:4 DO BEGIN NODENO:=J+J+J-2;
IF J>2 THEN LL:=2; N(NODENO):=M(1,1,KK)*M(2,1,LL);
NODM:=NODENO-1;  IF NODM<1 THEN NODM:=NODM+12;
N(NODENO):=N(NODENO)-N(NODENO+1)*2/3-N(NODM)*2/3-N(NODENO+2)/3-
N(NODM-1)/3; KK:=KK+INK; INK:=INK-1  END;
TAG:=S; NDER:=2; WRTSF(TAG,N,NODES,NDER,COORD);
COMMENT  develop shape functions for triangle elements;
PROCEDURE TLAN(K,CX,CXX);
BEGIN SCALAR LL; LL:=1; IF K=1 THEN GOTO ZZ; FOR J:=1:K-1 DO
    <<LL:=LL*(CXX-CX(J))/(CX(K)-CX(J))>>;     ZZ:RETURN LL   END;
NDER:=2;
COMMENT  set up locations matrix, to order nodes anti-clockwise;
FOR I:=1:36 DO BEGIN FOR J:=1:6 DO BEGIN LOC(J,I):=I END END;
LOC(2,4):=6;LOC(2,5):=4;LOC(2,6):=5;LOC(3,5):=9;LOC(3,6):=10;LOC(3,7):=5;
LOC(3,9):=6;LOC(3,10):=7;LOC(4,6):=12;LOC(4,7):=13;LOC(4,8):=14;LOC(4,9):=6;
LOC(4,10):=11;LOC(4,11):=15;LOC(4,12):=7;LOC(4,13):=10;LOC(4,14):=8;LOC(4,15):=9;
FOR MX:=1:4 DO BEGIN L3:=1-L1-L2; FOR J:=1:MX+1 DO <<CX(J):=(J-1)/MX>>; MM:=1;
FOR N3:=1:MX+1 DO BEGIN  FOR N2:=1:(MX-N3+2) DO BEGIN N1:=MX+3-N3-N2;
ZZ:=LOC(MX,MM); N(ZZ):=TLAN(N1,CX,L1)*TLAN(N2,CX,L2)*TLAN(N3,CX,L3); MM:=MM+1
END END;
TAG:=T; COORD(1):=L1; COORD(2):=L2; NODES:=(MX+2)*(MX+1)/2;
WRTSF(TAG, N, NODES, NDER, COORD) END; QUIT;
```

Appendix 1 cont

Listing of REDUCE input data

Appendix 2

REDUCE generated code for 6 node triangle element

```
      SUBROUTINE SF2T6(L1, L2 ,SF, ISF, SFDL, ISFDL, JSFDL)
C***SHAPE FUNCTION SUBROUTINE
C***(c) Peter and Jacqueline A. Bettess, 1987
C-------------------------------------------------------------
C PURPOSE
C      Forms element shape function and derivatives
C HISTORY
C      Written February 1987
C ARGUMENTS IN
C      L1     First local co-ordinate
C      L2     Second local co-ordinate
C      ISF    Dimension of shape function array, SF
C      ISFDL  1st dimension of shape function
C             derivative array, SFDL
C      JSFDL  2nd dimension of shape function
C             derivative array, SFDL
C ARGUMENTS OUT
C      SF     Shape function array
C      SFDL   Array of shape function derivatives
C             with respect to local co-ordinates
C*************************************************************
      DOUBLE PRECISION  SF, SFDL
      DOUBLE PRECISION L1, L2
      INTEGER  ISF, ISFDL, JSFDL
      DIMENSION  SF(ISF), SFDL(ISFDL,JSFDL)
C
C***form the element shape functions
C
      SF(1) = L1*(2.*L1-1.)
      SF(2) = 4.*L2*L1
      SF(3) = L2*(2.*L2-1.)
      SF(4) = 4.*L2*(-L2-L1+1.)
      SF(5) = 2.*L2**2+4.*L2*L1-3.*L2+2.*L1**2-3.*L1+1.
      SF(6) = 4.*L1*(-L2-L1+1.)
C
C***form the shape function derivatives
C
      SFDL(1,1) = 4.*L1-1.
      SFDL(1,2) = 4.*L2
      SFDL(1,3) = 0.
      SFDL(1,4) = -4.*L2
      SFDL(1,5) = 4.*L2+4.*L1-3.
      SFDL(1,6) = 4.*(-L2-2.*L1+1.)
      SFDL(2,1) = 0.
      SFDL(2,2) = 4.*L1
      SFDL(2,3) = 4.*L2-1.
      SFDL(2,4) = 4.*(-2.*L2-L1+1.)
      SFDL(2,5) = 4.*L2+4.*L1-3.
      SFDL(2,6) = -4.*L1
      RETURN
      END
```

THE GENERATION OF HYBRID STRUCTURED-UNSTRUCTURED GRIDS

N.P. Weatherill
Institute for Numerical Methods in Engineering
University College of Swansea, U.K.
and
Aircraft Research Association, Bedford, U.K.

SUMMARY

A method of generating hybrid structured-unstructured grids suitable for viscous flow calculations around high lift aerodynamic geometries is presented. The unstructured technique provides the flexibility to discretise the complex multiply connected domain whilst maintaining regular shaped triangles. The approach which will be described is based on the Voronoi construction and the dual, the Delaunay triangulation. In the near-wall regions, where viscous effects are dominant and where flow algorithm modifications, including the incorporation of a turbulence model may be required, it proves advantageous to construct a regular structured grid. In addition to details given on the grid generation, some attention will be focused on ways in which a flow algorithm based on an hybrid grid can be modified for increased efficiency.

1. INTRODUCTION

Many computational methods in engineering require a continuous domain, within which the problem is defined, to be subdivided into a set of control volumes or elements on which a numerical solution is constructed. This process is achieved by a suitable discretisation of the domain and an appropriate connectivity of these points defines the elements.

In view of its importance much effort has been expended in recent years to develop automatic methods for grid generation. Two apparently opposing strategies have emerged. One based on structured grids, in which each point has the same number of neighbouring points and favoured by the advocates of finite difference/finite volume methods, and secondly, unstructured grids favoured by finite element practitioners. Both approaches have their own advantages and disadvantages [1] and, therefore, an attractive option is to combine the two techniques.

As a first step in investigating the suitability of hybrid grids we will discuss an application of such an approach to the problem of simulation of the viscous compressible flow around high lift aerofoil systems.

2. HYBRID GRIDS

The motivation for combining structured and unstructured grids to produce hybrid grids is based on the inherent properties of the two approaches. Structured techniques are computationally efficient, readily programmed to utilise vector capabilities of supercomputers and they can take advantage of directional properties of the grid structure. In contrast, the strength of the unstructured approach is in the flexibility to handle arbitrary shaped computational domains and the ease of implementation of mesh point enrichment, either through apriori techniques, or in an adaptive manner. The latter technique involves the addition of points according to some measure of the local flow properties or to reduce some error in the transient solution.

The properties of these two techniques leads naturally to a basic strategy. In complicated geometrical regions and regions where grid embedding is required an unstructured approach is adopted. Where directional properties of a grid would be advantageous or in regions where a structured grid is readily constructed then a structured approach is adopted.

3. APPLICATION TO HIGH LIFT AERODYNAMIC GEOMETRIES

3.1 Strategy

Computational methods to simulate the flow, either inviscid or viscous, have long been of interest to the aerodynamicist. Inviscid methods, using the Full Potential or Euler equations are useful, but the requirement to predict the maximum lift attainable before the onset of flow separation, involves accurate modelling of the viscous effects. Boundary layer techniques are limited in their application at near flow separation conditions and hence attention is now focused on simulating the flow by the Reynolds-averaged Navier-Stokes equations [2]. In such an approach it is essential to ensure that the near-wall viscous effects are simulated accurately, which amongst other aspects can include the implementation of a suitable turbulence model.

To accommodate some of these requirements a regular

structured grid is constructed in the viscous dominated regions close to the aerofoils, with the global connectivity achieved by an unstructured triangular grid. The basic strategy is shown in Figure 1.

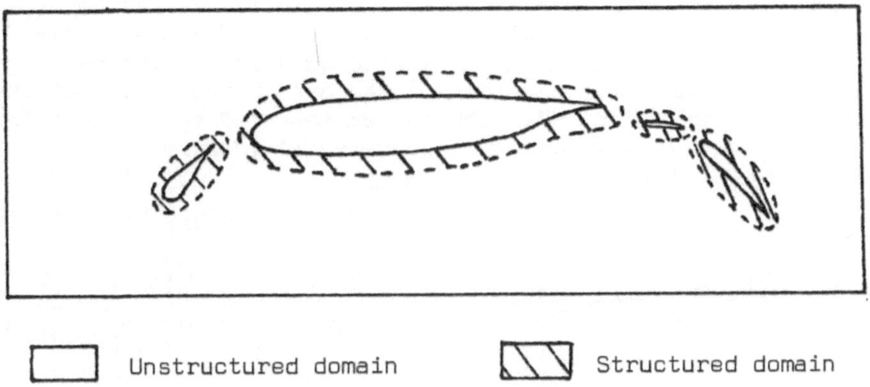

Figure 1. Subdomains of the grid

The grid generation technique is based on a combination of an analytic grid generator and a connectivity algorithm. Grid points, with a polar topology, are generated local to each aerofoil using a conformal mapping technique. Each aerofoil has a grid consisting of the data points ($1 \leqslant j \leqslant jmax_k$, $1 \leqslant i \leqslant i_{max_k}$, $k=1$, No. of aerofoils). Points are distributed near the bodies to ensure an adequate resolution of the boundary layers. Any point from one aerofoil grid which falls inside another aerofoil is automatically detected and rejected. The unstructured regions are obtained by applying an algorithm to construct the Delaunay triangulation to all points ($j_{min_k} < j \leqslant j_{max_k}$, $1 < i \leqslant imax_k$). The structured grid in the near-wall region is obtained by performing a direct triangulation technique to the points ($1 \leqslant j \leqslant jmin_k$, $1 \leqslant i \leqslant imax_k$).

3.2 Unstructured Grid Generation

The grid in the unstructured region is obtained by using the Delaunay criterion. This approach maximises the regularity of formed triangles. Dual to the Delaunay triangulation is the Voronoi diagram (or Dirichlet Tessellation). For a set of points p_m, it is possible to define the polygonal regions of the Voronoi diagram as

$$V_m = \{p: \|p-p_m\| < \|p - p_j\|, \forall_j \neq m\} \qquad (3.2.1)$$

i.e. the Voronoi region V_m is the set of all points p that are closer to p_m than to any other point. The sum of all points p forms a Voronoi polygon. From this definition it is clear that in two dimensions the territorial boundary, which forms the side of a Voronoi polygon, must be mid-way between the two points which it separates and is thus a segment of the perpendicular bisector of the line joining these two points. If all point pairs which have some segment of boundary in common are joined by straight lines the result is a triangulation of the data points. For each triangle there is an associated vertex of the Voronoi diagram which is at the circumcentre of the three points which form the triangle. These geometrical ideas are illustrated in Figure 2.

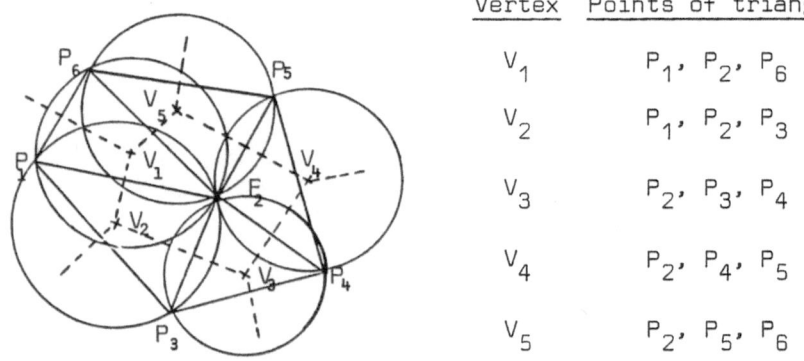

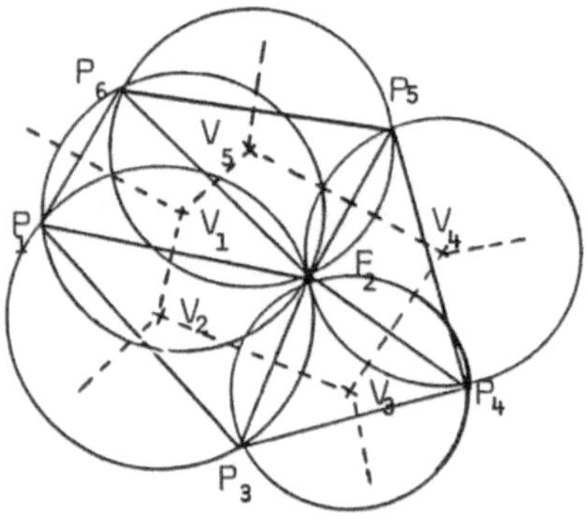

Vertex	Points of triangle
V_1	P_1, P_2, P_6
V_2	P_1, P_2, P_3
V_3	P_2, P_3, P_4
V_4	P_2, P_4, P_5
V_5	P_2, P_5, P_6

Figure 2. Geometrical concepts of Voronoi diagram

The algorithm to construct the triangulation is based on the method of Bowyer [3] and has been described elsewhere [4]. It is a sequential process, whereby, after the definition of 4 points which define the convex hull, together with the specification of the corresponding Voronoi structure, each point is input in turn. The addition of a point requires a local restructuring of the Voronoi polygons since the point must fall inside a circle associated with a vertex and thereby violates the Delaunay criterion [3.2.1]. Points associated with each of the vertices to be deleted are noted, since these form the contiguous points to the new point. Neighbours of the deleted Voronoi vertices are used since the points associated with these are used to determine valid combinations of old points with the new point. At the end of the process, which is illustrated diagrammatically in Figure 3, the vertex structure is reordered.

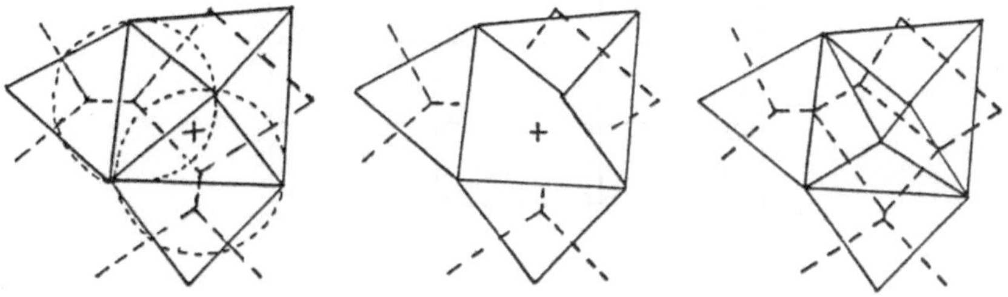

+ New Point

Figure 3. Diagramatic representation of the stages in the construction of the Delaunay triangulation

3.3 Structured Grid Generation

The connectivity of grid points within the inner near-wall structured regions ($1 \leqslant j \leqslant j_{min_K}$, $1 \leqslant i \leqslant i_{max_K}$, k=1, No. of aerofoils) is performed according to

$$P_{N1} \rightarrow P_{N2},\ P_{N2} \rightarrow P_{N3},\ P_{N2} \rightarrow P_{N4},$$

where

$$\begin{aligned} &P_{N1} \text{ is the point ordered } T_K + (j-1)\, i_{max_K} + i \\ &P_{N2} \text{ is the point ordered } T_K + (j-1)\, i_{max_K} + i+1 \\ &P_{N3} \text{ is the point ordered } T_K + (j)\, i_{max_K} + i \\ &P_{N4} \text{ is the point ordered } T_K + (j)\, i_{max_K} + i+1 \end{aligned} \qquad (3.3.1)$$

with

$$T_k = \sum_{k=2}^{\text{No. of aerofoils}} j_{min_{(k-1)}} \times i_{max_{(k-1)}} \text{ and } T_1 = 0.$$

A sketch of these connections is shown in Figure 4. It is clear that the definitions given in equation (3.3.1) completely define the data structure of points in the inner regions.

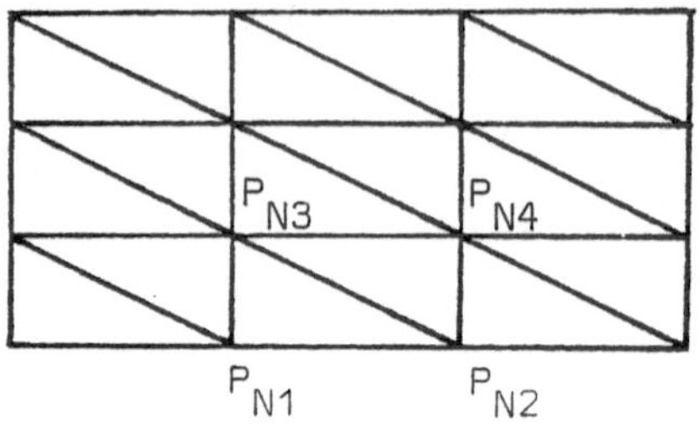

Figure 4. Connectivity of the points in the inner structured regions.

3.4 Example

Figure 5 shows the application of the ideas discussed to a two element slat-aerofoil geometry. Figure 5a shows the global features of the grid, in particular, how the component grid for the slat is local in nature, but the grid around the main aerofoil provides the global grid which extends to the outerboundary. Figure 5b and 5c show enlarged views of the gap region and the leading edge of the slat, respectively. The grid point spacings shown are appropriate to the simulation of laminar flow.

4. FLOW ALGORITHM CONSIDERATIONS

A flow algorithm designed to work on hybrid grids could incorporate two strategies. One, which is based on an unstructured framework, in which the limitations on speed and the difficulties in determining a particular direction in the flow field are accepted and secondly, in parallel, routines which can take advantage of the structured regions. Within such a framework, the same basic flow algorithm could utilise both direct and indirect addressing. In structured regions, local work arrays with a direct addressing format (i,j) could be used. Data stored in one-dimensional arrays could be copied to such work arrays then transferred back after updating. In this way, no interfacing problems between structured and unstructured regions would arise.

Some preliminary work has been undertaken using this approach. The speed-up in the direct structured regions has been found to be a factor of 5 over that obtained for the equivalent unstructured evaluation.

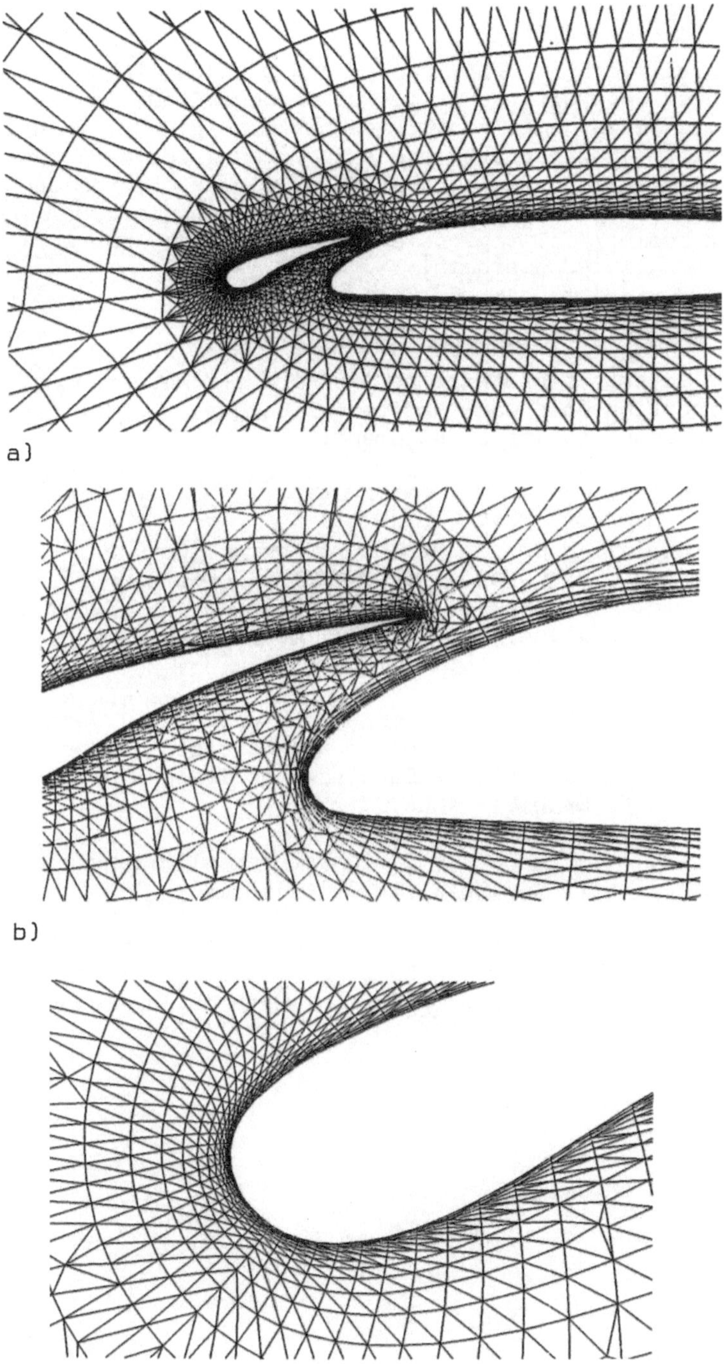

Figure 5. Hybrid structured-unstructured grid for a slat-aerofoil geometry.

5. CONCLUSIONS

A preliminary investigation into the use of hybrid structured-unstructured grids for viscous flow simulations over high lift aerodynamic geometries has been given. Techniques to generate the unstructured and structured grids have been highlighted. The resulting grids provide a compromise between the flexibility of the unstructured approach in modelling complex geometrical regions and the efficiency and applicability of the structured approach.

6. ACKNOWLEDGEMENTS

The author is grateful to the Aircraft Research Association for supporting this work and to colleagues who provided a valuable input to its development.

REFERENCES

1. Weatherill, N.P., 'Grid Generation: Structured, Unstructured or Both?', Proceedings of ~~IMA/~~SMAI Conference on Computational Methods in Aeronautical Fluid Dynamics, Reading, England, April 1987. To be published by Oxford University Press. Edited by P. Stow.

2. Weatherill, N.P., Johnston, L.J., Peace, A.J. and Shaw, J.A., 'Development of a Method for the Solution of the Reynolds-Averaged Navier-Stokes Equations on Triangular Grids', ARA Report 70, December 1986.

3. Bowyer, A., 'Computing Dirichlet Tessellations', Computer Journal, Vol, 24, No.2, (1981), 162-166.

4. Weatherill, N.P. 'A Method for Generating Irregular Computational Grids in Multiply Connected Planar Domains' to appear, Int. Jour. Numerical Methods in Fluids, 1987.